DRUG REPURPOSING IN CANCER THERAPY

DRUG REPURPOSING IN CANCER THERAPY

APPROACHES AND APPLICATIONS

Edited by

KENNETH K.W. TO

WILLIAM C.S. CHO

Academic Press is an imprint of Elsevier
125 London Wall, London EC2Y 5AS, United Kingdom
525 B Street, Suite 1650, San Diego, CA 92101, United States
50 Hampshire Street, 5th Floor, Cambridge, MA 02139, United States
The Boulevard, Langford Lane, Kidlington, Oxford OX5 1GB, United Kingdom

Library of Congress Cataloging-in-Publication Data
A catalog record for this book is available from the Library of Congress

British Library Cataloguing-in-Publication Data
A catalogue record for this book is available from the British Library

ISBN: 978-0-12-819668-7

For information on all Academic Press publications visit our website at https://www.elsevier.com/books-and-journals

Publisher: Stacy Masucci
Acquisitions Editor: Rafael E. Teixeira
Editorial Project Manager: Susan Ikeda
Production Project Manager: Selvaraj Raviraj
Cover Designer: Mark Rogers

Typeset by TNQ Technologies

Contents

5. Increasing opportunities of drug repurposing for treating breast cancer by the integration of molecular, histological, and systemic approaches

HARRAS J. KHAN, SAGAR O. ROHONDIA, ZAINAB SABRY OTHMAN AHMED, NIRAV ZALAVADIYA, AND Q. PING DOU

6. The success story of drug repurposing in breast cancer

SIDDHIKA PAREEK, YINGBO HUANG, ARITRO NATH, AND R. STEPHANIE HUANG

7. A personalized medicine approach to drug repurposing for the treatment of breast cancer molecular subtypes

ENRIQUE HERNÁNDEZ-LEMUS

8. Successful stories of drug repurposing for cancer therapy in hepatocellular carcinoma

YASMEEN M. ATTIA, HEBA EWIDA, AND MAHMOUD SALAMA AHMED

9. Stories of drug repurposing for pancreatic cancer treatment—Past, present, and future

MATTHIAS ILMER, MAXIMILIAN WENIGER, HANNO NIESS, YANG WU, CHUN ZHANG, C. BENEDIKT WESTPHALEN, STEPHAN KRUGER, MARTIN K. ANGELE, JENS WERNER, JAN G. D'HAESE, AND BERNHARD W. RENZ

10. Animal models and in vivo investigations for drug repurposing in lung cancer

HSUEN-WEN KATE CHANG AND VINCENT H.S. CHANG

11. Identification of chemosensitizers by drug repurposing to enhance the efficacy of cancer therapy

GE YAN AND THOMAS EFFERTH

12. Drugs repurposed to potentiate immunotherapy for cancer treatment

KENNETH K.W. TO AND WILLIAM C.S. CHO

13. Nanoparticle-based formulation for drug repurposing in cancer treatment

BEI CHENG AND PEISHENG XU

14. Nanotechnological approaches in cancer: the role of celecoxib and disulfiram

JOÃO BASSO, MARIA MENDES, ANA FORTUNA, RUI VITORINO, JOÃO SOUSA, ALBERTO PAIS, AND CARLA VITORINO

15. Clinical trials on combination of repurposed drugs and anticancer therapies

SÜREYYA ÖLGEN

Contributors

Mahmoud Salama Ahmed, Department of Internal Medicine, Division of Cardiology, University of Texas Southwestern Medical Center, Dallas, TX, United States

Martin K. Angele, Department of General, Visceral and Transplantation Surgery, Hospital of the University of Munich, Munich, Germany; German Cancer Consortium (DKTK), Partner Site Munich, German Cancer Research Center (DKFZ), Heidelberg, Germany

Yasmeen M. Attia, Department of Pharmacology, Faculty of Pharmacy, The British University in Egypt, El-Sherouk, Cairo, Egypt

João Basso, Faculty of Pharmacy, University of Coimbra, Azinhaga de Santa Comba, Coimbra, Portugal; Coimbra Chemistry Center, Department of Chemistry, University of Coimbra, Coimbra, Portugal

Gauthier Bouche, Anticancer Fund, Brussels, Belgium

Sohini Chakraborti, Molecular Biophysics Unit, Indian Institute of Science, Bangalore, Karnataka, India

Pushpaveni Chakravarthi, Department of Pharmaceutical Chemistry, T. John College of Pharmacy, Bangalore, Karnataka, India

Hsuen-Wen Kate Chang, Laboratory Animal Center, Taipei Medical University, Taipei, Taiwan

Vincent H.S. Chang, Department of Physiology, School of Medicine, College of Medicine, Taipei Medical University, Taipei, Taiwan; The Ph.D. Program for Translational Medicine, College of Medical Science and Technology, Taipei Medical University, Taipei, Taiwan

Bei Cheng, Department of Drug Discovery and Biomedical Sciences, College of Pharmacy, University of South Carolina, Columbia, SC, United States

William C.S. Cho, Department of Clinical Oncology, Queen Elizabeth Hospital, Hong Kong SAR, China

Jan G. D'Haese, Department of General, Visceral and Transplantation Surgery, Hospital of the University of Munich, Munich, Germany; German Cancer Consortium (DKTK), Partner Site Munich, German Cancer Research Center (DKFZ), Heidelberg, Germany

Q. Ping Dou, Barbara Ann Karmanos Cancer Institute, and Departments of Oncology, Pharmacology and Pathology, School of Medicine, Wayne State University, Detroit, MI, United States of America

Thomas Efferth, Department of Pharmaceutical Biology, Institute of Pharmacy and Biochemistry, Johannes Gutenberg University, Mainz, Germany

Heba Ewida, Department of Pharmacology and Biochemistry, Faculty of Pharmaceutical Sciences & Pharmaceutical Industries, Future University in Egypt, Cairo, Egypt

Ana Fortuna, Faculty of Pharmacy, University of Coimbra, Azinhaga de Santa Comba, Coimbra, Portugal; CIBIT/ICNAS–Coimbra Institute for Biomedical Imaging and Translational Research, University of Coimbra, Coimbra, Portugal

HemaSree GNS, Pharmacological Modelling and Simulation Centre, Faculty of Pharmacy, M.S. Ramaiah University of Applied Sciences, Bangalore, Karnataka, India

Enrique Hernández-Lemus, Research in Computational and Population Genomics, Computational Genomics Division, National Institute of Genomic Medicine, Ciudad de México, Mexico City, Mexico

Yingbo Huang, Department of Experimental and Clinical Pharmacology, University of Minnesota, Minneapolis, MN, United States

R. Stephanie Huang, Department of Experimental and Clinical Pharmacology, University of Minnesota, Minneapolis, MN, United States

Matthias Ilmer, Department of General, Visceral and Transplantation Surgery, Hospital of the University of Munich, Munich, Germany; German Cancer Consortium (DKTK), Partner Site Munich, German Cancer Research Center (DKFZ), Heidelberg, Germany

Harras J. Khan, Barbara Ann Karmanos Cancer Institute, and Departments of Oncology, Pharmacology and Pathology, School of Medicine, Wayne State University, Detroit, MI, United States of America

Mamatha Krishna Murthy, Pharmacological Modelling and Simulation Centre, Faculty of Pharmacy, M.S. Ramaiah University of Applied Sciences, Bangalore, Karnataka, India

Stephan Kruger, Department of Medicine III, University Hospital, LMU Munich, Munich, Germany

Hansaim Lim, The Ph.D. Program in Biochemistry, The Graduate Center, The City University of New York, New York, NY, United States

Swarna Mariam Jos, Pharmacological Modelling and Simulation Centre, Faculty of Pharmacy, M.S. Ramaiah University of Applied Sciences, Bangalore, Karnataka, India

Lydie Meheus, Anticancer Fund, Brussels, Belgium

Maria Mendes, Faculty of Pharmacy, University of Coimbra, Azinhaga de Santa Comba, Coimbra, Portugal; Coimbra Chemistry Center, Department of Chemistry, University of Coimbra, Coimbra, Portugal; Centre for Neurosciences and Cell Biology (CNC), University of Coimbra, Rua Larga, Faculty of Medicine, Coimbra, Portugal

Aritro Nath, Department of Medical Oncology and Therapeutics Research, City of Hope, Monrovia, CA, United States

Hanno Niess, Department of General, Visceral and Transplantation Surgery, Hospital of the University of Munich, Munich, Germany; German Cancer Consortium (DKTK), Partner Site Munich, German Cancer Research Center (DKFZ), Heidelberg, Germany

Süreyya Ölgen, Faculty of Pharmacy, Biruni University, Istanbul, Zeytinburnu, Turkey

Zainab Sabry Othman Ahmed, Barbara Ann Karmanos Cancer Institute, and Departments of Oncology, Pharmacology and Pathology, School of Medicine, Wayne State University, Detroit, MI, United States of America; Department of Cytology and Histology, Faculty of Veterinary Medicine, Cairo University, Giza, Egypt

Rachana R Pai, Pharmacological Modelling and Simulation Centre, Faculty of Pharmacy, M.S. Ramaiah University of Applied Sciences, Bangalore, Karnataka, India

Alberto Pais, Coimbra Chemistry Center, Department of Chemistry, University of Coimbra, Coimbra, Portugal

Pan Pantziarka, Anticancer Fund, Brussels, Belgium; The George Pantziarka TP53 Trust, London, United Kingdom

Siddhika Pareek, Department of Experimental and Clinical Pharmacology, University of Minnesota, Minneapolis, MN, United States

V Lakshmi PrasannaMarise, Pharmacological Modelling and Simulation Centre, Faculty of Pharmacy, M.S. Ramaiah University of Applied Sciences, Bangalore, Karnataka, India

Bernhard W. Renz, Department of General, Visceral and Transplantation Surgery, Hospital of the University of Munich, Munich, Germany; German Cancer Consortium (DKTK), Partner Site Munich, German Cancer Research Center (DKFZ), Heidelberg, Germany

Sagar O. Rohondia, Barbara Ann Karmanos Cancer Institute, and Departments of Oncology, Pharmacology and Pathology, School of Medicine, Wayne State University, Detroit, MI, United States of America

Klara Rombauts, Anticancer Fund, Brussels, Belgium

Ganesan Rajalekshmi Saraswathy, Pharmacological Modelling and Simulation Centre, Faculty of Pharmacy, M.S. Ramaiah University of Applied Sciences, Bangalore, Karnataka, India

João Sousa, Faculty of Pharmacy, University of Coimbra, Azinhaga de Santa Comba, Coimbra, Portugal; Coimbra Chemistry Center, Department of Chemistry, University of Coimbra, Coimbra, Portugal

Narayanaswamy Srinivasan, Molecular Biophysics Unit, Indian Institute of Science, Bangalore, Karnataka, India

Kenneth K.W. To, School of Pharmacy, Faculty of Medicine, The Chinese University of Hong Kong, Hong Kong SAR, China

Liese Vandeborne, Anticancer Fund, Brussels, Belgium

Rui Vitorino, Department of Medical Sciences and Institute of Biomedicine–iBiMED, University of Aveiro, Aveiro, Portugal

Carla Vitorino, Faculty of Pharmacy, University of Coimbra, Azinhaga de Santa Comba, Coimbra, Portugal; Coimbra Chemistry Center, Department of Chemistry, University of Coimbra, Coimbra, Portugal; Centre for Neurosciences and Cell Biology (CNC), University of Coimbra, Rua Larga, Faculty of Medicine, Coimbra, Portugal

Maximilian Weniger, Department of General, Visceral and Transplantation Surgery, Hospital of the University of Munich, Munich, Germany; German Cancer Consortium (DKTK), Partner Site Munich, German Cancer Research Center (DKFZ), Heidelberg, Germany

Jens Werner, Department of General, Visceral and Transplantation Surgery, Hospital of the University of Munich, Munich, Germany; German Cancer Consortium (DKTK), Partner Site Munich, German Cancer Research Center (DKFZ), Heidelberg, Germany

C. Benedikt Westphalen, Department of Medicine III, University Hospital, LMU Munich, Munich, Germany; Comprehensive Cancer Center Munich, Munich, Germany

Yang Wu, Department of General, Visceral and Transplantation Surgery, Hospital of the University of Munich, Munich, Germany; German Cancer Consortium (DKTK), Partner Site Munich, German Cancer Research Center (DKFZ), Heidelberg, Germany

Lei Xie, The Ph.D. Program in Biochemistry, The Graduate Center, The City University of New York, New York, NY, United States; Department of Computer Science, Hunter College, The City University of New York, New York, NY, United States; The Ph.D. Program in Computer Science & Biology, The Graduate Center, The City University of New York, New York, NY, United States; Helen and Robert Appel Alzheimer's Disease Research Institute, Feil Family Brain & Mind Research Institute, Weill Cornell Medicine, Cornell University, New York, NY, United States

Peisheng Xu, Department of Drug Discovery and Biomedical Sciences, College of Pharmacy, University of South Carolina, Columbia, SC, United States

Ge Yan, Department of Pharmaceutical Biology, Institute of Pharmacy and Biochemistry, Johannes Gutenberg University, Mainz, Germany

Nirav Zalavadiya, Barbara Ann Karmanos Cancer Institute, and Departments of Oncology, Pharmacology and Pathology, School of Medicine, Wayne State University, Detroit, MI, United States of America

Chun Zhang, Department of General, Visceral and Transplantation Surgery, Hospital of the University of Munich, Munich, Germany; German Cancer Consortium (DKTK), Partner Site Munich, German Cancer Research Center (DKFZ), Heidelberg, Germany

CHAPTER

1

Drug repurposing for cancer therapy—an introduction

Pan Pantziarka[1,2], *Lydie Meheus*[1], *Klara Rombauts*[1], *Liese Vandeborne*[1], *Gauthier Bouche*[1]

[1]Anticancer Fund, Brussels, Belgium; [2]The George Pantziarka TP53 Trust, London, United Kingdom

OUTLINE

Introduction

It incumbent upon us to seek new therapeutic options for cancer patients wherever they may be found—including seeking to extend the use of existing medications to the treatment of new diseases. This strategy, variously termed drug repurposing or drug repositioning, is in stark contrast to the primary drug development strategy based around the development of *de novo* molecules aimed at disease-specific molecular targets. In this review, we will introduce the field of oncological drug repurposing, outlining the rationale, briefly survey the current state of the field, and describe future prospects of progress.

Drug Repurposing in Cancer Therapy
https://doi.org/10.1016/B978-0-12-819668-7.00001-4

Definitions

While we refer throughout to "drug repurposing" there are in fact a number of competing terms with overlapping definitions. In addition to drug repurposing, other commonly used descriptive terms include "drug repositioning," "drug rescue," "drug reprofiling," and "drug rediscovery" [1]. For the purposes of this article, we adopt the following definition of drug repurposing: the application of an existing, licensed drug to a new disease for which it is not licensed or widely used as an off-label treatment. This definition, therefore, excludes the development of previously shelved compounds that have never been licensed as pharmaceuticals. Our definition does not exclude off-label use per se, rather it excludes those instances where drugs are widely used in clinical practice to treat diseases that are not included in the licensed indications for the drug [2,3]. However, the use of off-label prescribing for *new* uses is included within our scope of repurposing, although as we shall discuss later, it is not an ideal end point.

In the case of oncological repurposing, there is also a notable pragmatic distinction sometimes made between "soft" repurposing—the repurposing of existing oncological drugs to new cancer indications—and "hard" repurposing—the repurposing of noncancer drugs as cancer treatments [4,5]. In reality these strategies exist on a spectrum rather than being binary categories, as shown in Fig. 1.1.

In many respects, the "soft repurposing" side of the spectrum is the standard model in oncology—drugs are licensed for an initial cancer indication and then further developed for new indications while still within the patent-protected period—for example, a drug may be licensed as a second-line therapy and then developed for first-line or a drug developed for one cancer type may then be applied to other, often unrelated, malignancies. Such development benefits both from commercial sponsorship and from the degree of familiarity that treating physicians have with the use of these drugs in an oncological setting.

This process of soft repurposing can encompass all classes of anticancer drug at all stages of clinical and commercial development. Traditional cytotoxic chemotherapy drugs, the majority of which are generic, are still being explored for clinical use in sarcomas and other hard to

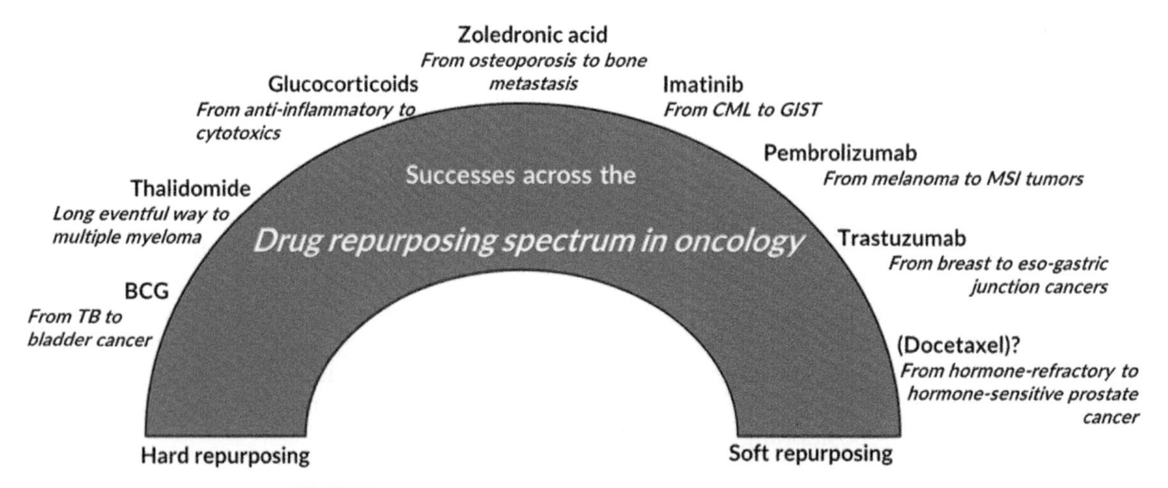

FIGURE 1.1 The spectrum of drug repurposing in oncology.

treat cancers in large academic-led trials such as the Euro Ewing 2012 trial [6]. In contrast to these efforts, which are not commercially supported, newer agents that are still within the market exclusivity period, such as immune checkpoint inhibitors, are being rapidly developed for multiple new indications in commercially supported trials. For example, pembrolizumab was first approved by the FDA in 2014 for patients with unresectable or metastatic melanoma and disease progression following ipilimumab and, if BRAF V600 mutation positive, a BRAF inhibitor. By January 2020, it has FDA approvals for over 20 cancer indications including in non-small cell lung cancer, small cell lung cancer, head and neck squamous cell carcinoma, classical Hodgkin lymphoma, and more. Indeed, it may be true to say that cancer drugs used in only a single indication are rare exceptions to a general trend of soft repurposing of anticancer agents.

However, the development of noncancer drugs as new cancer therapeutics is much rarer and more problematic, particularly in the case of generic medications. The most notable examples of hard repurposing have been thalidomide and all-trans retinoic acid (tretinoin), both drugs licensed for noncancer indications and then subsequently developed and licensed as treatments for multiple myeloma and acute promyelocytic leukemia, respectively. Hard repurposing may be more difficult to achieve in practice due to a lack of commercial incentives in the case of drugs that no longer benefit from IP protection, a topic we will discuss in more detail subsequently. A less tractable issue relates to a degree of skepticism from the medical community regarding the potential utility of noncancer drugs as oncological treatments. There may indeed be a rejection bias at work when assessing the evidence from repurposing trials, demanding a higher level of evidence for efficacy than would be expected for soft repurposing [7].

Increased interest in repurposing

However, despite these problems, there is little doubt that recent years have seen an increase in research—both clinical and preclinical—in drug repurposing in oncology. Fig. 1.2 shows the increase in Pubmed articles over the period 2008–19 using the following search term:

"drug repurposing" or "drug repositioning" and cancer.

The growth of the general oncology literature over the same period is also shown. It should be noted that "drug repositioning" was added as a MESH term in 2011.

Also important in the context of oncological drug repurposing is the attention from policymakers, regulators, patients, and other stakeholders. These, to be discussed in more detail in the appropriate sections of this chapter, attest both to the scientific and social impacts that repurposing may have in medicine in the future.

Why repurpose?

The rationale for seeking to repurpose existing medicines for cancer treatments rather than to develop novel agents rests on a number of perceived advantages that have been outlined in some detail [4,8,9]. To briefly recapitulate, repurposing offers a number of key benefits that may help to accelerate the development process:

- The availability of extensive data on safety and toxicity. For many widely used drugs, particularly generic medicines with decades of clinical use, we have an extensive pool of knowledge to drawn on in terms of common, rare, and very rare adverse events data
- There is existing data on pharmacokinetics and pharmacodynamics. There may be multiple dosing schedules in use for different existing indications, different populations, and for combinations with other drugs

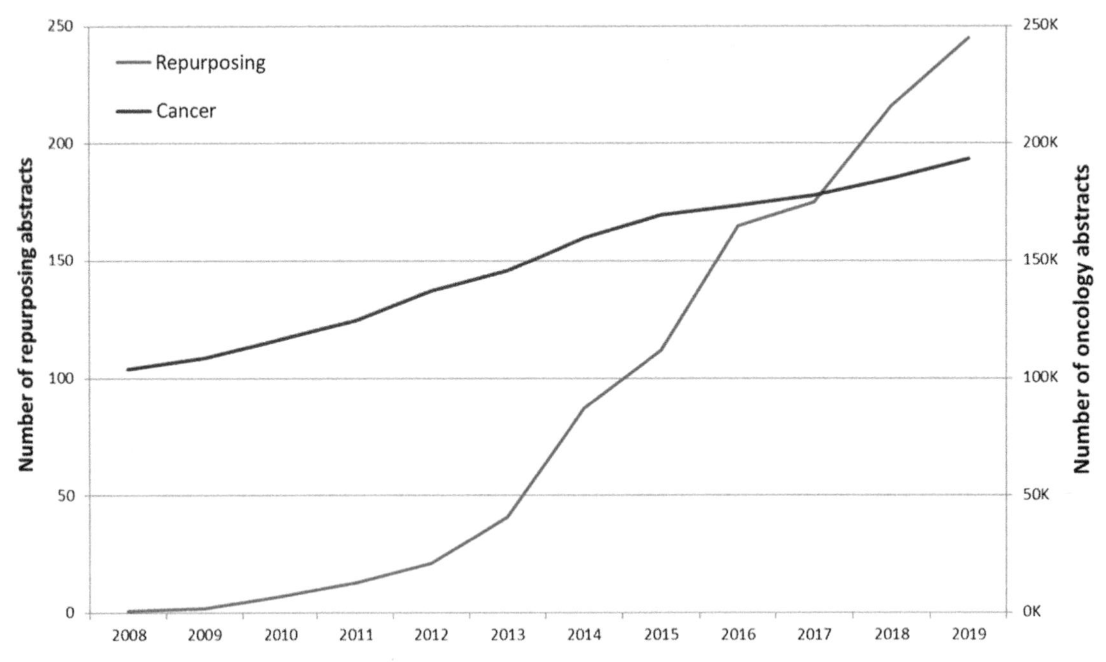

FIGURE 1.2 Growth in pubmed articles on drug repurposing in oncology and cancer (query performed on 20/01/20).

- Many of the repurposing candidates are widely available, often at low cost, particularly for generic medications in those markets where there are multiple competing manufacturers in place
- There is existing data on mechanisms of action, including molecular targets and pathways, which may be relevant in the anticancer activity of the candidate drug

This existing body of knowledge may mean that clinical development can proceed without having to first establish the initial safety and dosing information that a new drug, particularly a first-in-class drug, has to establish in animal models and in early phase clinical trials. Of course Phase I trials may still be required for repurposed drugs, for example, if it is part of a combination of drugs where the safety of the combination is not established despite the safety of the individual components or the repurposed drug is administered at a higher dose than in the original indication. However, even in these cases, the existing corpus of data have the potential to move the repurposed drug to clinical development significantly more quickly, and cheaply, than would a *de novo* drug candidate.

Economic incentives/disincentives

Essentially the head start that existing data availability provides for the development of a repurposed drug translates into a reduced cost of development. However, the economics of drug development are complex. Economic incentives are key drivers in medicine, just as they are in other fields of human endeavor. In the case of repurposing, many of the advantages outlined above may act as economic disincentives rather than as straightforward incentives [10].

De novo drug development may take many years longer (and therefore cost significantly more than repurposing), but for the pharmaceutical company funding the development, the end result will be a patent-protected asset able to generate a return on investment (ROI) [9]. However, for a generics manufacturer, the costs of further development of a drug for a new indication do not provide for the same potential ROI. Without a guarantee of market exclusivity, there is nothing to stop a competitor company reaping the benefit of a new indication for a generic drug (*free-riding* in economics parlance). This is borne out in analysis showing that the number of license extensions for new indications peaks in the period 6–3 years *prior* to the introduction of generic competitors, i.e., when the product originator retains market exclusivity, and thereafter declines sharply when generic competitors enter the market [11]. For example, in the case of the targeted therapy imatinib, the majority of repurposing trials occurred within the market exclusivity period, including all of the new FDA approvals [12].

A lack of economic incentives also applies to the development of new medicines for rare diseases, including some of the rare and ultra-rare cancers. In consequence, there is much interest in repurposing in the rare disease arena [13]. The small patient numbers suffering from many rare diseases means that the potential market for new drugs is relatively small and therefore unlikely to provide a positive ROI given the high costs associated with *de novo* drug development. Repurposing, therefore, may be viewed as providing a reservoir of successfully developed drugs that may be explored to discover new treatments that provide clinical benefits to rare disease sufferers.

In oncology, it is estimated that 24% of cancers diagnosed in the European Union in the period 2000–07 were rare cancers [14]. While this is a significant patient population, it is extremely heterogeneous, with over 600 different cancer diagnoses, of which more than 340 cancer types have fewer than 100 patients in Europe. Unmet needs in this population are high, with five year–relative survival for rare cancers at 48.5%, compared to 63.4% for the common cancers [14]. Drug repurposing, therefore, is an important strategy for the rare cancers field [15].

In recent years, a number of initiatives have brought together academic researchers, funders, and the pharmaceutical industry to address some of the institutional/financial challenges of repurposing [16,17]. One prominent example is the NIH National Center for Advancing Translational Sciences *Discovering New Therapeutic Uses for Existing Molecules* initiative, primarily focused on making shelved compounds available as repurposing candidates [18].

Drug candidates

Data to identify repurposing candidates can come from multiple sources [8], including

- Preclinical studies—*in vitro* and *in vivo*—showing the anticancer activity of the drug against one or more cancer cell lines
- Prospective drug screens to assess activity of a panel of drugs, including but not limited to licensed noncancer medicines, against one or more cancer cell lines
- Retrospective observational studies assessing cancer incidence or outcomes for patients prescribed a drug for noncancer indications
- Published case reports, including reports from off-label usage, describing unexpected or exceptional responses from cancer patients treated with the drug
- Data from current or previous repurposing clinical trials
- Data mining medical records, national drug/medical registries, or other data sources
- *In silico* studies or other forms of computational pharmacology

An approach adopted by the Repurposing Drugs in Oncology (ReDO) project has been to mine the biomedical literature and clinical trial registries to identify drug candidates [4,19]. This approach utilizes the existing peer-reviewed literature to assess the quality of evidence that indicates whether a given noncancer drug may have the potential to be repurposed as a cancer therapeutic. A number of drugs, listed in Table 1.1, were identified as candidates with high potential for further clinical development based on the availability of supporting data, particularly human data. The data supporting these candidates were summarized and published in review articles to bring it to the attention of the oncology community.

Notably absent from this list of candidates are such well-known candidates as aspirin, metformin, celecoxib, and sirolimus. These drugs are the subject of numerous clinical studies in a wide range of cancer types and were therefore assumed to be already well-known to the oncology community.

To date, the ReDO collaboration has identified 300 noncancer drugs with some peer-reviewed evidence of anticancer activity. An online database, ReDO_DB (http://www.redo-project.org/db/), has been made available as a resource to the oncology community [29]. The database lists information on each of the drugs, including whether the drug is included in the World Health Organization Essential Medicines List, patent status, and the types of information showing anticancer activity (e.g., in vitro, in vivo, case reports, etc.). The summary statistics for the drugs in the database are shown in Table 1.2.

The focus on licensed medications means that many of the herbal or folk medicines being investigated for anticancer uses, including well-known agents such as curcumin [30] or berberine [31], are not included in this definition of repurposing. While there is often very good preclinical evidence to support clinical investigation, such agents also suffer a number of disadvantages including a lack of standardized extracts and a lack of experience of clinical use. In many respects, developing these agents is more akin to developing novel drugs than it is repurposing. However, in cases where there are licensed drugs based on these agents—for example the artemisinin extract artesunate, used clinically as an antimalarial—the drugs are indeed relevant repurposing candidates [32,33].

An emerging area of repurposing research is the use of vaccines as anticancer therapeutics,

TABLE 1.1 ReDO repurposing candidate reviews.

Drug	Original indication	References
Mebendazole	Antihelminthic	[20]
Cimetidine	Antacid	[21]
Clarithromycin	Antibiotic	[22]
Itraconazole	Antifungal	[23]
Nitroglycerin	Angina	[24]
Diclofenac	Analgesia	[25]
Propranolol	Hypertension	[26]
Chloroquine/hydroxychloroquine	Antimalarial	[27]
PDE5 inhibitors	Erectile dysfunction	[28]

TABLE 1.2 Summary of data from ReDO_DB (January 01, 2020).

Item	Yes	No
Included in WHO EML?	98 (32.7%)	202(67.3%)
Are off-patent?	255 (85%)	37(12.3%)
Have in vitro evidence?	296 (98.7%)	4(1.3%)
Have in vivo evidence?	274 (91.3%)	26(8.7%)
Have cases reports?	91 (30.3%)	209(69.7%)
Have observational data?	37 (12.3%)	263(87.7%)
Included in clinical trials?	195 (65%)	105(35%)
Trial reports published?	124 (63.6%)	71(36.4%)
Human data exists?[a]	213 (71%)	87(29%)
WHO + off-patent + human data	78 (26%)	222(74%)
In vitro evidence only?	16 (5.3%)	284(94.7%)

[a] *Human data include case reports, observational studies, or clinical trials.*

particularly as adjuncts to existing oncoimmunology treatments. Bacillus Calmette–Guérin vaccine, a standard treatment to immunize against tuberculosis, is already used off-label as an intravesical treatment for noninvasive bladder cancer. Retrospective data from Sweden showed that colorectal cancer patients who received cholera vaccine postdiagnosis had a significantly reduced risk of death compared to patients who did not receive the vaccine [34]. Influenza, tetanus, and rotavirus vaccines are also potential repurposing candidates in oncology [35–37].

The ReDO approach, dependent as it is on the published literature, does not seek to identify novel candidates through the use of *in silico* modeling or other computational methods. This is an area of very active research, driven both by the development of new machine learning algorithms and the availability of large omics data sets. Some established repurposing candidates, such as mebendazole, have previously been identified in part through such approaches [38]. Unless validated, using *in vitro* or *in vivo* studies, candidate identification using computational pharmacological studies should be viewed as only the first stage of hypothesis generation. Recent work from the Broad Institute has described a process that incorporates genomic data and high-throughput screening to identify noncancer drugs able to reduce viability across hundreds of cancer cell lines [39].

Many repurposing candidates—for example aspirin, nitroglycerin, or propranolol—have decades of clinical use and were discovered and developed in the era of empirical drug development. These agents are sometimes viewed as "dirty drugs" in that they have multiple molecular targets and may affect many pathways, some of which have still not been fully elucidated. Polypharmacology seeks to exploit this multitargeting to enhance therapeutic effect, either by combining different drugs (i.e., seeking synthetic lethality) or by rationally designing new drugs that explicitly act on multiple pathways [40,41]. Many established anticancer drugs, such as those in the tyrosine kinase inhibitor class, have been analyzed to identify additional targets—therefore opening the door to repurposing for additional cancer indications [42]. While there are issues with our knowledge of the drug–target landscape [43], including off-target toxicity of cancer drugs [44], further work to enhance our knowledge of drug–protein interactions may present additional opportunities to identify noncancer drugs as repurposing candidates in the future.

A number of open access online databases that can aid in candidate selection are listed in Table 1.3.

A number of groups have worked to identify drug repurposing candidates in specific cancers, using both ReDO-type methods and other techniques, examples include acute myeloid leukemia [52], glioblastoma multiforme [53], hematological cancers [54], and colorectal cancer [55]. Such disease-centric approaches are able to utilize additional tools to analyze specific phenotypical features, molecular targets, and existing treatment modalities to identify repurposing candidates [55].

Clinical development

The ultimate aim of drug repurposing is to bring a drug to clinical use in a new disease area. The need to develop evidence of clinical efficacy is the same in repurposing as it is in any other form of drug development. Despite the many advantages of repurposing in terms of drug costs, speed of development, and so on, it is efficacy that is the primary criterion by which a treatment should be judged. In practice, this means well-designed and executed clinical trials with appropriate survival end points.

Given the existing medical uses for repurposing candidates, it may be thought that early phase clinical trials to establish safety and

TABLE 1.3 Open access drug repurposing databases.

Name	URL	Comment	References
DeSigN	design.cancerresearch.my/	A web-based tool for predicting drug efficacy against cancer cell lines using gene expression patterns.	[45]
Drug Repurposing hub	clue.io/repurposing	Extensive curated annotations for each drug, including details about commercial sources of all compounds.	[46]
DrugBank	www.drugbank.ca/	A comprehensive, freely accessible, online database containing information on drugs and drug targets.	[47]
DrugCentral	drugcentral.org/	A compendium of drug information, including indications, adverse events, and molecular targets.	[48]
DRUGSURV	www.bioprofiling.de/GEO/DRUGSURV	A computational tool to estimate the potential effects of a drug using patient survival information derived from clinical cancer expression data sets.	[49]
PRISM	depmap.org/repurposing/	Viability profiling of noncancer drugs against cancer cell lines.	[39]
ReDO_DB	www.redo-project.org/db	Curated database of noncancer drugs with published anticancer activity.	[29]
repoDB	apps.chiragjpgroup.org/repoDB/	A database of both approved and failed uses of drugs.	[50]
ProteomicsDB	www.ProteomicsDB.org/	Proteomics and drug sensitivity data and tools.	[51]

tolerability are not required. However, there may be some scenarios where such trials are necessary. For example, clinical trials where a combination of repurposed drugs are used and there are concerns about the cumulative toxicity of multiple drugs that are individually well-tolerated. One such example is the CUSP9* protocol for recurrent glioblastoma (NCT02770378), which combines nine repurposed drugs with metronomic temozolomide [56]. In other cases, there may be concern about the interaction of a repurposed drug not often used in oncology with standard of care chemotherapy or other anticancer therapeutics. And, in some cases, it may be that the repurposed drug is used at a higher dose than for other uses. However, it remains the case that many repurposing trials are in a position to benefit from the existing data on safety and tolerability.

Analysis of open clinical trials listed on ClinicalTrials.gov and the WHO International Clinical Trials Registry on September 10, 2019, for all the drugs on the ReDO database, identified 730 repurposing trials. Of these, 680 indicated a trial phase, from Phase 1 to Phase 4. The distribution of trials by phase is shown in Fig. 1.3.

The majority of active repurposing clinical trials in oncology are Phase 2 to Phase 4, with 119 in this data set being Phase 3 to Phase 4.

In the majority of cases, drug repurposing candidates are viewed as additions to standard of care therapies rather than as novel agents to be used as monotherapies. Metformin, celecoxib, and aspirin are the three most studied repurposing drugs in our data set, being included in 204 (28%) of the trials analyzed. The settings for these trials include combinations with existing neoadjuvant and adjuvant drug treatments and investigation as possible maintenance treatment following standard of care or, in a very few instances, as standalone treatments, as shown

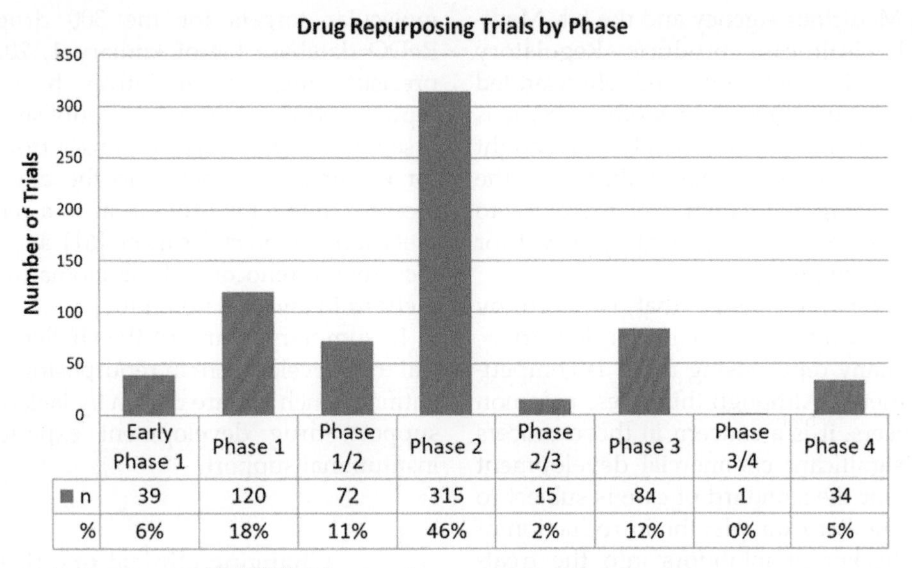

FIGURE 1.3 Oncology repurposing trials by phase.

in Fig. 1.4. Notable are the number of trials with biological or biomarker outcomes (17%). Often these are early phase trials, including window of opportunity trials prior to surgery, looking at biological correlates as preliminary evidence for future trials. Also notable are the 11% of clinical trials adding repurposed drugs to immunotherapy treatments, particularly with immune checkpoint inhibitors. Only 2% of the trials analyzed were testing repurposed drugs alone rather than in combination with other drugs or treatment modalities.

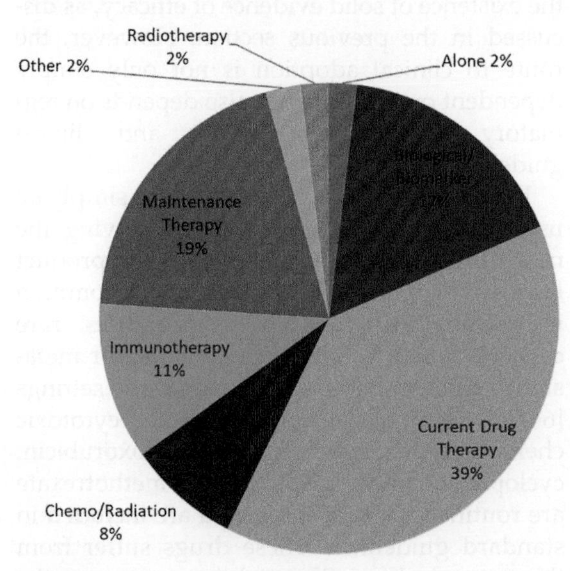

FIGURE 1.4 Drug repurposing combination therapies for metformin, celecoxib, and aspirin clinical trials.

A particular challenge for repurposing trials is the lack of support from commercial sources to fund trials. A previous analysis has shown that less than 4% of late-phase repurposing trials in oncology have commercial sponsorship [29]. The majority of funding for such trials comes from academic, institutional, philanthropic, or government sources. There is clearly a need to expand the availability of funding for repurposing trials—in the case of low-cost generic medications, the potential to address the financial toxicity associated with many new treatments may never be realized without such funding.

Commercial drug developers engage in regular dialog with regulators in order to design late-phase trials aimed at drug registration. One mechanism for doing this is the scientific advice route offered by regulators such as the

European Medicines Agency and the UK Medicines and Healthcare products Regulatory Agency [57]. For academic and clinician-led studies, as repurposing studies often are, it is highly recommended that such advice is sought in order to best design a trial to align with the concerns of drug regulators and therefore to maximize the chances of gaining approval for a new indication [58,59].

An additional challenge that is faced by many small academic or clinician-led trials, including many repurposing trials, is competition for patients. Although this is less common in rare cancers, it is a concern in those cancers in which significant commercial development is on-going or the standard of care is subject to rapid change. For example, the introduction of immune checkpoint inhibitors into the treatment of non-small cell lung cancer is creating challenges in trial recruitment for trials of other treatments. Competition for patients may be driven by financial factors in that commercial trials may generate additional income for recruiting centers, but may also be due to patient choice. When faced with a choice of a clinical trial of repurposed drugs or of a trial of new molecules, especially those with positive media speculation as to "breakthrough" status, many patients opt to go for the new drug rather than an "old-fashioned" repurposed drug.

While clinical trials are still the driving force for making changes to clinical practice, there is also an increasing interest in the use of precision medicine in oncology. The use of omics data to select treatments targeted at a patient's tumor is being used both in clinical practice and in the context of clinical trials. Some online databases for precision medicine, for example, OncoKB, do not include noncancer drugs [60]. By limiting the range of drugs included in precision medicine to licensed anticancer medications, the list of "actionable" targets is also being restricted. Here too, there may be a role for drug repurposing [5]. For example, there are 755 distinct molecular targets for the 300 drugs in the ReDO database (as of January 1, 2020). Some precision medicine initiatives have included repurposed drugs and have published credible case reports of notable responses from patients, for example, a response to the angiotensin II receptor antagonist irbesartan in a patient with metastatic colorectal cancer [61] and a case of recurrent adrenocortical carcinoma treated with metformin and melatonin [62].

In summary, many of the challenges of clinical trial development in repurposing are not scientific as such but are driven by lack of financial support, drug development experience, and institutional support.

Changing clinical practice

Drug repurposing is a strategy designed to move existing medicines into clinical practice to treat new diseases—in the case of oncology, our aim is to develop novel cancer treatments using existing licensed noncancer drugs. The route into clinical practice crucially depends on the existence of solid evidence of efficacy, as discussed in the previous section. However, the route to clinical adoption is not only simply dependent on evidence but also depends on regulatory issues, reimbursement, and clinical guidelines [63].

In theory, repurposed drugs could simply be used "off-label"—that is, without having the new cancer indication included in the product label. Off-label use is actually fairly common in oncology—particularly in pediatrics, rare cancers lacking standard treatments, for metastatic disease, and palliative care settings [64,65]. Many of the long-established cytotoxic chemotherapy drugs such as doxorubicin, cyclophosphamide, cisplatin, and methotrexate are routinely used off-label and are included in standard guidelines. These drugs suffer from the same lack of financial incentives as the repurposed noncancer drugs, and a lack of commercial sponsorship means that it is unlikely that the new uses will be added to the product labels.

Despite the common use of off-label prescribing, it is not an optimal solution for getting repurposed drugs into clinical use [66]. Off-label use may be subject to greater legal liability, therefore discouraging physicians from prescribing the treatments. Insurers or other payers may not support reimbursement of repurposed drugs. Outcomes data from off-label use is not routinely collected. Finally, there are cases where a drug is no longer used for its on-label indications and may disappear from a market, therefore becoming unavailable also for off-label use.

The ideal path for a drug developed through a repurposing strategy is to move from successful clinical trials to inclusion in standard treatment guidelines, gain marketing authorization/label extension, and be fully reimbursed by insurers or national health systems. In this way, "off-label" investigation in clinical trials can lead to "on-label" standard of care [66,67].

Future prospects

The number of noncancer drugs with evidence of anticancer activity, such as those included in the ReDO database, has shown a sustained increase in recent years, and there is little reason to think that it has reached any limit. DrugBank, for example, lists 3967 approved drugs (as of January 1, 2010), so less than 8% of them are repurposing in oncology candidates included in the ReDO database. A key driver of the increase in candidate identification is the preclinical work that is elucidating the molecular biology of cancer—and in doing so, new targets that can be accessed by existing drugs are also being uncovered.

Concomitant with the increase in candidate identification we can expect to see an increase in clinical trial activity. Our analysis of repurposing drug trials shows that 146 drugs are currently included in active clinical trials in cancer. Some drugs, such as metformin, aspirin, celecoxib, and sirolimus, are in multiple late-phase trials and will be reporting results in the near future. Positive results will necessarily focus attention on clinical adoption and the regulatory and institutional hurdles that stand in the way. However, it is likely that clinical adoption of one or two noncancer drugs as new cancer treatments will set important precedents and act as templates to be followed by other drugs.

Conclusions

Drug repurposing is a drug development strategy of increasing interest to the oncology community. By utilizing existing noncancer medications that have shown evidence of anticancer activity, clinical development can proceed more quickly than *de novo* medicines, particularly first-in-class drugs. The repurposing journey begins with candidate selection—a process that matches potential drugs to the specific cancer and setting that is being explored. The availability of online databases, such as the ReDO database, can help by making available a curated list of approved noncancer drugs with therapeutic potential for cancer treatment.

While much progress has been made in recent years, there is an urgent need to establish that a working repurposing pathway exists by moving from positive clinical trials to standard of care for one or more candidate drugs.

Acknowledgments

The authors wish to thank Andrea Leurs, KU Leuven, for assistance with the preparation of the data on drug repurposing clinical trials.

References

[1] Langedijk J, Mantel-Teeuwisse AK, Slijkerman DS, Schutjens MHDB. Drug repositioning and repurposing: terminology and definitions in literature. Drug Discov Today 2015;20(8):1027–34.

[2] Pulley JM, Jerome RN, Shirey-Rice JK, Zaleski NM, et al. Advocating for mutually beneficial access to shelved compounds. Future Med Chem 2018;10(12): 1395–8.

[3] Mullard A. Cancer charity sees success re-prioritizing industry's shelved compounds. Nat Rev Drug Discov 2014;13(5):319–21.

[4] Pantziarka P, Bouche G, Meheus L, Sukhatme V, Sukhatme VP. The repurposing drugs in oncology (ReDO) project. Ecancermedicalscience 2014;8:442.

[5] Pantziarka P, Bouche G, André N. 'Hard' drug repurposing for precision oncology: the missing link? Front Pharmacol 2018;9(June):637.

[6] Anderton J, Moroz V, Marec-Bérard P, Gaspar N, et al. International randomised controlled trial for the treatment of newly diagnosed Ewing sarcoma family of tumours - Euro Ewing 2012 Protocol. Trials 2020; 21(1):96.

[7] Gyawali B, Pantziarka P, Crispino S, Bouche G. Does the oncology community have a rejection bias when it comes to repurposed drugs? Ecancermedicalscience 2018;12:10–4.

[8] Bertolini F, Sukhatme VP, Bouche G. Drug repurposing in oncology–patient and health systems opportunities. Nat Rev Clin Oncol 2015;12(12):732–42.

[9] Ashburn TT, Thor KB. Drug repositioning: identifying and developing new uses for existing drugs. Nat Rev Drug Discov 2004;3(8):673–83.

[10] Simsek M, Meijer B, van Bodegraven AA, de Boer NKH, Mulder CJJ. Finding hidden treasures in old drugs: the challenges and importance of licensing generics. Drug Discov Today 2018;23(1):17–21.

[11] Langedijk J, Whitehead CJ, Slijkerman DS, Leufkens HGM, et al. Extensions of indication throughout the drug product lifecycle: a quantitative analysis. Drug Discov Today 2016;21(2):348–55.

[12] Carlisle BG, Zheng T, Kimmelman J. Imatinib and the long tail of targeted drug development. Nat Rev Clin Oncol 2020;17(1):1–3.

[13] Bloom BE. Recent successes and future predictions on drug repurposing for rare diseases. Expert Opin Orphan Drugs 2016;4(1):1–4.

[14] Gatta G, Capocaccia R, Botta L, Mallone S, et al. Burden and centralised treatment in Europe of rare tumours: results of RARECAREnet-a population-based study. Lancet Oncol 2017;18(8):1022–39.

[15] Pantziarka P, Meheus L. Omics-driven drug repurposing as a source of innovative therapies in rare cancers. Expert Opin Orphan Drugs 2018;6(9):513–7.

[16] Frail DE, Brady M, Escott KJ, Holt A, et al. Pioneering government-sponsored drug repositioning collaborations: progress and learning. Nat Rev Drug Discov 2015;14(12):833–41.

[17] Pushpakom S, Iorio F, Eyers PA, Escott KJ, et al. Drug repurposing: progress, challenges and recommendations. Nat Rev Drug Discov 2019;18(1): 41–58.

[18] Allison M. NCATS launches drug repurposing program. Nat Biotechnol 2012;30(7):571–2.

[19] Pantziarka P, Bouche G, Meheus L, Sukhatme V, Sukhatme VP. Repurposing drugs in your medicine cabinet: untapped opportunities for cancer therapy? Future Oncol 2015;11(2):181–4.

[20] Pantziarka P, Bouche G, Meheus L, Sukhatme V, Sukhatme VP. Repurposing drugs in oncology (ReDO)-mebendazole as an anti-cancer agent. Ecancermedicalscience 2014;8:443.

[21] Pantziarka P, Bouche G, Meheus L, Sukhatme V, Sukhatme VP. Repurposing drugs in oncology (ReDO)-cimetidine as an anti-cancer agent. Ecancermedicalscience 2014;8:485.

[22] Van Nuffel AM, Sukhatme V, Pantziarka P, Meheus L, et al. Repurposing drugs in oncology (ReDO)-clarithromycin as an anti-cancer agent. Ecancermedicalscience 2015;9:513.

[23] Pantziarka P, Sukhatme V, Bouche G, Meheus L, Sukhatme VP. Repurposing drugs in oncology (ReDO)-itraconazole as an anti-cancer agent. Ecancermedicalscience 2015;9:521.

[24] Sukhatme V, Bouche G, Meheus L, Sukhatme VP, Pantziarka P. Repurposing drugs in oncology (ReDO)-nitroglycerin as an anti-cancer agent. Ecancermedicalscience 2015;9:568.

[25] Pantziarka P, Sukhatme V, Bouche G, Meheus L, Sukhatme VP. Repurposing drugs in oncology (ReDO)-diclofenac as an anti-cancer agent. Ecancermedicalscience 2016;10:610.

[26] Pantziarka P, Bouche G, Sukhatme V, Meheus L, et al. Repurposing drugs in oncology (ReDO)-Propranolol as an anti-cancer agent. Ecancermedicalscience 2016;10:680.

[27] Verbaanderd C, Maes H, Schaaf MB, Sukhatme VP, et al. Repurposing drugs in oncology (ReDO)-chloroquine and hydroxychloroquine as anti-cancer agents. Ecancermedicalscience 2017;11:781.

[28] Pantziarka P, Sukhatme V, Crispino S, Bouche G, et al. Repurposing drugs in oncology (ReDO)-selective PDE5 inhibitors as anti-cancer agents. Ecancermedicalscience 2018;12:824.

[29] Pantziarka P, Verbaanderd C, Sukhatme V, Rica Capistrano I, et al. ReDO_DB: the repurposing drugs in oncology database. Ecancermedicalscience 2018;12:886.

[30] Giordano A, Tommonaro G. Curcumin and cancer. Nutrients 2019;11(10).

[31] Liu D, Meng X, Wu D, Qiu Z, Luo H. A natural isoquinoline alkaloid with antitumor activity: studies of the biological activities of berberine. Front Pharmacol 2019;10:9.

[32] Krishna S, Ganapathi S, Ster IC, Saeed MEM, et al. A randomised, double blind, placebo-controlled pilot study of oral artesunate therapy for colorectal cancer. EBioMedicine 2015;2(1):82–90.

[33] Augustin Y, Krishna S, Kumar D, Pantziarka P. The wisdom of crowds and the repurposing of artesunate as an anticancer drug. Ecancermedicalscience 2015;9: ed50.

[34] Ji J, Sundquist J, Sundquist K. Cholera vaccine use is associated with a reduced risk of death in patients with colorectal cancer: a population-based study. Gastroenterology 2018;154(1):86–92.e1.

[35] Tai L-H, Zhang J, Scott KJ, de Souza CT, et al. Perioperative influenza vaccination reduces postoperative metastatic disease by reversing surgery-induced dysfunction in natural killer cells. Clin Cancer Res 2013:5104–15.

[36] Mitchell DA, Batich KA, Gunn MD, Huang M-N, et al. Tetanus toxoid and CCL3 improve dendritic cell vaccines in mice and glioblastoma patients. Nature 2015; 519(7543):366–9.

[37] Shekarian T, Sivado E, Jallas A-C, Depil S, et al. Repurposing rotavirus vaccines for intratumoral immunotherapy can overcome resistance to immune checkpoint blockade. Sci Transl Med 2019;11(515).

[38] Dakshanamurthy S, Issa NT, Assefnia S, Seshasayee A, et al. Predicting new indications for approved drugs using a proteochemometric method. J Med Chem 2012;55(15):6832–48.

[39] Corsello SM, Nagari RT, Spangler RD, Rossen J, et al. Discovering the anticancer potential of non-oncology drugs by systematic viability profiling. Nature Cancer; 2020.

[40] Hopkins AL. Network pharmacology: the next paradigm in drug discovery. Nat Chem Biol 2008;4(11): 682–90.

[41] Reddy AS, Zhang S. Polypharmacology: drug discovery for the future. Expet Rev Clin Pharmacol 2013; 6(1):41–7.

[42] Klaeger S, Heinzlmeir S, Wilhelm M, Polzer H, et al. The target landscape of clinical kinase drugs. Science 2017;358(6367). New York, N.Y.

[43] Mestres J, Gregori-Puigjané E, Valverde S, Solé RV. Data completeness—the Achilles heel of drug-target networks. Nat Biotechnol 2008;26(9):983–4.

[44] Lin A, Giuliano CJ, Palladino A, John KM, et al. Off-target toxicity is a common mechanism of action of cancer drugs undergoing clinical trials. Sci Transl Med 2019;11(509).

[45] Lee BKB, Tiong KH, Chang JK, Liew CS, et al. DeSigN: connecting gene expression with therapeutics for drug repurposing and development. BMC Genomics 2017; 18(Suppl. 1):934.

[46] Corsello SM, Bittker JA, Liu Z, Gould J, et al. The Drug Repurposing Hub: a next-generation drug library and information resource. Nat Med 2017;23(4):405–8.

[47] Wishart DS, Feunang YD, Guo AC, Lo EJ, et al. DrugBank 5.0: a major update to the DrugBank database for 2018. Nucleic Acids Res 2018;46(D1):D1074–82.

[48] Ursu O, Holmes J, Bologa CG, Yang JJ, et al. DrugCentral 2018: an update. Nucleic Acids Res 2019;47(D1): D963–70.

[49] Amelio I, Gostev M, Knight RA, Willis AE, et al. DRUGSURV: a resource for repositioning of approved and experimental drugs in oncology based on patient survival information. Cell Death Dis 2014;5(2):e1051.

[50] Brown AS, Patel CJ. A standard database for drug repositioning. Sci data 2017;4:170029.

[51] Samaras P, Schmidt T, Frejno M, Gessulat S, et al. ProteomicsDB: a multi-omics and multi-organism resource for life science research. Nucleic Acids Res 2020;48(D1): D1153–63.

[52] Andresen V, Gjertsen BT. Drug repurposing for the treatment of acute myeloid leukemia. Front Med November 2017;4:211.

[53] Basso J, Miranda A, Sousa J, Pais A, Vitorino C. Repurposing drugs for glioblastoma: from bench to bedside. Canc Lett 2018;428:173–83.

[54] McCabe B, Liberante F, Mills KI. Repurposing medicinal compounds for blood cancer treatment. Ann Hematol 2015;94(8):1267–76.

[55] Nowak-Sliwinska P, Scapozza L, Ruiz I Altaba A. Drug repurposing in oncology: compounds, pathways, phenotypes and computational approaches for colorectal cancer. Biochim Biophys Acta Rev Cancer 2019; 1871(2):434–54.

[56] Kast RE, Karpel-Massler G, Halatsch M. CUSP9* treatment protocol for recurrent glioblastoma: aprepitant, artesunate, auranofin, captopril, celecoxib, disulfiram, itraconazole, ritonavir, sertraline augmenting continuous low dose temozolomide. Oncotarget 2014;5(18): 8052–82.

[57] Hofer MP, Jakobsson C, Zafiropoulos N, Vamvakas S, et al. Regulatory watch: impact of scientific advice from the European medicines agency. Nat Rev Drug Discov 2015;14(5):302–3.

[58] O'Connor D, McDonald K. Drug repurposing: innovation from the medicine cabinet. Regul Rapporteur January 2020;17:8–10.

[59] Pantziarka P. Scientific advice - is drug repurposing missing a trick? Nat Rev Clin Oncol 2017;14(8):455–6.

[60] Chakravarty D, Gao J, Phillips SM, Kundra R, et al. OncoKB: a precision oncology knowledge base. JCO Precis Oncol 2017;2017(1):1–16.

[61] Jones MR, Schrader KA, Shen Y, Pleasance E, et al. Response to angiotensin blockade with irbesartan in a patient with metastatic colorectal cancer. Ann Oncol 2016;27(5):801–6.

[62] Brown RE, Buryanek J, McGuire MF. Metformin and melatonin in adrenocortical carcinoma: morphoproteomics and biomedical analytics provide proof of concept in a case study. Ann Clin Lab Sci 2017;47(4):457–65.

[63] Verbaanderd C, Meheus L, Huys I, Pantziarka P. Repurposing drugs in oncology: next steps. Trends Cancer 2017;3(8):543–6.

[64] Goločorbin Kon S, Iliković I, Mikov M. Reasons for and frequency of off-label drug use. Med Pregl 2015; 68(1–2):35–40.

[65] Saiyed MM, Ong PS, Chew L. Off-label drug use in oncology: a systematic review of literature. J Clin Pharm Therapeut 2017;42(3):251–8.

[66] Pantziarka P, Verbaanderd C, Huys I, Bouche G, Meheus L. Repurposing drugs in oncology: from candidate selection to clinical adoption. Semin Cancer Biol 2020. https://doi.org/10.1016/j.semcancer.2020.01.008. pii: S1044-579X(20)30011-0.

[67] Verbaanderd C, Rooman I, Meheus L, Huys I. On-label or off-label? Overcoming regulatory and financial barriers to bring repurposed medicines to cancer patients. Front Pharmacol January 2020;10:1–11.

CHAPTER

2

A ligand-centric approach to identify potential drugs for repurposing: case study with aurora kinase inhibitors

Sohini Chakraborti[1], *Pushpaveni Chakravarthi*[2], *Narayanaswamy Srinivasan*[1]

[1]**Molecular Biophysics Unit, Indian Institute of Science, Bangalore, Karnataka, India;** [2]**Department of Pharmaceutical Chemistry, T. John College of Pharmacy, Bangalore, Karnataka, India**

OUTLINE

Introduction

Despite remarkable advancements in understanding the molecular basis of cancer, it remains one of the major causes of death worldwide. However, improved understanding over the decades has led to a paradigm shift in cancer therapy from treatment with cytotoxic chemotherapeutics to more selective targeted therapy focusing on the key molecular players in the

Drug Repurposing in Cancer Therapy
https://doi.org/10.1016/B978-0-12-819668-7.00002-6

cancer cell signaling pathways [1]. Protein kinases, which are the second most targeted group of drug targets after G protein–coupled receptors, play important role in cancer signaling pathways. More than 30 kinase inhibitors are currently available in the market as approved drugs for treatment of various types of cancer [2–4]. Eukaryotic genomes encode protein kinases, such as the Aurora kinases, which promote cell proliferation, survival, and migration. The hallmark of any form of cancer is rapid uncontrolled cell division and loss of apoptosis, which might originate in one part of the body and latter invade to adjoining parts to spread to other organs (metastasis). Therefore, targeting Aurora kinases would be an effective way toward management of many types of cancer as have been seen for the traditional antimitotic agents like taxol and its derivatives, which disrupt the mitotic spindle and thus arrest the cell division process [5–8].

The Aurora kinases (A, B, and C) are highly related serine/threonine protein kinases, which transfer the γ-phosphate group from ATP to serine/threonine residue of a substrate protein and play important role in mitosis. The ATP binding site is highly conserved within the Aurora kinase subfamily and across all Ser/Thr/Tyr kinases. Thus, targeting this site for protein inhibition is challenging as it leads to undesired off-target effects. However, reports are available where single amino acid difference in the ATP-binding pocket sequence has been exploited to achieve Aurora-Aselective inhibitor over its isoforms [9,10]. Aurora-A localizes to the centrosome from centrosome duplication through mitotic exit and largely plays a role in regulating the centrosome and forming mitotic spindle [11]. Dysregulation of Aurora-A has been linked to tumorigenesis, and studies have established it as a bona fide oncogene [12–15]. The active conformation of Aurora-A is induced by autophosphorylation, which is mediated by several cofactors [5,16]. One of the important autophosphorylation-mediating factors of Aurora-A is TPX2 (targeting protein for xenopus kinesin-like protein 2). Interaction between Aurora-A and TPX2 has been shown to involve three druggable hotspots. These three spots have been explored for allosteric inhibition of Aurora-A, which could potentially be helpful in achieving kinase selectivity (Fig. 2.1) [17]. Recent studies have shown that TPX2/Aurora-A signaling is a potential therapeutic target in genomically unstable cancer cells [18]. Aurora-B is a subunit of chromosomal passenger protein and controls accurate chromosome segregation and cytokinesis. The inner centromeric protein and survivin are the major substrates of activated Aurora-B. Experimental reports suggesting the significance of Aurora-B in tumorigenesis are available [19]. Aurora-C plays role in spermatogenesis and is specifically expressed in testis. The role of Aurora-C in oncogenesis was not very clear [16]. However, reports in the current decade hint the implication of aberrant expression of Aurora-C kinase in various forms of cancer like breast cancer, prostate cancer, etc. [20–22]. Aurora-A/B have been extensively exploited as attractive anticancer drug target since their discovery. A handful of inhibitors for these kinases have been reported in the past few years where some of those either specifically target Aurora-A kinase (MLN8054, MLN8237, etc.) [23–30] or Aurora-B kinase (Hesparadin, SU6668, etc.) [31–33], while many others have been found to target both Aurora-A and Aurora-B kinases (such as, AZD1 152) [34–39]. Reports of pan Aurora kinase inhibitors, i.e., Aurora-A, Aurora-B, and Aurora-C inhibitors (danusertib, PF-03824735, etc.) are also available [40–44]. These inhibitors and many other promising Aurora kinase inhibitors have been considered for clinical trial against various forms of cancer that have been elaborately reviewed in a number of articles [5,8,45,46]. However, currently, no approved drugs targeting Aurora kinases are available. This situation can be largely attributed to the most common challenge associated with any kinase drug

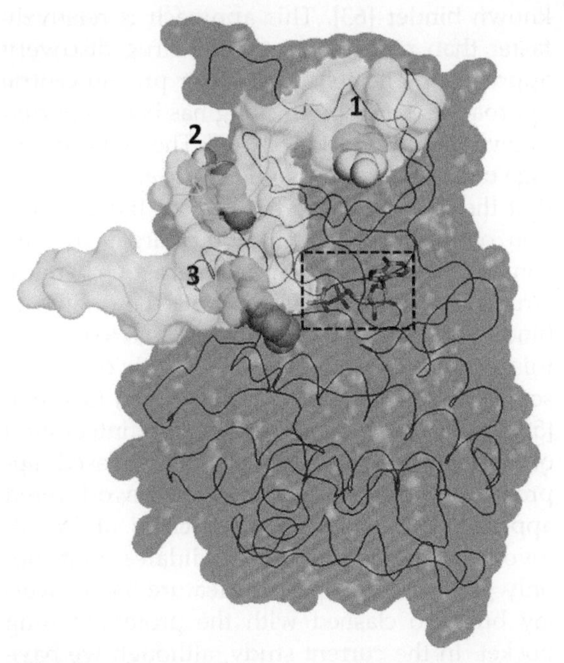

FIGURE 2.1 Structure of Aurora-A kinase complexed to TPX2 1-43 emphasizing the three allosteric druggable hotspots. This image has been generated by overlaying the structures corresponding to PDB entries 5DN3, 5ORN, and 5ORL on to 1OL5. The Aurora-A (pink) and TPX2: 1-43 residues (cyan) complex from 1OL5 has been shown in transparent surface representation, and the protein backbone is shown in ribbon representation. The sites marked as 1, 2, and 3 indicate the three allosteric druggable hotspots as discussed in the text, which correspond to allosteric site-1 (5DN3), allosteric site-2 (5ORN), and allosteric site-3 (5ORL), respectively. The three ligands shown in sphere representation at these three sites are 5DN (5DN3), A5E (5ORN), and A4W (5ORL). The ATP binding site on the other side of the protein is highlighted in black rectangle. The hydrogen atoms of all the ligands and the protein structure corresponding to 5DN3, 5ORN, and 5ORL have not been shown for visual clarity. The image has been generated using PyMOL (The PyMOL Molecular Graphics System, Schrödinger, LLC).

discovery programs, i.e., developing selective kinase inhibitors to minimize off-target mediated toxicity [3,20,47,48]. Efforts to identify more new allosteric inhibitors could be helpful in overcoming the problems of toxicity and achieving selectivity as the allosteric sites are evolutionary less conserved than the orthosteric ATP binding site [49]. Unfortunately, in general, anticancer drugs (irrespective of targets) have a very high attrition rate (~95%) [1] and the lengthy timescale involved in traditional drug discovery process further complicates the situation. Therefore, to meet the unmet medical need, a faster drug discovery process is desirable so that failures are expected to be detected quickly. In this regard, drug repurposing approaches to identify approved molecules that could potentially inhibit Aurora kinases would be helpful. Owing to the established pharmacokinetic and pharmacodynamic profiles of the molecules identified through drug repurposing approaches, the overall complexity in finding a right binding partner for the target of interest is generally lesser as compared to traditional drug discovery process, especially when available state-of-the art *in silico* techniques are rationally integrated with experimental findings [50,51]. Further, the established clinical safety profile of approved drugs reduces the risk of toxicity-related failures.

Broadly, computer-aided drug discovery/repurposing programs that use structural information can be either target-centric or ligand-centric or a combination of both [52]. Target-centric structure-based approaches commonly known as structure-based drug design (SBDD) use the information on three-dimensional (3D) structures of the host molecules, which are most often protein targets. Ligand-centric or ligand-based drug design (LBDD) approaches use the information on 2D/3D structures of the guest molecules, which are known to bind to the target of interest and are often small organic compounds [53]. The most popular SBDD approach involves molecular docking simulations, which focus on sampling the correct conformation of a given ligand in the protein-binding pocket. The predicted poses of each compound in the screening library are then ranked based on a scoring function that aims to calculate the energy of interaction between the binding partners. These scores are one of the determining factors to distinguish between probable weak and strong binders. However, scoring functions are formulated based on many approximations and the conformation sampling can also be inadequate. Hence, the results need to be interpreted cautiously [52, 54, 55]. In the absence of a reliable 3D structure of the host molecules, SBDD approaches cannot be employed. In such situations, LBDD may be the approach of choice. LBDD involves strategies that rely on using molecular fingerprints such as shape, functional groups, and/or other descriptors of known ligand/s as filters to screen databases of chemical compounds for searching new molecules that possess similar fingerprints. However, lack of consideration of target's structural information does not give insights on probable interactions between the binding partners [56,57]. Therefore, integration of LBDD and SBDD approaches are gaining popularity and have been found to perform better in certain instances [58,59]. Besides structure-centric approaches, there are many other computational drug repurposing approaches that are routinely used, such as network-based approaches, pathway mapping, and text mining as reviewed in some of the recent articles [50,60–62].

In this chapter, we have primarily demonstrated the use of a ligand-centric drug discovery/repurposing approach that exploits the information on the shape and stereo-electronic features of the known binders of Aurora kinases to identify new molecules of similar shapes and chemical features. The underlying idea for such an approach lies in the fact that a ligand which is known to bind to a target of interest is complementary to the shape and electrostatics features of the target binding site. Therefore, if another molecule with similar shape and stereo-electronic features as that of the known binder is identified, it is very likely to fit in the binding site of interest in the target protein and establish similar interactions with the protein as that of the known binder [63]. This approach is relatively faster than purely target-based drug discovery approaches, and its success over protein-centric approaches taken by docking has been reported in several instances [56,64–67]. The main advantage of shape-based ligand-centric approaches is that these methods do not require the information on the coordinates of the target and are comparatively faster. Availability of chemical structure of just one active molecule known to bind to the target of interest is enough to employ this technique. Nonetheless, in a study on virtual screening and lead hopping at Wyeth Research [59], it has been demonstrated that integration of target's information with shape-based approaches improve results. The two-layered approach employed by researchers at Wyeth involved elimination of candidates that not only gave good shape- and feature-based overlay but also clashed with the protein-binding pocket. In the current study, although we have primarily applied ligand-centric approach, at a later stage we have used the target's information by integrating molecular docking simulation to our pipeline in order to predict the binding affinity and binding mode of the selected sub-set of drugs against Aurora-A kinase. Our analysis aided in identifying number of potential Aurora kinase inhibitors from the repertoire of approved drugs, which could be considered for further investigation to examine their anticancer properties. Few selected cases have been presented and discussed here since elaborate discussion of all the findings is beyond the scope of this concise chapter.

Methodology

The overall workflow of the protocol that we have adopted in this study to identify potential Food and Drug Administration (FDA)–approved drugs which could be repurposed for treatment of cancer by targeting against Aurora-A and B kinases is summarized in Fig. 2.2.

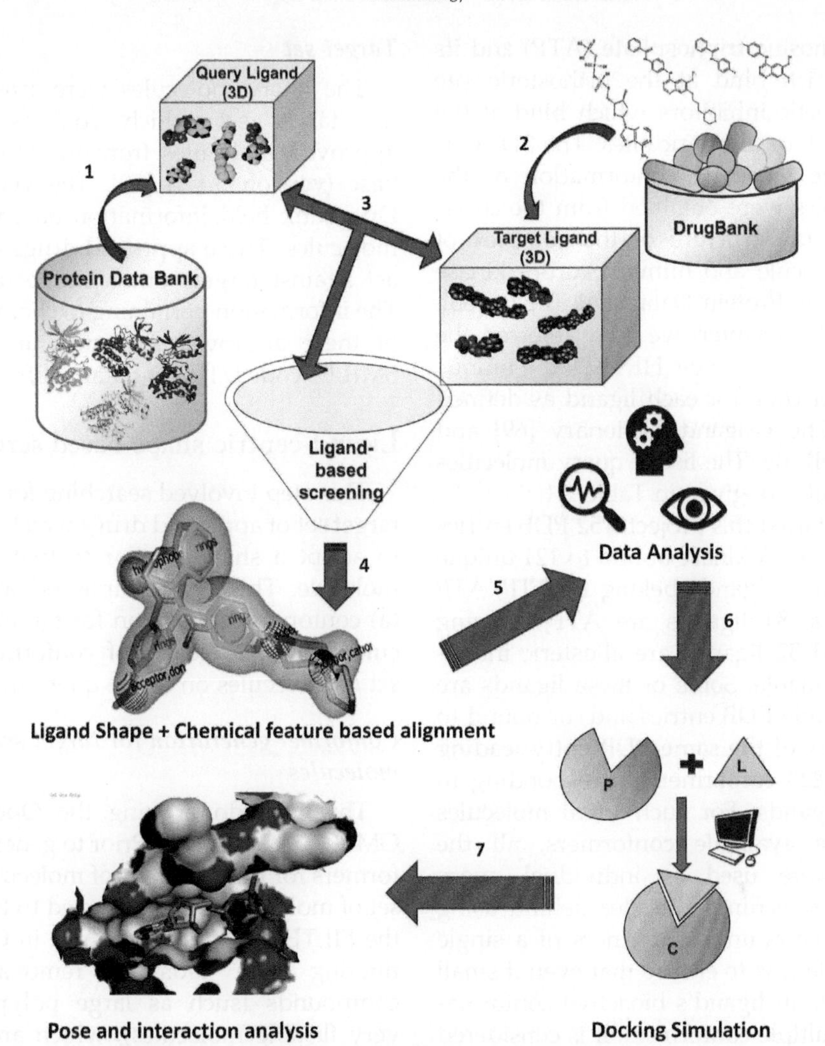

FIGURE 2.2 **Workflow used in the study.** In step 1, the coordinates of the bioactive conformation of the known ligands bound to the protein of interest are retrieved from the PDB. These ligands in their 3D conformation forms the query set. In step 2, the information of FDA-approved drugs is obtained followed by generation of their 3D conformers that forms the target set. The query molecules are then searched in the target database in step 3 using ligand-based screening approach (that involved ligand shape and field/chemical feature based alignment). Step 4 involves data analysis to shortlist high confidence reliable hits. Some of the hits are then subjected to docking studies in step 6 and 7 for in-depth analysis.

Data set

The data set for our study comprises of two parts: (a) query set and (b) target set, where the query set includes known binders of human Aurora kinases for which information on bioactive conformation are also available and the target set includes the search space where the former set of molecules are queried.

Query set

The query small molecules used in our study are known binders of Aurora kinases, for example, adenosine triphosphate (ATP) and its analogues which bind at the orthosteric site and the synthetic inhibitors which bind at the orthosteric and/or allosteric sites. The 3D coordinates of the bioactive conformation of the query molecules were obtained from the corresponding crystal structure of the complex of the query molecule and human Aurora kinase deposited in the Protein Data Bank (PDB) [68]. Throughout this chapter, we have referred the query molecules with their HET ID, the unique three character code for each ligand as defined in the PDBeChem ligand dictionary [69] and concerned PDB file. The list of query molecules and their details are given in Table 2.1.

When we started this project, 152 PDB entries of human Aurora-A kinase bound to 121 unique ligands (where 5 ligands belong to ATP/ATP analogue class, 84 ligands are ATP-replacing inhibitors, and 32 ligands are allosteric inhibitors) were available. Some of these ligands are found in multiple PDB entries and/or bound to multiple chains of the same PDB entry leading to a total of 223 conformers corresponding to 121 unique ligands. For such small molecules with multiple available conformers, all the conformers were used as individual query during shape screening. The idea behind using all the available bound conformers of a single query molecule was to ensure that even if small variation exists in ligand's bioactive conformation across multiple conformers, it is considered during the shape screening step for such variation might influence the shape of the molecule. For human Aurora-B kinase, only one ligand bound PDB entry (4AF3 bound to the inhibitor with HET ID: VX6) was available at the time of conducting this project. Therefore, only the bound conformation of VX6 in PDB entry 4AF3 was used to identify probable new binders of Aurora-B kinase. For Aurora-C kinase, there are no ligand bound experimental structure available. In this study, we have focused only on Aurora-A and Aurora-B kinases.

Target set

The query molecules were screened against the target set which comprised of FDA-approved molecules from the DrugBank database (version 5.1.3) [70]. The version 5.1.3 of DrugBank held information on 2556 approved molecules. These approved drugs are known to act against targets other than Aurora kinases. The information pertaining to chemical structure of these approved molecules in the form of SMILES code [71] was obtained from DrugBank.

Ligand-centric shape-based screening

This step involved searching for hits from the target set of approved drugs which could be able to adopt a shape similar to that of the query molecule. This is accomplished in two stages: (a) conformer generation for target set of molecules and (b) overlay of conformers of target set of molecules on to the query molecules.

Conformer generation for target set of molecules

This was done using the OpenEye tool—OMEGA 3.1.10.3 [72]. Prior to generation of conformers for the target set of molecules, the entire set of molecules were subjected to filtering using the FILTER program available in OMEGA. The filtering step aids in removal of those compounds (such as large polypeptides and very flexible molecules), which are unlikely to be useful for subsequent modeling applications. Therefore, early elimination of such compounds from a data set is beneficial. 3D conformers of the filtered set of compounds were then generated using OMEGA. 50 conformers per input molecules were generated.

Overlay of conformers of target set of molecules on to the query molecules

Each conformer of the target set of molecules generated using OMEGA was superposed on to the bioactive conformation (obtained from

TABLE 2.1 Details of query molecule used for shape screening and their best hit.

Serial Number	Query (LIGAND'S HET ID)	Ligand's Name	PDB ID	Ligand Binding Site	Best Reliable Hit (DRUGBANK ID)	Category of Hit	Tanimoto Combo score
1	ADP	ADENOSINE-5′-DIPHOSPHATE	1MQ4, 1OL5, 1OL7, 2WQE,4BN1, 4CEG, 4DEE, 5G1X, 5L8J, 5L8K, 5L8L, 5LXM, 5ODT, 5ORL, 5ORN, 5ORO, 5ORP, 5ORR, 5ORS, 5ORT, 5ORV, 5ORW, 5ORX, 5ORY, 5ORZ, 5OS0, 5OS1, 5OS2, 5OS3, 5OS4, 5OS5, 5OS6, 5OSD, 5OSE, 5OSF, 6R49, 6R4A, 6R4B, 6R4C, 6R4D	ATP	Inosine pranobex (DB13156)	Antiviral agent	1.251
2	ADN	ADENOSINE	1MUO, 4O0S, 4O0U, 4O0W	ATP	Inosine pranobex (DB13156)	Antiviral agent	1.774
3	ATP	ADENOSINE-5′-TRIPHOSPHATE	1OL6, 5DN3, 5DNR, 5DOS, 5DR2, 5DRD, 5DT3, 5DT4. 5OBJ	ATP	Vidarabine (DB00194)	Antiviral agent	1.088
4	ANP	PHOSPHOAMINOPHOSPHONIC ACID-ADENYLATE ESTER	2C6D, 2DWB	ATP	Inosine pranobex (DB13156)	Antiviral agent	1.214
5	ACP	PHOSPHOMETHYLPHOSPHONIC ACID ADENYLATE ESTER	4C3P, 4C3R, 5G15, 6C83, 6CPF	ATP	Inosine pranobex (DB13156)	Antiviral agent	1.098
6	MPY	(3E)-N-(2,6-DIETHYLPHENYL)-3-{[4-(4-METHYLPIPERAZIN-1-YL) BENZOYL]IMINO}PYRROLO[3,4-C]PYRAZOLE-5(3H)-CARBOXAMIDE	2BMC	ATP	N.A.	N.A.	N.A.
7	HPM	N-{5-[(7-{[(2S)-2-HYDROXY-3-PIPERIDIN-1-YLPROPYL]OXY}-6-METHOXYQUINAZOLIN-4-YL) AMINO]PYRIMIDIN-2-YL}BENZAMIDE	2C6E	ATP	Vandetanib (DB05294)	Antineoplastic agent (tyrosine kinase inhibitor)	1.029
8	626	4-(4-METHYLPIPERAZIN-1-YL)-N-[5-(2-THIENYLACETYL)-1,5-DIHYDROPYRROLO[3,4-C]PYRAZOL-3-YL]BENZAMIDE	2J4Z	ATP	N.A.	N.A.	N.A.

Continued

TABLE 2.1 Details of query molecule used for shape screening and their best hit.—cont'd

Serial Number	Query (LIGAND'S HET ID)	Ligand's Name	PDB ID	Ligand Binding Site	Best Reliable Hit (DRUGBANK ID)	Category of Hit	Tanimoto Combo score
9	627	N-[(3E)-5-[(2R)-2-METHOXY-2-PHENYLACETYL]PYRROLO[3, 4-C]PYRAZOL-3(5H)-YLIDENE]-4-(4-METHYLPIPERAZIN-1-YL)BENZAMIDE	2J50	ATP	Imatinib (DB00619)	Antineoplastic agent (tyrosine kinase inhibitor)	0.997
10	CC3	N-{3-[(4-{[3-(TRIFLUOROMETHYL) PHENYL]AMINO}PYRIMIDIN-2-YL)AMINO]PHENYL} CYCLOPROPANECARBOXAMIDE	2NP8	ATP	N.A.	N.A.	N.A.
11	L0C	4-{[2-(4-{[(4-FLUOROPHENYL) CARBONYL]AMINO}-1H-PYRAZOL-3-YL)-1H-BENZIMIDAZOL-6-YL]METHYL}MORPHOLIN-4-IUM	2WIC	ATP	Dacomitinib (DB11963)	Antineoplastic agent (tyrosine kinase inhibitor)	1.044
12	L0D	2-(1H-PYRAZOL-3-YL)-1H-BENZIMIDAZOLE	2W1D	ATP	Albendazole (DB00518)	Anthelmintic agent	1.317
13	L0E	4-[(2-{4-[(PHENYLCARBAMOYL)AMINO]-1H-PYRAZOL-3-YL}-1H-BENZIMIDAZOL-5-YL)METHYL]MORPHOLIN-4-IUM	2W1E	ATP	N.A.	N.A.	N.A.
14	L0F	N-[3-(1H-BENZIMIDAZOL-2-YL)-1H-PYRAZOL-4-YL]BENZAMIDE	2W1F	ATP	Praziquantel (DB01058)	Anthelmintic agent	1.115
15	L0G	2-{4-[(CYCLOPROPYLCARBAMOYL) AMINO]-1H-PYRAZOL-3-YL}-6-(MORPHOLIN-4-IUM-4-YLMETHYL)-1H-3,1-BENZIMIDAZOL-3-IUM	2W1G	ATP	N.A.	N.A.	N.A.
16	ZZL	4-{[9-CHLORO-7-(2,6-DIFLUOROPHENYL)-5H-PYRIMIDO[5,4-D][2]BENZAZEPIN-2-YL]AMINO}BENZOIC ACID	2WTV, 2WTW, 2X81	ATP	Midazolam (DB00683)	Hypnotic-sedative agent	1.057
17	X6D	6-BROMO-7-[4-(4-CHLOROBENZYL) PIPERAZIN-1-YL]-2-[4-(MORPHOLIN-4-YLMETHYL)PHENYL]-3H-IMIDAZO[4,5-B]PYRIDINE	2X6D	ATP	N.A.	N.A.	N.A.

18	YM4	6-BROMO-7-{4-[(5-METHYLISOXAZOL-3-YL)METHYL]PIPERAZIN-1-YL}-2-[4-(4-METHYLPIPERAZIN-1-YL)PHENYL]-1H-IMIDAZO[4,5-B]PYRIDINE	2X6E	ATP	N.A.	N.A.	N.A.
19	ASH	3-CHLORO-N-(4-MORPHOLIN-4-YLPHENYL)-6-PYRIDIN-3-YLIMIDAZO[1,2-A]PYRAZIN-8-AMINE	2XNE	ATP	Betrixaban (DB12364)	Anticoagulant agent	1.020
20	A0H	N-(3-{3-CHLORO-8-[(4-MORPHOLIN-4-YLPHENYL)AMINO]IMIDAZO[1,2-A]PYRAZIN-6-YL}BENZYL) METHANESULFONAMIDE	2XNG	ATP	N.A.	N.A.	N.A.
21	400	3-({[4-(4-METHYLPIPERAZIN-1-YL)PHENYL]CARBONYL}AMINO)-N-[(1R)-1-PHENYLPROPYL]-1H-THIENO [3,2-C]PYRAZOLE-5-CARBOXAMIDE	2XRU	ATP	N.A.	N.A.	N.A.
22	83H	8-ETHYL-3,10,10-TRIMETHYL-4,5,6,8,10,12-HEXAHYDROPYRAZOLO[4′,3′:6,7] CYCLOHEPTA[1,2-B]PYRROLO[2,3-F]INDOL-9(1H)-ONE	3COH	ATP	Pregnenolone (DB02789)	Hormone	1.144
23	VX6	CYCLOPROPANECARBOXYLIC ACID {4-[4-(4-METHYL-PIPERAZIN-1-YL)-6-(5-METHYL-2H-PYRAZOL-3-YLAMINO)-PYRIMIDIN-2-YLSULFANYL]-PHENYL}-AMIDE	3E5A, 4JBQ	ATP	N.A.	N.A.	N.A.
24	AK8	1-[3-METHYL-4-({3-[2-(METHYLAMINO) PYRIMIDIN-4-YL]PYRIDIN-2-YL}OXY)PHENYL]-3-[3-(TRIFLUOROMETHYL)PHENYL]UREA	3EFW	ATP	Sorafenib (DB00398)	Antineoplastic agent (tyrosine kinase inhibitor)	1.078
25	MMH	N-[3-(ACETYLAMINO)PHENYL]-5-{(2E)-2-[(4-METHOXYPHENYL) METHYLIDENE]HYDRAZINO}-3-METHYL-1H-PYRAZOLE-4-CARBOXAMIDE	3FDN	ATP	N.A.	N.A.	N.A.
26	48B	2-CHLORO-N-[4-({5-FLUORO-2-[(4-HYDROXYPHENYL) AMINO]PYRIMIDIN-4-YL}AMINO)PHENYL]BENZAMIDE	3H0Y	ATP	Rilpivirine (DB08864)	Anti-HIV agent	1.016
27	45B	4-{[2-({4-[2-(4-ACETYLPIPERAZIN-1-YL)-2-OXOETHYL]PHENYL}AMINO)-5-FLUOROPYRIMIDIN-4-YL]AMINO}-N-(2-CHLOROPHENYL)BENZAMIDE	3H0Z	ATP	N.A.	N.A.	N.A.

Continued

TABLE 2.1 Details of query molecule used for shape screening and their best hit.—cont'd

Serial Number	Query (LIGAND'S HET ID)	Ligand's Name	PDB ID	Ligand Binding Site	Best Reliable Hit (DRUGBANK ID)	Category of Hit	Tanimoto Combo score
28	97B	9-CHLORO-7-(2,6-DIFLUOROPHENYL)-N-{4-[(4-METHYLPIPERAZIN-1-YL) CARBONYL]PHENYL}-5H-PYRIMIDO[5,4-D][2] BENZAZEPIN-2-AMINE	3H10	ATP	N.A.	N.A.	N.A.
29	2JZ	N~2~-(3,4-DIMETHOXYPHENYL)-N~4~-[2-(2-FLUOROPHENYL)ETHYL]-N~6~-QUINOLIN-6-YL-1,3,5-TRIAZINE-2,4,6-TRIAMINE	3HA6	ATP	N.A.	N.A.	N.A.
30	PFQ	2-[(5,6-DIPHENYLFURO[2,3-D] PYRIMIDIN-4-YL)AMINO]ETHANOL	3K5U	ATP	Zolpidem (DB00425)	Hypnotic-sedative agent	1.096
31	OFI	N-[6-(4-HYDROXYPHENYL)-1H-INDAZOL-3-YL]BUTANAMIDE	3LAU	ATP	Estradiol valerate (DB13956)	Hormone	1.295
32	AKI	1-(4-{2-[(5,6-DIPHENYLFURO[2,3-D]PYRIMIDIN-4-YL)AMINO]ETHYL}PHENYL)-3-PHENYLUREA	3M11	ATP	N.A.	N.A.	N.A.
33	EML	2-{ETHYL[(5-{[6-METHYL-3-(1H-PYRAZOL-4-YL)IMIDAZO[1,2-A]PYRAZIN-8-YL]AMINO}ISOTHIAZOL-3-YL) METHYL]AMINO}-2-METHYLPROPAN-1-OL	3MYG	ATP	N.A.	N.A.	N.A.
34	NRM	N-(3-METHYLISOTHIAZOL-5-YL)-3-(1H-PYRAZOL-4-YL)IMIDAZO[1,2-A]PYRAZIN-8-AMINE	3NRM	ATP	Abacavir (DB01048)	Anti-HIV agent	1.088
35	LJE	N-{3-METHYL-4-[(3-PYRIMIDIN-4-YLPYRIDIN-2-YL)OXY]PHENYL}-3-(TRIFLUOROMETHYL)BENZAMIDE	3O50	ATP	Tolvaptan (DB06212)	Natriuretic agent	1.006
36	LJF	N-[4-({3-[5-FLUORO-2-(METHYLIDENEAMINO)PYRIMIDIN-4-YL]PYRIDIN-2-YL}OXY)PHENYL]-2-(PHENYLAMINO)BENZAMIDE	3O51	ATP	N.A.	N.A.	N.A.

37	P9J	4-[(5-METHYL-1H-PYRAZOL-3-YL)AMINO]-2-PHENYLPHTHALAZIN-1(2H)-ONE	3P9J, 3W16	ATP	Azelastine (DB00972)	Antiallergic agent	1.201
38	E9Z	5-CHLORO-N~4~-CYCLOPROPYL-N~2~-[4-(2-METHOXYETHOXY)PHENYL]PYRIMIDINE-2,4-DIAMINE	3QBN	ATP	Rilpivirine (DB08864)	Anti-HIV agent	1.200
39	D36	N-(2-AMINOETHYL)-N-{5-[(1-CYCLOHEPTYL-1H-PYRAZOLO[3,4-D]PYRIMIDIN-6-YL)AMINO]PYRIDIN-2-YL}METHANESULFONAMIDE	3R21	ATP	Palbociclib (DB09073)	Antineoplastic agent (kinase inhibitor)	1.065
40	D37	N-{5-[(1-CYCLOHEPTYL-1H-PYRAZOLO[3,4-D]PYRIMIDIN-6-YL)AMINO]PYRIDIN-2-YL}METHANESULFONAMIDE	3R22	ATP	Boscalid (DB12792)	Fungicide	1.159
41	0BZ	4-({4-[(2-FLUOROPHENYL)AMINO]PYRIMIDIN-2-YL}AMINO)BENZOIC ACID	3UNZ	ATP	Rilpivirine (DB08864)	Anti-HIV agent	1.282
42	0C0	4-{[4-(BIPHENYL-2-YLAMINO)PYRIMIDIN-2-YL]AMINO}BENZOIC ACID	3UO4	ATP	Rilpivirine (DB0864)	Anti-HIV agent	1.219
43	0BX	4-{[4-(PHENYLAMINO)PYRIMIDIN-2-YL]AMINO}BENZOIC ACID	3UO5	ATP	Paroxetine (DB00715)	Antidepressive agent	1.079
44	0BY	4-({4-[(2-CHLOROPHENYL)AMINO]PYRIMIDIN-2-YL}AMINO)BENZOIC ACID	3UO6	ATP	Rilpivirine (DB08864)	Anti-HIV agent	1.301
45	0C3	4-[(4-{[2-(TRIFLUOROMETHYL)PHENYL]AMINO}PYRIMIDIN-2-YL)AMINO]BENZOIC ACID	3UOD	ATP	Repaglinide (DB00912)	Antihyperglycemic agent	1.045
46	0C4	4-({4-[(2-BROMOPHENYL)AMINO]PYRIMIDIN-2-YL}AMINO)BENZOIC ACID	3UOH	ATP	Rilpivirine (DB08864)	Anti-HIV agent	1.250
47	0C5	4-({4-[(2-CYANOPHENYL)AMINO]PYRIMIDIN-2-YL}AMINO)BENZOIC ACID	3UOJ	ATP	Rilpivirine (DB08864)	Anti-HIV agent	1.213

Continued

TABLE 2.1 Details of query molecule used for shape screening and their best hit.—cont'd

Serial Number	Query (LIGAND'S HET ID)	Ligand's Name	PDB ID	Ligand Binding Site	Best Reliable Hit (DRUGBANK ID)	Category of Hit	Tanimoto Combo score
48	0C6	4-({4-[(2-CHLOROPHENYL)AMINO]-5-FLUOROPYRIMIDIN-2-YL}AMINO) BENZOIC ACID	3UOK	ATP	Rilpivirine (DB08864)	Anti-HIV agent	1.243
49	0C7	N~4~-(2-CHLOROPHENYL)-N~2~-[4-(1H-TETRAZOL-5-YL) PHENYL]PYRIMIDINE-2,4-DIAMINE	3UOL	ATP	Etravirine (DB06414)	Antiretroviral agent	1.030
50	0C8	4-[(4-{[2-(TRIFLUOROMETHOXY) PHENYL]AMINO}PYRIMIDIN-2-YL) AMINO]BENZOIC ACID	3UP2	ATP	N.A.	N.A.	N.A.
51	0C9	2-({2-[(4-CARBOXYPHENYL) AMINO]PYRIMIDIN-4-YL}AMINO) BENZOIC ACID	3UP7	ATP	Flufenamic acid (DB02266)	Antiinflammatory agent	1.106
52	0FY	3-(1-{2-[(3-FLUOROPYRIDINIUM-4-YL) AMINO]-2-OXOETHYL}-1H-PYRAZOL-4-YL)-6-METHYL-8-[(3-{[(1R,3R)-3-METHYLPIPERIDINIUM-1-YL]METHYL}-1,2-THIAZOL-5-YL) AMINO]IMIDAZO[1,2-A]PYRAZIN-1-IUM	3VAP	ATP	N.A.	N.A.	N.A.
53	RO9	1-(3-METHOXYPHENYL)-N-(5-METHYL-1H-PYRAZOL-3-YL)ISOQUINOLIN-3-AMINE	3W10	ATP	Orphenadrine (DB01173)	Muscle relaxant	1.163
54	N13	2-{3-[3-(1H-BENZIMIDAZOL-2-YL)-1H-INDAZOL-6-YL]-1H-PYRAZOL-5-YL}-N-(3-FLUOROPHENYL)ACETAMIDE	3W18	ATP	N.A.	N.A.	N.A.
55	N15	2-{4-[3-(1H-BENZIMIDAZOL-2-YL)-1H-INDAZOL-6-YL]-1H-PYRAZOL-1-YL}-N-(3-METHYLBUTYL) ACETAMIDE	3W2C	ATP	Pretomanid (DB05154)	Antitubercular agent	1.011
56	VEK	6-BROMO-2-(1-METHYL-1H-IMIDAZOLE-5-YL)-7-{4-[(5-METHYL-1,2-OXAZOL-3-YL)METHYL]PIPERAZIN-1-YL}-1H-IMIDAZO[4,5-B]PYRIDINE	4B0G	ATP	N.A.	N.A.	N.A.

57	FH3	(S)-N-((1-(6-CHLORO-2-(1,3-DIMETHYL-1H-PYRAZOL-4-YL)-3H-IMIDAZO[4,5-B]PYRIDIN-7-YL)PYRROLIDIN-3-YL) METHYL)ACETAMIDE	4BYI	ATP	Dexlansoprazole (DB05351)	Antiulcer agent (proton pump inhibitor)	1.064
58	FH5	(S)-N-(1-(6-CHLORO-2-(1,3-DIMETHYL-1H-PYRAZOL-4-YL)-3H-IMIDAZO[4,5-B]PYRIDIN-7-YL)PYRROLIDIN-3-YL)ACETAMIDE	4BYJ	ATP	Dexlansoprazole (DB05351)	Antiulcer agent (proton pump inhibitor)	1.105
59	NHI	4,4′-(PYRIMIDINE-2,4-DIYLDIIMINO) DIBENZOIC ACID	4DEA	ATP	Rilpivirine (DB08864)	Anti-HIV agent	1.123
60	NHJ	4-[(4-{[3-(TRIFLUOROMETHYL) PHENYL]AMINO}PYRIMIDIN-2-YL) AMINO]BENZAMIDE	4DEB	ATP	Rilpivirine (DB08864)	Anti-HIV agent	1.023
61	NHU	2-({2-[(4-CARBAMOYLPHENYL) AMINO]PYRIMIDIN-4-YL}AMINO) BENZAMIDE	4DED	ATP	Rilpivirine (DB08864)	Anti-HIV agent	1.296
62	0K6	7-CYCLOPENTYL-2-({1-METHYL-5-[(4-METHYLPIPERAZIN-1-YL) CARBONYL]-1H-PYRROL-3-YL}AMINO)-7H-PYRROLO[2,3-D] PYRIMIDINE-6-CARBOXAMIDE	4DHF	ATP	Ribociclib (DB11730)	Antineoplastic agent (CDK4/6 inhibitor)	1.144
63	CJ5	1-[4-[[4-[(5-CYCLOPENTYL-1 H-PYRAZOL-3-YL)AMINO]PYRIMIDIN-2-YL]AMINO]PHENYL]-3-[3-(TRIFLUOROMETHYL) PHENYL]UREA	4J8M	ATP	Sorafenib (DB00398)	Antineoplastic agent (tyrosine kinase inhibitor)	1.022
64	XU2	N-{4-[(6-OXO-5,6-DIHYDROBENZO [C][1,8]NAPHTHYRIDIN-1-YL)AMINO] PHENYL}BENZAMIDE	4JAI	ATP	Tolvaptan (DB06212)	Natriuretic agent	1.074
65	XU1	BENZO[C][1,8]NAPHTHYRIDIN-6(5H)-ONE	4JAJ	ATP	Dihydralazine (DB12945)	Antihypertensive agent	1.352
66	WPH	1-(4-{2-[(6-{4-[2-(DIMETHYLAMINO) ETHOXY]PHENYL}FURO[2,3-D]PYRIMIDIN-4-YL)AMINO]ETHYL} PHENYL)-3-PHENYLUREA	4JBO	ATP	N.A.	N.A.	N.A.
67	YPH	1-(4-{2-[(6-{4-[2-(4-HYDROXYPIPERIDIN-1-YL)ETHOXY]PHENYL} FURO[2,3-D]PYRIMIDIN-4-YL)AMINO] ETHYL}PHENYL)-3-PHENYLUREA	4JBP	ATP	N.A.	N.A.	N.A.

Continued

TABLE 2.1 Details of query molecule used for shape screening and their best hit.—cont'd

Serial Number	Query (LIGAND'S HET ID)	Ligand's Name	PDB ID	Ligand Binding Site	Best Reliable Hit (DRUGBANK ID)	Category of Hit	Tanimoto Combo score
68	2VU	N-[1-(3-CYANOBENZYL)-1H-PYRAZOL-4-YL]-6-(1H-PYRAZOL-4-YL)-1H-INDAZOLE-3-CARBOXAMIDE	4PRJ	ATP	Dapiprazole (DB00298)	Adrenergic antagonist (Neurotoxic agent)	1.103
69	Y3M	ETHYL (9S)-9-[5-(1H-BENZIMIDAZOL-2-YLSULFANYL)FURAN-2-YL]-8-HYDROXY-5,6,7,9-TETRAHYDRO-2H-PYRROLO[3,4-B]QUINOLINE-3-CARBOXYLATE	4UYN	ATP	N.A.	N.A.	N.A.
70	QMN	ETHYL (9S)-9-[3-(1H-BENZIMIDAZOL-2-YLOXY)PHENYL]-8-OXO-4,5,6,7,8,9-HEXAHYDRO-2H-PYRROLO[3,4-B] QUINOLINE-3-CARBOXYLATE	4UZD	ATP	N.A.	N.A.	N.A.
71	JVE	(4S)-4-(2-FLUOROPHENYL)-2,4,6,7,8,9-HEXAHYDRO-5H-PYRAZOLO[3,4-B][1,7]NAPHTHYRIDIN-5-ONE	4UZH	ATP	Pizotifen (DB06153)	Analgesic (antimigraine preparations)	1.213
72	4QV	5-HYDROXY-1′H-1,2′-BIBENZIMIDAZOL-2(3H)-ONE	4ZS0	ATP	Benzylparaben (DB14176)	Allergenic testing	1.266
73	4RM	(2Z,5Z)-2-[(4-ETHYLPHENYL) IMINO]-3-(2-METHOXYETHYL)-5-(PYRIDIN-4-YLMETHYLIDENE)-1,3-THIAZOLIDIN-4-ONE	4ZTQ	ATP	Praziquantel (DB01058)	Anthelmintic agent	1.015
74	4RJ	6-({4-[(Z)-{(2Z)-2-[(4-ETHYLPHENYL) IMINO]-3-METHYL-4-OXO-1,3-THIAZOLIDIN-5-YLIDENE}METHYL] PYRIDIN-2-YL}AMINO) PYRIDINE-3-CARBOXYLIC ACID	4ZTR	ATP	N.A.	N.A.	N.A.
75	4RK	(2Z,5Z)-2-[(4-ETHYLPHENYL)IMINO]-3-METHYL-5-[(2-{[4-(1H-TETRAZOL-5-YL)PHENYL]AMINO}PYRIDIN-4-YL) METHYLIDENE]-1,3-THIAZOLIDIN-4-ONE	4ZTS	ATP	N.A.	N.A.	N.A.
76	5GX	7-(1-BENZYL-1H-PYRAZOL-4-YL)-6-CHLORO-2-(1,3-DIMETHYL-1H-PYRAZOL-4-YL)-3H-IMIDAZO[4,5-B]PYRIDINE	5AAD	ATP	Ruxolitinib (DB08877)	Antineoplastic agent (tyrosine kinase inhibitor)	1.060

77	7HD	3-((4-(6-CHLORO-2-(1,3-DIMETHYL-1H-PYRAZOL-4-YL)-3H-IMIDAZO[4,5-B]PYRIDIN-7-YL)-1H-PYRAZOL-1-YL)METHYL)-5-METHYLISOXAZOLE	5AAE	ATP	Ruxolitinib (DB08877)	Antineoplastic agent (tyrosine kinase inhibitor)	1.135
78	NL4	3-((4-(6-CHLORO-2-(1,3-DIMETHYL-1H-PYRAZOL-4-YL)-3H-IMIDAZO[4,5-B]PYRIDIN-7-YL)-1H-PYRAZOL-1-YL)METHYL)-N,N-DIMETHYLBENZAMIDE	5AAF	ATP	N.A.	N.A.	N.A.
79	6F2	[3-[[4-[6-CHLORANYL-2-(1,3-DIMETHYLPYRAZOL-4-YL)-3H-IMIDAZO[4,5-B]PYRIDIN-7-YL]PYRAZOL-1-YL]METHYL]PHENYL]-(4-METHYLPIPERAZIN-1-YL)METHANONE	5AAG	ATP	N.A.	N.A.	N.A.
80	SKE	4-({5-AMINO-1-[(2,6-DIFLUOROPHENYL)CARBONYL]-1H-1,2,4-TRIAZOL-3-YL}AMINO)BENZENESULFONAMIDE	5DPV, 5DR6, 5DR9, 5DT0, 5OBR	ATP	Naratriptan (DB00952)	Analgesic (antimigraine preparations)	1.048
81	5VC	4-(3-CHLORANYL-2-FLUORANYL-PHENOXY)-1-[[6-(1,3-THIAZOL-2-YLAMINO)PYRIDIN-2-YL]METHYL]CYCLOHEXANE-1-CARBOXYLIC ACID	5EW9	ATP	N.A.	N.A.	N.A.
82	9YQ	4-(PROPANOYLAMINO)-~{N}-[4-[(5,8,11-TRIMETHYL-6-OXIDANYLIDENE-PYRIMIDO[4,5-B][1,4]BENZODIAZEPIN-2-YL)AMINO]PHENYL]BENZAMIDE	5ONE	ATP	N.A.	N.A.	N.A.
83	9A6	7-(4-METHYLPIPERAZIN-1-YL)-N-(5-METHYL-1H-PYRAZOL-3-YL)-2-[(E)-2-PHENYLETHENYL]QUINAZOLIN-4-AMINE	5ZAN	ATP	N.A.	N.A.	N.A.
84	EG7	(2R,4R)-1-[(3-CHLORO-2-FLUOROPHENYL)METHYL]-4-({3-FLUORO-6-[(5-METHYL-1H-PYRAZOL-3-YL)AMINO]PYRIDIN-2-YL}METHYL)-2-METHYLPIPERIDINE-4-CARBOXYLIC ACID	6C2R	ATP	N.A.	N.A.	N.A.

Continued

TABLE 2.1 Details of query molecule used for shape screening and their best hit.—cont'd

Serial Number	Query (LIGAND'S HET ID)	Ligand's Name	PDB ID	Ligand Binding Site	Best Reliable Hit (DRUGBANK ID)	Category of Hit	Tanimoto Combo score
85	EGJ	(2S,4R)-1-[(3-CHLORO-2-FLUOROPHENYL)METHYL]-2-METHYL-4-({3-[(1,3-THIAZOL-2-YL)AMINO]ISOQUINOLIN-1-YL}METHYL)PIPERIDINE-4-CARBOXYLIC ACID	6C2T	ATP	N.A.	N.A.	N.A.
86	35R	1-CYCLOPROPYL-3-{3-[5-(MORPHOLIN-4-YLMETHYL)-1H-BENZIMIDAZOL-2-YL]-1H-PYRAZOL-4-YL}UREA	6CPG	ATP	N.A.	N.A.	N.A.
87	F8Z	1-[(2~{R},3~{S})-2-[[1,3-BENZODIOXOL-5-YLMETHYL(METHYL)AMINO]METHYL]-3-METHYL-6-OXIDANYLIDENE-5-[(2~{S})-1-OXIDANYLPROPAN-2-YL]-3,4-DIHYDRO-2~{H}-1,5-BENZOXAZOCIN-8-YL]-3-(4-METHOXYPHENYL)UREA	6GRA	ATP	N.A.	N.A.	N.A.
88	G7T	~{N}-[4-(4-AZANYL-1-PROPAN-2-YL-PYRAZOLO[3,4-D]PYRIMIDIN-3-YL)-3-METHYL-PHENYL]-4-[4-FLUORANYL-3-(TRIFLUOROMETHYL)PHENYL]-4-OXIDANYLIDENE-BUTANAMIDE	6HJJ	ATP	N.A.	N.A.	N.A.
89	G7W	(~{E})-~{N}-[4-(4-AZANYL-1-PROPAN-2-YL-PYRAZOLO[3,4-D]PYRIMIDIN-3-YL)PHENYL]-4-[4-FLUORANYL-3-(TRIFLUOROMETHYL)PHENYL]-4-OXIDANYLIDENE-BUT-2-ENAMIDE	6HJK	ATP	N.A.	N.A.	N.A.
90	5DN	2-(3-BROMOPHENYL)-8-FLUOROQUINOLINE-4-CARBOXYLIC ACID	5DN3, 5DOS,5DPV, 5DT4,	Allosteric (1)	Diflunisal (DB00861)	Antiinflammatory agent	1.204
91	5E1	2-(3-BROMOPHENYL)QUINOLINE-4-CARBOXYLIC ACID	5DR2, 5DR6	Allosteric (1)	Lumiracoxib (DB01283)	Antiinflammatory agent	1.419

92	5E2	2-(3-BROMOPHENYL)-6-CHLOROQUINOLINE-4-CARBOXYLIC ACID	5DR9	Allosteric (1)	Lumiracoxib (DB01283)	Antiinflammatory agent	1.147
93	9QK	2-(3-FLUOROPHENYL) QUINOLINE-4-CARBOXYLIC ACID	5OBJ	Allosteric (1)	Diflunisal (DB00861)	Antiinflammatory agent	1.332
94	9QT	2-(3-CHLORANYL-5-FLUORANYL-PHENYL)QUINOLINE-4-CARBOXYLIC ACID	5OBR	Allosteric (1)	Diflunisal (DB00861)	Antiinflammatory agent	1.209
95	A5Q	4-[4-(TRIFLUOROMETHYL) PHENYL]-1,2,3-THIADIAZOL-5-AMINE	5ORR	Allosteric (1)	Edaravone (DB12243)	Free radical scavenger and neuroprotective agent	1.396
96	A5W	CYCLOBUTYL-[4-(2-METHOXYPHENYL)PIPERIDIN-1-YL]METHANONE	5ORS	Allosteric (1)	Benzoyl peroxide (DB09096)	Antiacne preparations	1.168
97	A5Z	[3-[2,6-BIS(CHLORANYL)PHENYL]-5-METHYL-1,2-OXAZOL-4-YL]METHANOL	5ORT	Allosteric (1)	Glycol salicylate (DB11323)	Antiinflammatory agent	1.351
98	A65	[3,5-BIS(METHYLSULFANYL)-1,2-THIAZOL-4-YL]METHANOL	5ORV	Allosteric (1)	Ethionamide (DB00609)	Antitubercular agent	1.393
99	A6E	3-(4-FLUORANYLPHENOXY)-1-THIOMORPHOLIN-4-YL-PROPAN-1-ONE	5ORW	Allosteric (1)	Pirfenidone (DB04951)	Antineoplastic /antiinflammatory agent	1.251
100	A6H	6-[2,6-BIS(CHLORANYL)PHENOXY] PYRIDIN-3-AMINE	5ORX	Allosteric (1)	Monobenzone (DB00600)	Depigmenting agent	1.482
101	AY4	2,4-BIS(FLUORANYL)-6-(1~{H}-PYRAZOL-3-YL)PHENOL	5ORY	Allosteric (1)	Pirfenidone (DB04951)	Antineoplastic/ antiinflammatory agent	1.511
102	A6W	METHYL 3-AZANYL-5-THIOPHEN-2-YL-THIOPHENE-2-CARBOXYLATE	5ORZ	Allosteric (1)	Methyl salicylate (DB09543)	Antirheumatic agent	1.473
103	A6Z	2-[4-(3-CHLOROPHENYL) PIPERAZIN-1-IUM-1-YL] ETHANENITRILE	5OS0	Allosteric (1)	Pirfenidone (DB04951)	Antineoplastic/ antiinflammatory agent	1.214
104	A7H	6-ETHOXY-2-METHYL-1,3-BENZOTHIAZOLE	5OS1	Allosteric (1)	Riluzole (DB00740)	Anticonvulsant	1.595

Continued

TABLE 2.1 Details of query molecule used for shape screening and their best hit.—cont'd

Serial Number	Query (LIGAND'S HET ID)	Ligand's Name	PDB ID	Ligand Binding Site	Best Reliable Hit (DRUGBANK ID)	Category of Hit	Tanimoto Combo score
105	A7K	[2-[4-(HYDROXYMETHYL) PIPERIDIN-1-YL]PHENYL] METHYLAZANIUM	5OS2	Allosteric (1)	Solriamfetol (DB14754)	Antidepressive agent	1.241
106	A7N	(1~{R})-1-(4-ETHOXYPHENYL) ETHANAMINE	5OS3	Allosteric (1)	Hydroxyamphetamine (DB09352)	Sympathomimetic agent (mydriatic)	1.482
107	A8H	(3~{A}~{R},5~{S},7~{A}~{S})-5-PHENYL-3~{A},4,5,6,7,7~{A} -HEXAHYDROISOINDOLE-1,3-DIONE	5OS4	Allosteric (1)	Metaxalone (DB00660)	Muscle relaxant	1.388
108	A8K	4-(4-HYDROXYPHENYL) SULFANYLPHENOL	5OS5	Allosteric (1)	Proflavine (DB01123)	Antiinfective agent	1.521
109	A7Q	(6-PHENOXYPYRIDIN-3-YL) METHANOL	5OS6	Allosteric (1)	Rufinamide (DB06201)	Anticonvulsant	1.276
110	JSB	(1~{S},10~{S})-12-CYCLOPROPYL-1-OXIDANYL-10-PROPAN-2-YL-9,12-DIAZATRICYCLO [8.2.1.0{2,7}]TRIDECA-2(7),3,5-TRIEN-11-ONE	6R49	Allosteric (1)	Meperidine (DB00454)	Antidepressive agent	1.075
111	JRT	2-(BENZIMIDAZOL-1-YL)-~{N}-(2-PHENYLETHYL) ETHANAMIDE	6R4A	Allosteric (1)	Azathioprine (DB00993)	Antineoplastic/ Antirheumatic agent	1.087
112	JSN	(6~{S})-6-[2,4-BIS (FLUORANYL) PHENYL]-~{N},~{N},4-TRIMETHYL-2-OXIDANYLIDENE-5,6-DIHYDRO-1~{H}-PYRIMIDINE-5-CARBOXAMIDE	6R4B	Allosteric (1)	Tipiracil (DB09343)	Metastatic colorectal cancer	1.290
113	JRQ	ETHYL 2-[(2~{R})-1-[(4-METHYLPHENYL) METHYL]-3-OXIDANYLIDENE-PIPERAZIN-2-YL]ETHANOATE	6R4C	Allosteric (1)	Diclofenac (DB00586)	Antiinflammatory/ antirheumatic/ antimigraine	1.087

114	JRW	(1 ~ {S},10 ~ {S})-12-CYCLOBUTYL-5-METHYL-1-OXIDANYL-10-PROPAN-2-YL-9,12-DIAZATRICYCLO [8.2.1.0{2,7}]TRIDECA-2(7),3,5-TRIEN-11-ONE	6R4D	Allosteric (1)	Venlafaxine (DB00285)	Antidepressive agent	1.117
115	A5E	3-THIOPHEN-2-YL-4,5-DIHYDRO-1 ~ {H}-PYRIDAZIN-6-ONE	5ORN	Allosteric (2)	Tegafur (DB09256)	Antineoplastic agent	1.471
116	A5H	3-(4-CHLOROPHENYL)-5, 6-DIHYDROIMIDAZO[2,1-B] [1,3]THIAZOLE	5ORO	Allosteric (2)	Anethole trithione (DB13853)	Gastrointestinal agent	1.295
117	A5K	1-[3-CHLORANYL-5-(TRIFLUOROMETHYL) PYRIDIN-2-YL]-1,4-DIAZEPANE	5ORP	Allosteric (2)	Betahistine (DB06698)	Antivertigo drug	1.371
118	A9B	5-(4-CHLOROPHENYL) FURAN-2-CARBOHYDRAZIDE	5OSD	Allosteric (2)	Edaravone (DB12243)	Free radical scavenger and neuroprotective agent	1.290
119	A98	METHYL ~ {N}-(5-ETHYLSULFANYL-1,3,4-THIADIAZOL-2-YL)CARBAMATE	5OSE	Allosteric (2)	Acetazolamide (DB00819)	Anticonvulsant	1.409
120	A9E	2-(4-ETHYLPHENOXY)-1-PIPERIDIN-1-YL-ETHANONE	5OSF	Allosteric (2)	Avobenzone (DB09495)	Sunscreen agent	1.315
121	A4W	~ {N}-(3-CHLORANYL-2-FLUORANYL-PHENYL)-3-SULFANYL-PROPANAMIDE	5ORL	Allosteric (3)	Pyridostigmine (DB00545)	Parasympathomimetic agent	1.533

P.S.: N.A. ("Not Applicable") indicates no reliable hits have been obtained based on the qualifying criteria used in the study.

crystal structure) of each of the query set molecule using the OpenEye tool—ROCS 3.3.0.3 [56]. ROCS is a shape-based superposition method which aligns 3D molecules by a solid-body optimization process by maximizing volume overlap between the query and the target molecule. This helps in identifying those compounds which are similar to the query molecule both in shape and chemistry. 500 hits per query molecule were identified from the target data set comprising approved drugs.

Data analysis

This step involves analysis of the output files from ROCS runs and selection of the most reliable hits for each query molecule. A cutoff of 0.99 for the TanimotoCombo score was set to identify the "reliable hits" from among the 500 hits obtained for each query molecule. TanimotoCombo score is the sum of shape and chemical similarity score of the query and target molecules in an optimized overlay (for details, see the original article in Ref. [56]). The reliable hits so obtained usually have molecular shape and chemical features highly similar to that of the corresponding query compound. Next, the details related to toxicity, pharmacodynamics, and description of each reliable hit were examined by consulting the DrugBank database to ensure that the subset of molecules we have selected for further in-depth study as potential anticancer agents are not known to cause serious side effects and/or do not belong to the class of "nutraceuticals" or "dietary supplement," which are unlikely to be good drug candidates. Molecules that were once approved but have been banned or voluntarily withdrawn from market due to safety concerns have also been discarded for subsequent analyses.

The hits for Aurora-A/B kinases were then segregated based on the type of query compound into three groups: (i) hits obtained when ATP/ATP analogues were queried (category-I; not applicable for Aurora-B kinase in this study), (ii) hits obtained when ATP-replacing inhibitors were queried (category-II), and (iii) hits obtained when allosteric inhibitors were queried (category-III; not applicable for Aurora-B kinase in this study). For the sake of brevity, only a few selected cases from each of these three categories were taken for further investigation to probe into their possibility of binding and inhibiting Aurora kinases, which has been discussed later in this chapter. For some cases, molecules have been selected for further study based on 2D chemical similarity/diversity as indicated by their Tanimoto similarity coefficient. Tanimoto similarity coefficients have been calculated using python RDKit package (RDKit open-source cheminformatics; http://www.rdkit.org).

Molecular docking

Molecular docking simulations for the selected set of hits were performed using AutoDock Vina [73] to predict the binding affinity between the molecules and Aurora-A kinase. This was done to ascertain whether the selected set of molecules could be favorably accommodated in the binding pocket of Aurora-A kinase to which the respective query molecule is known to bind. Additionally, possibility of forming crucial hydrogen bond interaction/s between the hinge residues (Glu211 and/or Ala213) of the protein and the predicted pose of the ligand was also checked. The importance of hinge residues in functioning of kinases has already been established [5,74]. Adenine ring of ATP and its analogues form hydrogen bonds with Glu211 and/or Ala213 of Aurora kinase (for example, PDB ID: 1OI5, 1MQ4, 2C6D, etc.). Also, the crystal structures of Aurora kinase with known ATP mimetic inhibitors show that the inhibitors form hydrogen bonds with the mentioned hinge residues (for example, PDB ID: 2WIC, 3VAP, 3NRM, etc.). The hinge region is structurally conserved across kinases, and topologically equivalent residues of Glu211 and Ala213 in other kinases also form hydrogen bond/s with adenine ring of ATP/ATP analogues or adenine-mimetic ring of respective ATP-replacing inhibitors. It has been shown earlier that purine and pyrimidine ring containing ATP mimetic agents which do not engage the important hinge residues in hydrogen bonding either fail to inhibit the kinase enzyme or do not show the desired extent of enzyme inhibition [75]. Therefore, performing docking studies to assess the possibilities of forming hydrogen bond interactions between the compounds shortlisted from our analysis and the important hinge residues of Aurora kinase is worthy to pursue in order to gain an understanding how likely a particular hit will be able to bind and inhibit the protein of our interest.

While selecting the protein structure for docking the compounds, it was ensured that the crystal structure of the protein has no missing residue and/or atom in the electron density map corresponding to the ligand binding site. Additionally, it was checked that the ligand and the binding site residues of interest in the selected protein crystal structures should be of good quality and this has been verified using the tool VHELIBS [76]. Further, it was also ensured that the protein structure of Aurora-A kinase selected for docking should be in the same conformational state (with respect to DFG motif and αC-helix) as that of the crystal structure to which the query molecule of the respective hit compounds selected for docking are bound. This was ensured by using the information fetched from KLIFS database [77]. Most (all but vandetanib, flibanserin, paroxetine, dimetotiazine, tolvaptan, midazolam, 4-(isopropylamino diphenylamine), and N-cyclohexyl-N′-phenyl-1,4-phenylenediamine) of the molecules that have been docked in the ATP binding site of Aurora-A kinase are the hits obtained against query molecules that are bound to crystal structures of Aurora-A kinase where the DFG and the αC-helix motif have been captured in "in" conformation. Rilpivirine has been obtained as hit against query molecules that are bound to protein structure of Aurora-A kinase crystallized in multiple conformation states (DFG in, αC-helix in; DFG in, αC-helix out-like; DFG out, αC-helix in) and deposited under different PDB entries. The protein structure belonging to PDB code 5G15 satisfies all our selection criteria and has the DFG as well as αC-helix in "in" conformation. Therefore, 5G15 was used for docking all the molecules in the ATP binding site except vandetanib. The query molecule for vandetanib is "HPM," which is bound to the protein structure with PDB code 2C6E. DFG and αC-helix of the protein structure in 2C6E are present in "out" conformation. Therefore, vandetanib was docked in the ATP-binding pocket of the protein structure with PDB code 2C6E as it satisfies other selection criteria used in our study. Based on similar criteria, the protein structure with PDB code 6R4A was selected to dock the hits obtained against known allosteric site-1 inhibitors of Aurora-A kinase.

The protein structures to which the corresponding query compounds of the hits: flibanserin, paroxetine, dimetotiazine, tolvaptan, midazolam, 4-(isopropylamino diphenylamine), and N-cyclohexyl-N′-phenyl-1,4-phenylenediamine, are bound did not qualify our selection criteria. Also, no other appropriate conformation of protein structures satisfying all our selection criteria for using in docking simulation were found for the mentioned cases. These molecules were docked in the ATP binding site of 5G15 to predict whether these molecules can be accommodated favorably by Aurora-A kinase with DFG-"in" and αC-helix-"in" conformation along with possibility of formation of crucial interactions.

In all the docking runs, only the flexibility of the ligands was considered. The binding site residues were considered as rigid. The docking protocols were validated through redocking experiment to ensure that the docking algorithm is able to reproduce the bound pose of the native ligand present in the respective crystal structures that are used in our docking study. The grid spacing was set at 1 Å. For docking molecules at the ATP binding site of 5G15 and allosteric site-1 of 6R4A, the x, y and z coordinates for the center of the grid boxes were chosen at (−23.334, 28.299, and −30.537) and (−12.208, 12.864, −20.887) respectively. A grid box measuring 16 and 12 Å in each direction was chosen for ATP site and allosteric site-1, respectively. For docking vandetanib in ATP binding site of 2C6E, the x, y, and z coordinates for the center of the grid box were chosen at (22.433, 90.693, and 70.795) and the dimension of the grid box was fixed at 12, 20, and 22 Å in x, y, and z direction, respectively. Only the best pose has been reported in this chapter, which was chosen based on consistency of binding mode along with retention of crucial interactions.

Results and discussion

Hits for Aurora-A kinase: For the ATP binding site, 535 unique reliable hits have been obtained based on our selection criteria. While 485 of them have been obtained on querying the ATP mimetic inhibitors, 13 of them have been obtained on querying ATP/ATP analogues. 37 hits are common between the query sets of ATP/ATP analogues and ATP mimetic inhibitors. For the allosteric sites, a total of 821 reliable hits have been obtained. The best reliable hit for each query (wherever applicable) has been presented in Table 2.1.

Category I: Hits belonging to this group have been obtained on querying ATP/ATP analogues against the target set of approved molecules. On close inspection of the hits, it was noted that majority of the hits in this group are known antiviral drugs. It is interesting to note that antiproliferative and cytotoxic properties of antiviral agents have been reported in the past, and their use in cancer treatment as adjuvant antiviral therapy is also being studied [78,79]. Mining of the ClinicalTrials.gov database, a resource provided by the US National Library of Medicine (available at https://clinicaltrials.gov/), reveals that few approved antiviral agents (such as ribavirin and acyclovir) are currently under clinical trials either as a monotherapy or as a combinatorial therapy in different types of cancer. Albeit anticancer properties of antivirals have been reported, but, in most cases, the molecular targets of these antivirals in cancer therapy have not been explored. Our results hint that Aurora kinases could be one of the potential targets of the antiviral agents through which these molecules mediate their anticancer effects. Therefore, we selected a subset of five antiviral drugs (vidarabine, ribavirin, didanosine, idoxuridine, and inosine) from the hit list of our shape-based screening experiment for further study to gain a better understanding of the possible molecular mechanism of action of antivirals in anticancer therapy by targeting Aurora kinases. Chemically, all these antiviral drugs have an adenine mimicking ring and target viral DNA/RNA or any protein that interferes with the function of DNA/RNA (information source: DrugBank) to exhibit their antiviral functions. Results of molecular docking simulations of the five selected antiviral drugs with Aurora-A kinase are discussed below.

Vidarabine: It is a nucleoside antibiotic isolated from *Streptomyces antibioticus*. It has antiviral activity against infections caused by a variety of viruses such as the herpes viruses, the vaccinia virus, and varicella zoster virus. Our shape-based screening experiment has yielded vidarabine as a hit against ATP and multiple ATP analogues bound to different crystal structures of Aurora-A kinase. Molecular docking simulation of vidarabine into the ATP binding pocket of Aurora kinase (PDB code: 5G15) reveals that vidarabine could be favorably accommodated by the protein as indicated by the predicted binding affinity of −6.2 kcal/mol for the best pose. Further, overlay of docked pose of vidarabine on to the bound pose of ACP (ATP analogue bound to the crystal structure 5G15) showed that the adenine ring of vidarabine and ACP are spatially proximal. Also, the adenine ring of vidarabine forms hydrogen bond with Ala213 (Fig. 2.3, Table 2.2). Additionally, the molecule is also engaged in hydrogen bonding with Lys162 (conserved catalytic lysine) and Asp274 (from conserved DFG motif). Earlier mutagenesis studies have shown that substitution of Lys with Arg at position 162 leads to loss of kinase activity [80]. Also, D274N mutation abolishes autophosphorylation of the kinase [81]. Thus, both Lys162 and Asp274 are important for the functioning of the protein.

Our results indicate that vidarabine could interact with the crucial functional residues of Aurora kinases and act as a potential inhibitor of these kinases contributing toward management of cancer conditions. Interestingly, literature survey reveals that vidarabine is known to possess antineoplastic property, and this has been demonstrated in several in vitro and cell-based studies [82–84]. It is worthy to mention here that fludarabine (DB01073), the 2-fluoro derivative of vidarabine, is an approved drug that is used in the treatment of hematological malignancies.

Ribavirin: It is a synthetic guanosine nucleoside that has broad spectrum activity against several RNA and DNA viruses. Like vidarabine,

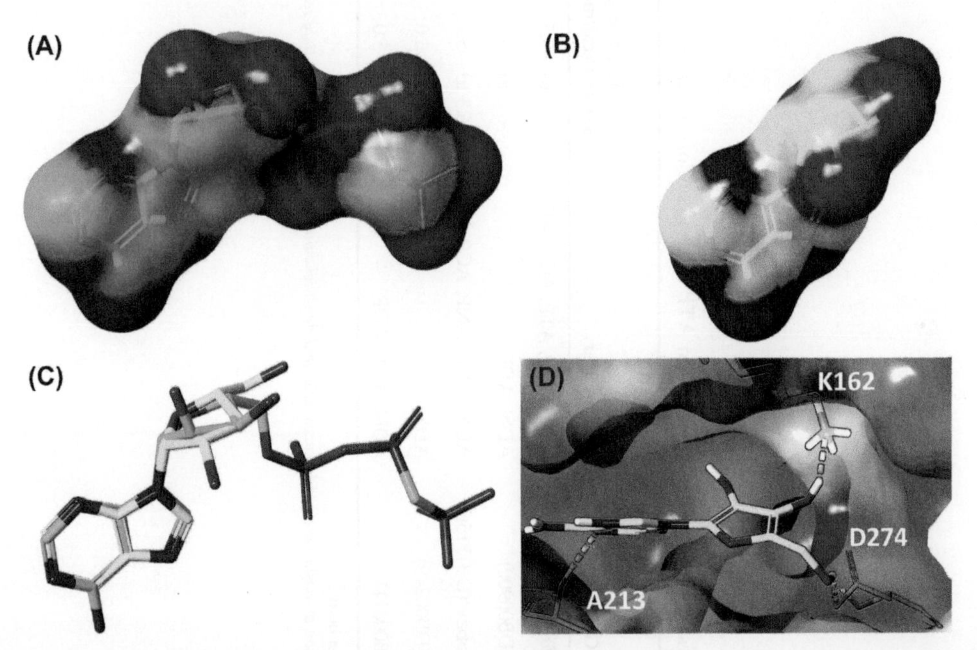

FIGURE 2.3 Analyses of shape screening and docking simulation of vidarabine. (A) The bioactive conformation of the ligand ACP as obtained from crystal structure 5G15 is shown in surface representation. (B) The 3D conformation of vidarabine predicted as hit against the query ACP from 5G15. (C) Overlay of (B: yellow) onto (A: green) shown in stick representation. (D) Docked pose of vidarabine (yellow carbon) in the ATP binding site of 5G15. Hydrogen bonds are shown as cyan *dashed lines* and the residues involved in hydrogen bonding are shown as stick (gray carbon) and labeled in white font. Nitrogen, hydrogen, oxygen, and sulfur atoms have been represented in standard color (blue, white, red, and yellow, respectively) in all the figures. None of the hydrogens in panel (A–C) as well as the nonpolar hydrogens in panel (D) were displayed while image generation for visual clarity. The images have been generated using free academic maestro graphical interface (Maestro, Schrödinger, LLC, NY).

TABLE 2.2 Docking results of antiviral compounds in ATP binding site of Aurora-A kinase.

Serial Number	Name (DRUGBANK ID)	Shape Query	Docking energy Range (kcal/mol)	Best pose	
				Energy (kcal/mol)	Hydrogen bonded residues
1	Vidarabine (DB00194)	ADP, ANP, ATP, ACP, ADN	−6.2 to −5.5	−6.2	**A213**, K162, D274
2	Didanosine (DB00900)	ADN, ADP	−6.2 to −5.5	−6.2	**A213**, D274
3	Inosine pranobex[a](DB13156)	ADN, ADP, ANP, ACP	−6.3 to −5.7	−6.3	**A213, E211**
4	Idoxuridine (DB00249)	ADN	−5.6 to −5.0	−5.6	**A213**, K162, D274
5	Ribavirin (DB00811)	ADN, ANP, ADP	−5.8 to −5.0	−5.8	**A213, E211**, K162, D274

Hinge residues are indicated in bold.

[a] *Inosine pranobex is a combination of inosine, acetamidobenzoic acid, and dimethylaminoisopropanol. For our study related to shape screening and docking, we have used only the inosine component.*

ATP and multiple query ATP analogues have yielded ribavirin as hit in the shape-based similarity screening experiments. Results of molecular docking simulations suggest that binding of ribavirin to the catalytic domain of Aurora-A kinase could be a favorable binding event as indicated by the predicted binding affinity of −5.8 kcal/mol for the best pose. Further, overlay of the docked pose of ribavirin on to bound pose of ACP in the crystal structure 5G15 showed that the triazole ring of ribavirin (which mimics adenine ring of ATP and its analogues) and adenine ring of ACP occupies spatially proximal regions. This facilitates establishment of hydrogen bonds between the triazole ring of ribavirin and the two hinge residues: Glu211 and Ala213, whose functional importance have been discussed earlier. Additionally, like vidarabine, ribavirin is also predicted to form hydrogen bonds with Lys162 and Asp274 (Fig. 2.4, Table 2.2). These findings potentiate the fact the ribavirin is likely to bind to Aurora-A kinase and thereby could lead to its inhibition. Kentsis et al. [85,86] showed that associations of ribavirin with eukaryotic translation initiation factor eIF4E leads to suppression of eIF4E-mediated oncogenic transformation in vitro and in vivo that leads to downregulation of a combination of oncogenes. Thus, experimental evidence for the anticancer property of ribavirin is already available. Based on our findings, it will be interesting to probe the effect of ribavirin on Aurora kinases alone and along with eIF4E to understand the subsequent mechanism of regulation of cancer signaling pathways.

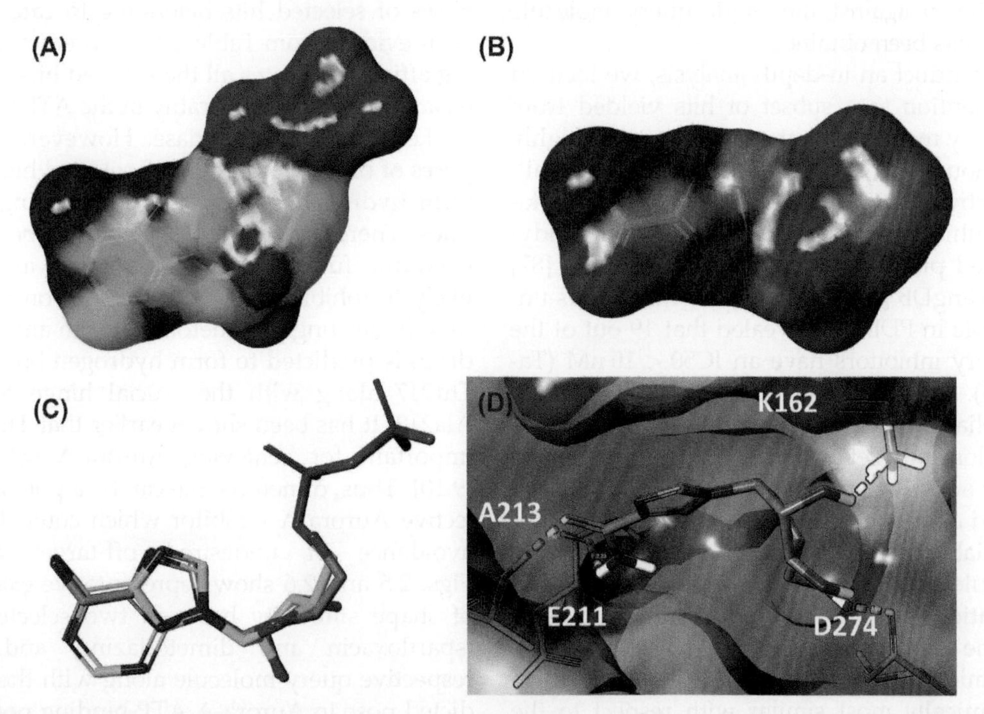

FIGURE 2.4 Analyses of shape screening and docking simulation of ribavirin. (A) The bioactive conformation of the ligand ADP as obtained from crystal structure 6R4B is shown in surface representation. (B) The 3D conformation of ribavirin predicted as hit against the query ACP from 6R4B. (C) Overlay of (B: pink) onto (A: green) shown in stick representation. (D) Docked pose of ribavirin (pink) in the ATP binding site of 5G15. Kindly refer to legend to Fig. 2.3 for further details.

Didanosine, idoxuridine, and inosine: In line with the findings for vidarabine and ribavirin, similar observations have been noted for didanosine, idoxuridine, and inosine. The docked poses of all these molecules show that hydrogen bond/s between the functionally important hinge residue/s of Aurora-A kinase and the respective molecule could be established (Table 2.2). Thus, inhibition of Aurora-A kinase upon treatment with these molecules is likely to be achieved.

Category II: Hits belonging to this category have been obtained on querying ATP-replacing inhibitors against the target set of approved molecules. As mentioned earlier, 485 unique reliable hits have been obtained exclusively against 84 unique known ATP replacing inhibitors of Aurora-A kinase. For Aurora-B kinase, only one reliable hit (viz. ribociclib, a known CDK4/6 inhibitor) against the single query molecule "VX6" has been obtained.

To conduct an in-depth analysis, we focused our attention to a subset of hits yielded from the query molecules that are known to be highly efficacious against Aurora-A kinase. The available activity data (IC50) of known Aurora A kinase inhibitors (used as query in our study) obtained primarily using PDBind database [87] or BindingDB [88] (in case when the IC50 is unavailable in PDBind) revealed that 19 out of the 84 query inhibitors have an IC50 < 10 nM (Table 2.3). However, 4 out of these 19 did not yield any reliable hits. Therefore, subsequent analyses were done on a set of 15 query molecules that are highly active against Aurora-A kinase and have yielded at least one reliable hit. Wherever multiple reliable hits are available, two hits per query molecule were chosen for molecular docking simulations that are representative of two extreme 2D chemical features, viz., a hit which is chemically most diverse and the other which is chemically most similar with respect to the query molecule. This was accomplished by calculating 2D Tanimoto similarity coefficients for each reliable hit obtained for the selected 15 query molecules. For query molecules like "HPM" and "ZZL," only one reliable hit has been obtained, therefore only that single hit has been considered in each of the mentioned two cases. Subsequently, 20 nonredundant hits were docked into the ATP binding site of Aurora-A kinase, and the results are summarized in Table 2.4. Interestingly, while some of the selected hits such as regorafenib and vandetanib are known antineoplastic agents, i.e., anticancer drugs, others mainly belong to the class of approved antibacterial drugs (sparfloxacin, linezolid, and sulphenazole)/antiviral agents (rilpivirine)/antiulcer agent (dexlansoprazole)/CNS-acting drugs (antianxiety, antidepressant, hypnotic, etc.)/antiinflammatory drugs (nabumetone), etc.

Similar analyses as have been done for category-I hits were also performed on docked poses of selected hits belonging to category-II. As is evident from Table 2.4, the predicted binding affinity indicates all the selected hits seem to be accommodated favorably in the ATP-binding pocket of Aurora-A kinase. However, docked poses of only 12 out of the 18 selected hits could form hydrogen bonds with crucial hinge residues. Therefore, these 12 hits could be prioritized for further probing as these are more likely to inhibit Aurora-A kinase upon binding to it. Interestingly, dimetotiazine (an antiallergic drug) is predicted to form hydrogen bond with Thr217 along with the crucial hinge residue, Ala213. It has been shown earlier that Thr217 is important for achieving Aurora-A selectivity [9,10]. Thus, dimetotiazine can be a potential selective Aurora-A inhibitor which could lead to avoidance of undesired off-target effects. Figs. 2.5 and 2.6 show representative examples of shape similarity between two selected hits (sparfloxacin and dimetotiazine) and their respective query molecule along with their predicted pose in Aurora-A ATP-binding pocket.

It is to be noted that four out of the six hits (tolvaptan, midazolam, 4-(isopropylamino diphenylamine), N-cyclohexyl-N′-phenyl-1,4-

TABLE 2.3 Details of query molecules belonging to category II and their selected hits for docking studies.

Serial Number	Shape query (HET ID)	PDB ID	DFG motif	α C-helix	H-bonded residues	IC50 (nM)	Reliable hits: Most similar ligand (Drugbank ID; 2D Tanimoto Coefficient)	Reliable hits: Most dissimilar ligand (Drugbank ID; 2D Tanimoto Coefficient)
1	HPM	2C6E	Out	Out	**A213**, K162	0.8	Vandetanib (DB05294; 0.697)	
2	L0F	2W1F	In	In	**A213**	5.9	Hexylcaine (DB00473; 0.502)	Sulphenazole (DB06729; 0.213)
3	L0G	2W1G	In	In	**E211, A213**	3	N.A.	
4	ZZL	2X81	–	–	R137, **A213**	5	Midazolam (DB00683; 0.502)	
5	83H	3COH	–	–	**A213**	4	Sparfloxacin (DB01208; 0.575)	Niclosamide (DB06803; 0.288)
6	AK8[a]	3EFW	In	In	K162, **A213**	4	Regorafenib (DB08896; 0.458)	Sorafenib[a] (DB00398; 0.448)
7	48B	3H0Y	Out	In	**A213**	6	Rilpivirine (DB08864; 0.539)	
8	45B	3H0Z	In	In	**A213**, T217, E260	4.3	N.A.	
9	NRM	3NRM	In	In	R137, **A213**	4	Nabumetone (DB00461; 0.424)	Linezolid (DB00601; 0.230)
10	E9Z	3QBN	–	–	**A213**	5.9	4-(Isopropylamino diphenylamine) (DB14195; 0.508)	N-Cyclohexyl-N′-phenyl-1,4-phenylenediamine (DB14196; 0.407)
11	0BZ	3UNZ	In	Out-like	**A213**	3.7	Rilpivirine (DB08864; 0.625)	Paroxetine (DB00715; 0.349)
12	0BY	3UO6	In	Out-like	**A213**	2.5	Rilpivirine (DB08864; 0.605)	Tolmetin (DB00500; 0.350)
13	0C4	3UOH	In	Out-like	R137, **A213**	2.1	Rilpivirine (DB08864; 0.618)	Paroxetine (DB00715; 0.320)
14	0C6	3UOK	In	Out-like	R137, **A213**	0.8	Rilpivirine (DB08864; 0.577)	Dimetotiazine (DB08967; 0.240)
15	0C9	3UP7	In	In	R137, K162, **A213**	6.1	Rilpivirine (DB08864; 0.677)	Alitretinoin (DB00523; 0.420)
16	0FY	3VAP	In	In	R137, **A213**, D274	4	N.A.	
17	FH3	4BYI	In	In	**A213**	9	Rilpivirine (DB08864; 0.428)	Dexlansoprazole (DB05351; 0.336)
18	XU2[a]	4JAI	Out-like	In	**A213**	4	Tolvaptan (DB06212; 0.367)	Flibanserin (DB04908; 0.317)
19	Y3M	4UYN	In	In	K162, E211, **A213**	2	N.A.	

N.A., "Not Applicable" indicates no reliable hits have been obtained based on the qualifying criteria used in the study. "–" indicates information regarding conformation of DFG motif and αC-helix is unavailable in KLIFS as it could not be ascertained due to many missing residues in these regions. Hinge residues are indicated in bold.

[a] *Only two reliable hits have been obtained. So, both were considered for docking studies. Selection based on 2D Tanimoto coefficient is not applicable here.*

TABLE 2.4 Docking results of selected hits against category II query molecules.

Serial Number	Name of hit (DRUGBANK ID)	Category of hit	Shape query (HET ID)	Energy Range of docking poses (kcal/mol)	Best pose	
					Energy (kcal/mol)	Hydrogen bonded residues
1	*Flibanserin (DB04908)*	Originally developed as antidepressant but currently repurposed to treat sexual disorder	XU2	−8.4 to −7.9	−8.1	**A213**
2	Regorafenib (DB08896)	Antineoplastic agent (protein kinase inhibitor)	AK8	−8.2 to −7.6	−8.0	**A213**
3	Rilpivirine (DB08864)	Antiviral agent	0BY,0BZ,0C4,0C6,0C9,48B, FH3	−8.3 to −7.8	−7.8	**A213, E211**
4	Sorafenib (DB00398)	Antineoplastic agent (protein kinase inhibitor)	AK8	−8.0 to −7.5	−7.7	**A213**
5	Sparfloxacin (DB01208)	Antibacterial agent	83H	−7.6 to −7.1	−7.5	**A213**, D274
6	*Paroxetine (DB00715)*	Antianxiety agent	0BZ,0C4	−7.8 to −6.9	−6.9	**A213**, D274
7	Linezolid (DB00601)	Antibacterial agent	NRM	−6.8 to −6.3	−6.8	**A213**, K162
8	*Dimetotiazine (DB08967)*	Antiallergic agent	0C6	−7.5 to −6.5	−6.6	**A213**, T217
9	Dexlansoprazole (DB05351)	Antiulcer agent (proton pump inhibitor)	FH3	−7.1 to −6.6	−6.6	**A213**
10	Niclosamide (DB06803)	Anthelmintic agent	83H	−7.1 to −5.8	−5.8	**A213, E211**
11	Tolmetin (DB00500)	Antiinflammatory agent	0BZ, 0BY	−6.8 to −6.2	−6.5	**A213**
12	Sulphenazole (DB06729)	Antibacterial agent	L0F	−6.9 to −6.2	−6.4	**A213, E211**
13	Nabumetone (DB00461)	Antiinflammatory agent	NRM	−7.1 to −5.8	−5.8	**A213, E211**
14	Hexylcaine (DB00473)	Local anesthetic agent	L0F	−6.0 to −5.6	−5.7	**A213**
15	*Tolvaptan (DB06212)*	Cardiac therapy fluid	XU2	−9 to −7.7	N.A. (L139, K143, K162, E260)	
16	*Midazolam (DB00683)*	Hypnotic and sedative	ZZL	−7.4 to −6.4	N.A. (D274)	

17	*4-(Isopropylamino diphenylamine) (DB14195)*	Diagnostic agent for allergic contact dermatitis	E9Z	−6.2 to −5.7	N.A. (L139, E260, D274)
18	*N-cyclohexyl-N′-phenyl-1,4-phenylenediamine (DB14196)*	Diagnostic agent for allergic contact dermatitis	E9Z	−7.1 to −6.6	N.A. (E260, D274)
19	Vandetanib[a] (DB05294)	Antineoplastic agent (tyrosine kinase inhibitor)	HPM	−9.1 to −8.2	N.A. (P214, E260)
20	Alitretinoin (DB00523)	Antineoplastic agent/Agents for dermatitis	DC9	−7.4 to −6.7	N.A. (K162, T217)

Shown in italics are the hits that have not been docked into structure with a conformation state identical to that of the corresponding query molecule. Hinge residues are indicated in bold. *N.A.*, Not Applicable; Best pose has not been assigned since no pose has been found to be H-bonded with hinge residues. However, indicated beside "N.A." within brackets () are the residue names that have been found to be hydrogen bonded with the docked poses of the molecules.

[a] *Docked into the ATP binding pocket of 2C6E; all the other molecules are docked into the ATP binding pocket of 5G15.*

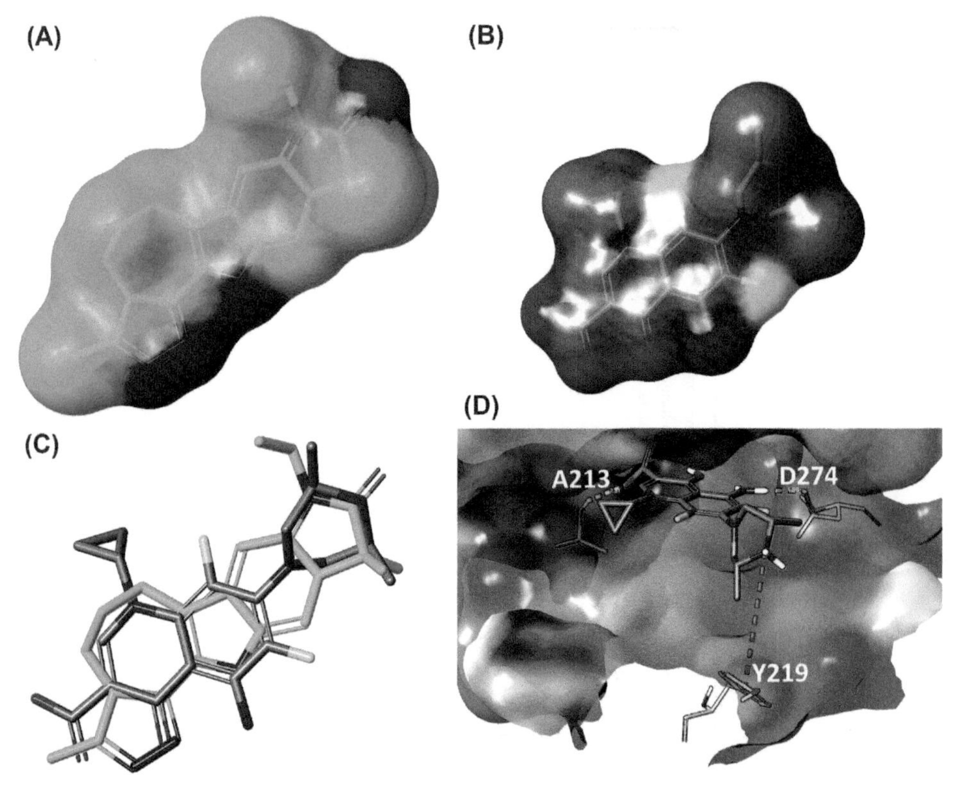

FIGURE 2.5 **Analyses of shape screening and docking simulation of sparfloxacin.** (A) The bioactive conformation of the ligand 83H as obtained from crystal structure 3COH is shown in surface representation. (B) The 3D conformation of sparfloxacin predicted as hit against the query 83H from 3COH. (C) Overlay of (B: purple) onto (A: green) shown in stick representation. (D) Docked pose of sparfloxacin (purple carbon) in the ATP-binding site of 5G15. The green dashed line represents Pi–cation interaction between the docked pose of sparfloxacin and Y219 of Aurora-A kinase. Kindly refer to legend to Fig. 2.3 for further details.

phenylenediamine, vandetanib, and altretinoin) that did not show any hydrogen bond interaction with the hinge residue/s have not been docked in a protein structure whose conformation is identical to that of the protein structure to which the respective query molecules are bound because of the reasons discussed in the methodology section. Given that all the ligands that have qualified as reliable hits could adopt a shape highly similar to that of the bioactive conformation of the respective query molecule, if the flexibility of the protein is considered, these ligands might be able to form crucial hydrogen bond interaction/s with the hinge residues of Aurora kinase, which is essential to elucidate the desired inhibition as seen in all the FDA-approved reversible kinase inhibitors except for type III allosteric inhibitors [89] as well as the corresponding query molecules for these hits (Table 2.3).

Category III: Hits of this category have been obtained against the query allosteric inhibitors of Aurora-A kinase. As mentioned earlier, these three allosteric sites form the interface of interaction between Aurora-A kinase and TPX2. While for allosteric site-1, 25 unique inhibitors are available as bound ligand fragments in crystal structures of Aurora-A kinase, only 6 and 1

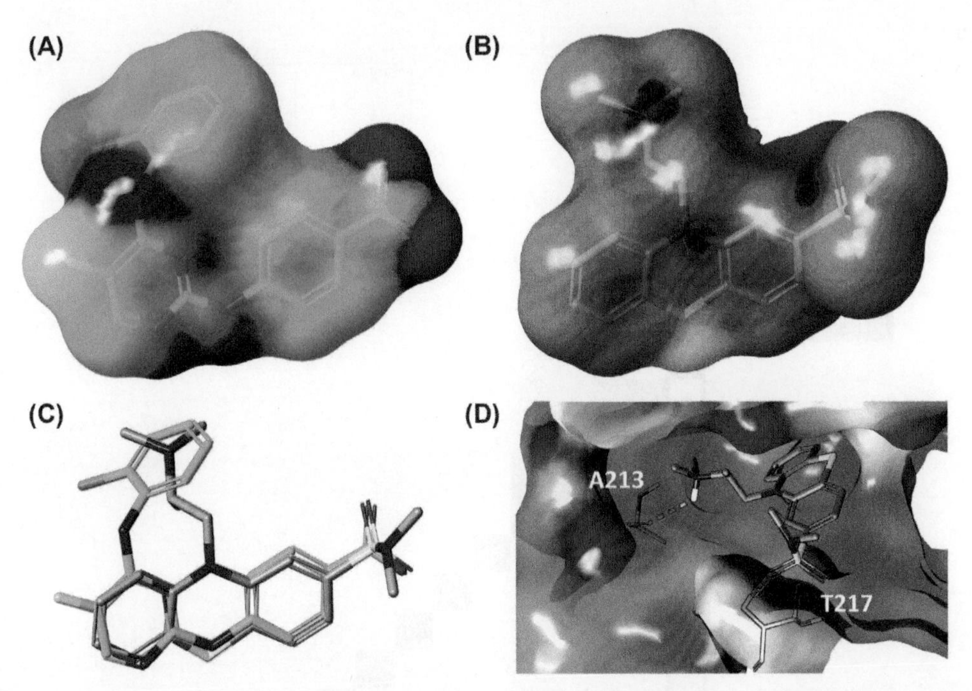

FIGURE 2.6 **Analyses of shape screening and docking simulation of dimetotiazine.** (A) The bioactive conformation of the ligand 0C6 as obtained from crystal structure 3UOK is shown in surface representation. (B) The 3D conformation of dimetotiazine predicted as hit against the query 0C6 from 3UOK. (C) Overlay of (B: orange) onto (A: green) shown in stick representation. (D) Docked pose of dimetotiazine (orange carbon) in the ATP binding site of 5G15. Kindly refer to legend to Fig. 2.3 for further details.

inhibitor are available for allosteric site-2 and 3, respectively (Table 2.1).

Ideally, a molecule which would bind only to the allosteric site (Aurora-A:TPX2 interaction interface) and not to the orthosteric ATP binding site is likely to be more selective to Aurora-A than any other kinases. Therefore, we focused on studying those hits that have been obtained only for allosteric site-1 and have not been identified as hits against any query molecule that binds to ATP binding site. From such a set, we chose three (valproic acid, paraldehyde, and bronopol) hits and performed docking studies. These three hits have been obtained upon querying the molecule A65 against the database of approved drugs. The ligand A65 is chemically most diverse among all the allosteric site-1 inhibitors used as query in our study as is indicated by its Tanimoto similarity coefficient with respect to rest of the molecules (Fig. 2.7). Therefore, the hits of A65 are expected to be chemically distinct from most of the hits obtained against other queries.

Analysis of the interaction profiles of all the allosteric site-1 inhibitors used as query in this study showed that most of these ligands are either hydrogen bonded to or are involved in Pi-cation/Pi-stacking with Arg179 of Aurora-A kinase in the respective crystal structures (Table 2.5). Arg179 is a surface residue of Aurora-A kinase, which interacts with TPX2 closely as could be seen in the structure with PDB ID 1OL5 [81]. Interestingly, two of the selected hits (valproic acid and bronopol) used in docking study also form hydrogen bond with Arg179 (Fig. 2.8, Table 2.6). It is exciting to note that

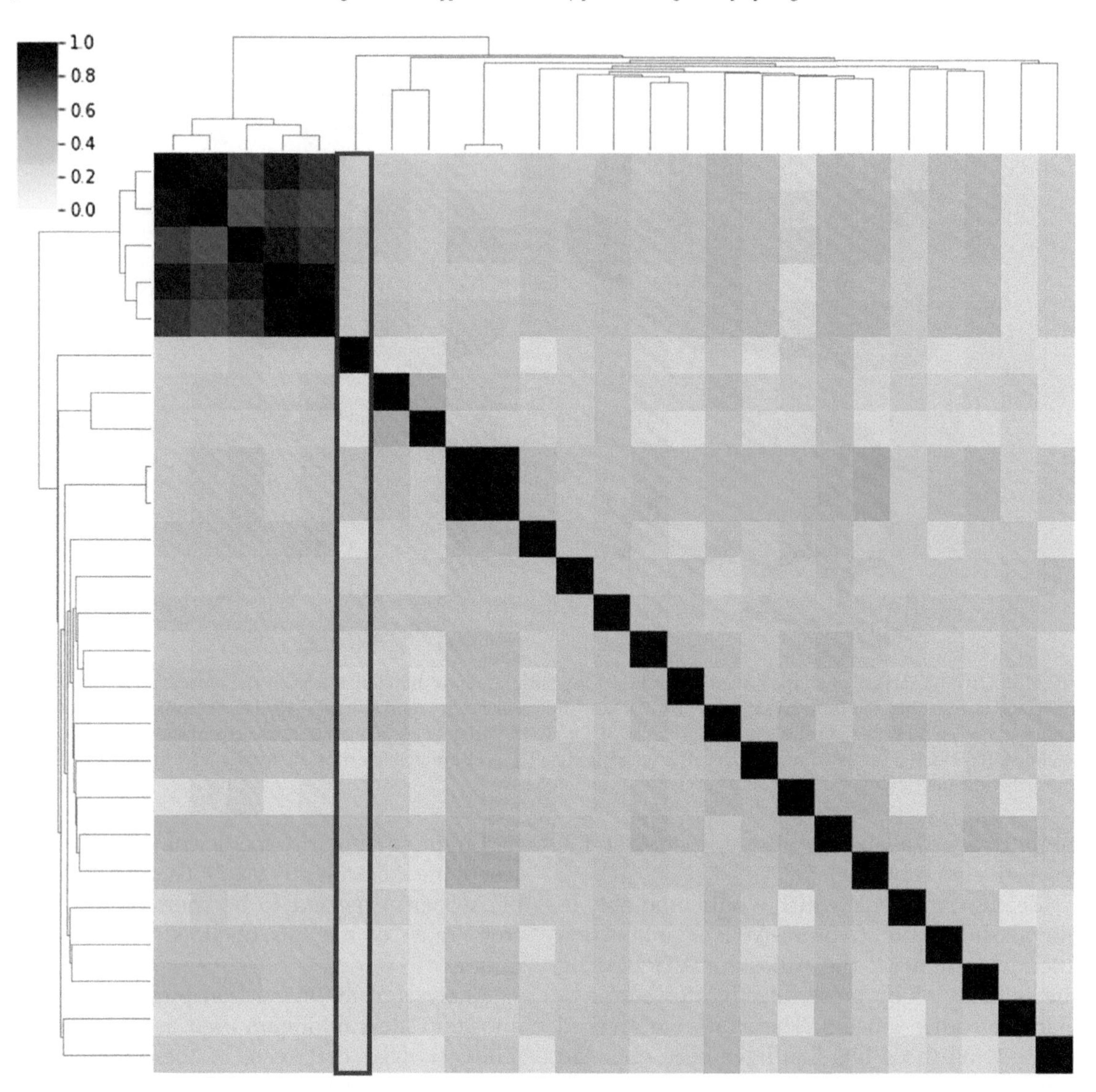

FIGURE 2.7 2D Tanimoto coefficient (TC) heatmap. Darker the shade, more is the chemical similarity between a pair of compounds and vice versa. The column within the red rectangle encloses the TC of A65 with respect to all other allosteric site-1 inhibitors used in the study. As could be seen, unlike other columns, most of the cells in the column corresponding to A65 are of lighter shade indicating that A65 is chemically not very similar to rest of the compounds.

TABLE 2.5 Interaction profiles (Hydrogen bonds and Pi-Cation/Pi-Stacking) of query molecules occupying allosteric site-1.

Serial Number	Query (LIGAND'S HET ID)	PDB ID	H-BONDED residues	Residues involved IN PI-cation/PI-stacking interaction
1	5DN	5DN3	K166	–
		5DOS	K166	R179
		5DPV	K166, H201	R179
		5DT4	K166	R179
2	5E1	5DR2	–	R179
		5DR6	K166	R179
3	5E2	5DR9	K166	R179
4	9QK	5OBJ	–	R179
5	9QT	5OBR	–	R179
6	A5Q	5ORR	–	R179
7	A5W	5ORS	K166	Y199
8	A5Z	5ORT	R179	R179
9	A65	5ORV	–	R179
10	A6E	5ORW	–	Y199
11	A6H	5ORX	–	Y199
12	AY4	5ORY	N.A.	
13	A6W	5ORZ	N.A.	
14	A6Z	5OS0	N.A.	
15	A7H	5OS1	N.A.	
16	A7K	5OS2	Y199, H201	H201
17	A7N	5OS3	N.A.	
18	A8H	5OS4	N.A.	

Continued

TABLE 2.5 Interaction profiles (Hydrogen bonds and Pi-Cation/Pi-Stacking) of query molecules occupying allosteric site-1.—cont'd

Serial Number	Query (LIGAND'S HET ID)	PDB ID	H-BONDED residues	Residues involved IN PI-cation/PI-stacking interaction
19	A8K	5OS5	E183	–
20	A7Q	5OS6	R179, E183	–
21	JSB	6R49	K166	–
22	JRT	6R4A	R179	R179
23	JSN	6R4B	–	H201
24	JRQ	6R4C	Y199	H201
25	JRW	6R4D	K166	H201

N.A.: These ligands are only held by hydrophobic contacts. "–" indicates absence of that particular interaction.

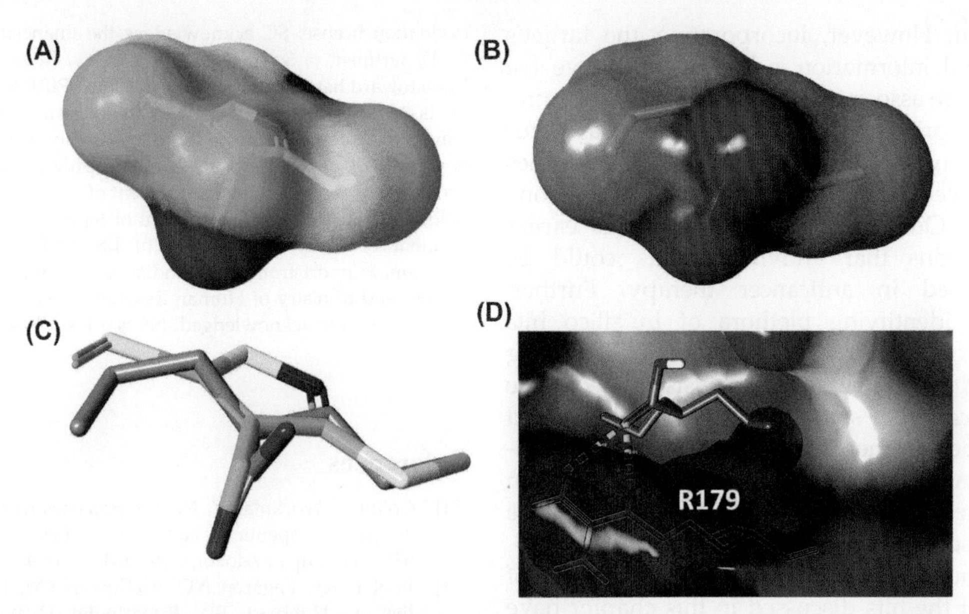

FIGURE 2.8 Analyses of shape screening and docking simulation of valproic acid. (A) The bioactive conformation of the ligand A65 as obtained from crystal structure 5ORV is shown in surface representation. (B) The 3D conformation of valproic acid predicted as hit against the query A65 from 3UOK. (C) Overlay of (B: gray) onto (A: green) shown in stick representation. (D) Docked pose of valproic acid (gray carbon) in the allosteric site-1 of 6R4A. Kindly refer to legend to Fig. 2.3 for further details.

TABLE 2.6 Docking results of selected hits for category III query molecules.

					Best pose	
Serial Number	**Name (DRUGBANK ID)**	**Category of hit**	**Shape query**	**Docking energy Range (kcal/mol)**	**Energy (kcal/mol)**	**Hydrogen bonded residues**
1	Valproic acid (DB00313)	Anticonvulsant	A65	−4.8 to −4.0	−4.3	R179
2	Paraldehyde (DB09117)	Anticonvulsant	A65	−4.4 to −3.8	N.A.	
3	Bronopol (DB13960)	Antimicrobial agent	A65	−3.4 to −3.0	−3.4	R179

N.A., Not Applicable; None of the poses were found to be hydrogen bonded to any residue of Aurora-A kinase. Therefore, best pose has not been assigned.

anticancer properties of valproic acid mediated via multiple pathways (such as downregulation of protein kinase C, inhibition of glycogen synthase kinase-3β, blocking histone deacetylase) have been reported [90–92], and it is also under clinical trial for anticancer therapy as mentioned in the DrugBank records and ClinicalTrial.gov database.

Conclusions

Our study has demonstrated the application of a simple pipeline for drug discovery/repurposing which is primarily a ligand-centric approach coupled with target's structural knowledge. Ideally, ligand-based approaches require only the information of a known binder as input. However, incorporating the target's structural information is likely to improve the confidence associated with the analysis ensuring rational prioritization of promising cases for further investigation (as noted in earlier studies and have been discussed in "Introduction" section). Our study resonates with the earlier observations that antiviral agents could be repurposed in anticancer therapy. Further, besides identifying plethora of in silico hits which could be targeted against the ATP binding site of Aurora kinases, our analysis also led to the identification of some *in silico* hits which could be probed further as potential allosteric inhibitors of Aurora-A kinase targeting the interaction interface between Aurora-A and TPX2 to achieve kinase selectivity (Table 2.1).

As mentioned earlier, anticancer properties of some of the hits discussed in this chapter have been reported by different experimental groups. However, in most cases, either the molecular target of such a compound has not been reported or the involvement of targets other than Aurora kinases have been indicated as in the case of ribavirin and valproic acid. Further experiments are required to validate our findings and dig deep into the molecular mechanism of actions of the identified hits as anticancer agents mediated through Aurora kinases alone and concurrently with their already reported anticancer targets. It is important to note that our study also paves the path for future rationale drug discovery using the chemical scaffolds of the identified hits as a medicinal chemistry starting point for design and development of new Aurora kinase inhibitors. Lastly, since drug repurposing is based on the principles of polypharmacology, the risks of undesired target mediated adverse effects need to be specially examined in any future analysis.

Acknowledgments

The authors are thankful to OpenEye Scientific Software, Santa Fe, NM, USA, for providing access to free academic evaluation license. SC acknowledges the financial support by Department of Science and Technology, Government of India toward her research through DST-INSPIRE fellowship. PC is thankful to All India Council of Technical Education, Government of India for her GPAT scholarship. This research is supported by Mathematical Biology program and FIST program sponsored by the Department of Science and Technology and also by the Department of Biotechnology, Government of India in the form of IISc-DBT partnership program. Support from UGC, India—Centre for Advanced Studies and Ministry of Human Resource Development, India, is gratefully acknowledged. NS is a J. C. Bose National Fellow.

References

[1] Collins I, Workman P. New approaches to molecular cancer therapeutics. Nat Chem Biol 2006;2(12): 689–700. https://doi.org/10.1038/nchembio840.

[2] Bhullar KS, Lagaron NO, McGowan EM, Parmar I, Jha A, Hubbard BP, Rupasinghe HPV. Kinase-targeted cancer therapies: progress, challenges and future directions. Mol Cancer 2018;17(1):48. https://doi.org/10.1186/s12943-018-0804-2.

[3] Ferguson FM, Gray NS. Kinase inhibitors: the road ahead. Nat Rev Drug Discov 2018;17(5):353–77. https://doi.org/10.1038/nrd.2018.21.

[4] Zhang J, Yang PL, Gray NS. Targeting cancer with small molecule kinase inhibitors. Nat Rev Cancer 2009;9(1):28–39. https://doi.org/10.1038/nrc2559.

[5] Borisa AC, Bhatt HG. A comprehensive review on Aurora kinase: small molecule inhibitors and clinical trial studies. Eur J Med Chem 2017;140:1–19. https://doi.org/10.1016/j.ejmech.2017.08.045.

[6] Carneiro BA, Meeks JJ, Kuzel TM, Scaranti M, Abdulkadir SA, Giles FJ. Emerging therapeutic targets in bladder cancer. Cancer Treat Rev 2015;41(2):170–8. https://doi.org/10.1016/j.ctrv.2014.11.003.

[7] Tang A, Gao K, Chu L, Zhang R, Yang J, Zheng J. Aurora kinases: novel therapy targets in cancers. Oncotarget 2017;8(14):23937–54. https://doi.org/10.18632/oncotarget.14893.

[8] Taylor S, Peters JM. Polo and Aurora kinases: lessons derived from chemical biology. Curr Opin Cell Biol 2008;20(1):77–84. https://doi.org/10.1016/j.ceb.2007.11.008.

[9] Bavetsias V, Faisal A, Crumpler S, Brown N, Kosmopoulou M, Joshi A, et al. Aurora isoform selectivity: design and synthesis of imidazo[4,5-b]pyridine derivatives as highly selective inhibitors of Aurora-A kinase in cells. J Med Chem 2013;56(22):9122–35. https://doi.org/10.1021/jm401115g.

[10] Bouloc N, Large JM, Kosmopoulou M, Sun C, Faisal A, Matteucci M, et al. Structure-based design of imidazo [1,2-a]pyrazine derivatives as selective inhibitors of aurora-A kinase in cells. Bioorg Med Chem Lett 2010; 20(20):5988–93. https://doi.org/10.1016/j.bmcl.2010.08.091.

[11] Crane R, Gadea B, Littlepage L, Wu H, Ruderman JV. Aurora A, meiosis and mitosis. Biol Cell 2004;96(3): 215–29. https://doi.org/10.1016/j.biolcel.2003.09.008.

[12] Bischoff JR, Anderson L, Zhu Y, Mossie K, Ng L, Souza B, et al. A homologue of Drosophila Aurora kinase is oncogenic and amplified in human colorectal cancers. EMBO J 1998;17(11):3052–65. https://doi.org/10.1093/emboj/17.11.3052.

[13] D'Assoro AB, Haddad T, Galanis E. Aurora-A kinase as a promising therapeutic target in cancer. Front Oncol 2016;5:1. https://doi.org/10.3389/fonc.2015.00295.

[14] Piszczatowski RT, Steidl U. Aurora kinase A inhibition: a mega-hit for myelofibrosis therapy? Clin Cancer Res 2019;25(16):4868–70. https://doi.org/10.1158/1078-0432.Ccr-19-1481.

[15] Zhou H, Kuang J, Zhong L, Kuo WL, Gray JW, Sahin A, et al. Tumour amplified kinase STK15/BTAK induces centrosome amplification, aneuploidy and transformation. Nat Genet 1998;20(2):189–93. https://doi.org/10.1038/2496.

[16] Carvajal RD, Tse A, Schwartz GK. Aurora kinases: new targets for cancer therapy. Clin Cancer Res 2006;12(23): 6869–75. https://doi.org/10.1158/1078-0432.CCR-06-1405.

[17] McIntyre PJ, Collins PM, Vrzal L, Birchall K, Arnold LH, Mpamhanga C, et al. Characterization of three druggable hot-spots in the aurora-A/TPX2 interaction using biochemical, biophysical, and fragment-based approaches. ACS Chem Biol 2017;12(11):2906–14. https://doi.org/10.1021/acschembio.7b00537.

[18] van Gijn SE, Wierenga E, van den Tempel N, Kok YP, Heijink AM, Spierings DCJ, et al. TPX2/Aurora kinase A signaling as a potential therapeutic target in genomically unstable cancer cells. Oncogene 2019;38(6): 852–67. https://doi.org/10.1038/s41388-018-0470-2.

[19] Ota T, Suto S, Katayama H, Han ZB, Suzuki F, Maeda M, et al. Increased mitotic phosphorylation of histone H3 attributable to AIM-1/Aurora-B overexpression contributes to chromosome number instability. Cancer Res 2002;62(18):5168–77.

[20] Quartuccio SM, Schindler K. Functions of Aurora kinase C in meiosis and cancer. Front Cell Dev Biol 2015;3. https://doi.org/10.3389/fcell.2015.00050.

[21] Tsou J, Chang K, Chang-Liao P, Yang S, Lee C, Chen Y, et al. Aberrantly expressed AURKC enhances the transformation and tumourigenicity of epithelial cells. J Pathol 2011;225(2):243–54. https://doi.org/10.1002/path.2934.

[22] Zekri A, Lesan V, Ghaffari SH, Tabrizi MH, Modarressi MH. Gene amplification and overexpression of aurora-C in breast and prostate cancer cell lines. Oncol Res 2012;20(5):241–50. https://doi.org/10.3727/096504013x13589503482978.

[23] Cervantes A, Elez E, Roda D, Ecsedy J, Macarulla T, Venkatakrishnan K, et al. Phase I pharmacokinetic/pharmacodynamic study of MLN8237, an investigational, oral, selective Aurora a kinase inhibitor, in patients with advanced solid tumors. Clin Cancer Res 2012;18(17):4764–74. https://doi.org/10.1158/1078-0432.CCR-12-0571.

[24] Dees EC, Cohen RB, von Mehren M, Stinchcombe TE, Liu H, Venkatakrishnan K, et al. Phase I study of Aurora A kinase inhibitor MLN8237 in advanced solid tumors: safety, pharmacokinetics, pharmacodynamics, and bioavailability of two oral formulations. Clin Cancer Res 2012;18(17):4775–84. https://doi.org/10.1158/1078-0432.CCR-12-0589.

[25] Dees EC, Infante JR, Cohen RB, O'Neil BH, Jones S, von Mehren M, et al. Phase 1 study of MLN8054, a selective inhibitor of Aurora A kinase in patients with advanced solid tumors. Cancer Chemother Pharmacol 2011;67(4):945–54. https://doi.org/10.1007/s00280-010-1377-y.

[26] Friedberg JW, Mahadevan D, Cebula E, Persky D, Lossos I, Agarwal AB, et al. Phase II study of alisertib, a selective Aurora A kinase inhibitor, in relapsed and refractory aggressive B- and T-cell non-hodgkin lymphomas. J Clin Oncol 2014;32(1):44–50. https://doi.org/10.1200/JCO.2012.46.8793.

[27] Kelly KR, Shea TC, Goy A, Berdeja JG, Reeder CB, McDonagh KT, et al. Phase I study of MLN8237–investigational Aurora A kinase inhibitor–in relapsed/refractory multiple myeloma, non-Hodgkin lymphoma and chronic lymphocytic leukemia. Invest N Drugs 2014;32(3):489–99. https://doi.org/10.1007/s10637-013-0050-9.

[28] Macarulla T, Cervantes A, Elez E, Rodriguez-Braun E, Baselga J, Rosello S, et al. Phase I study of the selective Aurora A kinase inhibitor MLN8054 in patients with advanced solid tumors: safety, pharmacokinetics, and pharmacodynamics. Mol Cancer Therapeut 2010; 9(10):2844–52. https://doi.org/10.1158/1535-7163.MCT-10-0299.

[29] Matulonis UA, Sharma S, Ghamande S, Gordon MS, Del Prete SA, Ray-Coquard I, et al. Phase II study of MLN8237 (alisertib), an investigational Aurora A kinase inhibitor, in patients with platinum-resistant or -refractory epithelial ovarian, fallopian tube, or primary peritoneal carcinoma. Gynecol Oncol 2012; 127(1):63–9. https://doi.org/10.1016/j.ygyno.2012.06.040.

[30] Mosse YP, Lipsitz E, Fox E, Teachey DT, Maris JM, Weigel B, et al. Pediatric phase I trial and pharmacokinetic study of MLN8237, an investigational oral selective small-molecule inhibitor of Aurora kinase A: a Children's Oncology Group Phase I Consortium study. Clin Cancer Res 2012;18(21):6058–64. https://doi.org/10.1158/1078-0432.CCR-11-3251.

[31] Godl K, Gruss OJ, Eickhoff J, Wissing J, Blencke S, Weber M, et al. Proteomic characterization of the angiogenesis inhibitor SU6668 reveals multiple impacts on cellular kinase signaling. Cancer Res 2005;65(15): 6919–26. https://doi.org/10.1158/0008-5472.CAN-05-0574.

[32] Nakamura T, Ozawa S, Kitagawa Y, Ueda M, Kubota T, Kitajima M. Antiangiogenic agent SU6668 suppresses the tumor growth of xenografted A-431 cells. Oncol Rep 2006;15(1):79–83.

[33] Sessa F, Mapelli M, Ciferri C, Tarricone C, Areces LB, Schneider TR, et al. Mechanism of Aurora B activation by INCENP and inhibition by hesperadin. Mol Cell 2005;18(3):379–91. https://doi.org/10.1016/j.molcel.2005.03.031.

[34] Boss DS, Witteveen PO, van der Sar J, Lolkema MP, Voest EE, Stockman PK, et al. Clinical evaluation of AZD1152, an i.v. inhibitor of Aurora B kinase, in patients with solid malignant tumors. Ann Oncol 2011; 22(2):431–7. https://doi.org/10.1093/annonc/mdq344.

[35] Dennis M, Davies M, Oliver S, D'Souza R, Pike L, Stockman P. Phase I study of the Aurora B kinase inhibitor barasertib (AZD1152) to assess the pharmacokinetics, metabolism and excretion in patients with acute myeloid leukemia. Cancer Chemother Pharmacol 2012; 70(3):461–9. https://doi.org/10.1007/s00280-012-1939-2.

[36] Kantarjian HM, Martinelli G, Jabbour EJ, Quintas-Cardama A, Ando K, Bay JO, et al. Stage I of a phase 2 study assessing the efficacy, safety, and tolerability of barasertib (AZD1152) versus low-dose cytosine arabinoside in elderly patients with acute myeloid leukemia. Cancer 2013;119(14):2611–9. https://doi.org/10.1002/cncr.28113.

[37] Kantarjian HM, Sekeres MA, Ribrag V, Rousselot P, Garcia-Manero G, Jabbour EJ, et al. Phase I study assessing the safety and tolerability of barasertib (AZD1152) with low-dose cytosine arabinoside in elderly patients with AML. Clin Lymphoma Myeloma Leuk 2013;13(5):559–67. https://doi.org/10.1016/j.clml.2013.03.019.

[38] Lowenberg B, Muus P, Ossenkoppele G, Rousselot P, Cahn JY, Ifrah N, et al. Phase 1/2 study to assess the safety, efficacy, and pharmacokinetics of barasertib (AZD1152) in patients with advanced acute myeloid leukemia. Blood 2011;118(23):6030–6. https://doi.org/10.1182/blood-2011-07-366930.

[39] Tsuboi K, Yokozawa T, Sakura T, Watanabe T, Fujisawa S, Yamauchi T, et al. A Phase I study to assess the safety, pharmacokinetics and efficacy of barasertib (AZD1152), an Aurora B kinase inhibitor, in Japanese patients with advanced acute myeloid leukemia. Leuk Res 2011;35(10):1384–9. https://doi.org/10.1016/j.leukres.2011.04.008.

[40] Jani JP, Arcari J, Bernardo V, Bhattacharya SK, Briere D, Cohen BD, et al. PF-03814735, an orally bioavailable small molecule Aurora kinase inhibitor for cancer therapy. Mol Cancer Therapeut 2010;9(4):883–94. https://doi.org/10.1158/1535-7163.MCT-09-0915.

[41] Meulenbeld HJ, Bleuse JP, Vinci EM, Raymond E, Vitali G, Santoro A, et al. Randomized phase II study of danusertib in patients with metastatic castration-resistant prostate cancer after docetaxel failure. BJU Int 2013;111(1):44–52. https://doi.org/10.1111/j.1464-410X.2012.11404.x.

[42] Schoffski P, Besse B, Gauler T, de Jonge MJ, Scambia G, Santoro A, et al. Efficacy and safety of biweekly i.v. administrations of the Aurora kinase inhibitor danusertib hydrochloride in independent cohorts of patients with advanced or metastatic breast, ovarian, colorectal, pancreatic, small-cell and non-small-cell lung cancer: a multi-tumour, multi-institutional phase II study. Ann Oncol 2015;26(3):598–607. https://doi.org/10.1093/annonc/mdu566.

[43] Schoffski P, Jones SF, Dumez H, Infante JR, Van Mieghem E, Fowst C, et al. Phase I, open-label, multicentre, dose-escalation, pharmacokinetic and pharmacodynamic trial of the oral aurora kinase inhibitor PF-03814735 in advanced solid tumours. Eur J Cancer 2011;47(15):2256–64. https://doi.org/10.1016/j.ejca.2011.07.008.

[44] Steeghs N, Eskens FA, Gelderblom H, Verweij J, Nortier JW, Ouwerkerk J, et al. Phase I pharmacokinetic and pharmacodynamic study of the aurora kinase inhibitor danusertib in patients with advanced or metastatic solid tumors. J Clin Oncol 2009;27(30):5094–101. https://doi.org/10.1200/JCO.2008.21.6655.

[45] Bavetsias V, Linardopoulos S. Aurora kinase inhibitors: current status and outlook. Front Oncol 2015;5:278. https://doi.org/10.3389/fonc.2015.00278.

[46] Falchook GS, Bastida CC, Kurzrock R. Aurora kinase inhibitors in oncology clinical trials: current state of the progress. Semin Oncol 2015;42(6):832–48. https://doi.org/10.1053/j.seminoncol.2015.09.022.

[47] Cohen P, Alessi DR. Kinase drug discovery—what's next in the field? ACS Chem Biol 2013;8(1):96–104. https://doi.org/10.1021/cb300610s.

[48] Goldenson B, Crispino JD. The aurora kinases in cell cycle and leukemia. Oncogene 2015;34(5):537–45. https://doi.org/10.1038/onc.2014.14.

[49] de Souza VB, Kawano DF. Structural basis for the design of allosteric inhibitors of the Aurora kinase A enzyme in the cancer chemotherapy. Biochim Biophys Acta Gen Subj 2020;1864(1):129448. https://doi.org/10.1016/j.bbagen.2019.129448.

[50] March-Vila E, Pinzi L, Sturm N, Tinivella A, Engkvist O, Chen H, Rastelli G. On the integration of in silico drug design methods for drug repurposing. Front Pharmacol 2017;8:298. https://doi.org/10.3389/fphar.2017.00298.

[51] Vanhaelen Q, Mamoshina P, Aliper AM, Artemov A, Lezhnina K, Ozerov I, et al. Design of efficient computational workflows for in silico drug repurposing. Drug Discov Today 2017;22(2):210–22. https://doi.org/10.1016/j.drudis.2016.09.019.

[52] Wang X, Song K, Li L, Chen L. Structure-based drug design strategies and challenges. Curr Top Med Chem 2018;18(12):998–1006. https://doi.org/10.2174/1568026618666180813152921.

[53] Leelananda SP, Lindert S. Computational methods in drug discovery. Beilstein J Org Chem 2016;12:2694–718. https://doi.org/10.3762/bjoc.12.267.

[54] Chen YC. Beware of docking! Trends Pharmacol Sci 2015;36(2):78–95. https://doi.org/10.1016/j.tips.2014.12.001.

[55] Huang SY, Grinter SZ, Zou X. Scoring functions and their evaluation methods for protein-ligand docking: recent advances and future directions. Phys Chem Chem Phys 2010;12(40):12899–908. https://doi.org/10.1039/c0cp00151a.

[56] Hawkins PC, Skillman AG, Nicholls A. Comparison of shape-matching and docking as virtual screening tools. J Med Chem 2007;50(1):74–82. https://doi.org/10.1021/jm0603365.

[57] Lee CH, Huang HC, Juan HF. Reviewing ligand-based rational drug design: the search for an ATP synthase inhibitor. Int J Mol Sci 2011;12(8):5304–18. https://doi.org/10.3390/ijms12085304.

[58] Prathipati P, Mizuguchi K. Integration of ligand and structure based approaches for CSAR-2014. J Chem Inf Model 2016;56(6):974–87. https://doi.org/10.1021/acs.jcim.5b00477.

[59] Rush 3rd TS, Grant JA, Mosyak L, Nicholls A. A shape-based 3-D scaffold hopping method and its application to a bacterial protein-protein interaction. J Med Chem 2005;48(5):1489–95. https://doi.org/10.1021/jm040163o.

[60] Park K. A review of computational drug repurposing. Transl Clin Pharmacol 2019;27(2). https://doi.org/10.12793/tcp.2019.27.2.59.

[61] Pushpakom S, Iorio F, Eyers PA, Escott KJ, Hopper S, Wells A, et al. Drug repurposing: progress, challenges and recommendations. Nat Rev Drug Discov 2019;18(1):41–58. https://doi.org/10.1038/nrd.2018.168.

[62] Xue H, Li J, Xie H, Wang Y. Review of drug repositioning approaches and resources. Int J Biol Sci 2018;14(10):1232–44. https://doi.org/10.7150/ijbs.24612.

[63] Kumar A, Zhang KYJ. Advances in the development of shape similarity methods and their application in drug discovery. Front Chem 2018;6:315. https://doi.org/10.3389/fchem.2018.00315.

[64] Koes DR, Camacho CJ. Shape-based virtual screening with volumetric aligned molecular shapes. J Comput Chem 2014;35(25):1824–34. https://doi.org/10.1002/jcc.23690.

[65] Muchmore SW, Debe DA, Metz JT, Brown SP, Martin YC, Hajduk PJ. Application of belief theory to similarity data fusion for use in analog searching and lead hopping. J Chem Inf Model 2008;48(5):941–8. https://doi.org/10.1021/ci7004498.

[66] Sastry GM, Dixon SL, Sherman W. Rapid shape-based ligand alignment and virtual screening method based on atom/feature-pair similarities and volume overlap scoring. J Chem Inf Model 2011;51(10):2455–66. https://doi.org/10.1021/ci2002704.

[67] Swann SL, Brown SP, Muchmore SW, Patel H, Merta P, Locklear J, Hajduk PJ. A unified, probabilistic framework for structure- and ligand-based virtual screening. J Med Chem 2011;54(5):1223–32. https://doi.org/10.1021/jm1013677.

[68] Berman HM, Westbrook J, Feng Z, Gilliland G, Bhat TN, Weissig H, et al. The protein data bank. Nucleic Acids Res 2000;28(1):235–42. https://doi.org/10.1093/nar/28.1.235.

[69] Dimitropoulos D, Ionides J, Henrick K. Using MSDchem to search the PDB ligand dictionary. Curr Protoc Bioinf 2006. https://doi.org/10.1002/0471250953.bi1403s15. Chapter 14, Unit14 13.

[70] Wishart DS, Knox C, Guo AC, Cheng D, Shrivastava S, Tzur D, et al. DrugBank: a knowledgebase for drugs, drug actions and drug targets. Nucleic Acids Res 2008;36(Database issue):D901–6. https://doi.org/10.1093/nar/gkm958.

[71] Weininger D. SMILES, a chemical language and information system. 1. Introduction to methodology and encoding rules. J Chem Inf Comput Sci 1988;28(1):31–6. https://doi.org/10.1021/ci00057a005.

[72] Hawkins PC, Skillman AG, Warren GL, Ellingson BA, Stahl MT. Conformer generation with OMEGA: algorithm and validation using high quality structures from the Protein Databank and Cambridge Structural Database. J Chem Inf Model 2010;50(4):572–84. https://doi.org/10.1021/ci100031x.

[73] Trott O, Olson AJ. AutoDock Vina: improving the speed and accuracy of docking with a new scoring function, efficient optimization, and multithreading. J Comput Chem 2010;31(2):455–61. https://doi.org/10.1002/jcc.21334.

[74] Xing L, Klug-Mcleod J, Rai B, Lunney EA. Kinase hinge binding scaffolds and their hydrogen bond patterns. Bioorg Med Chem 2015;23(19):6520–7. https://doi.org/10.1016/j.bmc.2015.08.006.

[75] Arris CE, Boyle FT, Calvert AH, Curtin NJ, Endicott JA, Garman EF, et al. Identification of novel purine and pyrimidine cyclin-dependent kinase inhibitors with distinct molecular interactions and tumor cell growth inhibition profiles. J Med Chem 2000;43(15):2797–804. https://doi.org/10.1021/jm990628o.

[76] Cereto-Massague A, Ojeda MJ, Joosten RP, Valls C, Mulero M, Salvado MJ, et al. The good, the bad and the dubious: VHELIBS, a validation helper for ligands and binding sites. J Cheminf 2013;5(1):36. https://doi.org/10.1186/1758-2946-5-36.

[77] van Linden OP, Kooistra AJ, Leurs R, de Esch IJ, de Graaf C. KLIFS: a knowledge-based structural database to navigate kinase-ligand interaction space. J Med Chem 2014;57(2):249–77. https://doi.org/10.1021/jm400378w.

[78] Alibek K, Bekmurzayeva A, Mussabekova A, Sultankulov B. Using antimicrobial adjuvant therapy in cancer treatment: a review. Infect Agents Cancer 2012;7(1):33. https://doi.org/10.1186/1750-9378-7-33.

[79] Shaimerdenova M, Karapina O, Mektepbayeva D, Alibek K, Akilbekova D. The effects of antiviral treatment on breast cancer cell line. Infect Agents Cancer 2017;12:18. https://doi.org/10.1186/s13027-017-0128-7.

[80] Katayama H, Sasai K, Kawai H, Yuan ZM, Bondaruk J, Suzuki F, et al. Phosphorylation by aurora kinase A induces Mdm2-mediated destabilization and inhibition of p53. Nat Genet 2004;36(1):55–62. https://doi.org/10.1038/ng1279.

[81] Bayliss R, Sardon T, Vernos I, Conti E. Structural basis of Aurora-A activation by TPX2 at the mitotic spindle. Mol Cell 2003;12(4):851–62. https://doi.org/10.1016/s1097-2765(03)00392-7.

[82] Cristalli G, Franchetti P, Grifantini M, Vittori S, Lupidi G, Riva F, et al. Adenosine deaminase inhibitors. Synthesis and biological activity of deaza analogues of erythro-9-(2-hydroxy-3-nonyl)adenine. J Med Chem 1988;31(2):390–3. https://doi.org/10.1021/jm00397a021.

[83] Cristalli G, Vittori S, Eleuteri A, Grifantini M, Volpini R, Lupidi G, et al. Purine and 1-deazapurine ribonucleosides and deoxyribonucleosides: synthesis and biological activity. J Med Chem 1991;34(7):2226–30. https://doi.org/10.1021/jm00111a044.

[84] Shen M, Asawa R, Zhang YQ, Cunningham E, Sun H, Tropsha A, et al. Quantitative high-throughput phenotypic screening of pediatric cancer cell lines identifies multiple opportunities for drug repurposing. Oncotarget 2018;9(4):4758–72. https://doi.org/10.18632/oncotarget.23462.

[85] Kentsis A, Topisirovic I, Culjkovic B, Shao L, Borden KL. Ribavirin suppresses eIF4E-mediated oncogenic transformation by physical mimicry of the 7-methyl guanosine mRNA cap. Proc Natl Acad Sci U S A 2004;101(52):18105–10. https://doi.org/10.1073/pnas.0406927102.

[86] Kentsis A, Volpon L, Topisirovic I, Soll CE, Culjkovic B, Shao L, Borden KL. Further evidence that ribavirin interacts with eIF4E. RNA 2005;11(12):1762–6. https://doi.org/10.1261/rna.2238705.

[87] Wang R, Fang X, Lu Y, Wang S. The PDBbind database: collection of binding affinities for protein-ligand complexes with known three-dimensional structures. J Med Chem 2004;47(12):2977–80. https://doi.org/10.1021/jm030580l.

[88] Liu T, Lin Y, Wen X, Jorissen RN, Gilson MK. BindingDB: a web-accessible database of experimentally determined protein-ligand binding affinities. Nucleic Acids Res 2007;35(Database issue):D198–201. https://doi.org/10.1093/nar/gkl999.

[89] Roskoski Jr R. Classification of small molecule protein kinase inhibitors based upon the structures of their drug-enzyme complexes. Pharmacol Res 2016;103:26–48. https://doi.org/10.1016/j.phrs.2015.10.021.

[90] Blaheta RA, Cinatl Jr J. Anti-tumor mechanisms of valproate: a novel role for an old drug. Med Res Rev 2002;22(5):492–511. https://doi.org/10.1002/med.10017.

[91] Brodie SA, Brandes JC. Could valproic acid be an effective anticancer agent? The evidence so far. Expert Rev Anticancer Ther 2014;14(10):1097–100. https://doi.org/10.1586/14737140.2014.940329.

[92] Michaelis M, Doerr HW, Cinatl Jr J. Valproic acid as anti-cancer drug. Curr Pharmaceut Des 2007;13(33):3378–93.

CHAPTER

3

Machine learning strategies for identifying repurposed drugs for cancer therapy

Hansaim Lim[1], *Lei Xie*[1,2,3,4]

[1]The Ph.D. Program in Biochemistry, The Graduate Center, The City University of New York, New York, NY, United States; [2]Department of Computer Science, Hunter College, The City University of New York, New York, NY, United States; [3]The Ph.D. Program in Computer Science & Biology, The Graduate Center, The City University of New York, New York, NY, United States; [4]Helen and Robert Appel Alzheimer's Disease Research Institute, Feil Family Brain & Mind Research Institute, Weill Cornell Medicine, Cornell University, New York, NY, United States

OUTLINE

Drug Repurposing in Cancer Therapy
https://doi.org/10.1016/B978-0-12-819668-7.00003-8

Introduction

The 2018 Nobel Prize in Physiology or Medicine was awarded to two researchers for their discovery of immune checkpoint proteins [1]. Although the discovery initiated the development of the revolutionary targeted immunotherapies, critical challenges remain to be addressed, making them far from perfection [2]. Limitations have been observed earlier in chemotherapies, where cytotoxic molecular compounds are administered to directly kill tumor cells and slow the progression of disease [3]. Other types of targeted therapies, where small molecules interfere with tumor-specific molecular abnormalities also has been actively studied for over half a century and used in parallel with other types of anticancer therapies. While advancements in our understanding of cancer biology and the treatment options made some types of cancers manageable, challenges remain in all types of anticancer therapies. First of all, the efficacy is often limited to a small subset of patients, and it is a difficult task to predict patients' responses to treatments [4]. Furthermore, it is now well known that many anticancer drugs, including both immunotherapeutic and other targeted therapeutic agents, cause serious side effects [5–7]. Natural redundancy and diversity in biological network, such as feedback mechanisms, which enhance robustness of phenotypical outcomes against perturbations, makes it difficult to pinpoint the targets that effectively stop tumor progression, and multidrug resistance develops over time, making the tumors immortal against the previously effective treatments [8–10]. More effective and safer drugs are needed to battle against and eventually conquer cancers.

Despite the urgent need, drug discovery is generally a lengthy and costly process with high risk of failure. A variety of risk factors delay new drugs entering the market and increases the chance of withdrawal, increasing the overall cost of drug development [11]. Drug-induced side effects and toxicity are one of the key issues relevant to the high rate of drug attrition [12]. The limited success of drug development suggests flaws in the long-standing paradigm: one drug–one target–one disease, where the goal is to design a molecule that inhibits a biological target (e.g., protein) known to be crucial for the disease. Under the paradigm, drug candidate molecules are optimized to interact with the intended target (on-target), without proactively understanding off-target interactions and leaving safety and toxicity tests to the later stage of pipeline. Off-target interactions, most of the time unexpected, may cause undesirable outcomes, even fatal ones in extreme cases, leaving irreparable damages to the business and patients [13]. On the other hand, the treatment of complex diseases may benefit from off-target interactions if both on- and off-target synergistically reverse the pathological processes. Indeed, it is now known that many approved drugs interact with more than one biological targets [14,15], and the importance of understanding the complex biological interactions for such multitargeting drug molecules is emphasized as a new paradigm of drug discovery, polypharmacology [15,16].

Instead of suffering from unexpected outcomes caused by off-target interactions, polypharmacology is an attempt to understand drug actions as results of multiple different interactions involving the drug molecule and maximize the benefit from the interactions. Indeed, the superiority of multitargeting drugs over highly selective single-target drugs is suggested [15]. Under the new paradigm, off-target interactions of existing drugs can be used to repurpose them for new indications. Protein–ligand interaction profiles for new ligands can be computationally predicted, and the chemical scaffolds active for multiple targets of interest can be integrated into a single molecule to maximize therapeutic effects and minimize adverse events [17]. New discoveries in biological studies reveal previously unattended anticancer targets, and new drugs can be designed to play multiple roles in the biological network [18,19]. Therefore, early understanding of drug–target interaction profile across whole genome space is essential for development of new, more effective and safer drugs. However, our knowledge of intermolecular interactions drug molecules cause is limited. It is prohibitive to experimentally evaluate all possible drug interactions. Drugs and drug candidate molecules are typically screened against a subset of potential biological targets, resulting in biased, noisy, sparse, and incomplete interaction profiles. At present, no experimental techniques are affordable and scalable enough to experimentally screen compounds for their complete bioactivity profile in humans.

Although not scalable yet to the whole human genome, high-throughput experimental methods have produced tremendous amount of compound bioactivity data, providing rich resources for data-driven knowledge discovery. To rapidly and systematically explore the data and discover hidden knowledge, various types of computational methods have been developed and applied to predict potential interactions of drug molecules. Among many classes of computational methods applicable for drug discovery, those specifically designed to predict unknown protein–ligand interactions are particularly suitable to fill in the sparse knowledge in bioactivity. This review aims to discuss the recent advancements in machine learning methods for the prediction of protein–ligand interactions as well as the efforts to collect and curate experimental data. While some methods directly aim to discover new anticancer therapies, other methods also provide opportunities to discover anticancer therapies when appropriate data sets and validation steps are incorporated. We try to guide readers who are interested in computer-aided drug discovery by providing information about collecting data, preprocessing the data, and methods that can take the processed data for inference. We discuss the major sources of biopharmaceutical databases that are frequently used to train and evaluate computational methods. Then, different types of computational approaches to predict intermolecular interactions are discussed with examples. Fig. 3.1 illustrates a general workflow for computational protein–ligand interaction prediction projects.

It should be emphasized that drug action is a complex process. The genome-wide protein–ligand interaction alone may be insufficient to predict clinical end points, such as therapeutic efficacies and side effects. A systems biology approach is needed to model the collective behavior of biomolecular interactions (e.g., DNA, RNAD, protein, metabolite, drug, etc.), which is beyond the scope of this chapter. Interested readers are referred to other publications [20–25].

Open-access databases for computational drug discovery projects

High-quality and large-scale compound activity data (e.g., protein–ligand interactions) are indispensable in artificial intelligence–based drug discovery projects. To train a computational prediction method, often called a computational model, known protein–ligand interactions are provided in the way the model can accept. Many public databases curating experimentally measured protein–ligand binding affinities can be used for such purpose. The protein–ligand pairs are numerically represented using fingerprinting techniques described below, and the model is trained and evaluated on the numerically processed data sets. Therefore, it is important to appropriately choose and utilize the databases. Databases that provide compound bioactivity data or proteomic information are especially useful for protein–ligand interaction prediction projects. While they are roughly divided into bioactivity-centric and proteomic databases, many current databases are actively

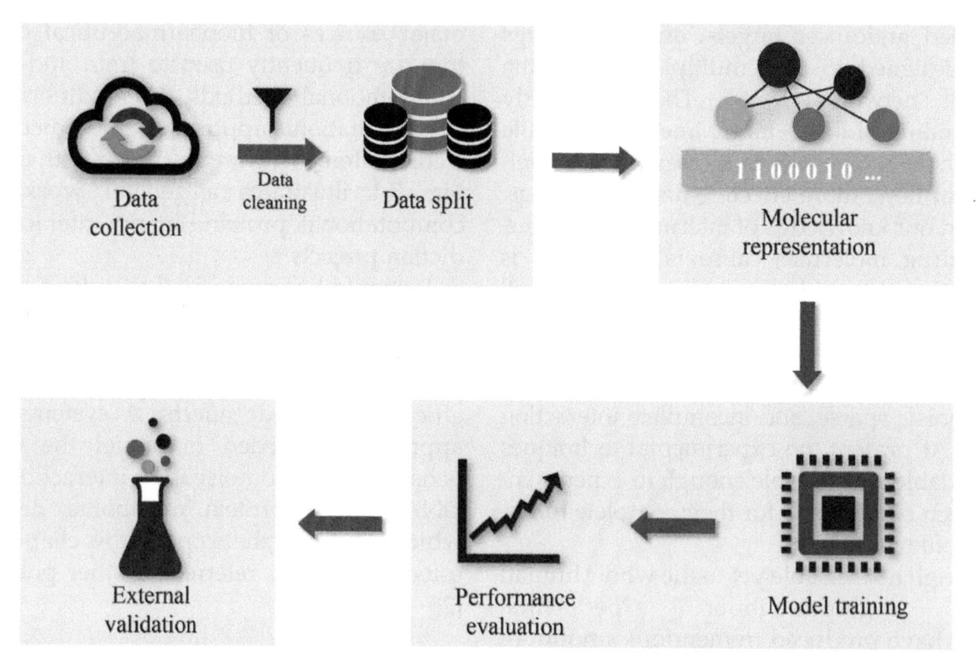

FIGURE 3.1 Illustration of a general workflow for computational protein–ligand interaction prediction projects.

maintained and updated to integrate data from multiple resources, providing more rich and comprehensive data sets. Also, biological network data can be integrated into computational models to help better prediction. The following sections are to describe the commonly used bioactivity and proteomics databases for artificial intelligence–based drug discovery processes and the standard training and evaluation methods (Table 3.1). Other types of large-scale omics data include genomics, transcriptomics, metabolomics, and biological pathway information, and relevant databases are discussed elsewhere [26].

Bioactivity-centric databases

Bioactivity-centric databases provide rich resources for protein–ligand interaction data, which can be used as the known protein–ligand interaction for computational models. While these provide large-scale data, the interaction profiles are not complete, and computational models are used to fill in the incomplete part of the interaction profiles. The databases introduced here are not comprehensive, but these have been updated frequently to maintain the quality and provide more comprehensive knowledge at the time. The bioactivity data that they provide are not necessarily mutually exclusive.

ZINC is a free database of commercially available compounds for virtual screening tasks [27]. Currently, it contains over 700 million purchasable compounds, among which over 200 million are with their 3D structures, making the database especially useful for virtual docking experiments. ChEMBL, a part of the European Molecular Biology Laboratory—European Bioinformatics Institute (EMBL-EBI), is a publicly available database of bioactivities from multiple sources [28]. ChEMBL database (version 22) contains approximately 14 million bioactivity values from more than 1 million assays, covering over 1.6 million unique compounds and 9000 proteins. While major sources of its bioactivity

TABLE 3.1 Commonly used databases for artificial intelligence–based drug discovery projects.

	Name (reference)	Features	Link
Bioactivity-centric	ZINC [27]	Large-scale compound library, compound 3D structures ready for docking	https://zinc.docking.org
	ChEMBL [28]	Continuous-valued bioactivities, large-scale bioassays from publications	https://www.ebi.ac.uk/chembl
	PubChem [29]	Large-scale compound library with activities	https://pubchem.ncbi.nlm.nih.gov
	BindingDB [30]	Continuous-valued bioactivities with focus on potential drug targets	https://www.bindingdb.org/bind/index.jsp
	LINCS [31]	Kinase-specific bioactivities	https://lincs.hms.harvard.edu
	STITCH [32]	Protein–ligand and ligand–ligand interaction data, including predicted activities	http://stitch.embl.de
	BioGRID [33]	Protein–ligand and protein–protein interaction data from publications	https://thebiogrid.org
	SIDER [34]	Drug-induced side effect data	http://sideeffects.embl.de
	KEGG [35]	Bioactivity and biological pathway data	https://www.genome.jp/kegg/kegg1.html
	DrugBank [36]	Drug–protein interactions for approved and investigational drugs	https://www.drugbank.ca
	ExCAPE-DB [37]	Example of systematic data integration	https://doi.org/10.5281/zenodo.2543724
	MUV [43]	Example of systematic data split	https://omictools.com/muv-tool
	DUD-E [44]	Example of systematic data split	http://dude.docking.org
Proteomic	UniProt [38]	Primary amino acid sequences and functional domain information	https://www.uniprot.org
	PDB [39]	Largest existing protein 3D structure database to date	https://www.rcsb.org
	STRING [40]	Protein–protein interaction with functional annotations, including predictions	https://string-db.org
	The Human Protein Atlas [41]	Human protein classifications based on functions and phenotypes	https://www.proteinatlas.org
	Harmonizome [42]	Multiple categories of data relevant to genes, proteins, cell lines, and pathways	http://amp.pharm.mssm.edu/Harmonizome

samples are publications, it also contains samples from both nonprofit and commercial organizations deposited data sets. PubChem, a chemical information database at the US National Center for Biotechnology Information, contains about 250 million bioactivities for over 200 million substances and 17,000 protein targets from over 30 million publications and 3 million patents [29]. BindingDB contains over 1.7 million measured protein–ligand binding affinities, which mainly focuses on small, druglike molecules, and potential druggable target proteins

[30]. The Library of Integrated Network–based Cellular Signatures (LINCS) is a database of cell-based perturbation-response signatures, which contains data samples for small molecules, cells, genes, and proteins categorized by the assay types [31]. STITCH is a protein–ligand interaction database containing over 400,000 chemicals and their interacting protein targets [32]. Its protein–ligand interaction data contain computationally predicted samples as well as samples from other databases. BioGRID contains protein and genetic interactions as well as chemical interactions with posttranslational modification information [33]. SIDER, which is also a part of EMBL, is a database containing known drug–side effect associations [34]. KEGG contains a large-spectrum biochemical and biomedical data sets, including protein–ligand interactions for approved drugs, gene–biological pathway associations, and biomolecular functions (KEGG Orthology) [35]. DrugBank provides rich resources for approved as well as investigational drugs with their known interactions with protein targets [36]. DrugBank is also a great source of drug–drug interactions. ExCAPE-DB is a relatively new database providing open access to the combination of bioactivities from ChEMBL and PubChem databases [37]. While its bioactivity samples are not new, ExCAPE-DB provides a unified set of samples from both databases with appropriate data integration steps.

Proteomic databases

UniProt is a large-scale database maintained by EMBL-EBI, containing over 120 million proteins across all branches of life, with their primary sequences [38]. RCSB Protein Data Bank (PDB) is a database containing 3D structures for biomolecules, including proteins and protein–ligand complexes [39]. STRING database allows researchers to connect these proteins to build protein–protein interaction networks [40]. The Human Protein Atlas project aims to map all human proteins in cells, tissues, and organs using various experimental and computational techniques [41]. It provides category information for genes and proteins based on their functions, compartments, and relevant diseases and drugs. Harmonizome contains data sets for genes and proteins with their associations with other biomolecules, expressions in cells and tissues, and knockout phenotypes [42]. Harmonizome is a collection of data sets from multiple sources, so users may also obtain other types of data sets.

Data preparation for training and evaluating computational models

The above databases provide rich data sets regarding chemical compound structures and their bioactivities, and their target information, including primary sequences and 3D structures of proteins and biomolecules. The differences and variety of data sets make it a critical and challenging step to appropriately prepare and divide samples for machine learning and artificial intelligence–based drug development projects. It may be too simplistic to train a computational model on a database and evaluate its performance on another. The databases mentioned above are regularly maintained and refined, and they often include samples from other databases for better coverage. As a result, for many databases of similar kinds, there are significant number of overlapping samples, making it necessary to properly combine and filter the databases. ExCAPE-DB is an example where appropriate merging and filtering steps are applied [37]. Once combined, a proper data split strategy must be applied so that the samples used to train the model are not included in the test sample set. A possible way of such data preparation step is to integrate multiple databases of interest (e.g., ChEMBL, PubChem, and ZINC for unique chemicals, proteins, and chemical–protein association pairs), filter out redundant samples, and randomly divide samples into training and evaluation sets. This naïve random split strategy often leads to overly optimistic evaluation of computational models.

A better data preparation strategy is to add property-matching filters in addition to the naïve random split so that the training or evaluation sample sets are not overrepresented by some factors that are unlikely to differentiate binding ligands and the others. Maximum Unbiased Validation (MUV) [43] and Database of Useful Decoys-Enhanced (DUD-E) [44] data sets are frequently used to avoid such overrepresentation issues and provide better-prepared data sets for computational drug development projects. MUV data set was designed to overcome artificial enrichment and analogue bias, two major biases in virtual screening data sets. Artificial enrichment stands for the bias when some simple molecular features separates the actives and decoys, such as molecular weight. Computational models suffer from analogue bias when they are trained with data sets having some substructures overrepresented in the actives. A computational model well-trained on such data sets is likely to make biased predictions based on the simple features or overrepresented chemical scaffolds. DUD-E adopts similar idea to the two biases while it focuses on protein–ligand pairs with 3D structures available, making it especially useful for virtual docking or structure-based models. The two data sets, however, have limitations. MUV excludes frequent hitters, the ligands that are active in most tested assays, making it less suitable for projects where the aim is to predict all unknown associations for drugs. DUD-E samples include activities for nonhuman homologs and exclude mutated targets, making it be subjective to species-specific or mutant-specific activities. Furthermore, despite the efforts, DUD-E samples are still not free from the biases as shown in a recent study regarding 3D structure–based protein–ligand binding prediction methods [45]. Both data sets are relatively small as well, partly because of the developed date and the structural coverage. As a result, the data sets and split strategies are often study-specific, and performances may be provided on these data sets for comparison. A gold-standard data set for artificial intelligence–based protein–ligand binding projects is needed.

Representation of molecules for artificial intelligence project

Machine learning and deep learning methods take numerical inputs. Thus, molecules, including small molecules and protein targets, must be numerically represented to be used as inputs (Fig. 3.1). Since the inputs are the information passed to the computational models, they should be meaningful rather than arbitrary numbers. Many efforts have been made to numerically represent small molecules and proteins. The methods to numerically represent biochemical molecules are often called molecular fingerprints or descriptors, and the numerical representations may be used to train computational models or calculate similarity (or distance) scores between molecules.

Chemical molecular fingerprints

Chemical molecular fingerprints are a class of methods that capture some information—commonly the atom types, bond types, topological distances, and chemical substructures—about the input molecules. Atom pair descriptor [46] and topological torsion descriptor [47] are both molecular fingerprint methods developed early in 1980s. While both of them take the atom types, heavy atom neighbors, and π-electrons to numerically represent molecules, they differ in that atom pair descriptor takes topological distances (e.g., how many bonds are in the shortest path between two atoms), while topological torsion descriptor does not. 2D pharmacophore fingerprint [48] is another method that takes the topological distances in addition to the appearances of the nine predefined classes of pharmacophores (e.g., hydrogen bond donor, halogens, etc.). While these fingerprint methods consider some atomic points in a molecule to form graphs connecting them, extended connectivity fingerprint (ECFP) [49] is another method that focuses on the appearance of substructures. ECFP is often used in two different versions: count vectors that count the appearance of substructures and hashed bits with "1"s for appeared substructures and "0"s for the rest. While 3D fingerprint methods exist, little to no benefit over ECFP is reported in virtual screening experiments [50]. Thus, ECFP has often been used to numerically represent small molecules. Many software tools exist for these conventional fingerprinting techniques, including a few open-source tools such as PaDEL descriptor [51], OpenBabel [52], and RDKit [53]. Users can choose appropriate methods for their cases.

Despite the extensive studies, there is no one best way to represent and compare small molecules in all scenarios. Ligand-based virtual screening studies suggest that the best-performing molecular fingerprint is dependent on the method and data [54–57]. Moreover, the conventional fingerprints introduced above are very sparse, containing mostly zeros, wasting the computational resources. It is also known that although with very low probability, bit collision may occur, where two different substructures are represented as the same bit information. Importantly, they are "not learnable" from the given tasks. In other words, the vector representation of a molecule will always be the same and are not adjusted during model training. Thus, ongoing efforts are to develop innovative, novel, yet systematic, and efficient descriptors. Several newer molecular fingerprint techniques have been developed using more advanced artificial intelligence methods.

Mol2vec is a natural language processing (NLP)–based approach to represent small molecule ligands into vectors of continuous values, overcoming the sparsity and bit collision issues [58]. Ligands are first represented as an ordered list of small substructures, like sentence of words in NLP, which are fed into the word2-vec model [59]. Word2vec is known to preserve the semantic similarity of the input words, which is also the case in mol2vec. In other words, the output of two similar ligands are closer than those less similar. The output vectors can then be used to train downstream tasks, such as molecular property prediction or protein–ligand interaction prediction. Neural molecular fingerprint (NeuMF) method is one of the newer chemical molecular fingerprints that uses convolutional neural network to represent small molecules [60]. NeuMF considers a molecule as a graph of atoms and substructures that are connected by varying degree of topological distances. Then, fully connected linear layers are applied to a set of atoms or substructures having certain degree. NeuMF is technically very similar to ECFP as they both split a molecule into a set of substructures comprising it according to the topological distances. The weights for NeuMF are, however, tuned particularly for the given tasks (e.g., solubility prediction or binding activity prediction), and the output vectors contain continuous values. Coley et al. demonstrated that incorporation of additional atomic and molecular features to the NeuMF can improve the performances in molecular property prediction tasks [61]. It is noted that although these newer fingerprint methods alleviate aforementioned issues, stereochemistry is mostly ignored. To date, only a limited number of ligands have known 3D conformation information. While there are ways to obtain the lowest-energy conformers, the lowest-energy conformation may not be the conformation when the ligands are in action (e.g., binding to a target protein). Also, a novel algorithm that learns accurate representations from 3D graphs is needed.

Representation of protein sequence and structure

Representing proteins as numeric vectors is often a critical step in data-driven drug activity modeling. Proteins are in general, however, larger and more complex than drugs: they are polymers of amino acid forming distinct folds and structures to perform biochemical activities. As a result, there are many ways to represent proteins at different scales and depth from whole primary sequence composition to more complex 3D interactions among atoms within binding pockets. Simple protein descriptors can be built based on the composition of amino acids. For example, how many times a particular amino acid appears in a protein can be counted and used as a descriptor for the protein. In the same way, the appearance of doublets or triplets of amino acids may be counted, which increases the complexity of the descriptor. These composition-based descriptors are however too simplistic to capture different properties of amino acids or positional dependences.

Physicochemical properties of amino acids have been studied, recorded, and continuously refined as more data and techniques become available [62,63]. Such physicochemical properties are called amino acid index, which have been utilized to design protein descriptors. For instance, conjoint triad descriptor is a variant of the composition-based descriptor for amino acid triplets [64]. It uses seven distinct amino acid classes based on polarity and volume, which reduces the dimension of the descriptor as well as takes some physicochemical properties into account. Moreau-Broto autocorrelation descriptor is defined as the average of products of a selected amino acid index between the ith and $(i+d)th$ amino acids along the sequence, where d is called the lag of autocorrelation [65]. Other variants of the autocorrelation descriptor can be obtained by replacing the averaging with some other statistics of the amino acid index values. Scale-based protein descriptors are a class of statistical method that attempt to capture the variability of amino acid features. A pioneering effort in the scale-based descriptor is to use a large number of physicochemical properties of the 20 natural amino acids and compress the features by principal component analysis [66]. Further efforts to improve the quality of scale-based descriptors have been made by adjusting the number of amino acid properties and/or including unnatural amino acids [67–69]. Various kinds of these protein descriptors are available [70].

Two key desired properties of protein descriptor are position dependency and long-range dependency. An amino acid may play completely different role in terms of biomolecular activities when its environment is changed, and two sequentially distant amino acids may closely interact with each other to render important properties (e.g., binding pocket). While the lag of autocorrelation somewhat accounts for the long-range dependency, it does not differentiate the amino acids that do interact from the rest. Moreover, above methods consider little, if any, about the positional differences. Profile-based descriptor is a position-dependent protein descriptor based on multiple sequence alignment [71]. Such a descriptor uses the observed and expected frequencies of amino acids at particular positions of aligned sequences, which instead focus less on long-range dependency. A recent advancement in NLP methods provide new opportunities to design protein descriptors that account for both of the dependencies. Based on the successful NLP model, Bidirectional Encoder Representations from Transformers (BERT) [72], Rives et al. showed that the model can learn biochemical properties of amino acids and proteins, such as charge, size, secondary structures, and residue–residue contacts [73]. Briefly, a large amount of protein sequence data from UniProt [30] is used to train the

BERT model in a self-supervised manner (i.e., hide some amino acids and let the model predict what was hidden) without preprocessing the sequences. This step is called pretraining of BERT, which is followed by fine-tuning steps with labeled data (e.g., protein family prediction, binding site prediction, etc.), using the BERT-calculated protein vectors as descriptors. The BERT-based protein descriptor is position-dependent and captures long-range dependency by design. BERT uses positional embedding that differentiates the representation of a word by its position. The long-range dependency is captured by one of the key components of BERT, the self-attention mechanism, which aligns the input embedding vectors by their impacts on each output embedding [74]. One drawback of BERT-based descriptor is the computational cost for training BERT. A recent development of a light version of BERT attempts to reduce the cost by improving the efficiency of parameters during BERT training stage [75]. Computational cost can also be alleviated by using more powerful computational devices or downsampling the input data by preprocessing (e.g., multiple sequence alignment and sequence identity filtering).

Computational methods for protein–ligand interaction prediction

In this chapter, we divide computational protein–ligand interaction prediction methods into four categories: ligand-based, structure-based, network-based, and deep learning–based methods (Table 3.2). Please note that many deep learning-based methods can also fall in one of the three other categories. Ligand-based methods rely on the assumption that the differences in chemical molecular structures can explain their different bioactivities. Small molecular compounds are numerically represented using molecular fingerprint techniques, and the fingerprint vectors are mathematically compared to other compounds having known protein targets. The closer the two fingerprint vectors are, the more likely that the two compounds share the same target proteins. Though it sounds simple, many ligand-based methods show high predictive performances, suggesting that the assumption is valuable. However, limitations exist. They cannot explain the activity cliff [76], and their performances are suboptimal for novel molecules. Structure-based methods also called protein–ligand docking methods take structural

TABLE 3.2 Strengths and weaknesses of protein–ligand interaction prediction methods.

Method type	Strength	Weakness
Ligand-based	Computational cost is generally low, high performance on some well-studied targets, simple to implement	Lack of high-resolution details in predicted interaction, performance limited for screening against targets with many known ligands
Structure-based	High-resolution details can be obtained, can make predictions for new targets (with structure available)	High computational cost, scalability also limited by structural coverage, often suffers from high false positive
Network-based	Systematically integrates knowledge from large amount of data across multiple aspects (multiomics)	Lack of high-resolution details, scalability depends on core algorithms and data requirement, comparing performances of multiple methods is not trivial
Deep learning–based	Often achieves higher performance compared to other types of methods, model flexibility to integrate assumptions from multiple types of methods	Computational cost is usually high, low interpretability of models, requires large amount of data to train models

information of both proteins and ligands and can partially overcome the drawbacks. The physicochemical interactions between atoms comprising the ligands and the binding pockets of proteins are calculated, and predefined energy functions are used to score the protein–ligand pairs of interest. While they can provide detailed pictures of how the given pair of molecules may interact, they suffer from high false positives. Also, high computational burden makes them inappropriate for screening protein–ligand interactions at a large scale. In network-based methods, protein–ligand interactions are treated as a bipartite graph where nodes are proteins and ligands and edges are known protein–ligand interactions. Nodes and edges may possess attributes, representing the characteristics of the molecules or the interactions. The missing edges (e.g., unknown interactions) are predicted using various kinds of mathematical models. Network-based methods often enjoy high scalability due to the mathematical models specially designed for large-scale networks. However, they rely on high-level features (e.g., only a few numbers representing a protein target) to represent the network, limiting their accuracy and interpretability. Deep learning–based methods use artificial neural networks to replace the mathematical models used in the methods of other kinds. With the great successes and explosive interest in deep learning, artificial neural networks, such as convolutional neural network, have been applied to computational protein–ligand interaction predictions. Deep learning–based fingerprints introduced above can be used to represent molecules. The flexibility of deep learning architecture makes the number of possible models infinite, and larger and more complex models are being developed. The field of deep learning is fast-evolving in both its software and hardware, continuously improving the performances on various tasks. While they enjoy the greatest flexibility and high performances, their interpretability has been criticized due to the "black-box" nature of complex artificial neural networks. Deep learning–based methods can also be classified into the other categories listed above.

Ligand-based virtual screening

A traditional assumption in ligand-based virtual screening is that ligands that are similar to each other are more likely to share common targets. For a given target protein, if a query ligand turns out to be very similar to a known active ligand, it is likely that the query ligand will also be active. A systematic approach of this is to decide the numerical representation of ligands (e.g., molecular fingerprints), define a similarity metric (e.g., Tanimoto coefficient or cosine distance), and evaluate the pairwise ligand similarity scores for each query ligand against each known active/inactive ligand. Then, the predicted protein–ligand interaction for the query ligand can be made simply based on the nearest neighbors [57] or based on the statistical significance of the similarity [54,55]. It was shown that with proper measurement, a set of weak ligand similarities (with statistical significance) may be an indication of common target. Moreover, the similarities between ligand pairs contain information relevant to the target proteins [54,77]. Due to the underlying assumption in the ligand-based virtual screening, however, an adequate number of known active/inactive ligands must be provided for each target protein. This makes ligand-based virtual screening inappropriate for cases where insufficient amount data are available. For instance, a suboptimal performance is expected if predictions are made for novel mutant target proteins, where the known active/inactive ligands are available only for the wild-type proteins. An approach taking advantage of the target features should be considered in such cases .

Riniker et al. performed an extensive comparison of 12 commonly used molecular fingerprints on DUD, MUV, and ChEMBL data sets [57]. For each test compound, its similarity score to all active molecules in training set was measured. The maximum similarity scores for each test compound are then used to rank them, and the efficiency of each fingerprint is evaluated using the rank. Circular fingerprints performed better when evaluated by early recognition metrics, whereas path-based fingerprints were better for AUC. It was also found that the performance varies more cross-target–same-fingerprint than the cross-fingerprint–same-target evaluations.

A popular method in ligand-based virtual screening is similarity ensemble approach (SEA), which applies statistical tests on top of the fingerprint-based pairwise ligand similarity search methods to rank the target proteins [54]. In SEA, each target protein is represented as a set of its known active ligands. For each query ligand, similarity to the known actives for each target is calculated, and the sum of the similarity scores above certain threshold are summed to derive raw scores for the query ligand–target protein association. The raw scores are then statistically evaluated against randomly populated set of ligands, and the statistical significance is used to rank the target proteins. In a later study, it was shown that the performance can be improved by using multiple SEA models, each with different types of fingerprints and by preparing for target-specific data sets (e.g., kinase data set) [56].

The SEA method is important in a few aspects. First, it uses all active ligands for a given target, instead of closest neighbors. The high performance of this method suggests that a set of weak similarities may indicate significant overall relatedness. Koutsoukas et al. proposed a probabilistic machine learning model, Parzen–Rosenblatt window method with similar ideas [78]. Each target protein is represented as a set of ECFP fingerprint features for the active ligands, and the probability of the query ligand being active is obtained by comparing how many of the fingerprint features they share. In other words, a target protein is a bag of fingerprint bits from its active ligands, and the query ligand may have many of the bits-in-bag if it is active. Second, the similarity scores are statistically tested against randomly populated set of ligands, instead of simply using the best similarity scores. Awale et al. measured pairwise ligand similarity scores based on multiple types of fingerprints, and the statistical significance of the distances between query and the closest neighbor in the active ligands for a target protein is used to rank the target proteins [55].

Lauria et al. developed DRUDIT, a ligand-based drug discovery tool that is freely accessible via web service [79]. Similar to an earlier work by Riniker et al. [57], ligands are represented as chemical molecular descriptor vectors, which are then used to build templates for target proteins. Here, the chemical descriptors contain not only the presence/absence of certain substructures but also some other physicochemical properties, such as the autocorrelation values. The target template represents the distribution of each descriptor from the active ligands. The input ligand's descriptors are then compared to the target template, where the score is higher if the input ligand's descriptor values are within a small range of the target template distribution. DRUDIT is strictly based on the ligand property and the known protein–ligand associations, and the easy-to-use web service (www.drudit.com) is freely available at the time of writing this review. Such a user-friendly service of computational drug discovery tool is rare and valuable.

Yang et al. used extreme gradient boost to predict the inhibition strength of compounds against JAK2. The authors collected active and inactive compounds for JAK2 from PubChem, BindingDB, and ZINC databases. Molecular fingerprints for the compounds were calculated using RDKit and fed into extreme gradient boost classification and regression models. It was shown that models using ECFP4 performed consistently better, and the extreme gradient boost classification model performed better than virtual docking methods in terms of early enrichment of active compounds [80].

Sadawi et al. reported that their multitask regression algorithm outperformed the single-task counterpart, and incorporation of evolutionary distances among protein targets further improved the performances in majority of the tasks [81]. In the multitask regression method, each target protein is a task, where the protein–ligand affinity data with IC_{50} measurements were obtained from ChEMBL. Using 1024-bit pharmacophore fingerprint (FCFP) as ligand representation, random forest regressor was used as the baseline single-task method, and two types of multitask methods, namely feature-based and instance-based methods, were compared. In feature-based multi-task method, FCFP was used to represent ligand, and the target proteins are simply grouped by their family hierarchy manually curated by ChEMBL, forming multitask data sets for related tasks. In addition to the feature-based method, instance-based method takes the target–target similarity profile as an additional input, where the similarity scores are based on amino acid sequence similarity. While the single-task baseline method has a flaw in that it does not include any target features, the study clearly showed that the target–target similarity scores help improve the predictive performances, i.e., instance-based performed better than feature-based multitask method. It is important to note that in the study, the evaluation data set contained same targets that were appeared in the training set.

Ligand 3D feature–based virtual screening

The limited success of ligand-based virtual screening may be partially attributed to the abstraction of chemical molecular features that cannot fully represent 3D structures of molecules. Therefore, virtual screening methods that utilize ligand 3D features have been studied to more accurately predict protein–ligand interactions. For the additional details in molecular representation, ligand 3D feature-based methods may provide more accurate predictions but may also suffer from higher computational costs. The requirement of ligand 3D structure also limits the scalability of models to chemical space.

Grisoni et al. used ligand 3D features, such as partial charges and 3D shape of compounds, to select a handful of ligands that are likely to show similar interaction profile to (-)-galantamine, a natural product approved for Alzheimer's disease [82]. Potential target proteins for these selected ligands were predicted, and some of them were experimentally validated. Although the proposed target is not cancer, it is a computational approach to discover multitarget drugs.

Hernandez et al. developed a 3D graph–based molecular similarity search method, which performed better than conventional fingerprints for classifying active ligands in DUDE data set [83]. The ligand molecules are represented as a graph of atoms and ring structures containing physical and pharmacophore features as node attributes. Then, maximum common substructures for pair of ligands were heuristically calculated to obtain ligand–ligand similarity scores. The similarity scores are then used to rank ligands in DUDE data sets to classify active and inactive ligands, and the early enrichment performances were generally higher than the conventional fingerprint-based methods. While the gain of performances in early enrichment is noticeable, the proposed 3D graph–based method did not outperform ECFP for all target classes. This may be due to that there are many ligands without known 3D conformation specific to the binding targets. As done in this study, the available standard conformers need to be used for the algorithm, which may not correctly represent the binding poses of the ligands. In cases where the ligands of interest are topologically similar but showing different activities, requiring high-resolution comparisons, such a 3D structure–based method can be more accurate than the 2D topology–based fingerprints.

Fan et al. proposed DStruBTarget, a ligand-based virtual screening method that utilizes 2D and 3D ligand information as well as protein–ligand binding affinity information [84]. In DStrubTarget, the prediction scores are calculated by finding a reference ligand that maximizes a combination of three values: 2D similarity between query and reference ligands, 3D similarity, and scaled binding affinity between the reference ligand and the target of interest. The evaluation demonstrated that incorporation of 3D similarity scores is helpful to improve the predictive performances.

Structure-based virtual screening

Structure-based virtual screening methods are a class of methods that uses 3D structures of the ligands and target proteins as input. Using the physicochemical properties of the 3D structures, the alignment between proteins and ligands are searched and scored. To date, molecular docking simulation methods are actively developed and applied to predict protein–ligand binding activities. Molecular docking simulation methods are frequently used to search for such binding poses by optimizing method–specific scoring functions. Scoring functions are metrics evaluating the protein–ligand docking poses based on physicochemical interactions (force field), statistics of atom pairs from known protein–ligand interactions, or some other types of variables designed to correlate with known binding affinities [85]. A usual workflow in studies based on molecular docking simulation is first to find the 3D structures for the ligands and targets of interest, where ligands often outnumber the targets. Using the structures as input, various molecular docking simulation approaches can be used to rank the ligands against each target with favorable (e.g., low-energy) protein–ligand binding complexes. Some of high-ranked binding complexes are obtained and filtered (e.g., discarding molecules predicted to interact with too many targets). Eventually, a handful of protein–ligand pairs are considered candidates and may be evaluated in vivo or in vitro for experimental validation of predicted activities [86,87]. A number of user-friendly graphic interfaces are available for molecular docking experiments, including AutoDock [88], UCSF Dock [89], MTiOpenScreen [90], HADDOCK [91], and SwissDock [92].

While traditional molecular docking experiments aim to screen a large library of ligands against a few particular targets of interest, inverse docking methods aim to screen a few ligands of interest against a large number of target proteins and therefore are "target fishing." Inverse docking is a useful scheme when a ligand with unknown targets is known for certain phenotypes. A potential target of a known tumor growth inhibitor was suggested by using an inverse docking method [93]. A general inverse docking approach was presented for multitargeting antibacterial drug design [94]. Recently, Wang et al. proposed a consensus inverse docking model by utilizing multiple different docking methods to predict consensus docking conformations [95]. It is available for web access (http://chemyang.ccnu.edu.cn/ccb/server/ACID/).

One of the key strengths of molecular docking methods is that the simulated protein–ligand complexes can be used to provide atom-level insights on how the binding occurs, generating hypotheses about the mechanism of action. For this reason, molecular docking methods can be used as a supplementary method for other classes of methods to enhance the predictions by providing finer-resolution view of the interactions [82,96,97]. However, due to the requirement of 3D structures and atom-level detailed computations, molecular docking methods suffer from computational cost and relatively low scalability. Thus, to date, such methods alone cannot fully exploit the abundance of archived data from high-throughput biochemical and biomedical experiments. When appropriate, other 3D structural features with lower resolution may be used to reduce the computational costs and thus increase scalability of the model [98]. More details of the strengths and drawbacks of molecular docking methods can be found in the reference [85].

Network-based virtual screening

In network-based virtual screening methods, protein–ligand interactions are viewed as a network, where proteins and ligands are nodes, and protein–ligand associations are edges connecting the nodes. A protein–ligand association network can be represented as an *m-by-n* rectangular matrix, where *m* and *n* are the number of unique proteins and ligands. In the matrix form, typically "1" indicates known protein–ligand associations and "0" indicates unknown ones, thereby representing the adjacency (or connectivity) between one type of nodes to another. While many databases can be integrated to build a large-scale protein–ligand association network, the protein–ligand adjacency matrix is usually sparse, noisy, biased, and incomplete. Due to the incomplete coverage of experiments, many interacting protein–ligand pairs have not been evaluated and thus are hidden in the ocean of "0"s in the adjacency matrix. Thus, the aim of many network-based virtual screening methods is to predict the hidden associations based on the known information. For example, the missing links between proteins and ligands can be predicted by evaluating the degree of shared neighbors within the protein–ligand interaction network [99]. Without using any other information than known protein–ligand associations, the assumption here is similar to that of ligand-based methods: similar nodes (in terms of shared neighbors) are likely to interact.

As drug activities, including protein–ligand interactions, happen within a complex and heterogeneous biological environment, many biological features are frequently used to help the protein–ligand interaction prediction, and the capability to utilize such side information is one of the key strengths of network-based methods. To be compatible with protein–ligand interaction data, side information may also be encoded in the form of adjacency matrix or similarity matrix. For example, drug-induced side effects may be in an *n-by-s* matrix, where *n* and *s* are the number of unique drugs and side effects, respectively. When a network consists of multiple types of nodes (e.g., proteins, ligands, diseases, side effects in one network), it is called heterogeneous network, where special treatments are often required to integrate and process the network for prediction. Also, similarities between nodes of same type can be measured and used as features, such as ligand–ligand similarity or protein–protein similarity. Within a network, the groups of nodes of same type are often referred as layers (e.g., ligand layer is a group of ligands in the network), edges connecting nodes of same type are intralayer (or within-layer) edges, and edges connecting nodes of different types are interlayer (or cross-layer) edges. The following examples are heterogeneous network-based protein–ligand interaction prediction methods.

BANDIT utilizes multiple biochemical and biomedical properties of compounds to identify target proteins of an orphan ligand [77]. BANDIT takes ligand chemical structure, gene expression profile upon chemical perturbation, cellular growth inhibition profile, drug-induced side effects, and other types of biochemical assay results to compare large number of ligands. Each pair of ligands are represented as a set of similarity scores, where each score is based on the aforementioned ligand properties. Then, a Bayesian statistic model is used to calculate the likelihood that two ligands share a target protein given the pieces of evidence. Using BANDIT, the researchers were able to predict the target protein of ONC201, an anticancer compound in clinical development whose biological target and mechanisms remained unknown. In addition to high performance and utility in clinical development, the performance of BANDIT suggests that it is beneficial to integrate multiple types of large-scale data sets to predict protein–ligand interactions. While the prediction procedure—that is to compare ligands based on their properties and shared targets are assumed for similar ligands—is similar to that of many ligand-based virtual screening methods introduced above, BANDIT is also capable of separating drugs based on their mechanisms of action, demonstrating its unique strength as a network-based method.

Chu et al. developed DTI-CDF, a network-based machine learning model that uses network representation of protein–ligand associations with intralayer similarity information to predict unknown drug–target associations [100]. In DTI-CDF, protein–protein, and ligand–ligand similarity scores are used to train different numbers of random forest [101] and XGBoost [102] models stacked multiple times to make final predictions. For each stack of the models, the input contains the multiple types of similarity information. From the second stack, the output from previous stack is additionally used as input. The model architecture resembles that of residual network, a popular CNN architecture that has been very successful in image recognition. DTI-CDF, however, is trained and evaluated only on KEGG data set [20], which contains a small subset of proteins and ligands, thus limiting the generalizability of the method to a larger protein–ligand space.

Luo et al. developed a computational framework to repurpose drugs by integrating heterogeneous network of drugs, diseases, side effects, and target proteins [103]. The heterogeneous network consists of both cross-layer (drug–protein, drug–side effect, drug–disease, and protein–disease associations) and within-layer (drug–drug interactions, drug–drug similarities, protein–protein interactions, and protein–protein similarities) edges, and random walk with restart (RWR) [104], a popular network propagation method, was used to extract the feature vectors for each nodes. A feature reduction technique was used to compress the size of the node vectors, and the drug vectors are projected to the target protein space, such that the projection approximates the known drug–protein interaction matrix. The method identified three drugs as potential cyclooxygenase inhibitors, and experimental validation confirmed the inhibitory activities of the drugs. The potential binding modes of the drugs were also predicted using a molecular docking simulation method.

Utilizing a convenient feature of adjacency matrices, Fu et al. attempted to computationally parse through the pathways that link proteins with ligands [105]. They first built more than 50 different semantic network relationships represented by commuting matrices—a kind of adjacency matrix representing the number of length–n pathways between nodes—from adjacency matrices of ligands, ligand substructures, side effects, diseases, protein targets, protein functional annotations, and biological pathways. The commuting matrices can be obtained by matrix multiplication of adjacency matrices. For example, if matrix $\mathbf{A}$ and $\mathbf{B}$ represent protein–ligand and ligand–disease associations, the matrix dot product of $\mathbf{A}$ and $\mathbf{B}$ (i.e., $\mathbf{A} \cdot \mathbf{B}$) represents the number of unique pathways connecting proteins and diseases after one intermediate ligand node. The commuting matrices represent the metapaths between proteins and ligands, which are used to train random forest and support vector machine models to predict binary protein–ligand associations. While the report did not include a repositioning of drugs, the high performance suggests that the incorporation of heterogeneous biological data sets is beneficial for computational predictions.

Notably, the "0"s in protein–ligand interaction network do not necessarily indicate noninteracting pairs. Zheng et al. showed that systematically sampling the true negative pairs can significantly improve performances of many machine learning algorithms for protein–ligand interaction prediction [106]. On a heterogeneous network of proteins, ligands, side effects, and protein functions, they applied guilt-by-association assumption—where protein–ligand pairs are likely negative if they are dissimilar to most of known positives—to obtain high-confidence negative protein–ligand interactions. The improved performances of machine learning models with the imputed negative data suggest that it is overly simplistic to use all "0"s as negative samples.

Nonnegative matrix factorization is also a popular network modeling method that has been applied to computational drug discovery. REMAP takes protein–ligand interaction network as well as the intralayer similarity networks to obtain compressed feature representations for all proteins and ligands in the network [107]. Liu et al. separately developed a similar method, called NRLMF [108], where the main difference from REMAP is that REMAP explicitly considers the possibility that some of "0"s are actually positives. To overcome the mathematical requirement that compressed feature vectors for proteins and ligands must be of the same length, a trifactorization version of REMAP was developed [109]. In both REMAP and TREMAP, ligand–ligand and protein–protein similarities can be calculated based on the 2D molecular structures and primary sequence comparisons, respectively. By the design of REMAP algorithm, once optimized, two structurally similar ligands are assigned similar compressed feature vectors, and so for proteins. This property of the algorithm provides a useful way to represent ligand molecules in a way that contains both its chemical structure and its global target profile in compressed vectors. Wang et al. used the compressed vectors from TREMAP to cluster ligands and applied RWR algorithm to repurpose diazoxide for triple-negative breast cancer [96]. Ayed et al. used the compressed vectors from REMAP to predict drug sensitivity of cancer cell lines, which showed superior performances compared to features based only on ligand structures [110]. A multilayered version of REMAP method was applied to predict drug-induced side effects and side effect–pathway relationships from heterogeneous biological network data [22].

Deep learning–based biochemical activity prediction

Zakharov et al. proposed a deep learning–based multitask protein–ligand interaction prediction model, named deep learning consensus architecture (DLCA) [111]. DLCA takes multiple types of molecular descriptors as inputs, each of which is processed separately by layers of fully connected artificial neurons, and the outputs from each of the separate layers are averaged to provide consensus predictions. The authors adopted both random split and scaffold out validation (i.e., training and test compounds do not share some predefined chemical substructures). The performance dropped significantly when scaffold out validation was performed. However, the consensus model performed better than every individual component model for both random split and scaffold out validation. DLCA is an example where the consensus models (or ensemble models) perform better than the individual component models. It was also shown no single type of molecular fingerprint is superior than the others in general.

Lee et al. developed DeepConv-DTI, a CNN-based binary classification tool for drug–target interaction prediction [112]. DeepConv-DTI takes compound 2D structure and target protein primary sequence as inputs, which are vectorized by ECFP and a simple lookup table (i.e., each amino acid has a unique integer identifier), respectively. The protein lookup values are then projected into embedding vectors, which are then pooled by convolutional layers before being concatenated with the compound vectors. The model was trained on a collection of binary protein–ligand associations in some of the databases mentioned above. Importantly, the convolutional filters applied to the protein sequences seem to highlight the parts of proteins interacting with the compound, suggesting the capability of convolutional neural network to capture the local features relevant to the physicochemical interactions.

Moridi et al. applied variational autoencoder and principal component analysis to extract features for small molecule drugs and diseases to repurpose approved drugs [113]. A small molecule drug is represented by its chemical structure, target protein sequences, relevant enzyme sequences, and gene expression profile under the treatment. These features are processed by variational autoencoders to reduce dimension and extract salient features, and then used to measure drug–drug similarity scores using cosine distance. On the other hand, a disease is represented by its phenotypes and genotypes (i.e., disease characteristics and involved genes), compressed by principal component analysis, which are used to measure disease–disease similarity scores. The predicted drug–disease association scores are then calculated by taking the square root of the maximum product of similarity scores within the drug–drug and disease–disease similarity network. When evaluated for each distinct disease, the model achieved high AUC values for most of the reported diseases. One shortcoming is that the model requires the target genes and relevant enzymes for the input drugs. While it is likely for an approved drug to have known target genes and enzymes, they are not necessarily the key players for the new indications, limiting the interpretability and applicability of the method.

Torng and Altman proposed a two-stage graph CNN method to predict protein–ligand binding affinity [114]. They first built an unsupervised deep autoencoder to represent the binding site pockets for druggable proteins. Each protein pocket is a graph of amino acid residues having any atoms within 6 Å to the bound ligands from PDB cocrystal structures. Residues within 7 Å are connected to form a binding pocket graph. The second stage comprises two separate graph CNN models that learn features from proteins and ligands, supervised by the binding classification labels. Ligand molecules are represented in the second stage using NeuMF graph CNN [60]. Since the two supervised graph CNNs run in parallel to predict the binding activity, the model does not require the cocomplex structure information. They applied negative sampling strategy to include both negative binding pockets and negative ligands. From DUD-E database, randomly chosen binding pockets dissimilar to the active pockets are included as negative pockets. From ChEMBL database, ligands having measured IC_{50} is greater than 50 μM are used as negative ligands. Such negative sampling technique can be critical for successful modeling as the benchmark data sets are often incomplete, sparse, and biased. Zhang et al. developed a deep learning–based protein–ligand binding affinity predictor, DeepBindRG [115]. DeepBindRG extracts interacting atom pairs from protein–ligand cocrystal structures similarly to the Torng and Altman method. The atom pairs are then featurized based on their chemical and force field–based atomic categories to prepare 2D picture-like input data. The popular ResNet architecture [116] was then trained to predict the protein–ligand affinity values in pK_d or pK_i.

Karimi et al. developed DeepAffinity, a protein–ligand affinity prediction tool based on RNN and CNN [117]. The inputs for DeepAffinity are the protein primary sequence and ligand 2D structures in SMILES string format, which are then processed by recurrent neural network (RNN)–based encoder [118] and pairwise attention mechanism. Before feeding into

RNN, protein primary sequences are used to represent their structural properties, such as secondary structure elements and physicochemical properties, while SMILES strings are k-hot encoded (i.e., binary bit vector indicating the individual characters). The RNN-encoded protein and ligand vectors are then further processed by pairwise attention mechanism, followed by CNN layers and fully connected layers to predict pIC_{50} values. It is noted that the authors split the protein–ligand affinity data for novel protein family prediction cases. In other words, the model performances were measured for protein targets that were not included in the training data. Also, it was shown that the attention mechanism assigned higher scores for protein segments that are closer to the binding sites although the input did not contain 3D structures. DeepAffinity does not require 3D structure data, making it more appropriate for large-scale, early stage predictions, where limited information is available for the proteins and ligands of interest.

Wan et al. proposed NeoDTI, a prediction tool for binary protein–ligand interactions based on graph convolutional neural network that is designed to learn from heterogeneous network, which showed superior performances compared to multiple types of methods [119]. Using the same data sets as Luo et al. [103], heterogeneous network was built to consist of intralayer and interlayer relationship edges between drugs, proteins, side effects, and diseases (nodes). Using graph convolutional neural network, similar to that of NeuMF [60], the information about nodes are collected to reflect the nodes themselves and their neighbor nodes connected by different edge types. The update of node information is governed by a loss function that is specially designed to drive the feature vectors of related nodes can be correctly identified after appropriate projection. The idea of the loss function is similar to that of matrix factorization methods—that is the feature vectors of proteins and ligands reproduce known protein–ligand network when multiplied. NeoDTI is a combination of graph-based neural network and matrix factorization method to efficiently represent and predict protein–ligand interactions.

Discussion

We discussed a few important subtypes of computational drug discovery approaches. A typical workflow of artificial intelligence–based drug discovery project contains a few steps, including problem definition, data set preparation, computational model building, training and evaluating the performance of the model, statistical analyses of the performance and predictions, and the inference and validation by assays (Fig. 3.1). Depending on the problem definition, a project can be classified as ligand-based (target), structure-based, network-based, or deep learning-based (Table 3.2). As reviewed in this chapter, ligand-based models mostly rely on the ligand homophily effect—ligands that are similar to each other are likely to share targets, which makes the ligand–ligand similarity measurement a key component for the models. Structure-based models consider more about the physicochemical interface of protein–ligand binding, often containing a submodel to represent binding pockets or interacting residues extracted from the 3D structures of protein–ligand interaction complexes. Network-based models treat protein–ligand binding as a network with edges (binding) connecting vertices (proteins and ligands) and aim to predict the unknown edges based on known edges.

There is no one class of models that are always the best. Ligand-based models are often simple and scalable, but they are heavily dependent on the ligand–ligand similarity measurement method, making it less suitable to predict and overcome the activity cliff—a small change in ligand structure leading to dramatic differences in bioactivity [76]. Structure-based methods are better suited for the activity cliff as they consider the physicochemical properties of both protein and ligand. However, they require the 3D structure of the target and suffer from higher computational complexity. Network-based models often attempt to integrate information from both proteins and ligands by building protein–protein and ligand–ligand intralayer networks in addition to the protein–ligand interlayer network. They often overcome the high computational complexity and the requirement of protein 3D structures by using sequence-based features. Network-based models, however, suffer from the sparsity, incompleteness, biases, and noises in the known protein–ligand interaction network. Also, higher level of feature abstraction may lead to loss of information. For instance, it is not guaranteed that the sequence-based protein features are sufficient to accurately predict protein–ligand interactions. It is desirable to combine multiple types of models to overcome the drawbacks for each class of methods as in a recent drug repositioning study [97].

To date, most cancer therapies focused on directly removing or killing the tumor cells. While such removal of growing seeds may exert some therapeutic effects, it is noted that less attention has been paid to the soil, called the tumor microenvironment (TME) [120]. So far, the inhibitors of aforementioned immune checkpoint proteins are one of few available TME-based therapeutic options whose inefficiency have not yet been fully understood [2]. Although incomplete, extensive biological research has revealed that there are subtypes of TMEs that either promote or inhibit tumor progression [121]. While the majority of computational drug discovery research in cancer therapies have focused on discovering molecules that can kill or inhibit the growth of cancer cells, a shift to target the specific subtypes of TMEs may result in novel therapeutic agents. The potential therapeutic targets that can possibly enhance quantity and quality of neutrophils—the most abundant circulating leukocytes that are responsible for innate immune response—have been recently reviewed [122]. Reported cases of drugs targeting cancer-associated fibroblasts (CAFs), a major component of TME in pancreatic cancer, have been discussed in a recent review [123]. While individual therapies have been of limited success, a number of potential CAF-related targeted therapies have been proposed.

Polypharmacology theorizes that inhibiting two or more of the intended targets will be more beneficial than single-targeting drugs. A dual-action agent inhibiting both mevalonate pathway and sterol regulatory element-binding protein activity may provide unprecedently efficient antitumor effects by avoiding the resistance mechanism, as suggested in Ref. [18]. A recent study demonstrated that PD-1 proteins on myeloid cells may have different roles from those on T cells [124], partially explaining the limited success of the immunotherapies. A hypothetical dual-action drug can be designed to inhibit both CAF-tumor cross talk (e.g., TGF-β inhibitor [125]) and CAF-produced tumor-promoting signals (e.g., interleukin-33, which induces tumor-promoting M2 macrophages [126]). An existing CAF-derived extracellular matrix-targeting drug, sonidegib [127], can be optimized to also intervene the abovementioned TME biological networks or cellular metabolic pathways. Another hypothetical, multitargeting drug that efficiently inhibits BCL-X_L and MCL-1 (BCL-2 family multidomain antiapoptotic proteins) may enjoy maximum benefits by reverting the adaptive resistance-based tumor cell survival with low toxicity [9]. Ligand-based screening methods can help identify chemical scaffolds that are potent against the multiple targets of interest, structure-based methods can help iteratively optimize the scaffolds, and network-based methods can identify the global bioactivity profiles at early stage. Biological interaction networks within and across cells have been actively revisited to reveal more potential therapeutic targets. Computational approaches should also be newly developed, revised, and fine-tuned to accelerate the discovery of drugs that can induce therapeutic effects from multiple interactions.

References

[1] Ledford H, Else H, Warren M. Cancer immunologists scoop medicine Nobel prize. Nature 2018;562(7725):20–1.

[2] Hegde PS, Chen DS. Top 10 Challenges in Cancer Immunotherapy. Immunity 2020;52(1):17–35.

[3] Schiller JH, Harrington D, Belani CP, Langer C, Sandler A, et al. Comparison of four chemotherapy regimens for advanced non-small-cell lung cancer. N Engl J Med 2002;346(2):92–8.

[4] Zugazagoitia J, Guedes C, Ponce S, Ferrer I, Molina-Pinelo S, et al. Current Challenges in Cancer Treatment. Clin Ther 2016;38(7):1551–66.

[5] Heinzerling L, Ott PA, Hodi FS, Husain AN, Tajmir-Riahi A, et al. Cardiotoxicity associated with CTLA4 and PD1 blocking immunotherapy. J Immunother Cancer 2016;4:50.

[6] Gronich N, Lavi I, Barnett-Griness O, Saliba W, Abernethy DR, et al. Tyrosine kinase-targeting drugs-associated heart failure. Br J Cancer 2017;116(10):1366–73.

[7] Santomasso B, Bachier C, Westin J, Rezvani K, Shpall EJ. The Other Side of CAR T-Cell Therapy: Cytokine Release Syndrome, Neurologic Toxicity, and Financial Burden. Am Soc Clin Oncol Educ Book 2019;39:433–44.

[8] Mansoori B, Mohammadi A, Davudian S, Shirjang S, Baradaran B. The Different Mechanisms of Cancer Drug Resistance: A Brief Review. Adv Pharm Bull 2017;7(3):339–48.

[9] Wood KC. Overcoming MCL-1-driven adaptive resistance to targeted therapies. Nat Commun 2020;11(1):531.

[10] Housman G, Byler S, Heerboth S, Lapinska K, Longacre M, et al. Drug resistance in cancer: an overview. Cancers (Basel) 2014;6(3):1769–17692.

[11] Dickson M, Gagnon JP. The cost of new drug discovery and development. Discov Med 2004;4(22):172–9.

[12] Waring MJ, Arrowsmith J, Leach AR, Leeson PD, Mandrell S, et al. An analysis of the attrition of drug candidates from four major pharmaceutical companies. Nature Reviews Drug Discovery 2015;14(7):475–86.

[13] Butler D, Callaway E. Scientists in the dark after French clinical trial proves fatal. Nature 2016;529(7586):263–4.

[14] Rask-Andersen M, Almen MS, Schioth HB. Trends in the exploitation of novel drug targets. Nat Rev Drug Discov 2011;10(8):579–90.

[15] Hopkins AL. Network pharmacology: the next paradigm in drug discovery. Nat Chem Biol 2008;4(11):682–90.

[16] Anighoro A, Bajorath J, Rastelli G. Polypharmacology: challenges and opportunities in drug discovery. J Med Chem 2014;57(19):7874–87.

[17] Ramsay RR, Popovic-Nikolic MR, Nikolic K, Uliassi E, Bolognesi ML. A perspective on multi-target drug discovery and design for complex diseases. Clin Transl Med 2018;7(1):3.

[18] Mullen PJ, Yu R, Longo J, Archer MC, Penn LZ. The interplay between cell signalling and the mevalonate pathway in cancer. Nat Rev Cancer 2016;16(11):718–31.

[19] Kowalik MA, Columbano A, Perra A. Emerging Role of the Pentose Phosphate Pathway in Hepatocellular Carcinoma. Front Oncol 2017;7:87.

[20] Xie L, Draizen EJ, Bourne PE. Harnessing Big Data for Systems Pharmacology. Annu Rev Pharmacol Toxicol 2017;57:245–62.

[21] Poleksic A, Xie L. Predicting serious rare adverse reactions of novel chemicals. Bioinformatics 2018;34(16):2835–42.

[22] Lim H, Poleksic A, Xie L. Exploring Landscape of Drug-Target-Pathway-Side Effect Associations. AMIA Jt Summits Transl Sci Proc 2017;2018:132–41.

[23] Chang RL, Xie L, Xie L, Bourne PE, Palsson BO. Drug off-target effects predicted using structural analysis in the context of a metabolic network model. PLoS Comput Biol 2010;6(9):e1000938.

[24] Nath A, Lau EYT, Lee AM, Geeleher P, Cho WCS, et al. Discovering long noncoding RNA predictors of anticancer drug sensitivity beyond protein-coding genes. Proc Natl Acad Sci U S A 2019;116(44):22020–9.

[25] Xie L, He S, Wen Y, Bo X, Zhang Z. Discovery of novel therapeutic properties of drugs from transcriptional responses based on multi-label classification. Sci Rep Aug 2, 2017;7(1):7136.

[26] Paananen J, Fortino V. An omics perspective on drug target discovery platforms. Brief Bioinform 2019;27.

[27] Sterling T, Irwin JJ. ZINC 15 – Ligand Discovery for Everyone. J Che Inform Model 2015;55(11):2324–37.

[28] Gaulton A, Hersey A, Nowotka M, Bento AP, Chambers J, et al. The ChEMBL database in 2017. Nucleic Acids Res 2017;45(D1):D945–54.

[29] Kim S, Chen J, Cheng T, Gindulyte A, He J, et al. PubChem 2019 update: improved access to chemical data. Nucleic Acids Res 2018;47(D1):D1102–9.

[30] Gilson MK, Liu T, Baitaluk M, Nicola G, Hwang L, et al. BindingDB in 2015: A public database for medicinal chemistry, computational chemistry and systems pharmacology. Nucleic Acids Res 2016;44(D1):D1045–53.

[31] Koleti A, Terryn R, Stathias V, Chung C, Cooper DJ, et al. Data Portal for the Library of Integrated

Network-based Cellular Signatures (LINCS) program: integrated access to diverse large-scale cellular perturbation response data. Nucleic Acids Res Jan 4, 2018; 46(D1):D558–66.

[32] Szklarczyk D, Santos A, von Mering C, Jensen LJ, Bork P, et al. STITCH 5: augmenting protein-chemical interaction networks with tissue and affinity data. Nucleic Acids Res 2016;44(D1):D380–4.

[33] Oughtred R, Stark C, Breitkreutz B-J, Rust J, Boucher L, et al. The BioGRID interaction database: 2019 update. Nucleic Acids Res 2018;47(D1):D529–41.

[34] Kuhn M, Letunic I, Jensen LJ, Bork P. The SIDER database of drugs and side effects. Nucleic Acids Res 2016; 44(D1):D1075–9.

[35] Kanehisa M, Furumichi M, Tanabe M, Sato Y, Morishima K. KEGG: new perspectives on genomes, pathways, diseases and drugs. Nucleic Acids Res 2017;45(D1):D353–61.

[36] Wishart DS, Feunang YD, Guo AC, Lo EJ, Marcu A, et al. DrugBank 5.0: a major update to the DrugBank database for 2018. Nucleic Acids Res 2018;46(D1): D1074–82.

[37] Sun J, Jeliazkova N, Chupakin V, Golib-Dzib JF, Engkvist O, et al. ExCAPE-DB: an integrated large scale dataset facilitating Big Data analysis in chemogenomics. J Cheminform 2017;9:17.

[38] Consortium TU. UniProt: a worldwide hub of protein knowledge. Nucleic Acids Res 2018;47(D1):D506–15.

[39] Burley SK, Berman HM, Bhikadiya C, Bi C, Chen L, et al. RCSB Protein Data Bank: biological macromolecular structures enabling research and education in fundamental biology, biomedicine, biotechnology and energy. Nucleic Acids Res 2018;47(D1):D464–74.

[40] Szklarczyk D, Gable AL, Lyon D, Junge A, Wyder S, et al. STRING v11: protein-protein association networks with increased coverage, supporting functional discovery in genome-wide experimental datasets. Nucleic Acids Res 2019;47(D1):D607–13.

[41] Uhlen M, Oksvold P, Fagerberg L, Lundberg E, Jonasson K, et al. Towards a knowledge-based Human Protein Atlas. Nat Biotechnol 2010;28(12): 1248–50.

[42] Rouillard AD, Gundersen GW, Fernandez NF, Wang Z, Monteiro CD, et al. The harmonizome: a collection of processed datasets gathered to serve and mine knowledge about genes and proteins. Database 2016:2016.

[43] Rohrer SG, Baumann K. Maximum Unbiased Validation (MUV) Data Sets for Virtual Screening Based on PubChem Bioactivity Data. Journal of Chemical Information and Modeling 2009;49(2):169–84.

[44] Mysinger MM, Carchia M, Irwin JJ, Shoichet BK. Directory of Useful Decoys, Enhanced (DUD-E): Better Ligands and Decoys for Better Benchmarking. Journal of Medicinal Chemistry 2019;55(14):6582–94.

[45] Chen L, et al. Hidden bias in the DUD-E dataset leads to misleading performance of deep learning in structure-based virtual screening. PLoS One 2019; 14(8):e0220113.

[46] Carhart RE, et al. Atom pairs as molecular features in structure-activity studies: definition and applications. J Chem Inf Comput Sci 1985;25(2):64–73.

[47] Nilakantan R, et al. Topological torsion: a new molecular descriptor for SAR applications. Comparison with other descriptors. J Chem Inf Comput Sci 1987; 27(2):82–5.

[48] Gobbi A, Poppinger D. Genetic optimization of combinatorial libraries. Biotechnol Bioeng 1998;61(1): 47–54.

[49] Rogers D, Hahn M. Extended-connectivity fingerprints. J Chem Inf Model 2010;50(5):742–54.

[50] Hu G, et al. Performance evaluation of 2D fingerprint and 3D shape similarity methods in virtual screening. J Chem Inf Model 2012;52(5):1103–13.

[51] Yap CW. PaDEL-descriptor: an open source software to calculate molecular descriptors and fingerprints. J Comput Chem 2011;32(7):1466–74.

[52] O'Boyle NM, et al. Open Babel: an open chemical toolbox. J Cheminf 2011;3(1):33.

[53] RDKit: Open-source cheminformatics. Available from: http://www.rdkit.org.

[54] Keiser MJ, et al. Relating protein pharmacology by ligand chemistry. Nat Biotechnol 2007;25(2):197–206.

[55] Awale M, Reymond JL. The polypharmacology browser: a web-based multi-fingerprint target prediction tool using ChEMBL bioactivity data. J Cheminf 2017;9:11.

[56] Wang Z, et al. Improving chemical similarity ensemble approach in target prediction. J Cheminf 2016;8:20.

[57] Riniker S, Landrum GA. Open-source platform to benchmark fingerprints for ligand-based virtual screening. J Cheminf 2013;5(1):26.

[58] Jaeger S, et al. Mol2vec: unsupervised machine learning approach with chemical intuition. J Chem Inf Model 2018;58(1):27–35.

[59] Distributed representations of words and phrases and their compositionality. In: Mikolov T, et al., editors. Advances in neural information processing systems; 2013.

[60] Convolutional networks on graphs for learning molecular fingerprints. In: Duvenaud DK, et al., editors. Advances in neural information processing systems; 2015.

[61] Coley CW, et al. Convolutional embedding of attributed molecular graphs for physical property prediction. J Chem Inf Model 2017;57(8):1757–72.

[62] Kawashima S, et al. AAindex: amino acid index database. Nucleic Acids Res 1999;27(1):368–9.

[63] Grantham R. Amino acid difference formula to help explain protein evolution. Science 1974;185(4154): 862–4.

[64] Shen J, et al. Predicting protein–protein interactions based only on sequences information. Proc Natl Acad Sci U S A 2007;104(11):4337.

[65] Moreau G, Broto P. The auto-correlation of a topological-structure-a new Molecular Descriptor. 120 Blvd Saint-Germain, 75280 Paris Cedex O6, France: Gauthier-Villars; 1980. p. 359–60.

[66] Sneath PHA. Relations between chemical structure and biological activity in peptides. J Theor Biol 1966; 12(2):157–95.

[67] Hellberg S, et al. Peptide quantitative structure-activity relationships, a multivariate approach. J Med Chem 1987;30(7):1126–35.

[68] Jonsson J, et al. Multivariate parametrization of 55 coded and non-coded amino acids. Quant Struct-Act Relat 1989;8(3):204–9.

[69] Sandberg M, et al. New chemical descriptors relevant for the design of biologically active peptides. A multivariate characterization of 87 amino acids. J Med Chem 1998;41(14):2481–91.

[70] Xiao N, et al. protr/ProtrWeb: R package and web server for generating various numerical representation schemes of protein sequences. Bioinformatics 2015;31(11):1857–9.

[71] Ye X, et al. An assessment of substitution scores for protein profile–profile comparison. Bioinformatics 2011;27(24):3356–63.

[72] Devlin J, et al. Bert: pre-training of deep bidirectional transformers for language understanding. arXiv preprint arXiv:181004805. 2018.

[73] Rives A, et al. Biological structure and function emerge from scaling unsupervised learning to 250 million protein sequences. bioRxiv 2019:622803.

[74] Attention is all you need. In: Vaswani A, et al., editors. Advances in neural information processing systems; 2017.

[75] Lan Z, et al. ALBERT: a lite BERT for self-supervised learning of language representations. arXiv preprint arXiv:190911942. 2019.

[76] Cruz-Monteagudo M, et al. Activity cliffs in drug discovery: Dr Jekyll or Mr Hyde? Drug Discov Today 2014;19(8):1069–80.

[77] Madhukar NS, et al. A Bayesian machine learning approach for drug target identification using diverse data types. Nat Commun 2019;10(1):5221.

[78] Koutsoukas A, et al. In silico target predictions: defining a benchmarking data set and comparison of performance of the multiclass naïve Bayes and Parzen-Rosenblatt Window. J Chem Inf Model 2013; 53(8):1957–66.

[79] Lauria A, et al. DRUDIT: web-based DRUgs DIscovery Tools to design small molecules as modulators of biological targets. Bioinformatics 2019;36(5): 1562–9.

[80] Yang M, et al. Machine learning models based on molecular fingerprints and eXtreme gradient boosting method lead to the discovery of JAK2 inhibitors. J Chem Inf Model 2019;59(12):5002–12.

[81] Sadawi N, et al. Multi-task learning with a natural metric for quantitative structure activity relationship learning. J Cheminf 2019;11(1):68.

[82] Grisoni F, et al. Design of natural-product-inspired multitarget ligands by machine learning. ChemMedChem 2019;14(12):1129–34.

[83] Hernandez M, et al. A quantum-inspired method for three-dimensional ligand-based virtual screening. J Chem Inf Model 2019;59(10):4475–85.

[84] Fan C, et al. DStruBTarget: integrating Binding Affinity with structure similarity for ligand binding protein prediction. J Chem Inf Model 2020;60(1):400–9.

[85] Guedes IA, et al. Empirical scoring functions for structure-based virtual screening: applications, critical aspects, and challenges. Front Pharmacol 2018;9:1089.

[86] Souza ML, et al. Discovery of potent, reversible and competitive Cruzain inhibitors with trypanocidal activity: a structure-based drug design approach. J Chem Inf Model 2020;60(2):1028–41.

[87] Kamsri P, et al. Discovery of new and potent InhA inhibitors as anti-tuberculosis agents: structure based virtual screening validated by biological assays and X-ray crystallography. J Chem Inf Model 2020;60(1): 226–34.

[88] Forli S, et al. Computational protein-ligand docking and virtual drug screening with the AutoDock suite. Nat Protoc 2016;11(5):905–19.

[89] Allen WJ, et al. Dock 6: impact of new features and current docking performance. J Comput Chem 2015; 36(15):1132–56.

[90] Labbe CM, et al. MTiOpenScreen: a web server for structure-based virtual screening. Nucleic Acids Res 2015;43(W1):W448–54.

[91] van Zundert GCP, et al. The HADDOCK2.2 web server: user-friendly integrative modeling of biomolecular complexes. J Mol Biol 2016;428(4):720–5.

[92] Grosdidier A, et al. SwissDock, a protein-small molecule docking web service based on EADock DSS. Nucleic Acids Res 2011;39(Web Server Issue): W270–7.

[93] Grinter SZ, et al. An inverse docking approach for identifying new potential anti-cancer targets. J Mol Graph Model 2011;29(6):795–9.
[94] Saenz-Méndez P, et al. Ligand selectivity between the ADP-ribosylating toxins: an inverse-docking study for multitarget drug discovery. ACS Omega 2017; 2(4):1710–9.
[95] Wang F, et al. ACID: a free tool for drug repurposing using consensus inverse docking strategy. J Cheminf 2019;11(1):73.
[96] Wang A, et al. ANTENNA, a multi-rank, multi-layered recommender system for inferring reliable drug-gene-disease associations: repurposing diazoxide as a targeted anti-cancer therapy. IEEE ACM Trans Comput Biol Bioinf 2018;15(6):1960–7.
[97] Lim H, et al. Rational discovery of dual-indication multi-target PDE/Kinase inhibitor for precision anti-cancer therapy using structural systems pharmacology. PLoS Comput Biol 2019;15(6): e1006619.
[98] Wang X, et al. Enhancing the enrichment of pharmacophore-based target prediction for the poly-pharmacological profiles of drugs. J Chem Inf Model 2016;56(6):1175–83.
[99] Lu Y, et al. Link prediction in drug-target interactions network using similarity indices. BMC Bioinf 2017; 18(1):39.
[100] Chu Y, et al. DTI-CDF: a cascade deep forest model towards the prediction of drug-target interactions based on hybrid features. Briefings Bioinf 2019.
[101] Breiman L. Random forests. Mach Learn 2001;45(1): 5–32.
[102] Chen T, Guestrin C. XGBoost: a scalable tree boosting system. San Francisco, California, USA: Association for Computing Machinery; 2016. p. 785–94.
[103] Luo Y, et al. A network integration approach for drug-target interaction prediction and computational drug repositioning from heterogeneous information. Nat Commun 2017;8(1):573.
[104] Tong H, et al. Random walk with restart: fast solutions and applications. Springer-Verlag; 2008. p. 327–46.
[105] Fu G, et al. Predicting drug target interactions using meta-path-based semantic network analysis. BMC Bioinf 2016;17:160.
[106] Zheng Y, et al. Old drug repositioning and new drug discovery through similarity learning from drug-target joint feature spaces. BMC Bioinf 2019;20(Suppl. 23):605.
[107] Lim H, et al. Large-scale off-target identification using fast and accurate dual regularized one-class collaborative filtering and its application to drug repurposing. PLoS Comput Biol 2016;12(10):e1005135.
[108] Liu Y, et al. Neighborhood regularized logistic matrix factorization for drug-target interaction prediction. PLoS Comput Biol 2016;12(2):e1004760.
[109] Lim H, Xie L. A new weighted imputed neighborhood-regularized tri-factorization one-class collaborative filtering algorithm: application to target gene prediction of transcription factors. IEEE ACM Trans Comput Biol Bioinf 2020.
[110] Ayed M, et al. Biological representation of chemicals using latent target interaction profile. BMC Bioinf 2019;20(Suppl. 24):674.
[111] Zakharov AV, et al. Novel consensus architecture to improve performance of large-scale multitask deep learning QSAR models. J Chem Inf Model 2019; 59(11):4613–24.
[112] Lee I, et al. DeepConv-DTI: prediction of drug-target interactions via deep learning with convolution on protein sequences. PLoS Comput Biol 2019;15(6): e1007129.
[113] Moridi M, et al. The assessment of efficient representation of drug features using deep learning for drug repositioning. BMC Bioinf 2019;20(1):577.
[114] Torng W, Altman RB. Graph convolutional neural networks for predicting drug-target interactions. J Chem Inf Model 2019;59(10):4131–49.
[115] Zhang H, et al. DeepBindRG: a deep learning based method for estimating effective protein-ligand affinity. PeerJ 2019;7:e7362.
[116] He K, et al., editors. Deep residual learning for image recognition. IEEE conference on computer vision and pattern recognition (CVPR); 2016 27–30 June 2016; 2016.
[117] Karimi M, et al. DeepAffinity: interpretable deep learning of compound-protein affinity through unified recurrent and convolutional neural networks. Bioinformatics 2019;35(18):3329–38.
[118] Sutskever I, et al. Sequence to sequence learning with neural networks. In: Proceedings of the 27th international conference on neural information processing systems - Volume 2; Montreal, Canada. 2969173: MIT Press; 2014. p. 3104–12.
[119] Wan F, et al. NeoDTI: neural integration of neighbor information from a heterogeneous network for discovering new drug-target interactions. Bioinformatics 2019;35(1):104–11.
[120] de Groot AE, et al. Revisiting seed and soil: examining the primary tumor and cancer cell foraging in metastasis. Mol Cancer Res 2017;15(4):361–70.
[121] Galon J, Bruni D. Tumor immunology and tumor evolution: intertwined histories. Immunity 2020;52(1): 55–81.
[122] Nemeth T, et al. Neutrophils as emerging therapeutic targets. Nat Rev Drug Discov 2020.

[123] Pereira BA, et al. CAF subpopulations: a new reservoir of stromal targets in pancreatic cancer. Trends Cancer 2019;5(11):724–41.

[124] Strauss L, et al. Targeted deletion of PD-1 in myeloid cells induces antitumor immunity. Sci Immunol 2020; 5(43).

[125] Pinho AV, et al. ROBO2 is a stroma suppressor gene in the pancreas and acts via TGF-beta signalling. Nat Commun 2018;9(1):5083.

[126] Andersson P, et al. Molecular mechanisms of IL-33-mediated stromal interactions in cancer metastasis. JCI Insight 2018;3(20).

[127] Cazet AS, et al. Targeting stromal remodeling and cancer stem cell plasticity overcomes chemoresistance in triple negative breast cancer. Nat Commun 2018; 9(1):2897.

CHAPTER

4

Unveiling potential anticancer drugs through in silico drug repurposing approaches

HemaSree GNS, V Lakshmi PrasannaMarise, Rachana R Pai, Swarna Mariam Jos, Mamatha Krishna Murthy, Ganesan Rajalekshmi Saraswathy

Pharmacological Modelling and Simulation Centre, Faculty of Pharmacy, M.S. Ramaiah University of Applied Sciences, Bangalore, Karnataka, India

OUTLINE

Drug Repurposing in Cancer Therapy
https://doi.org/10.1016/B978-0-12-819668-7.00004-X

List of abbreviations

ABPP Activity-based protein profiling
ADR Adverse drug reactions
AI Artificial intelligence
AML Acute myeloid leukemia
ANN Artificial neural networks
ATC Anatomical therapeutic classification
AUC Area under curve
AUROC Area under the receiver operating characteristic curve
BBB Brain—blood barrier
BSCE Basespace Correlation Engine
CancerDR Cancer drug resistance
CCLE Cancer Cell Line Encyclopedia
CCLP Cancer cell line profiler
CD Cancer drug
CDK2 Cyclin-dependent kinase 2
CDRscan Cancer drug response profile scan
CGC Cancer Gene Consensus
CGP Cancer Genome Project
CHAT Cancer Hallmarks Analytics Tool
cis-eQTL cis-Expression quantitative trait loci
CMap Connectivity Map
CNA Copy number alterations
CNN Convolutional neural network
COSMIC Catalog Of Somatic Mutations In Cancer
CPTAC Clinical Proteomic Tumor Analysis Consortium
CRC Colorectal cancer
CSTA Clinical and Translational Science Awards
CTD Comparative Toxicogenomics Database

CTRP Cancer Therapeutic Response Portal
CWR Cures Within Reach
DAPPLE Disease Association Protein–Protein Link Evaluator
DAVID Database for Annotation, Visualization, and Integrated Discovery
DEG Differentially expressed genes
DeSigN Differentially Expressed Gene Signatures
DIRAC Differential Rank Conservation
DNMT DNA methyl transferases
DP Disease proteins
Drug–SE Drug–side effect
EBI European Bioinformatics Institute
EHR Electronic Health Records
EMBL European Molecular Biology Laboratory
eMERGE Electronic Medical Records and Genomics
EMT Epithelial–mesenchymal transition
ER Estrogen receptor
ESSA Event sequence symmetry analysis
FAERS FDA Adverse Event Reporting System
FC Fold change
FDR False discovery rate
GC Gastric cancer
GDC Genomic Data Common
GDSC Genomics of Drug Sensitivity in Cancer
GEM Genome-scale metabolic model
GEO Gene Expression Omnibus
GPU Graphics processing unit
GRAIL Gene Relationships among Implicated Loci
GWAS Genome-wide association studies
HMDB Human Metabolome Database
HMP Human Metabolome Project
HNSCC Head and neck squamous cell carcinoma
HoC Hallmarks of cancer
HPA Human Protein Atlas
HTS High-throughput screening
IC Information components
IC_{50} Half maximal inhibitory concentration
ICGC International Cancer Genome Consortium
IE Information extraction
IntOGen Integrative Onco Genomics
IR Information retrieval
JMDC Japan Medical Data Center
KD Known drugs
KEGG Kyoto Encyclopedia of Genes and Genomes
KS Kolmogorov–Smirnov
LINCS Library of Integrated Network-Based Cellular Signatures
MDS Multidimensionality scaling
MedDRA Medical Dictionary for Regulatory Activities
MeSH Medical Subject Headings
MGI Mouse Genome Informatics
ML Machine learning
MRS Magnetic resonance spectroscopy
NCATS National Centre for Advancing Translational Sciences
NCBI National Center for Biotechnology Information
NCI National Cancer Institute
NER Named-entity recognition
NGS Next-generation sequencing
NHGRI National Human Genome Research Institute
NIH National Institutes of Health
NLP Natural language processing
NMR Nuclear magnetic resonance spectroscopy
PBMC peripheral blood mononuclear cell
PCA Principal component analysis
PCR Polymerase chain reaction
PDB Protein Data Bank
PharmGKB Pharmacogenomics Knowledge Base
PMID PubMed ID
POS Parts of speech
PPI Protein–protein interaction
ReDIReCT Repurposing of Drugs: Innovative Revision of Cancer Treatment
ReDO Repurposing Drugs in Oncology
REMC Roadmap Epigenomics Mapping Consortium
RGES Reversal Gene Expression Scores
RMSD Root mean square deviation
RMSE Root mean square error
ROC Receiving operating characteristics
ROR Reporting odds ratios
RPPA Reverse-phase protein microarrays
SD Synthetic derivative
SEA Similarity ensemble approach
SIDER Side Effect Resource
SMILES Simplified molecular-input line-entry system
SNP Single-nucleotide polymorphism
sscMap Statistically Significant Connection's Map
STITCH Search tool for interactions of chemical
SVM Support vector machine
T2DM Type 2 diabetes mellitus
TCGA The Cancer Genome Atlas
tINIT Integrative Network Inference for Tissues
TM Text mining
TMIC The Metabolomics Innovation Centre
TNBC Triple-negative breast cancer
TP Target proteins
TPU Tensor processing unit
TSDS Topological Score of Drug Synergy
TTD Therapeutic Target Database
UCSC University of California, Santa Cruz
UMLS Unified Medical Language System
VUMC Vanderbilt University Medical Center
WES Whole exome sequencing

Introduction

Cancer is a heterogeneous genetic disorder that causes overactivation of cell division signals, eventually precipitating uncontrolled cell proliferation, invasion, and metastasis [1,2]. It is one of the leading causes of death globally. According to GLOBOCAN [3] 2018, 18.1 million newly diagnosed cases and 9.6 million deaths related to cancer were reported.

Carcinogenesis is a result of interaction between environmental factors and genetic elements of an individual over a period of time, which could bring about an abnormal stimulation of protooncogenes or inhibition of tumor suppressor genes [4]. The exact mechanism of such genetic instabilities at the level of chromosome and nucleotides accountable for the progression and heterogeneity remains ambiguous [5]. Therefore, deep insights are sought from advanced bioinformatics and computational simulation techniques to unravel the molecular mechanisms lurking behind these invasive oncogenic processes.

Current cancer therapy: are we ready to battle cancer?

The current chemotherapeutic approaches are unsuccessful in targeting the stem cells from which cancer cells originate, and they merely focus on a limited number of genetic mutations that may not account for massive genetic variations linked with malignancies. Moreover, these therapies are imprecise as they presume normal somatic cells to possess malignant potential, thereby imposing cytotoxicity. On the other hand, deficient activation of certain enzymes that are responsible for conversion of prodrugs to their active forms, aberrant drug transporters or efflux pumps that undermine the drug concentrations within the cancer cells, irreparable DNA damage after a direct or indirect insult, and evasion of apoptosis are some of the factors that are likely to culminate in a drug gaining resistance. Although downsizing of tumor is evident after successful completion of chemotherapy cycles, there exists a plethora of cases where these agents fail to totally eliminate cancer stem cells with metastatic potential, thereby resulting in recurrence. This instigated the researchers to develop new drug regimens or combinational therapies for a successful treatment [6].

De novo drug discovery versus drug repurposing

Novel drug discovery is a complex process that consumes an average of 10 to 15 years for translation of a new molecule into an approved drug. The drug approval process is tedious and encounters higher attrition rates due to changing regulatory requirements. These rate-limiting steps in de novo synthesis of a drug necessitated a paradigm shift from conventional drug discovery pathways to contemporary drug repurposing research to expedite unraveling new indications for existing, banned, and investigational drugs (Fig. 4.1).

Drug repositioning bypasses the elaborative processes involved in conventional drug development and dramatically reduces the time required. It demands an investment of 1600 million USD in contrast to the 12,000 million USD required for traditional drug discovery. Furthermore, the failure rates are low as safety profiles of repurposable drugs are already established. The lower cost involved in drug repositioning research is advantageous for economically backward countries to satisfy their unmet medical needs [7–9]. Latest update on repurposed drugs in cancer therapy is listed in Table 4.1.

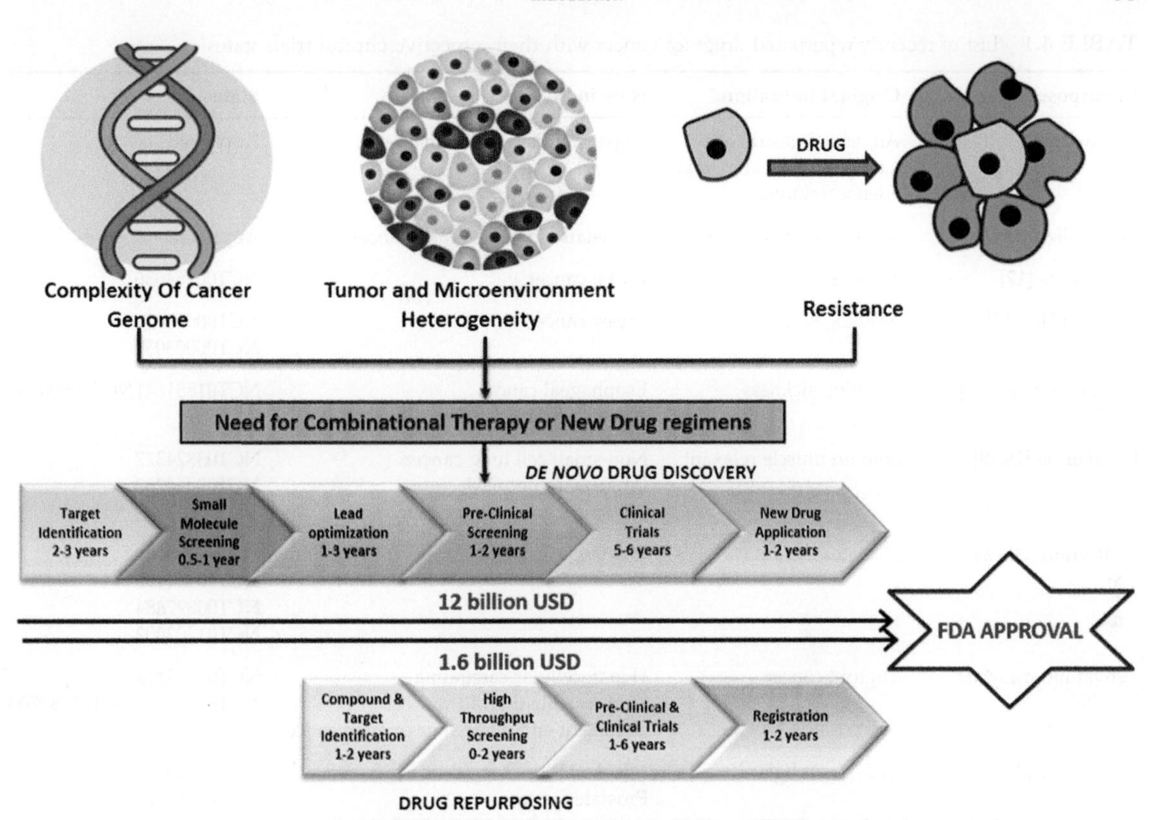

FIGURE 4.1 Evolution of drug repurposing.

Fundamental steps for a fruitful drug repurposing expedition

Drug repurposing mandates a thorough insight of polypharmacology, which facilitates the exploration of multitarget actions of a single drug and its involvement in other disease pathways. Latest information pertaining to unexplored pathways involved in disease pathogenesis and progression, along with their associated biomarkers, is required before initiating drug repurposing approaches. Drug repurposing investigations oriented toward genetic disorders demand supportive literature on the influence of environment and drugs on gene expression, transcription, translation, epigenome, and metabolism. The data acquired and accumulated over years are massive and too huge to be handled manually. This situation has enforced knowledge-based, signature-based, target-based, and network-based computational approaches to untangle the hidden relationships across drugs, targets, and diseases [9].

Inclusion of novel informatics approach, systems biology, and genomic information to reveal unknown targets or mechanisms of approved drugs improves drug repurposing methods by accelerating the timelines. The compounds derived from computational studies can be further validated through experimental testing. Hence, a combination of both computational and experimental assays is desirable to repurpose drugs for new indications [32,33] (Fig. 4.2).

Drugs are repurposed by employing omics data, such as genomics [34], transcriptomics [35], proteomics [36], epigenetics [37], and

TABLE 4.1 List of recently repurposed drugs for cancer with their respective clinical trials status.

Repurposed drug	Original indication	New indication	Status
Ramucirumab [10]	Advanced gastric or gastroesophageal junction adenocarcinoma	Hepatocellular carcinoma	Approved
Pembrolizumab [11]	Metastatic melanoma	Metastatic small cell lung cancer	Approved
Artesunate [12]	Malaria	Breast cancer	NCT00764036
Suramin [13–15]	Sleeping sickness	Breast cancer	NCT00054028 NCT00003038
Thalidomide [16,17]	Morning sickness	Esophageal cancer Advanced Colorectal cancer	NCT01551641NCT00890188
Papaverine [18,19]	Smooth muscle relaxant	Non-small cell lung cancer Prostatic hyperplasia treatment and cancer prevention	NCT03824327 NCT03064282
Metformin [20–24]	Diabetes mellitus	Prostate cancer Breast cancer	NCT03137186 NCT00984490 NCT00897884 NCT01302002
Lenvatinib mesylate [25–28]	Thyroid cancer	Hepatocellular carcinoma Unresectable thyroid cancer Recurrent endometrial or ovarian cancer	NCT03663114 NCT02430714NCT02788708
Quinacrine [29–31]	Malaria and giardiasis	Non-small cell lung cancer Prostate cancer	NCT01839955 NCT00417274

metabolomics [38]. Alongside omics databases, electronic health records and side effect data [39] also provide valuable hints to predict novel indications of existing drugs [40]. Further progress in this field has led to the construction and incorporation of mathematical algorithms and Machine Learning (ML) platforms for a rapid and accurate drug repurposing forecast analysis [41,42] (Fig. 4.3).

Global funding initiatives and evolving big data drug repurposing projects

In the wake of new horizon of drug repurposing in cancer, a number of funding schemes were initiated by both governmental and philanthropic agencies. The National Institute of Health (NIH), started the National Centre for Advancing Translational Sciences, which funds the development of novel therapeutic possibilities by various initial in silico predictions [43]. The Belgian initiative called the Anticancer Fund in collaboration with The Global Cures of USA, cofounded the Repurposing Drugs in Oncology [44] project, to screen and test the anticancer potential of the existing noncancer therapeutic armamentarium and redirect them for cancer therapy via drug repurposing. Apart from these, Repurposing of Drugs: Innovative Revision of Cancer Treatment [45], Clinical and Translational Science Awards [46], Findacure [47], Global Cures [48] and Cures Within Reach [49] are few other renowned funding agencies that are working toward addressing the challenges encountered in Cancer Drug (CD) discovery.

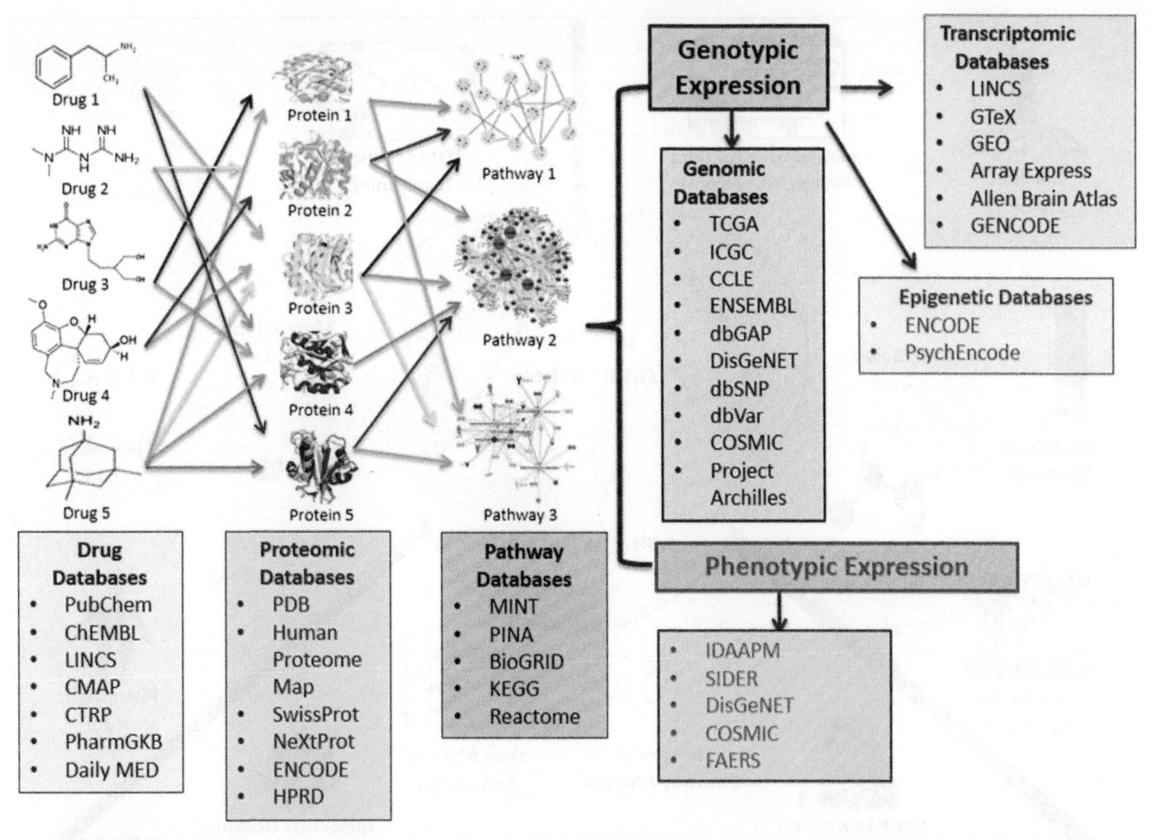

FIGURE 4.2 Polypharmacology concept exploring off-target actions of drugs and its downstream signaling.

Following the advent of funding programs, big data projects like The Cancer Genome Atlas (TCGA) [50], International Cancer Genome Consortium (ICGC) [51], NIH Library of Integrated Network-Based Cellular Signatures (LINCS) [52], Cancer Genome Project (CGP) [53], Clinical Proteomic Tumor Analysis Consortium [54], Cancer Drug Resistance database [55], Oncomine [56] etc., were constructed. The databases pertaining to cancer repurposing with their respective web links are tabulated in Table 4.2. These databases are used widely to extract data and generate hypotheses for repurposing through in silico methods.

Emergence of ML and Artificial Intelligence (AI) has increased the ease of drug repurposing predictions through integration of heterogeneous data using "garbage in, garbage out" method. Expanding supercomputing techniques like graphics processing unit and tensor processing unit has led to the generation of large diverse data and storage revolution. Many approaches utilize AI for predicting the pharmacological effects of chemicals or drugs by integrating medical data. Artificial Neural Networks (ANN), an outgrowth of AI, is employed to analyze the mechanism of action of drug molecules under the CD screening program initiated by National Cancer Institute (NCI) [108].

This chapter elaborates on various repurposing mediums with examples of their practical application. In addition, fusion of foregoing methods and their recent advancements toward

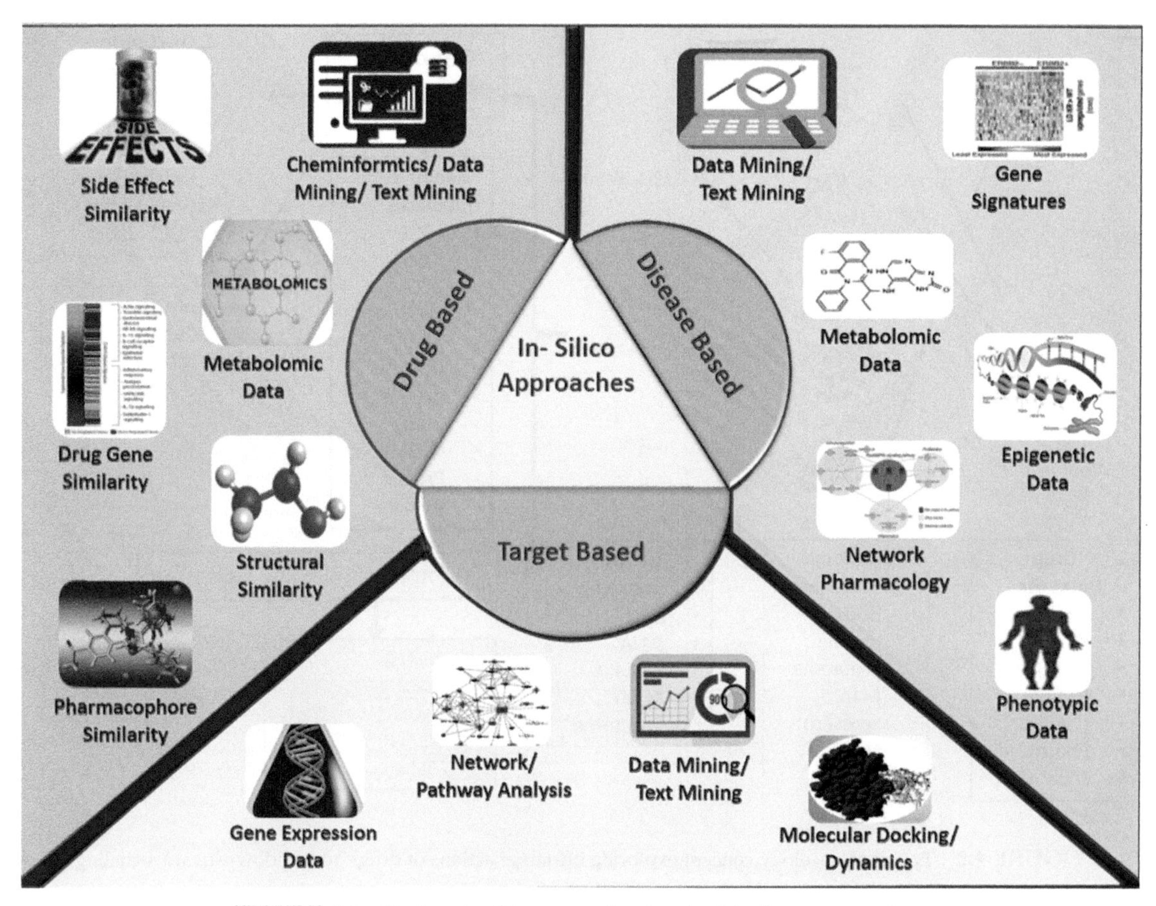

FIGURE 4.3 Various in silico approaches involved in drug repurposing.

TABLE 4.2 List of omics databases for cancer drug repurposing.

Serial number	Database	Link	References
1	TCGA	https://www.cancer.gov/about-nci/organization/ccg/research/structural-genomics/tcga	[50]
2	ICGC	https://dcc.icgc.org/	[51]
3	CCLE	https://portals.broadinstitute.org/ccle	[57]
4	Ensembl	https://asia.ensembl.org/index.html	[58]
5	dbGaP	https://www.ncbi.nlm.nih.gov/gap/	[59]
6	DisGeNET	https://www.disgenet.org/	[60]
7	dbSNP	https://www.ncbi.nlm.nih.gov/snp/	[61]

TABLE 4.2 List of omics databases for cancer drug repurposing.—cont'd

Serial number	Database	Link	References
8	dbVar	https://www.ncbi.nlm.nih.gov/dbvar/	[62]
9	COSMIC	https://cancer.sanger.ac.uk/cosmic	[63]
10	LINCS	http://www.lincsproject.org/	[52]
11	GTEx	https://gtexportal.org/home/	[64]
12	GEO	https://www.ncbi.nlm.nih.gov/pmc/	[65]
13	ArrayExpress	https://www.ebi.ac.uk/arrayexpress/	[66]
14	Allen Brain Atlas	https://portal.brain-map.org/	[67]
15	Protein Data Bank (PDB)	http://www.rcsb.org/	[68]
16	Human Proteome Project	https://hupo.org/human-proteome-project	[69]
17	SWISS-PROT	https://www.iop.vast.ac.vn/theor/conferences/smp/1st/kaminuma/SWISSPROT/index.html	[70]
18	neXtProt	https://www.nextprot.org/about/nextprot	[71]
19	GENCODE	https://www.gencodegenes.org/	[72]
20	ENCODE	https://www.genome.gov/Funded-Programs-Projects/ENCODE-Project-ENCyclopedia-Of-DNA-Elements	[73]
21	CCDS	https://www.ncbi.nlm.nih.gov/projects/CCDS/CcdsBrowse.cgi	[74]
22	Cancer Genome Interpreter	https://www.cancergenomeinterpreter.org/home	[75]
23	Roadmap	http://www.roadmapepigenomics.org/	[76]
24	PsychENCODE	https://www.nimhgenetics.org/resources/psychencode	[77]
25	BioGRID	https://thebiogrid.org/	[78]
26	KEGG	https://www.genome.jp/kegg/pathway.html	[79]
27	Reactome	https://reactome.org/	[80]
28	SIDER	http://sideeffects.embl.de/	[81]
29	HPRD	http://www.hprd.org/	[82]
30	MINT	https://mint.bio.uniroma2.it/	[83]
31	GPS-Prot	http://gpsprot.org/	[84]
32	PINA	https://omics.bjcancer.org/pina/	[85]
33	MPIDB	https://omictools.com/mpidb-tool	[86]
34	FAERS	https://open.fda.gov/data/faers/	[87]
35	IDAAPM	https://omictools.com/idaapm-tool	[88]
36	VigiAccess	http://www.vigiaccess.org/	[89]
37	Genomics of Drug Sensitivity in Cancer (GDSC)	http://www.cancerrxgene.org/	[90]

(Continued)

TABLE 4.2 List of omics databases for cancer drug repurposing.—cont'd

Serial number	Database	Link	References
38	Cancer Therapeutics Response Portal (CTRP)	https://portals.broadinstitute.org/ctrp/	[91]
39	STRING	https://string-db.org/cgi/input.pl	[92]
40	cBioPortal	https://www.cbioportal.org/	[93]
41	UniProt	https://www.uniprot.org/	[94]
42	CGP	https://icgc.org/icgc/cgp	[53]
43	CPTAC	https://cptac-data-portal.georgetown.edu/cptacPublic/	[54]
44	CancerDR	http://crdd.osdd.net/raghava/cancerdr/	[55]
45	Oncomine	https://www.oncomine.org/resource/login.html	[56]
46	GWAS	https://www.ebi.ac.uk/gwas/	[95]
47	Tumorscape	http://beroukhimlab.org/data-and-tools/	[96]
48	UCSC Cancer Genomics Browser	https://genome.ucsc.edu/	[97]
49	IntOGen	https://www.intogen.org/search	[98]
50	BioProfiling.de	http://www.bioprofiling.de/	[99]
51	CMap	https://www.broadinstitute.org/connectivity-map-cmap	[100]
52	DrugBank	https://www.drugbank.ca/	[101]
53	TTD	http://db.idrblab.net/ttd/	[102]
54	CTD	http://ctdbase.org/	[103]
55	PubChem	https://pubchem.ncbi.nlm.nih.gov/	[104]
56	ChEMBL	https://www.uniprot.org/database/DB-0174	[105]
57	HMDB	http://www.hmdb.ca/	[106]
58	PharmGKB	https://www.pharmgkb.org/	[107]

AI and ML world is brought to limelight to evince the upcoming cancer repurposing future.

Genomics: connecting genetics with drug repurposing in cancer

Genomics is an intriguing branch of omics sciences, composed of structural and functional genomics interlaced with elements of genetics [109]. Structural genomics utilizes Next-Generation Sequencing (NGS), whole exome sequencing, and Single-Nucleotide Polymorphism (SNP) microarray genotyping to identify tumor-specific mutations, copy number alterations, gene expression changes, gene fusions, and germline variants. On the other hand, functional genomics involves customizing the comparison between sequences of full-length complementary DNA and its respective genomic DNA to predict their corresponding transcriptomes and proteomes [110,111].

Owing to the labyrinthine genetic etiology of cancer, unearthing Differentially Expressed Genes (DEGs) using genomics is of immense assistance in identifying novel cancer targets for repositioning of drugs [96]. In addition, this approach can be further employed to monitor the treatment efficacy and deduce mechanisms of resistance.

Hitherto, the establishment of global cancer genome mapping projects like TCGA [50], ICGC [51] and the Genome-Wide Association Studies (GWAS) [95] has culminated in easy accessibility of information pertaining to genetic variations in cancer.

Besides, the dawn of user-friendly portals like Tumorscape [96], University of California, Santa Cruz, Cancer Genomics Browser [97], ICGC Data Portal [51], Catalog Of Somatic Mutations In Cancer (COSMIC) [63], cBioPortal [93], Integrative Onco Genomics [98], and BioProfiling.de [99] helps in retrieving and statistically analyzing oncogene data. The consortium of genomic data also aids in reducing heterogeneity bias since it collates the data from a cohort of patients. Integration of various genomic sources provides a pool of targets for the drugs to act on. Despite its inherent advantages, cohort data usage lacks specificity to an individuals' genome. This limitation can be outdone by individualized genomic N-of-one studies, which are the frontiers for personalized medication management and drug repurposing [112].

DeSigN tool: a web-based drug repurposing algorithm based on gene expression patterns

Lee et al. [113] developed a web-based algorithm or tool named "DeSigN" (Differentially Expressed Gene Signatures) to predict phenotypic characteristics of drugs in cancer cell lines by considering their half maximal Inhibitory Concentration (IC_{50}) values and individual gene expression patterns. This algorithm construction took place in three steps. Firstly, a reference database with sensitivity patterns of cell lines to drugs extracted from Genomics of Drug Sensitivity in Cancer (GDSC) [90] was created, which contained 140 drugs with their unique rank order–based gene signatures. The second step was generation of query inputs for DeSigN database using DEGs from microarray or RNA-Seq gene expression data of cell lines in tumor and control samples. In the third step, nonparametric modified Kolmogorov–Smirnov (KS) statistics, a rank order–based pattern-matching algorithm was implemented in Connectivity Map (CMap) [100] to correlate the query signatures with specific drug-associated gene expression profiles. Later, the drug candidates were prioritized by computing connectivity score obtained from modified KS test. Eventually, to demonstrate the validity of the tool in predicting candidate drugs, four datasets (two estrogen receptor positive breast cancer, one non-small cell lung cancer, one pancreatic cancer) from the National Center for Biotechnology Information Gene Expression Omnibus (NCBI GEO) [65] were selected. For all the included datasets, a DEG list was prepared to use as a query in DeSigN. The designed tool was experimentally validated among Oral Squamous Cell Carcinoma (OSCC) cell lines for identification of growth inhibitors. Thus obtained gene signatures containing 69 upregulated genes and 86 downregulated genes were used as a query in DeSigN that returned nine potential candidates namely, GSK-650394, pyrimethamine, RDEA 119, BIBW2992, CGP-082996, lapatinib, PF-562271, bosutinib, and PD-0325901. Among the abovementioned nine candidates, BIBW2992 and bosutinib were reported recently for their efficacy in head and neck squamous cell carcinoma cell lines. This in silico prediction for bosutinib was experimentally validated in ORL196, ORL-204, and ORL-48 OSCC cell lines, and the drug exhibited significant cytotoxicity at one micromolar concentration [113].

Anticancer drug repositioning through genome-wide association studies

Zhang et al. [34] designed an in silico pipeline that mapped 50 SNPs located in rectal mucosal cells obtained from National Human Genome Research Institute GWAS [95] catalog to 140 genes associated with Colorectal Cancer (CRC) using snp2gene algorithm. The mapped genes were prioritized based on (i) functional annotation using Database for Annotation, Visualization, and Integrated Discovery [114] bioinformatics resources, (ii) cis-expression quantitative trait loci effects using peripheral blood mononuclear cell data generated from 5311 European subjects, (iii) PubMed text mining (TM) via Gene Relationships among Implicated Loci tool [115], (iv) Protein–Protein Interaction (PPI) analyzed through Disease Association Protein—Protein Link Evaluator [116], (v) genetic overlaps with cancer somatic mutations by means of COSMIC database [63], (vi) genes mapped with knockout mouse phenotypes from Mouse Genome Informatics [117] database, and (vii) SNPs from linkage disequilibrium ($r^2 > 0.80$) that were annotated as missense variants using NIH Roadmap Epigenomics Mapping Consortium [76]. Thereafter, Pearson correlation method was used for gene scoring, which ranged between zero and seven, wherein 35 genes that scored ≥ two were considered as biological risk genes. These top priority genes were used as query signature inputs to predict repurposable drugs from Drugbank [101] and Therapeutic Target Database [102]. This study revealed anticancer potential of crizotinib, arsenic trioxide, vrinostat, dasatinib, estramustine, and tamibarotene against CRC.

Proteomics: proteins to pave way for cancer drug repurposing

Proteomics, a branch of biology, which encompasses a cluster of technologies such as activity-based protein profiling, reverse-phase protein microarrays, and magnetic resonance spectroscopy to investigate and characterize the total protein content of a cell, tissue, or organism. It also deals with the analysis of PPI and protein–nucleic acid interactions and post-translational modifications that influence the function of proteins. It provides comprehensive information pertaining to relative appraisal of normal and disease states, transcription/expression, side effects of drugs, and aids in biomarker identification. To add on, proteomics relies on polypharmacology concept that describes the role of a single protein in multiple pathways that might trigger multiple signaling mechanisms. These intricacies underpin the relevance of target-based proteomic approaches in novel drug discovery or repurposing [118].

Current strategies of oncotherapy, so far, are oriented toward inhibiting DNA synthesis and abnormal signaling mechanisms. However, these mechanisms are sabotaged by drug resistance. This accentuates the application of high-throughput proteomic screening methods that not only facilitate identification of protein molecules interlaced in cellular network cascades associated with cancer but also illuminates the underlying molecular mechanisms of pathogenesis and disease progression [119]. Designing these screening methods warrants structural information of target and drug. In due course, study of whole organism proteome, recognition of potential druggable binding sites, prediction of PPIs, and identification of interacting residues in the binding site are crucial in discerning potential drug candidates that can produce desired pharmacological effects. Further, innumerable target-based approaches like molecular docking, molecular dynamics, and pharmacophore modeling have been designed to explore potential druggable candidates.

Proteomic approach revealing synergistic cytotoxic effect of fluspirilene with 5-fluorouracil in hepato cellular carcinoma

Xi-Nan Shi et al. [120] perused Protein Data Bank (PDB) [68] and extracted 346 X-ray crystallographic structures of Cyclin-Dependent Kinase 2 (CDK2), a serine/threonine protein kinase that plays a potential role in cancer pathogenesis. Among the 346 PDB structures, 44 PDBs (resolution = 1.3–2.8 Å) with cocrystallized structures were selected for further analysis. These cocrystallized structures were then cross-docked against the selected 44 PDBs using Autodock [121] to shortlist only those PDBs possessing minimum root mean square deviation. Later, 4914 FDA-approved drugs from ZINC database [122] and Drugbank database [101] were docked against selected PDBs. The drugs that exhibited top binding free energies viz., nilotinib, LS-194959, estradiol benzoate, nandrolone phenylpropionate, vilazodone, fluspirilene, azelastine hydrochloride, latuda, and paliperidone, ranged between −9.95 and −10.46 kcal/mol were later evaluated in HepG2 and Huh7 hepatoma cell lines. Among the above nine drugs, fluspirilene demonstrated significant cytotoxicity by arresting cell cycle in G1 phase after 24 h. In addition, it was found to decrease the expression of CDK2, Rb, cyclin E, pho-CDK2, and pho-Rb proteins. To validate the in vitro results, in vivo studies were conducted in Huh7 hepatoma cells of BALB/c nude mice for 21 days, which displayed significant reduction in tumor weight and volume when treated with fluspirilene and 5-fluorouracil combination [120].

Synergistic effect of imatinib with vemurafenib in Triple-Negative Breast Cancer (TNBC): a network-based data integration approach

Francesca Vitali et al. [123] designed a target-based approach to predict repurposable drugs through Topological Score of Drug Synergy (TSDS) for TNBC. Initially, PPI network was constructed between 43 disease proteins (DPs) which were evidenced to possess potential role in pathogenesis of TNBC. In continuation, this PPI analysis revealed 554 nodes with 2602 edges, among which 110 were hub nodes while 139 were bridging nodes. Further, the DPs that exhibited high confidence score and significant interactions with drugs in search tool for interactions of chemicals and proteins [124] were deemed as Target Proteins (TPs). The aforementioned criteria yielded 33 TPs and around 6074 drugs with 180 proteins. The shortlisted TPs were then evaluated through TSDS calculations by taking the edge score of shortest path connecting TP with DP into consideration. This analysis revealed 134 PPI combinations of 16 TPs. The selected TPs were then screened for potential drug candidates from Drugbank [101] and Comparative Toxicogenomics Database (CTD) [103]. Thereafter, the employment of trifactorization algorithm (disease–gene, drug–target, and protein–protein data matrix) predicted eight drugs from 816 drug–target interactions. Out of the total, seven promising drugs i.e., imatinib, L-aspartic acid, vemurafenib, hydroxyurea, azacitidine, flucytosine, and trametinib were shortlisted as potential candidates. Further, a Boolean network was constructed between identified TP, drugs, and Kyoto Encyclopedia of Genes and Genomes (KEGG) [79] pathways. In conclusion, imatinib, a tyrosine kinase inhibitor implicated in 18 KEGG pathways, was found to be a top-notch potential drug candidate for TNBC. This promising candidate exhibited synergistic cytotoxicity in combination with vemurafenib in MCF7 TNBC cell line [123].

Transcriptomics: RNA expression road map for drug repositioning in cancer

Transcriptome, an intermediate between DNA and protein, refers to the sum of all RNA transcribed by genome in cells/tissues, while transcriptomics elucidates genome-wide expression in response to functional and environmental changes that are critical in comprehending the mechanisms relevant to disease development and/or drug perturbation [125,126].

Screening of individual transcriptional expressions against compound libraries remains inefficient due to existence of ambiguity in identifying new potential disease targets. Hence, contemporary techniques such as High-Throughput Screening (HTS), DNA microarrays, NGS, polymerase chain reaction, and RNA-Seq were developed to surmount traditional techniques, which in turn generated enormous omics data. These data are readily accessible for researchers globally to identify and formulate AI/ML approaches for investigating associations between disease—gene—drug. This attempt has opened a gateway for drug repositioning research parallel to unearthing polypharmacology facts [126].

In the recent era, computational approaches have gained tremendous attention in exploring potential repurposable drugs via transcriptome-based databases like CMap [100], GEO [65], Statistically Significant Connection's Map [127], NFFinder [128], and the European Molecular Biology Laboratory—European Bioinformatics Institute's Array Express [66].

Drug-induced transcriptional signature—based approach for drug repositioning in cancer

Lee et al. [129] designed a chemical genomics in silico drug repurposing approach for glioblastoma, lung cancer, and breast cancer using ~20,000 drug-induced expression profiles grounded on structural, target, and expression signatures that were generated across multiple cancer cell lines. This innovative approach mapped the transcriptional signatures, structural fingerprints, and target information of drugs or chemical compounds for the aforementioned cancer types. Initially, a gold standard set of Known Drugs (KD) for each disease was extracted from Drugbank [101], CTD [103], PubChem [104] and KEGG DRUG [79]. A total of 132, 216, and 256 drugs were obtained for glioblastoma, lung, and breast cancer, respectively. In addition to KD set, a compound database labeled as CD set comprising of 1155 active compounds was created from 243 PubMed literatures. All the drugs and compounds of KD and CD sets that had experimental proof of affinity toward their targets were mapped accordingly. Later, the upregulated and downregulated gene expression signatures of glioblastoma were gathered from TCGA [50], while lung and breast cancers from NCBI GEO [65] datasets. Subsequently, gene expressional signatures for CD and KD set were gained from LINCS [52] database. Simultaneously, fingerprint similarity (including isomers) was assessed for the above compounds. Upon assessment, 8860 unique entities were merged, of which core set of 2250 compounds possessing structural signatures (S), target annotations (T), and expression (E) patterns were used for drug repurposing predictions. Afterward, a series of classifiers describing S, T, E, ST, TE, SE, and STE were developed for identification of potential candidates. Based on the similarity among classifiers, a drug repurposing score (ranging zero—one) was assigned for each compound in core set by comparing with KD set using logistic regression. The compounds screened through drug repurposing score were evaluated for specificity, sensitivity, and Area Under Curve (AUC). Thereafter, the core set compounds that possessed structural similarity (Tanimoto index = 0.7) with KD set were shortlisted for HTS to assess anticancer potential against 29 human cancer cell line datasets obtained from NCI [130]. This in silico screening revealed outstanding performance of expression-based classifiers (E) over other classifiers. Finally, 14 FDA-approved CD possessing brain—blood

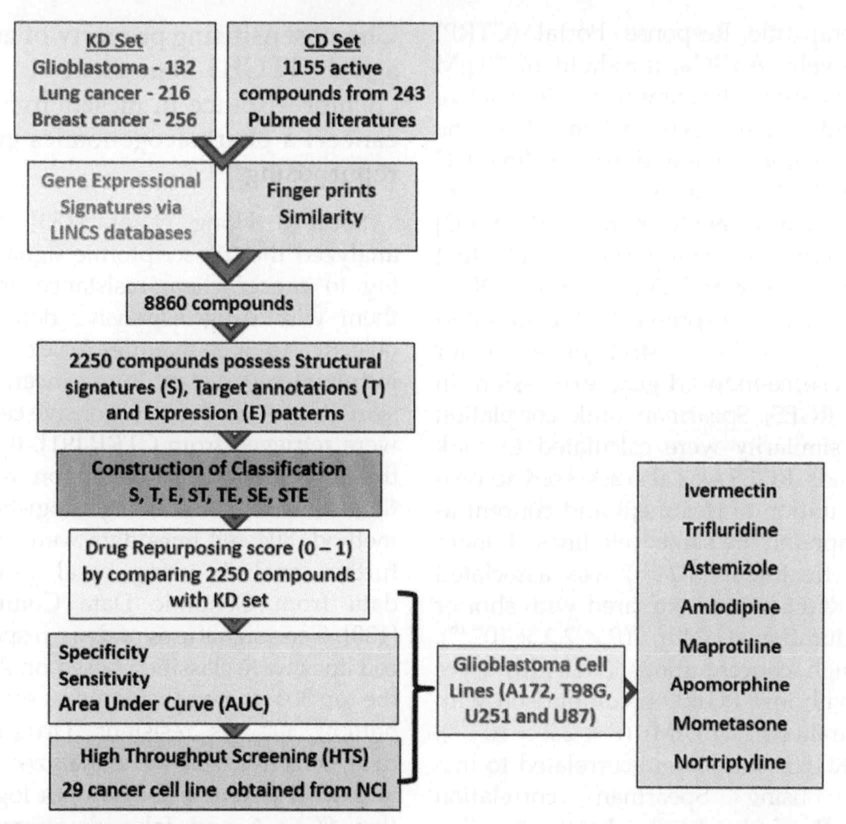

FIGURE 4.4 Workflow for chemical genomics in silico drug repurposing approach for glioblastoma.

barrier permeability and drug repurposing score >0.9 were validated experimentally in A172, T98G, U251, and U87 glioblastoma cell lines. The following drugs ivermectin, trifluridine, astemizole, amlodipine, maprotiline, apomorphine, mometasone, and nortriptyline among CD set were found to exhibit antitumor activity at 10 μM concentration [129] (Fig. 4.4).

Reversal gene expression profile–based drug repositioning for gastric cancer

In-Wha Kim et al. [131] developed a computational drug repositioning approach based on DEGs obtained from eight GEO datasets [65] for Gastric Cancer (GC) by calculating Reversal Gene Expression Scores (RGES). Among the 9113 significant DEGs retrieved from GEO datasets, 136 were upregulated and 53 were downregulated with $P < .001$ and log FC > 1.2. Similarities among gene expression profiles acquired from GEO were assessed using 40 GC cell lines data gained from Cancer Cell Line Encyclopedia database [57] using Spearman rank correlation test. The top 5000 genes were shortlisted according to interquartile range across all cell lines. In succession, LINCS database [52] was screened to extract repurposable drugs that possess similar gene profiles in concordance with the aforementioned top ranked genes. Only those molecules whose activity was established beforehand on the same cell lines were selected, and their median IC_{50} and AUC values were mined from ChEBI [132] and

Cancer Therapeutic Response Portal (CTRP) [91], respectively. An IC_{50} threshold of 10 μM was chosen to group the compounds into active (<10 μM) and inactive (>10 μM) moieties. The median IC_{50} values acquired by treating GC cell lines with 2025 compounds taken from ChEMBL [105], and 189 DEGs from LINCS [52] were considered for computing RGES that yielded a negative correlation between DEGs and compound gene expression. A compound with lower RGES has a stronger effect for reversal of disease-induced gene expression. In addition to RGES, Spearman rank correlation and cosine similarity were calculated to rank the compounds. RGES was also assessed according to the duration of treatment and concentration of compound in same cell lines. Longer duration of treatment (>24 h) was associated with lower RGES when compared with shorter treatment duration (<24 h) ($P < 2.2 \times 10^{-16}$). Similarly, high concentrations (>10 μM) were associated with low RGES on comparison with low concentrations (<10 μM) ($P < 2.2 \times 10^{-16}$). Significant RGES values were correlated to median IC_{50} using Spearman correlation (rho = 0.3, $P = 5.61 \times 10^{-3}$). Later, median RGES was calculated with multiple RGES found across several cell lines to represent summarized RGES (sRGES) values. Reversal genes were identified by comparing sRGES values of upregulated genes in active group with downregulated genes in inactive group and vice versa.

Ultimately, sensitivity analysis was performed for each compound. Genes that were found to be reversed in all trials with score of $P < .1$ were considered as reversal genes. Thus, this study identified COL4A1, PLOD3, UBE2C, MIF, and PRPF4 as reversal genes for GC. Fifteen drugs, cisplatin, cycloheximide, elesclomol, gefitinib, mitoxantrone, olaparib, paclitaxel, ponatinib, selumetinib, sirolimus, sorafenib, tanespimycin, temsirolimus, vincristine, and vorinostat were finally found to be active against GC [131].

Chemosensitizing property of atorvastatin against ITGB3-dependent chemoresistance in mesenchymal lung cancer: a pharmacogenomics guided drug repurposing

Soon-Ki Hong et al. [133] systematically analyzed the transcriptomic signatures pertaining to cancer chemoresistance and compared them with drug-responsive datasets, with an objective to open avenues for drug repurposing in treatment-resistant lung cancer. Gene expression data of 886 drug-responsive cancer cell lines were retrieved from CTRP [91]. By filtering out the low-quality profiles upon assessment of fitness score (<0.7) by logistic regression method, 804 cell lines data were considered for further analysis using basal gene expression data from Genomic Data Commons (GDC) [130]. Gene signatures of drug-responsive cancer cell lines were classified based on AUC, wherein the top 30% were considered as sensitive and the bottom 30% as resistant. Drug-resistant and drug-sensitive cells were analyzed subsequently to extract potential DEGs with log fold change (log FC) > 4 and false discovery rate (FDR) $P < .01$. This disclosed Epithelial–Mesenchymal Transition (EMT) to be the most upregulated pathways among the 32 chemotherapeutic drugs that exhibited significant resistance toward mesenchymal lung cancer cell line (A549TD). Thereafter, Spearman rank correlation test was used to calculate chemoresistant scores that explicated the frequency of appearance of a gene in upregulated DEGs. Amidst EMT genes, ITGB3 and CTGF were recurrently found in the upregulated DEGs. Finally, CMap database [100] was used to identify potential repurposable drugs for the prior mentioned chemoresistant genes. In addition, CMap also generated multiple gene signatures pertaining to each drug entity proposed against resistant genes. Following this, pairwise similarities among phenotypes and drugs were calculated using Jaccard index.

This in silico procedure highlighted the chemosensitizing property of FDA-approved antihyperlipidemic drug, atorvastatin. This was experimentally validated by pretreatment with atorvastatin that sensitized chemoresistant mesenchymal lung cancer cell line (A549TD) to doxorubicin treatment. Successively, a significant rise in apoptotic cell death with increased IκBα expression, followed by reduction in NF-κB reporter activity, BCL-xL, and ITGB3 expression, was observed, specifying the prominence of atorvastatin in overcoming resistance patterns [133].

Epigenomics: external factors that change the outlook on cancer drug repurposing

Epigenetics is an emerging omics science that describes the inherent modifications in gene expression without entailing alterations in DNA sequence, resulting in phenotypic changes by not affecting genotypic characteristics.

The factors that exert influence on aforesaid modifications include age, environment or lifestyle, and disease states. These numerous epigenetic changes are initiated and dynamically regulated by the processes such as (i) DNA methylation, (ii) histone modification, and (iii) noncoding RNA associated gene silencing [134,135].

Fundamentally, DNA methylation is a chemical process of inserting methyl groups by DNA methyl transferases at CpG sites where cytosine and guanine nucleotides are located adjacently with a phosphate linkage [136–139]. This methyl group addition renders structural and morphological changes in DNA, which ultimately results in modification of transcription, which is postulated to be involved in cancer pathophysiology [140]. Conversely, DNA demethylation is yet another process that aids in gene reprogramming [141].

Posttranslational histone modifications such as lysine acetylation, ubiquitination, and sumoylation, serine, threonine and tyrosine phosphorylation, lysine and arginine methylation, glutamate poly-ADP ribosylation, arginine deamination, and proline isomerization [142] can cause distortions in chromatin structure, DNA replication, transcription, DNA repair, and genomic stability.

Although noncoding RNAs do not encode proteins, nevertheless, on binding to DNA, they can change its conformation. Thereby, they regulate gene expression and finally affect the stability of mRNA after transcription [142]. These epigenetic modifications may be studied further to understand the heterogeneity in cancer cells and drug resistance. Therefore, recognizing the role of epigenetic changes is crucial in uncovering newer targets for drug repurposing.

KsRepo, a methylation-based drug repurposing method for acute myeloid leukemia

Brown et al. [143] designed a novel drug repositioning method that employed ksRepo, a modified KS test, to analyze differential DNA methylation patterns for AML. In this study, two genome-wide methylation datasets, which clearly described hyper- and hypomethylation status pertaining to AML were selected and investigated for differential methylation patterns using GEO2R. Subsequently, the resultant CpGs with FDR $P < .05$ were shortlisted and mapped to the nearest genes. The P-values of CpGs for each gene so obtained were combined via Empirical Brown's method and adjusted by Fisher's method. Significant genes (820 and 442) identified in the above datasets were found to be enriched in Wnt signaling pathways. In continuation, each dataset was individually analyzed through ksRepo [143], a methylation-based drug repurposing method. In this method, the compounds retrieved from Drugbank [101] and CTD [103] were subjected to KS test to establish compound–gene interactions which conceded

1075 drugs with at least one overlapping gene. Ultimately, four drugs, i.e., alitretinoin, cytarabine, panobinostat, and progesterone, were found to possess significant FDR *P*-value in both datasets. Out of which, cytarabine is an already existing FDA-approved drug for AML, while others are in investigational phase [143].

Metabolomics: following the footprints of cellular metabolites to uncover the treasure of novel therapy for cancer

Metabolomics is one of the latest branches of omics sciences emerged in the present decade that characterizes small molecule metabolites from biological samples (plasma, saliva, blood, tissues, and isolated cells) by commissioning innovative and sophisticated analytical instrumentation techniques (nuclear magnetic resonance spectroscopy or mass spectrometry, gas chromatography, liquid chromatography–mass spectrometry, or capillary electrophoresis–mass spectrometry in conjunction with statistical and computational tools) [144]. Variations in metabolite concentrations emerge as a consequence of diseased states, medical interventions, aberrant enzyme levels, environmental factors, and genetic influences [145]. Cancer is one such disease state, where cellular metabolism is altered in all cancer types regardless of their location, and this unique metabolic phenotype is exemplified by increased uptake of glutamate and glucose; biosynthesis of carbohydrates, fats, and proteins; and impaired mitochondrial activity. These altered pathways can be potentially recognized by metabolomic explorations [146,147].

Human Metabolome Project [148] funded by Canadian Institutes of Health Research, Canada Foundation for Innovation in 2004 rendered small molecule metabolites and biomarker discovery hassle-free. Later, The Metabolomics Innovation Centre [149] was launched with an objective to generate all detectable metabolites (>1 μM) in human body. Eventually, Human Metabolome Database [106] was launched in 2007, and it is currently considered as one of the world's largest organism specific, comprehensive metabolomics databases encompassing 351,754 experimentally validated metabolites, 25,770 illustrated metabolic pathways, with 5498 metabolite–disease relationships.

Ease of access to computational approaches and contemporary advances in statistical and technological aspects of affluent metabolomics data offers early detection and customization of personalized therapeutic regimen based on an individuals' metabolic profile.

Repurposing ifenprodil as a therapeutic agent for prostate cancer by adopting genome-scale metabolic modeling

Beste Turanli et al. [150] fabricated a prostate cancer–specific Genome-scale Metabolic Model (GEM) to identify potential repurposable drug candidates through drug–gene interactions derived via cancer-specific metabolomic data. Incipiently, a total of 14,778 protein coding genes were retrieved from transcriptome data (comprising of 495 tumor tissues and 52 nontumour tissues) and proteome data from NCBI GDC [130] and Human Protein Atlas [151]. Among which, 8558 genes were found to be overlapped between transcriptomic and proteomic data. In addition to the 8558 overlapped genes, 3328 proteome-specific and 2892 transcriptome-specific genes were also incorporated in GEM creation. Network analysis for overlapping proteins was constructed with Integrative Network Inference for Tissues algorithm using iCancer model [152] (a generic cancer metabolism model) as a template to identify prostate cancer–specific metabolic reactions. This, in turn, proclaimed 2655 genes and 6718 reactions. Amidst prostate cancer–specific metabolites, 86 were upregulated and 76 were downregulated. Functional enrichment analysis revealed the involvement of 23 genes in steroid biosynthesis pathways such as zymosterol biosynthesis (FDFT1, SQLE, LSS, CYP51A1, NSDHL, HSD17B7, and DHCR24), cholesterol production (DHCR24, DHCR7, SC5D, and EBP), terpenoid backbone biosynthesis, and mevalonate pathway (MVD, HMGCR, MVK, and PMVK). In continuation, Differential Rank Conservation analysis was used for network mapping of phenotypes with tumor and nontumour samples that uncovered lipid metabolism (especially glycosphingolipid biosynthesis, ether lipid metabolism, and steroid biosynthesis), riboflavin metabolism, pentose phosphate pathway, and thyroid cancer to be highly significant in tumor samples. KEGG pathways were analyzed to achieve genes intricated in the aforementioned lipid metabolism pathways. CMap2 was used for detecting repurposable drugs that can reverse these metabolic pathways using negative similarity scores. Initially, 81 drugs were shortlisted, of which 43 drugs were found to reverse metabolic signatures. Finally, sulfamethoxypyridazine, azlocillin, hydroflumethiazide, and ifenprodil were found to be noteworthy. Ifenprodil was then compared for cell viability between human prostate epithelial cell line (RWPE-1) and prostate cancer cell lines (LNCap, 22Rv1, PC3). Results revealed that 100 μM concentration of ifenprodil diminished cell viability of prostate cancer cell lines but had no influence on RWPE-1 cell line [150] (Fig. 4.5).

Side effects: providing prospective drug repositioning candidates for cancer using adverse drug event signals

Side effects or Adverse Drug Reactions (ADRs) are unintended phenotypic expressions of drugs that pose a considerable healthcare hurdle and may lead to drug withdrawal from the market. The unpredictability of ADRs uplifts their mysterious mechanisms as sources of interest to uncover off-targets of the respective drugs. In addition, the influence of drugs' structural features on incidence of ADR and comparability between ADR targets reinforce the fact that the side effect of one disease could possibly offer a solution for another disease. Besides, ADR mechanisms directed toward extended pharmacology of drugs correlate drug indications with side effect, thereby intriguing researchers to further upgrade their research inclining toward drug discovery and repositioning [153].

Although of immense advantage, the mechanism of ADRs is arduous to attain. The innovative Similarity Ensemble Approach [154] is a

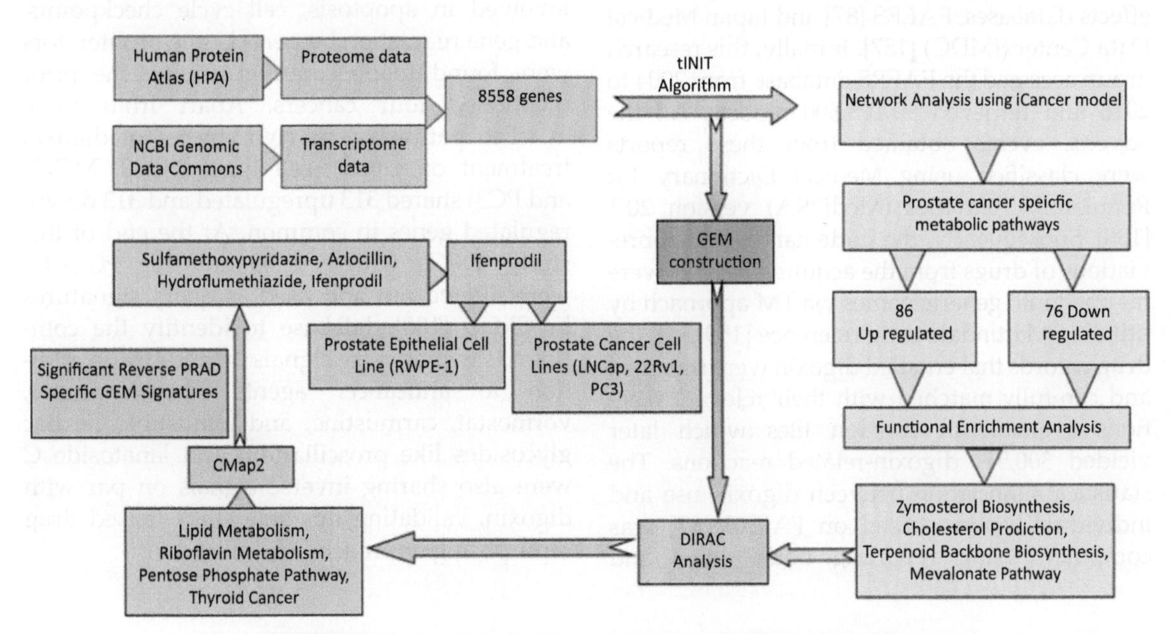

FIGURE 4.5 Outline of prostate cancer–specific GEM construction.

novel strategy that compares drug off-targets with similar actions of other drugs using simulations, thereby aiding in effective drug repositioning. Moreover, side effect databases like Side Effect Resource (SIDER) [81] or FDA Adverse Event Reporting System (FAERS) [87] yield frequently repeating ADRs as signals which on subsequent screening through genomics databases like Pharmacogenomics Knowledge Base [107] can abet in determining potential targets and diseases associated with those targets.

Furthermore, linking side effect profile with innately responsible TPs through ML and computational techniques accelerate and strengthen this approach. This in turn assists in better understanding of disease pathways and drug targets in order to retrieve novel indications [155].

Adverse effect database mining approach to elucidate anticancer potential of digoxin

Satoshi Yokoyama et al. [156] identified anticancer properties of digoxin by mining adverse effects databases: FAERS [87] and Japan Medical Data Center (JMDC) [157]. Initially, this research group accessed the FAERS database from 2004 to 2016 and retrieved 99,108,600 reports. All the adverse events obtained from these reports were classified using Medical Dictionary for Regulatory Activities (MedDRA) version 20.1 [158]. Subsequently, the trade names and abbreviations of drugs from the acquired reports were mapped into generic names via TM approach by utilizing Martindale drug reference [159]. All the drug records that entailed digoxin were screened and carefully matched with their relevant reactions available in reaction files which later yielded 300,541 digoxin-related reactions. The statistical association between digoxin use and individual cancers based on FAERS [87] was computed using reporting odds ratios and information components. The resultant analysis epitomized significant inverse signals associated between digoxin usage and melanoma, hematological malignancy, gastric, colorectal, pancreatic, breast, ovarian, prostate, and bladder cancers. Analogously, analysis of JMDC database [157] revealed 52,828 claims pertaining to digoxin. Alongside, Event Sequence Symmetry Analysis was performed on 3035 digoxin users to detect the association between digoxin use and cancer. This exposed inverse association between digoxin and diagnoses of esophageal, gastric, colorectal, lung, and prostate cancers as well as hematological malignancy. Upon comparison of results, among the two databases, overlapping inverse signals were seen between digoxin use and GC, CRC, prostate cancer, and hematological malignancy. In continuation, Basespace Correlation Engine database [160] was used to reclaim curated gene expression data of digoxin and the aforesaid cancer types. This was trailed by pathway enrichment analysis which revealed 197 upregulated and 71 downregulated pathways. Amidst the pathways, 17 upregulated and 8 downregulated pathways involved in apoptosis, cell cycle checkpoints, and gene regulation by peroxisome proliferators were found to be common among the prior mentioned four cancers. Apart from these in silico pathways, in vivo studies of digoxin treatment on cancer cell lines (HL60, MCF7, and PC3) shared 313 upregulated and 313 downregulated genes in common. At the end of this study, DEGs with $P < .05$ and log $FC > 1.5$ were filtered out and used as query signatures in CMap [100] database to identify the compounds with inverse signals. Incredibly, in addition to anticancer agents like etoposide, vorinostat, carmustine, and lomustine, cardiac glycosides like proscillaridin and lanatoside C were also sharing inverse signals on par with digoxin, validating this side effect–based drug repurposing approach [156] (Fig. 4.6).

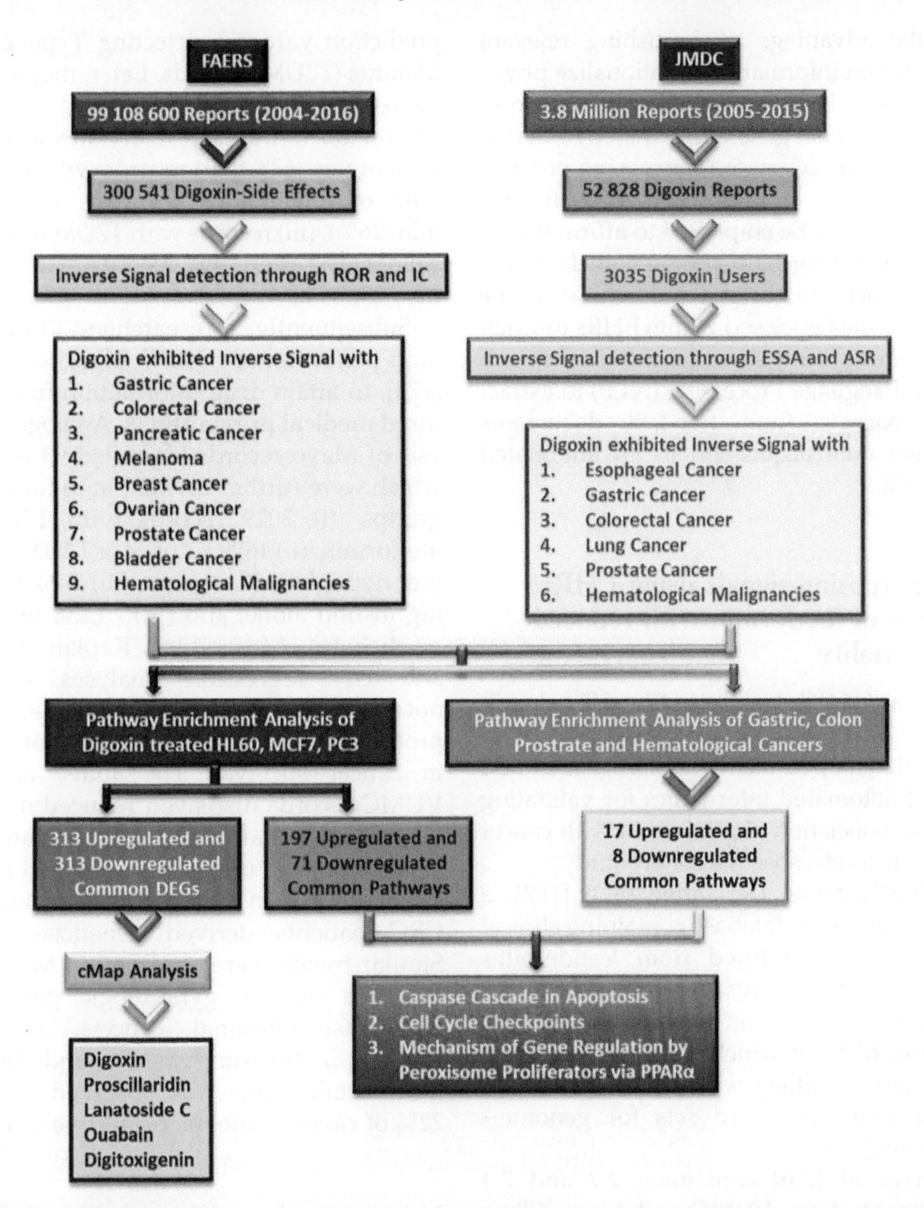

FIGURE 4.6 Workflow for adverse effect database mining approach to elucidate anticancer potential of digoxin.

Electronic health records: routine clinical practice data, a hidden treasure for cancer drug repurposing

EHR hoards extensive electronic data of patients' medical information, viz., demographics, phenotypes, family history, medical data coupled with diagnosis, clinical data, laboratory data, and other medical discrepancies, generated during their each consultation with healthcare delivery system [161]. Compilation of information reclaimed about drug exposure and the outcomes via EHR capacitates identification of various effects of the drug, thus promoting individualization of medication therapy, identification of risk factors, and drug–drug interactions [162].

Besides the advantage of furnishing relevant patient-centered information to rationalize physician's workflow, EHR can also serve as a platform that facilitates signal generation, thereby offering hints for drug discovery or repurposing projects. EHR is thus an efficient and cost-effective resource that could be employed to affirm the rationality of the drug repurposing signals generated from other sources as well. Evaluating the unstructured data enclosed within EHRs through the assistance of computational techniques such as Natural Language Processing (NLP) to extract medical concepts from free-text documents allows finer data acquisition in an automated fashion [163].

Drug repurposing signals using EHR: association of metformin with reduced cancer mortality

Xu H et al. [164] linked the EHRs of Vanderbilt University Medical Center (VUMC) [165] and Mayo Clinic [166] to their tumor registries and implicated automated informatics for validating the inverse association of metformin with cancer mortality via a retrospective cohort study.

Initially, Synthetic Derivative (SD) [167], a stand-alone research database containing clinical information, was captured from Vanderbilt's electronic medical records, which were altered to the point that they no longer resemble the original record from which they were derived. This SD resource along with BioVU [168] aids in identification of record sets for genome–phenome analysis.

SD image of EHR containing 2.2 and 7.4 million records from VUMC and Mayo Clinic were retrieved, respectively. Based on the eligibility criteria (cancer diagnosis should be as per IDC-O, exclusion of cases with skin cancer, congestive heart failure, and chronic kidney disease as they are contraindicated for metformin), 32,415 cancer records were selected from VUMC. These records were then assessed using electronic Medical Records and Genomics Network algorithm [169] which had a 98% post prediction value in detecting Type 2 Diabetes Mellitus (T2DM) records. Later, they subdivided the records in to four groups, i.e., (i) records using metformin as a primary medication along with others (63%), (ii) records with T2DM using other oral hypoglycemic agents but not metformin (26%), (iii) records with T2DM using only insulin (11%), and (iv) records with no T2DM (n = 28,917).

Subsequently, researchers implemented high-performance NLP algorithm, MedEx [170], to attain drug information from unstructured medical prescriptions. Analogously, analysis of Mayo records identified 93,169 records, which were further divided in to four exposure groups: (i) 3029 records with T2DM using metformin, (ii) 1629 records of T2DM with other oral hypoglycemic agents, (iii) 1462 records using insulin alone, and (iv) 73,138 records with no diabetes. After which, Kaplan–Meier plots and cox regression analyses were used not only to assess cancer detection survival probabilities but also the influence of metformin in cancer survival. The above analyses of VUMC records displayed reduced mortality in breast, colorectal, lung, and prostate cancers. Post metformin usage demonstrated maximum survival rates in breast cancer patients, while CRC patients derived beneficial outcomes. Similar trends were observed in Mayo records for bone marrow, gynecologic, genitourinary, and gastrointestinal cancers. Overall, an increase in the survival rate and decrease in the mortality rate was observed in 23% and 22% of cancer patients, respectively (Fig. 4.7).

Noncancer drug effects on survival of patients with cancer: a new-fangled hypothesis for drug repurposing via Electronic Health Records

Yonghui Wu et al. [171] exposed anticancer potential of 22 non-CDs by linking the cancer registry of VUMC [165] and Mayo clinic [166]. Initially, a total of 43,310 SD cancer patient records were identified by parsing pathology reports and billing codes. Medication information was then acquired using MedEx algorithm [170], which retrieved 301 drugs that were used by more than 5000 individuals. Among the total, short-term drugs, over-the-counter drugs, and cancer medications were excluded leaving out 146 noncancer medications. Multivariable Cox proportional hazards regression model was then applied to assess the survival rate of cancer patients after taking the noncancer medications for a minimum of 5 years. FDR <0.1 was used as a cutoff to select the top-ranking drugs for reducing cancer mortality and for excluding excessive false negatives. This model resulted

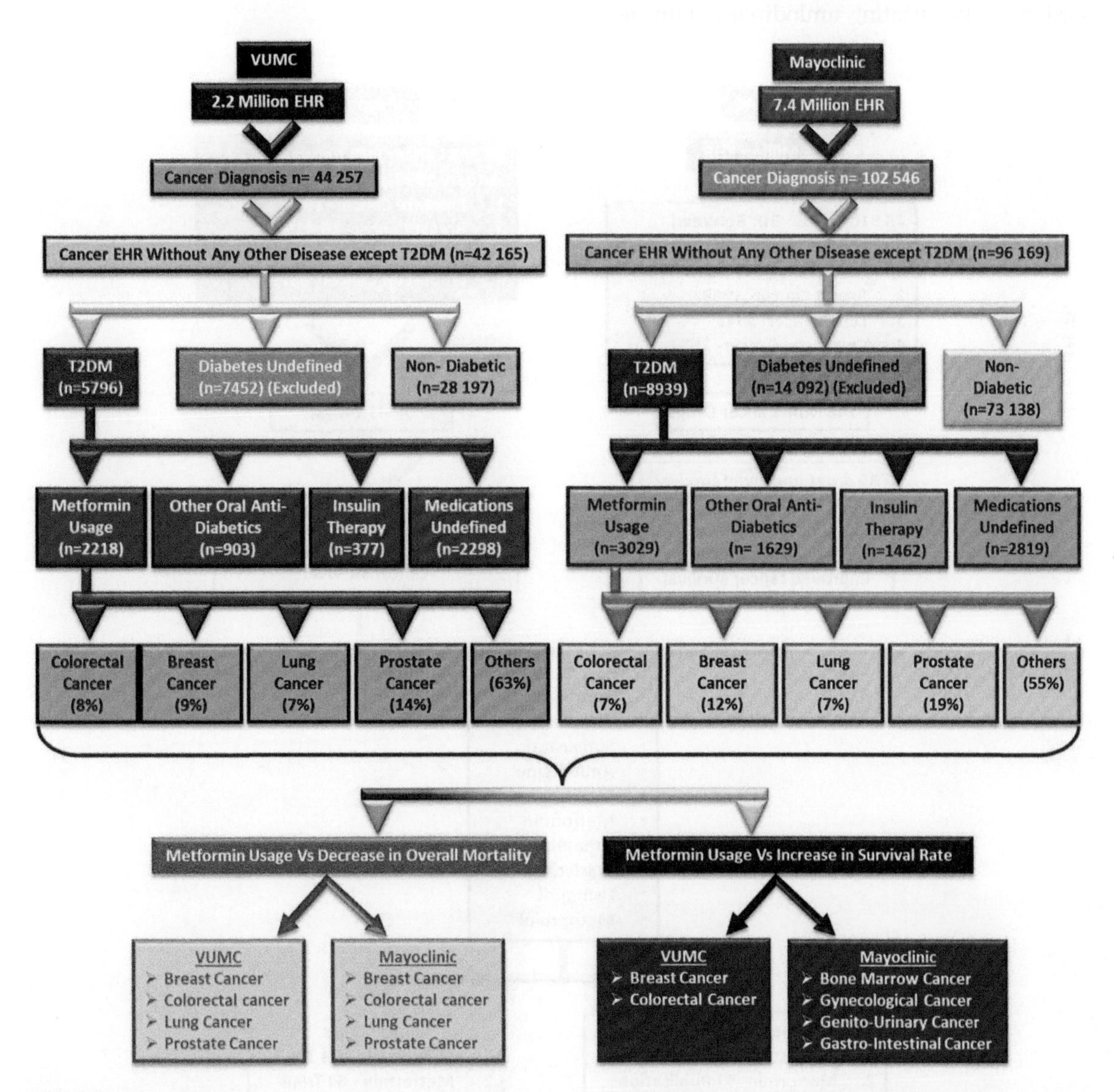

FIGURE 4.7 Workflow for detecting drug repurposing signals through EHR to explicate the association of metformin with reduced cancer mortality.

in 22 drugs belonging to the below six classes: statins, proton pump inhibitors, angiotensin-converting enzyme inhibitors, β-blockers, nonsteroidal antiinflammatory drugs, and alpha-1 blockers.

This study as such was replicated in Mayo clinic EHR database, which revealed nine drugs: rosuvastatin, simvastatin, amlodipine, tamsulosin, metformin, omeprazole, warfarin, lisinopril, and metoprolol. Eventually, literature reviews and clinical trial data for these nine drugs were reclaimed. This generated evidence revealed the antineoplastic effect of simvastatin in 18 studies and 23 trials and metformin in 40 studies and 64 trials (Fig. 4.8).

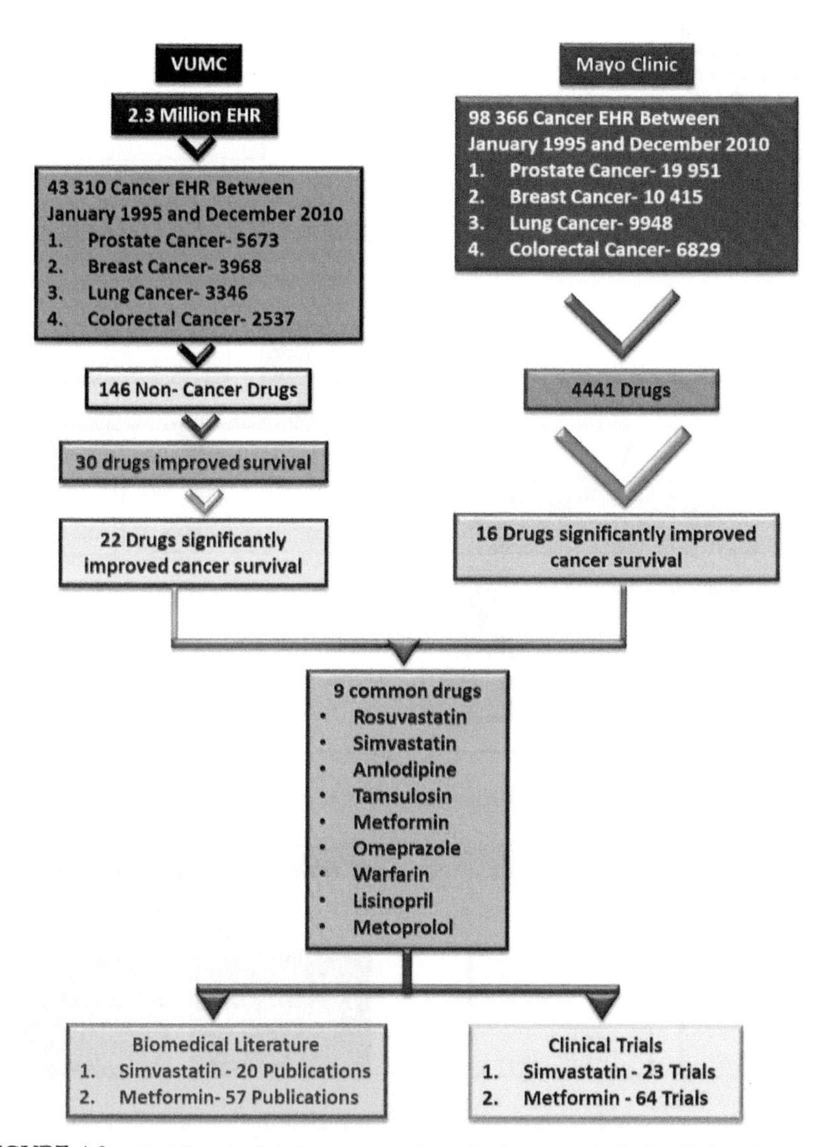

FIGURE 4.8 Workflow to detect noncancer drug effects on survival of patients with cancer.

Literature-based drug discovery: refreshing text records to optimize yields for drug repurposing in cancer

TM or literature mining, also called as text analytics, is an advanced technology that emerged recently, to unravel crucial information concealed in textual data sources. This multidisciplinary field encompasses Information Retrieval (IR), data mining, ML, statistics, and computational linguistics [172]. TM finds its application in exploration of significant entities of research interest from giant literature data, along with transformation of unstructured data to structured format. TM further translates the data in a pattern that is suitable for analysis and helps in framing a tangible research hypothesis.

NLP is one of the TM techniques that reads, analyses, and derives relationship between two or more keywords through Named-Entity Recognition (NER) in an unbiased manner based on their appearance in complex literature irrespective of language and theme [173]. NER retrieves unstructured data from Medical Subject Headings (MeSH), Scopus, Web of Science, PubMed, or MEDLINE and many other biomedical databases. It further captures and categorizes domain-specific concepts from text data and creates a dictionary of searched entities. This is followed by Information Extraction (IE) from the formed dictionary to establish relationships between specific attributes in order to conserve precise and accurate information within an internal database [172]. IE is succeeded by IR which derives patterns of attributes in accordance to their appearance and frames a meaningful relationship depending on the assigned weights. Later, algorithms like Boolean strategies, adhoc retrieval, inverted index, cosine similarity, term frequencies, term weightage, semantic search, and threading will independently trace the research gap between different entities, to ultimately formulate a valid research problem or hypothesis. Subsequently, clustering techniques (hierarchical, centroid, k means, and distributional) will be applied to identify the similarities among collected information in the database, to create groups, and to recognize unique patterns among the subgroups. Finally, the text summarization step elaborates the derived relations between entities in a logical manner using semantic graphs, fuzzy logic, regression models, neural networks, decision trees, etc. [172]. In the current decade, numerous hypotheses have been generated using TM techniques to foster the current research in the fields of drug discovery, drug repurposing, diagnosis, and disease–gene and drug–gene associations.

Availability of huge literature on cancer pathophysiology and molecular mechanisms poses an immense time stacking exercise necessitating computational approaches to streamline CD discovery research. This tedious task can be supported by the emerging TM techniques that paves path to organize and rediscover the desired information efficiently from sizable databases [174].

TM-centered construction of cancer drug toxicity knowledge base

Xu and Wang [175] developed a tool that offers a comprehensive and machine-understandable CD–side effect (drug–SE) relationship that forms a foreground for in silico CD and target discovery, drug repurposing, and toxicity prediction. This knowledge base was constructed from published biomedical literatures like MEDLINE records and clinical trials [176]. Initially, 21,354,075 MEDLINE records were parsed, and the abstracts and PubMed IDs were extracted to prepare the text corpus; subsequently, these abstracts were split into sentences. This was followed by creation of CD lexicon from clinical trials data which furnished 52,066 unique drug–disease pairs from 196,002 drug–disease pairs. Among the 52,066 drug–disease pairs, 17,386

CD pairs were found to be related to cancer semantic terms, according to Unified Medical Language System [177]. Thus, derived CD pairs were filtered depending on their cooccurrence in MEDLINE records, which revealed 358 pairs, out of which top ranked 100 drugs pairs were selected based on their appearance in clinical trials. Later, an SE lexicon comprising of 49,625 unique terms related to cancer was manually prepared out of 70,177 SE terms from MedDRA [158]. Each drug in top-ranked drug pairs were mapped with SE lexicon which were subjected to analysis of drug–SE cooccurrence in the text corpus. A total of 56,602 and 134,670 drug–SE pairs from sentences and abstracts were extracted, respectively. From the above search task, 44,816 unique drug–SE pairs were filtered and ranked based on their appearance in clinical trials and text corpus. Eventually, the top ranked drug–SE pairs were evaluated for precision and accuracy through an 11-point quality check scale to design a CD toxicity knowledge base. Precision, accuracy, and F1 score of this knowledge base were validated using irinotecan–SE as an example through SIDER [81] and manually created MEDLINE records, which yielded 30 common irinotecan–SE pairs. This study revealed that CDs that share similar SEs tend to possess overlapping gene targets and indications, thereby creating a platform for drug repurposing (Fig. 4.9).

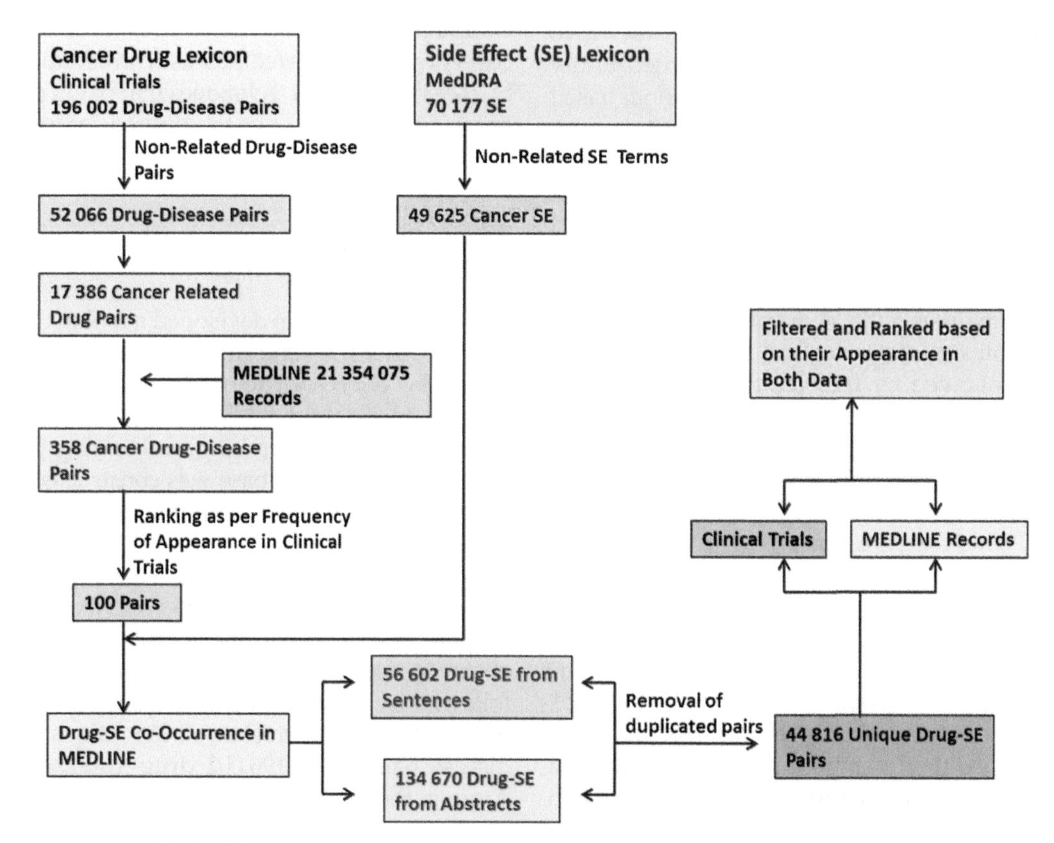

FIGURE 4.9 Workflow for construction of cancer drug toxicity knowledge base.

Cancer Hallmarks Analytics Tool: a text mining approach to organize and evaluate scientific literature on cancer

Simon Baker et al. [178] generated an extensive Hallmarks of Cancer (HoC) taxonomy and developed Cancer Hallmarks Analytics Tool (CHAT), an automatic TM tool capable of capturing and organizing heaps of cancer-related references from PubMed. This taxonomy was developed using 10 primary HoC and 37 secondary HoC. Later, a text corpus was created by retrieval of terms semantically similar with each HoC, from biomedical literature in PubMed database containing 150 million sentences from 24 million PubMed abstracts. Each sentence was screened for the presence of HoC and annotated with labels accordingly. The precision of created annotations was calibrated by assessing the interrater agreement. A supervised NLP pipeline was designed using seven semantic techniques for feature extraction, followed by usage of GENIA tagger [179] to execute the steps involved in NLP processing, viz., parts of speech tagging, NER, lemmatization, cosine similarity, and ANN. Feature selection was carried out through MeSH and Chem terms to classify the sentences and metadata were extracted from the created corpus. Thus mined features in metadata were represented in binary format based on their appearance (i.e., presence of feature is annotated as one and absence as zero). Support Vector Machines (SVM) and kernel methods (using scikit-learn) were applied to compare all the sentences in binary formats for their relevance in HoC taxonomy. The efficiency, accuracy, and F1 score of the tool were evaluated both intrinsically and extrinsically by case studies. Finally, during evaluation, CHAT identified the potentiality of aspirin in the treatment of CRC using invasion and metastasis as HoC. This tool will be supportive in identifying HoC associated with extrinsic factors, biomarkers, and therapeutic targets [178].

Integrating approaches to improve accuracy of prediction: the futuristic means to cancer drug repurposing

Computational approaches for drug repurposing has achieved great momentum in the recent years. However, there are numerous limitations in the methods opted. Most of these approaches depend on target-based, disease-based, and drug-based data. Target-based techniques rely on drug–protein interactions, whereas disease-based and drug-based approaches reckon on comparing the features and similarities of diverse diseases and drugs, respectively.

Target-based approaches work on reconnoitring target binding sites for new indications. Yet, till date, the existing algorithms are ineffective in accurately predicting the binding sites of new or unrelated proteins, thereby restricting their applicability in identification of novel compounds through drug repositioning. Network pharmacology, a background for target-based approaches, is involved in identification of gene signatures that predict the off-targets of drugs and recognize genetic modifications induced by drugs. However, the mechanisms underlying such gene expressional changes are yet to be explored to explicate off-target effects of drugs on genes. As a matter of fact, the noise perceived in gene expression data can lead to biased network predictions. Similar challenges are confronted in PPI network that maps PPI for data derived from experimental sources, as well as drug side effect data that correlates phenotypes with drug mechanisms, which in turn largely rely on individual genetic map, traits, and medication history.

The intricacies faced in drug-based approaches pose a threat of selection of erroneous chemical structures of drugs from available databases, due to which the entire process of drug discovery will be directed off the beam. Also, lack of relevant details owing to proprietary norms at some instances piles on further complications. Moreover, while adopting molecular docking technique, matching a ligand of interest with the target binding pocket is unrealistic as three-dimensional structures of countless proteins are unidentified. Though homology modeling builds up protein structures, the accuracy is still questionable. Another limitation of docking studies is that it provides flexibility only to the ligand, while the receptor is rigid to the residues around the binding site. Even though molecular dynamics can overcome this shortcoming by contributing flexibility to receptors, adding water molecules and salts, etc., in order to simulate the natural conditions, it demands supercomputing facilities. The pros and cons of above mentioned approaches are highlighted in Table 4.3. Considering these inadequacies in the aforementioned isolated approaches, it is always advisable to integrate the data obtained from multiple resources for better in silico predictions that offers a successful drug repurposing endeavor.

TABLE 4.3 Pros and cons of individual omics repurposing approaches.

Method	Advantages	Limitations
Literature Mining	• Time- efficient screening of multiple diseases • Helps screen a large data lasting over a longer period of time • Helps uncover serendipitous discoveries • Aids unveil rare events	• Publication bias • The vast data is prone to contain junk data and noise • False positives and false negatives need to be weeded out • Search strategy expertise required
Genomics	• Provides gene-specific data, thereby individual-specific treatment can be retrieved • Helps repurpose drugs for very rare or neglected diseases based on genetic similarity of targets • Aids in uncovering relation between gene and disease	• Post transcriptional and post translational changes are neglected • Epigenetic variations go unnoticed • Cohort effect overlooks herogeneity of cells • Obtaining genetic data is arduous due to the expensive and invasive techniques needed
Transcriptomics	• Helps uncover therapeutic possibilities based on differentially expressed genes • Aids in relating transcriptomic mutations with diseases	• Post translational modifications are not taken into account • Requires supercomputing facilities • Physiological simulations may not be accurately replicated
Proteomics	• Helps relate target proteins to disease mutations • Aids simulate protein structure from their amino acid sequence • Assists in determining ligand affinity to binding pocket	• Not all protein structures are available • Homology modelling of proteins may not be accurate • Physiological conditions cannot be accurately replicated • Supercomputing facilities required
Metabolomics	• Helps relate metabolite effects on disease phenotypes • Assists in regulating therapy based on metabolites	• Not all diseases have studied biomarkers • Metabolites need to be in larger quantities to get detected • Comparatively time consuming
Epigenomics	• Determines effect of external factors on the genomic activity	• Not all methylated regions in DNA get identified • Involves supercomputing • Challenging to link epigenetic data with gene expression • Lack of pre-identified epigenetic markers for many diseases

TABLE 4.3 Pros and cons of individual omics repurposing approaches.—cont'd

Method	Advantages	Limitations
Side effects	• Uncovers rare events • Helps strengthen side effect signals	• Side effects are individual-specific and may not be extrapolated to the larger populace • Safety profiles may not be available • Large data to be screened,- time consuming
Electronic Health Records	• Helps find serendipitous events • Aids to determine cost-effective regimen	• Rare effects may go unnoticed • Data recording bias • Diagnosis bias • The effect of concomitant medications go unnoticed

Deep learning algorithm integrating genomic and chemical fingerprints for drug repurposing

Chang et al. [180] developed Cancer Drug Response Profile scan (CDRscan), a deep learning algorithm by implementing five convolutional neural network–based models with varied architecture which can predict the anticancer efficacy by integrating large-scale screening data of 244 small molecule inhibitors and genomic profiles of 787 human cancer cell lines.

Initially, drug-responsive datasets and sequencing data of 1001 cancer cell lines spanning 30 cancer types as per TCGA criteria were retrieved from cancer cell line profiler [181], of which 787 cell lines corresponding to 25 cancer types were selected by excluding 5 cancer types that contained less than 10 cell lines. Subsequently, 265 anti-CDs were obtained from GDSC [90] based on the IC_{50} values, which were evaluated across the selected cell lines. Among these anti-CDs, 18 entities which lacked Simplified Molecular-Input Line-Entry System (SMILES) and rest 3 with high molecular weight were excluded. Thereafter, chemical features and structural similarities of 244 drugs were recovered from PubChem [104] using SMILES as an input. Compiling the above data, a deep learning dual convergent architecture model was constructed comprising 152,594 instances across 787 cell lines related to 25 cancer types with mutational information in 567 genes along with IC_{50} measurements of 244 chemical entities. This model relies on genomic fingerprint data collected from COSMIC database [63] and molecular fingerprint data from PubChem. Genomic fingerprints reclaimed from cancer cell lines were encoded in 28,328 binary digits (0 or 1), where each binary code represents the somatic mutation status as per Cancer Gene Consensus [182]. On the other hand, molecular fingerprints analogous to 244 drugs expressed as 3072 binary descriptors were extracted using PaDEL descriptor [183] via SMILES. Finally, a sum of 31,400 encoded binary digits of genomic and molecular fingerprints were used to develop five CDRscan models. These models were trained using 95% of 152,594 randomly selected instances from 25 cancer types and were validated with rest 5% instances which were designated as test data. The accuracy and precision of these models were evaluated by calculating R^2 (coefficient of determination), Root Mean Square Error (RMSE), and Area Under the Receiver Operating Characteristic Curve (AUROC). Drug-centric evaluation was then performed to classify the drugs as active ($lnIC_{50} < -2$) or inactive. An acceptable agreement was observed between IC_{50} values of drugs obtained from GDSC [90] and predictions of CDRscan algorithm across five models (R^2 ranging 0.838 to −53, RMSE of 1.069, and AUROC of 0.98). The above internal validation process was preceded by drug repurposing predictions for 1487 FDA-approved drugs taken from Drugbank [101]. Among 1487 drugs, 102 were already existing anticancer agents, of which 37 were found to possess potential for new cancer type indication. At the same time, 176 nononcology drugs were predicted for cancer indication, out of which 4 exhibited strong efficacy against 90% of cancer types (Fig. 4.10).

Antineoplastic potential of anthelmintic drugs via integrated drug–target–gene approach

Francesco Napolitano et al. [184] developed a novel approach for drug repurposing by integrating gene expression patterns, drug–drug similarity, and PPI. Initially, raw gene expression data of 1309 drugs from CMap were

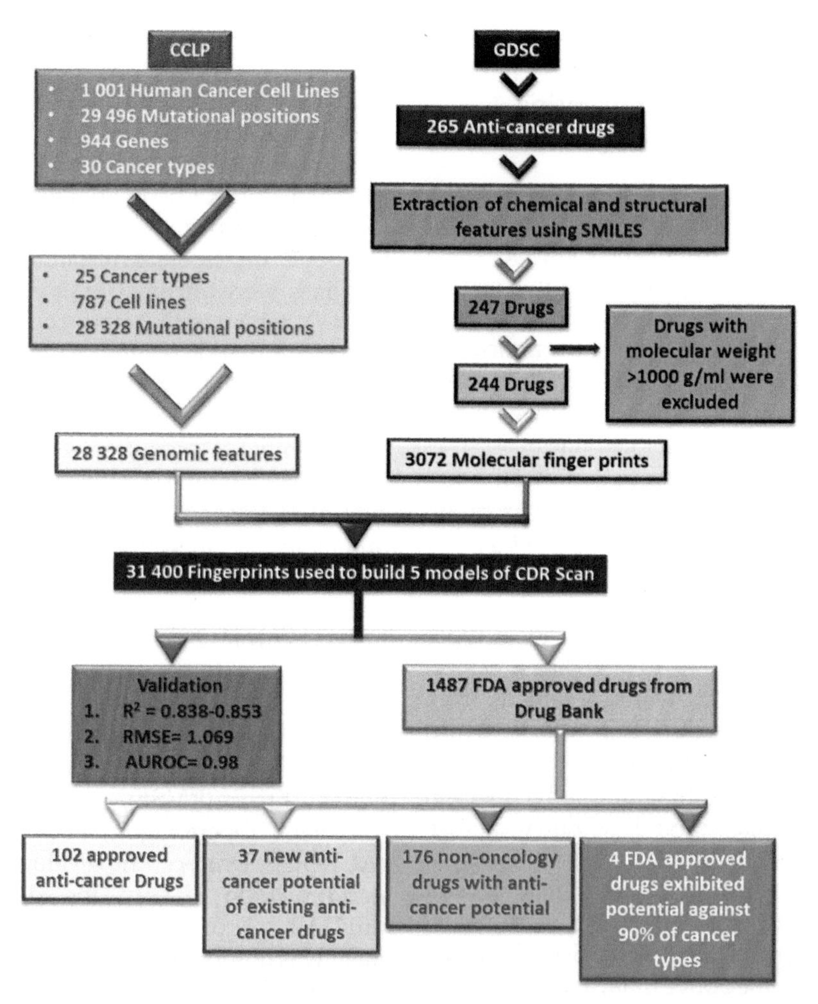

FIGURE 4.10 Workflow to design deep learning algorithm integrating genomic and chemical fingerprints for drug repurposing.

screened through stringent quality control criteria ($P < .05$) which yielded 1265 drugs' gene expression data. At the same time, structural similarities among fingerprints of 6594 FDA drugs were analyzed through Jaccard index, cosine similarities, and dice coefficient. Later, common targets of all the above drugs were compiled and subjected to PPI network analysis to asses target-based similarity, which revealed 1571 drugs sharing similar targets. The average similarity was calculated among DEGs, chemical finger prints, and targets using multi SVM, receiving operating characteristics, and kernel methods, which revealed 410 common drugs. Subsequently, drug similarity matrix was created for these drugs by incorporating average similarity scores and Anatomical Therapeutic Classification (ATC). Drugs that fall into small ATC classes (classes with ≤ two drugs) were then removed resulting in 281 drugs. These drugs were further processed for noise reduction using multidimensionality scaling and principal component analysis. Eventually, repositioning score was calculated for each drug based on their probable distribution patterns. This approach exemplified the antineoplastic potential of anthelmintic drugs niclosamide and oxamniquine (Fig. 4.11).

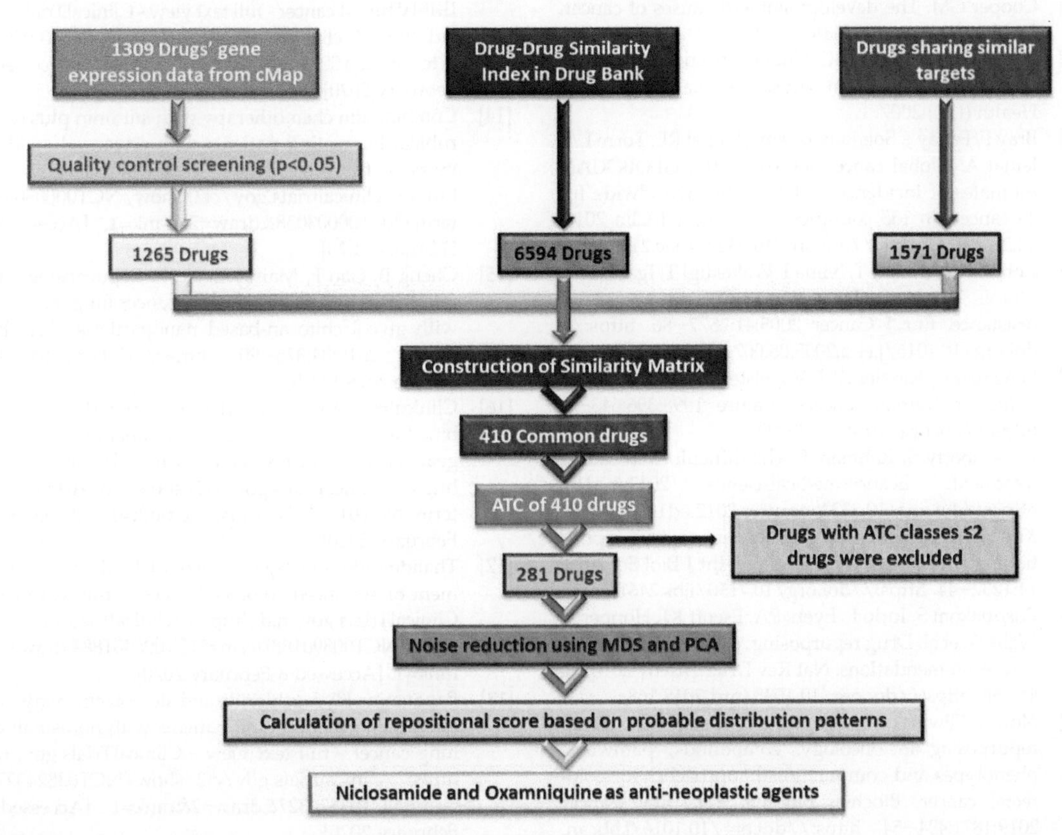

FIGURE 4.11 Workflow to explore antineoplastic potential of anthelmintic drugs via integrated drug–target–gene approach.

Conclusion

The aim of these computational approaches is not to forecast existing experimental results but instead to propose new concepts for drug repurposing, which requires experimental validation before extrapolating to clinical trials. These repurposing approaches aid in development of better pharmacotherapeutic regimens and offer quick solutions to tackle resistance and disease progression for a catastrophic disease like cancer.

References

[1] Cooper GM. The development and causes of cancer. 2nd ed. Sinauer Associates; 2000.
[2] (US) NI of H, Study BSC. Understanding cancer. NIH curriculum supplement series. National Institutes of Health (US); 2007.
[3] Bray F, Ferlay J, Soerjomataram I, Siegel RL, Torre LA, Jemal A. Global cancer statistics 2018: GLOBOCAN estimates of incidence and mortality worldwide for 36 cancers in 185 countries. CA Cancer J Clin 2018; 68:394–424. https://doi.org/10.3322/caac.21492.
[4] Kohno K, Uchiumi T, Niina I, Wakasugi T, Igarashi T, Momii Y, et al. Transcription factors and drug resistance. Eur J Cancer 2005;41:2577–86. https://doi.org/10.1016/j.ejca.2005.08.007.
[5] Lengauer C, Kinzler KW, Vogelstein B. Genetic instabilities in human cancers. Nature 1998;396:643–9. https://doi.org/10.1038/25292.
[6] Chakraborty S, Rahman T. The difficulties in cancer treatment. Ecancermedicalscience 2012;6:ed16. https://doi.org/10.3332/ecancer.2012.ed16.
[7] Xue H, Li J, Xie H, Wang Y. Review of drug repositioning approaches and resources. Int J Biol Sci 2018; 14:1232–44. https://doi.org/10.7150/ijbs.24612.
[8] Pushpakom S, Iorio F, Eyers PA, Escott KJ, Hopper S, Wells A, et al. Drug repurposing: progress, challenges and recommendations. Nat Rev Drug Discov 2018;18: 41–58. https://doi.org/10.1038/nrd.2018.168.
[9] Nowak-Sliwinska P, Scapozza L, Altaba AR i. Drug repurposing in oncology: compounds, pathways, phenotypes and computational approaches for colorectal cancer. Biochim Biophys Acta Rev Cancer 2019;1871:434–54. https://doi.org/10.1016/j.bbcan.2019.04.005.
[10] Lilly and Company E. Lilly's CYRAMZA® (ramucirumab) becomes first FDA-approved biomarker-driven therapy in patients with hepatocellular carcinoma. 2019.
[11] FDA approves pembrolizumab for metastatic small cell lung cancer | FDA; n.d. https://www.fda.gov/drugs/resources-information-approved-drugs/fda-approves-pembrolizumab-metastatic-small-cell-lung-cancer. [Accessed 8 February 2020].
[12] König M, Von Hagens C, Hoth S, Baumann I, Walter-Sack I, Edler L, et al. Investigation of ototoxicity of artesunate as add-on therapy in patients with metastatic or locally advanced breast cancer: new audiological results from a prospective, open, uncontrolled, monocentric phase i study. Cancer Chemother Pharmacol 2016;77:413–27. https://doi.org/10.1007/s00280-016-2960-7.
[13] Suramin and paclitaxel in treating women with stage IIIB-IV breast cancer - full text view - ClinicalTrials.gov; n.d. https://clinicaltrials.gov/ct2/show/NCT00054028?term=NCT00054028&draw=2&rank=1. [Accessed 8 February 2020].
[14] Combination chemotherapy with suramin plus doxorubicin in treating patients with advanced solid tumors - full text view - ClinicalTrials.gov; n.d. https://clinicaltrials.gov/ct2/show/NCT00003038?term=NCT00003038&draw=2&rank=1. [Accessed 8 February 2020].
[15] Cheng B, Gao F, Maissy E, Xu P. Repurposing suramin for the treatment of breast cancer lung metastasis with glycol chitosan-based nanoparticles. Acta Biomater 2019;84:378–90. https://doi.org/10.1016/j.actbio.2018.12.010.
[16] Clinicaltrialsgov. Phase II trial of thalidomide combined with concurrent chemoradiotherapy in esophageal cancer - full text view - ClinicalTrials.gov; n.d. https://clinicaltrials.gov/ct2/show/NCT01551641?term=NCT01551641&draw=2&rank=1. [Accessed 8 February 2020].
[17] Thalidomide and tegafur/uracil(UFUR) in the treatment of advanced colorectal cancer - full text view - ClinicalTrials.gov; n.d. https://clinicaltrials.gov/ct2/show/NCT00890188?term=NCT00890188&draw=2&rank=1. [Accessed 8 February 2020].
[18] Papaverine Hydrochloride and stereotactic body radiation therapy in treating patients with non-small cell lung cancer - full text view - ClinicalTrials.gov; n.d. https://clinicaltrials.gov/ct2/show/NCT03824327?term=NCT03824327&draw=2&rank=1. [Accessed 8 February 2020].

[19] Prostatic hyperplasia treatment and cancer prevention - full text view - ClinicalTrials.gov; n.d. https://clinicaltrials.gov/ct2/show/NCT03064282?term=NCT03064282&draw=2&rank=1. [Accessed 8 February 2020].

[20] Repurposing metformin as anticancer drug: in advanced prostate cancer - full text view - ClinicalTrials.gov; n.d. https://clinicaltrials.gov/ct2/show/NCT03137186?term=NCT03137186&draw=2&rank=1. [Accessed 8 February 2020].

[21] Metformin hydrochloride in treating women with stage I or stage II breast cancer that can be removed by surgery - full text view - ClinicalTrials.gov; n.d. https://clinicaltrials.gov/ct2/show/NCT00984490?term=NCT00984490&draw=2&rank=1. [Accessed 8 February 2020].

[22] Clinical and biologic effects of metformin in early stage breast cancer - full text view - ClinicalTrials.gov; n.d. https://clinicaltrials.gov/ct2/show/NCT00897884?term=NCT00897884&draw=2&rank=1. [Accessed 8 February 2020].

[23] The use of metformin in early breast cancer patients pre-surgery - full text view - ClinicalTrials.gov; n.d. https://clinicaltrials.gov/ct2/show/NCT01302002?term=NCT01302002&draw=2&rank=1. [Accessed 8 February 2020].

[24] Liu Q, Tong D, Liu G, Gao J, Wang LA, Xu J, et al. Metformin inhibits prostate cancer progression by targeting tumor-associated inflammatory infiltration. Clin Cancer Res 2018;24:5622–34. https://doi.org/10.1158/1078-0432.CCR-18-0420.

[25] A study of overall survival in participants with unresectable hepatocellular carcinoma - full text view - ClinicalTrials.gov; n.d. https://clinicaltrials.gov/ct2/show/NCT04008082?term=NCT03663114&draw=2&rank=2. [Accessed 8 February 2020].

[26] Post-marketing surveillance of lenvatinib mesylate (lenvima capsule) in patients with unresectable thyroid cancer (study LEN01T) - full text view - ClinicalTrials.gov; n.d. https://clinicaltrials.gov/ct2/show/NCT02430714?term=NCT02430714&draw=2&rank=1. [Accessed 8 February 2020].

[27] Lenvatinib and weekly paclitaxel for patients with recurrent endometrial or ovarian cancer - full text view - ClinicalTrials.gov; n.d. https://clinicaltrials.gov/ct2/show/NCT02788708?term=NCT02788708&draw=2&rank=1. [Accessed 8 February 2020].

[28] Merck & Co., Inc. - Anticancer Agent. LENVIMA® (lenvatinib mesylate) approved for additional indication of unresectable hepatocellular carcinoma (HCC) in Japan, first approval worldwide for LENVIMA for HCC; n.d. https://investors.merck.com/news/press-release-details/2018/Anticancer-Agent-LENVIMA-lenvatinib-mesylate-Approved-for-Additional-Indication-of-Unresectable-Hepatocellular-Carcinoma-HCC-in-Japan-First-Approval-Worldwide-for-LENVIMA-for-HCC/default.aspx. [Accessed 8 February 2020].

[29] Park S, Oh A-Y, Cho J-H, Yoon M-H, Woo T-G, Kang S-M, et al. Therapeutic effect of quinacrine, an antiprotozoan drug, by selective suppression of p-CHK1/2 in p53-negative malignant cancers. Mol Cancer Res 2018;16:935–46. https://doi.org/10.1158/1541-7786.MCR-17-0511.

[30] Erlotinib hydrochloride and quinacrine dihydrochloride in stage IIIB-IV non-small cell lung cancer - full text view - ClinicalTrials.gov; n.d. https://clinicaltrials.gov/ct2/show/NCT01839955?term=NCT01839955&draw=2&rank=1. [Accessed 8 February 2020].

[31] Quinacrine treatment in patients with androgen-independent prostate cancer - full text view - ClinicalTrials.gov; n.d. https://clinicaltrials.gov/ct2/show/NCT00417274?term=NCT00417274&draw=2&rank=1. [Accessed 8 February 2020].

[32] March-Vila E, Pinzi L, Sturm N, Tinivella A, Engkvist O, Chen H, et al. On the integration of in silico drug design methods for drug repurposing. Front Pharmacol 2017;8. https://doi.org/10.3389/fphar.2017.00298.

[33] Akhoon BA, Tiwari H, Nargotra A. Silico drug design methods for drug repurposing. In: Silico Drug Des; 2019. p. 47–84. https://doi.org/10.1016/b978-0-12-816125-8.00003-1. Elsevier.

[34] Zhang J, Jiang K, Lv L, Wang H, Shen Z, Gao Z, et al. Use of genome-wide association studies for cancer research and drug repositioning. PLoS One 2015;10:e0116477. https://doi.org/10.1371/journal.pone.0116477.

[35] Zhao H, Jin G, Cui K, Ren D, Liu T, Chen P, et al. Novel modeling of cancer cell signaling pathways enables systematic drug repositioning for distinct breast cancer metastases. Cancer Res 2013;73:6149–63. https://doi.org/10.1158/0008-5472.CAN-12-4617.

[36] Abbruzzese C, Matteoni S, Signore M, Cardone L, Nath K, Glickson JD, et al. Drug repurposing for the treatment of glioblastoma multiforme. J Exp Clin Cancer Res 2017;36. https://doi.org/10.1186/s13046-017-0642-x.

[37] Holder LB, Haque MM, Skinner MK. Machine learning for epigenetics and future medical applications. Epigenetics 2017;12:505–14. https://doi.org/10.1080/15592294.2017.1329068.

[38] Armitage EG, Southam AD. Monitoring cancer prognosis, diagnosis and treatment efficacy using metabolomics and lipidomics. Metabolomics 2016;12. https://doi.org/10.1007/s11306-016-1093-7.

[39] Lounkine E, Keiser MJ, Whitebread S, Mikhailov D, Hamon J, Jenkins JL, et al. Large-scale prediction and testing of drug activity on side-effect targets. Nature 2012;486:361–7. https://doi.org/10.1038/nature11159.

[40] Cha Y, Erez T, Reynolds IJ, Kumar D, Ross J, Koytiger G, et al. Drug repurposing from the perspective of pharmaceutical companies. Br J Pharmacol 2018;175:168–80. https://doi.org/10.1111/bph.13798.

[41] Yang X, Wang Y, Byrne R, Schneider G, Yang S. Concepts of artificial intelligence for computer-assisted drug discovery. Chem Rev 2019;119:10520–94. https://doi.org/10.1021/acs.chemrev.8b00728.

[42] Williams AM, Liu Y, Regner KR, Jotterand F, Liu P, Liang M. Artificial intelligence, physiological genomics, and precision medicine. Physiol Genomics 2018;50:237–43. https://doi.org/10.1152/physiolgenomics.00119.2017.

[43] Allison M. NCATS launches drug repurposing program. Nat Biotechnol 2012;30:571–2. https://doi.org/10.1038/nbt0712-571a.

[44] Pantziarka P, Bouche G, Meheus L, Sukhatme V, Sukhatme VP. The repurposing drugs in oncology (ReDO) project. Ecancermedicalscience 2014;8. https://doi.org/10.3332/ecancer.2014.442.

[45] Drug repurposing | Anticancerfund; n.d. https://www.anticancerfund.org/en/drug-repurposing. [Accessed 12 December 2019].

[46] Marusina K, Welsch DJ, Rose L, Brock D, Bahr N. The CTSA Pharmaceutical Assets Portal - a public-private partnership model for drug repositioning. Drug Discov Today Ther Strat 2011;8:77–83. https://doi.org/10.1016/j.ddstr.2011.06.006.

[47] Findacure | 7,000 rare diseases, 1 common goal; n.d. https://www.findacure.org.uk/. [Accessed 12 December 2019].

[48] Global cures; n.d. https://www.global-cures.org/. [Accessed 12 December 2019].

[49] Cures within reach - home; n.d. https://www.cureswithinreach.org/. [Accessed 12 December 2019].

[50] Tomczak K, Czerwińska P, Wiznerowicz M. The Cancer Genome Atlas (TCGA): an immeasurable source of knowledge. Współczesna Onkol 2015;(1A):A68–77. https://doi.org/10.5114/wo.2014.47136.

[51] Zhang J, Baran J, Cros A, Guberman JM, Haider S, Hsu J, et al. International cancer genome consortium data portal-a one-stop shop for cancer genomics data. Database 2011;2011. https://doi.org/10.1093/database/bar026.

[52] Koleti A, Terryn R, Stathias V, Chung C, Cooper DJ, Turner JP, et al. Data Portal for the Library of Integrated Network-based Cellular Signatures (LINCS) program: integrated access to diverse large-scale cellular perturbation response data. Nucleic Acids Res 2018;46:D558–66. https://doi.org/10.1093/nar/gkx1063.

[53] Hudson TJ, Anderson W, Aretz A, Barker AD, Bell C, Bernabé RR, et al. International network of cancer genome projects. Nature 2010;464:993–8. https://doi.org/10.1038/nature08987.

[54] Ellis MJ, Gillette M, Carr SA, Paulovich AG, Smith RD, Rodland KK, et al. Connecting genomic alterations to cancer biology with proteomics: the NCI clinical proteomic tumor analysis consortium. Cancer Discov 2013;3:1108–12. https://doi.org/10.1158/2159-8290.CD-13-0219.

[55] Kumar R, Chaudhary K, Gupta S, Singh H, Kumar S, Gautam A, et al. CancerDR: cancer drug resistance database. Sci Rep 2013;3. https://doi.org/10.1038/srep01445.

[56] Rhodes DR, Yu J, Shanker K, Deshpande N, Varambally R, Ghosh D, et al. ONCOMINE: a cancer microarray database and integrated data-mining platform. Neoplasia 2004;6:1–6. https://doi.org/10.1016/s1476-5586(04)80047-2.

[57] Barretina J, Caponigro G, Stransky N, Venkatesan K, Margolin AA, Kim S, et al. The Cancer Cell Line Encyclopedia enables predictive modelling of anticancer drug sensitivity. Nature 2012;483:603–7. https://doi.org/10.1038/nature11003.

[58] Birney E, Andrews TD, Bevan P, Caccamo M, Chen Y, Clarke L, et al. An overview of ensembl. Genome Res 2004;14:925–8. https://doi.org/10.1101/gr.1860604.

[59] Wong KM, Langlais K, Tobias GS, Fletcher-Hoppe C, Krasnewich D, Leeds HS, et al. The dbGaP data browser: a new tool for browsing dbGaP controlled-access genomic data. Nucleic Acids Res 2017;45:D819–26. https://doi.org/10.1093/nar/gkw1139.

[60] Piñero J, Queralt-Rosinach N, Àlex B, Deu-Pons J, Bauer-Mehren A, Baron M, et al. DisGeNET: a discovery platform for the dynamical exploration of human diseases and their genes. Database 2015;2015. https://doi.org/10.1093/database/bav028.

[61] Sherry ST, Ward M-H, Kholodov M, Baker J, Phan L, Smigielski EM, et al. dbSNP: the NCBI database of genetic variation. Nucleic Acids Res 2001;29:308–11. https://doi.org/10.1093/nar/29.1.308.

[62] MacDonald JR, Ziman R, Yuen RKC, Feuk L, Scherer SW. The database of genomic variants: a curated collection of structural variation in the human genome. Nucleic Acids Res 2014;42. https://doi.org/10.1093/nar/gkt958.

[63] Forbes SA, Beare D, Boutselakis H, Bamford S, Bindal N, Tate J, et al. COSMIC: somatic cancer genetics at high-resolution. Nucleic Acids Res 2017;45:D777–83. https://doi.org/10.1093/nar/gkw1121.

[64] Lonsdale J, Thomas J, Salvatore M, Phillips R, Lo E, Shad S, et al. The genotype-tissue expression (GTEx) project. Nat Genet 2013;45:580–5. https://doi.org/10.1038/ng.2653.

[65] Barrett T, Wilhite SE, Ledoux P, Evangelista C, Kim IF, Tomashevsky M, et al. NCBI GEO: archive for functional genomics data sets–update. Nucleic Acids Res 2013;41:D991–5. https://doi.org/10.1093/nar/gks1193.

[66] Parkinson H, Kapushesky M, Shojatalab M, Abeygunawardena N, Coulson R, Farne A, et al. ArrayExpress - a public database of microarray experiments and gene expression profiles. Nucleic Acids Res 2007;35. https://doi.org/10.1093/nar/gkl995.

[67] Sunkin SM, Ng L, Lau C, Dolbeare T, Gilbert TL, Thompson CL, et al. Allen Brain Atlas: an integrated spatio-temporal portal for exploring the central nervous system. Nucleic Acids Res 2012;41:D996–1008. https://doi.org/10.1093/nar/gks1042.

[68] Altunkaya A, Bi C, Bradley AR, Rose PW, Prli A, Christie H, et al. The RCSB protein data bank: integrative view of protein, gene and 3D structural information. Nucleic Acids Res 2017;45:271–81. https://doi.org/10.1093/nar/gkw1000.

[69] Legrain P, Aebersold R, Archakov A, Bairoch A, Bala K, Beretta L, et al. The human proteome project: current state and future direction. 2011. https://doi.org/10.1074/mcp.M111.009993.

[70] Bairoch A, Apweiler R. The SWISS-PROT protein sequence database and its supplement TrEMBL in 2000. Nucleic Acids Res 2000;28.

[71] Lane L, Argoud-Puy G, Britan A, Cusin I, Duek PD, Evalet O, et al. NeXtProt: a knowledge platform for human proteins. Nucleic Acids Res 2012;40. https://doi.org/10.1093/nar/gkr1179.

[72] Harrow J, Frankish A, Gonzalez JM, Tapanari E, Diekhans M, Kokocinski F, et al. GENCODE: The reference human genome annotation for The ENCODE Project; n.d. https://doi.org/10.1101/gr.135350.111.

[73] Davis CA, Hitz BC, Sloan CA, Chan ET, Davidson JM, Gabdank I, et al. The Encyclopedia of DNA elements (ENCODE): data portal update. Nucleic Acids Res 2018;46. https://doi.org/10.1093/nar/gkx1081.

[74] Pujar S, O'leary NA, Farrell CM, Loveland JE, Mudge JM, Wallin C, et al. Consensus coding sequence (CCDS) database: a standardized set of human and mouse protein-coding regions supported by expert curation. Nucleic Acids Res 2017;46:221–8. https://doi.org/10.1093/nar/gkx1031.

[75] Tamborero D, Rubio-Perez C, Deu-Pons J, Schroeder MP, Vivancos A, Rovira A, et al. Cancer Genome Interpreter annotates the biological and clinical relevance of tumor alterations. Genome Med 2018;10. https://doi.org/10.1186/s13073-018-0531-8.

[76] Bernstein BE, Stamatoyannopoulos JA, Costello JF, Ren B, Milosavljevic A, Meissner A, et al. The NIH roadmap epigenomics mapping consortium. Nat Biotechnol 2010;28:1045–8. https://doi.org/10.1038/nbt1010-1045.

[77] Akbarian S, Liu C, Knowles JA, Vaccarino FM, Farnham PJ, Crawford GE, et al. The PsychENCODE project the PsychENCODE consortium HHS public access author manuscript. Nat Neurosci 2015;18:1707–12. https://doi.org/10.1038/nn.4156.

[78] Oughtred R, Stark C, Breitkreutz B-J, Rust J, Boucher L, Chang C, et al. The BioGRID interaction database: 2019 update. Nucleic Acids Res 2018;47:529–41. https://doi.org/10.1093/nar/gky1079.

[79] Kanehisa M, Furumichi M, Tanabe M, Sato Y, Morishima K. KEGG: new perspectives on genomes, pathways, diseases and drugs. Nucleic Acids Res 2017;45:D353–61. https://doi.org/10.1093/nar/gkw1092.

[80] Croft D, O'Kelly G, Wu G, Haw R, Gillespie M, Matthews L, et al. Reactome: a database of reactions, pathways and biological processes. Nucleic Acids Res 2011;39:D691–7. https://doi.org/10.1093/nar/gkq1018.

[81] Kuhn M, Letunic I, Jensen LJ, Bork P. The SIDER database of drugs and side effects. Nucleic Acids Res 2016;44:D1075–9. https://doi.org/10.1093/nar/gkv1075.

[82] Peri S, Navarro JD, Kristiansen TZ, Amanchy R, Surendranath V, Muthusamy B, et al. Human protein reference database as a discovery resource for proteomics. Nucleic Acids Res 2004;32:D497–501. https://doi.org/10.1093/nar/gkh070.

[83] Chatr-aryamontri A, Ceol A, Palazzi LM, Nardelli G, Schneider MV, Castagnoli L, et al. MINT: the molecular INTeraction database. Nucleic Acids Res 2007;35. https://doi.org/10.1093/nar/gkl950.

[84] Fahey ME, Bennett MJ, Mahon C, Jäger S, Pache L, Kumar D, et al. GPS-Prot: a web-based visualization platform for integrating host-pathogen interaction data. BMC Bioinf 2011;12:1–13. https://doi.org/10.1186/1471-2105-12-298.

[85] Cowley MJ, Pinese M, Kassahn KS, Waddell N, Pearson JV, Grimmond SM, et al. PINA v2.0: mining interactome modules. Nucleic Acids Res 2012;40. https://doi.org/10.1093/nar/gkr967.

[86] Goll J, Rajagopala SV, Shiau SC, Wu H, Lamb BT, Uetz P. MPIDB: the microbial protein interaction database. Bioinformatics 2008;24:1743–4. https://doi.org/10.1093/bioinformatics/btn285. Appl NOTE.

[87] Fang H, Su Z, Wang Y, Miller A, Liu Z, Howard PC, et al. Exploring the FDA adverse event reporting system to generate hypotheses for monitoring of disease characteristics. Clin Pharmacol Ther 2014;95:496–8. https://doi.org/10.1038/clpt.2014.17.

[88] Legehar A, Xhaard H, Ghemtio L. IDAAPM: integrated database of ADMET and adverse effects of predictive modeling based on FDA approved drug data. J Cheminf 2016;8. https://doi.org/10.1186/s13321-016-0141-7.

[89] Randhawa G, Sharma R, Singh N, Sharma N. A qualitative and quantitative comparison of adverse drug reaction data in different drug information sources. Int J Appl Basic Med Res 2017;7:223. https://doi.org/10.4103/ijabmr.ijabmr_18_17.

[90] Yang W, Soares J, Greninger P, Edelman EJ, Lightfoot H, Forbes S, et al. Genomics of Drug Sensitivity in Cancer (GDSC): a resource for therapeutic biomarker discovery in cancer cells. Nucleic Acids Res 2013;41. https://doi.org/10.1093/nar/gks1111.

[91] Basu A, Bodycombe NE, Cheah JH, Price EV, Liu K, Schaefer GI, et al. An interactive resource to identify cancer genetic and lineage dependencies targeted by small molecules. Cell 2013;154:1151–61. https://doi.org/10.1016/j.cell.2013.08.003.

[92] Szklarczyk D, Morris JH, Cook H, Kuhn M, Wyder S, Simonovic M, et al. The STRING database in 2017: quality-controlled protein-protein association networks, made broadly accessible. Nucleic Acids Res 2017;45. https://doi.org/10.1093/nar/gkw937.

[93] Cerami E, Gao J, Dogrusoz U, Gross BE, Sumer SO, Aksoy BA, et al. The cBio Cancer Genomics Portal: an open platform for exploring multidimensional cancer genomics data. Cancer Discov 2012;2:401–4. https://doi.org/10.1158/2159-8290.CD-12-0095.

[94] Apweiler R, Bairoch A, Wu CH, Barker WC, Boeckmann B, Ferro S, et al. UniProt: the universal protein knowledgebase. Nucleic Acids Res 2004;32:D115–9. https://doi.org/10.1093/nar/gkh131.

[95] Welter D, MacArthur J, Morales J, Burdett T, Hall P, Junkins H, et al. The NHGRI GWAS Catalog, a curated resource of SNP-trait associations. Nucleic Acids Res 2014;42:D1001–6. https://doi.org/10.1093/nar/gkt1229.

[96] Lawrence MS, Stojanov P, Mermel CH, Robinson JT, Garraway LA, Golub TR, et al. Discovery and saturation analysis of cancer genes across 21 tumour types. Nature 2014;505:495–501. https://doi.org/10.1038/nature12912.

[97] Goldman M, Craft B, Swatloski T, Cline M, Morozova O, Diekhans M, et al. The UCSC cancer genomics browser: update 2015. Nucleic Acids Res 2015;43. https://doi.org/10.1093/nar/gku1073.

[98] Gonzalez-Perez A, Perez-Llamas C, Deu-Pons J, Tamborero D, Schroeder MP, Jene-Sanz A, et al. IntOGen-mutations identifies cancer drivers across tumor types. Nat Methods 2013;10:1081–4. https://doi.org/10.1038/nmeth.2642.

[99] Antonov AV. BioProfiling.de: analytical web portal for high-throughput cell biology. Nucleic Acids Res 2011;39. https://doi.org/10.1093/nar/gkr372.

[100] Lamb J, Crawford ED, Peck D, Modell JW, Blat IC, Wrobel MJ, et al. The connectivity map: using gene-expression signatures to connect small molecules, genes, and disease. Science 2006;313(80):1929–35. https://doi.org/10.1126/science.1132939.

[101] Law V, Knox C, Djoumbou Y, Jewison T, Guo AC, Liu Y, et al. DrugBank 4.0: shedding new light on drug metabolism. Nucleic Acids Res 2014;42:D1091–7. https://doi.org/10.1093/nar/gkt1068.

[102] Zhu F, Shi Z, Qin C, Tao L, Liu X, Xu F, et al. Therapeutic target database update 2012: a resource for facilitating target-oriented drug discovery. Nucleic Acids Res 2012;40:D1128–36. https://doi.org/10.1093/nar/gkr797.

[103] Davis AP, Grondin CJ, Johnson RJ, Sciaky D, McMorran R, Wiegers J, et al. The comparative toxicogenomics database: update 2019. Nucleic Acids Res 2019;47:D948–54. https://doi.org/10.1093/nar/gky868.

[104] Kim S, Thiessen PA, Bolton EE, Chen J, Fu G, Gindulyte A, et al. PubChem substance and compound databases. Nucleic Acids Res 2016;44:D1202–13. https://doi.org/10.1093/nar/gkv951.

[105] Gaulton A, Hersey A, Nowotka ML, Patricia Bento A, Chambers J, Mendez D, et al. The ChEMBL database in 2017. Nucleic Acids Res 2017;45:D945–54. https://doi.org/10.1093/nar/gkw1074.

[106] Wishart DS, Feunang YD, Marcu A, Guo AC, Liang K, Vázquez-Fresno R, et al. Hmdb 4.0: the human metabolome database for 2018. Nucleic Acids Res 2018;46:D608–17. https://doi.org/10.1093/nar/gkx1089.

[107] Thorn CF, Klein TE, Altman RB. PharmGKB: the pharmacogenomics knowledge base. Methods Mol Biol 2013;1015:311–20. https://doi.org/10.1007/978-1-62703-435-7_20.

[108] Explained: neural networks | MIT News; n.d. http://news.mit.edu/2017/explained-neural-networks-deep-learning-0414. [Accessed 12 December 2019].

[109] Genomics - an introduction to genetic analysis - NCBI bookshelf; n.d. https://www.ncbi.nlm.nih.gov/books/NBK21783/. [Accessed 12 December 2019].

[110] Karageorgos I, Mizzi C, Giannopoulou E, Pavlidis C, Peters BA, Zagoriti Z, et al. Identification of cancer predisposition variants in apparently healthy individuals using a next-generation sequencing-based family genomics approach. Hum Genom 2015;9:12. https://doi.org/10.1186/s40246-015-0034-2.

[111] Berger MF, Mardis ER. The emerging clinical relevance of genomics in cancer medicine. Nat Rev Clin Oncol 2018;15:353–65. https://doi.org/10.1038/s41571-018-0002-6.

[112] Marszalek RT. Cancer genomics just got personal. Genome Biol 2014;15. https://doi.org/10.1186/s13059-014-0464-5.

[113] Lee BKB, Tiong KH, Chang JK, Liew CS, Abdul Rahman ZA, Tan AC, et al. DeSigN: connecting gene expression with therapeutics for drug repurposing and development. BMC Genom 2017;18:1–11. https://doi.org/10.1186/s12864-016-3260-7.

[114] Dennis G, Sherman BT, Hosack DA, Yang J, Gao W, Lane HC, et al. DAVID: database for annotation, visualization, and integrated discovery. Genome Biol 2003;4. https://doi.org/10.1186/gb-2003-4-5-p3.

[115] Raychaudhuri S, Plenge RM, Rossin EJ, Ng ACY, Purcell SM, Sklar P, et al. Identifying relationships among genomic disease regions: predicting genes at pathogenic SNP associations and rare deletions. PLoS Genet 2009;5. https://doi.org/10.1371/journal.pgen.1000534.

[116] Rossin EJ, Lage K, Raychaudhuri S, Xavier RJ, Tatar D, Benita Y, et al. Proteins encoded in genomic regions associated with immune-mediated disease physically interact and suggest underlying biology. PLoS Genet 2011;7. https://doi.org/10.1371/journal.pgen.1001273.

[117] Eppig JT. Mouse genome informatics (MGI) resource: genetic, genomic, and biological knowledgebase for the laboratory mouse. ILAR J 2017;58:17–41. https://doi.org/10.1093/ilar/ilx013.

[118] Hewick RM, Lu Z, Wang JH. Proteomics in drug discovery. Adv Protein Chem 2003;65:309–42. https://doi.org/10.1016/S0065-3233(03)01024-6.

[119] Dias MH, Kitano ES, Zelanis A, Iwai LK. Proteomics and drug discovery in cancer. Drug Discov Today 2016;21:264–77. https://doi.org/10.1016/j.drudis.2015.10.004.

[120] Shi XN, Li H, Yao H, Liu X, Li L, Leung KS, et al. In silico identification and in vitro and in vivo validation of anti-psychotic drug fluspirilene as a potential CDK2 inhibitor and a candidate anti-cancer drug. PLoS One 2015;10. https://doi.org/10.1371/journal.pone.0132072.

[121] Forli S, Huey R, Pique ME, Sanner MF, Goodsell DS, Olson AJ. Computational protein-ligand docking and virtual drug screening with the AutoDock suite. Nat Protoc 2016;11:905–19. https://doi.org/10.1038/nprot.2016.051.

[122] Irwin JJ, Shoichet BK. Zinc – a free database of commercially available compounds for virtual screening. J Chem Inf Model 2005;45:177. https://doi.org/10.1021/CI049714.

[123] Vitali F, Cohen LD, Demartini A, Amato A, Eterno V, Zambelli A, et al. A network-based data integration approach to support drug repurposing and multi-target therapies in triple negative breast cancer. PLoS One 2016;11:e0162407. https://doi.org/10.1371/journal.pone.0162407.

[124] Kuhn M, von Mering C, Campillos M, Jensen LJ, Bork P. STITCH: interaction networks of chemicals and proteins. Nucleic Acids Res 2008;36:D684–8. https://doi.org/10.1093/nar/gkm795.

[125] Rani B, Sharma VK. Transcriptome profiling: methods and applications- A review. Agric Rev 2017;38. https://doi.org/10.18805/ag.r-1549.

[126] Toro-Domínguez D, Alarcón-Riquelme ME, Carmona-Sáez P. Drug repurposing from transcriptome data: methods and applications. Silico Drug Des 2019:303–27. https://doi.org/10.1016/b978-0-12-816125-8.00011-0. Elsevier.

[127] Zhang SD, Gant TW. sscMap: an extensible Java application for connecting small-molecule drugs using gene-expression signatures. BMC Bioinf 2009;10. https://doi.org/10.1186/1471-2105-10-236.

[128] Setoain J, Franch M, Martínez M, Tabas-Madrid D, Sorzano COS, Bakker A, et al. NFFinder: an online bioinformatics tool for searching similar transcriptomics experiments in the context of drug repositioning. Nucleic Acids Res 2015;43:W193–9. https://doi.org/10.1093/nar/gkv445.

[129] Lee H, Kang S, Kim W. Drug repositioning for cancer therapy based on large-scale drug-induced transcriptional signatures. PLoS One 2016;11:1–17. https://doi.org/10.1371/journal.pone.0150460.

[130] Jensen MA, Ferretti V, Grossman RL, Staudt LM. The NCI Genomic Data Commons as an engine for precision medicine. Blood 2017;130:453–9. https://doi.org/10.1182/blood-2017-03-735654.

[131] Kim IW, Jang H, Kim JH, Kim MG, Kim S, Oh JM. Computational drug repositioning for gastric cancer using reversal gene expression profiles. Sci Rep 2019;9. https://doi.org/10.1038/s41598-019-39228-9.

[132] Hastings J, Owen G, Dekker A, Ennis M, Kale N, Muthukrishnan V, et al. ChEBI in 2016: improved services and an expanding collection of metabolites. Nucleic Acids Res 2016;44:D1214–9. https://doi.org/10.1093/nar/gkv1031.

[133] Hong SK, Lee H, Kwon OS, Song NY, Lee HJ, Kang S, et al. Large-scale pharmacogenomics based drug discovery for ITGB3 dependent chemoresistance in mesenchymal lung cancer. Mol Cancer 2018;17:1–7. https://doi.org/10.1186/s12943-018-0924-8.

[134] Mariño-Ramírez L, Kann MG, Shoemaker BA, Landsman D. Histone structure and nucleosome stability. Expert Rev Proteomics 2005;2:719–29. https://doi.org/10.1586/14789450.2.5.719.

[135] Bannister AJ, Kouzarides T. Regulation of chromatin by histone modifications. Cell Res 2011;21:381–95. https://doi.org/10.1038/cr.2011.22.

[136] Shapiro O, Bratslavsky G. Genetic diseases. In: Brenner's, editor. Encyclopedia of Genetics. 2nd ed. Elsevier Inc.; 2013. p. 246–7. https://doi.org/10.1016/B978-0-12-374984-0.00614-8.

[137] Jones PA, Baylin SB. The fundamental role of epigenetic events in cancer. Nat Rev Genet 2002;3:415–28. https://doi.org/10.1038/nrg816.

[138] Robertson KD. DNA methylation and chromatin - unraveling the tangled web. Oncogene 2002;21:5361–79. https://doi.org/10.1038/sj.onc.1205609.

[139] Prachayasittikul V, Prathipati P, Pratiwi R, Phanus-umporn C, Malik AA, Schaduangrat N, et al. Exploring the epigenetic drug discovery landscape. Expet Opin Drug Discov 2017;12:345–62. https://doi.org/10.1080/17460441.2017.1295954.

[140] Horton JR, Gale M, Yan Q, Cheng X. Epigenetic targets and drug discovery part 2: histone demethylation and DNA methylation. Pharmacol Ther 2017;151: 121–40. https://doi.org/10.1007/978-3-319-59786-7.

[141] Dupont C, Armant DR, Brenner CA. Epigenetics: definition, mechanisms and clinical perspective. Semin Reprod Med 2009;27:351–7. https://doi.org/10.1055/s-0029-1237423.

[142] Franchini D-M, Schmitz K-M, Petersen-Mahrt SK. 5-Methylcytosine DNA demethylation: more than losing a methyl group. Annu Rev Genet 2012;46: 419–41. https://doi.org/10.1146/annurev-genet-110711-155451.

[143] Brown AS, Kong SW, Kohane IS, Patel CJ. ksRepo: a generalized platform for computational drug repositioning. BMC Bioinf 2016;17. https://doi.org/10.1186/s12859-016-0931-y.

[144] Wishart DS. Emerging applications of metabolomics in drug discovery and precision medicine. Nat Rev Drug Discov 2016;15:473–84. https://doi.org/10.1038/nrd.2016.32.

[145] Beger R. A review of applications of metabolomics in cancer. Metabolites 2013;3:552–74. https://doi.org/10.3390/metabo3030552.

[146] Yeung PK. Metabolomics and biomarkers for drug discovery. Metabolites 2018;8. https://doi.org/10.3390/metabo8010011.

[147] Serkova NJ, Glunde K. Metabolomics of cancer. Methods Mol Biol 2009;520:273–95. https://doi.org/10.1007/978-1-60327-811-9_20.

[148] Projects - wishart research group; n.d. http://www.wishartlab.com/projects/the-human-metabolome-project. [Accessed 12 December 2019].

[149] TMIC. The metabolomics innovation centre. 2013. https://www.metabolomicscentre.ca/.

[150] Turanli B, Zhang C, Kim W, Benfeitas R, Uhlen M, Arga KY, et al. Discovery of therapeutic agents for prostate cancer using genome-scale metabolic modeling and drug repositioning. EBioMedicine 2019;42:386–96. https://doi.org/10.1016/j.ebiom.2019.03.009.

[151] Thul PJ, Lindskog C. The human protein atlas: a spatial map of the human proteome. Protein Sci 2018;27:233–44. https://doi.org/10.1002/pro.3307.

[152] Uhlen M, Zhang C, Lee S, Sjöstedt E, Fagerberg L, Bidkhori G, et al. A pathology atlas of the human cancer transcriptome. Science 2017;(80):357. https://doi.org/10.1126/science.aan2507.

[153] Kuhn M, Campillos M, Letunic I, Jensen LJ, Bork P. A side effect resource to capture phenotypic effects of drugs. Mol Syst Biol 2010;6:1–6. https://doi.org/10.1038/msb.2009.98.

[154] Wang Z, Liang L, Yin Z, Lin J. Improving chemical similarity ensemble approach in target prediction. J Cheminf 2016;8. https://doi.org/10.1186/s13321-016-0130-x.

[155] Song M, Baek SH, Heo GE, Lee JH. Inferring drug-protein-side effect relationships from biomedical text. Genes 2019;10. https://doi.org/10.3390/genes10020159.

[156] Yokoyama S, Sugimoto Y, Nakagawa C, Hosomi K, Takada M. Integrative analysis of clinical and bioinformatics databases to identify anticancer properties of digoxin. Sci Rep 2019;9:3–4. https://doi.org/10.1038/s41598-019-53392-y.

[157] Kimura S, Sato T, Ikeda S, Noda M, Nakayama T. Development of a database of health insurance claims: standardization of disease classifications and anonymous record linkage. J Epidemiol 2010;20:413–9. https://doi.org/10.2188/jea.JE20090066.

[158] MedDRA; n.d. https://www.meddra.org/. [Accessed 12 December 2019].

[159] Martindale: The complete drug reference | Medicines-complete; n.d. https://about.medicinescomplete.com/publication/martindale-the-complete-drug-reference/. [Accessed 12 December 2019].

[160] BaseSpace Correlation Engine | A genomic data and decision tool library; n.d. https://sapac.illumina.com/products/by-type/informatics-products/base-space-correlation-engine.html. Accessed 12 December 2019].

[161] Robinson JR, Denny JC, Roden DM, Van Driest SL. Genome-wide and phenome-wide approaches to understand variable drug actions in electronic health records. Clin Transl Sci 2018;11:112–22. https://doi.org/10.1111/cts.12522.

[162] Electronic health records | CMS; n.d. https://www.cms.gov/Medicare/E-Health/EHealthRecords/index. [Accessed 12 December 2019].

[163] Sharma H, Mao C, Zhang Y, Vatani H, Yao L, Zhong Y, et al. Developing a portable natural language processing based phenotyping system. BMC Med Inf Decis Making 2019;19. https://doi.org/10.1186/s12911-019-0786-z.

[164] Xu H, Aldrich MC, Chen Q, Liu H, Peterson NB, Dai Q, et al. Validating drug repurposing signals using electronic health records: a case study of metformin associated with reduced cancer mortality. J Am Med Inf Assoc 2015;22:179–91. https://doi.org/10.1136/amiajnl-2014-002649.

[165] Roden DM, Pulley JM, Basford MA, Bernard GR, Clayton EW, Balser JR, et al. Development of a large-scale de-identified DNA biobank to enable personalized medicine. Clin Pharmacol Ther 2008;84: 362–9. https://doi.org/10.1038/clpt.2008.89.

[166] Sinsky CA, Trockel M, West CP, Nedelec L, Tutty MA, Shanafelt T. The association between perceived electronic health record usability and professional burnout among US physicians. Mayo Clin Proc 2019;1–12. https://doi.org/10.1016/j.mayocp.2019.09.024.

[167] Department of Biomedical Informatics. Synthetic derivative |; n.d. https://www.vumc.org/dbmi/synthetic-derivative. [Accessed 12 December 2019].

[168] BioVU | department of biomedical informatics; n.d. https://www.vumc.org/dbmi/biovu. [Accessed 12 December 2019].

[169] Ritchie MD, Verma SS, Hall MA, Goodloe RJ, Berg RL, Carrell DS, et al. Electronic medical records and genomics (eMERGE) network exploration in cataract: several new potential susceptibility loci. Mol Vis 2014;20:1281–95.

[170] Xu H, Stenner SP, Doan S, Johnson KB, Waitman LR, Denny JC. MedEx: a medication information extraction system for clinical narratives. J Am Med Inf Assoc 2010;17:19–24. https://doi.org/10.1197/jamia.M3378.

[171] Wu Y, Warner JL, Wang L, Jiang M, Xu J, Chen Q, et al. Discovery of noncancer drug effects on survival in electronic health records of patients with cancer: a new paradigm for drug repurposing. JCO Clin Cancer Informatics 2019;1–9. https://doi.org/10.1200/cci.19.00001.

[172] Talib R, Kashif M, Ayesha S, Fatima F. Text mining: techniques, applications and issues. Int J Adv Comput Sci Appl 2016;7. https://doi.org/10.14569/ijacsa.2016.071153.

[173] Agrawal R, Agrawal R, Batra M. A detailed study on text mining techniques. Int J Soft Comput Eng 2013;2: 118–21.

[174] Steinberger R. A survey of methods to ease the development of highly multilingual text mining applications; n.d. https://doi.org/10.1007/s10579-011-9165-9.

[175] Xu R, Wang QQ. Toward creation of a cancer drug toxicity knowledge base: automatically extracting cancer drug-side effect relationships from the literature. J Am Med Inf Assoc 2014;21:90–6. https://doi.org/10.1136/amiajnl-2012-001584.

[176] Clinicaltrials.gov. 2018. https://clinicaltrials.gov/ct2/show/NCT02002819?term=levetiracetam&cond=Alzheimer+Disease&rank=2. [Accessed 23 October 2018].

[177] Bodenreider O. The unified Medical Language system (UMLS): integrating biomedical terminology. Nucleic Acids Res 2004;32:D267–70. https://doi.org/10.1093/nar/gkh061.

[178] Baker S, Ali I, Silins I, Pyysalo S, Guo Y, Högberg J, et al. Cancer Hallmarks Analytics Tool (CHAT): a text mining approach to organize and evaluate scientific literature on cancer. Bioinformatics 2017;33: 3973–81. https://doi.org/10.1093/bioinformatics/btx454.

[179] Kulick S, Bies A, Liberman M, Mandel M, Mcdonald R, Palmer M, et al. Integrated annotation for biomedical information extraction; Proc. Hum. Lang. Technol. Conf. Annu. Meet. North Am. Chapter Assoc. Comput. Linguist., 2004, p. 61–8. n.d.

[180] Chang Y, Park H, Yang HJ, Lee S, Lee KY, Kim TS, et al. Cancer drug response profile scan (CDRscan): a deep learning model that predicts drug effectiveness from cancer genomic signature. Sci Rep 2018;8. https://doi.org/10.1038/s41598-018-27214-6.

[181] Cortes-Ciriano I, Murrell D, Chetrit B, Bender A, Malliavin T, Ballester P. Cancer Cell Line Profiler (CCLP): a webserver for the prediction of compound activity across the NCI60 panel. BioRxiv 2017. https://doi.org/10.1101/105478. 105478.

[182] Sondka Z, Bamford S, Cole CG, Ward SA, Dunham I, Forbes SA. The COSMIC Cancer Gene Census: describing genetic dysfunction across all human cancers. Nat Rev Cancer 2018;18:696–705. https://doi.org/10.1038/s41568-018-0060-1.

[183] Yap CW. PaDEL-descriptor: an open source software to calculate molecular descriptors and fingerprints. J Comput Chem 2011;32:1466–74. https://doi.org/10.1002/jcc.21707.

[184] Napolitano F, Zhao Y, Moreira VM, Tagliaferri R, Kere J, D'Amato M, et al. Drug repositioning: a machine-learning approach through data integration. J Cheminf 2013;5. https://doi.org/10.1186/1758-2946-5-30.

CHAPTER

5

Increasing opportunities of drug repurposing for treating breast cancer by the integration of molecular, histological, and systemic approaches

Harras J. Khan[1], *Sagar O. Rohondia*[1], *Zainab Sabry Othman Ahmed*[1,2], *Nirav Zalavadiya*[1], *Q. Ping Dou*[1]

[1]Barbara Ann Karmanos Cancer Institute, and Departments of Oncology, Pharmacology and Pathology, School of Medicine, Wayne State University, Detroit, MI, United States of America; [2]Department of Cytology and Histology, Faculty of Veterinary Medicine, Cairo University, Giza, Egypt

OUTLINE

Drug Repurposing in Cancer Therapy
https://doi.org/10.1016/B978-0-12-819668-7.00005-1

List of abbreviations

2-DG 2-Deoxyglucose
2-FDG 2-Fludeoxyglucose
3D Three direct-acting antivirals
ABC7 Adjuvant for breast cancer treatment using seven repurposed drugs
ABCG2 ATP-binding cassette superfamily G member 2
ADH Alcohol dehydrogenase
ADRB Adrenoreceptor beta
ALDH Aldehyde dehydrogenase
AMP Adenosine monophosphate
AMPK AMP-activated protein kinase
ART Artemisinin
ATP Adenosine triphosphate
BCRP Breast cancer–resistance protein
BCSCs Breast cancer stem cells
BLBC Basal-like breast cancer
BRCA-1 Breast cancer-1
CDDP Cisplatin
CDK2 Cyclin-dependent kinase-2
COX-2 Cyclooxygenase-2
CSCs Cancer stem cells
CuGlu Copper gluconate
CYP Cytochrome p
DCA Dichloroacetate
DCIS Ductal carcinoma in situ
DDI Drug–drug interaction
DDTC Diethyldithiocarbamate
DeCoST Drug Repurposing from Control System Theory
DMBA Dimethylbenz[a]anthracene
DNA Deoxyribonucleic acid
DNMT DNA methyltransferase
DOX Doxycycline
DSF Disulfiram
EGF Epidermal growth factor
EGFP Enhanced green fluorescent protein
EGFR Epidermal growth factor receptor
EMT Epithelial–mesenchymal transition
ENO1 Enolase-1

ER Estrogen receptor
ERK Extracellular receptor kinase
FDA Food and Drug Administration
FLN Flunarizine
GBM Glioblastoma
GLP-1 Glucagon like peptide-1
GSTP-1 Glutathione S-transferase P1
HCV Hepatitis C virus
HER2 Human epidermal growth factor receptor
HIV Human immunodeficiency virus
HK-2 Hexokinase-2
HR Hazard ratio
hTERT Human telomerase reverse transcriptase
IBC Inflammatory breast cancer
IGF-1 Insulin-like growth factor-1
IGFBP-3 Insulin-like growth factor binding protein-3
IL-6 Interleukin 6
IV Intravenous
JNK c-Jun N-terminal kinase
LC3 Light chain-3
LDHA Lactate dehydrogenase A
Lipo-DS Liposome-encapsulated disulfiram
LP-1 Lily polysaccharide-1
MAPK Mitogen-activated protein kinase
MATE Multidrug and toxin extrusion protein
MBC Metastatic breast cancer
MBZ Mebendazole
MGMT Methylguanine DNA methyltransferase
mM Millimolar
MMP Matrix metalloproteinases
mRNA Mitochondrial ribonucleic acid
MSCs Mesenchymal stem cells
MTD Maximum tolerated dose
MTF Metformin
mTOR Mammalian target of rapamycin
NAC N-acetylcysteine
NF-kB Nuclear factor kappa-light-chain-enhancer of activated B cells
NIH National Institutes of Health
nM Nanomolar
NS5B Nonstructural protein 5B
OATP Organic anion transporting polypeptide
OCT Organic cation transporter
OSR Oxidative stress response
P-gp P-glycoprotein
PAC Paclitaxel
PBPK Physiologically based pharmacokinetics
pCR Pathological complete response
PDK-1 Phosphoinositide-dependent kinase-1
PEG Polyethylene glycol
PET Positron emission tomography
PI3K Phosphoinositide 3-kinase
PLGA Poly-lactic-c-glycolic acid
PLK-1 Polo like kinase-1
PMAT Plasma membrane monoamine transporter
PR Progesterone receptor
PTEN Phosphatase and tensin homolog
qPCR Quantitative polymerase chain reaction
RANKL Receptor activator of nuclear factor kappa-beta ligand
RECK Reversion inducing cysteine rich protein with kazal motifs
ROS Reactive oxygen species
RPFNA Random periareolar fine-needle aspiration
SIL Silibinin
SOD Superoxide dismutase
TCGA The Cancer Genome Atlas
TGF-b Transforming growth factor beta
Th2 T helper type 2
TLR Toll-like receptor
TNBC Triple-negative breast cancer
Trp53 Transformation related protein 53
TTM Tetrathiomolybdate
TUNEL Terminal deoxynucleotidyl transferase dUTP nick end labeling
UGT Uridine diphosphate glucuronosyltransferase
UPS Ubiquitin proteasome system
VEGF Vascular endothelial growth factor
μM Micromolar

Introduction

Breast cancer is the most common type of cancer affecting women (https://www.cancer.gov/types/common-cancers; https://www.who.int/health-topics/cancer#tab=tab_1; https://www.wcrf.org/dietandcancer/cancer-trends/skin-cancer-statistics). The tremendous impact of breast cancer calls for any and every method possible to be used in its therapy and treatment. Drug repurposing (also known as repositioning, reprofiling, redirecting, or rediscovering [1]) means developing new uses for a drug beyond its original use or initially approved indication. Drug repurposing has attracted increasing attention in recent years as potentially inexpensive alternatives are needed urgently to compensate for the high costs and disappointing success rate associated with the drug discovery pipeline [2]. Repurposing can help identify new therapies for diseases at a lower cost and in a shorter time, particularly in those cases where preclinical safety studies have already been completed [3]. Repurposing an old drug for a new use, such as cancer therapy, is an attractive and exciting field [4,5]. During recent years, several reviews on drug repurposing have been published [2,5].

Drug repurposing is a possible alternative to the current therapies in breast cancer treatment. Multiple drugs have shown great promise in this aspect, from inhibition of breast cancer cell proliferation in general to blocking of pathways that specifically cause breast cancer. This chapter will focus on multiple old or currently used drugs that have shown potential promise in the therapy of breast cancer, by summarizing first their original uses or purposes and then their mechanisms of action responsible for the anti-breast cancer effects. This chapter will then comprehensively review both preclinical and clinical (as available) studiesusing these repurposed drugs and their multitude of effects on various subtypes of breast cancer. It will end by discussing the future of drug repurposing—both as a field in general and specifically in the therapy of breast cancer.

Metformin

Metformin (MTF; Fig. 5.1) is a well-acknowledged biguanide and a multiaction drug advertised under the name Glucophage as well as others. MTF was approved by the FDA in 1995 as an oral hypoglycemic drug in the management of diabetes mellitus and is the first-line medication for type 2 diabetes mellitus treatment [6] (Table 5.1). MTF could effectively reduce gluconeogenesis in the liver and improve insulin sensitivity by inducing peripheral glucose uptake and lowering the basal and postprandial plasma glucose [6]. It was reported that MTF has antiinflammatory [7], antiapoptotic, anticancer, hepatoprotective, cardioprotective [8], renoprotective [9], otoprotective [10], radioprotective, radiosensitizing, and antioxidant activities [10] (Table 5.1). There has been a recent development in MTF repositioning for anticancer therapy. It exhibited a great potential to change the metabolic reprogramming and acts as a candidate for cancer management [11]. In 2005, MTF was used for breast cancer treatment [12].

Metformin—mechanism of action of the original use

MTF is an orally bioavailable drug metabolized in the liver. It is transported into hepatocytes by organic cation transporter-1 (OCT1). It then accumulates in the mitochondrial inner membrane, inhibits complex 1 of the electron transport chain, and reduces ATP production. The increase in AMP and ADP causes AMPK activation. Activated AMPK inhibits fat synthesis and promotes fat oxidation instead, thus reducing hepatic lipid stores and enhancing hepatic insulin sensitivity. The increase in AMP: ATP ratio also inhibits fructose-1,6-bisphosphatase, resulting in the acute inhibition of gluconeogenesis. MTF also acts via other pathways to inhibit gluconeogenesis, glycogenolysis, and release of glucose in the blood (Table 5.1).

Metformin—preclinical studies

Metformin's anticancer effects

In vitro and in vivo data of MTF treatment revealed its ability to inhibit the growth of ovarian cancer stem cells (CSCs) [13], glioma-initiating cells [14], breast cancer cells [15], endometrial cancer cells [16], and non-small cell lung cancer cells [17].

Anticancer properties of MTF have been shown either through its direct effect on the cancer cells by activating AMPK/inhibiting mammalian target of rapamycin (mTOR) pathway [18] or via its indirect effect on the host by decreasing the blood glucose level in addition to its antiinflammatory effects [19,20] (Table 5.1). The inhibition of mTOR in tumor

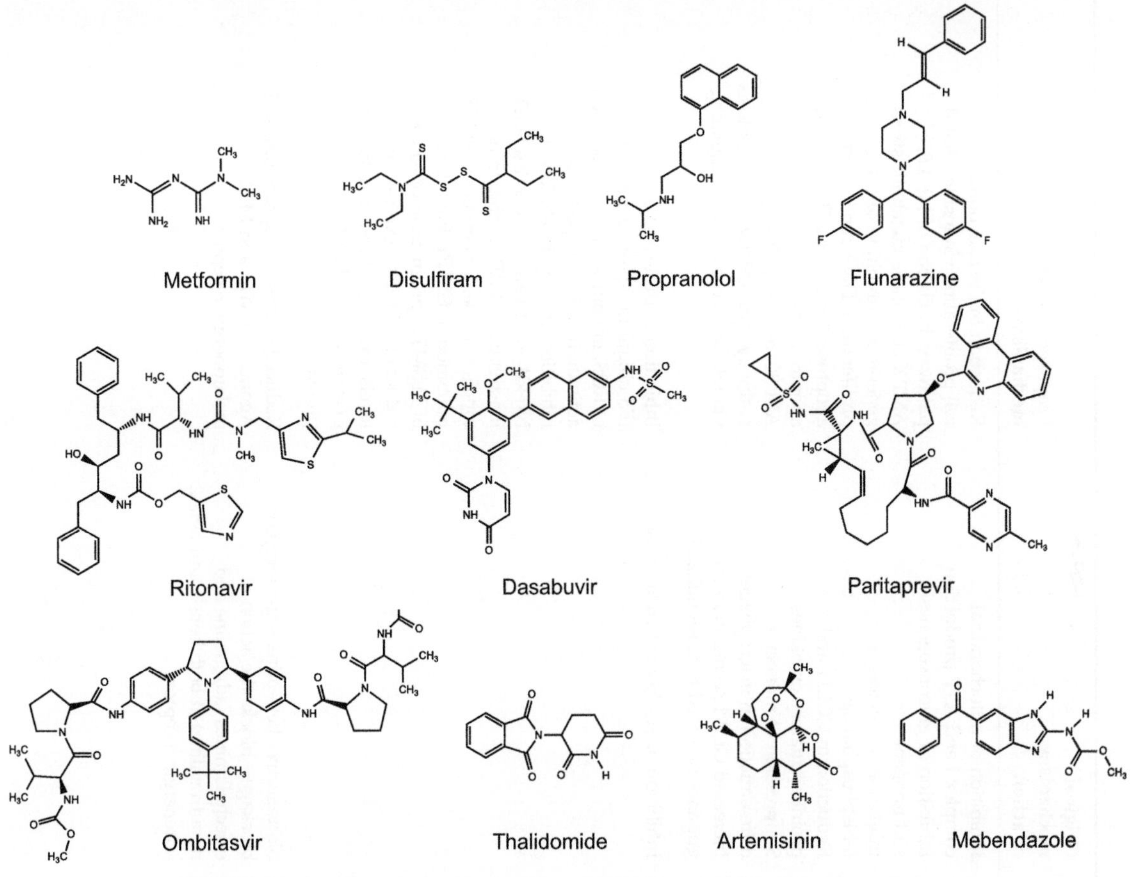

FIGURE 5.1 Chemical structures of repurposed drugs in breast cancer treatment.

cells was reported as one of the potential vital mechanisms of the anticancer properties of MTF [21]. AMPK could inhibit mTORC1 via phosphorylation of mTOR-binding raptor [22]. In addition, HER2 expression in human breast cancer cells treated with MTF was declined via direct inhibition of p70S6K1, which is a downstream effector of mTOR [23]. Several studies also suggested that MTF causes many biological and molecular effects on breast cancer cells in addition to its ability to reduce cellular response to the tumorigenesis critical factors as insulin, interleukin 6 (IL-6), and epidermal growth factor through downregulation of IGF-1R, p-Stat-3, and EGFR [24].

In MCF-7 breast cancer cells (Table 5.2), MTF reduced phosphorylation of S6 kinase, ribosomal protein S6, and eIF4E-binding protein, inhibited mTOR, and reduced translation initiation via AMPK activation [18].

Zordoky et al. reported that excess glucose amount in triple-negative MDA-MB-231 cells (Table 5.2) blocked the MTF-induced cell death, which suggested that high glucose levels resulted in the production of enough energy for cell proliferation via aerobic glycolysis [25]. In addition, Wahdan-Alaswad et al. reported that glucose increases the aggression of breast cancer cells and reduces the efficacy of MTF. Their preclinical studies on triple-negative breast cancer

TABLE 5.1 Summary of repurposed drugs in breast cancer.

Drug	History and use	Original mechanisms of action	Proposed anticancer mechanisms
MTF	FDA approved in 1995 as an oral hypoglycemic drug in the management of diabetes mellitus	Inhibition of mitochondrial complex I and ATP production Inhibition of gluconeogenesis and lipogenesis Inhibition of fructose-1, 6-bisphosphatase Reduction in CCL11 and proinflammatory cytokines Suppression of monocyte differentiation in macrophages Increased GLP-1 secretion in the gut and increased glucose utilization	Reduction in the carcinoma cell proliferation via insulin/IGF-1 pathway, inhibition of NF-kB, AMPK activation, metabolic stress generation, inhibition of cell proliferation, DNA replication apoptosis through the BAX/BCL-2 apoptotic pathway and AMPK/mTOR/p70S6 growth pathway, cytotoxicity, and apoptosis via toll-like receptor (TLR) signaling
DSF	FDA approved 1951 as Antabuse in alcohol deaddiction	Inhibition of alcohol dehydrogenase	Inhibition of proteasome (E3 ubiquitin ligase), generation of ROS, inhibition of SOD-1 activity, activation of MAPK pathway, inhibition of P-gp, inhibition of NF-kB, inactivation of Cu/Zn SOD, inactivation MMP, inhibition of DNA topoisomerase, inhibition of DNMT, inhibition of GSTP1, inhibition of MGMT, upregulation of RECK, reduction of CDK1 expression levels, causing G2 arrest, reduction in expression of PLK1 protein and mRNA
Propranolol	FDA approved in 1967 for the management of hypertension. It is used for other cardiovascular conditions cardiac arrhythmias, postmyocardial infarction to reduce mortality. Other noncardiovascular uses are for prophylaxis of migraine, essential tremors, anxiety, portal hypertension, hyperthyroidism, and pheochromocytoma	Nonselective beta-adrenergic receptor blockage—blocks the action of catecholamines (adrenaline and noradrenaline) at both beta-1 and beta-2 adrenergic receptors	Reduced expression of hexokinase-2 Agonism at MT1 and MT2 melatonergic receptors

Antivirals	Ombitasvir, dasabuvir, and paritaprevir were FDA approved in July 2016 for the treatment of hepatitis C Ritonavir was FDA approved in 1999 for the treatment of HIV AIDS	Ombitasvir—HCV NS5A inhibition Dasabuvir—HCV NS5B inhibition Paritaprevir—HCV NS3/4A inhibition Ritonavir—cytochrome P3A4 inhibition causing protease inhibition	Inhibition of P-gp, inhibition of breast cancer—resistance protein (BCRP), inhibition of transcriptional and translational factors, suppression of telomerase activity, and activation of human DNA polymerase, increase plasma exposure in drugs, binding to heat shock protein 90 (HSP90), and partial inhibition of its chaperone
Antipsychotics	Risperidone, 9-hydroxy-risperidone (paliperidone), olanzapine, quetiapine, clozapine, haloperidol, and chlorpromazine	Exact mechanisms are not known, but drugs of this class are believed to inhibit mainly dopaminergic D2 and serotonergic 5HT2A receptors and sometimes alpha-1 adrenergic, H-1 histaminic, and muscarinic receptors	Inhibition of cellular uptake of mitoxantrone, inhibition of RANKL-mediated MAPK and NF-kB signaling pathways, inhibition of breast cancer—resistance protein (BCRP)
Thalidomide	It was previously used in pregnant women to relieve morning sickness. However, its use was restricted because of its teratogenic effect on the developing fetus. Over the last decades, thalidomide showed a cytotoxic effect on different cancer cell lines, including breast cancer	The CD147 and MCT1 protein complex is stabilized and developed by the binding of the cereblon protein and this effect on the complex stimulates cell growth and facilitates the excretion of metabolic products like lactate. It is because of this that an increased abundance of this protein complex enables tumor cells to spread rapidly in diseases such as multiple myeloma. If such a cancer is treated with IMiDs, the complex is displaced from its binding to cereblon and therefore the protein complex of CD147 and MCT1 can no longer be activated, which in turn causes tumor cells to die.	The growth and progression of breast cancer cells including MCF-7 and MDA-MB-231 were inhibited by a small series of thalidomide-correlated compounds, which are very effective to induce cancer cell death via triggering TNFα-mediated apoptosis
Artemisinin	Artemisinin (ART) is a chemical that was isolated from the sweet wormwood by Youyou Tu at the Chinese Academy of Traditional Chinese Medicine in 1972. Originally used for the treatment of malaria, during the past two decades, studies revealed the anticancer activity of	Artemisinin and its semisynthetic derivatives were originally formulated to be used against malaria. These drugs contain endoperoxide bridges, which are needed for antimalarial activity. The mechanism of action for this original use was believed to be a two-step mechanism.	ARS and DHA were reported to inhibit TGF-β signaling that inactivates cancer-associated fibroblasts (CAFs) which play an important role in tumor growth and metastasis, such as stimulating angiogenesis, cell proliferation, migration, and invasion

Continued

TABLE 5.1 Summary of repurposed drugs in breast cancer.—cont'd

Drug	History and use	Original mechanisms of action	Proposed anticancer mechanisms
	artemisinin and its derivatives, which indicates the effectiveness of these compounds as cancer therapeutic drugs	First, ART is activated by intraparasitic heme iron which catalyzes the cleavage of the endoperoxide bridges which then causes a free radical intermediate to be formed and kill the parasite by alkylating and poisoning one or more essential malarial proteins	
Mebendazole	It was approved by the US FDA in 1974 for the treatment of nematode infections. Also, MBZ revealed efficacy against different types of solid tumors in vitro and in vivo, such as lung cancer, melanoma, colon cancer, glioblastoma multiforme, medulloblastoma, and head and neck squamous cell carcinoma	Mebendazole binds to the colchicine-sensitive site of tubulin and therefore causes degenerative alterations in the tegument and intestinal cells of the parasite. The parasite then has its glycogen stores depleted because this loss of cytoplasmic microtubules eventually leads to disruption in the glucose uptake by its larval and adult stages. This causes the parasites immobilization and eventual death	It was reported that MBZ was able to induce cell cycle arrest in the radiosensitive G2/M phase of the cell cycle in TNBC cells. This cell cycle arrest was followed by significant induction of apoptotic cell death in a dose- and time-dependent manner. Different studies revealed induction of apoptosis as the primary mode of cell death by MBZ in other tumor types, such as melanoma, lung cancer, and medulloblastoma. Also, tubulin depolymerization was reported to be the major target for benzimidazoles, including MBZ. MBZ exhibited induction of depolymerization of tubulin and inhibition of normal spindle formation in different cancer cell lines, resulting in mitotic arrest and apoptosis
Flunarizine	Not FDA approved. Not prescribed in the United StatesIn other countries, include prophylaxis of migraine, peripheral vascular disease, and vertigo	Selective calcium entry block with calmodulin-binding properties and histamine H1—blocking activity	N-Ras degradation, inhibition of colony formation through TG101348 and FLN combination, inhibition of growth and transforming activity of BLBC cells, induction of autophagy-like activity

TABLE 5.2 Inhibition of various subtypes of breast cancer cells by repurposed drugs.

Drug	Cancer cell line	ER	PR	HER2	BRCA1 mutation	Subtype	Source
Metformin	BT474	+	+	+	WT	LB	L
	MCF7, T47D	+	+	−	WT	LA	L
	ZR7530	+	−	+	WT	LB	L
	MDAMB453, SKBR3	−	−	+	WT	H	L
	MDAMB231, MDAMB157, MDAMB468, HCC70, BT20, BT549	−	−	−	WT	TNB	L
	HCC38, HCC1143, HCC1187, HCC1806	−	−	−	ND	TNB	L
	HCC1937, MDAMB436	−	−	−	MU	TNB	L
DSF	BT474	+	+	+	WT	LB	L
	MCF7, T47D	+	+	−	WT	LA	L
	SKBR3, SUM190	−	−	+	WT	H	L
	MDAMB157, MDAMB231, BT549	−	−	−	WT	TNB	L
	SUM149	−	−	−	MU	TNB	L
	HCC70, BT20	−	−	−	WT	TNA	L
	MDAMB436	−	−	−	MU	TNA	L
Propranolol	MCF7	+	+	−	WT	LA	L
	MDAMB231	−	−	−	WT	TNB	L
Antiviral drugs	MCF7, T47D	+	+	−	WT	LA	L
	MDAMB231	−	−	−	WT	TNB	L
	MDAMB436	−	−	−	MU	TNB	L
Antipsychotic drugs	MCF7	+	+	−	WT	LA	L
	MDAMB231	−	−	−	WT	TNB	L
Antimalarial drugs, Artemisinin (ART)	MCF7	+	+	−	WT	LA	L
	MDAMB231	−	−	−	WT	TNB	L
Mebendazole	MCF7, T47D	+	+	−	WT	LA	L
	MDAMB231, SUM159PT	−	−	−	WT	TNB	L
Thalidomide	MCF7	+	+	−	WT	LA	L
	MDAMB231	−	−	−	WT	TNB	L
Flunarizine	SUM102PT	−	−	−	WT	TNB	L
	SUM149PT	−	−	−	MU	TNB	L

H, HER2 positive; *LA*, luminal A; *LB*, luminal B; *MU*, mutant; *ND*, not decided; *TNA*, triple-negative A; *TNB*, triple-negative B; *WT*, wild type.

(TNBC) showed that MTF blocked several key enzymes necessary to glucose metabolism, but its effect was glucose-dependent. Thus, glucose monitoring for patients with breast cancer is suggested to be a crucial factor [26]. In vivo studies also revealed the antiproliferative effect of MTF in breast cancer [27,28].

Metformin enhances the anticancer activity of other drugs

In vitro and in vivo preclinical studies of MTF in combination with other drugs also revealed a synergistic effect to reduce cell proliferation in both breast cancer cell culture and animal models [29].

Metformin *in* combination with natural compounds

Curcumin with MTF exhibited dose-dependent cytotoxicity and antiproliferative activity, as shown by a declined hTERT expression, reduced vascular endothelial growth factor (VEGF) expression, induced transformation related protein 53–independent apoptosis, and triggered Th2 immune response [30]. In mice models, Falah et al. reported that the combination of MTF and curcumin in breast cancer caused angiogenesis inhibition, immune system modulation, and induction of p53-independent apoptosis [31].

A combination of ursolic acid with MTF at low concentrations significantly inhibited invasion and migration of transforming growth factor-β (TGF-β)–induced breast cancer in MDA-MB-231 and MCF-7 cells. Combination of ursolic acid and MTF also downregulated expression of CXCR4, uPA, vimentin, E-cadherin, N-cadherin, and MMP-2/9 proteins in breast cancer cells [32].

Vitamin D3 in combination with MTF increased levels of cleaved caspase-3, BAX, and AMPK and inhibited that of BCL-2, c-Myc, IGF-IR, mTOR, P70S6K, and S6, causing cell cycle arrest in breast cancer cells [33].

Silibinin in combination with MTF synergistically downregulated expression levels of hTERT and cyclin D1 and enhanced inhibition of breast cancer cell growth [34].

Chrysin with MTF synergistically caused breast cancer cell death by reducing cyclin D1 and hTERT gene [35]. Flavone with MTF decreased MDMX protein expression, regulated p53 downstream target gene Bcl-2, and cleaved caspase-3 [35].

Melatonin with MTF reduced tumor growth and size by increasing the expression of Bax and caspase-3 and inhibiting DMBA-induced breast cancer tumor growth [36].

Lily polysaccharide-1 with MTF downregulated Bcl-2 expression and upregulated Bax, causing enhanced antiproliferation and apoptosis in breast cancer cells [37].

Metformin in combination with other drugs used as anticancer agents

Recent studies revealed the ability of MTF to enhance the efficacy of chemotherapeutics in combination therapy [38]. It was reported that efficacies of several chemotherapeutics increased dramatically when given in combination with MTF [39].

A combination of MTF with doxorubicin inhibited NF-κB activity in breast tumor cells, which further decreased TNF-α and IL-6 expressions in breast tumor cells, suppressed tumor cell proliferation, and enhanced apoptosis. Moreover, the therapy enhanced nuclear doxorubicin accumulation and overcame drug resistance by downregulating P-gp (P-glycoprotein) and intracellular ATP content in a reduced dose [40,41]. Combination therapy of doxorubicin + 2-deoxyglucose (2-DG) with MTF showed substantial metabolic stress at low dose via inhibition of glucose uptake and suppression of lactate, fatty acid, and ATP production. This combination therapy was associated with decreased cell viability, increased the intracellular oxidation, induced apoptosis and autophagy, and completely inhibited colony formation by activation of AMPK [42]. Another combination experiment of MTF with doxorubicin and 2FDG increased phospho-AMPK but decreased phospho-Akt and phospho-ERK expressions [43].

Sahra et al. reported that the combination of MTF with 2-DG revealed inhibition of mitochondrial respiration and glycolysis via p53-dependent apoptosis through AMP pathway. Furthermore, MTF showed a synergistic effect when combined with paclitaxel (PAC), carboplatin, or doxorubicin in xenograft mouse models of breast, lung, and prostate cancer [44].

A combination of MTF with PAC exhibited significant inhibition of cell viability and induction of the G_2/M phase arrest [45]. In vitro studies revealed that MTF and PAC had a synergistic effect, and codelivery of the micelles induced higher cytotoxicity and apoptosis in 4T1 breast cancer cells than each free drug [46].

Tamoxifen with MTF synergistically inhibited cell proliferation and DNA replication and triggered apoptosis at reduced doses, through regulating both Bax/Bcl-2 and AMPK/mTOR/p70S6 pathways [47].

MTF in combination with 5-fluorouracil, epirubicin, and cyclophosphamide accelerated glucose consumption and lactate production in breast CSCs and significantly hampered intracellular ATP leading to a severe energy crisis and thus inducing DNA damage in breast cancer cells [48].

Combination of everolimus with MTF inhibited cell survival, clonogenicity, mTOR signaling activity, mitochondrial respiration, and abrogated S6 and 4E-BP1 phosphorylation in breast cancer cells [49,50]. In addition, the combination of MTF with erlotinib in breast cancer cells enhanced the reduction of EGFR, AKT, S6, and 4EBP1 phosphorylation, associated with inhibition of mammosphere outgrowth [29].

Aspirin and MTF combination enhanced breast cancer 4T1 cell apoptosis by inducing secretion of TGF-β1. The effect of this combination partly relied on cyclooxygenase-2 upregulation, without production of lipoxins [51]. In both immune-deficient and immune-competent breast cancer preclinical models, atenolol increased MTF activity against angiogenesis, local and metastatic growth of HER2+, and triple-negative BC. Aspirin increased the activity of MTF only in immune-competent HER2+ BC models. Both aspirin and atenolol, when added to MTF, significantly reduced the endothelial cell component of tumor vessels, whereas pericytes were reduced by the addition of atenolol but not aspirin. These data indicate that the combination of aspirin or atenolol with MTF might be beneficial for BC treatment and that this anti-BC activity is likely due to the inhibitory effects on both BC and microenvironment cells [52].

A combination of propranolol and MTF inhibited glucose metabolism via downregulation of ADRB2-dependent hexokinase-2, and posttranscription and activation of AMPK, along with the antioxidation activity [53].

Dichloroacetate and MTF combination revealed synergistic induction of caspase-dependent apoptosis via oxidative damage through PDK1 inhibition and decreased lactate production in breast cancer cells. Expression of glycolytic enzymes, including HK2, LDHA, and ENO1, were downregulated, associated with induction of cell death [54,55].

Topotecan with MTF was shown to activate AMPK, downregulate excision repair cross-complementation group 1, and suppress DNA replication by inhibiting nuclear enzyme topoisomerase I. This combination also depolarized mitochondrial membrane and induced cell cycle arrest in breast cancer cells [56]. MTF was also reported to prevent BRCA1 haploinsufficiency-driven RANKL gene overexpression and disrupt the autoregulatory feedback control of RANKL-addicted CSCs. The synergistic effect of MTF and denosumab decreased breast cancer–initiating cell (BCIC) population and self-renewal capacity [37].

Metformin—clinical trials

There are a total 45 registered clinical trials at different stages of development that have studied the efficacy of MTF as monotherapy or in combination with chemotherapy and/or radiotherapy exclusively for the management of breast cancer (Table 5.3). The studies also focus on the process of establishing the effects of MTF on markers of cellular proliferation, pathological response rate, progression-free survival, tolerated safe dose, and recurrence-free survival for breast cancer.

In vitro studies showed that MTF at 5–10 μM was able to activate AMPK [57]. However, clinically such low concentrations of MTF were not sufficient to cause AMPK activation, although metabolic alteration could be evoked [58]. Therefore, scientists have focused on using pharmacological combinations that can aid in producing anticancer effects at an effective and safe dose profile of MTF [59].

Other pharmacological agents may be used together with MTF for enhancing the therapeutic efficacy of the drug. Still, the effective and safe dose for breast cancer is a great challenge. A Phase I study was conducted using radioactive 400MBq 11C-MTF, injected in cubital veins of 10 males and females age >50 years with breast cancer, and PET images were taken after 120 min. Quantitative polymerase chain reaction confirmed the drug transportation pathway through OCT, specifically OCT1-3, MATE 1 and 2, and PMAT. PET imaging also confirmed the drug uptake by breast tissue [60]. In postmenopausal women with breast cancer, MTF therapy changed the levels of protein phosphorylation. In obese patients with the ordinal level of breast density, MTF decreased the risk of breast cancer [61].

MTF therapy prior to surgery was shown to reduce cancer cell proliferation and inhibit tumor growth in early stage breast cancer patients [62]. MTF was also shown to improve general body condition including weight loss in overweight or prediabetes mellitus–operable breast cancer patients [63]. In Stage I to III colorectal or breast cancer patients, MTF was able to decrease fasting insulin levels and change other insulin-related biomarkers like C-peptide, IGFR, IGFBP-3, adiponectin, and lectin [64].

In another trial, MTF was found to be able to inhibit cell proliferation and tumor growth in patients with early stage breast cancer before surgery. MTF before surgery in operable breast cancer patients improved general body condition. In a subset of breast cancer patients with tumor size >3 cm, 1500 mg extended-release MTF was shown to cause an alteration of S6K, 4E-BP-1, and AMPK via histochemical analysis, thus effectively altering breast cancer metabolism [65].

Phase I trial of MTF with exemestane and rosiglitazone was well tolerated in obese postmenopausal ER+ and PR+ breast cancer patients [66]. Another Phase I trial was conducted using MTF and temsirolimus to determine the maximum tolerated dose (MTD) for the combination in breast cancer patients [67]. A combination study of MTF with standard first-, second-, third- or fourth-line chemotherapy like anthracycline, taxane, platinum, capecitabine, or vinorelbine-based regimens showed improved survival and tumor response in women with metastatic breast cancer [68]. Another combination therapy of MTF and erlotinib showed good tolerance in TNBC patients [69]. MTF acted as a biomarker for tracing the mechanism of cancer cell resistance. MTF, along with myocet and cyclophosphamide in HER2-negative breast cancer patients, led to an increase in the progression-free survival as well as characterization of sensitivity in insulin levels [70].

In invasive breast cancer patients, when MTF was used before PAC therapy, there was a reduction in the occurrence of some adverse effects, specifically peripheral neuropathy [71]. In another subset of breast cancer patients, combination therapy of MTF and doxorubicin reduced cardiotoxicity by decreasing the

TABLE 5.3 Clinical trials of repurposed drugs in breast cancer.

	Clinical trial	Condition	N1	N2	Design	Primary outcome	Status
1	NCT00897884	Breast cancer	40	39	Metformin 500 mg tablet, TID X 2–3 weeks prior to surgery	Reduction in cell proliferation rates in tumor tissue	Completed
2	NCT01266486	Breast cancer	40	41	Extended release metformin 1500 mg OD X 14–21 days	Measure metformin induced effects in phosphorylation of S6K, 4e-BP-1 and AMPK	Completed
3	NCT02882581	Breast cancer	10	7	400MBq of 11 C-metformin is injected in the cubital vein, followed by PET scan	Metformin uptake in breast cancer	Completed
4	NCT01310231	Metastatic breast cancer	78	40	Metformin 850 mg BID + standard chemotherapy (anthracyclines, platinum, taxanes, or capecitabine; first or second line) Placebo + standard chemotherapy	Progression-free survival	Completed
5	NCT01650506	Breast cancer	20	8	Standard 3 + 3 dose escalation Metformin 850 mg BID to TID Erlotinib 150 mg daily	Maximum tolerated dose of metformin in combination with a fixed dose of 150 mg erlotinib daily	Completed
6	NCT01340300	Colorectal cancer Breast cancer	200	139	1. Exercise training 2. Exercise training + oral metformin QD X 2 weeks, then BID 3. Oral metformin QD X 2 weeks, then BID Control—educational information	Change in fasting insulin level	Completed
7	NCT01589367	Hormone receptor–positive malignant neoplasm of breast	208	208	Arm 1: Metformin experimental Arm 2: Letrozole alone/placebo	Clinical response rate comparing with RECIST 1.1 from baseline	Completed

Continued

TABLE 5.3 Clinical trials of repurposed drugs in breast cancer.—cont'd

	Clinical trial	Condition	N1	N2	Design	Primary outcome	Status
8	NCT01793948	Breast cancer Obesity	24	24	Experimental: Arm I: Metformin hydrochloride PO OD on days 1–30 in course 1 and BID on days 1–30 thereafter X 12 courses Arm II: Placebo	Changes in the phosphorylation of proteins after metformin exposure	Completed
9	NCT01885013	Human epidermal growth factor 2 negative carcinoma of breast	112	126	Arm A: Metformin (day 1–day 3, 1000 mg OD; day 4 to day 13, 1000 mg BID) + myocet 60 mg/m^2, IV on day 1/21 days + cyclophosphamide 600 mg/m^2 IV on day 1/21 days Arm B: Myocet + cyclophosphamide	Progression-free survival (PFS)	Completed
10	NCT00909506	Breast cancer	105	105	Placebo comparator: Placebo Active comparator: Metformin 500 mg/day OD X 1–2 weeks Active comparator: Metformin 1000 mg/day (dose-eascalate)	Weight loss	Completed
11	NCT00933309	Breast cancer	24	25	Group 1: Exemestane 25 mg PO OD Group 1: Exemestane 25 mg PO OD + avandamet (rosiglitazone 2 mg + metformin 500 mg PO OD)	Dose-limiting toxicity (DLT)	Completed

12	NCT00659568	Breast cancer Endometrial cancer Kidney cancer Lung cancer Lymphoma Unspecified adult solid tumor, protocol specific	28	28	Metformin hydrochloride OD/BID/TID on d1–d28 + temsirolimus IV d1, 8, 15, 22/28 days	Maximum tolerated dose and recommended Phase II dose of metformin hydrochloride when administered with temsirolimus	Completed
13	NCT02145559	Breast neoplasms Lung neoplasms Cancer of liver Lymphoma Cancer of kidney	64	24	Experimental: Sirolimus (3 mg daily) d1–7 + metformin XR (500 mg PO OD d8–d28; from d15 1000 mg PO OD) Experimental: Sirolimus (3 mg daily) d1–7, + delayed metformin XR (500 mg PO OD from d22; after 1 week 1000 mg PO OD)	Pharmacodynamic biomarker p70S6K	Completed
14	NCT02431676	Breast cancer Prostate cancer Lung cancer Colon cancer Melanoma of skin Endometrial cancer Liver cancer Pancreatic cancer Rectal cancer Kidney cancer Other solid malignant tumors	120	121	Active comparator: Self-directed behavioral self-control weight loss Experimental: Coach-directed behavioral weight loss Experimental: Metformin up to 2000 mg daily X 12 months	IGF-1 levels IGF-1 levels: IGFBP3 levels (ratio)	Completed
15	NCT02028221	Breast cancer	150	151	Metformin 850 mg PO OD X 4 weeks followed by 850 mg PO BID. Placebo	Change from baseline in breast density at 6 and 12 months	Active, not recruiting
16	NCT02488564	HER2-positive breast cancer	46	49	Day 1: Liposome-encapsulated doxorubicin, 50 mg/m^2 IV 1 h Day 2 and 9: Docetaxel, 30 mg/m^2 IV 1 h Day 2, 9, and 16: Trastuzumab	Pathologic complete response rate (pCR)	Active, not recruiting

Continued

TABLE 5.3 Clinical trials of repurposed drugs in breast cancer.—cont'd

	Clinical trial	Condition	N1	N2	Design	Primary outcome	Status
					4 mg/kg loading dose, 2 mg/kg/week for subsequent injections Day 13–11: Metformin 1000 mg OD; Day 10–0: Metformin 1000 mg BID until end of the study treatment.		
17	NCT01101438	Breast cancer	3582	3649	Experimental Arm I: Oral metformin HCL BID (OD in weeks 1–4) continue up to 5 years in ER+, PR+ Placebo Arm II: Placebo BID (OD in weeks 1–4) continue up to 5 years in ER+, PR+	Invasive disease-free survival in hormone receptor negative and positive subgroups	Active, not recruiting
18	NCT02278965	Stage 0—III breast carcinoma Breast neoplasms	20	19	Metformin 850 mg, oral, twice a day for 12 months Omega-3 fatty acids 2 capsules (560 mg each) oral, twice a day for 12 months	Assess the safety and feasibility of a 1-year intervention of metformin and omega-3 fatty acids in early stage breast cancer patients who completed adjuvant treatment	Active, not recruiting
19	NCT04143282	Metastatic breast cancer (nondiabetic)	50	50	Metformin 850 mg to 1 gm, BID	Disease progression through tumor size assessed by CT scan (chest–abdomen–pelvis), bone scan, and MRI	Recruiting
20	NCT04170465	Breast cancer female	60	60	1. Oral metformin HCl 850 mg BID X 6 months + AC-T chemotherapy regimen ([doxorubicin 60 mg/m^2 IV + cyclophosphamide 600 mg/m^2 IV[X 4 cycles, every 3 weeks + [paclitaxel 80 mg/m^2 IV] every	Evaluation of the effect on tumor proliferation as measured by Ki-67 immunohistochemical (IHC) assessment (%) Tissue level of Ki-67 expression in the excised tumor Evaluation of the effect on tumor apoptosis as measured by caspase-3 Chemotherapy toxicities	Recruiting

					week X 12 weeks) 2. AC-T chemotherapy regimen alone		
21	NCT03238495	HER2-positive breast cancer	100	100	1. Active comparator: Chemotherapy only—taxotere, carboplatin, herceptin + pertuzumab (TCH + P) 1. Experimental: Metformin 850 mg OD during the first cycle, then 850 mg twice daily for the remaining 5 cycles + (TCH + P)	Pathologic complete response	Recruiting
22	NCT01980823	Breast cancer Breast tumors	40	40	Metformin 1500 mg per day: Divided 500 mg in the morning and 1000 mg in the evening + atorvastatin 80 mg OD	Change in tissue levels of the proliferation marker Ki-67	Recruiting
23	NCT02506777	Breast cancer	96	96	1. Experimental: Metformin 850 mg BID + FDC (fluoruracil 500 mg/m^2, doxorubicin 50 mg/m^2, cyclophosphamide 500 mg/m^2) once every 21 days 2. Experimental: melatonin 3 mg before sleeping daily + FDC 3. Active comparator: FDC	Response rate evaluated by RECIST criteria Pathomorphological response assessed after surgery by miller and payne sscale	Recruiting
24	NCT02506790	Breast cancer	96	96	Experimental: Toremifene 60 mg daily + metformin 850 mg BID Experimental: Toremifene 60 mg daily with melatonin 3 mg before sleep daily Active comparator: Toremifene	Response rate evaluated by RECIST criteria Pathomorphological response assessed after surgery by miller and payne scale	Recruiting
25	NCT01929811	Breast cancer	200		Experimental: Metformin 500 mg PO TID (500 mg PO	Pathologic complete response rate (pCR)	Recruiting

Continued

TABLE 5.3 Clinical trials of repurposed drugs in breast cancer.—cont'd

	Clinical trial	Condition	N1	N2	Design	Primary outcome	Status
					OD for 1st cycle) + docetaxel + epirubicin + cyclophsophomide TEC: Docetaxel 75mg/m2 IV d1 q3w*6 + epirubicin 75mg/m2 IV d1 q3w* 6 + cyclophsophomide 500mg/m2 IV d1 q3w*6		
26	NCT01905046	Atypical ductal breast hyperplasia BRCA1, BRCA2 DCIS, LCIS	400	128	Arm I: Metformin hydrochloride PO (OD/BID X 24months; 850 mg BID 13–24months) Arm II: Placebo OD/BID X 12 months; patients may cross over to Arm I for months 13–24	Cytological atypia in unilateral or bilateral RPFNA aspirates	Recruiting
27	NCT03006172	Breast cancer Solid tumor	156	104	Experimental Stage I Arm A: GDC-0077 6 mg (escalating doses) d1, OD from d8/28 days cycle (35 days cycle 1) Experimental Stage I Arm B: GDC-0077 3 mg (escalating doses) OD, d1–d28 + palbociclib d1–d21 + letrozole d1–d28/28 days cycle Experimental Stage I Arm C: GDC-0077 (escalating doses) OD, d1–d28 + letrozole d1–d28/28 days cycle Experimental Stage II Arm B: GDC-0077 OD, d1-–28 + palbociclib d1–d21 + letrozole d1–d28/28 days cycle Experimental Stage II Arm C: GDC-0077 OD, d1–d28 + letrozole d1–	Dose-limiting toxicities Recommended Phase II dose of GDC-0077 Adverse events and serious adverse events	Recruiting

					d28/28days cycle Experimental Stage II Arm D: GDC-0077 d1−d28 + fulvestrant d1, d15/cycle 1, d1 from cycle 2/28 days Experimental Stage II Arm E: GDC-0077, d1−d28 + palbociclib d1−d21 + fulvestrant d1, d15/cycle 1, d1 from cycle 2/28 days Experimental Stage II Arm F: GDC-0077, d1−d28 + palbociclib d1−d21 + fulvestrant d1, d15/Cycle 1, d1 from cycle 2+ metformin d1−d28/28 days		
28	NCT03168880	Triple-negative breast cancer			A: Paclitaxel 100 mg/m2/week X 8 weeks + AC/EC (60/600 or 90/600)/3 weekly	Disease-free survival (DFS) Overall survival (OS)	Recruiting
29	NCT04001725	Brain metastases Melanoma Lung cancer Breast cancer	110		A: Dexamethasone 8 mg(minimum) PO/IM/IV OD/BID B: Dexamethasone + metformin (850 mg/day, progressively increased to 1700 mg/day on d4 and 2550 mg/day on d7, if well tolerated)	Prevention of precocious (14 days) dexamethasone-induced diabetes	Recruiting
30	NCT02874430	Breast carcinoma Endometrial clear cell adenocarcinoma Endometrial serous adenocarcinoma Uterine corpus cancer Uterine corpus carcinosarcoma	74	46	Metformin hydrochloride PO OD d1−3, BID from d4 + doxycycline PO BID/7 days	Change in the percent of stromal cells expressing caveolin-1 (CAV1) at an intensity of 1+ or greater assessed by immunohistochemistry analyzed using the wilcoxon signed-rank test	Recruiting

Continued

TABLE 5.3 Clinical trials of repurposed drugs in breast cancer.—cont'd

	Clinical trial	Condition	N1	N2	Design	Primary outcome	Status
31	NCT01042379	Breast neoplasms Breast cancer Breast tumors	800	1920	1. Active comparator: Paclitaxel, herceptin followed by doxorubicin, cyclophosphomide 2. AMG 386 with or without trastuzumab 3. AMG 479 (ganitumab) plus metformin 4. MK-2206 with or without trastuzumab 5. T-DM1 and pertuzumab 6. Pertuzumab and trastuzumab 7. Ganetespib 8. ABT-888 9. Neratinib 10. PLX3397 11. Pembrolizumab—4 cycles 12. Talazoparib plus irinotecan 13. Patritumab and trastuzumab 14. Pembrolizumab—8 cycles 15. SGN-LIV1A 16. Durvalumab plus olaparib 17. SD-101 + pembrolizumab 18. Tucatinib	Pathologic complete response (pCR) for experimental + standard neoadjuvant chemotherapy versus standard neoadjuvant chemotherapy	Recruiting
32	NCT02695121	Breast cancer Bladder cancer	NA		Dapagliflozin, insulin, metformin, sulfonylureas	Incidence of breast cancer and bladder cancer	Recruiting

33	NCT02201381	Cancer	2000	207	Experimental: Metabolic treatment Atorvastatin PO up to 80 mg OD + metformin up to 1000 mg PO OD, increased to BID after 2 weeks + Doxycycline 100 mg PO OD + mg PO OD.	Overall survival (OS)	Recruiting
34	NCT01302002	Breast cancer	30	0	Metformin 500 mg tablet, BID X 3 weeks	Determine the in situ effects of metformin on proliferation (Ki67) and apoptosis (TUNEL), phosphorylate AKT, CD1a CD83, CD68, F40/80, arginase iNOS and T cells CD4(+), CD45RA(+), CD 45RO, CD4, CD8, and FOXP3(+)	Withdrawn
35	NCT00984490	Breast cancer	30	5	Metformin 850 mg PO BID X 7–21 days	Change in Ki67 levels before and after treatment	Terminated (poor accrual)
36	NCT01627067	Breast cancer	40	23	Exemestane 25 mg OD + everolimus 10 mg OD + metformin 500 mg OD every 3 days—increase to 1000 mg BID if no toxicity	Progression-free survival (PFS)	Terminated (komen foundation funding terminated)
37	NCT01477060	Metastatic breast cancer	168	32	ARM A: Hormonal therapy + lapatinib (1250mg/die) ARM B: Hormonal therapy + metformin (1500 mg/die) ARM C: Hormonal therapy + lapatinib + metformin	Rate of patients free from disease progression	Terminated (poor accrual)
38	NCT02472353	Breast cancer Breast tumors	44	30	1. Active comparator—doxorubicin Experimental: Metformin + doxorubicin	Decrease in the incidence of change in left ventricle ejection fraction (LVEF)	Terminated (poor accrual)

Continued

TABLE 5.3 Clinical trials of repurposed drugs in breast cancer.—cont'd

	Clinical trial	Condition	N1	N2	Design	Primary outcome	Status
39	NCT02360059	Breast cancer	42	1	12 days prior to start of paclitaxel 1. Experimental arm: Metformin 500 mg OD X 5 days, followed by 500 mg BID X 5 days, followed by 1000 mg BID X 2 days Questionnaires; sensory and fine -motor tests 2. Placebo	Mean change in neuropathy	Terminated (poor accrual)
40	NCT00930579	Breast cancer	15	35	Metformin 1500 mg: divided 500 mg in the morning and 1000 mg in the evening, for at least 2 weeks prior to surgery	Effects of metformin on AMPK/mTOR signaling pathway	Unknown
41	NCT03192293	Breast cancer	28	28	Experimental: 7 days lead-in period: Metformin 850 mg OD + simvastatin 20 mg OD If well-tolerated, fulvestrant at standard doses: Cycle 1: 500 mg at day 1 and day 15/28 days cycle Cycle 2 and beyond: 500 mg at day 1/28 days cycle	Clinical benefit rate (CBR)	Unknown
42	NCT01666171	Breast cancer	458		Metformin hydrochloride clinical observation, diagnostic laboratory biomarker analysis, imaging biomarker analysis, medical chart review; procedure: Digital mammography	Change in percent mammographic breast density in contralateral (unaffected) breast from baseline to 1 year using two-sample t-test or wilcoxon rank-sum test	Unknown
43	NCT01566799	Locally advanced malignant neoplasm	60		Paclitaxel X 12 weeks followed by 4 cycles of FAC combined with metformin 500 mg PO OD X 24 weeks	Pathologic complete response (pCR)	Unknown

44	NCT01286233	Breast cancer Depression Fatigue Sleep disorders	454	394	Group 1: Metformin 850 mg PO BID X 5 years Group 2: Placebo PO BID X 5 years	Questionnaire scores about fatigue, stress, sleep, depression, general quality of life, and behavioral risks; biological correlates of fatigue DNA polymorphisms Changes in RNA gene expression	Unknown
45	NCT01302379	Breast neoplasms	340	333	1. Metformin: (week 1 - 500 mg PO OD, week 2–4–1000 mg PO OD, weeks 5+– 500 mg PO daily) + lifestyle intervention 2. Placebo + lifestyle intervention 3. Metformin + standard dietary guidelines 4. Placebo + standard dietary guidelines	Biological markers associated with breast cancer survival	Unknown
46	NCT03323346	Breast neoplasm Metastatic breast cancer	150	150	Disulfiram 400 mg PO daily + copper (2 mg elemental cu) PO daily	Clinical response rate (RR) Clinical benefit rate (CBR)	Recruiting
47	NCT01847001	Locally advanced malignant neoplasm Breast cancer	20	10	Propranolol (20 mg BID up to 40 mg BID) + neoadjuvant chemotherapy (regimen I: paclitaxel; Regimen II: doxorubicin and cyclophosphamide; if tumor is HER2 positive add trastuzumab and pertuzumab)	Percentage of patients compliant with taking > 80% take the drug while on chemotherapy	Active, not recruiting
48	NCT02596867	Breast cancer	30	2	Propranolol 1.5 mg/kg/day, BID X 3 weeks	Reduction the tumor proliferative index using Ki-67	Terminated (poor accrual)

Continued

TABLE 5.3 Clinical trials of repurposed drugs in breast cancer.—cont'd

	Clinical trial	Condition	N1	N2	Design	Primary outcome	Status
49	NCT00502684	Primary operable breast cancer	460	32	Experimental: propranolol 10 mg QID, starting on d-3 pre-op, X 6 days, till POD 2 + etodolac 400 mg BID, starting on d-3 pre-op, for 6 days, till POD 2 Placebo	Number and cytotoxic activity of NK cells, levels of NKT cells, lymphocytes, monocytes, and granulocytes; cytokine levels; in vitro cytokine secretion; levels of cortisol and VEGF. Cancer recurrence in 5 years	Unknown
50	NCT02013492	Male breast cancer, recurrent melanoma, Stage IV breast cancer, Stage IV melanoma, Stage IV ovarian epithelial cancer, Stage IV ovarian Germ cell tumor, unspecified adult solid tumor, protocol specific hepatocellular carcinoma	35	35	Propranolol PO BID X 4 months	Incidence of toxicity graded according to common criteria for adverse events (CTCAEs) V. 4.0 Change in vascular endothelial growth factor (VEGF) Effect of beta-adrenergic blockade on the tumor microenvironment and on the host immune system	Recruiting
51	NCT02649101	Metastatic breast cancer	60	60	Thalidomide tablet 100 mg QN, PO	Progression-free survival (PFS)	Unknown
52	NCT00193102	Breast cancer	40	40	Thalidomide	Overall response rate Time to disease progression	Terminated
53	NCT00049296	Cancer	26	26	Thalidomide PO, BID Docetaxel IV over 30 min once weekly	Determine maximum tolerated dose of docetaxel when administered with thalidomide	Completed
54	NCT00764036	Metastatic breast cancer Locally advanced breast cancer	18	23	Artesunate PO QD add-on therapy of 100, 150 or 200 mg	Dose-limiting adverse events with possible, probable, or definite relation with the respective dose level of the add-on therapy	

BID, twice a day; *FDC*, fixed dose combination; *HCL*, hydrochloride; *IGF*, insulin-like growth factor; *IM*, intramuscular; *IV*, intravenous; *N1*, Original estimated enrollment; *N2*, Estimated enrollment; *OD*, once daily; *PO*,per oral; *QD*, once daily; *QN*, every night; *TID*, three times a day; *XR*, extended release.

tendency of significant change in left ventricle ejection fraction [72].

An ongoing trial of MTF in postmenopausal obese women is studying the effect of MTF in the obesity-related breast cancer and breast density along with serum IGF-1 to IGFBP-3 ratio [73]. Another trial of MTF in women who have had a prior breast biopsy demonstrating atypical hyperplasia is studying the efficacy of MTF for prevention and alteration of RPFNA or blood biomarkers of atypical hyperplasia in unilateral or bilateral RPFNA aspirates of breast cancer [74].

The combined effect of MTF with 5-fluorouracil, doxorubicin, and cyclophosphamide is being studied in women with locally advanced breast cancer [75]. The pathomorphological and toxicological effect of MTF and toremifene is being studied in women above the age of 18 [76]. A presurgical trial of combination therapy of MTF and atorvastatin in women with operable DCIS breast cancer is studying the benefits of the combination as well as a change in the proliferation marker Ki-67 levels in the breast tissue of these patients [77]. In patients with ER+ breast cancer, the efficacy of neoadjuvant therapy of a combination of MTF and letrozole QD is currently being studied in a clinical trial. This trial also investigates the response rate of breast-conserving surgery and changes in Ki67 level after treatment with the combination [78].

An ongoing clinical trial with repurposed MTF along with taxotere, carboplatin, herceptin, and pertuzumab in patients with HER2+ breast cancer with cT1c-cT4a-d node without metastasis is studying the effect of the combination as adjuvant therapy [79].

An ongoing trial of MTF, doxorubicin, docetaxel, and trastuzumab focuses on the efficacy of the combination for operable and locally advanced HER2+ breast cancer. This trial evaluates pCR and systemic tolerance, with precise consideration to cardiac toxicity [80]. The clinical benefit rate and toxicity profile of combination therapy of MTF and simvastatin are currently being studied in a trial for ER+ breast cancer patients [81]. The pCR after combination therapy of MTF with docetaxel, epirubicin, and cyclophosphamide is also being studied in breast cancer patients with a life expectancy of >12 months [82].

An ongoing trial of a combination of MTF and doxycycline (DOX) is being studied in women with localized breast or uterine cancer. The main purpose of the study is to analyze the change in stromal cells expressing caveolin-1 at an intensity of 1+ by immunohistochemistry [83].

Metformin—summary

MTF has been off-patent protection since 2002 and can be easily synthesized with dimethylamine hydrochloride and 2-cyanoguanidine. It is known to have well-tolerated side effects making its compliance better than other anticancer medications. By reducing insulin, MTF reduces the anabolic capability of the body, which in turn reduces tumor growth. In the tumor cells, MTF alters the high energy metabolism, reducing their ability to survive and proliferate. The hydrophilic characteristics of the drug facilitate its direct transport through the cell membrane and its action via OCTs, plasma monoamine transporter, and multidrug and toxin extrusion protein.

Disulfiram

Disulfiram (DSF, tetraethylthiuram disulfide, or Antabuse; Fig. 5.1) is one of the most popular repurposed drugs. In the beginning, it was used in the rubber industry as an industrial catalyst [9], then it was introduced into medicine as scabicide and vermicide [84]. It was accidentally found to be a bad combination with alcohol when it was used in a study to treat stomach ailments. The US FDA-approved DSF in 1951 as Antabuse in alcohol deaddiction (Table 5.1). For about 50 years, DSF was used as an antialcoholism drug and now is being repurposed in cancer treatment [85,86] (Table 5.1). DSF reposition as a potential anticancer drug is based on its unique properties in the human body, e.g., low cost, fewer side effects, high selectivity against different cancers, and synergistic activity with other drugs [85]. Moreover, DSF showed proteasome inhibition and suppressed different cancer-associated pathways [9,87,88].

Disulfiram—mechanism of action of the original use

DSF is the best-known aldehyde dehydrogenase (ALDH) irreversible inhibitor [4]. Inhibition of ALDH by DSF is through the formation of an intramolecular disulfide bond by two mechanisms: (i) between DSF and an active site thiol in the enzyme or (ii) between the active site thiol and thiol of another cysteine residue via unstable mixed disulfide adduct [89]. When alcohol enters the body, it is converted into acetaldehyde by alcohol dehydrogenase and then rapidly metabolized into acetic acid by the action of ALDH in the liver. DSF was reported to block the action of ALDH that causes an increased serum acetaldehyde concentration, which is toxic and results in an unpleasant DSF-ethanol reaction to the individual (Table 5.1).

Disulfiram—preclinical studies

Since the 1970s, scientists have noticed the tumor-suppressive effect of DSF [87]. Many anticancer properties of DSF have been demonstrated in different preclinical models of breast, prostate, myeloma, leukemia, lung cancer, cervical adenocarcinoma, melanoma, neuroblastoma, and colorectal cancer [85] (Table 5.1).

Disulfiram complexes with metals

DSF, as a thiuram disulfide of dithiocarbamate, could form complexes with metals. The interaction of DSF with copper (II) chloride in solution reveals a rapid formation of the bis (N, N-diethyldithiocarbamate) copper (II) complex in situ [89]. The anticancer activity of DSF is copper-dependent [90]. Up to 10 μM, DSF alone did not exhibit cytotoxic effect in cancer cells [91], while DSF/Cu at lower concentrations exerted anticancer activity [90] and suppressed cell proliferation, tumor growth, invasion, and migration [84,90], in addition to eliminating tumor-initiating cells and inhibiting colony formation and tumor formation in different animal models [92]. The antiangiogenic effect was also seen following DSF/Cu treatment via reducing microvessel density and VEGF expression [9]. DSF with exogenous copper had better antitumor effect when compared to DSF alone in a breast cancer xenograft model [93].

DSF–Cu complex in breast cancer stem cells

CSCs are a subpopulation of cells that have the ability of self-renewal and differentiation. Levels of CSCs correlate with tumorigenesis, metastasis, radio-/chemoresistance, and cancer recurrence clinically [94,95]. DSF exhibited a highly cytotoxic effect on breast cancer stem cells (BCSCs), which are considered a significant cause of chemoresistance that leads to the failure of breast cancer chemotherapy.

DSF–Cu complex exhibited proteasome inhibition and induction of apoptosis in breast cancer cell cultures. It suppressed the growth of breast cancer xenografts, which is independent of the status of PIK3CA, the gene encoding α form of class IA PI3Ks, that heightened and overexpressed in a variety of human cancers, including breast cancer. Treatment of a DSF–Cu mixture decreased the expression of PTEN protein in a time- and dose-dependent manner and activated AKT in cell lines regardless of the presence of PIK3CA mutations, which are thought to lead to the activation of the AKT signaling pathway. Therefore, DSF appears to trigger two conflicting signaling pathways in breast cancer cells. First, DSF–Cu inhibits the proteasome, thus bringing about the death pathway and inducing apoptosis. Second, DSF–Cu activates the PI3K/PTEN/AKT survival signaling pathway [96].

HER2 is an important determinant of survival for BCSCs that are associated with a high risk of tumor recurrence [97,98]. HER2 overexpression [99] is able to alter the apoptotic and molecular signaling pathways [100]. DSF/Cu combination was reported to inhibit HER2/Akt signaling and eliminated BCSCs, suggesting the potential effectiveness of DSF for HER2-positive breast cancer treatment [101].

DSF–Cu complex for TNBC therapy

TNBC represents an aggressive subtype, for which radiation and chemotherapy are the only options [102]. Acquired chemoresistance remains the primary cause of therapeutic failure of TNBC [103]. In the clinic, the relapsed TNBC is commonly pan-resistant to various drugs with entirely different resistance mechanisms. Investigation of the resistance mechanisms and development of new drugs to target pan-chemoresistance will potentially improve the therapeutic outcomes of TNBC patients [104].

The MDA-MB-231$_{PAC10}$ cell line is made up of a high population of cells expressing stem cell markers that may play a vital role in the pan-resistance. These cells express high ALDH activity and a panel of embryonic stem cell–related proteins (e.g., Oct4, Sox2, Nanog) and nuclearization of HIF2a and NF-kBp65. These cells are highly cross-resistant to PAC, cisplatin (CDDP), docetaxel, and doxorubicin. DSF was reported to abolish CSC characters and completely reverse PAC and CDDP resistance in MDA-MB-231$_{PAC10}$ cells (Table 5.2). In addition, DSF/Cu exposure for 4 h leads to inhibition of both ALDH activity and expression of Sox2 and Nanog in the resistant cells. In combination with DSF/Cu, the cytotoxicity of PAC and CDDP in MDA-MB-231$_{PAC10PAC10}$ cells was significantly higher than PAC, CDDP, or DSF/Cu single-drug exposure. The cytotoxicity of DSF/Cu plus PAC was synergistic in a wide range of concentrations [104].

DSF/Cu combination induced the expression of Bax while inhibited the expression of Bcl2 in MDA-MB-231$_{PAC10}$ cells, which lead to a significant increase in Bax/Bcl2 ratio in the resistant cell line. Expression of p21 and p53 proteins was also enhanced by DSF/Cu combination, which did not effect on that of CDK2 and cyclin D1 and E [104].

Analysis of the impact of various treatments on CSCs population in MDA-MB-231 and T47D cell lines revealed that formation of mammosphere from both cell lines was blocked entirely by exposure to the combination of DSF (1 μM)/Cu (1 μM) plus PAC (40 nM) for 48 h but not affected by PAC, DSF, or Cu alone. In addition, DSF/Cu, but not DSF or Cu, treatment significantly inhibited the ALDH-positive population in mammospheres [88].

DSF–Cu complex for the treatment of inflammatory breast cancer

Although there have been advances in multimodality treatment, inflammatory breast cancer (IBC) is a discrete, advanced BC subtype characterized by high rates of residual disease and recurrence [105,106]. Metagene analysis of patient samples revealed significantly higher oxidative stress response (OSR) scores in IBC tumor samples when compared to normal or non-IBC tissues. These higher OSR scores were responsible for the inadequate response of IBC tumors to standard treatment strategies. DSF–Cu antagonized NF-kB signaling [107], ALDH activity, and antioxidant levels [88], causing the induction of oxidative stress–mediated apoptosis in multiple IBC cellular models [93].

Under in vivo condition, DSF–Cu caused apoptosis and inhibited growth exclusively in tumor cells without significant toxicity. It has been suggested that strong redox adaption of IBC tumors may contribute to their resistance to ROS-inducing therapies. DSF, through redox modulation, may be advantageous in the manner of enhancing chemo- and/or radiosensitivity for extreme BC subtypes that showed therapeutic contention. DSF's potency was significantly enhanced by the addition of 10 μM Cu (DSF–Cu), with a decrease of roughly 100-fold in IC_{50} values in SUM149 and rSUM149 (Table 5.2). Remarkably, DSF-Cu caused cell death in redox-adapted rSUM149 cells at levels comparable to parental redox-sensitive SUM149 cells [93].

DSF–Cu complex is nontoxic, at up to 20 μM , to normal, immortalized breast cells (MCF10A). Similar to SUM149 cells, DSF-induced, Cu-dependent cell death was observed in other IBC cell lines tested, which included MDA-IBC-3 (HER2-overexpressing), SUM190 (HERer2-overexpressing, ROS sensitive), and rSUM190 (an isogenic derivative of SUM190 with therapeutic resistance, redox adaptation) (Table 5.2). SUM190 cells, even at extremely reduced concentrations of DSF, were exceedingly susceptible to DSF–Cu treatment. In the presence of DSF–Cu, the lessening of IBC cell viability corresponded with a drop off in levels of XIAP, the most potent mammalian caspase inhibitor and antiapoptotic protein. This is coherent with the induction of oxidative stress triggering the intrinsic apoptotic pathway. XIAP overexpression in IBC cells was identified to be correlated with resistance to therapeutic apoptosis. DSF–Cu-mediated cell death was also related to the decreased expression of eIF4G1 (a disease pathogenesis factor identified in IBC tumors) and increased PARP cleavage. These findings suggested the induction of apoptotic cell death being a fundamental mode of action of DSF–Cu [93].

Addition of bathocuproine disulfonate or tetrathiomolybdate (TTM), two high affinity chelators to sequester free Cu, hindered DSF–Cu-induced cell death altogether in SUM149 and SUM149 cells, thus showing the importance of Cu binding in the intensification of DSF's cytotoxic effects. Although TTM is able to induce cancer cell death through sequestering Cu, the TTM–Cu mixture showed no effect in these models of IBC. These data confirmed that the activity of DSF is not related to Cu sequestration, but rather a gain of function that results in increased ROS production [93].

Han et al. reported that without copper supplementation, exogenous SOD potentiated subtoxic DSF toxicity antagonized by subtoxic TTM or by the antioxidant N-acetylcysteine. Exogenous glucose oxidase, another H_2O_2 generator, paralleled exogenous SOD in potentiating subtoxic DSF. It was confirmed that potentiation of sublethal DSF toxicity by extracellular H_2O_2 against the human tumor cell lines only needed basal Cu and heightened ROS production. This is critical as these findings accentuate the importance of extracellular H_2O_2 as a novel mechanism to improve the anticancer effects of DSF while minimizing copper toxicity [108]. The authors also documented that DSF could significantly suppress the TGF-β–induced upregulation of vimentin and N-cadherin as well as downregulation of E-cadherin in a dose-dependent manner. These findings suggested that DSF can inhibit TGF-β–induced EMT by modifying the expression of EMT-related proteins.

DSF–Cu complex as a BC proteasome inhibitor

DSF was able to affect 20S proteasome activity at low micromolar concentrations, but through a mechanism different from that of the authentic proteasome inhibitors as bortezomib or MG132 that block the active site of the proteasome. DSF acts at some other sites on the proteasome as a slow-binding partial noncompetitive inhibitor [109]. However, DSF and its combination with copper exhibited a strong inhibitory effect on 26S proteasomes [110], and the inhibition was copper-dependent [84].

MDA-MB-231 BC cells with high levels of cellular copper showed proteasome inhibition in the presence of DSF alone, which suggested that copper acts as an endogenous element in the cell, which can form a complex with DSF and induce cell death. Similar findings were observed in mice bearing MDA-MB-231 tumor xenografts treated with DSF that inhibited tumor growth and proteasome activity and induced apoptosis in vivo [84] (Table 5.2).

DSF–cadmium complex

A study reported that breast cancer MCF-10 DCIS cells were more sensitive to DSF–Cd-induced 20S proteasome inhibition and apoptosis induction than nontumorigenic human breast MCF-10A cells [111,112]. Thus, the DSF–Cd complex could selectively induce proteasome inhibition and apoptosis in human breast tumor cells [113].

DSF and zinc availability

The availability of extracellular zinc was suggested to influence DSF efficacy significantly. Live cell confocal microscopy using fluorescent endocytic probes and the zinc dye FluoZin-3 showed that DSF elevated zinc levels selectively and speedily in endolysosomes. DSF was reported to cause the spatial disorganization of late endosomes and lysosomes, suggesting that they could be used as novel targets for DSF. Moreover, DSF was shown to increase the intracellular zinc levels in breast cancer cells specifically. It was reported that 10–100 μM DSF greatly heightened intracellular zinc levels in both MCF-7 and MDA-MB-231 cell lines (Table 5.2), while zinc levels in the noncancerous MCF-10A cells remained unaffected by the same treatment. In serum-free media containing low zinc and copper, neither sodium pyrithione nor DSF induced a statistically significant increase in intracellular zinc in MCF-7 cells. Supplementation of serum-free media with 20 μM zinc was enough to restore completely, and exaggerate, the ionophore ability of both DSF and sodium pyrithione, demonstrating that this ionophore activity is dependent on extracellular zinc levels [114].

Disulfiram combination with other drugs

DSF combination with doxorubicin, daunorubicin, mitoxantrone, colchicine, or paclitaxel

Robinson et al. reported that DSF, as well as the related compound thiram, were identified as the most potent growth inhibitors against multiple TNBC cell lines. Combination treatment with four different drugs commonly used to treat TNBC revealed that DSF synergizes most effectively with doxorubicin to inhibit cell growth of TNBC cells. DSF and doxorubicin cooperated to induce cell death as well as cellular senescence and targeted the ESA+/CD24−/low/CD44+ CSC population. The results suggested that DSF may be repurposed to treat TNBC in combination with doxorubicin [102].

Studies performed on HCC70, MDA-MB-231, MDA-MB-436, and Bt549 cell lines revealed that DSF was more effective against each cell line than doxorubicin, daunorubicin, mitoxantrone, colchicine, or PAC. Notably, MDA-MB-436 cells were resistant to the mitotic inhibitors, colchicine, and PAC but highly susceptible to DSF. In a panel of 13 human-derived TNBC lines, both DSF and thiram effectively suppressed the growth of TNBC cells, with an average IC_{50} across all lines of 300 and 360 nM, respectively (Table 5.2). Similar effect was seen when these drugs were used in basal-A or basal-B TNBC cell lines. These and other results all indicate that DSF/Cu is cytotoxic and nonspecific to breast cancer cells and can kill all subtypes of breast cancer.

There was a strong synergistic effect between DSF/Cu and PAC over a wide range of concentrations (IC_{50}– IC_{90}). DSF/Cu significantly enhanced (3.7- to 15.5-fold) the cytotoxicity of PAC in BC cell lines. Contrary to the slight induction of apoptosis at low concentration of PAC alone (1 nM), the proportion of apoptotic cells was greatly elevated by DSF/Cu (DSF 100–150 nM/Cu 1 μM) and PAC in combination. DSF/Cu-induced cytotoxicity in BC cell lines was caused by ROS activation and was reversed by the addition of NAC, a ROS inhibitor, in the culture [88].

DSF/Cu combination was also seen to inhibit NF-kB activity in BC cell lines. The NF-kB is a ROS-induced transcription factor with substantial antiapoptotic activity, which in turn dampens the proapoptotic effect of ROS [115]. Therefore, the blocking of NF-kB activation enhances ROS-induced cytotoxicity. Both PAC and DSF/Cu inhibited NF-kB DNA-binding activity in BC cell lines. The most potent inhibition was observed in the BC cells treated with PAC/DSF/Cu in combination.

The clonogenicity of BC cell lines was significantly inhibited by DSF/Cu and eradicated by exposure to PAC plus DSF/Cu [88]. Furthermore, inhibition of ALDH activity by DSF has been suggested to contribute to reduce CSCs and overcome drug resistance [116].

DSF–Cu combination with sunitinib

The combination of DSF and antiangiogenic agent sunitinib was studied in more detail. DSF–sunitinib combination induced apoptosis and reduced androgen receptor protein expression more than either of the compounds alone. Moreover, combinatorial exposure reduced metastatic characteristics such as cell migration and 3D cell invasion, as well as induced epithelial differentiation (e.g., elevated E-cadherin expression). Androgenic and antioxidative compounds were reported to antagonize the DSF effect. However, inhibitors of receptor tyrosine kinase, proteasome, topoisomerase II, glucosylceramide synthase, or cell cycle were among compounds sensitizing prostate cancer cells to DSF [117].

DSF with cisplatin

DSF was reported to enhance the cytotoxicity of CDDP via modulation of ROS accumulation in breast cancer [118]. DSF inhibits ALDH activity and the expression of stemness-related transcription factors (Sox, Nanog, Oct) in CSC derived from breast cancer cell lines. It also modulates intracellular generation of ROS. ALDH+ stem-like cells were involved in resistance to the conventional chemotherapeutic agent CDDP. DSF was reported to enhance the cytotoxic effect of CDDP through inhibiting the stemness and overcoming CDDP resistance in ALDH+ stem-like cells [118].

Disulfiram delivery approach

The anticancer application of DSF is bound because it has a very short half-life in the bloodstream, which prompted to develop a liposome-encapsulated DSF (Lipo-DS). Lipo-DS was shown to be able to block NF-κB activation and specifically target CSCs in vitro and in vivo with strong anticancer efficacy. Lipo-DS/Cu was highly cytotoxic to mesenchymal stem cells (MSCs), which are resistant to conventional anti-BC drugs. Further, it induced Bax expression and inhibited Bcl2 expression. The ability of sphere formation was also completely eradicated after 4 h of exposure to Lipo-DS/Cu but not Lipo-DS nor Cu alone. The ALDH+ and CD24 low/CD44 high CSC population in the mammospheres were eliminated by Lipo-DS/Cu but not Lipo-DS, Cu, or anti-BC drugs. In addition, Lipo-DS/Cu induced ROS activity in MSCs, which was reversed by NAC, a ROS inhibitor. After a 4-h exposure to Lipo-DS/Cu, the major MAPK pathway elements, e.g., phosphorylated JNK, phosphorylated C-JUN, and phosphorylated p38, but not ERK, were greatly induced. However, in contradiction with the MAPK pathway, Lipo-DS/Cu inhibited IκBα degradation and blocked NF-κB p65 nuclear translocation in both MCF7 and T47D CSCs [119].

Biodegradable and controllable drug delivery carriers have been designed to overcome some disadvantages, such as poor solubility and rapid metabolism, in gastrointestinal fluid by providing sustained release of DSF [120]. FDA-approved poly-lactic-c-glycolic acid (PLGA) is one such a biodegradable polymer. Encapsulation of DSF by PLGA could protect DSF from degradation in blood and therefore enhance its anticancer activity [121,122]. PLGA–polyethylene glycol nanoparticles were developed for encapsulation and delivery of DSF into breast cancer cells [123,124].

Encapsulation of DSF with nanoparticles could protect the thiol groups of DSF and extend the half-life of DSF in the serum, thus delivering intact DSF to cancer tissues [125]. Nanoencapsulated DSF, combined with oral administrated Cu gluconate, exhibited significant anticancer activity against breast, liver, lung, and brain tumors in the mouse model [126,127]. The FDA has approved liposomes as one of the nanoplatforms for cancer therapy due to its biodegradability, biocompatibilities, and low toxicity [128]. Lipo-DS was able to target the breast CSC population, exhibiting potent anticancer activity via inhibition of the NF-kB pathway [119]. The oral administration of DS/copper gluconate (CuGlu) was compared with intravenous administration of Lipo-DS plus oral administration of CuGlu. In comparison with the control group, both Lipo-DS/CuGlu and Lipo-DS significantly inhibited tumor growth but Lipo-DS/CuGlu demonstrated highest anticancer efficacy. Oral administration of DS/CuGlu also showed a tumor-inhibiting effect, although it was not statistically significant. Lipo-DS/CuGlu, Lipo-DS, and DS/CuGlu induced Bax, inhibited Bcl-2, and induced TUNEL staining. Also, the ALDH expression in tumor tissues was inhibited when the animals were treated with Lipo-DS/CuGlu and Lipo-DS [119].

In vivo, acute, and chronic toxicity of nanoparticles and their efficacy on inhibition of breast cancer tumor growth were investigated in another study. The results showed that the folate receptor–targeted nanoparticles were internalized into the cells, induced ROS formation, induced apoptosis, and inhibited cell proliferation more efficiently compared to the nontargeted nanoparticles. The maximum dose of DSF equivalent of nanoparticles for intravenous injection is 6 mg/kg while showed a significant decrease in the tumor growth rate. It is believed that the developed formulation could be used as a potential vehicle for the successful delivery of DSF, an old and inexpensive drug, into breast cancer cells, and other solid tumors [129].

DSF is metabolized to active diethyldithiocarbamate (DDTC) in the body by reduction. A redox-sensitive DDTC–polymer conjugate was developed for targeted cancer therapy. It was found that the DDTC–polymer conjugate modified with a β-D-galactose receptor targeting ligand can self-assemble into LDNP nanoparticle and efficiently enter cancer cells by receptor-mediated endocytosis. Upon cellular uptake, the LDNP nanoparticle degrades and releases DDTC due to the cleavage of disulfide bonds and subsequently forms a copper (II)–DDTC complex to kill a broad spectrum of cancer cells. 3D cell culture revealed that this nanoparticle has much stronger tumor mass penetrating and destructive capacity. Furthermore, LDNP nanoparticles exhibited great potency in inhibiting tumor growth in a peritoneal metastatic ovarian tumor model [130].

Disulfiram—clinical studies

A nationwide epidemiological study demonstrated that patients who continuously used DSF have a lower mortality risk from cancer compared to those who stopped using DSF at their diagnoses [90].

In one epidemiological study, researchers studied whether the use of DSF was associated with a change in cancer mortality at a population level due to the relative scarcity of cancer-related clinical trials with DSF. The hazard ratios (HR) of cancer-specific mortality associated with DSF use among first-time cancer patients during 2000–13 were estimated using the Danish nationwide demographic and health registries. DSF users were put into two groups, as (i) "previous users" for patients who were only prescribed DSF for alcohol dependence before their cancer diagnosis and (ii) "continuing users" for patients prescribed DSF both before and after the diagnosis. As expected from the harmful effects of alcohol abuse on cancer risk and prognosis, cancer-specific mortality was higher among previous DSF users than among never users of DSF. Notably, the authors also found reduced cancer-specific mortality for cancer overall, as well as for cancers of the colon, prostate, and breast among continuing users compared to previous DSF users. It was revealed by stratification by the clinical stage that reduced cancer-specific mortality with ongoing use of DSF was reduced, even among patients with metastatic disease. While not allowing conclusions about causality, these findings supported the hypothesis that DSF may exert anticancer effects among patients suffering from common cancers, prompting researchers to perform preclinical analyses [90].

Multiple clinical trials have studied the effect DSF in various cancers. There are currently 19 clinical trials of DSF in various cancers of which 8 have been completed. However, there is only one clinical trial that is studying the effect of DSF in breast cancer, which is currently in its Phase II stage and is still recruiting. The purpose of this study is to establish clinical evidence for introducing DSF and Cu as an active therapy for metastatic breast cancer upon failure of conventional systemic and/or locoregional therapies [131] (Table 5.3).

DSF has earlier been studied in cancers like glioblastoma, melanoma, prostate cancer, and lung cancer [132]. DSF has been shown to potentiate the effect of several chemotherapeutic agents in vitro, including CDDP, gemcitabine, temozolomide, PAC, docetaxel, cyclophosphamide, 5-fluorouracil, doxorubicin, sunitinib, and BCNU [4]. DSF has shown good tolerance in Phase I trials involving solid tumors. DSF in combination with chemotherapeutics like CDDP and vinorelbine was well tolerated and appeared to prolong survival in patients with newly diagnosed non-small cell lung cancer [133].

In glioblastoma (GBM) patients who received standard chemoradiotherapy, DSF monotherapy and its combination with temozolomide was studied. Although this combination was safe, a few reversible neurological toxicities were observed [110].

The safety and efficacy of combining DSF with CDDP and vinorelbine were also studied in a Phase II, multicenter, randomized, double-blinded trial. In the DSF group, there was prolonged survival. But, there were only two long-term survivors [134].

Disulfiram—summary

DSF is the most popular repurposed drug for cancer therapy; this is because it shows great potential for cancer treatment due to its multiple mechanisms. The mechanisms and targets of cancer therapy proposed for DSF include inhibition of the UPS, alteration of intracellular ROS, and suppression of cancer cell stemness (Table 5.1). In addition, DSF also induces tumor cell apoptosis, inhibits proliferation of cancer cells, and suppresses cancer cell metastasis. These advantages make DSF quite an ideal and useful repurposed drug for breast cancer therapy.

Some limitations of DSF include that, as a repurposed drug, the odds of DSF being widely used in cancer therapy are very low as there is no patent protection on the drug, and therefore the pharmaceutical companies practically will not push DSF into mass use.

Propranolol

Propranolol—mechanism of action of the original use

Since its FDA approval in 1967 for the management of hypertension, propranolol (Fig. 5.1) is a noncardioselective β-blocker used for the treatment of high blood pressure, irregular heartbeats, shaking (tremors), and other conditions (Table 5.1). It is used after a heart attack to improve the chance of survival. It is also used to prevent migraine headaches and chest pain (angina) [135].

Propranolol is highly lipophilic and an orally bioavailable drug. It is completely absorbed in the GI tract and metabolized in the liver. Only 25% of the drug reaches the systemic circulation. It is a nonselective β-blocker that blocks the action of catecholamines (adrenaline and noradrenaline) at both beta-1 and beta-2 adrenergic receptors. By blocking the beta-adrenergic sites, propranolol inhibits sympathetic effects that act through these receptors.

Propranolol—preclinical studies

Breast cancer tissue expresses beta-adrenergic receptors [136], particularly type 2 beta-adrenergic receptors [137]. In vitro and in vivo studies have shown the ability of β-blockers to inhibit cancer cell migration and angiogenesis [138]. Preclinical studies have documented the inhibitory effects of propranolol on several pathways involved in breast cancer progression and metastasis [139].

MDA-MB-231, MCF-7, and 4T1 breast cancer cells (Table 5.2) were positive for ADRB1/2 expression. After propranolol treatment, HK-2 expression and its posttranscriptional level were significantly decreased in the BC cells. In addition, mice treated with propranolol showed decreased levels of tumor/nontumor values and brown adipose tissue as well as reduced HK-2 expression, compared to the control group [140]. Breast cancer metastases in mice were increased by 30-folds after exposure to chronic stress with neuroendocrine activation. This spread was inhibited by propranolol treatment [141].

ABC7 regimen (adjuvant for breast cancer treatment using seven repurposed drugs) was developed as a new approach for metastatic breast cancer. It aims to defeat aspects of epithelial–mesenchymal transition (EMT) that lead to the dissemination of breast cancer to bone. As an add-on to current standard treatment with capecitabine, ABC7 uses ancillary attributes of seven already-marketed noncancer treatment drugs to stop both the natural EMT process inherent to breast cancer and the added EMT occurring as a response to current treatment modalities. Chemotherapy, radiation, and surgery provoke EMT in cancer generally and in breast cancer specifically. ABC7 uses standard doses of capecitabine as used in treating breast cancer today. In addition, the mechanisms of action of ABC7 includes (1) using an older psychiatric drug, quetiapine (QUE), to block RANK signaling; (2) using pirfenidone, an antifibrosis drug to block TGF-beta signaling; (3) using rifabutin, an antibiotic to block beta-catenin signaling; (4) using MTF, a first-line antidiabetic drug to stimulate AMPK and inhibit mTOR; (5) using propranolol, a β-blocker to block beta-adrenergic signaling; (6) using agomelatine, a melatonergic antidepressant to stimulate M1 and M2 melatonergic receptors; and (7) using ribavirin, an antiviral drug to prevent eIF4E phosphorylation. Agonism at MT1 and MT2 melatonergic receptors has been shown to inhibit both breast cancer EMT and growth. All of these block the signaling pathways—RANK, TGF-beta, mTOR, beta-adrenergic receptors, and phosphorylated eIF4E—that have been shown to trigger EMT and enhance breast cancer growth and so are worthwhile targets to inhibit. This ensemble was designed to be safe and augment capecitabine efficacy. Given the expected outcome of metastatic breast cancer as it stands today, ABC7 warrants a cautious trial [142].

Propranolol—clinical trials

Barron et al. queried the Linked National Cancer Registry and prescription dispensing data to identify women with a diagnosis of Stage I to IV invasive breast cancer between January 1, 2001, and December 31, 2006 (Table 5.3). They compared women taking propranolol or atenolol in the year before breast cancer diagnosis to women not taking a β-blocker and assessed the risk of local tumor invasion at diagnosis (T4 tumor), nodal or metastatic involvement at diagnosis (N2/N3/M1 tumor), and time to breast cancer–specific mortality. It was found that propranolol users had a lower risk of developing a T4 (odds ratio [OR], 0.24; 95% confidence interval [CI], 0.07 to 0.85) or N2/N3/M1 (OR, 0.20; 95% CI, 0.04 to 0.88) tumor compared with matched nonusers. Propranolol users also had a lower cumulative probability of breast cancer–specific mortality compared with matched nonusers (HR, 0.19; 95% CI, 0.06 to 0.60). There was no difference in T4 or N2/N3/M1 tumor incidence or breast cancer–specific mortality between atenolol users and matched nonusers. The study concluded with statistics in favor of using propranolol to lower the risk of breast cancer progression and mortality [136].

Using the National Cancer Data Repository in England, Cardwell et al. conducted a nested case–control study of breast cancer patients. Their aim was to investigate the association between postdiagnostic β-blocker usage and risk of cancer-specific mortality. 18.9% of 1435 breast cancer–specific deaths and 19.4% of their 5697 matched controls were identified to be using β-blockers postdiagnosis. This showed that there is lower association between β-blocker use and breast cancer-specific mortality (OR 1/4 0.97; 95% CI 0.83, 1.13). There was also little evidence of an association when analyses were restricted to cardio nonselective β-blockers (OR 1/4 0.90; 95% CI 0.69, 1.17). Similar results were observed in analyses of drug dosage frequency and duration and β-blocker type [142].

In addition, another clinical study investigated whether breast cancer patients who used propranolol or other nonselective β-blockers had reduced breast cancer–specific or all-cause mortality in eight European cohorts. In 55,252 and 133,251 breast cancer patients, there was no association between propranolol use after diagnosis of breast cancer and breast cancer–specific or all-cause mortality (fully adjusted HR = 0.94; 95% CI, 0.77, 1.16 and HR = 1.09; 95% CI, 0.93, 1.28, respectively). There was also no association between propranolol use before breast cancer diagnosis and breast cancer–specific or all-cause mortality (fully adjusted HR = 1.03; 95% CI, 0.86, 1.22 and HR = 1.02; 95% CI, 0.94, 1.10, respectively) (Table 5.3). This study concluded that in this large pooled analysis of breast cancer patients, the use of propranolol or nonselective β-blockers was not associated with improved survival [139].

Sorenson et al. studied the association between use of β-blockers and breast cancer recurrence in a large Danish cohort of 18,733 women diagnosed with nonmetastatic breast cancer between 1996 and 2003. In any β-blocker users, the recurrence HR in unadjusted models was 0.91 (95% CI, 0.81–1.0) and in adjusted models was 1.3 (95% CI, 1.1–1.5). Besides metoprolol and sotalol, which showed increased recurrence rates (adjusted metoprolol HR 1.5; 95% CI, 1.2–1.8; adjusted sotalol HR 2.0; 95% CI, 0.99–4.0), most β-blockers had no association with recurrence. Thus, their analysis concluded that β-blockers have no association with attenuation of breast cancer recurrence risk [143].

Propranolol—summary

Preclinical studies showed that propranolol inhibits several pathways that are involved in breast cancer progression and metastasis. However, in the mentioned clinical studies, there was little to no improved survival or improved mortality in breast cancer patients who used propranolol or other nonselective β-blockers. This shows that although propranolol showed promise in inhibiting breast cancer–involved pathways in preclinical studies, once actually applied in clinical studies, there was virtually no impact on the health of patients.

Antiviral drugs

Antivirals—mechanism of action of original use

Drugs such as ombitasvir, paritaprevir, dasabuvir, and ritonavir are among antiviral drugs (Fig. 5.1). Ombitasvir, dasabuvir, and paritaprevir were FDA approved in July 2016 for the treatment of hepatitis C (Table 5.1). Ombitasvir is a direct-acting antiviral medication used as part of combination therapy to treat chronic hepatitis C, an infectious liver disease caused by infection with the hepatitis C virus (HCV). Paritaprevir is a direct-acting antiviral medication used as part of combination therapy to treat chronic hepatitis C. Ritonavir was FDA approved in 1999 for the treatment of HIV AIDS.

Dasabuvir is a nonnucleoside inhibitor of the HCV nonstructural protein 5B (NS5B), an RNA-dependent RNA polymerase, with potential activity against HCV. It is primarily metabolized by cytochrome P450 (CYP) 2C8, with a minor contribution from CYP3A, and is a substrate of P-gp and breast cancer–resistance protein (BCRP), as well as an inhibitor of BCRP and uridine diphosphate glucuronosyltransferase (UGT) 1A1. Ritonavir is an HIV protease inhibitor that interferes with the reproductive cycle of HIV.

Antiviral drugs—preclinical studies

Antiviral agents like ribavirin have shown antiproliferative activity in cancers through various mechanisms like inhibition of transcriptional and translational factors, suppression of telomerase activity, and activation of human DNA polymerase [144] (Table 5.1).

Ombitasvir, a hepatitis C NS5B polymerase inhibitor, was reported to be "repurposed" for control and prevention of beta-tubulin–driven breast cancers [145]. Dasabuvir was reported to inhibit P-gp and BCRP (Table 5.1). However, its net effect on transporter inhibition occurs when combined with paritaprevir and ritonavir, which are also P-gp and BCRP inhibitors [146].

To assess drug–drug interaction (DDI) potential for the three direct-acting antiviral (3D) regimen of ombitasvir, dasabuvir, and paritaprevir, in vitro studies profiled drug-metabolizing enzyme and transporter interactions. DDI potential was predicted for CYP3A, CYP2C8, UGT 1A1, organic anion-transporting polypeptide (OATP) 1B1/1B3, BCRP, and P-gp. When perpetrator interactions were assessed, ritonavir was responsible for the substantial increase in exposure of sensitive CYP3A substrates, whereas paritaprevir (an OATP1B1/1B3 inhibitor) significantly increased the exposure of sensitive OATP1B1/1B3 substrates. The 3D regimen drugs were UGT1A1 inhibitors and were predicted to moderately increase plasma exposure of sensitive UGT1A1 substrates. Plasma exposures of the 3D regimen were reduced by potent CYP3A inducers (paritaprevir and ritonavir; major CYP3A substrates) but were not affected by strong CYP3A4 inhibitors, since ritonavir (a CYP3A inhibitor) is already present in the regimen. Potent CYP2C8 inhibitors increased plasma exposure of dasabuvir (a primary CYP2C8 substrate), OATP1B1/1B3 inhibitors increased plasma exposure of paritaprevir (an OATP1B1/1B3 substrate), and P-gp or BCRP inhibitors (all compounds are substrates of P-gp and/or BCRP) increased plasma exposure of the 3D regimen. Overall, the comprehensive mechanistic assessment of compound disposition along with mechanistic and PBPK approaches to predict victim and perpetrator

DDI liability may enable better clinical management of several drug combinations with the 3D regimen [147].

MCF7 and T47D are ER-positive, estradiol-dependent lines, while MDA-MB-231 is ER-negative line (Table 5.2); all of the BC cell lines were sensitive to ritonavir treatment, suggesting that antiviral drugs are nonspecific to breast cancer cells and can kill all subtypes of breast cancer. Ritonavir-induced G_1 arrest in addition to depletion of cyclin kinases 2, 4, and 6 and cyclin D1 but not cyclin E. It also caused depletion of phosphorylated Rb and ser473 Akt. MDAMB231 xenograft and intratumoral Akt activity were suppressed after ritonavir treatment. Ritonavir was also reported to bind to heat shock protein 90 (HSP90) and inhibit its chaperone partially. It could block the association of HSP90 with Akt and could notably deplete HSP90. Sustained exposure to ritonavir downregulated ER in ER-positive, estradiol-dependent lines. Nontransformed, ER-positive breast epithelial line, MCF10A, was less sensitive to ritonavir than ER-positive breast cancer cells [148].

Although ritonavir showed inhibition of 20S proteasome in vitro, recent studies revealed its ability to enhance 26S activity in vitro, which suggests that ritonavir may be a proteasome modulator rather than a direct inhibitor [149].

Antiviral drugs—clinical studies

There have been no clinical studies for ombitasvir, dasabuvir, or paritaprevir. However, ritonavir has three clinical trials, of which one is complete. This study was a Phase I/II trial, which studied the best dose of ritonavir and the impact it has on biomarkers in women who are undergoing surgery for newly diagnosed breast cancer. In this Phase I trial, dose escalation was used with three levels of (200, 400, and 600 mg BID) ritonavir. These three levels were applied to the following groups: (i) ER+, HER2−; (ii) ER+, HER2+; (iii) ER−, HER2+; (iv) ER−, PR+, HER2−; and (v) ER−, PR−, HER2−. In Phase II, the dose used was the MTD that was used in Phase I. Also, once the MTD of ritonavir was established, 19 ER+, HER2− patients were enrolled at MTD during the Phase II component along with conventional therapeutic surgery [150].

Antiviral—summary

Karuppasamy et al. showed the advantages of antiviral repurposed drugs to be the control and prevention of beta-tubulin–driven breast cancers. Antiviral drugs, such as ombitasvir, are one of the only types of repurposed drugs that treat cancer with this type of interaction. Ombitasvir was also seen to be very promiscuous. It showed high chemical and pharmacophore similarity with PAC, which makes it very promising in the future of chemotherapy. However, one fallback of this drug is the complications arising when determining dosage. The interindividual variation in both pharmacokinetic and pharmacodynamic factors affecting the drug dose–response relationship makes it extremely difficult to estimate dosage levels correctly. It requires collaborations between biologists and bioinformaticians, which can be a very complicated procedure.

Antipsychotic drugs

Antipsychotics—mechanism of action of the original use

Antipsychotic medications are broadly divided into typical and atypical antipsychotics, although this distinction does not necessarily consider the individuality in receptor profiles of the individual antipsychotic medications. Most typical/first-generation antipsychotic medications block dopamine receptors (D2). Many atypical/second-generation antipsychotics block serotonin receptors (5HT2A) along with dopamine receptors (D2). Additionally, the antipsychotic drugs also produce side effects through their action on cholinergic, histaminic, and adrenergic receptors in the body (Table 5.1).

Antipsychotics—preclinical studies

BCRP, the ATP-binding cassette superfamily G member 2 transporter, has been suggested to play an essential role in the disposition of many drugs and environmental toxins [151]. The effects of different antipsychotic drugs, risperidone, 9-hydroxy-risperidone (paliperidone), olanzapine, QUE, clozapine, haloperidol, and chlorpromazine, along with a positive control inhibitor Ko143, were investigated on functions of BCRP in MCF7 and BCRP-overexpressing MCF7/MX100 BC cell lines using a BCRP prototypical substrate mitoxantrone (Table 5.2). The antipsychotic drugs exhibited inhibition of cellular uptake of mitoxantrone in MCF7 and MCF7/MX100 cells. The intracellular fluorescence of mitoxantrone in the wild-type MCF7 cells was 8.1-fold higher than that in the BCRP overexpression cell line MCF7/MX100, indicating that BCRP was overexpressed in the MCF7/MX100 cells but not in MCF7 cells. Treatment of the MCF7/MX100 cells with the positive control inhibitor of BCRP, Ko143 (0.01–1 μM), significantly increased the intracellular fluorescence of mitoxantrone in the MCF7/MX100 cells. While in the wild-type MCF7 cells, Ko143 increased the intracellular fluorescence of mitoxantrone to a significant but lesser degree compared to those in the MCF7/MX100 cells. There was no statistically significant difference between MCF7/MX100 cells and MCF7 cells regarding their intracellular fluorescence of mitoxantrone in the presence of Ko143 [151].

Risperidone at concentrations ranging from 1 to 100 μM significantly inhibited the intracellular accumulation of oestrone-3-sulfate in Caco-2 cell monolayers, a human colon carcinoma cell line, suggesting that a potential source of pharmacokinetic interactions exists between BCRP substrates and several antipsychotics.

RAW 264.7 cells and bone marrow–derived macrophages (BMMs) were used to detect the inhibitory effect of the antipsychotic drug QUE on osteoclastogenesis in vitro. Mouse model of breast cancer metastasis to bone was used also to test the suppressive effect of QUE on breast cancer–induced bone loss in vivo. The findings revealed that QUE has the ability to inhibit RANKL-induced osteoclast differentiation from RAW 264.7 cells and BMMs without showing any signs of cytotoxicity. In addition, QUE was able to reduce the occurrence of MDA-MB-231 cell–induced osteolytic bone loss by suppressing the differentiation of osteoclasts. Suppression of the osteoclast differentiation was suggested to be through inhibiting RANKL-mediated MAPK and NF-kB signaling pathways [49].

Antipsychotics—clinical studies

Dalton et al. compared cancer risks in a population-based cohort study of 25,264 neuroleptic (antipsychotic) medication users (2 or more prescriptions) with those who did not receive such prescriptions. They found that there was a reduced risk of rectal cancer among antipsychotic medication users (adjusted incidence rate [IRR] of 0.61; 95% CI 0.41–0.91, among women, and adjusted IRR of 0.82; 95% CI 0.56–1.19, among men) compared to nonusers. They also observed decreased colon cancer risk among female users (adjusted IRR 0.78; 95% CI 0.62–0.98). However, the results for breast cancer and antipsychotic use association were not statistically significant. The authors also adjusted for reproductive history and use of hormone replacement to assess the risk of breast cancer. They found that for breast cancer, there was only a slight reduction in the risk among neuroleptic users (adjusted IRR 0.93; 95% CI 0.74–1.17). The results from this study provide some support for anticarcinogenic effects of antipsychotic medications [152].

According to a retrospective cohort study using nested case–control analysis within the General Practice Research Database in the United Kingdom, atypical antipsychotics were compared to typical antipsychotics with regard to breast cancer risk. The cohort consisted of females who received at least one antipsychotic prescription between January 1, 1988, and December 31, 2007, with follow up until December 2010. They identified 1237 breast cancer cases and matched each case with up to 10 controls. It was observed that exclusive users of atypical antipsychotics had reduced risk of breast cancer compared to users of typical antipsychotics (rate ratio 0.81; 95% CI 0.63–1.05). The authors further categorized and compared those who were only prescribed risperidone (most commonly prescribed atypical antipsychotic) and found similar results (rate ratio0.86; 95% CI 0.60–1.25). It cannot be concluded whether these effects are due to anticancer properties of atypical antipsychotics or higher carcinogenicity of typical antipsychotics. The results need to be further confirmed by a study comparing antipsychotic users with nonusers in the general population [153].

Antipsychotic drugs—summary

Antipsychotic drugs showed some reduced risk of certain cancer types, partly supporting for the anticarcinogenic hypothesis of antipsychotics. However, there was no overall decrease in cancer incidence found, and there is no strong causation between antipsychotic use and cancer therapy. In summary, antipsychotics have not been proved to be a strong repurposed class of drugs for cancer therapy but instead have slight anticarcinogenic effects.

Thalidomide

Thalidomide (a(N-phthalimido)-glutarimide, a derivative of glutamic acid with two rings and two optically active forms; Fig. 5.1) is an odorless, white crystalline compound with low water solubility [154, 155]. Thalidomide has been shown to be clinically useful in several conditions due to its ability to inhibit TNF-α synthesis. At the beginning, it was used in pregnant women to relieve the morning sickness. However, its use was restricted because of its teratogenic effect on the developing fetus [156, 157]. Furthermore, insolubility may cause serious problems in terms of systemic bioavailability. The major solvent of thalidomide is dimethyl sulfoxide, which however can mimic the cytotoxic effects of thalidomide [158]. Over the last decades, thalidomide was shown to have a cytotoxic effect on different cancer cell lines, including breast cancer [154, 155].

Thalidomide—mechanism of action of the original use

Due to the ability to modify body's immune response, drugs such as thalidomide, lenalidomide, and pomalidomide are known as immunomodulatory drugs (IMiDs). It has been established that the cellular protein cereblon plays an important role in the function of ImiDs, but the detailed mechanisms have only recently been understood. It is known that CD147 and MCT1 proteins present in blood building and immune cells are important for proliferation, metabolism, and angiogenesis (Table 5.1). In order to perform these functions, CD147 and MCT1 will form a complex which requires cereblon. Binding of cereblon stabilizes the CD147 and MCT1 protein complex, which stimulates cell growth and facilitates the excretion of metabolic products like lactate. Also, an increased abundance of this protein complex enables tumor cells to spread rapidly in diseases such as multiple myeloma. If such a cancer is treated with IMiDs, the CD147 and MCT1 complex will be displaced because binding to cereblon will be inhibited by IMiDs. This will lead to inactivation of the protein complex of CD147 and MCT1 and consequent tumor cell death. Interestingly, this mechanism is identical to that which causes the birth defects thalidomide was known for in the past. This supports the hypothesis of thalidomide-induced deformities being related to the reduction and abnormal formation of new blood vessels.

Thalidomide—preclinical studies

Anticancer mechanism of action

The growth and progression of breast cancer cells including MCF-7 and MDA-MB-231 were inhibited by a small series of thalidomide-correlated compounds, which are very effective to induce cancer cell death via triggering TNFα-mediated apoptosis (Table 5.2). The most active compounds can drastically reduce the migration of breast cancer cells by regulation of the two major proteins involved in EMT: vimentin and E-cadherin. Moreover, these compounds diminish the intracellular biosynthesis of VEGF, which is primarily involved in the promotion of angiogenesis, sustaining tumor progression [157].

Thalidomide's anticancer effects

Thalidomide has immunomodulatory and antiangiogenic properties that may underlie its activity in cancer. It showed its success in myeloma, and it has also been investigated in other types of cancers including renal cell carcinoma, advanced breast cancer, and colon cancer [154].

Thalidomide dithiocarbamate analogs revealed a significant antiproliferative effect on HUVECs and MDA-MB-231 cells without cytotoxicity, in addition to a powerful antiangiogenicity in wound healing, migration, tube formation, and NO assays (Table 5.1). Thalidomide analogues exhibited more suppressive effect against IL-6, IL-8, TNF-α, VEGF165, and MMP-2 than thalidomide [159]. In addition, thalidomide treatment resulted in apoptosis induction in a time- and dose-dependent manner when treated mouse breast cancer cells, and the apoptosis induction was demonstrated by caspase-3 enzyme activity [160]).

The antitumor activity of carboplatin was increased after combining carboplatin with thalidomide in mouse 4T1 breast cancer models. The systemic administration of carboplatin and thalidomide significantly decreased tumor growth through increased tumor cell apoptosis compared with either control group. These findings suggested the synergistic suppressive effect of this combination [161].

Thalidomide—clinical studies

A Phase I trial of docetaxel and thalidomide was conducted in 26 patients with solid tumors including breast cancer. Thalidomide 100 mg BID with escalating doses of docetaxel from 10 to 30 mg/m2/week was given for 12 consecutive weeks per cycle of therapy. Dose-limiting toxicities included bradycardia, fatigue, fever, hyperbilirubinemia, leukopenia, myocardial infarction, and neutropenia. The dose of thalidomide with docetaxel that caused $\leq$ grade 1 nonhematologic or $\leq$ grade 2 hematologic toxicity for cycle one was considered as the MTD. Among the evaluable patients, prolonged freedom from disease progression was 44.4%. The recommended phase II dosing schedule was thalidomide 100 mg BID, with docetaxel 25 mg/m2/week. This combination was well tolerated and showed beneficial clinical activity (Table 5.3).

However, another study of capecitabine in combination with thalidomide did not show beneficial clinical activity in patients with metastatic breast cancer. In 24 previously treated metastatic breast cancer patients, this combination was poorly tolerated with multiple adverse effects like grade 3/4 neutropenia, grade 3 nausea, vomiting, diarrhea, and grade 2/3 hand-foot syndrome. The partial response was 13%, stable disease was seen in 17%, and progressive disease at first evaluation was seen in 35%. The median time to progression and overall survival was 2.7 and 11.0 months, respectively. These results did not support further investigation of this combination for MBC (Table 5.3).

An ongoing clinical trial of thalidomide plus chemotherapy versus only chemotherapy for advanced breast cancer is currently being conducted and is in its recruitment stage. Thalidomide 100 mg orally every night will be used along with the standard chemotherapy prescribed for the breast cancer patients. Efficacy and safety of the chemotherapy–thalidomide combination as well as the progression-free survival and overall survival will be evaluated (Table 5.3). Obtaining the pre- and posttreatment serum specimens will provide essential information about the mechanisms by which VEGF inhibition affects tumor growth and helps evaluate the molecular effects of thalidomide on breast tumor.

Thalidomide—summary

Thalidomide was originally marketed as a sedative in a postwar era and therefore was used quite often with reports indicating prevalence of this drug was as much as aspirin. One out of seven Americans had taken the drug regularly that was considered completely safe. It was also found to alleviate morning sickness and therefore was prescribed to pregnant mothers; this later caused the birth defect tragedy of "flipper babies" as they were referred to because the drug caused phocomelia, resulting in shortened, absent, or flipper-like limbs. However, after being banned for some time, the drug had recently been seen to promote tumor reduction due to the disruption of the CD147-MCT1 protein complex and is now being tested preclinically as well as clinically as used in combination with chemotherapy in advanced breast cancer as well as on its own. Thalidomide was also seen to reduce tumor growth in other cancers such as multiple myeloma.

Artemisinin

Artemisinin (ART; Fig. 5.1) is a chemical that was isolated from the sweet wormwood (qinghao, *Artemisia annua* L., Asteraceae) by Youyou Tu at the Chinese Academy of Traditional Chinese Medicine in 1972 [1, 2]. Its unique chemical structure is a sesquiterpene lactone that has a potent antimalarial effect. Recently, via inserting new groups to the parent structure of ART, three generations of compounds were synthesized in order to improve the efficacy and tolerability [3] that showed better bioactivity or solubility, including artemether (ARM), artesunate (ARS) and dihydroartemisinin (DHA) [4]. ART and its derivatives are currently considered as the most effective drugs for treatment of cerebral malaria and chloroquine-resistant falciparum malaria [5, 6]. During the past two decades, studies revealed the anticancer activity of ART and its derivatives that indicates the effectiveness of these compounds as cancer therapeutic drugs [4].

Artemisinin—mechanism of action of original use

ART and its semisynthetic derivatives were originally formulated to be used against malaria. These drugs contain endoperoxide bridges that are needed for antimalarial activity. The mechanism of action for this original use was believed to be a two-step mechanism. First, ART is activated by intraparasitic heme iron which catalyzes the cleavage of the endoperoxide bridges which then causes a free radical intermediate to be formed and kill the parasite by alkylating and poisoning one or more essential malarial proteins (Table 5.1).

Artemisinin—preclinical studies

Anticancer mechanism of action

ARS and DHA were reported to inhibit TGF-β signaling that inactivates cancer-associated fibroblasts (CAFs). CAFs play an important role in tumor growth and metastasis, by stimulating angiogenesis, cell proliferation, migration, and invasion. CAFs have the ability to release growth factors and cytokines, produce metalloproteinases (MMPs), and downregulate CD36 [7–9]. In addition, CAFs are able to affect extracellular matrix (ECM) stiffness at primary tumors, enhancing cancer cell migration and invasion by inducing EMT (7, 10, 11). Therefore, ART derivatives could be potential therapeutic agents for the treatment of breast cancer. CAF inactivation by ART and its derivatives was suggested to decrease the interaction between tumor and its ECM (Table 5.1). Thus, it could be a promising therapeutic strategy for breast cancer treatment [4].

Artemisinin's anticancer effects

The migration and invasion of 4T1 cells were decreased by ART derivatives. In addition, ART derivatives were able to reduce the cross talk between CAFs and cancer cells, leading to inhibition of tumor metastasis [4].

Treatment of L-929-CAFs and CAFs with ARS and DHA resulted in dramatic decrease in the migration and invasion of 4T1 cells when compared with untreated CAFs. However, ART and ARM were not able to inhibit the migration and invasion of 4T1 cells. These results indicated the promoting functions of CAFs on migration and invasion were impaired by ARS and DHA. It was also studied whether ART, ARM, ARS, and DHA could affect CAF-induced tumor burden and metastatic effect in vivo. ARS and DHA were reported to significantly prolong the life span of the breast cancer orthotopic model in BALB/c nude mice when compared with the other groups. The probabilities of metastasis induced by CAFs were downregulated by ARS (50%) and DHA (37.5%), while that of PAC, ART, and ARM were 62.5%, 87.5% and 87.5%, respectively.

ARS and DHA treatment resulted in reduction of tumor volume more than PAC, ART, and ARM. In the orthotopic model, cancer cells were able to metastasize to different organs, including brain, lung, liver, and bone. ARS and DHA were able to inhibit metastases in lung and liver.

It was reported that ART, ARM, ARS, and DHA did not induce tumor apoptosis or inhibit cell proliferation in vivo. However, ARS and DHA reduced collagen in the tissue, and the expression of specific markers such as MMP-9, fibronectin, vimentin, and α-SMA. Also, ARS and DHA reduced the expression of p-Smad3 (Ser423/425) and TGF-β1, but ART and ARM did not exhibit any effect on the protein expression [4]. In addition, the expression of 84 genes involved in cell motility of MCF-7 breast cancer cells were reported to be altered after ART treatment [12].

Artemisinin—clinical studies

ART was tried as an add-on drug in patients with metastatic breast cancer to their guideline-based oncological therapy for a study period of 4 weeks with frequent clinical and laboratory monitoring until 4–8 weeks thereafter. The dose of either 100 or 150 or 200 mg oral daily ART was tried in 23 patients. Three patients experienced drug related toxicities including leukopenia, neutropenia, asthenia, and anemia during the actual trial period of 4 ± 1 weeks. Up to 200 mg/day (2.2–3.9 mg/kg/day) oral ART was safe and well tolerated; therefore, 200 mg/day is recommended for Phase II/III trials (Table 5.3). The study also suggested that regular monitoring should include reticulocytes, NTproBNP, and audiological and neurological exploration to ensure safety.

Artemisinin—summary

ART was originally developed in order to combat malaria and has saved millions of lives and is also seen as one of the largest contributions of China to global health. The 2015 Nobel Prize in Physiology or Medicine was awarded to Professor Youyou Tu for her contribution in the discovery of this drug. ART and its derivatives, however, have now been seen to have some anticancer activity, indicating that these drugs may be effective cancer therapeutic agents. Clinical studies have shown that the cytotoxic effects of ART can cause patients to experience drug related toxicities.

Mebendazole

A broad-spectrum anthelminthic mebendazole (MBZ) (methyl 5-benzoyl-2-benzimidazole-carbamate; Fig. 5.1) is part of a class of structurally related, tubulin-disrupting drugs (benzimidazoles) that have been used for treatment of helminthic disease in humans. MBZ was approved by the US FDA for the treatment of nematode infestations [162] (Table 5.1). Also, MBZ revealed efficacy against different types of solid tumors in vitro and in vivo, such as lung cancer [163], melanoma [164], colon cancer [165], glioblastoma multiforme [166], medulloblastoma [167], and head and neck squamous cell carcinoma [112].

Mebendazole—mechanism of action of original use

MBZ binds to the colchicine-sensitive site of tubulin and therefore causes degenerative alterations in the tegument and intestinal cells of the parasite (Table 5.1). The parasite then has its glycogen stores depleted because this loss of cytoplasmic microtubules eventually leads to disruption in the glucose uptake by its larval and adult stages. The parasites ATP production is drained because of the degenerative changes in the endoplasmic reticulum, germinal layer mitochondria, and release of lysosomes. This causes the parasites immobilization and eventual death.

Mebendazole—preclinical studies

Anticancer mechanism of action

It was reported that MBZ was able to induce cell cycle arrest in the radiosensitive G_2/M phase of the cell cycle in TNBC cells. This cell cycle arrest was followed by significant induction of apoptotic cell death in a dose- and time-dependent manner [162] (Table 5.1). Different studies revealed induction of apoptosis as the primary mode of cell death by MBZ in other tumor types, such as melanoma [168, 169], lung cancer [170, 171], and medulloblastoma [172]. Also, tubulin depolymerization was reported to be the major target for benzimidazoles, including MBZ [173, 174, 175]. MBZ exhibited induction of depolymerization of tubulin and inhibition of normal spindle formation in different cancer cell lines, resulting in mitotic arrest and apoptosis [168, 171, 172, 176, 177, 178]. Single treatment of TNBC cells with MBZ resulted in a significant increase in double-strand breaks (DSBs) after exposure to MBZ by 24 h. The increase in the percentage of cells that arrest in the G_2/M phase may be contributing to the increased levels of DSB observed at this time point [162].

Mebendazole's anticancer effects

MBZ has exhibited anticancer effect in several preclinical models of cancer [179]. MBZ was identified as a "hit" that efficiently inhibits the radiation-induced dedifferentiation and induces cell death in TNBC cells [162]. In addition,

MBZ was reported to prevent the occurrence of radiation-induced reprogramming and improve the effect of radiation therapy in patients with TNBC. It efficiently depletes BCIC pool and prevents the ionizing radiation (IR)–induced conversion of breast cancer cells into therapy-resistant the BCICs, which are thought to contribute to disease recurrence. MBZ was reported to arrest cells in the G_2/M phase of the cell cycle and causes DSBs and apoptosis. Also, MBZ sensitizes TNBC cells to IR in vitro and in vivo, resulting in improved tumor control in a human xenograft model of TNBC [162]. Exposure to IR can dedifferentiate surviving nontumorigenic BC cells into BCICs [180]. These findings were confirmed in 4 BC lines representing different subsets of BC: luminal (MCF7 and T47D), basal (MDA-MB-231), and claudin low (SUM159PT), indicating MBZ as a nonspecific drug to all subtypes of breast cancer. [162]

Treatment of the tumor-bearing mice with MBZ (10 mg/kg or 20 mg/kg) for 3 weeks (5 days on, 2 days off) resulted in modest delay of tumor growth on its own, but a single dose of 10 Gy significantly delayed tumor growth. The combination of 10 Gy with 10 mg/kg of MBZ did not have any added effect on tumor growth compared with 10 Gy alone. However, the combination of a single 10 Gy treatment with 20 mg/kg of MBZ resulted in delayed tumor growth compared with 10 Gy alone. These results suggested that MBZ can also enhance the effect of RT on TNBC tumors in vivo. Importantly, no toxicity was observed from in vivo administration of 20 mg/kg MBZ for the duration of the experiment. MBZ was reported to have a modest effect on tumor growth in vivo in a human TNBC tumor model [162]. In addition, other studies have shown that MBZ potentiates the effect of different chemotherapeutics in different tumor models [169, 181].

Mebendazole—clinical studies

An ongoing clinical trial aims at studying the safety, tolerability, and efficacy of metabolic combination treatments on cancer, including breast cancer. Patients will be treated with the four metabolic drugs—oral MBZ 100 mg OD, oral atorvastatin up to 80 mg OD, oral MTF up to 1000 mg OD/BID, and oral DOX 100 mg OD for study duration. Every 3 months they will bring in their medical records and data will be collected to determine the effect of the intervention. Efficacy and safety of this regimen as well as the progression-free survival and overall survival will be evaluated.

Mebendazole—summary

MBZ was originally formulated to treat helminthic disease in humans and has shown efficacy against solid tumors both in vitro and in vivo. In preclinical studies, MBZ has shown promise as it both inhibits the radiation-induced dedifferentiation and serves as a potent inducer of TNBC cell death. However, MBZ has only one clinical study relevant to breast cancer but has yet to be completed, and therefore its long-term clinical use as well as clinical efficiency is unknown.

Flunarizine

Flunarizine—mechanism of action of original use

Flunarizine (FLN) (Fig. 5.1) is a selective calcium entry blocker with calmodulin binding properties and histamine H1 blocking activity. It is effective in the prophylaxis of migraine, occlusive peripheral vascular disease, vertigo of central and peripheral origin, and as an adjuvant in the therapy of epilepsy [182].

FLN physically plugs the extracellular calcium channel, therefore inhibiting the influx of extracellular calcium through myocardial and vascular membrane pores. The decrease in intracellular calcium inhibits the contractile processes of smooth muscle cells, causing dilation of the coronary and systemic arteries, increases oxygen delivery to the myocardial tissue, decreases total peripheral resistance, decreases systemic blood pressure, and decreases afterload (Table 5.1).

Flunarizine—preclinical studies

FLN which approved for treating migraine and epilepsy [182] was reported to induce an autophagy pathway to selectively degrade N-Ras, thus blocking the growth of basal-like breast cancer (BLBC cells) in vitro and in vivo [182], as N-Ras plays an important role in the growth and transforming activity of the aggressive BLBC [183]. Ras GTPases are powerful drivers for tumorigenesis, but directly targeting Ras for treating cancer remains challenging [182].

In BLBC, FLN-induced N-Ras degradation was affected by autophagy inhibitors rather than a 26S-proteasome inhibitor. Furthermore, N-Ras was seen colocalized with active autophagosomes upon FLN treatment, suggesting that FLN alters the autophagy pathway to degrade N-Ras. The observed decrease in N-Ras levels was suggested to be mainly due to increased protein degradation. Importantly, FLN treatment recapitulated the effect of N-Ras silencing in vitro by selectively inhibiting the growth of BLBC cells, but not that of breast cancer cells of other subtypes. BLBC cells were more sensitive to FLN, and among BLBC cell lines, SUM149PT and SUM102PT cells were more sensitive to FLN. In addition, they showed great sensitivity to N-Ras inhibition by shRNA3 [182].

In glioblastoma cells, FLN has been reported to induce autophagylike activity [184]. FLN was reported to activate LC3-containing autophagosomes in BLBC cells and decrease p62 levels in SUM149PT cells (Table 5.2). EGFP-LC3 formed a pronounced coarse punctate pattern indicative of strong autophagy activation. Likewise, instead of mostly being concentrated at the plasma membrane and Golgi, FLN-induced mCherry-N-Ras also accumulated in such structures, leading to substantial colocalization in the coarse punctate pattern with EGFP-LC3. Also, instead of activating the conventional Raf effector pathway, N-Ras in BLBC cells has been shown to act through JAK2, and colony formation efficiency of these cells in soft agar can be inhibited by a JAK2 inhibitor, TG1013483. TG101348 and FLN, when used separately, had little effect on soft agar colony formation by another BLBC cell line, SUM102PT; however, when the two drugs were combined, colony formation was greatly inhibited. These results suggested that FLN, like N-RAS silencing, can efficiently inhibit growth and transforming activity of BLBC cells, and this inhibition can be further enhanced by combining with drugs targeting additional components in this N-Ras pathway [182]. In vivo studies exhibited inhibition of tumor growth of a BLBC xenograft model by using FLN. SUM102PT xenograft mouse model was chosen because its growth is N-Ras dependent. The data showed that when FLN was added at levels comparable to those used in humans, tumor growth was efficiently inhibited, mimicking the effect of DOXDOX-inducible N-RAS silencing [182]. There are currently no clinical studies being done on FLN on the treatment of breast cancer or any other type of cancer.

Flunarizine—summary

Directly targeting Ras for cancer treatment is a difficult task; as before-mentioned, growth and transforming activity of aggressive BLBC is driven by N-Ras. FLN's ability to induce N-Ras degradation as well as target the N-Ras pathway in a multitude of ways, including altering the autophagy pathway to degrade N-Ras. FLN also inhibited the growth of BLBC cells as well as has low toxicity making it a good candidate for a repurposed drug for cancer treatment. However, FLN is very new in repurposing for cancer therapy with the first preclinical study being done in 2018, and there are yet to be any clinical studies. Therefore, its clinical uses are still unknown as well as its long-term impact.

Future of drug repurposing in breast cancer therapy

Drug repurposing is a huge industry and has a promising future as it is one of the most active areas in pharmacology in the last decade. Drugs are constantly being tested in order to be used for new purposes, of which they were not originally intended. Drug repurposing may allow for more systematic and substantially less expensive methods in the discovery of new treatments for diseases when compared to traditional drug development. Even though drug repurposing methods have moved from "one: disease—gene—drug" to "multi: gene—drug" and from "lazy guilt-by-association" to "systematic model—based pattern matching," a mathematical system and control system has not been generally applied to model the system biology connectivity among drugs, genes, and diseases. One paper proposed a DeCoST, Drug Repurposing from Control System Theory, a framework to apply a control system model for drug repurposing purposes. In this system, the DeCoST framework being proposed, which is among the first methods in drug repurposing with the control theory system, applies biological and pharmaceutical knowledge to quantify rich connective data sources among drugs, genes, and diseases to construct a disease-specific mathematical model. They use a linear-quadratic regulator control technique to assess the therapeutic effect of a drug in disease-specific treatment. DeCoST framework could classify between FDA-approved drugs and rejected/withdrawn drugs, which is the foundation to apply DeCoST in recommending potentially new treatment [185].

Another paper guided prioritizing and integrating drug repositioning methods for specific drug repositioning pipelines. Low-risk and low-cost drug repositioning strategies have been widely used to identify new clinical opportunities for old drugs. Accordingly, numerous strategies have been developed from drug repositioning studies. They had used a fishbone flowchart to present the existing methods with preclinical and clinical validation. In the era of precision medicine, it is crucial to delineate disease mechanisms, such as signaling pathways, or treatment mechanisms, such as off-targets and targeted pathways, to explain the mechanisms of action of drugs. This also leads to the application of drug repositioning to new indications for individual patients. Mechanism-based repositioning approaches can consider the heterogeneity and complexity of patients fully while reducing the inefficacy and toxicity caused by patient variability. This paper emphasized that drug repositioning studies have to be solidly grounded on science to be successful. Toward better drug repositioning, the field needs better development of more in-depth mechanistic computational methods or models that can readily be customized into drug repositioning pipelines that integrate computational and experimental methods seamlessly to ensure high success rates of repositioned drugs [186].

Summary

Drug repurposing is a valuable tool in the treatment for many diseases. Although in this chapter, we looked into one disease, that being breast cancer, this type of therapy and treatment can be used for a variety of diseases. Drug repurposing provides an extra "weapon" in the treatment of diseases and allows healthcare professionals to be able to give the option of the best quality of life possible to patients.

In this chapter, we specifically looked into the repurposed drugs that target breast cancer. Many drugs have shown great promise, and each drug has a mechanism of action (while some have multiple) in order to target breast cancer in many different ways, some of which include inhibiting the pathways that cause breast cancer specifically or inhibiting cell proliferation. There are many ways in which breast cancer is formed or spreads, and each repurposed drug serves its function in dealing with the variety of pathways and cell functions. Some drugs are beneficial and show great potential (such as DSF); however, they have not had many clinical trials for breast cancer specifically and, therefore, will need to be addressed further in the future. Other drugs have had clinical trials but have not shown significant positive results as compared to when they were in the preclinical stage, such as propranolol. This can be attributed to the clinical trials for propranolol being used for a primary condition other than breast cancer and not yet through the traditional phases of clinical trials for treating breast cancer specifically. Therefore drug dosage and appropriate population selection could be the primary reason for the lack of significant positive results for propranolol. Overall, many drugs show great potential and could have a significant impact on breast cancer therapy.

References

[1] Simsek M, et al. Finding hidden treasures in old drugs: the challenges and importance of licensing generics. Drug Discov Today 2018;23(1):17–21.

[2] Ashburn TT, Thor KB. Drug repositioning: identifying and developing new uses for existing drugs. Nat Rev Drug Discov 2004;3(8):673–83.

[3] Baker NC, et al. A bibliometric review of drug repurposing. Drug Discov Today 2018;23(3):661–72.

[4] Viola-Rhenals M, et al. Recent advances in antabuse (disulfiram): the importance of its metal-binding ability to its anticancer activity. Curr Med Chem 2018; 25(4):506–24.

[5] Klug DM, Gelb MH, Pollastri MP. Repurposing strategies for tropical disease drug discovery. Bioorg Med Chem Lett 2016;26(11):2569–76.

[6] Zhou M, Xia L, Wang J. Metformin transport by a newly cloned proton-stimulated organic cation transporter (plasma membrane monoamine transporter) expressed in human intestine. Drug Metab Dispos 2007;35(10):1956–62.

[7] Scheen AJ, Esser N, Paquot N. Antidiabetic agents: potential anti-inflammatory activity beyond glucose control. Diabetes Metab 2015;41(3):183–94.

[8] Calvert JW, et al. Acute metformin therapy confers cardioprotection against myocardial infarction via AMPK-eNOS-mediated signaling. Diabetes 2008; 57(3):696–705.

[9] Nasri H, Rafieian-Kopaei M. Metformin: current knowledge. J Res Med Sci 2014;19(7):658–64.

[10] Mujica-Mota MA, et al. Safety and otoprotection of metformin in radiation-induced sensorineural hearing loss in the guinea pig. Otolaryngol Head Neck Surg 2014;150(5):859–65.

[11] De A, Kuppusamy G. Metformin in breast cancer: preclinical and clinical evidence. Curr. Probl Cancer 2020; 44(1):100488.

[12] Evans JM, et al. Metformin and reduced risk of cancer in diabetic patients. BMJ 2005;330(7503):1304–5.

[13] Shank JJ, et al. Metformin targets ovarian cancer stem cells in vitro and in vivo. Gynecol Oncol 2012;127(2): 390–7.

[14] Sato A, et al. Glioma-initiating cell elimination by metformin activation of FOXO3 via AMPK. Stem Cells Transl Med 2012;1(11):811–24.

[15] Song CW, et al. Metformin kills and radiosensitizes cancer cells and preferentially kills cancer stem cells. Sci Rep 2012;2:362.

[16] Cantrell LA, et al. Metformin is a potent inhibitor of endometrial cancer cell proliferation—implications for a novel treatment strategy. Gynecol Oncol 2010; 116(1):92–8.

[17] Storozhuk Y, et al. Metformin inhibits growth and enhances radiation response of non-small cell lung cancer (NSCLC) through ATM and AMPK. Br J Cancer 2013;108(10):2021–32.

[18] Dowling RJ, et al. Metformin inhibits mammalian target of rapamycin-dependent translation initiation in breast cancer cells. Cancer Res 2007;67(22):10804–12.

[19] Fidan E, et al. The effects of rosiglitazone and metformin on inflammation and endothelial dysfunction in patients with type 2 diabetes mellitus. Acta Diabetol 2011;48(4):297–302.

[20] Dowling RJ, Goodwin PJ, Stambolic V. Understanding the benefit of metformin use in cancer treatment. BMC Med 2011;9:33.

[21] Chae YK, et al. Repurposing metformin for cancer treatment: current clinical studies. Oncotarget 2016; 7(26):40767–80.

[22] Gwinn DM, et al. AMPK phosphorylation of raptor mediates a metabolic checkpoint. Mol Cell 2008; 30(2):214–26.

[23] Vazquez-Martin A, Oliveras-Ferraros C, Menendez JA. The antidiabetic drug metformin suppresses HER2 (erbB-2) oncoprotein overexpression via inhibition of the mTOR effector p70S6K1 in human breast carcinoma cells. Cell Cycle 2009;8(1): 88–96.

[24] Wahdan-Alaswad R, et al. Glucose promotes breast cancer aggression and reduces metformin efficacy. Cell Cycle 2013;12(24):3759–69.

[25] Zordoky BN, et al. The anti-proliferative effect of metformin in triple-negative MDA-MB-231 breast cancer cells is highly dependent on glucose concentration: implications for cancer therapy and prevention. Biochim Biophys Acta 2014;1840(6):1943–57.

[26] Wahdan-Alaswad RS, et al. Metformin targets glucose metabolism in triple negative breast cancer. J Oncol Transl Res 2018;4(1).

[27] Hadad S, et al. Evidence for biological effects of metformin in operable breast cancer: a pre-operative, window-of-opportunity, randomized trial. Breast Cancer Res Treat 2011;128(3):783−94.

[28] Niraula S, et al. Metformin in early breast cancer: a prospective window of opportunity neoadjuvant study. Breast Cancer Res Treat 2012;135(3): 821−30.

[29] Lau YK, et al. Metformin and erlotinib synergize to inhibit basal breast cancer. Oncotarget 2014;5(21): 10503−17.

[30] Thomas F, et al. Fibronectin confers survival against chemotherapeutic agents but not against radiotherapy in DU145 prostate cancer cells: involvement of the insulin like growth factor-1 receptor. Prostate 2010;70(8):856−65.

[31] Falah RR, Talib WH, Shbailat SJ. Combination of metformin and curcumin targets breast cancer in mice by angiogenesis inhibition, immune system modulation and induction of p53 independent apoptosis. Ther Adv Med Oncol 2017;9(4):235−52.

[32] Zheng G, et al. Synergistic chemopreventive and therapeutic effects of Co-drug UA-Met: implication in tumor metastasis. J Agric Food Chem 2017;65(50): 10973−83.

[33] Guo LS, et al. Synergistic antitumor activity of vitamin D3 combined with metformin in human breast carcinoma MDA-MB-231 cells involves m-TOR related signaling pathways. Pharmazie 2015;70(2):117−22.

[34] Chatran M, et al. Synergistic anti-proliferative effects of metformin and silibinin combination on T47D breast cancer cells via hTERT and cyclin D1 inhibition. Drug Res (Stuttg) 2018;68(12):710−6.

[35] Rasouli S, Zarghami N. Synergistic growth inhibitory effects of chrysin and metformin combination on breast cancer cells through hTERT and cyclin D1 suppression. Asian Pac J Cancer Prev 2018;19(4): 977−82.

[36] Bojková B, et al. Metformin and melatonin inhibit DMBA-induced mammary tumorigenesis in rats fed a high-fat diet. Anticancer Drugs; 2017.

[37] Cuyàs E, et al. Metformin inhibits RANKL and sensitizes cancer stem cells to denosumab. Cell Cycle 2017; 16(11):1022−8.

[38] Hirsch HA, et al. Metformin selectively targets cancer stem cells, and acts together with chemotherapy to block tumor growth and prolong remission. Cancer Res 2009;69(19):7507−11.

[39] Chatterjee S, Thaker N, De A. Combined 2-deoxy glucose and metformin improves therapeutic efficacy of sodium-iodide symporter-mediated targeted radioiodine therapy in breast cancer cells. Breast cancer, vol. 7. Dove Med Press; 2015. p. 251−65.

[40] Phan LM, Yeung SC, Lee MH. Cancer metabolic reprogramming: importance, main features, and potentials for precise targeted anti-cancer therapies. Cancer Biol Med 2014;11(1):1−19.

[41] Lu Z, et al. A size-shrinkable nanoparticle-based combined anti-tumor and anti-inflammatory strategy for enhanced cancer therapy. Nanoscale 2018;10(21): 9957−70.

[42] Xue C, et al. Targeting P-glycoprotein function, p53 and energy metabolism: combination of metformin and 2-deoxyglucose reverses the multidrug resistance of MCF-7/Dox cells to doxorubicin. Oncotarget 2017; 8(5):8622−32.

[43] Cooper AC, et al. Changes in [18F]Fluoro-2-deoxy-D-glucose incorporation induced by doxorubicin and anti-HER antibodies by breast cancer cells modulated by co-treatment with metformin and its effects on intracellular signalling. J Cancer Res Clin Oncol 2015;141(9):1523−32.

[44] Ben Sahra I, et al. Targeting cancer cell metabolism: the combination of metformin and 2-deoxyglucose induces p53-dependent apoptosis in prostate cancer cells. Cancer Res 2010;70(6):2465−75.

[45] Rocha GZ, et al. Metformin amplifies chemotherapy-induced AMPK activation and antitumoral growth. Clin Cancer Res 2011;17(12):3993−4005.

[46] Xiao Y, et al. Co-delivery of metformin and paclitaxel via folate-modified pH-sensitive micelles for enhanced anti-tumor efficacy. AAPS PharmSciTech 2018;19(5):2395−406.

[47] Ma J, et al. Metformin enhances tamoxifen-mediated tumor growth inhibition in ER-positive breast carcinoma. BMC Cancer 2014;14:172.

[48] Menendez JA, et al. Metformin is synthetically lethal with glucose withdrawal in cancer cells. Cell Cycle 2012;11(15):2782−92.

[49] Wang H, et al. Quetiapine inhibits osteoclastogenesis and prevents human breast cancer-induced bone loss through suppression of the RANKL-mediated MAPK and NF-κB signaling pathways. Breast Cancer Res Treat 2015;149(3):705−14.

[50] Ariaans G, et al. Anti-tumor effects of everolimus and metformin are complementary and glucose-dependent in breast cancer cells. BMC Cancer 2017; 17(1):232.

[51] Amaral MEA, et al. Pre-clinical effects of metformin and aspirin on the cell lines of different breast cancer subtypes. Invest New Drugs 2018;36(5):782−96.

[52] Talarico G, et al. Aspirin and atenolol enhance metformin activity against breast cancer by targeting both neoplastic and microenvironment cells. Sci Rep 2016; 6:18673.

[53] Rico M, et al. Metformin and propranolol combination prevents cancer progression and metastasis in different breast cancer models. Oncotarget 2017;8(2): 2874–89.
[54] Haugrud AB, et al. Dichloroacetate enhances apoptotic cell death via oxidative damage and attenuates lactate production in metformin-treated breast cancer cells. Breast Cancer Res Treat 2014;147(3): 539–50.
[55] Hong SE, et al. Targeting HIF-1α is a prerequisite for cell sensitivity to dichloroacetate (DCA) and metformin. Biochem Biophys Res Commun 2016; 469(2):164–70.
[56] Banala VT, et al. Synchronized ratiometric codelivery of metformin and topotecan through engineered nanocarrier facilitates in vivo synergistic precision levels at tumor site. Adv Healthc Mater 2018;7(19): e1800300.
[57] Owen MR, Doran E, Halestrap AP. Evidence that metformin exerts its anti-diabetic effects through inhibition of complex 1 of the mitochondrial respiratory chain. Biochem J 2000;348(3):607–14.
[58] Zhou G, et al. Role of AMP-activated protein kinase in mechanism of metformin action. J Clin Invest 2001; 108(8):1167–74.
[59] Chen S, et al. Combined cancer therapy with nonconventional drugs: all roads lead to AMPK. Mini Rev Med Chem 2014;14(8):642–54.
[60] Hospital AU. Metformin in breast cancer, visualized with positron emission tomography. 2016. https://ClinicalTrials.gov/show/NCT02882581.
[61] Storniolo AM, Institute NC, University I. Metformin hydrochloride vs. Placebo in overweight or obese patients at elevated risk for breast cancer. 2013. https://ClinicalTrials.gov/show/NCT01793948.
[62] Mount Sinai Hospital C, Princess Margaret Hospital C. Clinical and biologic effects of metformin in early stage breast cancer. 2008. https://ClinicalTrials.gov/show/NCT00897884.
[63] Hospital SNU. Efficacy and safety of adjuvant metformin for operable breast cancer patients. 2009. https://ClinicalTrials.gov/show/NCT00909506.
[64] Institute D-FC. Exercise and metformin in colorectal and breast cancer survivors. 2011. https://ClinicalTrials.gov/show/NCT01340300.
[65] Trust OUHN, UK CR. Effect of metformin on breast cancer metabolism. 2011. https://ClinicalTrials.gov/show/NCT01266486.
[66] Center MDAC, Foundation SGKBC. The impact of obesity and obesity treatments on breast cancer. 2009. https://ClinicalTrials.gov/show/NCT00933309.
[67] Centre LHS, Institute NC. Metformin and temsirolimus in treating patients with metastatic or unresectable solid tumor or lymphoma. 2008. https://ClinicalTrials.gov/show/NCT00659568.
[68] Inc OR, Foundation BCR. A trial of standard chemotherapy with metformin (vs placebo) in women with metastatic breast cancer. 2011. https://ClinicalTrials.gov/show/NCT01310231.
[69] University C, Foundation SGKBC, Inc AP. Study of erlotinib and metformin in triple negative breast cancer. 2012. https://ClinicalTrials.gov/show/NCT01650506.
[70] Tumori ISRplSelcd. Myocet + cyclophosphamide + metformin vs myocet + cyclophosphamide in 1st line treatment of HER2 neg. Metastatic breast cancer patients. 2010. https://ClinicalTrials.gov/show/NCT01885013.
[71] Center MDAC. Metformin for reduction of paclitaxel-related neuropathy in patients with breast cancer. 2015. https://ClinicalTrials.gov/show/NCT02360059.
[72] Hospital AM, Center UH. Use of metformin to reduce cardiac toxicity in breast cancer. 2014. https://ClinicalTrials.gov/show/NCT02472353.
[73] Arizona Uo, Institute NC. Phase II study of metformin for reduction of obesity-associated breast cancer risk. 2014. https://ClinicalTrials.gov/show/NCT02028221.
[74] Oncology AfCTi, Institute NC. Metformin hydrochloride in preventing breast cancer in patients with atypical hyperplasia or in situ breast cancer. 2013. https://ClinicalTrials.gov/show/NCT01905046.
[75] Oncology NNPNMRCo. Neoadjuvant FDC with melatonin or metformin for locally advanced breast cancer. 2015. https://ClinicalTrials.gov/show/NCT02506777.
[76] Oncology NNPNMRCo. Neoadjuvant toremifene with melatonin or metformin in locally advanced breast cancer. 2015. https://ClinicalTrials.gov/show/NCT02506790.
[77] University C. Pre-surgical trial of the combination of metformin and atorvastatin in newly diagnosed operable breast cancer. 2013. https://ClinicalTrials.gov/show/NCT01980823.
[78] Hospital SNU. Neoadjuvant letrozole plus metformin vs letrozole plus placebo for ER-positive postmenopausal breast cancer. 2012. https://ClinicalTrials.gov/show/NCT01589367.
[79] Khan Q, Center UoKM. Randomized trial of Neoadjuvant chemotherapy with or without metformin for HER2 positive operable breast cancer. 2017. https://ClinicalTrials.gov/show/NCT03238495.

[80] Tumori ISRplSelcd. A study of liposomal doxorubicin + docetaxel + trastuzumab + metformin in operable and locally advanced HER2 positive breast cancer. 2014. https://ClinicalTrials.gov/show/NCT02488564.

[81] National University Hospital S. Metformin and simvastatin in addition to fulvestrant. 2017. https://ClinicalTrials.gov/show/NCT03192293.

[82] Medicine SJTUSo. NeoMET study in neoadjuvant treatment of breast cancer. 2013. https://ClinicalTrials.gov/show/NCT01929811.

[83] University SKCCaTJ, University TJ. Metformin hydrochloride and doxycycline in treating patients with localized breast or uterine cancer. 2016. https://ClinicalTrials.gov/show/NCT02874430.

[84] Chen D, et al. Disulfiram, a clinically used anti-alcoholism drug and copper-binding agent, induces apoptotic cell death in breast cancer cultures and xenografts via inhibition of the proteasome activity. Cancer Res 2006;66(21):10425–33.

[85] Yang Y, et al. Disulfiram chelated with copper promotes apoptosis in human breast cancer cells by impairing the mitochondria functions. Scanning 2016;38(6):825–36.

[86] Zha J, et al. Disulfiram targeting lymphoid malignant cell lines via ROS-JNK activation as well as Nrf2 and NF-kB pathway inhibition. J Transl Med 2014;12:163.

[87] Wattenberg LW. Inhibition of dimethylhydrazine-induced neoplasia of the large intestine by disulfiram. J Natl Cancer Inst 1975;54(4):1005–6.

[88] Yip NC, et al. Disulfiram modulated ROS-MAPK and NFκB pathways and targeted breast cancer cells with cancer stem cell-like properties. Br J Cancer 2011; 104(10):1564–74.

[89] Lewis DJ, et al. On the interaction of copper(II) with disulfiram. Chem Commun 2014;50(87):13334–7.

[90] Skrott Z, et al. Alcohol-abuse drug disulfiram targets cancer via p97 segregase adaptor NPL4. Nature 2017;552(7684):194–9.

[91] Liu P, et al. Cytotoxic effect of disulfiram/copper on human glioblastoma cell lines and ALDH-positive cancer-stem-like cells. Br J Cancer 2012;107(9): 1488–97.

[92] Jin N, et al. Disulfiram/copper targets stem cell-like ALDH. J Cell Biochem 2018;119(8):6882–93.

[93] Allensworth JL, et al. Disulfiram (DSF) acts as a copper ionophore to induce copper-dependent oxidative stress and mediate anti-tumor efficacy in inflammatory breast cancer. Mol Oncol 2015;9(6): 1155–68.

[94] Siddique HR, Saleem M. Role of BMI1, a stem cell factor, in cancer recurrence and chemoresistance: preclinical and clinical evidences. Stem Cell 2012;30(3): 372–8.

[95] Carnero A, et al. The cancer stem-cell signaling network and resistance to therapy. Cancer Treat Rev 2016;49:25–36.

[96] Zhang H, et al. Disulfiram treatment facilitates phosphoinositide 3-kinase inhibition in human breast cancer cells in vitro and in vivo. Cancer Res 2010;70(10): 3996–4004.

[97] Korkaya H, et al. HER2 regulates the mammary stem/progenitor cell population driving tumorigenesis and invasion. Oncogene 2008;27(47):6120–30.

[98] Duru N, et al. Breast cancer adaptive resistance: HER2 and cancer stem cell repopulation in a heterogeneous tumor society. J Cancer Res Clin Oncol 2014;140(1): 1–14.

[99] Knuefermann C, et al. HER2/PI-3K/Akt activation leads to a multidrug resistance in human breast adenocarcinoma cells. Oncogene 2003;22(21): 3205–12.

[100] Takahashi T, et al. Cyclin A-associated kinase activity is needed for paclitaxel sensitivity. Mol Cancer Ther 2005;4(7):1039–46.

[101] Kim JY, et al. Disulfiram targets cancer stem-like properties and the HER2/Akt signaling pathway in HER2-positive breast cancer. Cancer Lett 2016; 379(1):39–48.

[102] Robinson TJ, et al. High-throughput screen identifies disulfiram as a potential therapeutic for triple-negative breast cancer cells: interaction with IQ motif-containing factors. Cell Cycle 2013;12(18): 3013–24.

[103] Borst P. Cancer drug pan-resistance: pumps, cancer stem cells, quiescence, epithelial to mesenchymal transition, blocked cell death pathways, persisters or what? Open Biol 2012;2(5):120066.

[104] Liu P, et al. Disulfiram targets cancer stem-like cells and reverses resistance and cross-resistance in acquired paclitaxel-resistant triple-negative breast cancer cells. Br J Cancer 2013;109(7):1876–85.

[105] Saigal K, et al. Risk factors for locoregional failure in patients with inflammatory breast cancer treated with trimodality therapy. Clin Breast Cancer 2013; 13(5):335–43.

[106] Rueth NM, et al. Underuse of trimodality treatment affects survival for patients with inflammatory breast cancer: an analysis of treatment and survival trends from the National Cancer Database. J Clin Oncol 2014;32(19):2018–24.

[107] Wang W, McLeod HL, Cassidy J. Disulfiram-mediated inhibition of NF-kappaB activity enhances cytotoxicity of 5-fluorouracil in human colorectal cancer cell lines. Int J Cancer 2003;104(4):504–11.

[108] Han D, et al. Disulfiram inhibits TGF-β-induced epithelial-mesenchymal transition and stem-like

features in breast cancer via ERK/NF-κB/Snail pathway. Oncotarget 2015;6(38):40907–19.

[109] Hasinoff BB, Patel D. Disulfiram is a slow-binding partial noncompetitive inhibitor of 20S proteasome activity. Arch Biochem Biophys 2017;633:23–8.

[110] Huang J, et al. A phase I study to repurpose disulfiram in combination with temozolomide to treat newly diagnosed glioblastoma after chemoradiotherapy. J Neuro Oncol 2016;128(2):259–66.

[111] Cvek B, et al. Ni(II), Cu(II), and Zn(II) diethyldithiocarbamate complexes show various activities against the proteasome in breast cancer cells. J Med Chem 2008;51(20):6256–8.

[112] Li L, et al. Disulfiram promotes the conversion of carcinogenic cadmium to a proteasome inhibitor with pro-apoptotic activity in human cancer cells. Toxicol Appl Pharmacol 2008;229(2):206–14.

[113] Zhang Z, et al. Organic cadmium complexes as proteasome inhibitors and apoptosis inducers in human breast cancer cells. J Inorg Biochem 2013;123:1–10.

[114] Wiggins HL, et al. Disulfiram-induced cytotoxicity and endo-lysosomal sequestration of zinc in breast cancer cells. Biochem Pharmacol 2015;93(3):332–42.

[115] Nakano H, et al. Reactive oxygen species mediate crosstalk between NF-kappaB and JNK. Cell Death Differ 2006;13(5):730–7.

[116] Clark DW, Palle K. Aldehyde dehydrogenases in cancer stem cells: potential as therapeutic targets. Ann Transl Med 2016;4(24):518.

[117] Ketola K, Kallioniemi O, Iljin K. Chemical biology drug sensitivity screen identifies sunitinib as synergistic agent with disulfiram in prostate cancer cells. PloS One 2012;7(12). e51470.

[118] Yang Z, et al. Disulfiram modulates ROS accumulation and overcomes synergistically cisplatin resistance in breast cancer cell lines. Biomed Pharmacother 2019;113:108727.

[119] Liu P, et al. Liposome encapsulated Disulfiram inhibits NFκB pathway and targets breast cancer stem cells in vitro and in vivo. Oncotarget 2014;5(17):7471–85.

[120] Yang Q, et al. An Updated Review of Disulfiram: Molecular Targets and Strategies for Cancer Treatment. Current pharmaceutical design 2019;25(30):3248–56.

[121] Makadia HK, Siegel SJ. Poly lactic-co-glycolic acid (PLGA) as biodegradable controlled drug delivery carrier. Polymers (Basel) 2011;3(3):1377–97.

[122] Wang K, et al. FoxM1 inhibition enhances chemosensitivity of docetaxel-resistant A549 cells to docetaxel via activation of JNK/mitochondrial pathway. Acta Biochim Biophys Sin (Shanghai) 2016;48(9):804–9.

[123] Song W, et al. Stable loading and delivery of disulfiram with mPEG-PLGA/PCL mixed nanoparticles for tumor therapy. Nanomedicine 2016;12(2):377–86.

[124] Chen W, et al. Disulfiram copper nanoparticles prepared with a stabilized metal ion ligand complex method for treating drug-resistant prostate cancers. ACS Appl Mater Interfaces 2018;10(48):41118–28.

[125] Zhou L, et al. Membrane loaded copper oleate PEGylated liposome combined with disulfiram for improving synergistic antitumor effect in vivo. Pharm Res 2018;35(7):147.

[126] Najlah M, et al. Development and characterisation of disulfiram-loaded PLGA nanoparticles for the treatment of non-small cell lung cancer. Eur J Pharm Biopharm 2017;112:224–33.

[127] Madala HR, et al. Brain- and brain tumor-penetrating disulfiram nanoparticles: sequence of cytotoxic events and efficacy in human glioma cell lines and intracranial xenografts. Oncotarget 2018;9(3):3459–82.

[128] Agarwal R, et al. Liposomes in topical ophthalmic drug delivery: an update. Drug Deliv 2016;23(4):1075–91.

[129] Fasehee H, et al. Delivery of disulfiram into breast cancer cells using folate-receptor-targeted PLGA-PEG nanoparticles: in vitro and in vivo investigations. J Nanobiotechnol 2016;14:32.

[130] He H, et al. Repurposing disulfiram for cancer therapy via targeted nanotechnology through enhanced tumor mass penetration and disassembly. Acta Biomater 2018;68:113–24.

[131] Marian Hajduch MD, et al. Phase II trial of disulfiram with copper in metastatic breast cancer. Ph.D. 2017. https://ClinicalTrials.gov/show/NCT03323346.

[132] Ekinci E, et al. Repurposing disulfiram as an anti-cancer agent: updated review on literature and patents. Recent Pat Anticancer Drug Discov 2019;14(2):113–32.

[133] Organization HM, Augusta Hospital B. Initial Assessment of the Effect of the Addition of disulfiram (antabuse) to standard Chemotherapy in lung cancer. 2006. https://ClinicalTrials.gov/show/NCT00312819.

[134] Nechushtan H, et al. A phase IIb trial assessing the addition of disulfiram to chemotherapy for the treatment of metastatic non-small cell lung cancer. Oncologist 2015;20(4):366–7.

[135] Al-Majed AA, et al. Propranolol. Profiles Drug Subst Excip Relat Methodol 2017;42:287–338.

[136] Barron TI, et al. Beta blockers and breast cancer mortality: a population- based study. J Clin Oncol 2011;29(19):2635–44.

[137] Powe DG, et al. Alpha- and beta-adrenergic receptor (AR) protein expression is associated with poor clinical outcome in breast cancer: an immunohistochemical study. Breast Cancer Res Treat 2011;130(2):457–63.

[138] Wilson JM, et al. β-Adrenergic receptors suppress Rap1B prenylation and promote the metastatic phenotype in breast cancer cells. Cancer Biol Ther 2015;16(9):1364–74.

[139] Cardwell CR, et al. Propranolol and survival from breast cancer: a pooled analysis of European breast cancer cohorts. Breast Cancer Res 2016;18(1):119.

[140] Kang F, et al. Propranolol inhibits glucose metabolism and 18F-FDG uptake of breast cancer through post-transcriptional downregulation of hexokinase-2. J Nucl Med 2014;55(3):439–45.

[141] Sloan EK, et al. The sympathetic nervous system induces a metastatic switch in primary breast cancer. Cancer Res 2010;70(18):7042–52.

[142] Cardwell CR, et al. Beta-blocker usage and breast cancer survival: a nested case-control study within a UK clinical practice research datalink cohort. Int J Epidemiol 2013;42(6):1852–61.

[143] Sørensen GV, et al. Use of β-blockers, angiotensin-converting enzyme inhibitors, angiotensin II receptor blockers, and risk of breast cancer recurrence: a Danish nationwide prospective cohort study. J Clin Oncol 2013;31(18):2265–72.

[144] Alibek K, et al. Using antimicrobial adjuvant therapy in cancer treatment: a review. Infect Agent Cancer 2012;7(1):33.

[145] Karuppasamy R, et al. An integrative drug repurposing pipeline: switching viral drugs to breast cancer. J Cell Biochem 2017;118(6):1412–22.

[146] King JR, et al. Clinical pharmacokinetics of dasabuvir. Clin Pharmacokinet 2017;56(10):1115–24.

[147] Shebley M, et al. Mechanisms and predictions of drug-drug interactions of the hepatitis C virus three direct-acting antiviral regimen: paritaprevir/ritonavir, ombitasvir, and dasabuvir. Drug Metab Dispos 2017;45(7):755–64.

[148] Srirangam A, et al. Effects of HIV protease inhibitor ritonavir on Akt-regulated cell proliferation in breast cancer. Clin Cancer Res 2006;12(6):1883–96.

[149] Gaedicke S, et al. Antitumor effect of the human immunodeficiency virus protease inhibitor ritonavir: induction of tumor-cell apoptosis associated with perturbation of proteasomal proteolysis. Cancer Res 2002;62(23):6901–8.

[150] Masonic Cancer Center, U.o.M., S.G.K.B.C. Foundation. Ritonavir and its effects on biomarkers in women undergoing surgery for newly diagnosed breast cancer. 2010. https://ClinicalTrials.gov/show/NCT01009437.

[151] Wang JS, et al. Antipsychotic drugs inhibit the function of breast cancer resistance protein. Basic Clin Pharmacol Toxicol 2008;103(4):336–41.

[152] Dalton SO, et al. Cancer risk among users of neuroleptic medication: a population-based cohort study. Br J Cancer 2006;95(7):934–9.

[153] Azoulay L, et al. The use of atypical antipsychotics and the risk of breast cancer. Breast Cancer Res Treat 2011;129(2):541–8.

[154] Singhal S, Mehta J. Thalidomide in cancer. Biomed Pharmacother 2002;56(1):4–12.

[155] Stebbing J, et al. The treatment of advanced renal cell cancer with high-dose oral thalidomide. Br J Cancer 2001;85(7):953–8.

[156] Marriott JB, et al. CC-3052: a water-soluble analog of thalidomide and potent inhibitor of activation-induced TNF-alpha production. J Immunol 1998; 161(8):4236–43.

[157] Iacopetta D, et al. Old Drug Scaffold, New Activity: Thalidomide-Correlated Compounds Exert Different Effects on Breast Cancer Cell Growth and Progression. C hem Med Chem 2017;12(5):381–9.

[158] Eter N, Spitznas M. DMSO mimics inhibitory effect of thalidomide on choriocapillary endothelial cell proliferation in culture. Br J Ophthalmol 2002;86(11): 1303–5.

[159] El-Aarag BY, et al. In vitro anti-proliferative and anti-angiogenic activities of thalidomide dithiocarbamate analogs. Int Immunopharmacol 2014;21(2):283–92.

[160] Simsek Öz E, Aydemir E, Fışkın K. DMSO exhibits similar cytotoxicity effects to thalidomide in mouse breast cancer cells. Oncology letters 2012;3(4):927–9.

[161] de Souza CM, et al. Combination therapy with carboplatin and thalidomide suppresses tumor growth and metastasis in 4T1 murine breast cancer model. Biomed Pharmacother 2014;68(1):51–7.

[162] Zhang L, et al. Mebendazole Potentiates Radiation Therapy in Triple-Negative Breast Cancer. Int J Radiat Oncol Biol Phys 2019;103(1):195–207.

[163] Liedtke C, et al. Response to neoadjuvant therapy and long-term survival in patients with triple-negative breast cancer. J Clin Oncol 2008;26(8):1275–81.

[164] Al-Hajj M, et al. Prospective identification of tumorigenic breast cancer cells. Proc Natl Acad Sci U S A 2003;100(7):3983–8.

[165] Ginestier C, et al. ALDH1 is a marker of normal and malignant human mammary stem cells and a predictor of poor clinical outcome. Cell Stem Cell 2007;1(5): 555–67.

[166] Phillips TM, McBride WH, Pajonk F. The response of CD24(-/low)/CD44+ breast cancer-initiating cells to radiation. J Natl Cancer Inst 2006;98(24):1777–85.

[167] Woodward WA, et al. WNT/beta-catenin mediates radiation resistance of mouse mammary progenitor cells. Proc Natl Acad Sci U S A 2007;104(2):618–23.

[168] Doudican N, et al. Mebendazole induces apoptosis via Bcl-2 inactivation in chemoresistant melanoma cells. Mol Cancer Res 2008;6(8):1308–15.

[169] Simbulan-Rosenthal CM, et al. The repurposed anthelmintic mebendazole in combination with trametinib suppresses refractory NRASQ61K melanoma. Oncotarget 2017;8(8):12576–95.
[170] Mukhopadhyay T, et al. Mebendazole elicits a potent antitumor effect on human cancer cell lines both in vitro and in vivo. Clin Cancer Res 2002;8(9):2963–9.
[171] Sasaki J, et al. The anthelmintic drug mebendazole induces mitotic arrest and apoptosis by depolymerizing tubulin in non-small cell lung cancer cells. Mol Cancer Ther 2002;1(13):1201–9.
[172] Larsen AR, et al. Repurposing the antihelmintic mebendazole as a hedgehog inhibitor. Mol Cancer Ther 2015;14(1):3–13.
[173] Laclette JP, Guerra G, Zetina C. Inhibition of tubulin polymerization by mebendazole. Biochem Biophys Res Commun 1980;92(2):417–23.
[174] Lacey E, Watson TR. Structure-activity relationships of benzimidazole carbamates as inhibitors of mammalian tubulin, in vitro. Biochem Pharmacol 1985;34(7):1073–7.
[175] Gull K, et al. Microtubules as target organelles for benzimidazole anthelmintic chemotherapy. Biochem Soc Trans 1987;15(1):59–60.
[176] Bai RY, et al. Antiparasitic mebendazole shows survival benefit in 2 preclinical models of glioblastoma multiforme. Neuro Oncol 2011;13(9):974–82.
[177] Pinto LC, et al. The anthelmintic drug mebendazole inhibits growth, migration and invasion in gastric cancer cell model. Toxicol In Vitro 2015;29(8):2038–44.
[178] De Witt M, et al. Repurposing Mebendazole as a Replacement for Vincristine for the Treatment of Brain Tumors. Mol Med 2017;23:50–6.
[179] Pantziarka P, et al. The Repurposing Drugs in Oncology (ReDO) Project. Ecancer medicals cience 2014;8:442.
[180] Lagadec C, et al. Radiation-induced reprogramming of breast cancer cells. Stem Cells 2012;30(5). 833-544.
[181] Zhang F, et al. Anthelmintic mebendazole enhances cisplatin's effect on suppressing cell proliferation and promotes differentiation of head and neck squamous cell carcinoma (HNSCC). Oncotarget 2017; 8(8):12968–82.
[182] Zheng ZY, et al. Induction of N-Ras degradation by flunarizine-mediated autophagy. Sci Rep 2018;8(1): 16932.
[183] Zheng ZY, et al. Wild-type N-Ras, overexpressed in basal-like breast cancer, promotes tumor formation by inducing IL-8 secretion via JAK2 activation. Cell Rep 2015;12(3):511–24.
[184] Zhang L, et al. Small molecule regulators of autophagy identified by an image-based high-throughput screen. Proc Natl Acad Sci U S A 2007; 104(48):19023–8.
[185] Nguyen TM, et al. DeCoST: a new approach in drug repurposing from control system theory. Front Pharmacol 2018;9:583.
[186] Jin G, Wong ST. Toward better drug repositioning: prioritizing and integrating existing methods into efficient pipelines. Drug Discov Today 2014;19(5): 637–44.

CHAPTER

6

The success story of drug repurposing in breast cancer

Siddhika Pareek[1], *Yingbo Huang*[1], *Aritro Nath*[2], *R. Stephanie Huang*[1]

[1]Department of Experimental and Clinical Pharmacology, University of Minnesota, Minneapolis, MN, United States; [2]Department of Medical Oncology and Therapeutics Research, City of Hope, Monrovia, CA, United States

OUTLINE

Drug Repurposing in Cancer Therapy
https://doi.org/10.1016/B978-0-12-819668-7.00006-3

Introduction

Breast cancer is the most prevalent cancer in women worldwide, and despite tremendous research efforts, it remains the second leading cause of cancer-related mortalities among women (https://seer.cancer.gov/data/). According to the 2018 GLOBOCAN report, over 2 million new breast cancer incidences and nearly 600,000 deaths were reported globally [1]. Breast cancer is vastly heterogeneous owing to differences in the underlying molecular architecture among tumor subtypes. These molecular alterations serve as molecular biomarkers that are useful for diagnosing, staging and grading, therapeutic intervention, prognosis and clinical management of recurrent and metastatic cases. The commonly assessed and clinically validated pharmacodiagnostic markers are estrogen receptor (ER), progesterone receptor (PR), and human epidermal growth factor receptor 2 (HER2, also called HER2/neu) [1].

Breast cancer classification

Several factors influence the prognosis of breast cancer patients, such as the stage of the disease, size of the tumor, molecular subtype, and age at the time of first diagnosis. Breast cancer staging system primarily includes anatomic factors based on TNM classification, tumor size (T), the extent of spread to nearby lymph nodes (N), and metastases of the malignant growth to distant organs (M). Not long ago, American Joint Committee on Cancer recognized and incorporated the prognostic influence of the biologic factors to refine the precision of the breast cancer staging system. This includes accounting for tumor expression of ER and PR, HER2 expression, proliferation (Ki67), and expression-based classification panels like PAM50 into the staging system [2,3]. Based on these markers, breast cancer has been classified broadly into three major clinical subtypes:

(i) Luminal like, which is further categorized into luminal A: high ER/PR, low proliferation and luminal B: low ER/PR, high proliferation.
(ii) HER2 like: ER/PR negative, HER2 positive.
(iii) Basal like: ER/PR negative, HER2 negative, also known as triple-negative breast cancer (TNBC).

Overall, the 5-year survival rates for breast cancer are 91%, 86% after 10 years, and 80% after 15 years. Relative survival rates, however, do not predict individual prognosis as it fails to take into account the genetic variability between patients and the characteristics of the tumor. For instance, metastatic breast cancer (MBC) 5-year survival rates are only 22% [6]. Depending on the stage of breast cancer, the overall 5-year relative survival rate is 99% for localized disease, 85% for regional disease, and 27% for distant-stage disease. Furthermore, within each stage, the survival varies by tumor size; for regional disease, the 5-year relative survival is 95% for tumors less than or equal to 2.0 cm, 85% for tumors 2.1–5.0 cm, and 72% for tumors greater than 5.0 cm. Among the clinical subtypes, TNBCs accounts for 15%–20% of breast cancers, are often difficult to treat due to lack of surface markers, and are associated with a high rate of recurrence [4]. TNBC is more common in premenopausal women and those with a *BRCA1* mutation. Less than 30% of women with metastatic TNBC survive past 5 years, and almost all die due to their disease [5,6]. Moreover, the remarkable amount of heterogeneity of TNBCs concerning molecular alterations and tumor microenvironment makes it very challenging to target or treat disease.

Current therapies for breast cancer

Generally, breast cancer diagnosis and management rely on clinical assessment of its pathophysiology, tissue sampling, and examining the markers for molecular subtyping of the tumor. Besides, information on the patient's age and general health is also crucial to assess the other comorbidities that might impact choices of therapy. Based on the clinicopathological factors of the tumor and hormone receptor status, there is a range of modalities employed toward treatment including surgery, radiation therapy, systemic pharmacotherapies that include chemotherapy and endocrine therapy, and targeted immune therapies (antibodies against tumors antigens) [7,8]. Breast cancer surgery may be preceded by systemic neoadjuvant therapies to shrink the tumor for effective surgery. For instance, HER2+ breast cancers that are aggressive kind are administered with trastuzumab (Herceptin) and pertuzumab (Perjeta) as neoadjuvant therapy where trastuzumab is continued postsurgery [9,10]. Postsurgery patients undergo radiotherapy and/or chemotherapy to ensure the destruction of remnant micrometastatic cancer cells, thereby reducing chances of remission and thus increasing the overall patient survival [11].

Pharmacotherapies relies on the status of hormone receptors or target protein−targeted adjuvant (additional) therapies. In case of early stage ER/PR+ breast tumors that have not spread to the lymph nodes, patients are given long-term endocrine therapies such as tamoxifen or aromatase inhibitors together or alone with the surgical resection, depending on the patient's menopausal status [12]. In case of TNBCs, the common treatment strategies include chemotherapy with taxane and adjuvant treatment with anthracycline and taxane, sometimes in combination with PARP inhibitors [13] (discussed in detail in Section 6.6.2).

It is noteworthy that while these adjuvant systemic chemotherapies are routinely used for early stage ER/PR+ breast cancer patients, not all of them offer long-term survival benefits when compared to the cost of side effects due to overtreatment. To overcome these inadequacies of the conventional methods, genomic assays such as Oncotype DX have been successfully combined with traditional methods that allow clinicians to predict behavior of the cancer, its recurrence, and whether or not chemotherapy will be beneficial for the early stage breast cancer patients. Initially described and introduced into clinical practice in 2004, Oncotype DX is a multigene prognosis assay that predicted recurrence of tamoxifen-treated, node-negative breast cancer [14]. It analyzes the expression profiles of 21 genes and predicts the disease recurrence by providing a recurrence score ranging between 0 and 100 to early stage breast cancer patients. Oncotype DX assays were validated in ER+ and node-negative breast cancer and have been included in the guidelines of major oncology organization such as National Comprehensive Cancer Network, American Society of Clinical Oncology, and the National Institute for Health and Care Excellence. Several other commercial genomic tests are available to analyze breast cancer tumors, including EndoPredict (Myriad Genetics, Salt Late City, UT, USA), Prosigna Breast Cancer Prognostic Gene Signature Assay (PAM50; NanoString Technologies, Seattle, WA, USA), and MammaPrint (Agendia, Amsterdam, the Netherlands), with the goal to categorize patients based on their potential benefit from continued endocrine therapy or chemotherapy [15].

In addition to the abovementioned therapeutic approaches, recent years have seen a prominent reemergence of immunotherapy as biological therapeutics, the idea where the patient's immune system is used to fight cancer cells [16]. As of 2019, two immunotherapies have been approved to treat breast cancer treatment. The first is atezolizumab (Tecentriq) along with protein-bound paclitaxel (Abraxane), which is approved for locally advanced inoperable and metastatic TNBC that tested positive for the PD-L1 protein. Another approved candidate is pembrolizumab (Keytruda) for metastatic cancers that are microsatellite instability high or have DNA mismatch repair deficiency. These treatments are likely to work in patients with higher levels of PD-L1 or gene mutations but with considerable side effects. Nonetheless, there are more than 250 clinical trials underway extensively exploring immunotherapy for breast cancer treatment. Especially given the lack of efficient treatment option of TNBCs, there has been a constant need to develop novel drugs with much better efficacy than the existing ones offered for breast cancer therapy.

Drug repurposing

Drug repositioning, also referred to as drug repurposing or drug reprofiling, is the process of uncovering new indications of the approved or failed/abandoned compounds for use in a different disease [17]. This approach benefits from the fact that such compounds have already undergone extensive rigor of elaborate phases of new drug discovery that often includes detailed information on their safety, efficacy, formulation, dose, and potential toxicity [18] and in many instances, Phase I clinical trials. Fundamentally, drug repositioning is being made feasible because drugs perturb a multitude of biological processes that are common to many different diseases. Few of the prominent and obvious advantages of drug repositioning are a significant reduction in the cost and development time as their safety in humans is usually well established [17]. Advances in genomics, proteomics, transcriptomics, and metabolomics have provided tremendous insights into the molecular and metabolic alterations that manifest in cancers. The fundamentals of drug repurposing through the integration of systems biology and bioinformatics depend on basic concepts, i.e., activity-based and in silico drug discovery [19]. The former approach relies on utilizing experimental approaches to evaluate the anticancer activity directly. Drug candidates, which are structural similar, are likely to share biological activity and indications [20]. Likewise, if the same metabolic pathway is affected in the two different conditions, then the drug candidates targeting the specific pathway can be utilized as therapeutics for both diseases despite their structural dissimilarity [21]. The tendencies of the drugs with a strong side effect in certain disease can be explored further to see if these "off-target" effects for one disease could be relevant and novel for the treatment in some other disease [22,23].

In silico drug repositioning offers a new paradigm by integrating the wide range of high-throughput biological data for the discovery of novel indications for the existing drugs. Big data emerging from molecular interaction studies such as genome-wide association studies (GWAS), together with text-based searches using online Mendelian inheritance in man and PubMed databases can be utilized to perform systematic and coordinated bioinformatics analysis for repositioning studies. For example, researchers analyzed GWAS data together with proteomics and metabolomics and identified 992 proteins as potential antidiabetic targets, for which nine drugs were repositioned [24,25]. A recent study showcased a new approach, "Drug Repurposing from Control System theory," which provides a comprehensive framework using control theory paradigm for drug repurposing that considers various limitations in the previous databases such as variation in gene of interest copy numbers, mutations, and lack of reference for normal range of gene expression in different diseases and generates a disease-specific mathematical models [26].

Success stories of drug repurposing in breast cancer

Nobel laureate pharmacologist Sir James Black (1988) has stated that "the most fruitful basis for the discovery of a new drug is to start with an old drug." The drug repurposing efforts echo this wisdom and attempt to utilize the unexplored therapeutic potential of the existing drugs. However, since usually the information about mechanism of action of these drugs is limited and narrow in addressing the question around the disease of intended use, most of the success stories of repositioned drugs are serendipitous to date. Often they were discovered during later stages of clinical trials as unexpected yet beneficial findings. Even though the drugs have established bioavailability and safety information, they cannot be extrapolated for new diseases due to unknown or complex mechanism of drug actions [27,28]. This scenario is rapidly changing as with the ever-growing knowledge emerging from bioinformatics approaches, more rational and systematic pipelines are generated/being generating to narrow down the pharmacologically relevant biomolecules for repurposing efforts. In this section, we have expanded on the stories of the drugs that have been successfully repositioned for each clinical subtype of breast cancer. While the term drug repurposing is often used for drugs that are now being used for diseases other than their original intended use, in this section, we also discuss oncology drugs that were originally approved for other cancer types besides breast cancer. For example, some of the repurposing candidates for TNBCs that are either in the preclinical or clinical trials stages were previously approved for other cancers. Finally, in addition to these success stories, we have elaborated on the potential drug candidates for repurposing, which have shown promises in the animal studies or are part of ongoing clinical trials in Section 6.4. The approved repositioned drugs for breast cancer are listed in Table 6.1, grouped according to subtypes.

Hormone receptor—positive breast cancer

Nearly 70% of the breast tumors are receptors positive, which are treated with antiestrogen therapy. These nuclear hormone receptors belong to large family of nuclear receptors that acts as transcription factors and are modulated by the steroid hormones estrogen and progesterone [29,30]. A series of epigenetic events contribute to ER signaling following stimulation of ER by estrogen. Therapeutics have effectively targeted ER signaling; for instance, tamoxifen antagonizes the binding of estrogen to the ER or downregulation of ER by fulvestrant (Faslodex) [31].

Tamoxifen for treatment and prevention for HR+ breast cancer

Tamoxifen is selective estrogen receptor modulator (SERM) that directly acts on ER and mediates effects specific to tissue or organ. SERMs may concomitantly activate the ER pathway in certain tissues, while inhibiting the pathway in others. Inhibition of the hormone receptor has been pursued since the 1960s, although the initial goal was to develop them as contraceptives [32]. The association between hormones and cancer was recognized as early as 1916 [33], but their use to treat cancer was achieved during the 1930s by Lacassagne who showed in mice studies a direct link between estrogens and breast cancer appearance. Subsequent work by Charles Huggins at the University of Chicago employed synthetic compound stilbestrol [34], which had estrogenic properties and was used to treat prostate cancer. Another compound with estrogenic activity to be analyzed was triphenylethylene. Years of extensive studies by ICI laboratories (now part of AstraZeneca) of triphenylethylene derivative led to the synthesis of dimethylamino ethoxy compound ICI 46,474 or Tamoxifen (brand name Nolvadex) in 1962. Tamoxifen was pursued as a part of the oral contraception program in ICI and was designed to serve as the contraceptive pill. Initial clinical trial results and isolation of ERs in 1968 [35] accelerated the pipeline development of tamoxifen. However, in 1971 a clinical trial conducted in the Karolinska Institute in Stockholm pulled the plug on tamoxifen's future as a contraceptive pill in women. The results showed that contrary to studies in rats, rather than suppression the ovulation, tamoxifen

TABLE 6.1 Advanced/FDA approved repositioned drugs in different breast cancer subtypes.

Breast cancer subtype	Drugs	Properties	Original use	Current status
Hormone receptor positive (estrogen receptor/progesterone receptor; ER/PR)	Tamoxifen	Estrogen receptor antagonist	Fertility drug (ovulation induction) Osteoporosis	FDA approved for breast cancer prevention and treatment
	Raloxifene	Estrogen receptor antagonist	Osteoporosis treatment in menopausal and postmenopausal women	FDA approved for breast cancer prevention
HER2 positive	Nelfinavir/Viracept	HIV protease inhibitor	HIV treatment	Preclinical study
	Propranolol	Beta-1 adrenergic receptors antagonist	Infantile hemangioma, hypertension, anxiety, cardiac arrhythmia, hyperthyroidism, infarction, thyrotoxicosis	Clinical trial
Triple-negative breast cancer and advanced metastatic breast cancer	Clofazimine/Lamprene	Antibacterial	Multibacillary leprosy treatment	Preclinical study
	Penfluridol	Synthetic nucleoside analog	Antipsychotic drug for schizophrenia treatment	Preclinical study
	Paclitaxel (PTX)/Taxol	Suppression of microtubule dynamics	Atrial restenosis, ovarian cancer	FDA approved for metastatic breast cancer combination therapy
	Gemcitabine	Cell division	Antiviral drug	FDA approved
	Goserelin	Luteinizing hormone–releasing hormone (LHRH) agonist	Prostate cancer, uterine fibroids, assisted reproduction	**Phase II clinical trial standard neoadjuvant therapy for TNBC**

resulted in its stimulation. While ICI initially considered closing the program, parallel studies and data from its clinical trial initiated a global interest in its potential as an antibreast cancer drug. In 1973, Nolvadex was launched in the UK markets as both fertility drug and palliative treatment for breast cancer. In between 1973 and 1975, tamoxifen was upgraded from palliative to adjuvant therapy and finally to chemopreventive. Tamoxifen was first of a kind due to its origin as a contraceptive drug. Therefore, its administration was oral. The series of clinical trials that underwent during the journey of tamoxifen from a contraceptive to cancer pill can be further referred to in the article by Refs. [36,37]. Over the years, studies have demonstrated a clear reduction in mortality with use in the adjuvant setting. Metaanalyses of 15 years follow-up data from randomized trials of patients with ER-positive disease given 5-year adjuvant tamoxifen after 6 months of anthracycline-based chemotherapy have demonstrated an approximately 50% reduction in breast cancer mortality. One of beneficial effects of tamoxifen is it preserves the bone mineral density in postmenopausal patients; however, owing to its estrogenic activity in other tissues, patients are at high risk of experiencing some serious side effects including vaginal bleeding, endometrial hyperplasia, and an increased risk of invasive endometrial cancer and thromboembolism [38,39]. Therefore, other SERMs have been developed, with the hope to achieve more effective mode of action with lower or negligible adverse effects.

Raloxifene for treatment and prevention of HR+ breast cancer

Another drug belonging to SERMS is raloxifene, which is a benzothiaphene type of second-generation SERM that received its approval in the United States for breast cancer prevention in postmenopausal women in 2007. However, in many countries, its use remains limited [40]. Initially approved for the osteoporosis treatment in menopausal and postmenopausal women, raloxifene acts essentially by mimicking the action of estrogen in bone tissue. It regulates the proliferation activity of bone cells, including both osteoblasts and osteoclasts by modulating expression levels of matrix proteins and collagen and IL-6, respectively [41]. However, in the breast and endometrium tissue, raloxifene exhibits antiestrogen effects, unlike tamoxifen, and thus may prove a safer alternative. It binds to the ER's ligand-binding domain, specifically AF-2 region that is the binding site for coactivators and corepressors that are crucial for the binding of the ER to estrogen-responsive elements, thereby activating the transcription of the target genes. To investigate the effects of raloxifene, the MORE trial [42]was conducted, which is a large multicenter randomized placebo-controlled study conducted across 180 clinical centers in 25 countries with 7705 postmenopausal women with osteoporosis enrolled. The trial's primary aim was to evaluate the risk of fracture in the postmenopausal women with osteoporosis after 3 years of raloxifene administration at a daily dose of either 60 or 120 mg. The studies showed that candidates on raloxifene had increased bone mineral density and a reduced risk of vertebral fractures compared to the placebo group. As a secondary endpoint of the study, the occurrence of breast cancer was evaluated, which showed that raloxifene reduced the risk of invasive breast cancer by 76% after 3 years compared to the placebo group. The benefits were most significant in women with ER-positive cancers who showed a reduced risk of invasive breast cancer by 90% and 86% after 3 [43]and 4 years [44]. Unlike tamoxifen, which was reported to show an enhanced risk of endometrial cancer, the MORE trial showed no effect of the treatment with raloxifene on the endometrium. After 8 years, the patients from the MORE trial were included in a follow-up study, referred to as the CORE trial. This trial reported a continued protective effect of raloxifene for breast cancer incidence. The risk of invasive breast cancer and ER-positive breast cancer in the patients showed a staggering reduction by 66% and 76%, respectively, compared to the placebo group [45]. Raloxifene continued to prove its efficacy toward breast cancer prevention in the subsequent clinical studies. It was not only as effective as tamoxifen in terms of risk reduction but also showed a decreased risk of endometrial cancer in addition to reduced side effects like pulmonary embolism and cataracts compared to tamoxifen [40,46]. Currently, drug repositioning efforts focus mostly on efficacy. In the future, drug associated side effects will inevitably be a major determinant of success for drug repositioning as well.

HER2 + breast cancer

HER2 or HER2/neu gene was identified in early 1980s by Robert Weinberg group. HER2 is a tyrosine kinase receptor (RTK) that is involved in the regulation of cellular growth and differentiation signaling pathways. HER2 expression is highly correlated with the amplification of the c-ERBB2 proto-oncogene [47]. Amplification of HER2/neu oncogene and associated genetic elements on chromosome 17 results in its significant overexpression (up to 100-fold) in breast tumor cells compared to normal tissues. Prior to the introduction of targeted therapy, HER2-positive breast cancers had a very poor prognosis, with high risk invasiveness and recurrence. The monoclonal antibody trastuzumab (Herceptin) introduced in 1998, in combination with chemotherapy, has been the standard treatment for the patients with metastatic HER2-positive breast cancer for the past two decades. Herceptin brings about its action by binding and blocking the Her2 receptors present on cell surface, thus preventing the cell from receiving the signal for its growth and division. However, this treatment suffered from the challenge of de novo and acquired resistance in which patients fail to show objective response and thus develop disease progression while continuing on trastuzumab therapy.

Nelfinavir

Nelfinavir, also known as Viracept, is an aspartyl protease inhibitor that was originally approved by the FDA for the treatment of HIV in 1997. For HIV patients, the initial oral dosage was 750 mg three times a day, which was revised to a simpler dose regimen of 1250 mg twice a day based on clinical trials showing improved efficacy [48]. However, both regimen have proven to be quite effective with a few of the common side effects that include insulin resistance, hyperglycemia, and lipodystrophy [49]. Around early 2000, studies showing anticancer effects of nelfinavir began to emerge, which subsequently led to an increasing interest to identify the underlying mechanism of its action [46]. It was reported independently by several groups that nelfinavir inhibited AKT signaling in cancer cells. This is highly pertinent because insulin resistance is brought about by the inhibition of phosphatidylinositol-3-kinase (PI3K)/AKT signaling pathway, which is also an important mediator for cancer cell survival and resistant to chemo- and radiation therapy.

In a breast cancer study, Srirangam et al. [50] showcased that ritonavir, an HIV protease inhibitor that is a structural relative of nelfinavir, selectively inhibited the growth of HER2-positive breast cancer cells by binding to HSP90. This binding between nelfinavir and HSP90 inhibited HSP90 and AKT interaction resulting in the degradation of HER2. HSP90 is a molecular chaperone that functions as a stabilizer for several cellular proteins and facilitates protein folding posttranslation, and under stress, it has been of interest in cancer [51]. Several inhibitors such as geldanamycin and novobiocin target HSP90 by binding to its ATP binding pocket and other chaperons such as HSP70 or by binding its dimerization domain thus inducing conformation changes. While nelfinavir does enhance interaction between HSP90 and HSP70, it does not affect the process of dimerization. Additionally, it was shown that nelfinavir reduced the levels of HSP90 client proteins, CDK4 and CDK6, thus inhibiting AKT signaling. Additionally, it showed an impressive growth inhibition in HER2-positive breast cancer cells that were resistant to either trastuzumab and/or lapatinib. In vivo analysis showed that nelfinavir brought about selective inhibition of HER2-positive breast cancer cell growth. Emerging studies have proposed an additional mechanism of nelfinavir, endoplasmic reticulum stress induction, and downregulation of hypoxia-inducible factor 1α resulting in cancer cell autophagy and angiogenesis inhibition [52]. This implies that nelfinavir interacts with a myriad of cellular proteins that potentially results in the inhibition of multiple pathways of significance for cancer cell growth and survival. Ongoing Phase I/II clinical trial data would help in substantiating the use of nelfinavir as a successfully repositioned breast cancer drugs.

Propranolol

The nonselective beta-blocker propranolol, targeting beta-adrenergic receptors, has emerged as one of the prominent drug repurposing candidates for its antimetastatic effect, particularly in breast cancer. It was originally developed in the early 1960s in ICI and was effectively administered for the treatment of myriads of conditions including infantile hemangioma, hypertension, anxiety, cardiac arrhythmia, and hyperthyroidism infarction thyrotoxicosis. In the case of HER2-positive breast cancer, propranolol has been shown to potentially revert resistance to trastuzumab. In this retrospective study, patients with HER2 overexpressing MBCs showed improved overall survival rates of 38.71% in the group given trastuzumab and concurrent chemotherapy with β-blocker propranolol versus 28.57% in the control group treated with trastuzumab and chemotherapy without propranolol after 36 months of treatment [53].

TNBC

Despite the established complexities of breast cancer, endocrine receptors ER and PR and aberrant expression of HER2 remain markers for clinical assessments. By definition, TNBC lacks the expression of all these classical markers that serve as targets of therapeutic agents. Traditionally, systemic chemotherapy has proven to provide higher response rates for breast cancer patients with early stage as well as advanced stages; however, fewer than 30% of metastatic TNBC patients survive 5 years after diagnosis, with none of them surviving the disease eventually [54]. About 60%–70% of patients with TNBC fail to respond fully to chemotherapy; however, nearly 90% of them carry alterations in the pathways that are being targeted by clinically tested agents such as PARP inhibitors, PI3K inhibitors, MEK inhibitors, heat shock protein 90 inhibitors, and histone deacetylase inhibitors [REF]. Clinical trials with platinum agents in TNBC have demonstrated clinical benefits but have not proven to show overall survival benefits [55]. It is suggested that this tumor cell population, which survived chemotherapy, constitutes the subpopulations of cells with both tumor-initiating and mesenchymal features. Inhibitors of the dynamic cancer stem cell population, also referred to as tumor-initiating cells, have not yet been established. Studies have separately reported different signaling pathways such as Wnt [56], JAK/STAT [57], Notch [58], and TGF-β [59] pathways that can potentially contribute to the maintenance of breast cancer stem cells (BCSCs) and therefore constitute as strong targets for novel therapeutics [60].

Standard of care drugs: paclitaxel and goserelin for TNBC and metastatic breast cancer

Paclitaxel is classified as a taxane, an antimicrotubule agent that promotes tubulin dimerization and inhibits depolymerization of the microtubules. Originally isolated from the Pacific yew's bark in the 1960s, taxol's antitumor activity in the mouse melanoma was confirmed by 1977. In 1992, taxol was FDA approved for the treatment for ovarian cancer. Subsequent testing and clinical trials showed Taxol as an effective candidate for advanced breast cancer, and in 1994, the FDA approved Taxol for use against breast cancer.

Gemcitabine was initially developed in early 1980 as antiviral drug; however, preclinical studies showed its cytotoxicity in leukemia cells in vitro. It is a synthetic pyrimidine nucleoside analog that replaces cytidine during DNA replication resulting in arrest of tumor growth and induces apoptosis. Gemcitabine and paclitaxel combination therapy was approved for MBC therapy around 2004.

Goserelin, a luteinizing hormone–releasing hormone agonist, was used for the treatment of prostate cancer and later also approved for the treatment of premenopausal women with hormone-sensitive breast cancers in 1989 [55]. In a study with 257 premenopausal females with ER-negative breast cancer, groups administered with goserelin in combination with chemotherapy also showed improved survival rates compared to chemotherapy alone [56]. Subsequently, an a Phase II clinical trial was initiated to assess its effectiveness in addition to standard neoadjuvant therapy for TNBC patients and is expected to complete by 2023 [57].

Clofazimine

In the case of TNBC and several other cancers such as liver and colon cancer, Wnt signaling pathway, which is active during embryogenesis, is emerging as a candidate for targeted therapeutics. Its reactivation in adults, possibly due to mutations or epigenetic modifications, results in erroneous growth signals that results in tumor development. Wnt signaling is overactivated in TNBC and several other cancers, and its suppression may be an effective anticancer treatment. However, drugs targeting the Wnt pathway do not currently exist on the market or under advanced clinical trial.

Clofazimine, first introduced in 1957, is a riminophenazine derivative antibiotic. It was initially approved for the treatment of tuberculosis but has been used mainly for the treatment of multibacillary leprosy, owing to its lower efficacy toward former. Researchers at the University of Lausanne in 2014 showed that clofazimine exerts inhibitory effect on Wnt signaling pathway in TNBC cell line HTB19 [61]. Subsequently, the preclinical anticancer efficacy and effect of clofazimine was recently validated in vivo by the same research group using xenograft mice models. Clofazimine administration resulted in significant reduction of tumor growth with no adverse side effects. The drug was even more effective when administered in combination with the cytotoxic drug doxorubicin [62]. Considering low toxicity profile of clofazimine and proven success of this above-discussed drug combination in Phase II clinical trial of primary hepatocellular carcinoma [63], clofazimine is a promising candidate of repositioning for TNBC.

Penfluridol

Penfluridol is a first-generation antipsychotic drug used to treat schizophrenia that exerts antiproliferative effects, presumably by interference with integrin signaling. Its antitumor effects have been demonstrated in vivo, both in localized disease and in an experimental model of breast cancer metastasis to the brain [40]. In schizophrenia patients, penfluridol is approved for oral intake. One recent study showed antiproliferative and antimetastatic effect of penfluridol treatment on TNBC in vitro and in vivo. In a breast tumor orthotopic mice model, oral administration of penfluridol led to significant suppression of tumor growth and brain metastases, potentially through the inhibition of integrin $\alpha6\beta4$ signaling axis. There were no significant side effects for the dose showing significant tumor growth suppression, thus making it a highly favorable candidate for repurposing for TNBC [64].

Current candidates for TNBC repositioning

This section focuses on examples of drug candidates currently under consideration for repurposing, which are targeting the intractable issues that make finding suitable treatment modalities for the cancer challenging, specifically tumor heterogeneity and cancer stem cells. Given the poor prognosis of TNBC and challenges in establishing the treatment modalities, different approaches have tried to capture its complex molecular landscape. Although not uniform across TNBC, the somatic mutations for TNBC average up to 1.68 mutations per Mb of the coding region, thereby amounting for nearly 60 mutations in each tumor. The most frequently mutated gene is TP53 (60%–70%), followed by PIK3CA (~10%) and others (<1%) including *ERBB2* and $BRAF^{V600E}$. It is noteworthy to mention that the presence of germline mutations in BRCA1/2 increases the lifetime risk of breast cancer to 60%–70% [65] and is observed in about 10% of patients with TNBC [66,67]. These molecular alterations present a formidable challenge for ad hoc drug development. Redefining the trial designs and including genomic screening in the clinics [68] would allow focusing on the scope of these novel genetic mutations for their candidacy as druggable candidates.

Ongoing research has shown emergence of few promising drug candidates to be repositioned for TNBC. Ras has been shown to drive the progression of basallike breast cancer, an aggressive molecular subtype of TNBC. Ras is an important drug target because it mediates the growth factor signaling that promotes tumorigenesis. To target Ras-driven cancers, inhibitors such as PLX4032/vemurafenib targeting the effector kinase B-Raf in melanoma were used but exhibited drug resistance later on. Thus, it may be effective to target the Ras protein directly in order to overcome such challenges. In this league, FDA-approved drugs that degrade Ras were screened and flunarizine, an N-Ras inhibitor, approved for migraine or vertigo has shown promise in basal-like subtype of TNBC. In xenograft mice models, flunarizine was shown to induce degradation of N-Ras GTPases that are responsible for driving tumor growth by inducing autophagy [69].

Targeting metabolic pathways is yet another direction that has gained attention for cancer treatment. Especially in the case of TNBCs, which suffer from several disadvantages including enormous heterogeneity, lack of effective targets, and drug resistance, metabolic inhibitors could prove to be a highly effective option. A recent study [70] at the University of Chicago identified BACH1 as a prominent regulator of mitochondrial metabolism by controlling transcription of its electron transport chain genes and the key determinant of TNBC response to the treatment with metformin. The BACH1 protein is often highly expressed in the TNBC patients and is required for metastasis. The researchers have provided mechanistic evidence for repurposing two old drugs, metformin and heme, to treat resistant tumors in xenograft mice models. Discovered in 1922, metformin has been clinically used to treat type-2 diabetes and has been shown to have an anticancer effect by repressing the proliferation of tumor cells [71]. Heme or Panhematin was initially crystallized from blood in 1853 and has been used to treat porphyrias and related ailments. The study found that heme targets BACH1, thus forcing the BACH1-depleted breast cancer cells to change their metabolic pathways. Subsequent treatment with metformin caused mitochondrial suppression of the susceptible cancer cells. The researchers proposed that BACH1 levels could serve as biomarkers to predict the metformin resistance in the patients and the addition of heme treatment would sensitize them to metformin. Thus, reprogramming the metabolic network offers a novel approach to enhance the efficiency of the inhibitor drugs.

One of the formidable challenges for ongoing repurposing efforts is heterogeneity of breast cancer especially toward intractable subtypes such as TNBCs and aggressive MBC. This is important because heterogeneity and complexity of breast cancer have a major influence on the efficacy of drug treatment. This variability of response between individuals is partly due to genetic polymorphisms in drug-metabolizing enzymes, drug transporters, receptors, and other drug targets [72]. For example, studies have shown that the CYP2D6 genotype in patients determines the potency of tamoxifen treatment for ER+ breast cancer [73]. Thus, drugs that showed poor efficacy in initial trials that were conducted in patient cohorts with mix genomic profiles may still hold promise for certain individuals. For example, the lack of efficacy of sunitinib malate, a multitarget oral RTK inhibitor, led to failed Phase III trials for metastatic and TNBC [74,75]. But the drug was later approved for imatinib-resistant or imatinib-intolerant gastrointestinal stromal tumors, advanced kidney cancer, and advanced pancreatic neuroendocrine tumors by the FDA. Recent ongoing clinical trials are again investigating sunitinib's efficacy in combination with other drugs such as crizotinib, paclitaxel, doxorubicin, or cyclophosphamide [76].

All of the therapeutic strategies aim to achieve effective remission-free survival for patients yet nearly 30% lymph node–negative and 70% lymph node–positive breast cancer relapse after 5–20 years from their initial diagnosis. This is a result of a period of dormancy ascribed to dormant disseminated tumor cells (DTCs) that contribute to the micrometastases even before the primary treatment began and therefore are inherently chemoresistant. These nonproliferative cells could persist for extended periods in the patients before reemerging as aggressive secondary lesions [77]. Similar to DTC, BCSCs share similarities with respect to intrinsic pathways regulating the survival, state of quiescence, and interaction with their microenvironment [78]. Elevated activation of autophagy is one of the many prosurvival strategies and has emerged as primary focus of 32 ongoing human clinical trials as detailed by Flynn et al. [77]. Antimalarial agents chloroquine (CQ) and its closely related molecule hydroxychloroquine block autophagosome–lysosome fusion causing inhibition of autophagy and are extensively explored as repurposing candidates for breast cancer. Mice implanted with metastasizing 4T1 mouse breast cancer cells showed reduction in the primary tumor volume and enhanced survival time following administration of 25 and 50 mg/kg of CQ [79]. In TNBC, CQ was shown to eliminate BCSC through reduced expression of Janus-activated kinase 2 and DNA methyl transferase 1 [80]. More recently, researchers have shown in TNBC cell lines, HCC38 and MB-MDA-231, that CQ could potentially boost the effectiveness of the novel glutaminase-inhibiting drug via combined effect of inhibition of autophagy in response to accelerated lipid catabolism and oxidative stress brought about by glutaminolysis inhibition [81].

Challenges and future directions in breast cancer drug repurposing

Drug repurposing certainly offers scope of faster access of drugs to patients at reduced economic burden, yet its success is limited by challenges of patient and disease heterogeneity, reemergence, and metastasis of cancer. In this section, in addition to challenges, we have also discussed the potential of integrative approaches of identifying novel drug targets, utilizing machine learning computational methods, and including combination drug therapy in order to predict and validate successful disease–drug pairs.

Despite improved and targeted pharmacotherapies, intratumoral heterogeneity and drug resistance continued to play a major part in therapeutic failure for many patients. Drug resistance is common in all breast cancer types irrespective of the treatment modalities applied, due to complicated underlying mechanisms. Repurposed drugs effects are similar to the drugs that are specifically developed to target the cancer considering that tumor cells acquire resistance to drugs later on in the therapy. Now since the molecular pathways involved in any disease condition are interconnected, action of a combination of drug could attack the cancer from multiple sides and thus offer better susceptibility. For instance, combining repurposed drugs (nitroglycerin) with chemotherapeutic agents (vinorelbine and cisplatin) has shown promising outcome in one of the randomized Phase II trials for nonsquamous cell lung cancer [82]. Another study looked into utilizing 5-fluorouracil (5-FU) in combination with the itraconazole, which was one of the 6 drug candidates predicted for repurposing in breast cancer (ReDO) project, for breast cancers that have developed chemoresistance toward 5-FU. Itraconazole, combined with 5-FU, showed better responses (increased reduction of cellular viability) compared to both individual drug treatment in cancer cell lines.

Another important point of consideration in addressing therapeutic failures is identifying the correct patient cohorts that are critical in the success in drug development. This is particularly true for the highly heterogeneous disease (i.e., TNBC). Heterogeneity of TNBCs is certainly a major challenge in drug repositioning efforts. Due to large number of potential subtypes within TNBCs, discovery of candidate drugs would require validation in rare cohorts representing these subtypes. This hurdle may prevent pharmaceutical companies to invest in developing new candidates. On the other hand, this provides an excellent opportunity for future repurposing efforts that may eventually reduce the cost and time of developing new therapeutic regimens for targeted cohorts. One mean of doing so is by further classifying disease using well-defined gene signatures (either at mutation or gene expression level). For example, the traditional drug development pipeline where drug leads are identified from chemical screening of a large number of candidate compounds towards a clearly defined target, therefore enables the development of a companion diagnosis kit. One such recent example is alpelisib for PI3K-mutated, progressive ER+ breast cancers. Besides general hormone receptor status, only a limited number of companion diagnostics are being used to make treatment decisions to treat certain breast cancers. The exact target of a repositioned drug however in the new disease setting is not always apparent. Therefore, although cancer genomic data are certainly propelling development of new targeted drugs to treat breast cancers, there has not been examples of companion diagnosis kit developed for repurposed drug yet. The commercial genomic tests in breast cancer mentioned in the chapter for the most part are used to characterize the disease prognosis risk and subsequently assist in determining the aggressiveness of treatment. To our knowledge, none of them are used to guide the selection of specific therapeutic agent. Nonetheless, the future of drug development either through traditional pipeline or reposition would require identification of proper patient population.

As discussed in the previous section, autophagy and its markers have gained significant attention as druggable and biomarker candidate status, respectively, especially, since autophagy is involved in both tumor suppression and promotion and therefore requires careful considerations before targeting it in the clinical settings. This is in fact evident that none of FDA approved drug targeting autophagy has been approved for breast cancer yet. Studies using breast cancer cell lines showed a differential pattern of survival in response to inhibition of autophagy. These effects were driven by signal transducer and activator of transcription 3 activity and interleukins IL-6 secretion. Utilizing dependencies on such biomarkers during clinical trials could allow identification of the responders from nonresponders for the autophagy targeting drugs. For instance, a glioblastoma (GBM) combination Phase I trial is evaluating potential of EGFRvIII as biomarker to assess response of administering CQ together with chemo- and radiation therapy (NCT02378532) [83]. Specifically, GBM patients show very poor prognosis with median survival of only 14.6 months following chemo- and radiation therapy. Amplification/mutation of epidermal growth factor receptor *EGFR* is common in GBM and is thought to cause resistance from radiotherapy. 50%–60% of patients with EGFR amplification have EGFRvIII mutation. This mutation enhances the ability of cancer cells to survive via autophagy during stresses such as hypoxia and starvation. So while trial primarily attempts to establish the maximum tolerated dose for CQ in combination with chemo- and radiotherapy, it will also provide data that can shed light on utilizing these markers to drive targeted therapy in GBM patient with or without EGFRvIII mutations. In case of breast cancer, for instance, AKT inhibitor, ipatasertib, together with paclitaxel showed 56% reduction in the risk of disease progression in patients with AKT E17 mutant TNBC [84]. It will be pertinent to include serial assessment of biomarkers in clinical trials while pursuing the drug repurposing candidates.

Similar to the regular pipeline of drug development, the earlier drug repositioning efforts were mainly focused on identifying drugs that aimed at preventing breast cancer. However, with the increasing threat of MBC, more efforts are needed to identify drugs that can prevent the spread of the disease in addition to controlling localized disease. We expect to see more candidate drugs show up in the near future combating metastatic disease through a combination of known biology-driven discovery and machine learning approach. As in new drug development, the newly repurposed drugs may be used as first- or second-line treatment either alone or with the existing standard of care agents. However, these remained to be evaluated through prospective clinical trials.

Emerging data from cell line, mice, and clinical studies have been actively leveraged in association with bioinformatics explorations to identify the drug repurposing possibilities. For instance, the Connectivity Map (CMap) provides a repository of gene expression profiles for drug phenotype in the human cells. The analysis of the potential of drugs to alter gene expression profiles in cancer cells provides insights into their mechanism of action. One such example is repositioning studies for antidepressant trifluoperazine, initially approved for schizophrenia. It showed synergistic tendencies with the current standard of care drug gefitinib in lung cancer. Another study identified several phenothiazines using CMap analysis, with potential efficacy in tamoxifen-resistant breast cancer cell lines. In addition to CMap, another gene signature based bioinformatics tool is Differentially Expressed Gene Signatures—Inhibitors (DeSigN). DeSigN identified bosutinib (originally used for chronic myeloid leukemia), which was shown in vitro to be effective toward oral squamous cell carcinoma.

With the enhanced understanding of molecular resistance mechanisms and the growing arsenal of novel therapeutic agents, one can imagine a future of truly personalized medicine in which therapeutic cocktails of targeted and other agents are administered according to a customized protocol planned based on the molecular characterization of the tumor.

Conclusion

In this chapter, we have elaborated on the advantages, success stories, and future directions of drug repurposing in breast cancer treatment. Based on the clinical subtypes, we have provided a comprehensive narrative of drugs that were successfully repositioned for breast cancer treatment.

Drug sensitivity prediction approaches have been assessed and benchmarked in vitro data [85] and are continually extended to obtain response predictions in patients. Combining biomarker discovery and drug repurposing together will open new avenues for cancer therapy. Considering that developing new drug is a highly expensive and time-consuming venture [86,87], identifying new clinical indications for existing drugs, or for those that are in the pipeline, substantially cuts down the risk, costs, and wait period for drug discovery. Recently, computational drug repositioning has shown promises in accelerating drug discoveries for cancer. Also, the combined use of drugs has long been recognized to improve treatment efficacy. However, in vitro and in vivo screening of all possible drug combinations is not feasible. Therefore, computational methods to effectively explore this space of drug combinations are highly desirable. These methods could significantly impact the clinical evaluation and subsequent utility of the validated drug combination in patient care.

References

[1] Prat A, Pineda E, Adamo B, Galvan P, Fernandez A, Gaba L, et al. Clinical implications of the intrinsic molecular subtypes of breast cancer. Breast 2015;24(Suppl. 2):S26–35.

[2] Banin Hirata BK, Oda JM, Losi Guembarovski R, Ariza CB, de Oliveira CE, Watanabe MA. Molecular markers for breast cancer: prediction on tumor behavior. Dis Markers 2014;2014:513158.

[3] Parker JS, Mullins M, Cheang MC, Leung S, Voduc D, Vickery T, et al. Supervised risk predictor of breast cancer based on intrinsic subtypes. J Clin Oncol 2009;27(8): 1160–7.

[4] Dent R, Trudeau M, Pritchard KI, Hanna WM, Kahn HK, Sawka CA, et al. Triple-negative breast cancer: clinical features and patterns of recurrence. Clin Cancer Res 2007;13(15 Pt 1):4429–34.

[5] Hurvitz S, Mead M. Triple-negative breast cancer: advancements in characterization and treatment approach. Curr Opin Obstet Gynecol 2016;28(1):59–69.

[6] Pal SK, Childs BH, Pegram M. Triple negative breast cancer: unmet medical needs. Breast Cancer Res Treat 2011;125(3):627–36.

[7] Nounou MI, ElAmrawy F, Ahmed N, Abdelraouf K, Goda S, Syed-Sha-Qhattal H. Breast cancer: conventional diagnosis and treatment modalities and recent patents and Technologies. Breast Cancer: Basic Clin Res 2015;9(Suppl. 2):17–34.

[8] Conway A, McCarthy AL, Lawrence P, Clark RA. The prevention, detection and management of cancer treatment-induced cardiotoxicity: a meta-review. BMC Cancer 2015;15:366.

[9] Swain SM, Im Y-H, Im S-A, Chan V, Miles D, Knott A, et al. Safety profile of pertuzumab with trastuzumab and docetaxel in patients from Asia with human epidermal growth factor receptor 2-positive metastatic breast cancer: results from the phase III trial CLEOPATRA. Oncologist 2014;19(7):693–701.

[10] Tsang RY, Finn RS. Beyond trastuzumab: novel therapeutic strategies in HER2-positive metastatic breast cancer. Br J Cancer 2012;106(1):6–13.

[11] Matsen CB, Neumayer LA. Breast cancer: a review for the general surgeon. JAMA Surgery 2013;148(10): 971–9.

[12] Puhalla S, Brufsky A, Davidson N. Adjuvant endocrine therapy for premenopausal women with breast cancer. Breast 2009;18(Suppl. 3):S122–30.

[13] Lebert JM, Lester R, Powell E, Seal M, McCarthy J. Advances in the systemic treatment of triple-negative breast cancer. Curr Oncol 2018;25(Suppl. 1):S142–50.

[14] Paik S, Shak S, Tang G, Kim C, Baker J, Cronin M, et al. A multigene assay to predict recurrence of tamoxifen-treated, node-negative breast cancer. N Engl J Med 2004;351(27):2817–26.

[15] Siow ZR, De Boer R, Lindeman G, Mann GB. Spotlight on the utility of the Oncotype DX® breast cancer assay. Int J Wom Health 2018;Volume 10:89–100.

[16] Aggarwal S, Verma SS, Aggarwal S, Gupta SC. Drug repurposing for breast cancer therapy: old weapon for new battle. Semin Cancer Biol 2019;19:30294–9.

[17] Oprea TI, Bauman JE, Bologa CG, Buranda T, Chigaev A, Edwards BS, et al. Drug repurposing from an academic perspective. Drug Discov Today Ther Strateg 2011;8(3–4):61–9.

[18] Ashburn TT, Thor KB. Drug repositioning: identifying and developing new uses for existing drugs. Nat Rev Drug Discov 2004;3(8):673–83.

[19] Li J, Zheng S, Chen B, Butte AJ, Swamidass SJ, Lu Z. A survey of current trends in computational drug repositioning. Brie Bioinf 2016;17(1):2–12.

[20] Oprea TI, Tropsha A, Faulon JL, Rintoul MD. Systems chemical biology. Nat Chem Biol 2007;3(8):447–50.

[21] Gupta SC, Sung B, Prasad S, Webb LJ, Aggarwal BB. Cancer drug discovery by repurposing: teaching new tricks to old dogs. Trends Pharmacol Sci 2013;34(9): 508–17.

[22] Oprea TI, Mestres J. Drug repurposing: far beyond new targets for old drugs. AAPS J 2012;14(4):759–63.

[23] Napolitano F, Zhao Y, Moreira VM, Tagliaferri R, Kere J, D'Amato M, et al. Drug repositioning: a machine-learning approach through data integration. J Cheminf 2013;5(1):30.

[24] Zhang M, Luo H, Xi Z, Rogaeva E. Drug repositioning for diabetes based on 'omics' data mining. PLoS One 2015;10(5):e0126082.

[25] Akhoon BA, Tiwari H, Nargotra A. Silico drug design methods for drug repurposing. 2019. p. 47–84.

[26] Nguyen TM, Muhammad SA, Ibrahim S, Ma L, Guo J, Bai B, et al. DeCoST: a new approach in drug repurposing from control system theory. Front Pharmacol 2018; 9.

[27] Yella JK, Yaddanapudi S, Wang Y, Jegga AG. Changing trends in computational drug repositioning. Pharmaceuticals 2018;11(2).

[28] Pushpakom S, Iorio F, Eyers PA, Escott KJ, Hopper S, Wells A, et al. Drug repurposing: progress, challenges and recommendations. Nat Rev Drug Discov 2019; 18(1):41–58.

[29] Evans RM. The steroid and thyroid hormone receptor superfamily. Science 1988;240(4854):889–95.

[30] Yasar P, Ayaz G, User SD, Gupur G, Muyan M. Molecular mechanism of estrogen-estrogen receptor signaling. Reprod Med Biol 2017;16(1):4–20.

[31] Lim E, Metzger-Filho O, Winer EP. The natural history of hormone receptor-positive breast cancer. Oncology 2012;26(8):688–94. 96.

[32] Nasrazadani A, Thomas RA, Oesterreich S, Lee AV. Precision medicine in hormone receptor-positive breast cancer. Front oncol 2018;8:144.

[33] Lathrop AE, Loeb L. Further investigations on the origin of tumors in mice. III. On the part played by internal secretion in the spontaneous development of tumors. J Cancer Res 1916;1(1):1–19.

[34] Dodds EC, Goldberg L, Lawson W, Robinson R. Oestrogenic activity of certain synthetic compounds. Nature 1938;141(3562):247–8.

[35] Gorski J, Toft D, Shyamala G, Smith D, Notides A. Hormone receptors: studies on the interaction of estrogen with the uterus. Recent Prog Horm Res 1968;24:45–80.

[36] Quirke VM. Tamoxifen from failed contraceptive pill to best-selling breast cancer medicine: a case-study in pharmaceutical innovation. Front Pharmacol 2017;8: 620.

[37] Jordan VC. Tamoxifen (ICI46,474) as a targeted therapy to treat and prevent breast cancer. Br J Pharmacol 2006; 147(Suppl. 1):S269–76.

[38] Perez EA. Safety profiles of tamoxifen and the aromatase inhibitors in adjuvant therapy of hormone-responsive early breast cancer. Ann Oncol 2007; 18(Suppl. 8):viii26–35.

[39] Fisher B, Dignam J, Bryant J, Wolmark N. Five versus more than five years of tamoxifen for lymph node-negative breast cancer: updated findings from the National Surgical Adjuvant Breast and Bowel Project B-14 randomized trial. J Natl Cancer Inst 2001;93(9):684–90.

[40] Sleire L, Forde HE, Netland IA, Leiss L, Skeie BS, Enger PO. Drug repurposing in cancer. Pharmacol Res 2017;124:74–91.

[41] Clemett D, Spencer CM. Raloxifene: a review of its use in postmenopausal osteoporosis. Drugs 2000;60(2): 379–411.

[42] Ettinger B, Black DM, Mitlak BH, Knickerbocker RK, Nickelsen T, Genant HK, et al. Reduction of vertebral fracture risk in postmenopausal women with osteoporosis treated with raloxifene: results from a 3-year randomized clinical trial. Multiple Outcomes of Raloxifene Evaluation (MORE) Investigators. JAMA 1999;282(7): 637–45.

[43] Cummings SR, Eckert S, Krueger KA, Grady D, Powles TJ, Cauley JA, et al. The effect of raloxifene on risk of breast cancer in postmenopausal women: results from the MORE randomized trial. Multiple Outcomes of Raloxifene Evaluation. JAMA 1999;281(23):2189–97.

[44] Cauley JA, Norton L, Lippman ME, Eckert S, Krueger KA, Purdie DW, et al. Continued breast cancer risk reduction in postmenopausal women treated with raloxifene: 4-year results from the MORE trial. Multiple outcomes of raloxifene evaluation. Breast Cancer Res Treat 2001;65(2):125–34.

[45] Martino S, Cauley JA, Barrett-Connor E, Powles TJ, Mershon J, Disch D, et al. Continuing outcomes relevant to Evista: breast cancer incidence in postmenopausal osteoporotic women in a randomized trial of raloxifene. J Natl Cancer Inst 2004;96(23):1751–61.

[46] Vogel VG, Costantino JP, Wickerham DL, Cronin WM, Cecchini RS, Atkins JN, et al. Effects of tamoxifen vs raloxifene on the risk of developing invasive breast cancer and other disease outcomes: the NSABP Study of Tamoxifen and Raloxifene (STAR) P-2 trial. JAMA 2006;295(23):2727–41.

[47] Slamon DJ, Clark GM, Wong SG, Levin WJ, Ullrich A, McGuire WL. Human breast cancer: correlation of relapse and survival with amplification of the HER-2/neu oncogene. Science 1987;235(4785):177–82.

[48] Marzolini C, Buclin T, Decosterd LA, Biollaz J, Telenti A. Nelfinavir plasma levels under twice-daily and three-times-daily regimens: high interpatient and low intrapatient variability. Ther Drug Monit 2001; 23(4):394–8.

[49] Shim JS, Liu JO. Recent advances in drug repositioning for the discovery of new anticancer drugs. Int J Biol Sci 2014;10(7):654–63.

[50] Srirangam A, Mitra R, Wang M, Gorski JC, Badve S, Baldridge L, et al. Effects of HIV protease inhibitor ritonavir on Akt-regulated cell proliferation in breast cancer. Clin Cancer Res 2006;12(6):1883–96.

[51] Trepel J, Mollapour M, Giaccone G, Neckers L. Targeting the dynamic HSP90 complex in cancer. Nat Rev Cancer 2010;10(8):537–49.

[52] Koltai T. Nelfinavir and other protease inhibitors in cancer: mechanisms involved in anticancer activity. F1000Research 2015;4:9.

[53] Liu D, Yang Z, Wang T, Yang Z, Chen H, Hu Y, et al. beta2-AR signaling controls trastuzumab resistance-dependent pathway. Oncogene 2016;35(1):47–58.

[54] Bonotto M, Gerratana L, Poletto E, Driol P, Giangreco M, Russo S, et al. Measures of outcome in metastatic breast cancer: insights from a real-world scenario. Oncologist 2014;19(6):608–15.

[55] Cortazar P, Zhang L, Untch M, Mehta K, Costantino JP, Wolmark N, et al. Pathological complete response and long-term clinical benefit in breast cancer: the CTNeoBC pooled analysis. Lancet 2014;384(9938): 164–72.

[56] DiMeo TA, Anderson K, Phadke P, Feng C, Perou CM, Naber S, et al. A novel lung metastasis signature links Wnt signaling with cancer cell self-renewal and epithelial-mesenchymal transition in basal-like breast cancer. Cancer Res 2009;69(13):5364–73.

[57] Marotta LLC, Almendro V, Marusyk A, Shipitsin M, Schemme J, Walker SR, et al. The JAK2/STAT3 signaling pathway is required for growth of CD44+CD24− stem cell−like breast cancer cells in human tumors. J Clin Invest 2011;121(7):2723−35.

[58] Harrison H, Farnie G, Howell SJ, Rock RE, Stylianou S, Brennan KR, et al. Regulation of breast cancer stem cell activity by signaling through the Notch4 receptor. Cancer Res 2010;70(2):709−18.

[59] Bhola NE, Balko JM, Dugger TC, Kuba MG, Sánchez V, Sanders M, et al. TGF-β inhibition enhances chemotherapy action against triple-negative breast cancer. J Clin Invest 2013;123(3):1348−58.

[60] Creighton CJ, Li X, Landis M, Dixon JM, Neumeister VM, Sjolund A, et al. Residual breast cancers after conventional therapy display mesenchymal as well as tumor-initiating features. Proc Natl Acad Sci U S A 2009;106(33):13820−5.

[61] Koval AV, Vlasov P, Shichkova P, Khunderyakova S, Markov Y, Panchenko J, et al. Anti-leprosy drug clofazimine inhibits growth of triple-negative breast cancer cells via inhibition of canonical Wnt signaling. Biochem Pharmacol 2014;87(4):571−8.

[62] Ahmed K, Koval A, Xu J, Bodmer A, Katanaev VL. Towards the first targeted therapy for triple-negative breast cancer: repositioning of clofazimine as a chemotherapy-compatible selective Wnt pathway inhibitor. Cancer Lett 2019;449:45−55.

[63] Falkson CI, Falkson G. A phase II evaluation of clofazimine plus doxorubicin in advanced, unresectable primary hepatocellular carcinoma. Oncology 1999;57(3):232−5.

[64] Ranjan A, Gupta P, Srivastava SK. Penfluridol: an antipsychotic agent suppresses metastatic tumor growth in triple-negative breast cancer by inhibiting integrin signaling Axis. Cancer Res 2016;76(4):877−90.

[65] Antoniou A, Pharoah PDP, Narod S, Risch HA, Eyfjord JE, Hopper JL, et al. Average risks of breast and ovarian cancer associated with BRCA1 or BRCA2 mutations detected in case series unselected for family history: a combined analysis of 22 studies. Am J Hum Genet 2003;72(5):1117−30.

[66] Comprehensive molecular portraits of human breast tumours. Nature 2012;490(7418):61−70.

[67] Foulkes WD. Germline BRCA1 mutations and a basal epithelial phenotype in breast cancer. Cancer Spectr Knowl Environ 2003;95(19):1482−5.

[68] Jameson JL, Longo DL. Precision medicine — personalized, problematic, and promising. N Engl J Med 2015;372(23):2229−34.

[69] Zheng ZY, Li J, Li F, Zhu Y, Cui K, Wong ST, et al. Induction of N-Ras degradation by flunarizine-mediated autophagy. Sci Rep 2018;8(1):16932.

[70] Lee J, Yesilkanal AE, Wynne JP, Frankenberger C, Liu J, Yan J, et al. Effective breast cancer combination therapy targeting BACH1 and mitochondrial metabolism. Nature 2019;568(7751):254−8.

[71] Leone A, Di Gennaro E, Bruzzese F, Avallone A, Budillon A. New perspective for an old antidiabetic drug: metformin as anticancer agent. Cancer Treat Res 2014;159:355−76.

[72] Evans WE, Relling MV. Pharmacogenomics: translating functional genomics into rational therapeutics. Science 1999;286(5439):487−91.

[73] Goetz MP, Kamal A, Ames MM. Tamoxifen pharmacogenomics: the role of CYP2D6 as a predictor of drug response. Clin Pharmacol Ther 2008;83(1):160−6.

[74] Crown JP, Diéras V, Staroslawska E, Yardley DA, Bachelot T, Davidson N, et al. Phase III trial of sunitinib in combination with capecitabine versus capecitabine monotherapy for the treatment of patients with pretreated metastatic breast cancer. J Clin Oncol 2013;31(23):2870−8.

[75] Bergh J, Bondarenko IM, Lichinitser MR, Liljegren A, Greil R, Voytko NL, et al. First-line treatment of advanced breast cancer with sunitinib in combination with docetaxel versus docetaxel alone: results of a prospective, randomized phase III study. J Clin Oncol 2012;30(9):921−9.

[76] Wragg JW, Heath VL, Bicknell R. Sunitinib treatment enhances metastasis of innately drug-resistant breast tumors. Cancer Res 2017;77(4):1008−20.

[77] La Belle Flynn A, Schiemann WP. Autophagy in breast cancer metastatic dormancy: tumor suppressing or tumor promoting functions? J Cancer Metastasis Treat 2019;2019.

[78] Hen O, Barkan D. Dormant disseminated tumor cells and cancer stem/progenitor-like cells: similarities and opportunities. Semin Cancer Biol 2020;60:157−65.

[79] Jiang P-D, Zhao Y-L, Deng X-Q, Mao Y-Q, Shi W, Tang Q-Q, et al. Antitumor and antimetastatic activities of chloroquine diphosphate in a murine model of breast cancer. Biomed Pharmacother 2010;64(9):609−14.

[80] Choi DS, Blanco E, Kim Y-S, Rodriguez AA, Zhao H, Huang TH-M, et al. Chloroquine eliminates cancer stem cells through deregulation of Jak2 and DNMT1. Stem Cell 2014;32(9):2309−23.

[81] Halama A, Kulinski M, Dib SS, Zaghlool SB, Siveen KS, Iskandarani A, et al. Accelerated lipid catabolism and autophagy are cancer survival mechanisms under inhibited glutaminolysis. Cancer Lett 2018;430:133−47.

[82] Bayat Mokhtari R, Homayouni TS, Baluch N, Morgatskaya E, Kumar S, Das B, et al. Combination therapy in combating cancer. Oncotarget 2017;8(23):38022−43.

[83] Levy JMM, Towers CG, Thorburn A. Targeting autophagy in cancer. Nat Rev Cancer 2017;17(9):528–42.

[84] Hyman DM, Smyth LM, Donoghue MTA, Westin SN, Bedard PL, Dean EJ, et al. AKT inhibition in solid tumors with AKT1 mutations. J Clin Oncol 2017;35(20):2251–9.

[85] Costello JC, Heiser LM, Georgii E, Gonen M, Menden MP, Wang NJ, et al. A community effort to assess and improve drug sensitivity prediction algorithms. Nat Biotechnol 2014;32(12):1202–12.

[86] Dickson M, Gagnon JP. The cost of new drug discovery and development. Discov Med 2004;4(22):172–9.

[87] Dickson M, Gagnon JP. Key factors in the rising cost of new drug discovery and development. Nat Rev Drug Discov 2004;3(5):417–29.

CHAPTER

7

A personalized medicine approach to drug repurposing for the treatment of breast cancer molecular subtypes

Enrique Hernández-Lemus

Research in Computational and Population Genomics, Computational Genomics Division, National Institute of Genomic Medicine, Ciudad de México, Mexico City, Mexico

OUTLINE

Introduction

Breast cancer is a major public health concern and a relevant mortality issue in young women worldwide [1]. Mammary tumors are extremely heterogeneous at the histological, molecular, and systemic levels. To handle such variability, oncology researchers and clinicians have

Drug Repurposing in Cancer Therapy
https://doi.org/10.1016/B978-0-12-819668-7.00007-5

developed a number of prognostic and therapeutic approaches ranging from classifications based on clinical parameters or histopathologic markers (such as estrogen, progesterone, and epidermic human growth factor receptors) [2] to classifications based on gene expression profiling [3] as well as various attempts to patient-based phenotyping.

The main objective of these methods is to allow the design of improved therapeutic procedures. However, the lack of such options for certain subtypes—in particular for triple-negative tumors—represents an important source of frustration challenge for clinical oncologists who often have to resort to cytotoxic therapies with a large number of adverse side effects.

The development of new anticancer drugs, although extremely relevant, is a very slow and high-cost endeavor. In contrast, the repurposing of many approved drugs (both anticancer and non–anti-cancer) has become an effective strategy to broaden the options in the oncologic therapeutic spectrum with the undeniable advantages of being faster, cheaper, and faster to go through preclinical and clinical stages of validation protocols, tier studies, and clinical trials [4]. Of particular relevance has been the strategy to develop tailor-made drug cocktails based on personalized medicine studies.

There is indeed a large body of evidence to support the claim that combination therapies are more effective against late-stage neoplastic tumors than single agents or sequential drugs combinations, given the large inter- and intratumoral heterogeneity among patients [5,6]. Tailor-made combination therapy is not without caveats, in particular given the fact that the development and testing procedures in the pharma industry are not, in general, designed with multitherapy in mind [6a].

Another important caveat for the rational design of multidrug repurposing approaches lies in the fact that it is a quite interdisciplinary task, even more reliant on computational biology, bioinformatics, and artificial intelligence (AI) than "usual" oncopharmacology. Clinical practitioners and pharmaceutical company officers need to become aware of this fact and adapt their current practices accordingly, take for instance, the wealth of information on chemical, pharmacological, and genomic databases.

Let us briefly analyze one (of many) instance in which data mining approaches become relevant. Most compound and drugs currently used in the clinic have a large number of off-target effects aside of its main therapeutic mechanism of action. Such off-target actions are indeed the basis for a lot of drug repurposing strategies. Computer-aided (CA) interrogations of the many large databases on drugs and mechanisms of action may allow the identification of (additional) specific targets [8,9].

Other computational strategies of drug repurposing include the combination of knowledge discovery in databases (KDD) with molecular profiling and modelization to identify novel drug–target interactions. Often machine learning (ML) algorithms are used to screen enormous catalogs of molecules to search for drug–target interactions. The combination of KDD/ML with high-throughput in vitro assay screening has shown to be an extremely effective strategy in multifactorial diseases such as cancer largely outperforming single-drug approaches [11]. Aside from "single-shot" drug–target interactions, molecular and phenotypic heterogeneity in cancer tumors needs to be taken into account for the design of effective anticancer therapeutic interventions, for instance, when dealing with immunotherapy [12]. Although highly publicized and extremely successful in some cases, the majority of cancer patients do not benefit from immunotherapy, quite likely by effects related to the immunosuppressive nature of their individual tumor microenvironment. Designs considering a personalized approach need to focus on deciphering the individually affected metabolic pathways. Li and coworkers have discussed quite extensively the way metabolic circuits regulate/deregulate intrinsic antitumor immunity pathways and how some (many, indeed) of these relationships have even reached the clinical trial stage (see, for instance, Table 1 in Ref. [12]). Along the same lines, repurposing of immunomodulatory drugs such as thalidomide, lenalidomide, and pomalidomide has been highly enhanced by the extensive validation of computationally predicted biomarkers in patient-diverse subpopulations [13]. There is of course the concern that the therapeutic efficacy of these treatments is highly heterogeneous [148]. For this reason, these drugs are usually part of a polypharmacological treatment. For instance, Shen and coworkers have reported that the use of thalidomide increases delivery and efficacy of cisplatin [149]. On the other hand, the main challenges for their wider use rely on the fact that there have been many reports of adverse drug effects, these include peripheral neurotoxicity [150], as well as teratogenic and dermatological, among other effects [151].

Of course, the challenges of anticancer drug repurposing do not end with the (many) molecular targets and off-target prediction technicalities. Barriers to repurposing often start with the availability of better diagnostic tools able to predict and stratify patients response to therapy. This is another field in which CA and AI approaches may result helpful [14–17]. Aside from CA/AI tools, systems biology modeling may also allow for better phenotyping and prognostics, leading to better-suited drug repurposing designs [18,19].

The use of patient-derived genomic information to study how genetic alterations influence the routes to tumor progression and cell survival is key to uncover tumor-specific vulnerabilities that may lead to the development of narrowly targeted therapeutic interventions, including knowledge-driven drug repurposing [20–22].

Once novel repurposing ideas have been found, there is the need to develop strategies to make these ideas reach the patients [23]. Among the many caveats, three main questions have been considered particularly relevant [4]:

1. How to establish the recommended dose to achieve anticancer activity, especially when repurpose drugs were not initially intended as antitumor drugs. In this regard, novel computational approaches based on a systems biology philosophy could be applied [24–26].
2. How to deal with intellectual property, patent, and licensing issues, both in generic and proprietary treatments?
3. Since cancer-related clinical trials are usually more expensive, need longer follow-ups, and are very prone to failure than those of noncancer drugs, there may be sufficient financial incentives for the pharmaceutical stakeholders. This is so since most repurposed drugs, aside from being clinically significant, need to demonstrate higher cost-effectiveness ratios than newer treatments.

In this chapter, we will outline and discuss some of the major themes in this regard, particularly related to the development of translational bioinformatics strategies to guide clinical oncologist in the design of more effective and personalized therapies using repurposed drugs to treat breast cancer subtypes at the individual (personalized) level.

Mutation-specific therapies as an approach to personalized medicine in cancer: pros and cons

Ever since the discovery of the first cancer-associated mutations and oncogenes, one important tenet of anticancer therapy has been the discovery of such cancer "causal" mutations (in particular tumor drivers), to later try to figure out the structure of a "silver-bullet" drug, a molecule that may target tumors on an extremely specific fashion leaving nontumor cells unaffected. It actually seemed like a pretty good idea.

In Fig. 7.1, a simplified view of a generalistic mutation profile–based approach to personalized drug repurposing to treat breast cancer tumors is presented. First, by means of high-precision DNA sequencing, a tumor-specific mutation is found in the genome of a patient. In the best scenario, it may be a mutation that is already known and annotated in a "cancer panel" so that its intrinsic molecular characteristics are available.

With this knowledge and after assessing that these mutation is absent in the germline, a targeted therapy may be developed, being this the finding of a monoclonal antibody able to recognize the effect of the mutation at the protein level [27–29], the composition of an antibody–drug conjugate complex [30–33], or the synthesis of a small molecule drug [34,35].

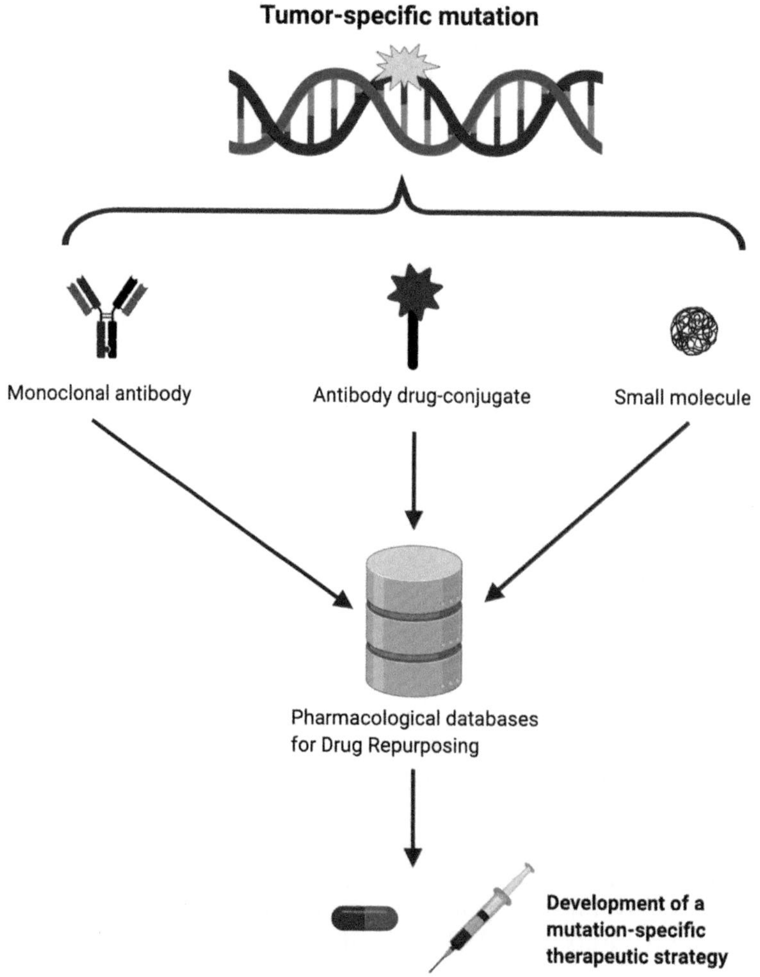

FIGURE 7.1 Schematic depiction of a drug repurposing strategy to personalized therapy based on mutation-specific profiling. *Image created with BioRender https://Biorender.com.*

Once this therapy or combination of given therapies is known, one must interrogate the pharmacological databases for drug repurposing to look up for the existence of such pharmaceuticals and in due case its possible off-targets and side effects [36–43].

Unfortunately, except on a few exceptional cases of highly penetrant mutations for which the response to such silver bullets has been equally exceptional, most cancer patients did not benefit by using these approaches [44,45].

Indeed, it has been argued that "...*Precision oncology promises to pair individuals with cancer with drugs that target the specific mutations in their tumor, in the hope of producing long-lasting remission and extending their survival. The basic idea is to use genetic testing to link patients with the drugs that will work best for them, irrespective of the tissue of origin of their tumor. Enthusiasm has been fueled by reports of exceptional or super responders — individuals for whom experimental therapies seem to work spectacularly well* ... [yet] ... *Few patients benefit from precision oncology. Data from some 2600 people enrolled in a sequencing program at the MD Anderson Cancer Center in Houston, Texas, showed that just 6.4% were paired with a targeted drug for identified mutations*" [46,47].

The field is moving fast in recent times, and the general trends however are not apparently changing yet. A recent large-scale survey intended to evaluate the benefits of genome-driven oncology, the MOSCATO study [152,153], suggests that a genome-driven strategy for cancer therapy is able to improve clinical outcomes in a significant minority of patients who undergo molecular screening. The increase in progression-free survival rate as compared with the group with no screening is larger than 1.3 (30% increase) in about 33% of the patients with a targeted therapy matched to a genomic alteration. This seems to be good news and they are indeed, but only, for a fraction of patients. According to the MOSCATO study "...although these results are encouraging, only 7% of the successfully screened patients benefited from this approach...". So, apparently figures go up from 6.4% to about 7% of success. This situation may nonetheless improve in the upcoming future with the advent of better, more precise tumor variation assessment and improved therapeutic designs.

These facts have eroded the optimism on mutation-centered personalized medicine that was prevailing years ago [48,49]. Mutational heterogeneity is indeed key to understand the challenges of this conceptually appealing approach. In recent times, mutational variability in breast cancer tumors has been unveiled at an unprecedented scale [50].

It has been revealed that mutation induced by pharmacological treatment increases the severity and the drug resistance of metastatic tumors. This is so, since mutational burden enhances up to the double the relative contributions that secondary breast cancer mutational signatures present [51].

In particular, the mutation frequency of well-known driver genes increases in metastatic breast cancer. Such changes in frequency are indeed significantly associated with previous pharmacological treatment [52,53]. It is actually known that the APOBEC family of APO enzymes plays a relevant role in these mutational heterogeneity [54,55].

To make things still more complex, the genomic region of APOBEC enzymes is one of abnormally high sequence heterogeneity [56]. Breast cancer tumors in particular (and most epithelial tumors indeed) are also characterized by a complex subclonal structure [57], a fact that may cause that even drugs perfectly targeting the mutational profile of a given clone may be unable to have a sustained therapeutic effect on these tumors.

Strategy: combining pathway analysis, network approaches, and data mining

An alternative to mutation-based therapeutic design that has been gaining relevance in recent years is the one based on functional pathway studies based on gene expression profiling. Here, we will discuss at some detail an approach of this kind that was developed within our research group [76]. This approach combines pathway enrichment [74], pathway crosstalk [117], and pathway deregulation analysis [10] with network approaches [58] with probabilistic approaches to data mining [59].

Fig. 7.2 presents a simplified view of this approach. We begin with a tumor biopsy sample from one patient from which mRNA is extracted and purified. Then gene expression levels from the sample are measured either by RNA sequencing (RNASeq, the current standard) or by gene expression arrays, a Luminex panel such as the L1000 [11], etc. The sample is subtyped by comparison with other patient tumor samples within our databases (in our case, in a corpus mainly composed by the TCGA and METABRIC breast cancer transcriptomic databases).

Once the individual patient gene expression sample profile is analyzed in the context of this large data corpus, the next step is a KDD mining from pathway databases such as KEGG [60], Reactome [61,62], and Pathway Commons [63]. We can opt to consider a specific set of pathways (e.g., for the design of metabolic or immunotherapy) or on a data-driven approach by considering all currently annotated pathways and look up for the more deregulated ones, the ones with more associated pharmacological targets in the market, those with more stringent targets or less side effects, etc.

By using the Pathifier algorithm, pathway deregulation scores (PDSs) can be calculated for every patient (including our test case) by comparing pathway-specific gene expression profiles with other tumors and with normal samples to account for gene expression variability [10]. It is to be noticed that the nature of the Pathifier algorithm approach is agnostic to whether gene expression heterogeneity comes from mutational tumor heterogeneity or other phenotypic or environmental origins.

The set of deregulated scores allow for another round of KDD, this time involving the joint analysis of pathway deregulation and differential gene expression data on the one hand and of drug–target interaction as well as off-target and side effects databases such as PharmGKB [64,65], DrugBank ([66]), the Therapeutic Target Database [67], and others. Our strategy made use of the automated query capabilities of a metadatabase, the Drug-Gene Interaction Database (www.dgidb.org). This platform integrates information of more than 15 of the main pharmacological databases.

Queries at DGIDB include information about drugs, pharmacological targets, type of drug-target interaction, data sources, and other characteristics, and it is available as an R-package RDGIdb in the Bioconductor suite [68].

As a result of this joint computational analysis, it is feasible to generate a data table for each patient (or groups of patients) including its deregulation score for every pathway, for a given group of pathways of interest or for the set of high Z-scored deregulated pathways.

Aside from pathway information, the table includes a list of differentially expressed genes in such pathways that are known targets of drugs already in the market (i.e., available for repurposing) either as anticancer drugs or not (in which case, dose reassessment studies may apply), further information includes possible off-targets and side effects of these drugs and a note on whether the drug is aimed at either "normalizing" cell behavior (to "cure" the cells)—called homeostatic effect—or further deregulating such pathway (to "kill" the cells)—called antihomeostatic effect for controlled cytotoxicity.

Examples of such tables can be found in Table 3 and recommendation charts for patients (MB.3058) and (MB.5387) in Ref. [76]. At this stage, such tables are considered not as actual

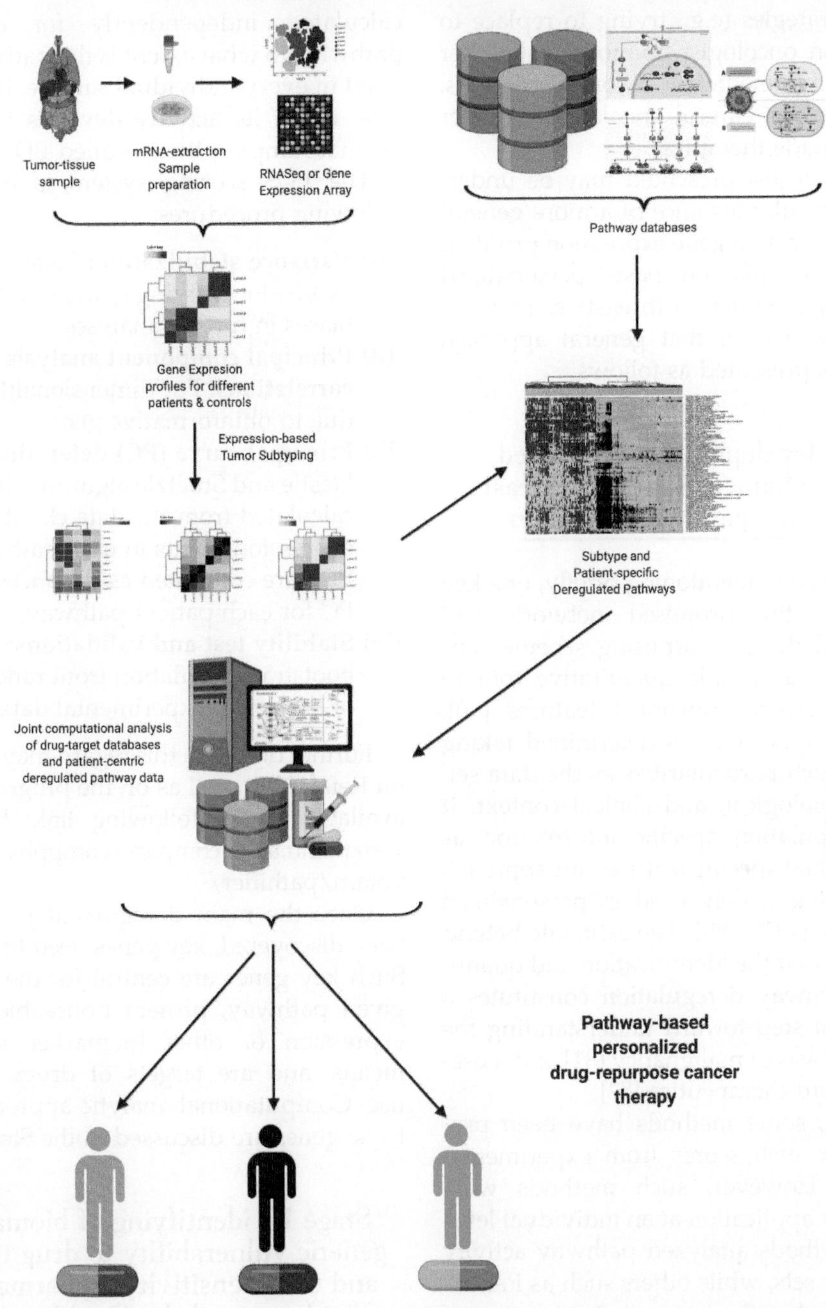

FIGURE 7.2 Schematic depiction of a drug repurposing strategy to personalized therapy based on patient-specific pathway deregulation inferred from gene expression profiling. *Image created with BioRender https://Biorender.com.*

therapeutic strategies (e.g., trying to replace to some extent an oncologist opinion), but rather as personalized medicine recommendations, tools for the expert clinical oncologist to better design tailor-made therapies.

The approach just presented may be understood as a particular instance of a more general four-stage approach to gene expression profiling and pathway deregulation–based personalized therapeutic repurposing to (breast) cancer. An outline of how would that general approach may be built is presented as follows.

Stage I: developing a personalized functional characterization of breast tumors—the pathifier algorithm

As we have been mentioning briefly, one key constituent of the proposed network and pathway-based drug repurposing scheme consists in having a reliable quantitative tool to assess patient-centric functional features [10]. The devised metric, PDS is determined taking into account such particularities as the data set, cancer type, biological, and clinical context. It combines, population specific information as well as individual specific features, an approach that is becoming widely used in personalized medicine designs [77–80]. The rationale behind this method is that the identification and quantification of pathway deregulation constitutes a very important step toward understanding the biomolecular basis of malignancy [81] and a useful guide toward therapeutics [82].

In the past, some methods have been proposed to infer such scores from experimental data [83,84]. However, such methods were limited in their application at an individual level since most methods analyzed pathway activity for full sample sets, while others such as PARADIGM [84] relied on the connectivity structure of the pathways, hence are not applicable for pathways that being relevant are of undetermined functional connectivity, some of which are central to cancer biology. The pathifier algorithm calculates, independently for every given pathway, to what extent is the pathway deregulated in every individual sample, by measuring how much its activity deviates from a set of normal samples, the so-called PDS.

The PDS scoring system is based on the following procedures:

(i) **Variance stabilization**: To avoid highly expression varying genes to introduce biases in further analyses.
(ii) **Principal component analysis of correlations:** For dimensionality reduction due to uninformative genes.
(iii) **Principal curve (PC) determination**: Using Hastie and Stuetzle algorithm [85], the PC is calculated from the data cloud of gene expression points in each pathway space. PDS are computed as distances along this PC for each patient pathway.
(iv) **Stability test and validations**: Using bootstrap calculation from randomization of the actual experimental data.

Further details in this regard may be consulted on Ref. [10] as well as on the program web page available at the following link: http://www.weizmann.ac.il/complex/compphys/software/yotam/pathifier/

Once the main deregulated pathways have been discovered, key genes need to be unveiled. Such key genes are central for the activity of a given pathway, present noticeable changes of expression or other biomarker identifiability means, and are targets of drugs currently in use. Computational analytic approaches to find these genes are discussed in the Stage II.

Stage II: identifying of biomarkers of genetic vulnerability to drug treatments and drug sensitivity—pharmacological databases and the DepMap approach

Identifying molecular or phenotypic markers of tumor vulnerability is an extremely difficult task. This is particularly so in the case of epithelial tumors, since these are known to carry a large number of genomic alterations largely scattered upon many pathways thus veiling the access to know which are the genes needed for the tumors to survive. As we have already mentioned, this is one of the main limitations of mutation-driven strategies to anticancer therapies.

However, it is also known that a large number of such dependencies fall into a fairly bounded number of cases so that up to 82% of them can be related to biomarkers based on gene expression profiles [11]. Strikingly, only 16% of the dependencies can be identified by mutational signatures and just 2% can be associated to DNA copy number alterations.

The fact that (1) molecular dependencies arise in the context of functional pathways and (2) most of these dependencies may be associated with gene expression signatures has been of enormous relevance for the development of functionally founded strategies to therapeutic designs. Perhaps the best known approach of this kind is the Cancer Dependency Map (DepMap), developed by a large group of highly competent molecular oncologists and computational biologists at the Broad Institute and some of Harvard Medical School associated hospitals, most notably the Dana-Farber Cancer Institute [11].

The main goal of the DepMap approach is the integration of a large corpus of genome-scale loss-of-function and/or pharmacological perturbation experiments aimed at understanding the molecular basis of tumor functions and vulnerabilities on living cells. The large pro is that the DepMap approach has generated a number of the largest gene expression—drug perturbation databases available, which are extremely well annotated: To date, a set of at least 769 well-validated interactions in 501 cell lines has been curated by the DepMap. These 769 targets include several classes of "acting" proteins, like transcription factors or kinases from which at least 20% (around 152) are potential drug targets in tumor cells.

However, targeting transcription factors has the downside that it may induce cell death also in normal cells expressing those factors. Hence, the greater utility is focus on cell lineage—specific transcription factors like TEAD1, ESR1, GATA3, or FOXA1 and others that are unfortunately not so abundant. The main downside with the DepMap approach is that precisely due to lineage-specific features some of the findings made on cell lines may not be easily translated to the clinical practice, due mainly to off-target effects [69,70]. In this regard, recent approaches have looked at loss-of-function assays relying on CRISPR-Cas9 techniques to complement RNAi and somehow alleviate this unspecificity given by the use of cell lines [71,72].

Another way to circumvent these issues, one that we consider to be to some extent more promissory in terms of early delivering to the clinical settings, is the use of the vast wealth of information in these high-throughput databases with computational intelligence and probabilistic modeling of patient data. However, some issues have to be considered.

In the first place, the advent of high-throughput technologies capable of retrieving a large amount of biological data from single, highly standardized and well-controlled experiments as is the case of cDNA expression microarrays or, recently, RNA-seq techniques that allow us to approach the study of the transcriptome on an unprecedented scale with full coverage and relatively large data sample sizes.

Data themselves are not enough; however, important advances in functional genomics and systems biology have clarified our conception that generally speaking, a biological function of the living cell can be understood as resulting from a large number of molecular interactions. Hence, a complex phenotype often does not arise as the consequence of the action or malfunction of a single gene or a single molecule. Instead biological functions can be abstracted as a set of genes in a genome forming a dense network of mutually interacting molecules in the cell. Hence, a pathway is a representation of a set of high-order biological functions [60].

With this in mind, and analyzing the success in the development of areas such as pharmacogenomics that probes on a global scale, all the genes are involved in the response to a certain perturbation or drug under a broad variety of conditions [73]. Pharmacogenomics has evolved from gene-centered pharmacogenetics to a whole gene- and/or pathway-based approach [74] considering information from high-throughput biomolecular technologies including genomics, transcriptomics, proteomics, and others, as well as from the use of manually and computationally curated databases to improve its standards to generate and test hypotheses and, more importantly, of being able to translate basic molecular knowledge right to the clinical practice [75].

Even considering the outstanding advances in this field, too many breast cancer patients still have to face low survival rates and poor quality of life (which is indeed further diminished by the presence of side and off-target effects of the pharmacological therapy itself). We are thus in the search for better design practices, able to reach low-harm yet effective therapeutic options. Optimized therapeutic approaches are particularly relevant to treat certain breast cancer subtypes—such as the basal or triple-negative tumors—or those neoplasms that develop pharmacological resistance to standard therapeutic schemes [2,72,154](Friese et al., 2017).

As previously discussed, the lack of better therapeutic options are of varied origins. Aside from the molecular and phenotypic heterogeneity and other intricacies inherent to cancer biology, there are other conundrums of a more technical, logistic, or even economic nature present not only in traditional drug development (i.e., the path from concept to drug's approval) but also in the clinical evaluation or drugs to be repurposed. An accurate selection of optimized pharmacological targets (i.e., the ones with "sharper" direct effects and small off-target and adverse effects) is crucial to achieve both technical and financial viability and thus a higher rate of success.

One possible avenue of improvement for personalized but financially sound breast cancer therapy is by the use of individual functional profiling (instead of the more difficult to standardize personal mutation profiling) to define a route to drug repurposing. We need thus to ask ourselves if it is possible to identify stringent associations between pharmacological targets with drugs currently available in the market and their respective pathways of action to cover up for every breast cancer molecular phenotype, even better combinations tailored to an individual patient's molecular, and functional profile.

To address this question, one proposal recently made consists in the implementation of a CA analysis based on the systematic inquiry of all possible deregulated pathways—as proxied by their gene expression profiles—specific for each breast cancer subtype, and the assessment of target genes (within said pathways), which are susceptible to pharmacological modulation. This approach relies on large, well-curated gene expression datasets from high-throughput transcriptome experiments on patients (as opposed to the DepMap approach relying on cancer cell lines) [76].

Stage III: generating a consensus approach to personalized breast cancer treatment

As discussed in the previous sections, one important challenge for the development of personalized drug repurposing approaches to anticancer therapy lies in the fact that there is a large molecular and phenotypic heterogeneity between cancer patients—and even molecular heterogeneity among the different cell populations of a single patient tumor(s). This high variability imposes constrains both at the purely technical and the logistical dimensions of therapeutic designs. The use of large-scale databases spanning all over different breast tumors, such as the ones in large research consortia such as The Cancer Genome Atlas—now Genomic Data Commons—[86–89] and METABRIC [90,91], by means of integrated computational analyses has made possible to discern the commonalities and differences in the expression traits, the phenotypes, and survival for thousands of cancer patients. This knowledge is making possible in turn to develop dynamic maps (much in the spirit of the DepMap) of tumor features and vulnerabilities by classes.

Hence, by recognizing that individuals vary in multiple ways that are however constrained by the physiology and functionality of tumor and normal cells, it has been possible to tackle variability by defining functional subclasses, in a similar way to the manner in which breast cancer molecular subtypes were defined [92,93]. For instance, Shah and coworkers were able to track down functional and molecular evolution features of the elusive triple-negative breast cancer by means of genomic analysis to develop hints about the way in which unique mutation patterns give rise to gene expression and functional tumor commonalities [94]. Similar approaches have led to the establishment of novel biomarkers useful in the clinical setting [95].

The combination of computational analysis of gene expression profiles in hundreds/thousands of patients with probabilistic modeling and a networked approach has allowed a significant improvement of diagnostics and prognostics [96,97], opening the way to more insightful designs to breast cancer therapeutics [98]. ML studies on these large databases have also lead to targeted assay proposals to determine the efficacy of competing therapies such as chemotherapy and hormone-guided designs [99] or the effects of combinatorial immune therapies [100].

One additional way in which we can go from purely descriptive gene expression profiling to actual biological function is by studying the patterns in which molecular pathways are deregulated. The best way to actually do this would be by using large-scale phosphoproteomic and metabolomic data. Since these databases are still on their infancy, an alternative approach has been the use of gene expression profiles of the genes within a given pathway to evaluate—by means of differential gene expression—to what extent any given pathway of interest is deregulated [76,101]. Pathway deregulation analysis as already discussed is an extremely useful way to understand cellular function from molecular profiling [10]. In the case of breast cancer, pathway deregulation analysis is already paving the way to the design of a personalized approach to cancer therapeutics.

As an outstanding example, Livshits and collaborators [102] have identified nine breast tumor subtypes of which a new one comprising approximately 7% of the cases considered (on a study of around 2000 tumors and 144 controls) is highly deregulated on a striking 38 PKA pathways. Of enormous relevance for therapeutics is the fact that although this is a very large number of protein kinase–driven pathways with enormous consequences for the patients, these pathways are all inducible by a single molecule: *PRKACB* which is indeed a druggable gene. These researchers also found that there are two classes of basal tumors that are actually distinguishable by a cluster of respectively lowly and highly deregulated immune system pathways. The highly deregulated group is characterized by a noticeable presence of tumor infiltrating lymphocytes and is of much better prognosis (and potentially a target of immunotherapy). PRKACB is a target for staurosporine, a drug belonging to the class of p-glycoprotein/abcb1 inhibitors. The activity of ABC transporters in cancer metabolism is well established. Staurosporine has been proved to induce cell death in MCF7 human breast cancer cells [145] and is know to also disrupt HUNK, a cell cycle associated kinase in Her2+ tumor derived cells [146]. So PRKACB inhibitor staurosporine is able to treat two different breast cancer subtypes (luminal and Her2+) by different mechanisms, in both cases by inhibiting proliferation.

The same large-scale study also identified 9 EGFR-related pathways that can be targeted by drugs already available in the market such as anlotinib [103,104], a drug that has been used also in other aggressive, drug-resistant tumors such as glioblastoma [105]; poziotinib [106–108], and dacomitinib [109], as well as by the novel cationic polyamidoamine dendrimers already under clinical investigation [110]. For this class of tumors, it is also possible to modulate EGFR pathways by the action of glucocorticoids (also available for repurposing) [111], caution has to be taken however, since the hormone-mediated mechanisms of action are much less specific than other EGFR modulators known.

23 PAK1-mediated pathways were identified, which can be useful since a number of PAK1 inhibitors are susceptible of being repurposed [112] and their actual mechanisms of action—e.g., inducing PUMA-mediated cell death and p21-mediated cell cycle arrest—are currently starting to be understood [113].

Although EGFR targeting therapies have been found to be less effective than initially expected in breast tumors in general, recent studies have found that targeting EGFR may enhance chemosensitivity of triple-negative breast cancer tumors by inducing a "rewiring" of apoptotic signaling networks. Hence, anti-EGFR therapy may be used as a chemosensitizer or to prevent metastases in such (difficult to treat) cases [147].

By looking at the downstream functional effects, that as we have discussed are shared by individuals with different mutational profiles, it has been possible to reach *functional consensus* rather than *genome variation consensus*. Since the ultimate action of pharmacological therapy is the modification of the functional and physiological traits of the disease (the so-called pathophenotype), strategies based on functional features (as proxied by pathway deregulation patterns and gene expression profiles) are believed to be more therapeutically effective by being less prone to off-target effects and also more financially efficient by alleviating the extraordinary burden of human and tumor genomic heterogeneity.

Stage IV: coping with pharmacological resistance: the role of pathway crosstalk and secondary targets

Once a therapeutic scheme has been tailor-designed for a given individualized profile, one still has to consider the fact that the actual efficacy of a drug (considered not as a silver bullet but rather as a systemic perturbation) to "control" a given biological process via a transduction network depends heavily on the connectivity structure of said network [114,115]. It has been long known that intermediate signaling steps and the presence of crosstalk events are able to induce pharmacological resistance [116,117] and that this fact has indeed to be taken into account in the optimization of therapeutic strategies [118–120].

To this end, database and computational resources are available [121–123]. Based on these resources, it is advisable to develop crosstalk inhibition studies [124–127], evaluate drug synergism [128–131], and analyze crosstalk-induced resistance [117,132–134].

Related to the crosstalk phenomenon, there is the semimechanistic effects due to the presence of secondary targets, i.e., any molecular target whose mode of action and/or effects are not related to the desired therapeutic target of the drug, that need to be taken into account. Although the analysis of therapeutic targets has been recognized as valuable for a long time and in spite of the availability of genomic technologies to measure their effects [135–137], trustworthy resources for high-throughput assessment of secondary targets are still being developed [138,139].

This has led to drug resistance and secondary effects databases such as the one by the COSMIC consortium (https://cancer.sanger.ac.uk/cosmic/drug_resistance) [140,141] that while being extremely comprehensive and well curated is restricted to analyze somatic mutations that, as we have stated, provide a narrow, often limited view of cancer biology. More general approaches (however based on less comprehensive databases) are currently being developed [142], and some of them have already make it into the pharmaceutical development stages, as is the case of pembrolizumab (commercial name Keytruda), an immune checkpoint inhibitor whose secondary targets provide synergistic effects [143].

Once this last stage of the proposed repurposing schema has been taken, one will be in a position to provide the clinical decision makers with therapeutic designs, either in the form of a single multipurpose drug or more often a personalized drug cocktails that need of course to be evaluated via highly controlled trials. Of course, the presented approach seems to be quite complex an endeavor. However, since most of these studies are based on experimental measurements through well-established high-throughput techniques (expression arrays, RNASeq) and most computational analyses may be automated to a certain degree, once a reliable pipeline has been established, it may represent 1 week work for a small team of translational bioinformaticians, i.e., it can be made with resources available in most medium and high-end cancer hospitals. In this regard, the present proposal, although ambitious, is still realistic. Let us consider some plausible scenarios.

As previously reported by our research group [76], breast cancer subtypes phenotypic similarities are due at least partially to similar biochemical and functional features. As it can be seen from Fig. 3 in Ref. [76], pathway deregulation profiles for hundreds of individuals in all KEGG-annotated pathways cluster together by subtype. For instance, basal breast tumors generally share deregulation of the cell cycle, oocyte meiosis, and steroid biosynthesis pathways. Our approach has also identified actionable targets specific to these pathways in basal tumors. By analyzing approved drugs available for this targets, we are able to propose a therapeutic scheme. A druggable gene able to influence the cell cycle pathway is CDK1, which is a target of dinaciclib. In the case of oocyte meiosis, there are two actionable kinases, AURKA1 (a target of alisertib) and also CDK1. An actionable gene in the steroid biosynthesis pathway is SQLE, for which terbinafine is a known therapeutic agent.

It is know that basal tumors are highly proliferative, and although they usually do not express estrogen and progesterone receptors, their steroid biogenesis pathway is commonly deregulated, leading to cholesterol mediated proliferation [144]. With these facts in mind, the drug repurposing approach described will point out to suggest a combination therapy for basal breast tumors consisting dinaciclib, alisertib, and terbinafine. By interrogating off-target effects and drug–drug interactions in pharmacological databases (our approach makes use of DGIDB, a metadatabase reporting the major drug databases available), we have seen that these drugs do not have reported evidence of interactions or crossed effects.

The approach just described is intended to be used at a cancer subtype–specific level. The method is actually intended to be able to make recommendations at a personalized level. Let us consider the case of two patients, both with basal breast tumors presenting high deregulation scores of the cell cycle, oocyte meiosis, and steroid biosynthesis pathways as before.

However, one of the patients (basal patient 1) presents a high deregulation of the apoptosis pathway, whereas the other one (basal patient 2) does not.

In order to decrease tumor growth by preventing apoptosis evasion, patient 1 will be suggested to receive a drug cocktail combination including either alvocidib or paclitaxel (in addition to dinaciclib, alisertib, and terbinafine), whereas patient 2 will not receive these apoptosis enhancers because the adverse effects of these drugs will not be justified by the marginal benefits given the low apoptosis deregulation.

The previous example highlights the importance of the combination of molecular profiling (comprising differential gene expression and pathway deregulation profiles) with knowledge discovery in pathway and drug–target and drug–drug interaction databases: given a patient-specific expression and pathway profile, our approach makes use of databases and computational intelligence to find the best drugs to modify the state of the abnormal pathways by targeting key genes, while at the same time avoiding side and off-target effects as much as possible.

In practice, the application of the four-stage approach just described would be as follows:

Stage I: Developing a personalized functional characterization of breast tumors—the pathifier algorithm:

- Starting with a biopsy tumor sample for a patient we perform RNA extraction, it is not actually needed to know in advance to which molecular subtype does the tumor belong (though the information can be used for validation purposes if available).
- RNA is purified and gene expression profiling is performed either via RNASeq or expression microarrays.
- Gene expression profile is used to subtype the tumor (via the PAM50 approach) and expression data are loaded into a database containing our training set of (several hundred) expression profiles of the given tumor subtype as well as expression profiles for (also hundreds of) "controls" (normal tumor-adjacent tissue).
- Differential gene expression and pathway deregulation (via Pathifier) scoring is performed for the whole training set (including the test patient sample).
- The end product of this stage is a list of score-ranked deregulated pathways for each sample, including our test patient as well as a list of differentially expressed genes.

Stage II: Identifying of biomarkers of genetic vulnerability to drug treatments and drug sensitivity—pharmacological databases and the DepMap approach:

- Starting from the data coming from Stage I, KDD is used to look up for actionable targets for the most deregulated/relevant pathways (in the test patient). Once these targets are known, another KDD instance looks up to find whether such actionable targets belong to the "druggable genome," which drug(s) target these genes, whether these drugs are FDA approved, and what are their known side effects.
- A preliminary list of drugs is generated as recommended therapy for the test patient.
- The list of drugs is ranked according to the DepMap database scoring system for action on known cancer vulnerabilities.
- The ranked list of drugs is analyzed via KDD to look up for off-target effects and drug–drug interactions.

Stage III: Generating a consensus approach to personalized breast cancer treatment:

- Once the ranked list of suggested drugs is available, a consensus procedure is carried out to choose between drugs with similar mechanisms of action, the one with the best rank, the lowest side effects, and so on.

- For a patient-specific (personalized) design, the resulting list will move on to Stage IV.
- If a subtype-specific therapeutic design is desired (say for establishing a protocol), then a patient consensus approach list must be generated. This list will move on to Stage IV.

Stage IV: Coping with pharmacological resistance—the role of pathway crosstalk and secondary targets:

- Once a therapeutic scheme has been devised, it is useful to determine in advance what kind of pharmacological resistance events may develop. In order to do this, crosstalk analysis should be made to assess what are the biological pathways that are crossing the pathways targeted by our therapy, downstream of the intervention point (i.e., the targeted molecules), and what is the state of activity of such pathways either in patients that have developed resistance or, if possible, in the patient for whom this specific therapy has been devised.

Pharmacological resistance is indeed a very complex phenomenon that may or may not be caused by pathway crosstalk. Pathway crosstalk scenarios may be also quite complex (for the case of tamoxifen resistance, you may want to look at our paper on the subject: [117].).

As a working example, however, consider the well-documented case of tamoxifen resistance generated by crosstalk with the insulin signal transduction pathway [154]. In this case, the gene that needs to be blocked is IGF2, which is a target of dusigitumab, an FDA-approved human monoclonal antibody commercialized by AstraZeneca. So if a given patient is developing tamoxifen resistance and its insulin pathway is abnormally upregulated, dusigitumab may be an elective drug for treatment.

Concluding remarks

Along this chapter, we have discussed an approach (or rather, a close family of approaches) to personalized therapy for breast cancer using a drug repurposing strategy. The rationale behind is that based on the following assumptions: (i) due to large mutation heterogeneity, the design of tailor-made therapies would induce the need for a much larger space of drug—target interactions than is currently available, thus preventing for repurposing approaches, (ii) gene expression profiles provide a much closer picture to cellular functional features, providing a basis for process-based strategies, i.e., pathway deregulation as a proxy for pathogenic phenotype, (iii) once the deregulated pathways and genes have been unveiled at an individual level, it is possible to look for approved drugs targeting such genes (and affecting their associated pathways) available to be repurposed, (iv) from the set of drug—targets prescreened, we can assess their mechanisms of action, off-target and secondary targets, and interactions thus formulating a validated proposal for clinical oncologists.

We are aware that this concept is still under development, but we believe the essential ideas have been presented, discussed, and supported by evidence in such a way that this document may serve not only as an introduction to these ideas but also, to trigger, as a source of inspiration, the development of further refinements that in the (hopefully, not so) long run may help patients and clinicians to take the best possible therapeutic decisions. Although most of our efforts have been in the context of breast cancer therapeutics, the general schema may serve as a basis to develop similar approaches to other maladies, oncologic, and other.

References

[1] Ferlay J, Soerjomataram I, Dikshit R, Eser S, Mathers C, Rebelo M, et al. Cancer incidence and mortality worldwide: sources, methods and major patterns in GLOBOCAN 2012. Int J Cancer 2015; 136(5):E359–86.

[2] Prat A, Pineda E, Adamo B, Galván P, Fernández A, Gaba L, et al. Clinical implications of the intrinsic molecular subtypes of breast cancer. Breast 2015;24: S26–35.

[3] Parker JS, Mullins M, Cheang MC, Leung S, Voduc D, Vickery T, Quackenbush JF, et al. Supervised risk predictor of breast cancer based on intrinsic subtypes. J Clin Oncol 2009;27(8):1160.

[4] Bertolini F, Sukhatme VP, Bouche G. Drug repurposing in oncology—patient and health systems opportunities. Nat Rev Clin Oncol 2015;12(12):732.

[5] Yap TA, Gerlinger M, Futreal PA, Pusztai L, Swanton C. Intratumor heterogeneity: seeing the wood for the trees. Sci Transl Med 2012;4(127). 127ps10-127ps10.

[6] Gerlinger M, Rowan AJ, Horswell S, Larkin J, Endesfelder D, Gronroos E, et al. Intratumor heterogeneity and branched evolution revealed by multiregion sequencing. N Engl J Med 2012;366(10): 883–92.

[6a] Robert C, Karaszewska B, Schachter J, Rutkowski P, Mackiewicz A, Stroiakovski D, et al. Improved overall survival in melanoma with combined dabrafenib and trametinib. New Eng. J. Med 2015;372(1):30–9.

[7] Deleted in review

[8] Barratt MJ, Frail DE, editors. Drug repositioning: bringing new life to shelved assets and existing drugs. John Wiley & Sons; 2012.

[9] Lamb J. The connectivity map: a new tool for biomedical research. Nat Rev Cancer 2007;7(1):54.

[10] Drier Y, Sheffer M, Domany E. Pathway-based personalized analysis of cancer. Proc Natl Acad Sci U S A 2013;110(16):6388–93.

[11] Tsherniak A, Vazquez F, Montgomery PG, Weir BA, Kryukov G, Cowley GS, et al. Defining a cancer dependency map. Cell 2017;170(3):564–76.

[12] Li X, Wenes M, Romero P, Huang SCC, Fendt SM, Ho PC. Navigating metabolic pathways to enhance antitumour immunity and immunotherapy. Nat Rev Clin Oncol 2019;1.

[13] Lopez-Girona AEA, Mendy D, Ito T, Miller K, Gandhi AK, Kang J, et al. Cereblon is a direct protein target for immunomodulatory and antiproliferative activities of lenalidomide and pomalidomide. Leukemia 2012;26(11):2326.

[14] Bera K, Schalper KA, Rimm DL, Velcheti V, Madabhushi A. Artificial intelligence in digital pathology—new tools for diagnosis and precision oncology. Nat Rev Clin Oncol 2019;16(11):703–15.

[15] Bejnordi BE, Veta M, Van Diest PJ, Van Ginneken B, Karssemeijer N, Litjens G, et al. Diagnostic assessment of deep learning algorithms for detection of lymph node metastases in women with breast cancer. JAMA 2017;318(22):2199–210.

[16] Steiner DF, MacDonald R, Liu Y, Truszkowski P, Hipp JD, Gammage C, et al. Impact of deep learning assistance on the histopathologic review of lymph nodes for metastatic breast cancer. Am J Surg Pathol 2018;42(12):1636.

[17] Liu Y, Kohlberger T, Norouzi M, Dahl GE, Smith JL, Mohtashamian A, et al. Artificial intelligence–based breast cancer nodal metastasis detection: insights into the black box for pathologists. Arch Pathol Lab Med 2018.

[18] Yurkovich JT, Tian Q, Price ND, Hood L. A systems approach to clinical oncology uses deep phenotyping to deliver personalized care. Nat Rev Clin Oncol 2019: 1–12.

[19] Kaissis G, Ziegelmayer S, Lohöfer F, Steiger K, Algül H, Muckenhuber A, et al. A machine learning algorithm predicts molecular subtypes in pancreatic ductal adenocarcinoma with differential response to gemcitabine-based versus FOLFIRINOX chemotherapy. BioRxiv 2019:664540.

[20] Perales-Patón J, Domenico TD, Fustero-Torre C, Piñeiro-Yáñez E, Carretero-Puche C, Tejero H, et al. vulcanSpot: a tool to prioritize therapeutic vulnerabilities in cancer. Bioinformatics 2019;35(22):4846–8.

[21] Brunen D, Bernards R. Drug therapy: exploiting synthetic lethality to improve cancer therapy. Nat Rev Clin Oncol 2017;14(6):331.

[22] Lord CJ, Ashworth A. Targeted therapy for cancer using PARP inhibitors. Curr Opin Pharmacol 2008;8(4): 363–9.

[23] Pantziarka P, Bouche G, Meheus L, Sukhatme V, Sukhatme VP, Vikas P. The repurposing drugs in oncology (ReDO) project. Ecancermedicalscience 2014;8:442.

[24] Barbolosi D, Ciccolini J, Lacarelle B, Barlési F, André N. Computational oncology—mathematical modelling of drug regimens for precision medicine. Nat Rev Clin Oncol 2016;13(4):242.

[25] Powathil GG, Swat M, Chaplain MA. Systems oncology: towards patient-specific treatment regimes informed by multiscale mathematical modelling. In: Seminars in cancer biology, vol. 30. Academic Press; 2015. p. 13–20.

[26] Agur Z, Elishmereni M, Kheifetz Y. Personalizing oncology treatments by predicting drug efficacy, side-effects, and improved therapy: mathematics, statistics, and their integration. Wiley Interdiscip Rev: Syst Biol Med 2014;6(3):239–53.

[27] Pento JT. Monoclonal antibodies for the treatment of cancer. Anticancer Res 2017;37(11):5935–9.

[28] Scott AM, Allison JP, Wolchok JD. Monoclonal antibodies in cancer therapy. Cancer Immun Arch 2012; 12(1):14.

[29] Weiner LM, Dhodapkar MV, Ferrone S. Monoclonal antibodies for cancer immunotherapy. Lancet 2009; 373(9668):1033–40.

[30] Chau CH, Steeg PS, Figg WD. Antibody–drug conjugates for cancer. Lancet 2019;394(10200):793–804.

[31] Walko CM, West HJ. Antibody drug conjugates for cancer treatment. JAMA Oncol 2019;5(11). 1648-1648.

[32] Coats S, Williams M, Kebble B, Dixit R, Tseng L, Yao NS, et al. Antibody drug conjugates: future directions in clinical and translational strategies to improve the therapeutic index. Clin Cancer Res 2019. clincanres-0272.

[33] Pegram MD, Miles D, Tsui CK, Zong Y. HER2-Overexpressing/Amplified breast cancer as a testing ground for antibody–drug conjugate drug development in solid tumors. Clin Cancer Res 2019.

[34] Zhang T, Li J, He Y, Yang F, Hao Y, Jin W, et al. A small molecule targeting myoferlin exerts promising anti-tumor effects on breast cancer. Nat Commun 2018;9(1):3726.

[35] Sakoff J, Gilbert J, McCluskey A. 100 Small molecules selectively targeting breast cancer cells. Eur J Cancer 2014;50:36.

[36] Gonzalez-Fierro A, Dueñas-González A. Drug repurposing for cancer therapy, easier said than done. In: Seminars in cancer biology. Academic Press; 2019.

[37] Shuptrine CW, Surana R, Weiner LM. Monoclonal antibodies for the treatment of cancer. In: Seminars in cancer biology, vol. 22. Academic Press; 2012. p. 3–13.

[38] Van Nuffel AM, Sukhatme V, Pantziarka P, Meheus L, Sukhatme VP, Bouche G. Repurposing drugs in oncology (ReDO)—clarithromycin as an anti-cancer agent. Ecancermedicalscience 2015;9.

[39] Dan N, Setua S, Kashyap V, Khan S, Jaggi M, Yallapu M, Chauhan S. Antibody-drug conjugates for cancer therapy: chemistry to clinical implications. Pharmaceuticals 2018;11(2):32.

[40] Flemming A. Antibody engineering: fine-tuning antibody–drug conjugates. Nat Rev Drug Discov 2014;13(3):178.

[41] Banerji U, van Herpen CM, Saura C, Thistlethwaite F, Lord S, Moreno V, et al. Trastuzumab duocarmazine in locally advanced and metastatic solid tumours and HER2-expressing breast cancer: a phase 1 dose-escalation and dose-expansion study. Lancet Oncol 2019;20(8):1124–35.

[42] Nowak-Sliwinska P, Scapozza L, i Altaba AR. Drug repurposing in oncology: compounds, pathways, phenotypes and computational approaches for colorectal cancer. Biochim Biophys Acta Rev Cancer 2019;871(2):434–54.

[43] Chen H, Wu J, Gao Y, Chen H, Zhou J. Scaffold repurposing of old drugs towards new cancer drug discovery. Curr Top Med Chem 2016;16(19):2107–14.

[44] Jeibouei S, Akbari ME, Kalbasi A, Aref AR, Ajoudanian M, Rezvani A, Zali H. Personalized medicine in breast cancer: pharmacogenomics approaches. Pharmacogenomics Pers Med 2019;12:59.

[45] Chan C, Law B, So W, Chow K, Waye M. Novel strategies on personalized medicine for breast cancer treatment: an update. Int J Mol Sci 2017;18(11):2423.

[46] Prasad V. Perspective: the precision-oncology illusion. Nature 2016;537(7619):S63.

[47] Meric-Bernstam F, Brusco L, Shaw K, Horombe C, Kopetz S, Davies MA, et al. Feasibility of large-scale genomic testing to facilitate enrollment onto genomically matched clinical trials. J Clin Oncol 2015;33(25): 2753.

[48] Tan SH, Lee SC, Goh BC, Wong J. Pharmacogenetics in breast cancer therapy. Clin Cancer Res 2008; 14(24):8027–41.

[49] Huang RS, Duan S, Shukla SJ, Kistner EO, Clark TA, Chen TX, et al. Identification of genetic variants contributing to cisplatin-induced cytotoxicity by use of a genomewide approach. Am J Hum Genet 2007; 81(3):427–37.

[50] Angus L, Smid M, Wilting SM, van Riet J, Van Hoeck A, Nguyen L, et al. The genomic landscape of metastatic breast cancer highlights changes in mutation and signature frequencies. Nat Genet 2019:1–9.

[51] Ng CK, Bidard FC, Piscuoglio S, Geyer FC, Lim RS, De Bruijn I, et al. Genetic heterogeneity in therapy-naive synchronous primary breast cancers and their metastases. Clin Cancer Res 2017;23(15):4402–15.

[52] Nik-Zainal S, Morganella S. Mutational signatures in breast cancer: the problem at the DNA level. 2017.

[53] Nik-Zainal S, Davies H, Staaf J, Ramakrishna M, Glodzik D, Zou X, et al. Landscape of somatic mutations in 560 breast cancer whole-genome sequences. Nature 2016;534(7605):47.

[54] Swanton C, McGranahan N, Starrett GJ, Harris RS. APOBEC enzymes: mutagenic fuel for cancer evolution and heterogeneity. Cancer Discov 2015;5(7): 704–12.

[55] Burns MB, Lackey L, Carpenter MA, Rathore A, Land AM, Leonard B, et al. APOBEC3B is an enzymatic source of mutation in breast cancer. Nature 2013;494(7437):366.

[56] Hakata Y, Landau NR. Reversed functional organization of mouse and human APOBEC3 cytidine deaminase domains. J Biol Chem 2006;281(48):36624–31.
[57] Savas P, Teo ZL, Lefevre C, Flensburg C, Caramia F, Alsop K, et al. The subclonal architecture of metastatic breast cancer: results from a prospective community-based rapid autopsy program "CASCADE". PLoS Med 2016;13(12):e1002204.
[58] Espinal-Enriquez J, Fresno C, Anda-Jauregui G, Hernández-Lemus E. RNA-Seq based genome-wide analysis reveals loss of inter-chromosomal regulation in breast cancer. Sci Rep 2017a;7(1):1760.
[59] Espinal-Enríquez J, Mejía-Pedroza RA, Hernández-Lemus E. Computational approaches in precision medicine. In: Progress and challenges in precision medicine. Academic Press; 2017b. p. 233–50.
[60] Kanehisa M, Goto S. Kegg: kyoto encyclopedia of genes and genomes. Nucleic Acids Res 2000;28:27–30.
[61] Croft D, O'Kelly G, Wu G, Haw R, Gillespie M, Matthews L, et al. Reactome: a database of reactions, pathways and biological processes. Nucleic Acids Res 2010;39(Suppl. 1):D691–7.
[62] Fabregat A, Jupe S, Matthews L, Sidiropoulos K, Gillespie M, Garapati P, et al. The reactome pathway knowledgebase. Nucleic Acids Res 2017;46(D1): D649–55.
[63] Cerami EG, Gross BE, Demir E, Rodchenkov I, Babur Ö, Anwar N, et al. Pathway Commons, a web resource for biological pathway data. Nucleic Acids Res 2010;39(Suppl. 1):D685–90.
[64] Hewett M, Oliver DE, Rubin DL, Easton KL, Stuart JM, Altman RB, Klein TE. PharmGKB: the pharmacogenetics knowledge base. Nucleic Acids Res 2002;30(1):163–5.
[65] Thorn CF, Klein TE, Altman RB. PharmGKB. Pharmacogenomics. Humana Press; 2005. p. 179–91.
[66] Wishart DS, Feunang YD, Guo AC, Lo EJ, Marcu A, Grant JR, et al. DrugBank 5.0: a major update to the DrugBank database for 2018. Nucleic Acids Res 2017;46(D1):D1074–82.
[67] Zhu F, Shi Z, Qin C, Tao L, Liu X, Xu F, et al. Therapeutic target database update 2012: a resource for facilitating target-oriented drug discovery. Nucleic Acids Res 2011;40(D1):D1128–36.
[68] Wagner AH, Coffman AC, Ainscough BJ, Spies NC, Skidmore ZL, Campbell KM, et al. DGIdb 2.0: mining clinically relevant drug–gene interactions. Nucleic Acids Res 2015;44(D1):D1036–44.
[69] Marcotte R, Sayad A, Brown KR, Sanchez-Garcia F, Reimand J, Haider M, Virtanen C, Bradner JE, Bader GD, Mills GB, et al. Functional genomic landscape of human breast cancer drivers, vulnerabilities, and resistance. Cell 2016;164:293–309.
[70] Cheung HW, Cowley GS, Weir BA, Boehm JS, Rusin S, Scott JA, et al. Systematic investigation of genetic vulnerabilities across cancer cell lines reveals lineage-specific dependencies in ovarian cancer. Proc Natl Acad Sci U S A 2011;108(30):12372–7.
[71] Aguirre AJ, Meyers RM, Weir BA, Vázquez F, Zhang CZ, Ben-David U, Cook A, Ha G, Harrington WF, Doshi MB, et al. Genomic copy number dictates a gene-independent cell response to CRISPR/Cas9 targeting. Cancer Discov 2016;6: 914–29.
[72] Munoz DM, Cassiani PJ, Li L, Billy E, Korn JM, Jones MD, Golji J, Ruddy DA, Yu K, McAllister G, et al. CRISPR screens provide a comprehensive assessment of cancer vulnerabilities but generate false-positive hits for highly amplified genomic regions. Cancer Discov 2016;6:900–13.
[73] Pirmohamed M. Pharmacogenetics and pharmacogenomics. Br J Clin Pharmacol 2001;52:345–7.
[74] García-Campos MA, Espinal-Enríquez J, Hernández-Lemus E. Pathway analysis: state of the art. Front Physiol 2015;6:383.
[75] Wang L. Pharmacogenomics: a systems approach. Wiley Interdiscp. Rev. Syst. Biol. Med. 2010;2:3–22.
[76] Mejía-Pedroza RA, Espinal-Enríquez J, Hernández-Lemus E. Pathway-based drug repositioning for breast cancer molecular subtypes. Front Pharmacol 2018;9:905.
[77] Mc Cord KA. Int J Publ Health 2019;64:1255. https://doi.org/10.1007/s00038-019-01293-2.
[78] Khoury MJ, Gwinn M, Glasgow RE, Kramer BS. A population perspective on how personalized medicine can improve health. Am J Prev Med 2012;42(6):639.
[79] Bachtiar M, Ooi BNS, Wang J, Jin Y, Tan TW, Chong SS, Lee CG. Towards precision medicine: interrogating the human genome to identify drug pathways associated with potentially functional, population-differentiated polymorphisms. Pharmacogenomics J 2019;19(6):516–27.
[80] Juengst ET, McGowan ML. Why does the shift from "personalized medicine" to "precision health" and "wellness genomics" matter? AMA J Ethics 2018; 20(9):881–90.
[81] Chin L, Hahn WC, Getz G, Meyerson M. Making sense of cancer genomic data. Genes Dev 2011;25(6):534–55.
[82] Bild AH, Yao G, Chang JT, Wang Q, Potti A, Chasse D, et al. Oncogenic pathway signatures in human cancers as a guide to targeted therapies. Nature 2006;439(7074):353–7.
[83] Emmert-Streib F, Glazko GV. Pathway analysis of expression data: deciphering functional building blocks of complex diseases. PLoS Comput Biol 2011; 7(5).

[84] Vaske CJ, Benz SC, Sanborn JZ, Earl D, Szeto C, Zhu J, et al. Inference of patient-specific pathway activities from multi-dimensional cancer genomics data using PARADIGM. Bioinformatics 2010;26(12):i237–45.

[85] Hastie T, Stuetzle W. Principal curves. J Am Stat Assoc 1989;84(406):502–16.

[86] Cancer Genome Atlas Network. Comprehensive molecular portraits of human breast tumours. Nature 2012;490(7418):61.

[87] Weinstein JN, Collisson EA, Mills GB, Shaw KRM, Ozenberger BA, Ellrott K, Cancer Genome Atlas Research Network. The cancer genome atlas pan-cancer analysis project. Nat Genet 2013;45(10):1113.

[88] Ciriello G, Gatza ML, Beck AH, Wilkerson MD, Rhie SK, Pastore A, et al. Comprehensive molecular portraits of invasive lobular breast cancer. Cell 2015;163(2):506–19.

[89] Berger AC, Korkut A, Kanchi RS, Hegde AM, Lenoir W, Liu W, et al. A comprehensive pan-cancer molecular study of gynecologic and breast cancers. Cancer Cell 2018;33(4):690–705.

[90] Curtis C, Shah SP, Chin SF, Turashvili G, Rueda OM, Dunning MJ, et al. The genomic and transcriptomic architecture of 2,000 breast tumours reveals novel subgroups. Nature 2012;486(7403):346.

[91] Pereira B, Chin SF, Rueda OM, Vollan HKM, Provenzano E, Bardwell HA, et al. The somatic mutation profiles of 2,433 breast cancers refine their genomic and transcriptomic landscapes. Nat Commun 2016;7:11479.

[92] Milioli HH, Vimieiro R, Tishchenko I, Riveros C, Berretta R, Moscato P. Iteratively refining breast cancer intrinsic subtypes in the METABRIC dataset. BioData Mining 2016;9(1):2.

[93] Mukherjee A, Russell R, Chin SF, Liu B, Rueda OM, Ali HR, et al. Associations between genomic stratification of breast cancer and centrally reviewed tumour pathology in the METABRIC cohort. NPJ Breast Cancer 2018;4(1):5.

[94] Shah SP, Roth A, Goya R, Oloumi A, Ha G, Zhao Y, et al. The clonal and mutational evolution spectrum of primary triple-negative breast cancers. Nature 2012;486(7403):395.

[95] Milioli HH, Vimieiro R, Riveros C, Tishchenko I, Berretta R, Moscato P. The discovery of novel biomarkers improves breast cancer intrinsic subtype prediction and reconciles the labels in the metabric data set. PLoS One 2015;10(7):e0129711.

[96] Wang M, Klevebring D, Lindberg J, Czene K, Grönberg H, Rantalainen M. Determining breast cancer histological grade from RNA-sequencing data. Breast Cancer Res 2016;18(1):48.

[97] Dadiani M, Ben-Moshe NB, Paluch-Shimon S, Perry G, Balint N, Marin I, et al. Tumor evolution inferred by patterns of microRNA expression through the course of disease, therapy, and recurrence in breast cancer. Clin Cancer Res 2016;22(14):3651–62.

[98] Kim YH, Jeong DC, Pak K, Goh TS, Lee CS, Han ME, et al. Gene network inherent in genomic big data improves the accuracy of prognostic prediction for cancer patients. Oncotarget 2017;8(44):77515.

[99] Mucaki EJ, Baranova K, Pham HQ, Rezaeian I, Angelov D, Ngom A, et al. Predicting outcomes of hormone and chemotherapy in the molecular taxonomy of breast Cancer international consortium (METABRIC) study by biochemically-inspired machine learning. F1000Research 2016.

[100] Liu C, Xiao Y, Tao Z, Hu XC. Identification of immune microenvironment subtypes of breast cancer in TCGA set: implications for immunotherapy. 2019.

[101] Peña-Chilet M, Oltra SS, Martinez MT, Fores J, Ayala G, Ribas G. Pathway deregulation networks in breast cancer young patients: own data with METABRIC and TCGA databases. Eur J Cancer 2016;61:S173.

[102] Livshits A, Git A, Fuks G, Caldas C, Domany E. Pathway-based personalized analysis of breast cancer expression data. Mol Oncol 2015;9(7):1471–83.

[103] Sun Y, Niu W, Du F, Du C, Li S, Wang J, et al. Safety, pharmacokinetics, and antitumor properties of anlotinib, an oral multi-target tyrosine kinase inhibitor, in patients with advanced refractory solid tumors. J Hematol Oncol 2016;9(1):105.

[104] Lu J, Shi Q, Zhang L, Wu J, Lou Y, Qian J, et al. Integrated transcriptome analysis reveals KLK5 and L1CAM predict response to anlotinib in NSCLC at 3rd line. Front oncol 2019;9:886.

[105] Lv Y, Zhang J, Liu F, Song M, Hou Y, Liang N. Targeted therapy with anlotinib for patient with recurrent glioblastoma: a case report and literature review. Medicine 2019;98(22):e15749.

[106] Kim JY, Lee E, Park K, Jung HH, Park WY, Lee KH, et al. Molecular alterations and poziotinib efficacy, a pan-HER inhibitor, in human epidermal growth factor receptor 2 (HER2)-positive breast cancers: combined exploratory biomarker analysis from a phase II clinical trial of poziotinib for refractory HER2-positive breast cancer patients. Int J Cancer 2019;145(6):1669–78.

[107] Park YH, Lee KH, Sohn JH, Lee KS, Jung KH, Kim JH, et al. A phase II trial of the pan-HER inhibitor poziotinib, in patients with HER2-positive metastatic breast cancer who had received at least two prior HER2-directed regimens: results of the NOV120101-203 trial. Int J Cancer 2018;143(12):3240–7.

[108] Robichaux JP, Elamin YY, Vijayan RSK, Nilsson MB, Hu L, He J, et al. Pan-Cancer landscape and analysis of ErbB2 mutations identifies Poziotinib as a clinically active inhibitor and enhancer of T-DM1 activity. Cancer Cell 2019;36(4):444–57.
[109] Kalous O, Conklin D, Desai AJ, O'Brien NA, Ginther C, Anderson L, et al. Dacomitinib (PF-00299804), an irreversible Pan-HER inhibitor, inhibits proliferation of HER2-amplified breast cancer cell lines resistant to trastuzumab and lapatinib. Mol Cancer Therapeut 2012;11(9):1978–87.
[110] Akhtar S, Al-Zaid B, El-Hashim AZ, Chandrasekhar B, Attur S, Yousif MH, Benter IF. Cationic polyamidoamine dendrimers as modulators of EGFR signaling in vitro and in vivo. PLoS One 2015;10(7):e0132215.
[111] Lauriola M, Enuka Y, Zeisel A, D'Uva G, Roth L, Sharon-Sevilla M, et al. Diurnal suppression of EGFR signalling by glucocorticoids and implications for tumour progression and treatment. Nat Commun 2014;5:5073.
[112] Semenova G, Chernoff J. Targeting PAK1. Biochem Soc Trans 2017;45(1):79–88.
[113] Woo TG, Yoon MH, Hong SD, Choi J, Ha NC, Sun H, Park BJ. Anti-cancer effect of novel PAK1 inhibitor via induction of PUMA-mediated cell death and p21-mediated cell cycle arrest. Oncotarget 2017;8(14): 23690.
[114] Yin N, Ma W, Pei J, Ouyang Q, Tang C, Lai L. Synergistic and antagonistic drug combinations depend on network topology. PLoS One 2014;9(4).
[115] Logue JS, Morrison DK. Complexity in the signaling network: insights from the use of targeted inhibitors in cancer therapy. Genes Dev 2012;26(7): 641–50.
[116] Sun X, Bao J, You Z, Chen X, Cui J. Modeling of signaling crosstalk-mediated drug resistance and its implications on drug combination. Oncotarget 2016; 7(39):63995.
[117] de Anda-Jáuregui G, Mejía-Pedroza RA, Espinal-Enríquez J, Hernández-Lemus E. Crosstalk events in the estrogen signaling pathway may affect tamoxifen efficacy in breast cancer molecular subtypes. Comput Biol Chem 2015;59:42–54.
[118] Camidge DR, Pao W, Sequist LV. Acquired resistance to TKIs in solid tumours: learning from lung cancer. Nat Rev Clin Oncol 2014;11(8):473.
[119] Ivanov M, Barragan I, Ingelman-Sundberg M. Epigenetic mechanisms of importance for drug treatment. Trends Pharmacol Sci 2014;35(8):384–96.
[120] Behar M, Barken D, Werner SL, Hoffmann A. The dynamics of signaling as a pharmacological target. Cell 2013;155(2):448–61.
[121] Chen X, Yan CC, Zhang X, Zhang X, Dai F, Yin J, Zhang Y. Drug–target interaction prediction: databases, web servers and computational models. Briefings Bioinf 2016a;17(4):696–712.
[122] Chen X, Ren B, Chen M, Wang Q, Zhang L, Yan G. NLLSS: predicting synergistic drug combinations based on semi-supervised learning. PLoS Comput Biol 2016b;12(7).
[123] Kirouac DC, Du JY, Lahdenranta J, Overland R, Yarar D, Paragas V, et al. Computational modeling of ERBB2-amplified breast cancer identifies combined ErbB2/3 blockade as superior to the combination of MEK and AKT inhibitors. Sci Signal 2013;6(288). ra68-ra68.
[124] Huang TX, Guan XY, Fu L. Therapeutic targeting of the crosstalk between cancer-associated fibroblasts and cancer stem cells. Am J Cancer Res 2019;9(9): 1889.
[125] Liang F, Ren C, Wang J, Wang S, Yang L, Han X, et al. The crosstalk between STAT3 and p53/RAS signaling controls cancer cell metastasis and cisplatin resistance via the Slug/MAPK/PI3K/AKT-mediated regulation of EMT and autophagy. Oncogenesis 2019;8(10):1–15.
[126] Dhanasekaran R, Baylot V, Kim M, Kuruvilla S, Bellovin DI, Adeniji N, et al. MYC and Twist1 cooperate to drive metastasis by eliciting crosstalk between cancer and innate immunity. eLife 2020;9: e50731.
[127] Chen HT, Liu H, Mao MJ, Tan Y, Mo XQ, Meng XJ, et al. Crosstalk between autophagy and epithelial-mesenchymal transition and its application in cancer therapy. Mol Cancer 2019;18(1):101.
[128] Sidorov P, Naulaerts S, Ariey-Bonnet J, Pasquier E, Ballester P. Predicting synergism of cancer drug combinations using NCI-ALMANAC data. Front Chem 2019;7:509.
[129] Li H, Li T, Quang D, Guan Y. Network propagation predicts drug synergy in cancers. Cancer Res 2018; 78(18):5446–57.
[130] Celebi R, Don't Walk OB, Movva R, Alpsoy S, Dumontier M. In-silico prediction of synergistic anti-cancer drug combinations using multi-omics data. Sci Rep 2019;9(1):1–10.
[131] Tosi D, Pérez-Gracia E, Atis S, Vié N, Combès E, Gabanou M, et al. Rational development of synergistic combinations of chemotherapy and molecular targeted agents for colorectal cancer treatment. BMC Cancer 2018;18(1):812.
[132] Norouzi S, Valokala MG, Mosaffa F, Zirak MR, Zamani P, Behravan J. Crosstalk in cancer resistance and metastasis. Crit Rev Oncol Hematol 2018;132: 145–53.

[133] Choe C, Shin YS, Kim C, Choi SJ, Lee J, Kim SY, et al. Crosstalk with cancer-associated fibroblasts induces resistance of non-small cell lung cancer cells to epidermal growth factor receptor tyrosine kinase inhibition. Onco Targets Ther 2015;8:3665.

[134] Furth N, Bossel N, Pozniak Y, Geiger T, Domany E, Aylon Y, Oren M. Tumor suppressor crosstalk: hippo and p53. Eur J Cancer 2016;61:S50.

[135] Marton MJ, DeRisi JL, Bennett HA, Iyer VR, Meyer MR, Roberts CJ, et al. Drug target validation and identification of secondary drug target effects using DNA microarrays. Nat Med 1998;4(11):1293–301.

[136] O'Donnell III JJ, Somberg JC, O'Donnell JT. Introduction to drug discovery and development. In: Drug discovery and development. 3rd ed. CRC Press; 2019. p. 1–13.

[137] Bourdeau V, Deschênes J, Laperrière D, Aid M, White JH, Mader S. Mechanisms of primary and secondary estrogen target gene regulation in breast cancer cells. Nucleic Acids Res 2008;36(1):76–93.

[138] Whitebread S, Dumotier B, Armstrong D, Fekete A, Chen S, Hartmann A, et al. Secondary pharmacology: screening and interpretation of off-target activities—focus on translation. Drug Discov Today 2016;21(8):1232–42.

[139] de Anda-Jáuregui G, Espinal-Enriquez J, Hur J, Alcalá-Corona SA, Ruiz-Azuara L, Hernández-Lemus E. Identification of Casiopeina II-gly secondary targets through a systems pharmacology approach. Comput Biol Chem 2019;78:127–32.

[140] Tate JG, Bamford S, Jubb HC, Sondka Z, Beare DM, Bindal N, et al. COSMIC: the catalogue of somatic mutations in cancer. Nucleic Acids Res 2019;47(D1):D941–7.

[141] Jubb HC, Saini HK, Verdonk ML, Forbes SA. COSMIC-3D provides structural perspectives on cancer genetics for drug discovery. Nat Genet 2018;50(9):1200–2.

[142] de Anda-Jáuregui G, Guo K, Hur J. Network-based assessment of adverse drug reaction risk in polypharmacy using high-throughput screening data. Int J Mol Sci 2019;20(2):386.

[143] Dang TO, Ogunniyi A, Barbee MS, Drilon A. Pembrolizumab for the treatment of PD-L1 positive advanced or metastatic non-small cell lung cancer. Expet Rev Anticancer Ther 2016;16(1):13–20.

[144] dos Santos CR, Domingues G, Matias I, Matos J, Fonseca I, de Almeida JM, Dias S. LDL-cholesterol signaling induces breast cancer proliferation and invasion. Lipids Health Dis 2014;13(1):16.

[145] Xue LY, Chiu SM, Oleinick NL. Staurosporine-induced death of MCF-7 human breast cancer cells: a distinction between caspase-3-dependent steps of apoptosis and the critical lethal lesions. Exp Cell Res 2003;283(2):135–45.

[146] Zambrano JN, Williams CJ, Williams CB, Hedgepeth L, Burger P, Dilday T, et al. Staurosporine, an inhibitor of hormonally up-regulated neu-associated kinase. Oncotarget 2018;9(89):35962.

[147] Masuda H, Zhang D, Bartholomeusz C, Doihara H, Hortobagyi GN, Ueno NT. Role of epidermal growth factor receptor in breast cancer. Breast Cancer Res Treat 2012;136(2):331–45.

[148] Iacopetta D, Carocci A, Sinicropi MS, Catalano A, Lentini G, Ceramella J, et al. Old drug scaffold, new activity: thalidomide-correlated compounds exert different effects on breast cancer cell growth and progression. ChemMedChem 2017;12(5):381–9.

[149] Shen Y, Li S, Wang X, Wang M, Tian Q, Yang J, et al. Tumor vasculature remolding by thalidomide increases delivery and efficacy of cisplatin. J Exp Clin Cancer Res 2019;38(1):427.

[150] Islam B, Lustberg M, Staff NP, Kolb N, Alberti P, Argyriou AA. Vinca alkaloids, thalidomide and eribulin-induced peripheral neurotoxicity: from pathogenesis to treatment. J Peripher Nerv Syst 2019;24:S63–73.

[151] Tseng S, Pak G, Washenik K, Pomeranz MK, Shupack JL. Rediscovering thalidomide: a review of its mechanism of action, side effects, and potential uses. J Am Acad Dermatol 1996;35(6):969–79.

[152] Schram AM, Hyman DM. Quantifying the benefits of genome-driven oncology. Cancer Discov 2017;7(6):552–4.

[153] Massard C, Michiels S, Ferté C, Le Deley MC, Lacroix L, Hollebecque A, et al. High-throughput genomics and clinical outcome in hard-to-treat advanced cancers: results of the MOSCATO 01 trial. Cancer Discov 2017;7(6):586–95.

[154] Hawsawi Y, El-Gendy R, Twelves C, Speirs V, Beattie J. Insulin-like growth factor—oestradiol crosstalk and mammary gland tumourigenesis. Biochim Biophys Acta (BBA)-Rev Cancer 2013;1836(2):345–53.

CHAPTER

8

Successful stories of drug repurposing for cancer therapy in hepatocellular carcinoma

Yasmeen M. Attia[1], *Heba Ewida*[2], *Mahmoud Salama Ahmed*[3]

[1]Department of Pharmacology, Faculty of Pharmacy, The British University in Egypt, El-Sherouk, Cairo, Egypt; [2]Department of Pharmacology and Biochemistry, Faculty of Pharmaceutical Sciences & Pharmaceutical Industries, Future University in Egypt, Cairo, Egypt; [3]Department of Internal Medicine, Division of Cardiology, University of Texas Southwestern Medical Center, Dallas, TX, United States

OUTLINE

Drug Repurposing in Cancer Therapy
https://doi.org/10.1016/B978-0-12-819668-7.00008-7

Hepatocellular carcinoma

Hepatocellular carcinoma (HCC) is a major health problem with devastating consequences associated with treatment failure. It has an estimated global incidence of more than 850,000 new cases annually [1]. HCC is currently the second leading cause of cancer-related death worldwide and accounts for 90% of cases with primary liver cancer [2]. Several risk factors contribute to HCC development such as liver cirrhosis, viral hepatitis including hepatitis B virus (HBV) and hepatitis C virus (HCV) infections, fatty liver, and alcohol abuse. Smoking and the fungal carcinogen, aflatoxin B1, are also well-known contributors to HCC [3]. Recent advances in HBV vaccination and antiviral therapeutics have remarkably contributed to a decrease in incidence. However, nonalcoholic fatty liver disease (NAFLD) and its progressive form, nonalcoholic steatohepatitis (NASH), at its current pace of growing prevalence approaching epidemic proportions is projecting as the most common underlying etiology of HCC, presented in almost 60% of cases, which made NAFLD-associated HCC an emerging indication for liver transplantation [4–6]. Additionally, HCC has a notable gender predilection where incidence in men is threefold that in women [1,7].

Molecular pathogenesis of HCC

Over the past decade, substantial progress has been achieved in understanding how HCC develops and progresses in an attempt to improve treatment options [8]. HCC is a very heterogeneous disease in terms of both phenotype and genotype. This heterogeneity could be attributed, in part, to the divergent nature of the contributing factors, the complexity of the liver microenvironment, and the stage at which HCC turns to be clinically evident/detectable.

Malignant transformations in liver cells are driven by several factors such as chronic injury or inflammation due to oxidative stress that may lead to genetic and epigenetic modification. These modifications consequently lead to disrupted cellular signaling pathways leading to an overexpression in several growth factors and their receptors. This inevitably results in cell resistance to apoptotic signals, stimulation of angiogenesis, and uncontrollable proliferation besides the acquisition of a metastatic phenotype [9,10].

Since HCC almost exclusively develops in patients with chronic liver diseases, injury of liver cells can promote the progression to HCC over a long period of time [11] driven by a number of cytokines and inflammatory mediators along with aberrant activity of several signaling pathways, as shown in Fig. 8.1 (will be discussed later).

Therapeutic management of HCC

HCC mostly exhibits resistance to conventional chemotherapy. Besides, patients with HCC are usually intolerant to treatment due to an underlying hepatic dysfunction.

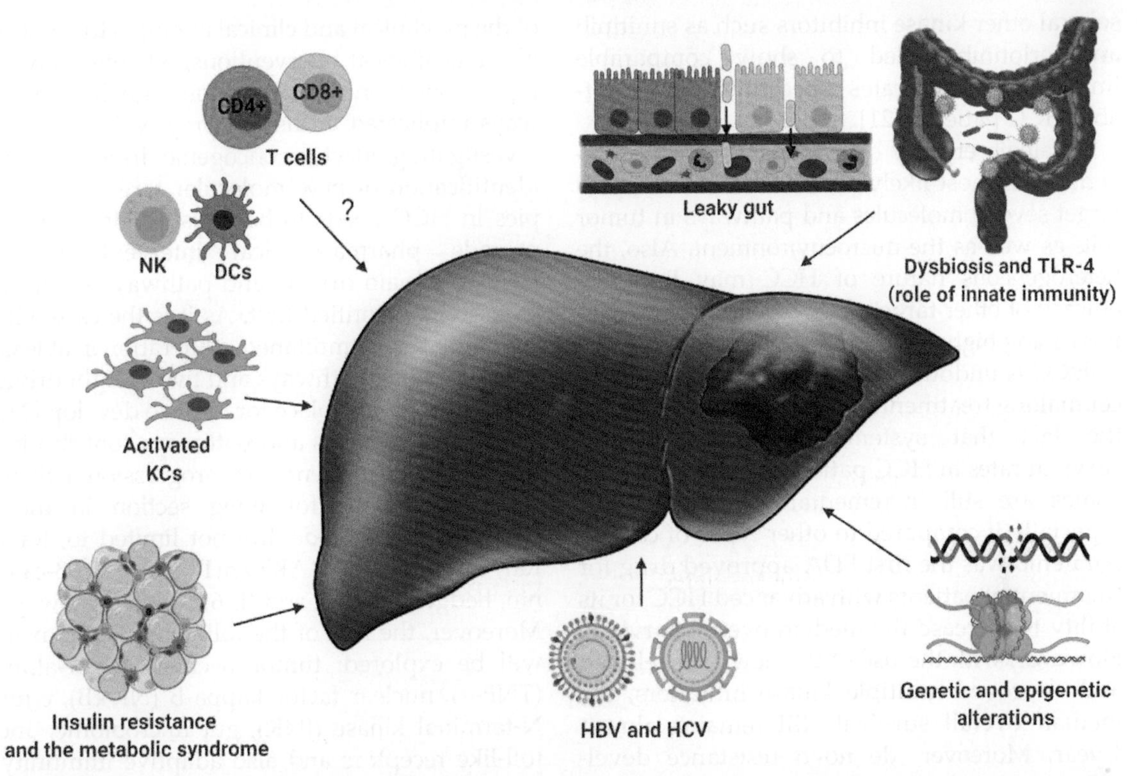

FIGURE 8.1 Proposed mechanisms for the progression of HCC. These mechanisms include viral infections by hepatitis B and C viruses (HBV and HCV), genetic and epigenetic alterations, dysbiosis and leaky gut, role of innate immunity through Toll-like receptors-4 (TLR4) and Kupffer cells (KCs), metabolic syndrome and its hepatic manifestation (nonalcoholic fatty liver disease), and adaptive immune cells (CD4+ and CD8+ T cells) whose role is still quite debatable. Also, dendritic cells (DCs) and natural killer T cells (NK T cells) bridge innate with adaptive immune systems.

However, almost half of HCC patients still receive chemotherapy at some point during the course of the disease [12,13]. Chemotherapeutic intervention mainly relies on the multi-targeted tyrosine kinase, sorafenib, that acts by effectively blocking the Ras/Raf/MAPK pathway impeding cancer cell ability to circumvent apoptotic signals and induce angiogenesis, proliferation, and invasion. Sorafenib was found to extend the median survival in advanced HCC patients for up to 3 months with manageable adverse effects; however, it lacks predictive biomarkers to reflect on responsiveness [14]. Sorafenib remained the only approved systemic treatment for HCC between 2007 and 2016. Yet, promising outcomes were reported in randomized phase III trials using other multiple kinase inhibitors such as lenvatinib [15], regorafenib [16], cabozantinib [17], and ramucirumab [18,19], where regorafenib has received FDA approval in the second-line setting. Moreover, nivolumab, a monoclonal immunotherapy-based antibody targeting the immune checkpoint programmed cell death protein 1, showed also positive response rates and mean overall survival durations in a phase I–II trials performed on patients who were formerly treated with sorafenib [20] for which it has been successfully granted an accelerated FDA approval. On the other hand, several other kinase inhibitors such as sunitinib and erlotinib failed to show comparable improved survival rates especially in unrespectable HCC patients [21].

Sorafenib efficacy over other proposed interventions is most likely attributed to its ability to target several molecules and pathways in tumor cells as well as the microenvironment. Also, the heterogeneous nature of HCC may limit the efficacy of other targeted therapeutic approaches possessing higher selectivity [22].

HCC is undoubtedly a resistant type of cancer making treatment more challenging. Despite the fact that systemic therapy enhanced survival rates in HCC patients, therapeutic outcomes are still incremental and inadequate, especially if compared to other types of cancer. Sorafenib was the first FDA-approved drug for treatment of patients with advanced HCC for its ability to increase the median overall survival. However, with the use of the newly developed and approved multiple kinase inhibitors, the median overall survival still remains almost 1 year. Moreover, de novo resistance developing to sorafenib has been recently heavily reported impeding its beneficial clinical applications [23–26]. Resistance to sorafenib involves a cross talk between several pathways such as Janus kinase/signal transducer and activator of transcription 3 (STAT3), phosphatidylinositol-3-kinase (PI3K)/AKT/mammalian target of rapamycin (mTOR), and hypoxia-inducible pathways beside others [27]. Providing new treatment options for HCC, therefore, still remains an unmet medical need. Accordingly, further insights into the molecular targets affiliated with the pathophysiology of HCC will be dissected in the following section.

HCC drivers and affiliated molecular targets

To date, we still have very few identified "druggable" drivers for HCC. However, most of the preclinical and clinical attempts to identify pharmacological interventions, whether new or repositioned ones, focused on signaling pathways implicated in disease progression. Hence, investigating effect on oncogenic drivers for the identification of new molecular targeted therapies in HCC needs to be revisited. In order to provide pharmacological interventions for HCC, the main drivers and pathways involved have to be identified first. As it is the case with solid tumors, a simultaneous alteration in at least three signaling pathways and five to eight driver genes should take place for HCC to develop [28].

The main drivers and pathways contributing to HCC development and progression will be discussed in the following section in more details. These include, but not limited to, Ras/Raf/MAPK, PI3K/AKT/mTOR, Wnt/β-catenin, hedgehog (Hh), and IL-6/STAT3 pathways. Moreover, the role of the following key players will be explored: tumor necrosis factor-alpha (TNF-α), nuclear factor kappa-B (NF-κB), c-Jun N-terminal kinase (JNK), gut microbiome, and toll-like receptors and also adaptive immunity. All these molecular targets and signaling pathways represent potential candidates for targeting HCC, as shown in Fig. 8.2.

Ras/Raf/MAPK pathway

Among the most extensively investigated pathways implicated in HCC development and progression is the Ras/Raf/MAPK pathway. Cell surface receptor tyrosine kinases such as insulin-like growth factor (IGF) receptor, endothelial and vascular epidermal growth factor receptor (EGFR), c-Met, and platelet-derived growth factor receptor send signals that are transmitted to the nucleus via this pathway to control cell survival, growth, and differentiation. Upregulation of the Ras/Raf/MAPK pathway in liver cells induces cell growth augmenting antiapoptotic signals leading ultimately to HCC [29]. Aberrant upstream IGF and EGFR signaling, suppression of Raf kinase inhibitor protein,

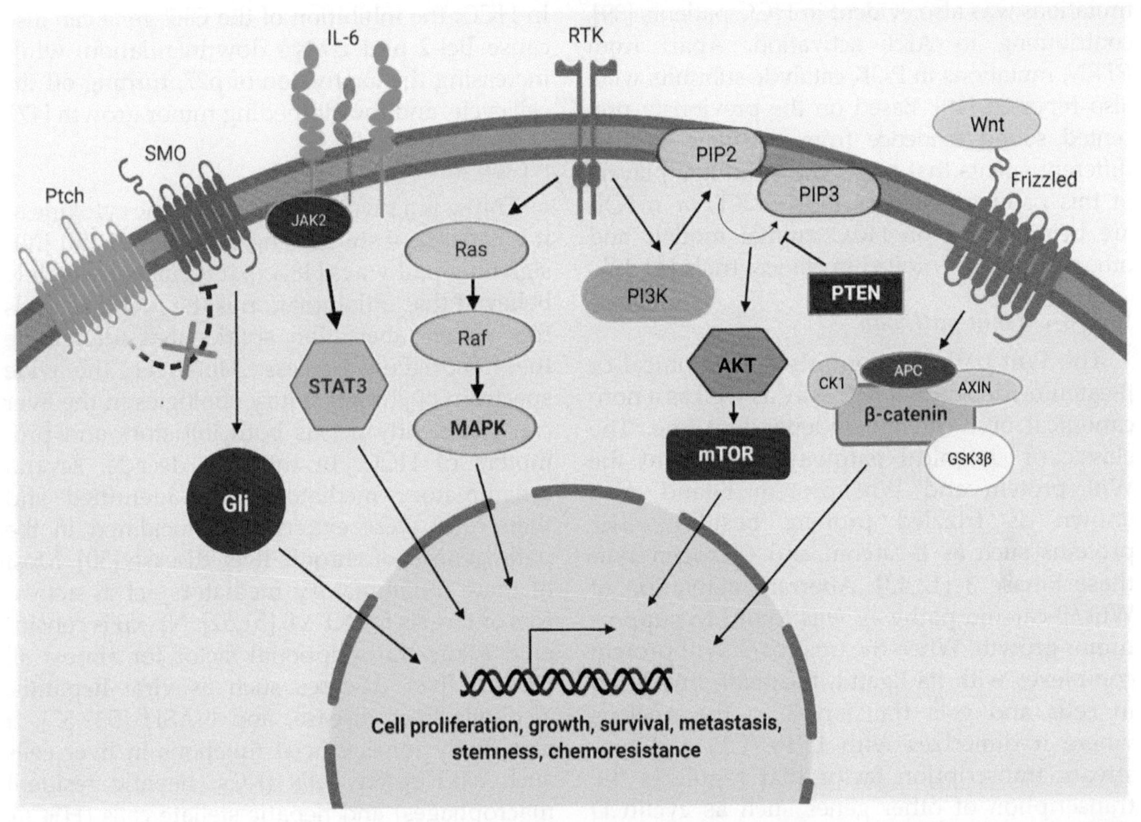

FIGURE 8.2 Molecular targets and pathways affiliated with HCC. Ras/Raf/MAPK, IL-6/STAT3,Wnt/β-catenin, PI3K/AKT/mTOR, and hedgehog signaling pathways are among the most important pathways implicated in HCC, which are presented as potential targets for treatment. *APC*, adenomatosis polyposis coli; *CK1*, caseinkinase 1; *JAK2*, janus kinase-2; *mTOR*, mammalian target of rapamycin; *PI3K*, phosphoinositide3-kinases; *PIP2*, phosphatidylinositol 4,5-bisphosphate; *PIP3*, phosphatidylinositol (3,4,5)-trisphosphate; *RTK*, receptor tyrosine kinase; *STAT3*, signal transducer and activator of transcription.

and HBV- and HCV-related proteins induction are among the mechanisms by which the Ras/Raf/MAPK pathway can be activated toward HCC development [30].

PI3K/AKT/mTOR pathway

Mounting evidence over the past years suggested the aberrant upregulation of PI3K/AKT/mTOR pathway in HCC where mTOR phosphorylation and a subsequent upregulation of its downstream effector, p70S6k, were reported in almost 50% of HCC patients [31]. Likewise, almost 40% of patients encountered an activated mTOR [32]. Aberrant PI3K/Akt/mTOR pathway was also correlated with poor prognosis, especially in advanced-stage HCC [33]. The underlying mechanism by which PI3K/AKT/mTOR pathway is activated in HCC is not yet fully deciphered, yet overly expressed upstream IGF, c-Met, or EGFR is likely [34–36]. Hepatitis viral infections are also capable of activating PI3K/AKT/mTOR pathway in liver [37], where HCV infection was found to increase neuroblastoma (N)-Ras expression in HCC, which turns on the PI3K/AKT/mTOR pathway [38]. Loss of *PTEN*, a tumor suppressor gene, by mutations was also evident in HCC patients [39], contributing to AKT activation. Apart from *PTEN*, mutations in PI3K catalytic subunits were also reported [40]. Based on the previously presented sound evidence from previous studies, different agents that target the main key players of this pathway such as PI3K, AKT, or mTOR are being tested on HCC animal models and others currently enrolled in clinical trials [41,42].

Wnt/β-catenin pathway

The Wnt pathway comprises a canonical or β-catenin–dependent pathway as well as a noncanonical or β-catenin–independent one. The classic or canonical pathway consists of the Wnt protein and Wnt protein ligand, also known as frizzled protein, besides other proteins such as β-catenin and glycogen synthase kinase 3 [12,43]. Aberrant activation of Wnt/β-catenin pathway was found to support tumor growth. When the upstream Wnt protein complexes with its ligand, β-catenin builds up in cells and gets transferred to the nucleus where it dimerizes with LEF/TCF, a downstream transcription factor that regulates the transcription of other genes such as cyclin D [44]. Abnormal upregulation of Wnt/β-catenin pathway was linked to HCC development and cancer stem cell maintenance. Aberrant β-catenin was detected in almost 90% of liver cancers [12].

Hedgehog pathway

Hh signaling pathway is among the most important pathways implicated in HCC. It consists of Hh ligand, Ptch and Smo (two transmembrane receptors), and Glib nuclear transcription factor along with downstream genes. Once activated, Hh ligands bind to Ptch receptors blocking Ptch inhibitory effect on Smo. Smo then moves to the cytoplasm activating Gli, to induce specific genes upregulation, thereby controlling growth and division. The Hh pathway is rarely activated in normal liver cells; however, in HCC, it is abnormally active [45,46]. In HCC, the inhibition of the Gli2 gene can also cause Bcl-2 and c-Myc downregulation, while increasing the expression of p27, turning off the cell cycle, and thus impeding tumor growth [47].

NF-κB and JNK

TNF-α is a pivotal protumorigenic cytokine as it is capable of stimulating both NF-κB and JNK signaling pathways [48,49]. It is currently widely believed that inflammation is the fuel that feeds the genetic aberration sparks that inaugurate the tumorigenic process. Moreover, the wide spectrum of chronic injury etiologies in the liver can sufficiently act as both initiators and promoters of HCC. In this last decade, several inflammatory mediators were identified and their roles were extensively elucidated in the pathogenesis of chronic liver disease [50]. Most of these inflammatory mediators act as activators or targets for NF-κB [51,52]. NF-κB is considered a key transcriptional factor for almost all chronic liver diseases such as viral hepatitis, alcoholic liver disease, and NASH [53–57]. It also finely tunes crucial functions in liver cells such as Kupffer cells (KCs; hepatic resident macrophages) and hepatic stellate cells (HSCs). Inhibition of NF-κB–related signaling cascades, however, may lead to liver fibrosis and tumorigenesis and that is why NF-κB is considered very essential in maintaining homeostasis and wound-healing processes in the liver [58]. Accordingly, NF-κB is thought of as a two-edged sword since inhibition may be sometimes beneficial; however, it may also influence other essential processes pertaining to liver homeostasis.

Regarding JNK, two isoforms are expressed in liver cells, namely, JNK1 and JNK2, where the former being linked to carcinogenesis. JNK1 protumorigenic potential is defined by its ability to induce the proliferative potential of HCC [28]. It also regulates the proliferation of HCC cells via downregulation of p21 and upregulation.of c-Myc [59]. Moreover, apoptosis induced by caspase-8 was found to activate JNK and hence promote proliferation of liver cells [60]. These findings propose that hepatocyte apoptosis is what triggers JNK activation. Although, the role of JNK in HCC is controversial, JNK1 and 2 knockout attenuated HCC in an animal model of DEN, exhibited p21 upregulation, and decreased c-Myc [61]. Overall, TNFα, NF-κB, and JNK pathways can either have prosurvival or proapoptotic potentials augmenting HCC proliferation and growth. Thus, treatment approaches involving NF-κB and JNK inhibition should act moderately to avoid a complete shut-down in hepatocytes leading to liver injury and subsequent HCC.

IL-6/STAT3 pathway

IL-6, as a mediator of STAT3 activation, is an essential driver of liver cell proliferation, which may consequently lead to HCC [62]. Moreover, overactive STAT3 coinciding with high levels of IL-6 was found in patients with HCC [63]. Likewise, STAT3 upregulation had promoted DEN-induced HCC experimentally [64]. IL-6 autocrine production is essential for malignant transformations in HCC, which once develops, IL-6 paracrine production from KCs start initiating growth and proliferation of HCC cells [65]. Activation of STAT3 may also take place via IL-22, produced by Th17 cells, that is overly expressed in HCC patients. Similar to IL-6, IL-22 also plays a role in promoting DEN-induced HCC in mice via STAT3 pathway [66]. Besides IL-6 and IL-22, IL-17 can also activate STAT3. This action, however, is IL-6-dependent [67] Collectively, these findings propose a role for IL-6/STAT3 pathway in promoting hepatic carcinogenesis suggesting it as a potential therapeutic target for treatment of HCC.

Innate and adaptive immunity

Overwhelming evidence from the past few decades have underlined the role of dysbiosis in the development of chronic liver disease and HCC. Moreover, the role of activating the innate immune system–related receptors in general and toll-like receptors (TLRs) in particular, in HCC development, and progression was also explored [68]. Due to its anatomical location, the liver is considered the first organ exposed to gut-derived microbial products translocated through the portal vein that in turn leads to TLR activation in the liver [68]. TLR-4 is expressed in different liver cells such as KCs, HSCs, and hepatocytes. It was previously reported that activation of TLR-4 on liver cells by bacterial lipopolysaccharides, a gram-negative cell wall component, leads to subsequent fibrotic and carcinogenic events [69]. The immune surveillance hypothesis suggests that the immune system may protect against nascent tumors by destroying malignant cells early on, before they develop into detectable tumors. The T-cell antitumor role is facilitated through immune surveillance by CD4+ and CD8+ T cells. Interestingly, an animal model of NASH-associated HCC was found to cause a depletion of CD4+ T cells impairing immune surveillance [70]. Other studies, however, reported a protumorigenic potential of CD8+ T, natural killer T cells, and T-helper (Th) 17 cells in an animal model of choline-deficient high fat diet [71]. Yet still debatable, these findings suggest a role for adaptive immunity in HCC development and progression. Thus, targeting it in the appropriate context may offer a chemopreventive strategy for HCC patients.

Drug repositioning: drug discovery tool

Drug discovery process involves multidisciplinary integrating experiences starting with in silico computational modeling to design scaffolds of interest based on the affiliated therapeutic molecular targets to different clinical diseases. Typically, this is followed by synthetic methodologies optimization to assemble final ligands ready for biological evaluation [72]. Preclinical biological evaluations involving in vitro and in vivo assessments create significant portion for getting the potential ligands to be considered as potential drug candidates or active pharmaceutical ingredients; however, there are different discovery stories that failed to make it into the market due to intellectual property, efficacy, safety, toxicity, biopharmaceutical compatibility, cost/benefit feasibility, regulatory affairs, etc. The whole discovery process typically takes 10–15 years to get a drug into the market at a cost close to 1 billion US dollars.

In the last decade (2009–19), drug repurposing/repositioning approach has been an emerging drug discovery tool to overcome the associated challenges faced by the pharmaceutical industry. Conceptually, the whole drug–receptor binding theory focuses on two approaches, whether key-lock theory or induced-fit theory [73], where most of the pharmaceutical ingredients exhibit on-target effects targeting receptors/proteins of interest in a drug positioning fashion, as shown in Fig. 8.3. Therefore, existing drugs might show potential binding interactions along with other molecular targets, where drug repositioning shows off-target effects to pronounce newly alternative therapeutic applications. Drug repositioning offered excellent opportunity to offer more pharmaceutical candidates to the market with established postmarketing surveillance safety data, toxicity, and pharmacokinetics profile [74]. This approach was adopted by the industrial, funding, and academic bodies to evaluate assets on the shelves by having a trilateral funding mechanism integrating universities along with pharmaceutical industry through National Institute of Health to support preclinical/clinical investigations for already existing drugs/assets with established efficacy and safety for novel clinical applications. Most recently, the outbreak of the new strain of COVID-19 virus has prompted the researchers to adopt the repurposing strategy to offer rapid therapeutic regimens targeting polymerase, protease, or spike protein.

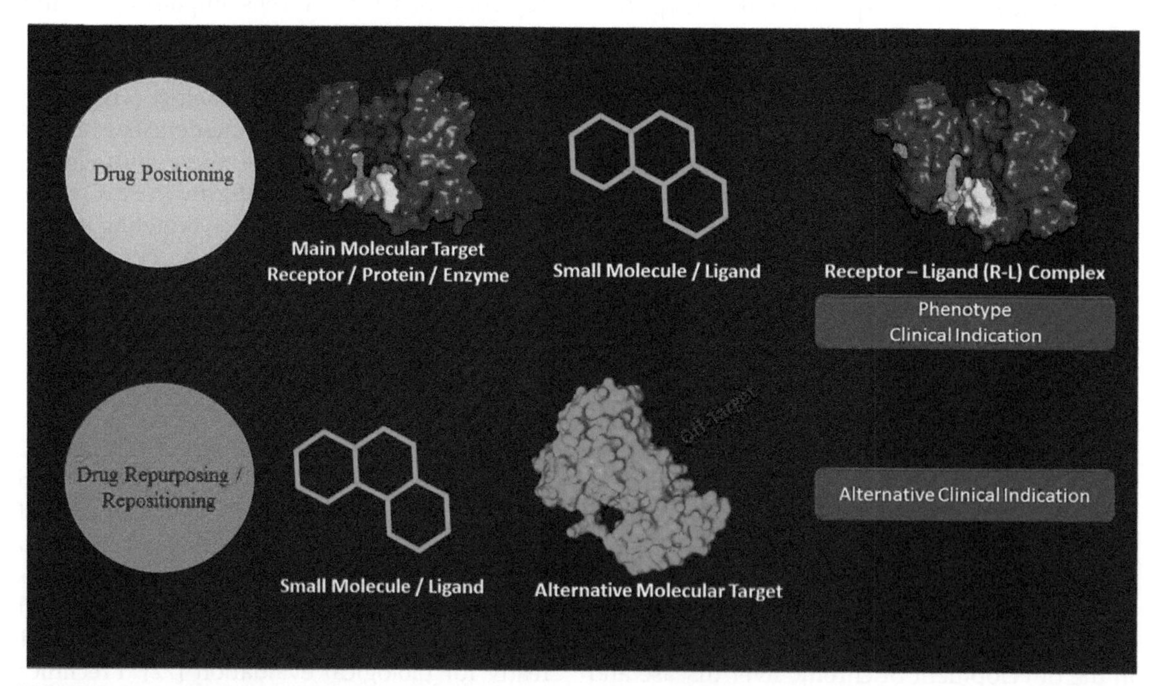

FIGURE 8.3 Drug positioning approach, where the receptor–ligand complex is known for certain clinical indication. Drug epositioning approach, where already existing molecule is repurposed for alternative molecular target (off-target mechanism) whether structurally or biochemically to reveal alternative therapeutic application.

Drug repositioning strategies development

The biological system is complicated enough to repurpose different pharmaceutical drugs for various clinical indications blindly based on off-targets screening, adverse effects reporting, serendipity, or clinical trials. In this section, there is a general chronological highlight for the drug development strategies development, as shown in Fig. 8.4. The whole arise of the drug repositioning integration started with **blinded serendipity–based repositioning** stories, such as sildenafil, minoxidil, and everolimus [75,76]. Consequently, the strategic progress led to **disease network–based repositioning** where certain diseases share same histological, pathological, pharmacological, or biochemical phenotypes; this certainly could be potential opportunity for testing the already established drugs for identified diseases to be evaluated for other different diseases having the same histopathological/biochemical features. The advancement of biological sciences made a huge transition from -ology era to -omics era; consequently, this showed huge impact on the transformation of the drug repositioning strategy; where the integration of proteomics, metabolomics, transcriptomics, and genomics heavily participated in the evolution of the drug repositioning. Ultimately, this gave the opportunity for the **high-throughput screening–based repositioning** to evolve; where the phenotypic assessment using cell lines, fluorescence assays, or enzyme kinetics and the genomic/proteomic-based assays by analyzing multigenomics profiling, mRNA, and protein sequencing offered molecular signature for different molecular targets [77]. This strategy helped potential drugs, such as metformin, digoxin, and statins to be repurposed for different types of cancer. The integration of computational tools has the adequate

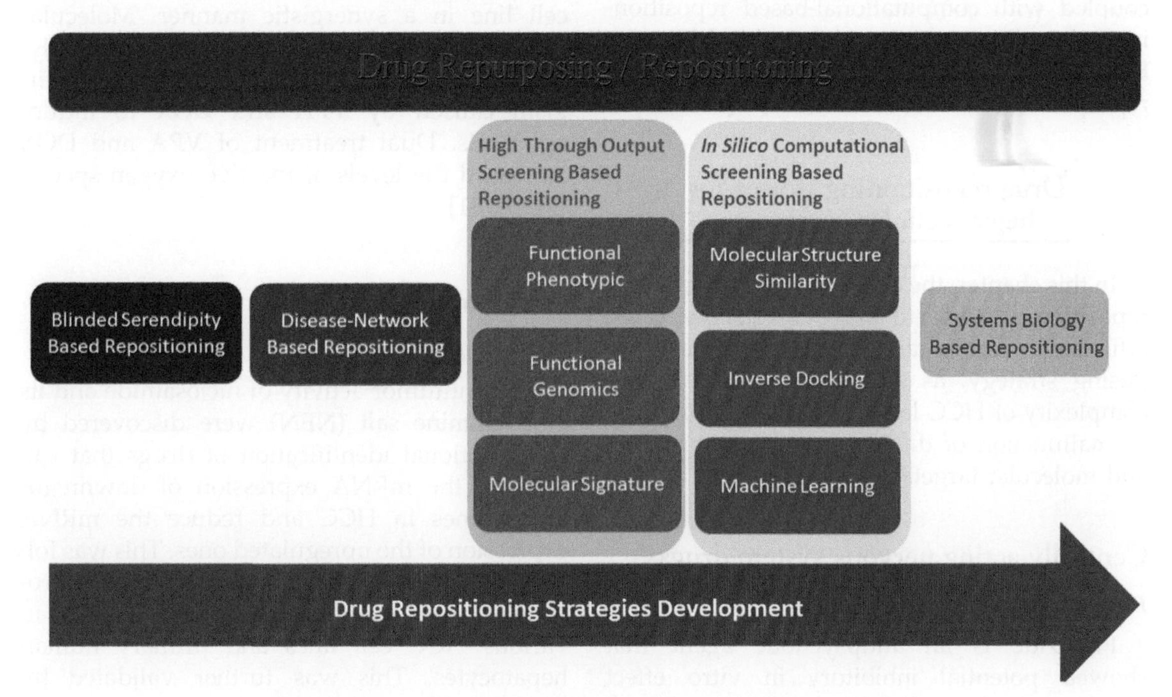

FIGURE 8.4 Drug repositioning strategies development categories.

share for the advancement and validation of the drug repositioning strategy to evolve in silico **computational screening–based repositioning** [72]. This strategy helped the repositioning of nonsteroidal antiinflammatory drugs and proton-pump inhibitors for different types of cancer [78,79]. The main in silico screening involves different tactics:

(1) Molecular structure similarity assessment using Jaccard index or Tanimoto coefficient calculations.
(2) Molecular docking to validate the drug–receptor(s) interactions with respect to binding mode/energy with respect to enthalpy and entropy.
(3) Dealing with huge consortium of data on the molecular and structural levels applying artificial intelligence network to create algorithm connecting the structural input along with molecular output.

The association of omics-based repositioning coupled with computational-based repositioning led to the appearance of **systems biology–based repositioning**, for instance, digoxin was repurposed for medulloblastoma [76].

Drug repositioning stories for hepatocellular carcinoma

In this chapter, the authors will focus on drug repositioning stories to target HCC, showing the affiliated molecular targets identifying the repurposing strategy, as shown in Table 8.1. The complexity of HCC leads to multiple mutations or malfunction of different receptors, proteins, and molecular targets.

Centrally acting nervous system drugs

Pimozide

Pimozide is an antipsychotic agent that showed potential inhibitory in vitro effect against cell proliferation of HCC cell lines via induction of apoptosis at G0/G1 phase. In addition, pimozide exhibited inhibitory profile for HCC stem-like cells, particularly the CD133-positive cells side population. Pimozide was found to target STAT3 expression via luciferase assay activity along with downregulation of the transcription levels of downstream oncogenes for STAT3 signaling. The antitumor activity for pimozide was further validated in vivo in nude mice [80].

Valproate

Valproic acid (VPA), a potent and specific histone deacetylase (HDAC) inhibitor, is widely used antiepileptic drug. HDAC has been recognized for its significant role as identified target with known molecular signature in the progression of different types of cancer. The mono- and adjuvant therapeutic in vitro effects of VPA and doxorubicin (DOX) managed to identify specific, efficient, and antiproliferative profiles of VPA and DOX combination against HepG2 cell line in a synergistic manner. Molecular levels of caspase-3 and poly (ADP-ribose) polymerase (PARP) activation validated the synergism caused by VPA and DOX to induce apoptosis. Dual treatment of VPA and DOX increased the levels of reactive oxygen species (ROS) [81].

Anthelmintic drugs

Niclosamide ethanolamine

The antitumor activity of niclosamide and its ethanolamine salt (NEN) were discovered by computational identification of drugs that can increase the mRNA expression of downregulated genes in HCC and reduce the mRNA expression of the upregulated ones. This was followed by in vitro evaluation to show the antiproliferative activity of niclosamide and NEN in various HCC cell lines and primary human hepatocytes. This was further validated by in vivo evaluation against two mouse models

TABLE 8.1 List of repurposed drugs targeting hepatocellular carcinoma.

Drug	Original therapeutic indication	HCC molecular target	Repositioning strategy	Investigation status
Pimozide	Antipsychotic	STAT3	Blinded screening	In vivo
Valproate	Antiepileptic	HDAC	Molecular signature	In vitro
Niclosamide ethanolamine	Anthelmintic	STAT3	Systems biology	In vivo
Amiodarone	Antiarrhythmic	mTOR	Disease network	In vivo preclinical
Lanatoside C	Antiarrhythmic	MMP, PKCδ, AIF	Systems biology	In vivo
Simvastatin	Antihypercholesterolemia	AMPK, STAT3	Phenotype screening Artificial intelligence	In vivo
Guanabenz acetate	Antihypertensive	DNA damage-inducible, p34, eukaryotic initiation factor 2α	Phenotype screening	In vitro
Fenofibrate	Antihypercholesterolemia	PPARa, AKT, CTMP	Serendipity	In vitro
Metformin	Oral hypoglycemic	KLF6/p21, AMPK	Systems biology	In vitro
Canagliflozin	Oral hypoglycemic	ERK, p38, AKT	Blinded screening	In vivo
Linagliptin	Oral hypoglycemic	ADORA3	Structure-based docking	In vitro
Atovaquone	Pneumonia	DNA double-stranded breaks	Blinded screening Phenotype screening	In vivo
Ketoconazole	Antifungal	PTGS2	Disease network Phenotype screening	In vivo
Obeticholic acid	Primary biliary cholangitis	IL-6/STAT3 pathway	Blinded screening Phenotype screening	In vivo

(genetically induced liver tumors and patient-derived xenografts [PDXs]) for HCC to show significant reduction in the tumor growth after oral administration of NEN compared to niclosamide. The dual administration of NEN and sorafenib also showed significant improvement to reduce the progression of PDX compared to either sole drug treatment. In HepG2 cell lines and PDX models, administration of niclosamide or NEN showed unique molecular signature in terms of gene expression compared to HCC molecular signature. Administration of NEN to PDX model reduced expression of proteins in the Wnt/β-catenin, STAT3, AKT/mTOR EGFR/Ras/Raf signaling pathways, and disruptive interactions along with heat shock protein 90 [82].

Cardiovascular system-acting drugs

Amiodarone

Amiodarone, a class III antiarrhythmic agent and a potent mTOR inhibitor, was found to suppress liver tumor formation through induction of autophagy activity in the rat orthotropic model and in the mouse xenograft model. Furthermore, a big data analysis of 32,625 case–control provided by Taiwan's National Health Insurance program revealed that long-term regular amiodarone usage significantly decreases the risk of HCC. Amiodarone, as a repurposed drug, has antitumor potential to suppress liver tumor formation and prevent HCC incidence through induction of the autophagy activity [83].

Lanatoside C

Lanatoside C is an antiarrhythmic agent, naturally occurring compound extracted from *Digitalis lanata*. Integration of systems biology repositioning approach showed the potential of lanatoside C to tackle HCC at the in vitro and in vivo levels with significant reduction in tumor growth. The molecular mechanistic investigation revealed the ability of lanatoside C to trigger mitochondrial membrane potential (MMP) loss, followed by activation of apoptotic markers to induce cell death. Inhibition of Thr505 phosphorylation for protein kinase delta (PKCδ) reversed lanatoside C–induced MMP loss and apoptosis validating the molecular mechanism for lanatoside C, where AKT/mTOR pathway is involved via modulation of PKCδ activation [84].

Statins (simvastatin)

Statins are HMG-CoA reductase inhibitors, to intervene with the cholesterol synthesis via inhibiting mevalonate pathways targeting different cardiovascular system complications. Statins were identified as potential drugs tackling various types of cancer, including breast cancer, prostate, head, and HCC using high-throughput activity–based screens of disease phenotypes and in silico data-driven algorithms. Simvastatin induced cell cycle arrest at G0/G1 phase accompanied by series of molecular changes via promoting AMP-activated protein kinase (AMPK), leading to induction of p21 and p27 accumulation. Using the transcriptomics, simvastatin showed significant reduction in the Skp2 expression, resulting in p27 accumulation by preventing proteasomal degradation, mediated by STAT3 inhibition. In addition, simvastatin significantly decreased tumor growth in HepG2 xenograft mice [85].

Guanabenz acetate

Guanabenz acetate, an antihypertensive drug, was screened as a potential candidate for HCC via a phenotypic screening assay by in vitro antiproliferative activity against HCC cell lines using high-throughput screening–based repositioning approach. Guanabenz acetate reduced HCC cell viability via inhibition of growth and induction of DNA damage, leading to increased phosphorylation of eukaryotic initiation factor 2α, increased activation of transcription factor 4, and induction of apoptosis [86].

Fenofibrate

Fenofibrate is known for its lipid lowering capability; however, it has shown huge potential to target various types of cancer recently. Fenofibrate showed significant reduction in the viability of human HepG2 cells via necrosis, but not apoptosis. The mechanistic investigation for the fenofibrate revealed an increase in the levels of ROS and a decrease in glutathione (GSH, an important cellular antioxidant) accompanied with impairment for the mitochondrial function in HepG2 cell lines. In other studies, fenofibrate showed cell cycle arrest for Huh-7 cell lines at G2/M phase through downregulation of cyclins group with upregulation of p27. Fenofibrate also activated endogenous peroxisome proliferator-activated receptor (PPAR)α in Huh7, HepG2, and Li7 cell lines, but the antiproliferative activity induced by fenofibrate was not affected by the PPARα inhibitor GW6471 or the knockdown of the expression of PPARα by siRNA. Moreover, fenofibrate suppressed AKT phosphorylation and increased the expression of C-terminal modulator protein, which binds specifically to AKT [87,88].

Oral hypoglycemic drugs

Metformin

Metformin is the drug of choice toward treatment of type II diabetes. The in vitro antiproliferative activity for metformin showed multiple phenotypic events starting with induction of cell cycle arrest at G0/G12 phase accompanied with significant reduction in cell growth with elevated levels of KLF6/p21 protein content in HepG-2 cell line. Metformin played a unique role via modulating the microenvironment of tumors by reducing cellular lipid accumulation and promoting AMPK activity. Metformin downregulated the expression of IGF I and II [89].

Canagliflozin

Canagliflozin, a sodium–glucose cotransporter 2 inhibitor (SGLT2-I), is known as an antidiabetic agent. Canagliflozin was selected for the phenotypic screening based on the given fact of high expression levels of SGLT2 in HCC cell lines. Initially, the antiproliferative ability of canagliflozin against HCC cell lines was assessed to show inhibitory profile in a dose-dependent manner along with induction of apoptosis at G2/M phase with elevated levels of caspase-3 and inhibition of ERK. This suggested that canagliflozin can inhibit glycolytic metabolism including glucose uptake, lactate, and intracellular ATP production. This was validated by in vivo screening for tumor growth assay after oral administration of canagliflozin (10 mg/kg/day) to show significant reduction in the tumor size and attenuated intratumor vascularization in HepG2-derived xenograft tumors in BALB/c nude mice [90].

Linagliptin

Linagliptin is known as antidiabetic class targeting dipeptidyl peptidase-4, where the scaffold of 3,7 dihydro-1H-purine-2,6-dione functional group was an attractive scaffold to be structurally repositioned in a trial to target HCC by modulation of adenosine 3 receptors (ADORA3) via in silico molecular modeling simulations coupled with in vitro analysis. Linagliptin and its degradation product showed antiproliferative activity against HepG-2 and Huh-7 cell lines inducing cell cycle arrest at G2/M phase with elevated levels of apoptotic marker, caspase-3. This was validated by monitoring the expression levels of ADORA3 [91].

Respiratory system targeting drugs

Atovaquone

Atovaquone, an FDA-approved drug for pneumocystis pneumonia, significantly inhibited hepatoma cell proliferation via S phase cell cycle arrest and induced both extrinsic and intrinsic apoptotic pathways associated with upregulation of p53 and p21. Molecular investigations demonstrated that atovaquone inhibits hepatoma cell proliferation by inducing double-stranded DNA breaks, leading to sustained activation of ataxia telangiectasia mutated and its downstream molecules such as cell cycle checkpoint kinase-2 and H2AX. In addition, atovaquone also induced apoptosis, inhibited both cell proliferation and angiogenesis in vivo, and prolonged the survival time of tumor-bearing mice, without any obvious side [92].

Others

Ketoconazole

Ketoconazole is a broad-spectrum antifungal agent, exhibiting antiproliferative activity against HCC cell lines by worsening mitophagy in vitro and in vivo. Ketoconazole-induced suppression of prostaglandin-endoperoxide synthase 2 led to PINK1-PRKN–mediated mitophagy, resulting in mitochondrial dysfunction and induction of apoptosis [93].

Obeticholic acid

Obeticholic acid (OCA) was granted accelerated approval from the FDA in 2016 for the treatment of primary biliary cholangitis. OCA is a synthetically modified bile acid agonist for the nuclear farnesoid X receptor (FXR). A cross talk between IL-6/STAT3 pathway and the hepatic FXR in HCC was previously reported [94]. In this study, activating FXR using OCA interfered with HCC cell growth by downregulating IL-6/STAT3 pathway in vitro. These effects were hampered in the presence of an FXR antagonist.

Future opportunities and limitations

Drug repositioning has become a significant emerging shortcut approach to deliver chemical entities into the pharmaceutical market cutting cost with an eye on the established toxicity profile, pharmacokinetics behavior, and postmarketing safety surveillance. However, repurposing novel clinical indications for already existing molecules should not grab our attention from investing in assembly of novel chemical entities to different molecular targets for different pathological conditions. Conceptually, the repositioning approach should be good platform for novel scaffold repurposing to synthesize novel ligands and pharmacophores.

In this chapter, we have outlined the repositioning as an excellent opportunity to bypass the toxicity issues, where the FDA-approved drugs show clinical safety profiles including cardiac, hepatic, and metabolic safety profiles. The authors tried to cover the serious repurposing trials in that field, where most of the repositioned drugs for HCC were revealed via phenotypic screening or systems biology without rational approaches with respect to structural or ligand-based drug design. However, the literature is full of random repositioning studies without solid understanding for fundamental basics for in vitro/in vivo biological assessment coupled with in silico molecular modeling simulations with no revealing for molecular target validation. The authors believe that despite the huge efforts for drug repositioning targeting HCC, amiodarone can be considered as a potential therapeutic drug for HCC due to its repositioning strategy counting on disease network, where data analysis of 32,625 case–control provided by Taiwan's NHI program revealed that long-term regular amiodarone usage significantly decreases the risk of HCC. Technically, the integration of artificial intelligence to get business solutions, servers, and healthcare platforms, such as Watson developed by IBM, can create potential opportunity for drug repositioning to offer robust methodology driving more drugs to clinical trials for novel therapeutic applications.

References

[1] Torre LA, et al. Global cancer statistics, 2012. CA Cancer J Clin 2015;65:87–108.

[2] Sartorius K, Sartorius B, Aldous C, Govender PS, Madiba TE. Global and country underestimation of hepatocellular carcinoma (HCC) in 2012 and its implications. Cancer Epidemiol 2015;39:284–90.

[3] El-Serag HB. Hepatocellular carcinoma. N Engl J Med 2011;365:1118–27.

[4] Wong RJ, Cheung R, Ahmed A. Nonalcoholic steatohepatitis is the most rapidly growing indication for liver transplantation in patients with hepatocellular carcinoma in the U.S. Hepatology 2014;59:2188–95.

[5] Baffy G, Brunt EM, Caldwell SH. Hepatocellular carcinoma in non-alcoholic fatty liver disease: an emerging menace. J Hepatol 2012;56:1384–91.

[6] Sanyal A, Poklepovic A, Moyneur E, Barghout V. Population-based risk factors and resource utilization for HCC: US perspective. Curr Med Res Opin 2010; 26:2183–91.

[7] Yang D, et al. Impact of sex on the survival of patients with hepatocellular carcinoma: a surveillance, epidemiology, and end results analysis. Cancer 2014;120: 3707–16.

[8] Zucman-Rossi J, Villanueva A, Nault JC, Llovet JM. Genetic landscape and biomarkers of hepatocellular carcinoma. Gastroenterology 2015;149:1226–39. e4.

[9] Llovet JM, Bruix J. Novel advancements in the management of hepatocellular carcinoma in 2008. J Hepatol 2008;48.

[10] Yu MC, Yuan JM. Environmental factors and risk for hepatocellular carcinoma. In: Gastroenterology, vol. 127. W.B. Saunders; 2004.

[11] Nakamoto Y, Guidotti LG, Kuhlen CV, Fowler P, Chisari FV. Immune pathogenesis of hepatocellular carcinoma. J Exp Med 1998;188:341–50.

[12] Bruix J, Sherman M. Management of hepatocellular carcinoma: an update. Hepatology 2011;53:1020–2.

[13] Llovet JM, et al. Hepatocellular carcinoma. Nat Rev Dis Prim 2016;2.

[14] Llovet JM, Montal R, Sia D, Finn RS. Molecular therapies and precision medicine for hepatocellular carcinoma. Nat Rev Clin Oncol 2018;15:599–616.

[15] Kudo M, et al. Lenvatinib versus sorafenib in first-line treatment of patients with unresectable hepatocellular carcinoma: a randomised phase 3 non-inferiority trial. Lancet 2018;391:1163–73.

[16] Bruix J, et al. Regorafenib for patients with hepatocellular carcinoma who progressed on sorafenib treatment (RESORCE): a randomised, double-blind, placebo-controlled, phase 3 trial. Lancet 2017;389:56–66.

[17] Abou-Alfa GK, et al. Cabozantinib in patients with advanced and progressing hepatocellular carcinoma. N Engl J Med 2018;379:54–63.

[18] Zhu AX, et al. REACH-2: a randomized, double-blind, placebo-controlled phase 3 study of ramucirumab versus placebo as second-line treatment in patients with advanced hepatocellular carcinoma (HCC) and elevated baseline alpha-fetoprotein (AFP) following first-line sorafenib. J Clin Oncol 2018;36. 4003–4003.

[19] Zhu AX, et al. Ramucirumab after sorafenib in patients with advanced hepatocellular carcinoma and increased α-fetoprotein concentrations (REACH-2): a randomised, double-blind, placebo-controlled, phase 3 trial. Lancet Oncol 2019;20:282–96.

[20] El-Khoueiry AB, et al. Nivolumab in patients with advanced hepatocellular carcinoma (CheckMate 040): an open-label, non-comparative, phase 1/2 dose escalation and expansion trial. Lancet 2017;389:2492–502.

[21] Llovet JM, Hernandez-Gea V. Hepatocellular carcinoma: reasons for phase III failure and novel perspectives on trial design. Clin Cancer Res 2014;20:2072–9.

[22] Llovet JM, Villanueva A, Lachenmayer A, Finn RS. Advances in targeted therapies for hepatocellular carcinoma in the genomic era. Nat Rev Clin Oncol 2015;12:408–24.

[23] Méndez-Blanco C, Fondevila F, García-Palomo A, González-Gallego J, Mauriz JL. Sorafenib resistance in hepatocarcinoma: role of hypoxia-inducible factors. Exp Mol Med 2018;50.

[24] Lai S-C, et al. DNMT3b/OCT4 expression confers sorafenib resistance and poor prognosis of hepatocellular carcinoma through IL-6/STAT3 regulation. J Exp Clin Cancer Res 2019;38:474.

[25] Suk F-M, et al. Treatment with a new benzimidazole derivative bearing a pyrrolidine side chain overcomes sorafenib resistance in hepatocellular carcinoma. Sci Rep 2019;9:17259.

[26] Lai Y, et al. Non-coding RNAs: emerging regulators of sorafenib resistance in hepatocellular carcinoma. Front Oncol 2019;9.

[27] Zhu YJ, Zheng B, Wang HY, Chen L. New knowledge of the mechanisms of sorafenib resistance in liver cancer. Acta Pharmacol Sin 2017;38:614–22.

[28] Sakurai T, Maeda S, Chang L, Karin M. Loss of hepatic NF-κB activity enhances chemical hepatocarcinogenesis through sustained c-Jun N-terminal kinase 1 activation. Proc Natl Acad Sci USA 2006;103:10544–51.

[29] Galuppo R, Ramaiah D, Ponte OM, Gedaly R. Molecular therapies in hepatocellular carcinoma: what can we target? Dig Dis Sci 2014;59:1688–97.

[30] Galuppo R, et al. Synergistic inhibition of HCC and liver cancer stem cell proliferation by targeting RAS/RAF/MAPK and WNT/β-catenin pathways. Anticancer Res 2014;34:1709–14.

[31] Sahin F, et al. mTOR and P70 S6 kinase expression in primary liver neoplasms. Clin Cancer Res 2004;10:8421–5.

[32] Sieghart W, et al. Mammalian target of rapamycin pathway activity in hepatocellular carcinomas of patients undergoing liver transplantation. Transplantation 2007;83:425–32.

[33] Zhou L, Huang Y, Li J, Wang Z. The mTOR pathway is associated with the poor prognosis of human hepatocellular carcinoma. Med Oncol 2010;27:255–61.

[34] Tavian D, et al. u-PA and c-MET mRNA expression is co-ordinately enhanced while hepatocyte growth factor mRNA is down-regulated in human hepatocellular carcinoma. Int J cancer 2000;87:644–9.

[35] Daveau M, et al. Hepatocyte growth factor, transforming growth factor α, and their receptors as combined markers of prognosis in hepatocellular carcinoma. Mol Carcinog 2003;36:130–41.

[36] Tovar V, et al. IGF activation in a molecular subclass of hepatocellular carcinoma and pre-clinical efficacy of IGF-1R blockage. J Hepatol 2010;52:550–9.

[37] Lee YI, Kang-Park S, Do SI, Lee YI. The hepatitis B virus-X protein activates a phosphatidylinositol 3-kinase-dependent survival signaling cascade. J Biol Chem 2001;276:16969–77.

[38] Mannová P, Beretta L. Activation of the N-Ras-PI3K-Akt-mTOR pathway by hepatitis C virus: control of cell survival and viral replication. J Virol 2005;79:8742–9.

[39] Su R, et al. Associations of components of PTEN/AKT/mTOR pathway with cancer stem cell markers and prognostic value of these biomarkers in hepatocellular carcinoma. Hepatol Res 2016;46:1380–91.

[40] Cao H, et al. Functional role of SGK3 in PI3K/Pten driven liver tumor development. BMC Cancer 2019;19.

[41] Sun L, et al. microRNA-1914, which is regulated by lncRNA DUXAP10, inhibits cell proliferation by targeting the GPR39-mediated PI3K/AKT/mTOR pathway in HCC. J Cell Mol Med 2019. https://doi.org/10.1111/jcmm.14705.

[42] Dai N, et al. Capsaicin and sorafenib combination treatment exerts synergistic anti-hepatocellular carcinoma activity by suppressing EGFR and PI3K/Akt/mTOR signaling. Oncol Rep 2018;40:3235–48.

[43] Calvisi DF, Factor VM, Loi R, Thorgeirsson SS. Activation of beta-catenin during hepatocarcinogenesis in transgenic mouse models: relationship to phenotype and tumor grade. Cancer Res 2001;61:2085–91.

[44] Thompson MD, Monga SPS. WNT/β-catenin signaling in liver health and disease. Hepatology 2007:45 1298–1305.

[45] Arzumanyan A, et al. Hedgehog signaling blockade delays hepatocarcinogenesis induced by hepatitis B virus X protein. Cancer Res 2012;72:5912–20.

[46] Wang Y, Han C, Lu L, Magliato S, Wu T. Hedgehog signaling pathway regulates autophagy in human hepatocellular carcinoma cells. Hepatology 2013;58: 995–1010.

[47] Kim Y, et al. Selective down-regulation of glioma-associated oncogene 2 inhibits the proliferation of hepatocellular carcinoma cells. Cancer Res 2007;67: 3583–93.

[48] Ringelhan M, Pfister D, O'Connor T, Pikarsky E, Heikenwalder M. The immunology of hepatocellular carcinoma review-article. Nat Immunol 2018:19 222–232.

[49] Taniguchi K, Karin MNF-B. inflammation, immunity and cancer: coming of age. Nat Rev Immunol 2018:18 309–324.

[50] Yang YM, Kim SY, Seki E. Inflammation and liver cancer: molecular mechanisms and therapeutic targets. Semin Liver Dis 2019;39:26–42.

[51] Xiao C, Ghosh S. NF-kappaB, an evolutionarily conserved mediator of immune and inflammatory responses. In: Advances in experimental medicine and biology, vol. 560; 2005. p. 41–5.

[52] Ghosh S, Karin M. Missing pieces in the NF-kappaB puzzle. Cell 2002;109(Suppl. l):S81–96.

[53] Hösel M, et al. Not interferon, but interleukin-6 controls early gene expression in hepatitis B virus infection. Hepatology 2009;50:1773–82.

[54] Boya P, et al. Nuclear factor-kB in the liver of patients with chronic hepatitis C: decreased RelA expression is associated with enhanced fibrosis progression. Hepatology 2001;34:1041–8.

[55] Kosters A, Karpen SJ. The role of inflammation in cholestasis: clinical and basic aspects. Semin Liver Dis 2010;30:186–94.

[56] Mandrekar P, Szabo G. Signalling pathways in alcohol-induced liver inflammation. J Hepatol 2009;50: 1258–66.

[57] Ruggie J. Lipid metabolism and liver inflammation II. Hum Rights Counc 2008:1–27. Eighth Ses.

[58] Bettermann K, et al. TAK1 suppresses a NEMO-dependent but NF-κB-Independent pathway to liver cancer. Cancer Cell 2010;17:481–96.

[59] Hui L, Zatloukal K, Scheuch H, Stepniak E, Wagner EF. Proliferation of human HCC cells and chemically induced mouse liver cancers requires JNK1-dependent p21 downregulation. J Clin Invest 2008; 118:3943–53.

[60] Vucur M, et al. RIP3 inhibits inflammatory hepatocarcinogenesis but promotes cholestasis by controlling caspase-8- and JNK-dependent compensatory cell proliferation. Cell Rep 2013;4:776–90.

[61] Das M, Garlick DS, Greiner DL, Davis RJ. The role of JNK in the development of hepatocellular carcinoma. Genes Dev 2011;25:634–45.

[62] Kao JT, et al. IL-6, through p-STAT3 rather than p-STAT1, activates hepatocarcinogenesis and affects survival of hepatocellular carcinoma patients: a cohort study. BMC Gastroenterol 2015;15.

[63] Zhou M, Yang H, Learned RM, Tian H, Ling L. Non-cell-autonomous activation of IL-6/STAT3 signaling mediates FGF19-driven hepatocarcinogenesis. Nat Commun 2017;8.

[64] He G, et al. Hepatocyte IKKβ/NF-κB inhibits tumor promotion and progression by preventing oxidative stress-driven STAT3 activation. Cancer Cell 2010;17: 286–97.

[65] He G, et al. XIdentification of liver cancer progenitors whose malignant progression depends on autocrine IL-6 signaling. Cell 2013;155:384.

[66] Jiang R, et al. Interleukin-22 promotes human hepatocellular carcinoma by activation of STAT3. Hepatology 2011;54:900–9.

[67] Hu Z, et al. IL-17 activates the IL-6/STAT3 signal pathway in the proliferation of hepatitis B virus-related hepatocellular carcinoma. Cell Physiol Biochem 2017;43:2379–90.

[68] Roh YS, Seki E. Toll-like receptors in alcoholic liver disease, non-alcoholic steatohepatitis and carcinogenesis. J Gastroenterol Hepatol 2013:28 38–42.

[69] Dapito DH, et al. Promotion of hepatocellular carcinoma by the intestinal microbiota and TLR4. Cancer Cell 2012;21:504–16.

[70] Ma C, et al. NAFLD causes selective CD4+ T lymphocyte loss and promotes hepatocarcinogenesis. Nature 2016;531:253–7.

[71] Gomes AL, et al. Metabolic inflammation-associated IL-17A causes non-alcoholic steatohepatitis and hepatocellular carcinoma. Cancer Cell 2016;30:161–75.

[72] Moridi M, Ghadirinia M, Sharifi-Zarchi A, Zare-Mirakabad F. The assessment of efficient representation of drug features using deep learning for drug repositioning. BMC Bioinf 2019;20:577.

[73] Spyrakis F, BidonChanal A, Barril X, Luque FJ. Protein flexibility and ligand recognition: challenges for molecular modeling. Curr Top Med Chem 2011;11:192–210.

[74] Sahu UN, Kharkar SP, Computational Drug Repositioning. A lateral approach to traditional drug discovery? Curr Top Med Chem 2016;16:2069–77.

[75] Kochar P, et al. Exploring the potential of minoxidil tretinoin liposomal based hydrogel for topical delivery in the treatment of androgenic alopecia. Cutan Ocul Toxicol 2019:1–30. https://doi.org/10.1080/15569527.2019.1694032.

[76] Kim JY, Son JY, Lee BM, Kim HS, Yoon S. Aging-related repositioned drugs, donepezil and sildenafil citrate, increase apoptosis of anti-mitotic drug-resistant KBV20C cells through different molecular mechanisms. Anticancer Res 2018;38:5149–57.

[77] Varbanov HP, Kuttler F, Banfi D, Turcatti G, Dyson PJ. Screening-based approach to discover effective platinum-based chemotherapies for cancers with poor prognosis. PLoS One 2019;14:e0211268.

[78] He J, et al. Proton pump inhibitors can reverse the YAP mediated paclitaxel resistance in epithelial ovarian cancer. BMC Mol cell Biol 2019;20:49.

[79] Saxena P, Sharma PK, Purohit P. A journey of celecoxib from pain to cancer. Prostag Other Lipid Mediat 2020;147.

[80] Chen JJ, et al. The neuroleptic drug pimozide inhibits stem-like cell maintenance and tumorigenicity in hepatocellular carcinoma. Oncotarget 2017;8:17593–609.

[81] Saha SK, et al. Valproic acid induces endocytosis-mediated doxorubicin internalization and shows synergistic cytotoxic effects in hepatocellular carcinoma cells. Int J Mol Sci 2017;18.

[82] Chen B, et al. Computational discovery of niclosamide ethanolamine, a repurposed drug candidate that reduces growth of hepatocellular carcinoma cells in vitro and in mice by inhibiting cell division cycle 37 signaling. Gastroenterology 2017;152:2022–36.

[83] Lan S-H. The prophylactic effect of amiodarone on HCC occurrence. Clin Oncol 2018;3. OPEN ACCESS. Remedy Publications LLC, http://clinicsinoncology.com.

[84] Chao M-W, et al. Lanatoside C, a cardiac glycoside, acts through protein kinase Cδ to cause apoptosis of human hepatocellular carcinoma cells. Sci Rep 2017;7:46134.

[85] Wang S-T, Ho HJ, Lin J-T, Shieh J-J, Wu C-Y. Simvastatin-induced cell cycle arrest through inhibition of STAT3/SKP2 axis and activation of AMPK to promote p27 and p21 accumulation in hepatocellular carcinoma cells. Cell Death Dis 2017;8:e2626.

[86] Kang HJ, et al. Guanabenz acetate induces endoplasmic reticulum stress-related cell death in hepatocellular carcinoma cells. J Pathol Transl Med 2019;53:94–103.

[87] Jiao H, Zhao B. Cytotoxic effect of peroxisome proliferator fenofibrate on human HepG2 hepatoma cell line and relevant mechanisms. Toxicol Appl Pharmacol 2002;185:172–9.

[88] Yamasaki D, et al. Fenofibrate suppresses growth of the human hepatocellular carcinoma cell via PPARα-independent mechanisms. Eur J Cell Biol 2011;90:657–64.

[89] Vacante F, et al. Metformin counteracts HCC progression and metastasis enhancing KLF6/p21 expression and downregulating the IGF Axis. Internet J Endocrinol 2019:7570146.

[90] Kaji K, et al. Sodium glucose cotransporter 2 inhibitor canagliflozin attenuates liver cancer cell growth and angiogenic activity by inhibiting glucose uptake. Int J Cancer 2018;142:1712–22.

[91] Ayoub BM, Attia YM, Ahmed MS. Structural re-positioning, in silico molecular modelling, oxidative degradation, and biological screening of linagliptin as adenosine 3 receptor (ADORA3) modulators targeting hepatocellular carcinoma. J Enzym Inhib Med Chem 2018;33.

[92] Gao X, Liu X, Shan W, Liu Q, Wang C, Zheng J, Yao H, Tang R, Zheng. J. Anti-malarial atovaquone exhibits anti-tumor effects by inducing DNA damage in hepatocellular carcinoma. Am J Cancer Res 2018;9:1697–711.

[93] Chen HN, Chen Y, Zhou ZG, Wei Y, Huang C. A novel role for ketoconazole in hepatocellular carcinoma treatment: linking PTGS2 to mitophagy machinery. Autophagy 2019:15 733–734.

[94] Attia YM, Tawfiq RA, Ali AA, Elmazar MM. The FXR agonist, obeticholic acid, suppresses HCC proliferation & metastasis: role of IL-6/STAT3 signalling pathway. Sci Rep 2017;7.

CHAPTER

9

Stories of drug repurposing for pancreatic cancer treatment—Past, present, and future

Matthias Ilmer[1,2], Maximilian Weniger[1,2], Hanno Niess[1,2], Yang Wu[1,2], Chun Zhang[1,2], C. Benedikt Westphalen[3,4], Stephan Kruger[3], Martin K. Angele[1,2], Jens Werner[1,2], Jan G. D'Haese[1,2], Bernhard W. Renz[1,2]

[1]Department of General, Visceral and Transplantation Surgery, Hospital of the University of Munich, Munich, Germany; [2]German Cancer Consortium (DKTK), Partner Site Munich, German Cancer Research Center (DKFZ), Heidelberg, Germany; [3]Department of Medicine III, University Hospital, LMU Munich, Munich, Germany; [4]Comprehensive Cancer Center Munich, Munich, Germany

OUTLINE

Drug Repurposing in Cancer Therapy
https://doi.org/10.1016/B978-0-12-819668-7.00009-9

Introduction

Cancer in general is one of the leading causes of mortality worldwide [1]. Pancreatic ductal adenocarcinoma (PDAC), as the most frequent malignant tumor of the pancreas, has not a high incidence on a global level. However, it is projected to become the second leading cause of cancer-related death in particular in the United States and in Europe by 2030 [2–4]. At the time of diagnosis, only about 10% of all PDACs are local disease with a 32% 5-year survival rate and 29% are regional disease with a 12% 5-year survival rate [5] Some of the latter group will become eligible for surgery after upfront chemotherapy. Unfortunately, the majority of patients (52%) are diagnosed with distant stage disease and will undergo relatively inefficient systemic therapy regimens, such as modified FOLFIRINOX (FFX; fluorouracil, folinic acid, irinotecan, oxaliplatin) or gemcitabine (GEM) plus nanoparticle albumin-bound paclitaxel (nab-paclitaxel; GnP) with a 5-year survival rate of currently only 3% at the moment [5,6]. Although lately, some progress can be seen, due to the mentioned facts, PDAC patients usually still encounter a dismal prognosis at the time of diagnosis, and thus, the 5-year overall survival (OS) of all PDAC patients is about 8% [7]. Prevention is difficult in this neoplasm, where only few factors associated with higher incidence are known and only a small number of specific risk factors are tightly linked to its occurrence. Randomized screening of patients at high risk, such as those with a family history of PDAC, by MRI or endoscopic ultrasound has not yielded the expected success [8]. Similarly, current screening methods do not permit early discovery, which is in part due to the difficult accessibility of the pancreas, although tremendous efforts are invested into approaches that might help to delineate the disease at earlier, and therefore, treatable stages of disease, such as analysis of exosomes or other liquid biomarkers [9].

At the same time, remarkable energy is being invested in drug discovery, not only for PDAC but also anticancer strategies in general. Research spending of private and public foundations (e.g., National Institutes of Health, Deutsche Forschungsgemeins, etc.) is rising, biotechnology companies or spin-offs are being launched, and established pharmaceutical companies are investing increasingly higher budgets on drug development. Many of those ventures have not translated into more or better drugs to the extent of the invested resources. Estimates of different studies show that research and development expenditures for a new drug might accumulate up to $2.6 billion or higher [10]. Moreover, from basic research to the eventual FDA filing, it is a lengthy process and might take anywhere from 9 to 20 years (Fig. 9.1). Apart from the economic and long-lasting efforts, respectively, high failure rates add up to a very interminable route, and so, overall, only about 5%–12% of all drugs that enter Phase I clinical trials ultimately receive FDA approval (Figs. 9.1 and 9.2) [10,11].

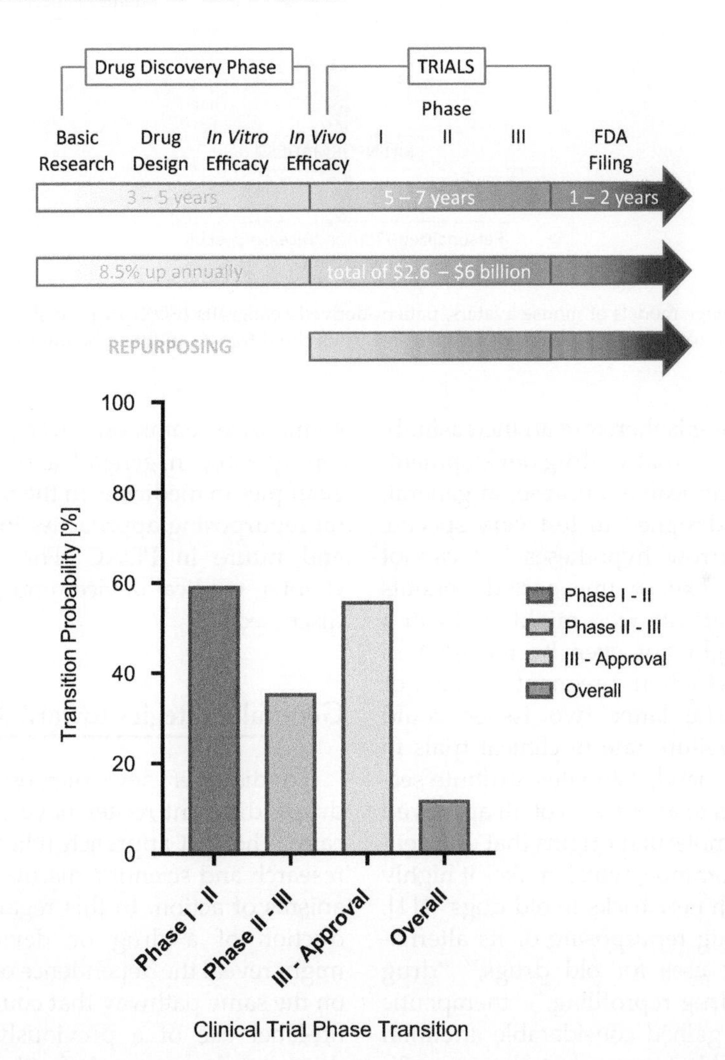

FIGURE 9.1 Illustration of the estimated costs and time that is currently estimated to be needed for the development of a drug from bench (basic research) to bedside (FDA approval). Repurposing significantly shortens that trajectory (upper image). Transition probability for clinical trials based on different phases as well as overall transition (modified according to DiMasi and colleagues, 2016 [10]).

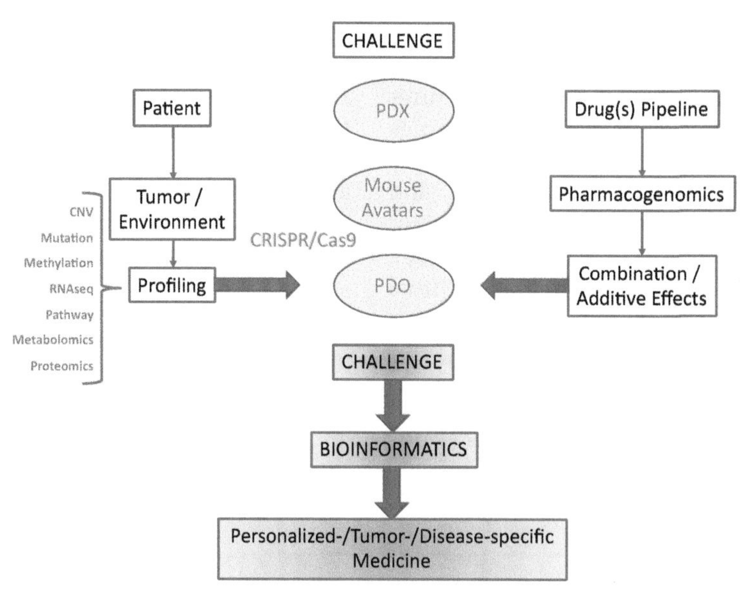

FIGURE 9.2 Challenge models of mouse avatars, patient-derived xenografts (PDX), or patient-derived organoids (PDO) with mutation profiling of patients as well as drugs. Together, this could transform into a bioinformatics pipeline with gold standard methods [20].

Drug repurposing is therefore an increasingly attractive field for alternative drug development. This is especially appealing because, in general, clinical trials are designed to test very specific questions and narrow hypotheses but cannot focus on and/or expose unexpected benefits [12]. Moreover, patients who might profit in a different way might not even be included in initial trials, e.g., children, pregnant women, or elderly people. The latter two issues could explain the high failure rate in clinical trials in part. On the other hand, estimates attribute secondary indications to about 90% of all approved drugs because of molecular origins that different diseases have in common, which makes it highly attractive to "teach new tricks to old dogs" [11]. In this regard, drug repurposing or its alternative terms ("new uses for old drugs," "drug repositioning," "drug reprofiling," "therapeutic switching") have gained considerable attention not only in the field of cancer but also in medicine in general. In summary, this terminology in principal refers to identifying or developing new uses for existing or abandoned pharmacotherapies [13]. In this chapter, we will briefly summarize common strategies toward drug repurposing in general and discuss successful examples in medicine. In the main part, we focus on repurposing approaches in the past, present, and future in PDAC. For the scope of this chapter, medical device repurposing will not be discussed.

General strategies toward drug repurposing

To discover new purposes for established drugs, different routes have been taken historically. The first approach relies on ongoing *basic* research and scientific discovery of novel mechanisms of action. In this regard, after the introduction of a drug or device, investigations might reveal the dependence of a different tumor on the same pathway that could then lead to the off-label use of a previously well-established drug. A prime example for that is the expansion of imatinib, originally developed for the treatment of chronic myelogenous leukemia (CML) to other cancer types on the basis of common principal pathways [13,14].

Second, a *translational* way of discovery depends on the understanding of the pathophysiologic basis of how a drug works. With that, scientists can widen the search for additional applications of a drug to completely different conditions with yet the same mechanism. For instance, beta-adrenergic antagonists or beta-blockers, originally developed to treat to treat angina and cardiac arrhythmias [15,16], have also been applied to alleviate performance anxiety or essential tremor [17].

Third, *clinical* observations might help to understand additional and unrecognized (positive) side effects of FDA-approved drugs. For example, bupropion use could be linked to easing smoking cessation but was originally developed as an antidepressant drug [13]. In a similar way, sildenafil was developed to treat angina pectoris and arterial hypertension. During early phase studies, it was discovered that erectile dysfunction could be significantly improved by its application [13].

Combining the abovementioned pathways with cutting-edge techniques, such as high-throughput screening as well as computational approaches, might lead to reasonable and modern ways to repurpose drugs to other diseases or vice versa. Ideally, Connectivity Maps of drugs or even drug pipelines, diseases and their most abundant pathways, and genes of the respective tumor would be created. The obtained data could be challenged against each other, evaluated with phenotypic screening in modern models of disease (among others, patient-derived xenografts [PDX], mouse avatars or patient-derived organoids [PDO]), and personalized candidate drugs would be proposed by bioinformatics and/or artificial intelligence (AI) (Fig. 9.2). In this regard, PDOs have the advantage of accelerated assessment of individual neoplastic cell dependencies and drug responses of the pipeline. Also, they can be transplanted into mice with various genetic backgrounds to investigate interactions with the TEM [18]. Genetically engineered mouse models, including CRISPR/Cas9-edited animals, could serve as avatars to novelties of single and combination therapies as suggested by the drug pipeline [19]. Genomic alterations often prioritize driver pathways on which the tumor becomes dependent. This dependence can—in theory—be exploited for significant cancer viability inhibition. Moreover, genomic data or gene expression signatures could help to refine data and hence classify additive effects of potential combinations of clinically useful treatment options by prioritizing therapeutic vulnerabilities or helping to identify suitable biomarkers.

Perspectives of drug repurposing and successful examples

Recent history of drug repurposing is rich in examples of successfully repurposed drugs including some with initial failure or even severely harmful side effects.

In the field of cardiology, several blockbuster drugs have been developed to help people with different ailments such as clogged coronary arteries, arrhythmia, or arterial hypertension. In this regard, the calcium channel blocker verapamil was introduced in 1963 for the treatment of angina pectoris; however, through basic and translational strategies as described in Section General strategies toward drug repurposing, pathologies in calcium channel functioning were shown to be involved in many other medical conditions, such as esophageal spasms, peripheral vascular disease, preterm labor, hypertension, and even migraine. For all of those ailments, verapamil is in use nowadays in a repurposed manner [12]. Another prime example in the field of cardiology is the widened use of beta-blockers, initially introduced to treat tachycardic conditions, but with remarkable effects for patients with performance anxiety or essential tremor [17]. Not only that, but as we will discuss later, beta-blockers might also be very useful in the field of oncology to treat tumors that thrive on sympathetic, adrenergic signaling networks [21].

Last, the phosphodiesterase inhibitor sildenafil (Viagra) was originally developed for coronary artery disease and angina pectoris, which was ineffective. However, clinical testing showed relevant side effects that were fully exploited by repurposing of the drug in order to improve male erectile disfunction and pulmonary arterial hypertension. It has become a blockbuster drug of its own ever since its release in 1998 [22].

A notable example of initial failure having at first had teratogenic side effects to unborn babies is thalidomide. Initially developed as a sedative drug in the early 1960s, it proved to cause severe malformations if taken during pregnancy [23]. Nevertheless, basic, translational, and clinical scientific approaches showed its clinical benefits in treating leprosy (FDA approval in 1998) and mechanistically through inhibition of TNF-α as an antitumorigenic drug for multiple myeloma as well as rheumatoid arthritis [12].

Many other interesting examples of successful drug repurposing are summarized in Table 9.1 below.

TABLE 9.1 Successfully repurposed blockbuster drugs and their original purpose.

Drug	Original purpose	Repurpose	References
Beta-blockers	Tachycardia	Performance anxiety or essential tremor	
Bupropion (Wellbutrin)	Depression	Smoking cessation	[13]
Diclofenac	Rheumatoid arthritis, osteoarthritis	Analgesia, primary dysmenorrhea, ankylosing spondylitis	
Duloxetine (Cymbalta)	Depression	Stress urinary incontinence	
Everolimus	mTOR, Renal, astrocytoma immunosuppression		
Imatinib (Gleevec, Glivec)	CD117 = cKIT Chronic myelogenous leukemia (CML)	Gastrointestinal stroma tumors (GIST)	[14,24]
Sildenafil	Coronary artery disease	Erectile dysfunction Pulmonary arterial hypertension	[25]
Thalidomide	Sedative drug in the early 1960s → teratogenic effects	Leprosy, 1998 inhibiton of TNF-α → tumors (multiple 2006 myeloma), rheumatoid arthritis	[12] [22] https://www.fda.gov/AboutFDA/CentersOffices/CDER/ucm095651.htm.
Verapamil	Ca^{2+} channel blocker verapamil (introduced in 1963) for angina pectoris	Esophageal spasms, peripheral vascular disease, preterm labor, hypertension, prophylaxis to migraine	[12]

Drug repurposing strategies in pancreatic cancer (PDAC)

This section reviews strategies that are being applied to the specific purpose of identifying helpful drugs for developing novel approaches toward PDAC treatment. We include trends of both academic institutions and pharmaceutical industry and try to corroborate our statements with prominent case studies of past and present.

Interestingly, exploration of scientific search engines revealed only a small number of hits using the search terms "drug repurposing" and "pancreatic cancer" or "drug repositioning" and "pancreatic cancer" (Fig. 9.3), which emphasizes the enormous potential that these approaches might still promise. However, even in the field of pancreatic cancer, an increase in publications over the last 3–4 years could be noticed and we have reviewed some of those trends before (Fig. 9.3) [26,27].

Last, this section will include a subjective outlook on future repurposing strategies and fields that we consider promising with substantial opportunities of improving medical treatment for patients with PDAC.

Case studies of the past

Permanent intense scientific efforts are undertaken and try to seek more curative treatment options for PDAC in adjuvant, neoadjuvant, or primary metastatic settings. However, most efforts remain unsatisfactory, making PDAC one of the leading causes for cancer-related death in Europe and the United States [28,29]. Still, resection of the primary tumor is the only curative option that is possible in roughly 20% of all PDAC patients, and surgery increases the 5-year survival rate significantly [30,31]. Conversely, locally unresectable tumors,

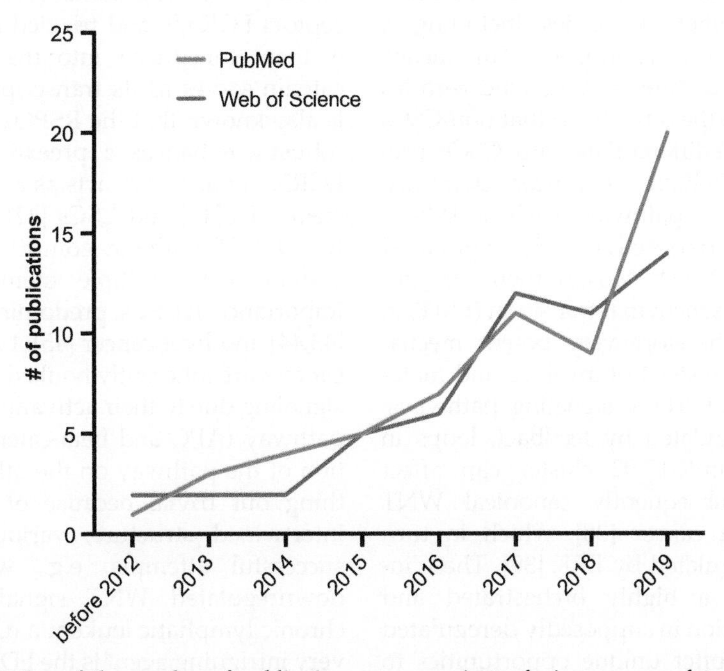

FIGURE 9.3 Illustration of the number of publications per year. A steady increase was detected with most of the publications published after 2016. In purple numbers found in Web of Science and in gray numbers found in PubMed databases are illustrated. Terms for the search were as follows: "drug repurposing" and "pancreatic cancer."

concurrent distant metastasis, or local recurrence render patient prognosis devastating [29].

The search for further and more potent therapeutic approaches is imperative and should therefore include both the search for novel drugs as well as repurposing compounds that have already shown antitumoral activity in other cancers [26].

In this section, we aim to describe case studies of drugs repurposed in the past, their original mechanism of action, and their implication in PDAC therapy and reveal how they performed.

CSC pathway inhibition

The concept of cancer stem cells (CSCs) postulates that in heterogeneous tumors, a subgroup of cells that are not necessarily predefined but endowed with CSC characteristics, could sit at the top of hierarchy. Initially portrayed as a subclass of indefinitely propagating neoplastic cells that produce an overt cancer (self-renewal and tumorigenesis) [32], current literature ascribes many other properties including a higher grade of plasticity to CSCs. This means on the one hand that different CSC subcategories could exist and on the other hand that non-CSCs can potentially dedifferentiate into CSCs presumably by activation of cancer stemness-associated signaling pathways such as WNT/RSPO [33,34], Nodal/Activin [35], miR-based pathways [36], or Notch. Most of them can control epithelial–mesenchymal transition (EMT) in PDAC and are therefore very potent mechanisms that regulate steps of invasive and metastatic cascades [37]. These signaling pathways, moreover, are regulated by feedback loops in CSCs, e.g., the miR-17-92 cluster can affect p38alpha and subsequently canonical WNT signaling in lung cancer [38], which in turn regulates or is regulated by ERK [39]. These interplays suggest a highly orchestrated and controlled interaction in supposedly deregulated cancer cells that offer unique opportunities to specifically target those CSCs with previously established FDA-approved drugs. Cycling between the different states of stemness and differentiation means that cells must be able to change their functional properties. Apart from the mentioned signaling pathways, metabolic dependencies seem to profoundly impact the plasticity of CSCs'. Preclinical models of CSCs in PDAC were able to demonstrate that FDA-approved drugs could possess inhibitory properties to the abovementioned CSC-specific capabilities. Here, we will summarize mostly preclinical data around repurposing FDA-approved drugs that could confront PDAC and its CSCs. Although the concept of CSCs is not completely novel anymore, given the ongoing challenges of a rising PDAC incidence, repurposing drugs might emerge as a considerable opportunity with new perspectives and angles in the course of PDAC treatment.

Canonical WNT signaling

Canonical WNT signaling is activated by the binding of glycoproteins (WNTs) to their coreceptors LGR5/6 and frizzled (FZD). This signal is then transduced into the cell where beta-catenin serves as its transcription factor [40]. It is also known that the RSPO/LGR complex can robustly enhances a preexisting WNT signal. LGR5, in particular, acts as a marker for normal stem cells [41] and CSCs [42]. Thus, the WNT/RSPO/LGR pathway comprises a very intricate system with multiple members and critical importance in CSCs, predominantly in colorectal [43,44] and liver cancer [45]. Liver and colorectal cancers are inherently hooked to canonical WNT signaling due to their activating mutations in the pathway (APC and beta-catenin genes). Inhibition of the pathway on the other hand is everything but trivial because of its complex and intertwined structure; various studies report successful attempts, e.g., with salinomycin-downregulated WNT signaling in CSCs of chronic lymphatic leukemia (CLL) [46]. Another very intriguing agent is the FDA-approved small molecule inhibitor aprepitant, which is directed against the NK1-receptor (NK1-R). Its primary clinical use is for chemotherapy-induced nausea and vomiting (CINV), but it also showed strong and efficient inhibitory activity in hepatoblastoma in vitro and in vivo [47]. Aprepitant as well as similar NK1-R inhibitors robustly decreased the viability of CSCs in colon and pediatric liver cancer (hepatoblastoma) [48] by potently suppressing both AKT/mTOR and canonical WNT [49] through the disruption of the FoxM1/beta-catenin complex [50]. Intriguingly, FoxM1 is often deregulated in PDAC and exerts metabolic control over tumor growth influencing oxidative glycolysis (Warburg effect) [51]. Last, low-dose ketamine induced ADNP expression, which in turn acted as a WNT repressor in a colorectal cancer model. This ultimately inhibited tumor growth and prolonged survival of tumor-bearing mice in vivo [52].

Disulfiram

Disulfiram is a compound that is used to treat patients with chronic alcoholism. However, it is also known to modify a range of cellular functions, some of which could be useful to target in PDAC. Notably, disulfiram seems to have an effect on pancreatic CSCs by elimination of the highly tumorigenic and therapy-resistant Aldefluor-positive cells in vitro [53]. Additionally, EMT was significantly inhibited by inhibition of ERK and NF-κB signaling in breast cancer cells [54].

In a Phase 2 clinical pilot study (NCT03714555) involving metastatic PDAC, disulfiram and copper gluconate will be added to chemotherapy (nab-paclitaxel plus GEM, FOLFIRINOX, or GEM mono) [55]. Additionally, disulfiram is a potential inhibitor of muscle degeneration and as such seems to ameliorate muscle wasting in animal models. In this regard, it could exert supplementary effects in patients with cancer cachexia or sarcopenia, which are conditions that not only debilitate the PDAC patients early on in the course of the disease but also result in chemotherapy-precluding frailty. Moreover, the use of disulfiram was supposed to sensitize cancer cells to GEM hydrochloride (NCT02671890) [56].

Quinomycin A

The quinoxaline antibiotic quinomycin A was attributed with a potential to influence several important CSC hallmarks, among them, inhibition of spheroid formation of PDAC, downregulation of several CSC markers, and suppression of the expression of multiple members of the Notch pathway in vitro. Quinomycin A treatment significantly reduced tumor burden in vivo, which suggested that it could be useful as a potent drug to target CSCs in PDAC [57].

Modulation of epithelial–mesenchymal transition

EMT is highly conserved and was shown to be mechanistically involved in regulating CSC plasticity of different tumor types, e.g., breast cancer, colorectal cancer, and PDAC [58,59]. Given that background, it was hypothesized that inhibition of EMT should then diminish the CSC population of a tumor and, in consequence, decrease the frequency of tumor relapse and metastasis. In light of this, the development of strategies to inhibit EMT has been pursued. In a systematic drug screen for EMT inhibitors, Meidhof and colleagues were able to show that mocetinostat, a class I histone deacetylase inhibitor, can act as an epigenetic drug by interfering with ZEB 1 function. This restored miR-203 expression and ultimately repressed EMT as well as CSC properties in PDAC and prostate cancer. In addition, mocetinostat was also able to resensitize CSCs for treatment with conventional chemotherapy [60]. In PDAC, therapy resistance, e.g., against GEM as an established chemotherapeutic agent for systemic treatments, seems to depend heavily on EMT mechanisms. It was shown that inhibition of EMT increased the expression of two types of human nucleoside transporters, namely the concentrative nucleoside transporter and the equilibrative nucleoside transporter (ENT) proteins, which are important for the molecular transport of hydrophilic GEM into the cells [61]. However, invasive or metastatic behaviors of PDAC were not significantly changed by the same treatment. Together, that might suggest that targeting EMT in PDAC could be of therapeutic benefit affecting various CSC hallmarks downstream of EMT [62].

Plerixafor

CXCR4, a receptor to CXCL12 (SDF1) or MIF, was initially found to be mainly expressed on hematopoietic stem cells, and plerixafor (Mozobil) was developed to inhibit CXCR4. In this regard, plerixafor is in use to mobilize stem cells, e.g., before harvest for transplantation therapies. Besides from that, CXCR4 was also found to be highly expressed in cells with increased invasive and metastatic capabilities in PDAC and other malignancies [63,64]. Moreover, CXCR4 conferred drug resistance in some cell populations [65]. With that, plerixafor could be useful in targeting cancer, especially in the context of highly CXCL12-producing surroundings. Last, current studies showed that it might actually be supportive in combination with immunotherapeutic drugs such as α-PD-1/α-CTLA4 [66].

Hedgehog inhibitors and hyaluronidase (HA, PegPH20) as stromal depleting strategies

Due to the desmoplastic microenvironment of PDAC, many drugs might not even reach their target, the cancer cell itself. Hence, substantial work is being spent on identifying means to beneficially alter the microenvironmental structure. One major component of the extracellular matrix or stroma of PDAC is hyaluronic acid, a glycosaminoglycan, which can be degraded by HA [67]. This knowledge was translated into a recombinant human drug produced by Halozyme Therapeutics (PegPH20) and is heavily tested in clinical trials of different phases. In theory, increased hyaluronic acid content of the PDAC stroma is associated with increased interstitial fluid pressure leading to reduced tissue perfusion and therefore collapse of supplying vessels. Vice versa, reduction of the stromal components then should increase vascularization and ease distribution of cytotoxic drugs [7]. In preclinical mouse models, PegPH20 in combination with GEM significantly remodeled the tumor stroma as seen by different alpha-SMA and SHG staining patterns [68]. A Phase I study was able to demonstrate that application of PegPH20 twice a week reduced HA up to 60% [69]. Later on, partial response rates of 67% in tumors with high HA levels could be detected in a Phase Ib trial [70]. Due to the promising data, several trials ensued, among them the HALO-109-202 Phase II trial. Here, higher rates of thromboembolic events in the PegPH20-containing arm were reported during the first stage of enrollment; these events could be successfully handled by the use of enoxaparin. While this trial was still successful and led to the initiation of a Phase III trial, the latter one failed to show improvement in median OS of the treatment arm (PegPH20 in combination with GEM and nab-paclitaxel [Abraxane]) compared to GEM and nab-paclitaxel alone—11.2 months compared to 11.5 months ($HR = 1.00$, $P = .9692$), neither did it translate into an improvement in duration of response or progression-free survival (PFS). This Phase III trial was evaluating PegPH20 as a first-line therapy for treatment of Stage IV PDAC patients (metastatic pancreas cancer). Although patients in both treatment arms of the HALO-301 trial surpassed the published median OS rates from the fundamental study of nab-paclitaxel plus GEM as first-line therapy for Stage IV PDAC [71], the study had to be discontinued based on the lack of improvement in comparison to the standard-of-care arm.

The reasons why this approach has not translated into a benefit for PDAC patients remain to be fully elucidated; however, previous studies on the desmoplastic stroma in PDAC and its manipulation could be deducted from hedgehog inhibition stories. Initially, drugs that could alter the stroma were thought to ease the way for chemotherapeutics into the fortress of the cancer [72]. On the other hand, several studies suggested that the stroma might also act as a physical barrier to restrain the tumor from growing and progression into adjacent tissues. Depletion of the stroma then led to a more aggressive PDAC phenotype [73].

Although abovementioned studies were not successful, the field keeps its hopes high to find use for HA modifiers. In PDAC, use of checkpoint inhibitors (CPI) has not been successful as of yet with reported response rates of 0% [74]. However, clinical studies are currently testing whether combination of CPI with TME modifiers, such as PegPH20, could increase the effects of CPI. In this regard, one Phase I study (Table 9.2. NCT03481920) tests the hypothesis whether elimination of HA by PegPH20 in the PDAC microenvironment affects stromal remodeling with subsequent increase of immune infiltrates as well as improved tumor vascularization and vessel patency, which could then facilitate the activity of CPIs. Both increased immune infiltrates and improved drug delivery could be fruitful in combination with CPIs such as avelumab, a specific monoclonal antibody for programmed death ligand 1 (PD-L1). In Phase II trials, the combination of PegPH20 with pembrolizumab, a humanized antibody (IgG4) targeting programmed cell death protein 1 (PD-1), is being investigated in metastatic PDAC patients to evaluate its capacity to interfere with the ability of PDAC cells to grow and further spread (NCT04058964 and NCT03634332).

TABLE 9.2 Clinical trials investigating repurposing potential for anticancer therapies.

Study drug	Original use	Mechanism of action	Repurposed target	ClinicalTrials.gov (NCT) identifier	Study phase
Atovaquone	Antimalaria [157,158]	Mitochondrial inhibition	Breast cancer cell	NCT02628080	1
Beta-blockers	Hypertension	Beta-2 adrenergic pathway	Cancer cell	(PROSPER)	2
Chloroquine	Malaria	Autophagy, hedgehog	Cancer cell	NCT01777477	1 [159]
Disulfiram	Alcoholism	CSCs Reduction of tumor-induced muscle wasting	ALDH inhibitor, EKR/NF-κB/EMT [53,54]	NCT02671890 NCT03714555	1 [55]
Hedgehog inhibitors	Vismodegib (GDC-0449) Saridegib (IPI-926) [160] LDE225	Binding SMO and prevention of GLI activation [161]	Basal cell carcinoma [162] Glioblastoma, prostate cancer, RCC [163–165]	NCT01088815 NCT01195415 NCT01130142 NCT01485744 NCT01713218 NCT01064622 NCT00878163	
Hydroxychloroquine	Malaria	Autophagy, hedgehog	Cancer cell	NCT04132505 NCT03825289 NCT01273805	[166]

(Continued)

TABLE 9.2 Clinical trials investigating repurposing potential for anticancer therapies.—cont'd

Study drug	Original use	Mechanism of action	Repurposed target	ClinicalTrials.gov (NCT) identifier	Study phase
Losartan	Hypertension	TGF-β1, CCN2, ET-1	Desmoplasia	NCT01494155 NCT03344172 NCT01978184 NCT01128296 NCT01506973 NCT01821729 NCT03563248 NCT01276613	1, 2 [75,167]
Metformin	Type 2 diabetes	ETC complex I, mTOR/AMPK-axis, pancreatic CSCs	PDAC	NCT01210911 NCT01167738 NCT01954732 NCT02336087 NCT01666730 NCT03889795 NCT02153450 NCT02005419 NCT01971034 NCT02048384 NCT04033107 NCT04207944 NCT02431676 NCT02294006 NCT02978547 NCT02201381	1–3
Mocetinostat	Class I HDAC inhibitor	EMT inhibition	Cancer cell (PDAC, prostate)	NCT02805660, NCT00372437	1, 2
Plerixafor	Mobilization of HSCs	PAUF/CXCR4	Cancer cell	NCT03277209 NCT02695966 NCT02179970	1
Recombinant human hyaluronidase: PEGylated-rHuPH20 (PegPH20)			Tumor microenvironment	NCT03481920 NCT04058964 NCT01839487 (HALO-109-202) NCT01959139 NCT01453153 NCT03634332 NCT02910882 NCT02045589 NCT02715804 (HALO-109-301) NCT02045602	1, 2
Tigecycline	Broad-spectrum antibiotic	Mitochondrial inhibition [168,169]	Cancer cell (AML, breast, prostate)	NCT01332786	1
Vitamin D (calcitriol)	Vitamin D deficiency	AKT pathway		NCT03472833 NCT00238199	2, 3

Losartan

Experimental evidence

As already mentioned in PDAC, the desmoplastic reaction comprises mainly collagen fibers and hyaluronan and these components are a significant barrier, limiting the therapeutic efficacy of cytotoxic agents [68,75,76]. In PDAC models, stromal modifiers, which reduce desmoplasia, can improve vascular perfusion, drug delivery, and the effectiveness of cytotoxic agents [68,75,76]. In this regard, activation of the renin–angiotensin system in fibroblasts increases tumor fibrosis and desmoplasia, a mentioned key feature of PADC. This activation is mediated by the transforming growth factor β (TGF-β) pathway. The primary effector of the renin–angiotensin system is angiotensin II. Inhibition of the renin–angiotensin system activity is achieved by angiotensin I receptor blockers (ARBs), such as losartan. These drugs have the potential to both reduce the malignant potential of cancer cells and alter the tumor microenvironment, activating immunity and normalizing the extracellular matrix by reducing the levels of collagen and hyaluronan in orthotopic PDAC models to allow enhanced delivery of cytotoxic chemotherapy. In addition, losartan increased the fraction of perfused vessels and the delivery and efficacy of 5-fluorouracil (5FU) [75]. Losartan also increased the efficacy of the nanotherapeutic Doxil [77]. In this regard, it has to be mentioned that administration of 5FU, Doxil, or losartan alone did not affect the growth of pancreatic tumors, but tumors were significantly smaller in mice treated with losartan combined with either Doxil or 5FU [75]. In summary, losartan has potential as an additional drug to safely enhance the intratumoral penetration and efficacy of small and large therapeutics in patients with PDAC.

Clinical trials

In the past retrospective, observational cohort studies suggested that PDAC patients who were already taking angiotensin-converting enzyme inhibitors or ARBs because of preexisting cardiovascular disease had longer survival [78]. More importantly, in a recently published prospective Phase II trial, losartan in combination with neoadjuvant FOLFIRINOX followed by individualized chemoradiotherapy patients with locally advanced PDAC was evaluated [79]. The primary end point was the margin-negative (R0) resection rate, with secondary end points of safety, PFS, OS), and circulating biomarkers of response. Of the 49 patients, 39 completed eight cycles of FOLFIRINOX and losartan; 10 patients had fewer than eight cycles due to progression (5 patients), losartan intolerance (3 patients), or toxicity (2 patients). Overall median PFS was 17.5 months and median OS was 31.4 months. Among patients who underwent resection, median PFS was 21.3 months, and median OS was 33.0 months. This prospective evaluation of FOLFIRINOX and losartan therapy followed by individualized radiochemotherapy showed a high rate of R0 resection (61%) and prolonged survival rates in locally advanced PDAC. This study also has a significant limitation: Due to the lack of randomization in a single-arm Phase 2 clinical trial, the specific role of losartan contributing to the benefit seen in locally advanced PDAC cannot be assessed. The favorable outcomes of this study lead to the design of an additional multicenter randomized Phase 2 study using these agents alone or with immunotherapy, which is now underway (NCT03563248).

Metabolism: (hydroxy)chloroquine, metformin, antibiotics

PDAC is a disease with numerous alterations, most prominently in the KRAS pathway [80], which can be found mutated in more than 90% of all PDAC cases [81]. Together with several other oncogenes (e.g., AKT) or tumor suppressors (e.g., p53), KRAS is known to regulate many different processes in the malignant cell including, among others, also metabolic pathways [82]. Recently, Daemen et al. showed that PDAC cells can be subdivided into glycolytic or lipogenic subtypes by metabolite profiling depending on their preferred catabolic pathway [83]. Furthermore, Hardie and colleagues were able to identify several novel mutations in mitochondrial DNA (mtDNA)—encoded subunits of the electron transport chain (ETC) [82]—and intriguingly, at least one such mutation was found in every patient cell line tested. Due to the pivotal function of mitochondria cancer cell metabolism, therapeutic agents could be developed or repurposed to fight special metabolic traits of PDAC.

Several diseases, such as malaria, rheumatoid arthritis, or systemic lupus erythematous, can be treated with the FDA-approved drug hydroxychloroquine [84]. This drug has been found to impact various signaling pathways, among others, Toll-like receptor signaling on immune cells (plasmacytoid dendritic cells) [85], with the latter also promoting tumor cell growth and survival. Another pivotal mechanism is autophagy, a machinery by which cells can digest unnecessary and dysfunctional cellular components (proteins, organelles, and macromolecules) with subsequent lysosomal degradation and recycling of bioenergetic damaged pieces in order to maintain the energy levels of the cell, especially in circumstances of extreme starvation. In the case of PDAC, it is known that the process of autophagy is activated and moreover, correlates with poor patient survival as shown by immunohistochemical staining of the autophagy marker LC3 in PDAC specimens [86]. Autophagy per se in PDAC is a process with opposing roles; in the early stages of cancer development, autophagy seems to have antitumor effects during tumorigenesis, while later on, cancer progression heavily relies on the supportive energy supply of fuel provided by autophagic digestion [87]. Moreover, autophagy of stromal bystander cells, especially pancreatic stellate cells, seems to support and maintain metabolic homeostasis of malignant PDAC cells [88]. This finding is especially intriguing because it shows foremost that PDAC cells could recruits stromal cells for their purposes and second, that these more abundant cells in a pancreatic tumor could be a decent aim for therapeutic targeting. Autophagy inhibitors have been developed and tested, however with mixed clinical results so far. As already stated, (hydroxy) chloroquine (HQ) has been shown to inhibit autophagy and, consequently, cell growth in pancreatic cancer, at least in preclinical models [87,89]. However, a recent publication showed that OS did not improve upon hydroxychloroquine use in patients with metastatic PDAC [90]. Further research needs to decipher whether biomarker use could improve the application of HQ or whether blockade of potential escape routes needs to be added to this treatment option. On another note, its use in neoadjuvant settings is further being investigated in trials (see Table 9.2, e.g., NCT01128296).

More antimalarial compounds have been tested for their modulatory effects on metabolic behavior in tumors and use in anticancer strategies. In this regard, as well as further aspects, chloroquine, for instance, was also shown to have significant effects on pancreatic CSCs not only by potentially suppressing autophagy and metabolic addiction of PDAC [91] but also mainly due to inhibition of the CXCL12/CXCR4 axis [92]. Regardless of all the exciting preclinical pathway deciphering work in vitro and in vivo, clinical success for the PDAC patient is still pending.

Metformin

Metformin (1,1-dimethylbiguanide hydrochloride), a biguanide derivate, is a drug most commonly used to treat hyperglycemia in non—insulin-dependent diabetes type 2; pharmacologically, metformin reduces hepatic gluconeogenesis and induces glucose uptake in peripheral tissues with subsequently decreased systemic blood glucose levels [93]. As far as its mechanism of action is understood, metformin inhibits the mitochondrial respiratory chain in complex I, cyclic adenosine monophosphate, and glycerophosphate dehydrogenase and activates AMP-activated protein kinase (AMPK) among others; also, these effects seem to depend on the tissue type and drug concentrations [94]. Apart from this well-investigated use and benefit, recent research shows that metformin might delay age-related diseases due to its anti-inflammatory effects. To understand whether metformin can actually impact age-related ailments, completed and forthcoming trials are underway, most prominently studies with acronyms like MILES (Metformin in Longevity Study NCT02432287), MASTERS (Metformin to Augment Strength Training Effective Response in Seniors NCT02308228), and TAME (Targeting Aging with Metformin). Targeting inflammatory states in the body might be useful to fight not only aging but also cancer, especially because cancer can be regarded as a condition of chronic inflammation; tumor-promoting inflammation per se has been considered one of the hallmarks of cancer in a myriad of different neoplasms, among them PDAC [95]. Two types of inflammation, namely chronic inflammation caused by pancreatitis, obesity, or other genetic and environmental factors as well as tumor-associated inflammation contribute to PDAC initiation and progression. Oncogenic KRAS in PDAC cells leads to attraction of proinflammatory cells, which in turn secrete inflammatory mediators that in a positive feedback loop stimulate PDAC cells [96]. Moreover, defective autophagy presumably mediated by oncogenic KRAS [87] promotes inflammation through diverse mechanisms, among others, failure to remove damaged mitochondria with subsequently increased levels of reactive oxygen species [97]. The growing epidemic of glucose intolerance, hypertension, dyslipidemia, and obesity, particularly central obesity, which, taken together, have been termed metabolic syndrome [98], leads to dysglycemia, hyperinsulinism, and ultimately diabetes type 2; these conditions also activate AKT/mTOR signaling, thereby inhibiting autophagy [99,100]. Loss of autophagy can then cause insulin resistance again. In the long term and through different pathomechanisms, these conditions are fertile soils for the growing cancer with plenty of necessary nutrients in circulation.

Metformin could tackle and disrupt many of the abovementioned mechanisms, most importantly, exerting its effects through manipulation of the mTOR/AMPK-dependent and AMPK-independent pathways. These pathways have been shown to play pivotal roles in PDAC tumorigenesis, progression, and cancer stemness, e.g., Matsubara and colleagues demonstrated that mTOR regulates and maintains PDAC stem cell behavior via downstream phosphorylation of its effectors 4E-BP1 and S6K and mTOR inhibition by rapamycin robustly inhibited CSC functions in PDAC [101]. Similarly, repurposing aprepitant, as mentioned above, an antiemetic drug and inhibitor to the NK1R system, was found to decrease mTOR/AKT signaling in various cancer types, thereby disrupting tumor growth mechanisms [47,49]. Metformin on the other hand though was shown to possess convincing properties in fighting CSCs of PDAC, especially by promoting significant mitochondrial toxicity through decreasing their transmembraneous potential and mitochondrial protein synthesis after mTOR inhibition [102]. Pancreatic CSC defining CD44+ subpopulations was also robustly decimated in in vivo mouse models. CSCs seem prone to depend on OXPHOS (oxidative phosphorylation), which in the aftermath of its metformin-mediated inhibition leads to energy crisis and apoptosis of PDAC cells, because CSCs were portrayed as less flexible in their metabolic consumptive pathway usage [103]. Notwithstanding, some CSCs were able to escape those treatment forms and transformed into metformin-resistant cells. Last, metformin could also exert its antitumor action by reducing systemic glucose and inhibition of complex I within the mitochondrial ETC [104].

Alone as well as in combination, these drugs could be useful for prevention and treatment of PDAC [105]. With the caveats mentioned, most probably longstanding use and combinatorial use will be necessary for clinical success. Especially, metformin has already been heavily tested in clinical trials (Table 9.2) with much less success than promised by many preclinical papers; however, further trials are ongoing, and given the heterogeneity of PDAC, subgrouping by metabolic profiling might be necessary for positive accomplishments.

Vitamin D

Vitamin D3 (VD) is mainly (90%) obtained from cutaneous 7-dehydrocholesterol cholesterol synthesis induced by ultraviolet radiation (UVR) in the skin and partly (10%) from dietary intake [106]. 1,25-dihydroxyvitamin D (calcitriol), the normal active form of VD, is predominantly produced in the kidney and has the highest biological activity. Beyond the traditional role in the regulation of bone metabolism and calcium phosphate homeostasis, VD has recently been shown to regulate aggressive behavior of cancer cells, such as cell growth, differentiation, apoptosis, and angiogenesis [107]. Epidemiologic studies have also indicated that low VD status is associated with increased risk of cancer in different kinds of tumors, such as colorectal, breast, bladder, prostate, and ovarian cancer [108–112]. With regard to the most agreed criteria for serum vitamin D levels, a concentration of less than 50 nmol/L (20 ng/mL) has been considered "deficient," a concentration of 50–74 nmol/L (20–29 ng/mL) "insufficient," and a concentration of not less than 75 nmol/L (30 ng/mL) "sufficient" [113,114]. Here, we will synopsize the potential therapeutic role of VD in PDAC based on recent clinical and experimental data.

Epidemiological outcomes

The association between VD and the risk for occurrence, prognosis, and incidence/mortality of PDAC remains unclear because data from different trials are at times inconsistent and conflicting.

Epidemiologic studies in the United States and Australia have indicated that the mortality rate of PDAC inversely correlates with exposure to UVR, a circumstance that was attributed to the presence of VD [115,116]. In addition, an inverse correlation between UVR and the incidence rate of PDAC has been reported by the Naval Health Research Center in the United States [117]. An Australian case–control study also showed an inverse correlation between PDAC risk and UVR, possibly mediated by VD synthesis [118]. A recent investigation conducted by Garland et al. has shown that residents of countries with low UVB irradiance have approximately 6 times the incidence rates of PDAC as those of countries with high UVB irradiance, and there was an inverse association of UVR irradiance with the incidence of PDAC [119].

However, trials based on dietary intake of VD have exhibited inconsistent outcomes. No associations were found between VD intake and PDAC risk in a pooled analysis of 14 prospective cohort studies [120]. Surprisingly, a pooled analysis of 9 case–control studies has indicated an increased risk of PDAC with higher dietary intake of VD.

[121]. On the other hand, a recent metaanalysis on VD intake and PDAC risk including 25 correlative studies (with a total of 1.214.995 individuals) revealed that VD intake can decrease the risk of PDAC [122].

Recent studies measuring circulating concentrations of 25-hydroxyvitamin D (25(OH)D), a parameter taking into account both UVR and vitamin D intake, have also indicated conflicting results. The National Cancer Institute has reported that a higher 25(OH)D concentration ($\geq$100 nmol/L) was correlated with a twofold increase of PDAC risk based on a pooled case–control study of eight cohorts [123]. In contrast, another pooled analysis of five cohorts has indicated that higher serum concentrations of 25(OH)D were related to a lower risk for

PDAC [124]. In addition, an Egyptian investigation has revealed that low serum 25(OH)D is not a risk factor for PDAC in Egyptian patients [125]. In European patient cohorts, serum VD concentrations could not be associated with PDAC risk [126]. This discrepancy may be attributed to different study populations, interference of confounding factors, and the respective methodology for VD measurement.

As to the role of VD in PDAC prognosis, Wang and colleagues have found that a lower expression of the VD receptor (VDR) in tumor is associated with a poor prognosis of PDAC [127]. Cho et al. reported that lower circulating levels (less than 20 ng/mL) of VD correlated with poorer prognosis [128]. A Japanese study has shown that a higher plasma level of VD is a major factor for extended distant metastasis-free survival in PDAC patients after preoperative chemoradiation therapy [129]. Van Loon et al. found no relationship between circulating VD and PDAC prognosis in a randomized trial of 256 patients [130]. However, Yuan et al. observed longer OS in PDAC patients with sufficient plasma levels of VD (24.6 ng/mL) in five prospective US cohorts of 493 patients [131]. A metaanalysis conducted by Zhang et al. revealed that high plasma VD levels were significantly associated with improved survival in PDAC patients [132].

In conclusion, VD may play a role in the development and prognosis of PDAC according to epidemiological data. However, due to the inconsistent results, more studies are needed in the future to obtain a better understanding of the association between VD and PDAC.

Experimental evidence

Many experimental studies have provided evidence that VD and its analogues could serve as potential therapeutic drugs in PDAC treatment. For instance, it was shown that VD analogues have the capability to suppress proliferation and growth by inducing cell cycle arrest in the G1/S phase as well as apoptosis, inhibiting migration and invasion of PDAC in cell lines, tumor tissue, and PDAC xenografts [133–140]. These capabilities are most likely to be mediated by the AKT pathway via upregulation of cyclin-dependent kinase inhibitors p21 and p27 and downregulation of cyclin D3 and CDK 4 and 5 [141–146]. It was also reported that other pathways were affected, such as Smo/hedgehog, the cadherin switch, F-actin synthesis, decreased secretion of matrix metalloproteinases-2 (MMP-2) and MMP-9, and ultimately EMT [145,147].

The antitumor abilities of VD analogues were reported to be mediated by VDRs expressed in PDAC cell lines and tumor tissue [148]. Yu et al. showed that a combination of calcitriol and GEM increased caspase-dependent apoptosis, thereby enhancing antitumor activity compared to single drug treatment in a mouse model of PDAC [149]. Interestingly, a hallmark study by Sherman et al. has revealed that pancreatic stellate cells in the desmoplastic stroma of PDAC highly express VDR. VD analogue treatment then significantly reprogrammed the tumor microenvironment and enhanced the effects of GEM chemotherapy in PDAC [150].

In summary, the VD system plays an important role in PDAC development and VD analogues show a potential in PDAC therapy according to current experimental evidence. Future studies are also needed to explore the mechanism in more detail, by which VD analogues exhibit their antitumorigenic functions or target the microenvironment remodeling of PDAC.

Clinical trials

Four published and eight ongoing clinical trials investigating the potential therapeutic role of VD analogues in PDAC treatment are listed in Table 9.2.

In 2002, Evans et al. published the first clinical trial in England. Only 14 patients with inoperable PDAC successfully completed an 8-week treatment with EB1089, while the other 22 patients were withdrawn due to tumor progression or other reasons [151]. Although EB1089 (daily dose: 10–15 μg) was well tolerated, no antitumor activity was observed in inoperable PDAC with hypercalcemia as the only side effect.

A study evaluating the safety of arachitol, a VD analogue, in patients with operable PDAC was conducted by Barreto et al. in India. The outcomes revealed that VD supplement by arachitol intramuscular injection did not influence the overall morbidity and mortality rate but induced hyperglycemia [152].

In 2009, Blanke et al. reported in a phase II trial in Canada, which evaluated the efficacy of a combination of calcitriol and docetaxel in 25 patients with inoperable PDAC. Three patients attained showed a partial response and seven achieved stable disease [153]. Although side effects like hyperglycemia and fatigue existed, patients treated with calcitriol and docetaxel showed a modest increase in time to progression when compared to a historical cohort receiving single docetaxel treatment only [153].

Currently, a phase II study (NCT02754726) investigating the combination of nivolumab, cisplatin, nab-paclitaxel, paricalcitol, and GEM (NAPPCG) in untreated metastatic PDAC treatment is underway, and preliminary data from 24 patients were encouraging. The objective response rate (ORR) was 83%, with a median PFS of 8.17 months and a median OS of 15.3 months [154].

A pilot study (NCT02030860), evaluating the combination of GEM/abraxane and paricalcitol prior to surgery for resectable PDAC, has been completed recently (no data published yet). Further clinical trials (NCT03472833, NCT03331562, NCT03520790, NCT03300921 NCT03883919, NCT03519308, NCT03415854) investigating the effects of oral supplementation with VD analogues (mainly paricalcitol) and different chemotherapy drugs in PDAC treatment are also in progress.

Outlook

Many clinical and laboratory results have highlighted the potential role of VD in PDAC development as well as the promising role of VD analogues in PDAC treatment.

Results from epidemiological studies on sun exposure and risk of PDAC are relatively consistent with an inverse association of UVR and incidence of PDAC. As reported by Garland et al. incidence rates of residents in countries with low UVR are approximately six times higher than those in countries with high UVR [119]. Therefore, increasing UVR may be a good way to increase serum VD and ultimately reduce the risk for PDAC. Although epidemiological studies on VD dietary intake and risk of PDAC showed conflicting results, a metaanalysis revealed that VD intake can decrease PDAC risk [122]. Circulating VD is a better indicator to reflect VD level in the human body. However, most original investigations as well as metaanalysis have shown that there is no correlation between circulating VD level and PDAC risk. In addition, several studies have indicated that plasma VD levels are significantly associated with improved survival in PDAC patients [129,131,132]. Since inconsistent data exist, more and well-designed studies are needed in the future before adding routine VD supplement to PDAC treatment.

Experimental data have supported the idea that VD analogues might repress aggressiveness of PDAC in cell lines, tumor tissue, and xenografts [133–136,138–140,155]. Possible mechanisms include AKT pathway regulation, Smo/hedgehog pathway inhibition, EMT reduction, and regulation of matrix MMP-2 and MMP-9 [141–147]. Recently, studies have revealed an important role of VD analogues in reprogramming the tumor microenvironment and enhancing chemotherapy in PDAC [129,150].

This has provided new concepts for PDAC treatment with a focus on the role of VD in remodeling of the tumor microenvironment in PDAC. Further studies are needed to investigate mechanisms by which VD analogues inhibit PDAC or its TME.

A series of clinical studies have been conducted and published data have shown promising results. The clinical trial by Barreto et al. has proved the safety of calcitriol analogues in patients with operable PDAC [152]. In an early clinical trial, EB1089 showed no antitumor activity in inoperable PDAC, with hypercalcemia as the side effect [151]. Considering the synergistic effect shown in experimental settings [149], follow-up trials have applied the combination of VD analogues and chemotherapeutic drugs. Encouragingly, a Phase II trial has indicated that patients treated with calcitriol and docetaxel showed a modest increase in time to progression of PDAC [153]. Preliminary results of a recent phase II study (NCT02754726) have demonstrated that patients treated with NAPPCG exhibit a 83% ORR with a PFS of 8.17 months and OS of 15.3 months [154].

Haloperidol

Jandaghi et al. conducted microarray analysis on PDAC patients as well as nontumor pancreatic tissue specimens in the search for new therapeutic targets. The dopamine receptor DRD2 mRNA and to an even higher extent its protein level were elevated in samples of PDAC and chronic pancreatitis. Gene expression analyses showed that the DRD2 signaling pathway was activated. DRD2 itself is involved in many psychotic diseases, such as schizophrenia, as a key molecule and can be targeted by numerous established drugs. Dopamine antagonists that inhibit the function of DRD2 were discovered in the 1950s, e.g., haloperidol. Blockade of DRD2 by knockdown experiments or administration of haloperidol had antiproliferative effects, activated ER stress, and displayed antitumor activity in mouse models of PDAC. These results highlight the possibilities that DRD2 antagonists could be useful in patients with PDAC and susceptibility could be tested based on DRD2 expression [156].

Current development and case studies

Today, approaches such as the ones outlined in Section General strategies toward drug repurposing might become less relevant with the emerging importance of machine learning (ML) and AI. Archives of existing publicly available databases are rapidly growing, and the fact that sequencing of tumors, samples, or even single cells is much more affordable today than ever will further emphasize the need for machine-based evaluations. Various computational methods already exist, for example, a tool called PanDrugs (http://www.pandrugs.org). It uses a large database of drug–target associations that ranges from drugs in experimental preclinical phases over FDA-approved drugs for ailments other than cancer to drugs in clinical use for cancer treatment. For this database mining, AI is applied and will certainly be further integrated. Moreover, these approaches of prioritized evidence-based anticancer therapies were applied to TCGA patients and validated in xenograft mouse models [170] (Fig. 9.4). Phenotypic screening could also be carried out in PDOs embedded in collaborative networks.

Another interesting approach was recently presented by Datta et al. by screening compounds of different libraries (LOPAC with 1.280 pharmacologically active compound and NPC with 3.300 compounds for clinical use) in their pursuit to inhibit intercellular communication via exosomes of advanced cancers [171]. Among others, inhibitors of exosome production could be identified, such as antifungal agents (neticonazole, climbazole, ketoconazole) and others (triadimenol, tipifarnib).

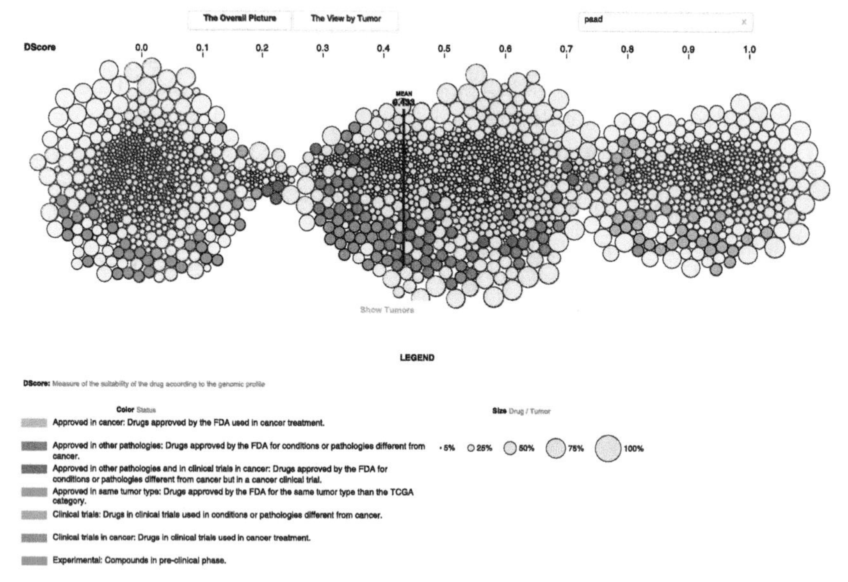

FIGURE 9.4 Bubble plot adapted from www.PanDrugs.org [170]. It shows the drugs for all tumor types (in light colors) of the TCGA database as well as PDAC-specific suggested drugs (in bright colors). The size of the bubble is representative for the percentage of TCGA patients and color code of the bubbles is descriptive of the drug approval status by the FDA as outlined in the legend.

Aspirin

Aspirin has long been suspected to confer anticancer effects. Although there is broad evidence from epidemiological observational studies and preclinical animal models for the efficacy of aspirin to prevent, contain, or even treat different types of cancer [172], evidence from Phase III studies or even recommendations for its use in cancer treatment or prevention guidelines is sparse. Data for its effects on PDAC, both positive and negative, are even more limited. However, since the anticancer effects of aspirin are suspected to be mediated via universal, cancer-type independent targets, most notably via platelets, this might allow us to extrapolate data from other adenocarcinomas on putative effects in PDAC.

Mechanisms of action against cancer

Several of the known actions of aspirin in the human body could explain the potential anticancer effects, either by acting indirectly via the tissue/tumor microenvironments or directly on tumor cells. In general, these effects can be divided into two mechanisms of action: (1) Cox-dependent and (2) Cox-independent pathways. Regarding the former, aspirin is an irreversible inhibitor of cycloxygenase-1 (Cox-1) and Cox-2, thus reducing the conversion of arachidonic acid to downstream, tissue-specific signaling lipids known as prostanoids [173]. These include the prostaglandins (PG)D_2, PGE2, and $PGF_{2\alpha}$, prostacyclin (PGI_2), and thromboxane A (TXA_2). Prostanoids are important biologically active signaling lipids that influence processes in several cells and tissues, such as angiogenesis, apoptosis, cell proliferation and migration, inflammatory response, and thrombosis [174]. Besides the fact that these processes are also of importance in cancer pathophysiology, the hypothesis that anticancer effects of aspirin are conferred by Cox inhibition is deducted from several observations: (1) cancers express a higher amount of PGs than their surrounding tissue [175], (2) PG expression can be driven by oncogenes and growth factors [176], and (3) Cox-2 overexpression has been shown in several tumor types, including PDAC [177]. Furthermore, in mice carrying a Cox knockout, the intestinal neoplasia formation by induction of an *Apc* mutation is reduced by 80% [178].

However, the half-life of aspirin in the human body is only 15–20 min, which cannot convey a sustained inhibitory effect on Cox in nucleated cells when aspirin is taken once per day. Thus, it is unclear how long-lasting effects of the drug on cancer cells directly can be explained. Instead, the predominant mechanism of action against cancer appears to be by influencing platelet function.

The main indication for aspirin use is in patients with cardiovascular disease, where the inhibition of TXA_2 synthesis is the desired mechanism of action. TXA_2, which is the main metabolite in platelets, not only causes their activation and aggregation but also leads to vasoconstriction and smooth muscle cell proliferation. Due to the irreversible nature of the Cox enzyme inhibition by aspirin through acetylation of a critical serine residue, cells require de novo synthesis of the enzymes in order to recover from aspirin exposure. This makes mature platelets within the blood stream, which lack a cell nucleus and are thus incapable of resynthesis of the enzyme, especially susceptible to long-lasting aspirin effects. Platelets on the other hand are believed to be of importance in the process of metastasis formation as they are encroached by circulating tumor cells (CTC) in the blood stream to aid in their survival by shielding them from NK-cell mediated killing, sheer stress, and supplying them with survival signals after loss of their stromal environment [179]. Thus, they facilitate the otherwise extremely insufficient cascade of metastasis formation, in which the vast majority of CTCs succumbs after exiting the primary tumor into the circulation. In PDAC, thrombocytosis is associated with poor prognosis and inhibition of platelets reduces metastasis formation in an animal model [180,181].

Regarding Cox-independent pathways of aspirin action, modulation of the NF-κB pathway to induce apoptosis in neoplastic cells has been proposed as a likely mode of action [182]. Other molecules and pathways important for tumorigenesis that aspirin has been shown to interact with, independently of Cox inhibition, are β-catenin, WNT signaling, tumor necrosis factor (TNF), polyamine metabolism, tumor angiogenesis, and DNA mismatch repair [174,183].

Evidence for efficacy in human trials of nonpancreatic cancer

The largest body of evidence for an anticancer effect of aspirin has been established in colorectal cancer. Here, randomized controlled trials have proven aspirin to prevent recurrence of colorectal adenomas [184] and lower the incidence of and mortality from colorectal cancer [185]. This effect is even more prominent in patients with a high hereditary risk for cancer development due to Lynch syndrome, where the risk for development of colorectal cancer was reduced by 59% [186]. According to a large observational study, aspirin appears to be an effective treatment for colorectal cancer even after the diagnosis is made. In a study of 1279 patients diagnosed with stage I–III colorectal cancer and a median follow-up of 11.8 years, the colorectal cancer-specific mortality was reduced by 29% in patients that regularly took aspirin [187]. Patients whose primary tumors overexpressed Cox-2 showed the largest benefit (HR 0.39, 95% CI: 0.2–0.76).

According to a large metaanalysis of case–control and cohort studies by Bosetti et al. the chemopreventive effect of aspirin has also been shown for other cancer entities, such as esophageal (relative risk [RR]: 0.72), gastric (RR: 0.84), breast (RR: 0.91), and lung (RR: 0.94) [188]. In another metaanalysis of seven randomized controlled trials of aspirin use in cardiovascular disease with an average treatment period of at least 4 years, a highly relevant reduction in all cancer-related mortality (HR 0.66, 95% CI 0.50–0.87) with an even more profound effect in gastrointestinal cancers (HR: 0.65, 95% CI 0.54–0.78) was observed [185]. For PDAC (HR: 0.25, 95% CI 0.07–0.92), along with esophageal, brain, and lung cancer, the effect on cancer-related mortality was seen after a latent period of 5 years. For stomach, colorectal, and prostate cancer, the latent period was even longer. Across all tumor entities, the greatest benefit was observed for adenocarcinoma. Intriguingly, in consequent studies by Rothwell et al. the authors were able to show a marked reduction in metastatic cancer by aspirin intake (HR 0.64, 95%,CI 0.48–0.84), which explains the observed mortality reduction of previous studies [189]. This effect was most clinically relevant and significant in cases of adenocarcinoma (HR 0.54, 95% CI 0.38–0.77) and nonsignificant for other solid cancers. Additionally, even in patients that developed nonmetastatic adenocarcinoma during the study period, subsequent metastasis formation was reduced by approximately 70%. This speaks for a short-term effect of aspirin, specifically in the process of metastasis formation, which stands in contrast to the long time periods of aspirin intake that are necessary for the primary prevention of cancer.

On the other hand, there are several studies that were negative with regards to a therapeutic effect of aspirin in patients with a cancer diagnosis. For example, there was no effect observed when aspirin was combined with conventional chemotherapy in patients with small cell lung cancer or renal cell carcinoma [190,191].

Aspirin PDAC treatment and research

Concerning the chemopreventive effect of aspirin, the evidence for pancreatic cancer is not conclusive. Besides the aforementioned positive study by Rothwell et al. large cohort studies of 141,940 patients with 1122 cases of pancreatic cancer showed no association between the incidence of pancreatic cancer between aspirin and nonaspirin groups, even after a long latent period [192].

To our knowledge, no randomized Phase III studies that studied aspirin use in an adjuvant or palliative setting in pancreatic cancer patients have previously been or are currently conducted (according to ClinicalTrials.gov). However, efficacy of aspirin against PDAC has been asserted in preclinical studies by counteracting CSC features, desmoplasia, and increasing GEM resistance [193].

Furthermore, aspirin has been shown to influence the immune response to PDAC in a genetically engineered mouse model of PDAC by reducing Foxp3+ regulatory T cells, thus prolonging survival in treated animals versus control animals [194].

Unselective beta-blockers (e.g., propranolol)

The sympathetic nervous system (SNS) is an anatomical and functional network of nerves, the adrenal medulla, and their respective neurotransmitters and receptors. Catecholamines, including epinephrine and norepinephrine, act on effector tissues through alpha- and beta-adrenergic receptors, which are widely expressed in both normal and neoplastic tissues including PDAC cell lines [195]. Catecholamines are stress molecules, produced by the SNS and linked to PDAC growth via beta-adrenergic signaling demonstrated via in vitro [196,197] and in vivo studies [198]. It is known from epidemiologic studies that stress can promote tumor growth [199] and that PDAC patients have the highest level of psychological stress of all investigated types of cancer [200]. In addition, cancer mortality is significantly increased by high levels of psychological stress [201]. On the other hand, it is suggested that beta-blocker treatment of patients suffering from colorectal cancer [199], breast cancer [202], ovarian cancer [203], melanoma [204], and PDAC [205,206] may lead to improved survival. Moreover, systemic stress has been shown to accelerate growth of cancer cells grown in immunocompromised mice, including ovarian cancer and prostate cancer [207]. A recent preclinical study demonstrated that stress-dependent sympathetic signaling can induce PDAC in preneoplastic lesions (i.e., PanINs) and thus cancer development in a susceptible host could be inhibited by ADRB2 antagonists. ADRB2 signaling likely promotes tumorigenesis through effects on both the epithelial and the stromal compartment [208]. Although the effects of stress in promoting cancer have been linked in the past to suppression of the immune response [209] or to recruitment of M2 macrophages [210,211], it could be shown recently that many of the effects of circulating catecholamines are indeed direct, with stimulation of ADRB2-dependent pancreatic epithelial growth [21].

As aforementioned, at an in vitro level, the positive effects of beta blockade on PDAC rely on ADRB2 inhibition. However, ADRB1-selective beta-blockers (SBBs) are more commonly prescribed than nonselective beta-blockers (NSBB), and populations with greater SBB use are unlikely to demonstrate a benefit from their beta-blocker use. Nevertheless, there are several studies that have investigated the impact of beta-blocker use on cancer survival. Among the 2394 patients in a very recent study [206], only 21 NSBB are in their medication. However, patients who used beta-blockers (selective and nonselective) (n = 522) had a lower cancer-specific mortality rate than nonusers (NBB). No clear rate differences were observed by beta-blocker receptor selectivity owing to very few patients on NSBB. A SEER-Medicare analysis also showed improved survival in patients with PDAC on beta-blocker therapy [205].

The role of ADRB2 in PDAC progression was also supported by a retrospective clinical study from our group, which demonstrated a significantly improved OS in patients on NSBB, although the analysis was limited by the uncertain duration and dosage of beta-blockers. Beta-1–selective drugs provided no clinical advantage in this analysis [212], consistent with in vitro data showing that ADRB1 blockers did not suppress proliferation of *Kras* mutant pancreatic cells.

Taken together, as there is mounting evidence of the potential impact of beta-blockers on the outcomes of patients with cancer, a prospective clinical trial is warranted to identify those patients who would benefit most from beta-blocker use and to identify the best beta-blocker for a specific tumor type based on adrenergic receptor expression. Tumor cell expression of ADRB2 could be used as a biomarker for selecting those patients who would benefit from a specific beta-blocker. Beta-blockers could then be used as an adjuvant therapy during surgical recovery and chemotherapy to decrease tumor angiogenesis, tumor growth, delays in wound healing, and metastasis [209]. Beta-blockers also may reduce cancer-related psychological distress in patients newly diagnosed with cancer [213]. Therefore, beta-blockers have the potential to impact not only cancer biology and immunology but also the psychological well-being of patients with cancer.

In this regard, the PROSPER trial (Pancreatic Resection with perioperative Off-label Study of Propranolol and Etodolac—a phase II Randomized trial) studies a combination therapy of propranolol (NSBB) and etodolac in the perioperative setting in PDAC. It is hypothesized that this combination might attenuate psychological, surgical, and inflammatory stress responses that facilitate metastasis of the tumor [214]. This approach is interesting because it uses that critical period of surgery as a window for cancer directed therapy that is currently largely unexploited. Moreover, it does not interfere with current practice or future implementation of new adjuvant chemotherapy regimens (https://www.anticancerfund.org/en/approval-pancreatic-cancer-trial-perioperative-use-β-blocker-and-anti-inflammatory-drug).

Bethanechol

Emerging evidence has pointed to innervation as a critical factor in the microenvironment that contributes to the growth of a variety of malignancies [215,216]. Multiple studies have suggested a marked increase in neural density and nerve size in solid tumors [217–219]. In addition, experimental models systems have demonstrated a direct contribution of nerves to the development of several cancers [208,220]. Furthermore, studies have revealed an increased cross talk between tumor cells and nerves, with tumors able to induce active axonogenesis [218,221]. In many cases, nerves appear to promote strongly the growth of tumors, but the result of neural input is likely site-specific, influenced largely by the unique nerve–tumor interactions. Unlike findings in other tumor types, in PDAC, some reports indicate that vagus nerve signaling may have the potential to actually slow tumor progression. In this regard, vagotomy was reported to promote tumor growth and shortened OS [222] in orthotopic and syngeneic PDAC mouse models. Furthermore, treatment of these mice with the cholinergic agonist, bethanechol, improved survival of tumor-bearing mice.

Bethanechol chloride is a cholinergic agent that is a synthetic ester that is structurally and pharmacologically related to acetylcholine. It is currently FDA approved with an indication as the treatment of acute postoperative and postpartum nonobstructive (functional) urinary retention and for neurogenic atony of the urinary bladder with retention.

Preclinical data have indicated that blockade of parasympathetic signaling in genetically engineered mouse models of PDAC leads to increased carcinogenesis and that this is mediated by signaling via the cholinergic muscarinic receptor 1 [223]. Furthermore, in related experiments utilizing a genetically engineered mouse model of PDAC (LSL-*Kras*$^{+/LSL-G12D}$; LSL-*Trp53*$^{+/R172H}$; *Pdx1*-Cre mice, KPC), it has been demonstrated that treatment of tumor-bearing animals with GEM + bethanechol leads to significant improvement in OS with a corresponding decrease in CD44-positive pancreatic CSCs. In a genetically engineered mouse that models the multistep development and progression of PDAC (LSL-*Kras*$^{+/LSL-G12D}$; *Pdx1*-Cre mice, KC), parasympathetic blockade by subdiaphragmatic vagotomy resulted in increased incidence of tumor formation. This increase in tumor formation was accompanied by an increased in F4/80+ macrophages and an increase in the serum TNFα levels. Treatment of these mice with single agent bethanechol attenuates tumor formation and decreases F4/80+ macrophages and serum TNFα. Taken together, this preclinical data suggest that bethanechol may be an attractive target in PDAC therapy and may lead to a novel clinical trial in the near future.

Future directions of drug repurposing in PDAC

Future treatments for pancreatic cancer—the microbiome

With the advent of affordable sequencing technology for bacterial sequences, a plethora of publications has linked the microbiome to multiple, at first sight not overtly connected conditions. Current research has related disruptions of the composition of the microbiome, referred to as dysbiosis, to metabolic [231], neurologic [232,233], and cardiovascular diseases [234].

Furthermore, fecal microbiota transplantation is already being utilized for *Clostridium difficile* colitis and in clinical trials for inflammatory bowel diseases [235]. Cancer, in comparison, has been linked to dysbiosis both by novel sequencing approaches and traditional means of biomedical research. As many as 20% of all cancers worldwide are believed to be linked to alterations of the microbiome [236], with well-known examples such as *Helicobacter pylori* [237], *Fusobacterium nucleatum* [238], and Ebstein–Barr virus [239]. Strikingly, vaccinations against human papilloma viruses are already used to prevent cancers [240], and it is intriguing to speculate how the emerging knowledge on the influence of the microbiome in pancreatic cancer may lead to novel therapeutic approaches. Herein, we discuss how microbiota affects oncogenesis, response to therapy, and how modulation of the microbiome may be exploited for therapeutic benefit.

Interestingly, human and murine pancreatic cancers have been shown to harbor a more abundant microbiome than normal pancreas [241]. Moreover, qPCR analyses of human pancreatic cancer specimen from upfront resections showed an about 1000-fold increased amount of bacterial DNA in comparison to healthy controls, with significant differences in their microbiological composition [241]. Not only do these results demonstrate that pancreatic tissue is not entirely sterile, migration experiments with WT mice and fluorescently labeled *Escherichia coli* and *Escherichia faecalis* showed that these bacteria can migrate into the pancreas [241]. Similar results were reported for fungi, where human pancreatic cancer specimen showed an about 3000-fold increase of fungal DNA in comparison to healthy controls and GFP-labeled *Saccharomyces cerevisiae* migrated into pancreata of WT and LSL-$Kras^{+/LSL-G12D}$; *p48*-Cre (KC) [242]. Accordingly, while it seems clear that PDAC has an increased propensity to be populated with microbiota, it is important to determine whether these organisms influence tumor biology. In this regard, KC mice treated with a microbiologically ablative oral treatment consisting of vancomycin, neomycin, metronidazole, and amphotericin showed delayed tumor progression in terms of delayed acinar effacement, reduced dysplasia, reduced fibrosis, and lower overall pancreas weights [241]. Additionally, WT C57BL/6J implanted subcutaneously with pancreatic cancer cells showed lower tumor volumes and weights after ablative antibiotic treatment [243]. Importantly, ablative antibiotic treatment remained without effect in Rag1 knockout mice lacking mature T- and B lymphocytes, indicating that the effects of the microbiome on tumor growth are crucially dependent on the immune response [243]. More precisely, ablative antibiotic treatment results in increased numbers of interferon gamma producing Th1 cells and reduced numbers of IL-17 producing T cells [243]. While an increased Th1 to Th2 ratio has been linked to improved survival in pancreatic cancer [244], IL-17 accelerates tumor progression in KC mice [245].

Further evidence not only links the microbiome to an unfavorable phenotype of the immune reaction but also establishes a connection to poor success rates with immune CPI seen in PDAC. Orthotopic tumor models show that after ablative antibiotic treatment immunosuppressive $CD206^+$ M2 macrophages are reduced, more cytotoxic $CD8^+$ cells are present and PD1 expression is increased in intratumoral $CD4^+$ and $CD8^+$ cells [241]. While the aforementioned evidence suggests that antibiotic treatment is favorable in PDAC or could even render immune CPI effective, studies on patients with lung cancer (non-small cellular lung cancer, NSCLC) indicate that a more sophisticated approach might be necessary. NSCLC patients treated with commonly used antibiotics (beta-lactam, quinolones, macrolides) within 2 months before or 1 month after initiation of PD-1 or PD-L1 treatment had worse OS than those who did not receive antibiotic treatment (8.3 vs. 15.3 months, $P = .01$) [246]. Furthermore,

NSCLC patients not responding to immune CPI were shown to have low fecal abundance levels of *Akkermansia muciniphila*, which when supplemented to mice with NSCLC and treated with antibiotics could restore response to PD-1 inhibitors [246]. Similarly, a nuanced interaction between antibiotic use and risk of colorectal cancer was found when more than 150,000 patients from the United Kingdom were retrospectively analyzed. While penicillins increased the risk for colon cancer, the risk for rectal cancer was decreased for patients who had received tetracyclines [247].

In addition to modulating responses to immune CPI, the microbiome has also been shown to be an important factor in response to chemotherapy in pancreatic cancer. In this regard, human pancreatic cancer specimen have been shown to harbor gammaproteobacteria, which by expressing the long version of the enzyme cytidine deaminase can render cancer resistant to GEM [248]. Cytidine deaminase has been shown to deaminate GEM in cell cultures positive for *Mycoplasma hyorhinis* [249]. Interestingly, antitumoral activity of GEM is restored in colon cancers positive for gammaproteobacteria after treatment with ciprofloxacin [248].

Accordingly, the abovementioned evidence suggests that outcomes in PDAC could be improved by modulating the microbiome. However, no clinical data indicate which means of antimicrobial or probiotic therapy are suitable for patients with pancreatic cancer. While experimental studies show that there might be a rationale for reshaping the microbiome by using antibiotics, significant doubts remain that removal of an overly broad spectrum of microbiota might do more harm than good. Ideally, targeted therapies would remove species identified as harmful. Such treatment strategies are already applied for recurrent infections with *C. difficile*, where the *C. difficile* toxin B targeting monoclonal antibody Bezlotoxumab has been shown to lower recurrent *C. difficile* infections in comparison to placebo in two recent clinical trials (MODIFY I: 17% vs. 28%, MODIFY II: 16% vs. 26%) [250]. As an alternative to eliminating harmful species, adding beneficial microbiota using fecal microbiota transplantation may be a viable option to improve outcomes. Fecal microbiota transplants have been shown to be enormously effective in recurrent *C. difficile* infections, with clinical trials reporting resolution of diarrhea in as many as 90% of patients [251]. Interestingly, murine orthotopic tumor models of PDAC show that after ablative antibiotic treatment and fecal microbiota transplant with stool either from human long- or short-term survivors of PDAC (median survival 10.1 vs. 1.6 years), mice transplanted with stool from long-term survivors show reduced tumor volumes and higher numbers of $CD8^+$ cells as well as reduced infiltration with immunosuppressive $CD4^+FOXP3^+$ and myeloid-derived suppressor cells [252]. Furthermore, reshaping of the microbiome could also be achieved using dietary measures. Studies on colon cancer have shown that the gut microbiome shows rapid responses to dietary changes with subsequent cellular effects on the gut mucosa [253]. In this respect, a study with healthy African Americans and rural South Africans, two populations that have a remarkably different incidence of colon cancer (65:100,000 vs. <5:100,000) showed that dietary switches can induce favorable changes of the microbiome within 14 days. More precisely, African Americans were put on an African-style low-fat, high-fiber and South Africans on an American-style high-fat, low-fiber diet. Biopsies taken before and after dietary switch showed increased numbers of butyrate producers, which is a critical fecal metabolite reducing colon cancer risk [253,254]. Moreover, dietary switch did not only alter the microbiome favorably, low-fat, high-fiber nutrition was also associated with reduced Ki67 indices of the colonic mucosa [253].

Accordingly, it is clear that the microbiome plays a major role in carcinogenesis, immune response and chemoresistance in pancreatic cancer, and examples from other conditions show multiple avenues of potential future therapies. However, despite a torrent of basic research evidence being published currently, clinical evidence regarding utilization of the microbiome in pancreatic cancer is lacking. Future studies, especially clinical trials, will have to determine how antibiotics, targeted therapies, fecal microbiota transplantation, and also dietary interventions can help shape pancreatic cancer risk and therapy outcomes.

Future treatments for PDAC—lifestyle interventions

In light of dietary modifications to influence the body's microbiome, this idea could be also used as a repurposed and personalized drug. It has been shown that a ketogenic diet (KD) consisting of a very low-carbohydrate, but high-fat, composition may improve the effectiveness of certain classes of cancer drugs, e.g., drugs that target the insulin-activated enzyme phosphatidylinositol-3 kinase (PI3K). The underlying hypothesis for that study was that blocking PI3K cannot be effective, if the intrinsic insulin levels are kept at high levels; however, dietary modifications targeting hyperglycemia with KD could show a much improved response to the drug [255]. In a pancreatic xenograft model, it was also suggested that KD could increase radiation sensitivity [256]. Although a healthy dietary pattern with dietary fibers [257] or Mediterranean-style diet alone in nonalcoholic patients could not significantly reduce the risk for PDAC, a BMI > 25 kgm^2 and smoking were independently associated with it [258]. Diet and obesity, however, have significant influence on chronic inflammation as outlined in our remarks about metformin, and obesity itself as an expression of an increased inflammatory burden of the whole body also correlates with a heightened risk for PDAC [100]. Taken together, dietary modifications, such as KD on the one hand in conjunction with certain drug treatments or Mediterranean-style diet with a plant-based focus on fruits and vegetables, whole grains, and fish, might be highly beneficial and a future target for repurposing algorithms.

At the same time, dosed physical exercise to foster improvements of perioperative results or as additional adjunct therapeutic modifications for improving postoperative chemotherapeutic regimens is an exciting topic to "repurpose" in this regard.

As reported before, neuropsychological stress can cause a deleterious feedforward loop in some patients prone for pancreatic cancer and seems to accelerate pancreatic tumorigenesis [21]. It might also cause sleep disorders, thereby contributing to chronic inflammation with ultimately increased risk for cancer [259].

Finally, it should be mentioned that all dietary modifications as well as preventive use of antibiotics, nonsteroidal antiinflammatory drugs or other drugs can influence many physiological homeostatic balances within the body, e.g., the normally healthy population of microorganisms in the gut with unknown consequences.

Future directions

Out of the armory of compounds out of which we only presented a small part in this chapter, many have not been tested in clinical trials on patients with PDAC, yet (Table 9.3). Hence, our tables are far from complete with plenty of drugs potentially still being out there to be discovered. To generate more promising preclinical data and translate them into meaningful treatment options for PDAC patients, first of all, valid clinical studies need to be conceptualized, carried out, and analyzed.

Results from these studies need to be readily available for researchers, physicians, caretakers, and, last but not least, the affected patient. Data should go into a drug and disease archives with public availability as it is already the case for tumor-specific databases, such as the ones

TABLE 9.3 Compounds with repurposing potential for therapies in PDAC.

Study drug	Original use	Mechanism of action	Cancer type	Reference
Aprepitant	Antiemetic	AKT/mTOR, WNT inhibition	Hepatoblastoma, colorectal cancer, PDAC	[47–49]
Azithromycin	Macrolide	Unknown	Various: e.g., NSCLC, PDAC	[169,224]
Crocetinic acid	Saffron compound	Hedgehog inhibition	PDAC	[225]
Haloperidol	Psychosis	DRD2 antagonism	PDAC	[156]
Itraconazole	Antifungal	Angiogenesis, autophagy, hedgehog	PDAC	[226]
Ketamine	NMDA receptor antagonist	WNT inhibition	Colorectal cancer	[52]
Nigericin	Ionophore	EMT inhibition	Colorectal cancer	[227]
Pirfenidone	Pulmonary fibrosis	Stellate cells	PDAC	[228]
Quinomycin A	Quinoxaline antibiotic	Notch inhibition	PDAC	[57]
Salinomycin	Ionophore	WNT inhibition, K-ras	Breast cancer; CLL	[229,230]

AML, acute myeloid leukemia; *CLL*, chronic lymphatic leukemia; *N/A*, not available; *NSCLC*, non-small cell lung cancer; *PDAC*, pancreatic ductal adenocarcinoma; referenced trials were searched on https://clinicaltrials.gov/ct2/home.

provided by the The Cancer Genome Atlas (TCGA) [260]. In the future, drug scores derived from drug databases (see also www.PanDrugs.org [170]), tumor database, and the gene score of sequenced PDAC tumors might be combined by AI-driven algorithms, and results could then suggest completely novel and personalized treatment options. Before applying these therapies, however, fast preclinical testing would be desirable, such as in vitro in PDOs [261] or in vivo in mouse avatars to verify actionable mechanisms [20]. Final approval should then be discussed in established tumor boards at the respective institution (Fig. 9.5).

On a different note, the process of translational research will have to be accelerated and new technologies including AI algorithms will certainly assist to cut down time restraints. Different approaches will help to shorten bench to bedside turnovers. Intriguing compounds with negligible side effects could be identified earlier by AI or better tailored to the respective patient. More sophisticated preclinical models will make testing easier and enhance transition into clinical trials or direct application. In patients after first-line therapy, case-by-case decisions might apply, and molecular profiling could contribute to shorter transition time as well. Ultimately, predictive markers for repurposed compounds could then further enhance the usefulness of their application, e.g., high expression of NK1-R full length versus truncated in the case of aprepitant treatment [262,263].

Conclusion

In this chapter, we tried to shed light on studies of the past, present, and future that investigate potential antineoplastic regimens of various established compounds that are in clinical use for different diseases and conditions.

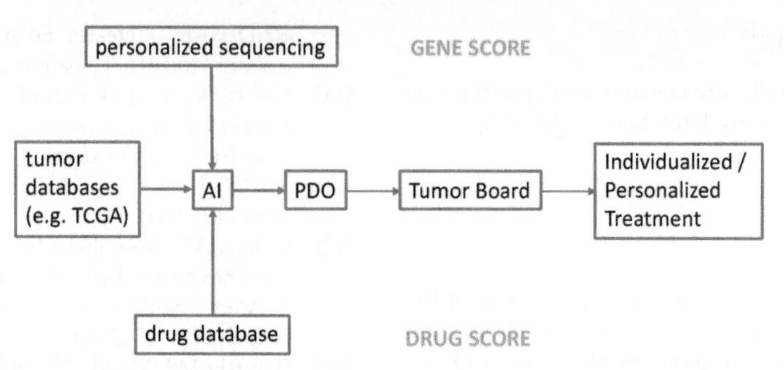

FIGURE 9.5 Schematic illustration of an outlook for future personalized therapeutic applications that include standard treatments as well as repurposed compounds tailored to tumor subtype with preclinical testing. *AI*, artificial intelligence; *PDO*, patient-derived organoids; *TCGA*, The Cancer Genome Atlas.

Although 5-year survival numbers are slowly increasing and reaching double digits in 2020 according to the Cancer Facts & Figures report of the American Cancer Society PDAC, is still a devastating disease and on the downside more people than ever before will be diagnosed.

While the real Achilles heel of PDAC, if it exists, is yet to be found, thanks to a growing field of researchers and increased federal funding, plenty of different approaches have been extensively investigated over the last decades to improve outcome and survival of patients with PDAC. Also, grassroots advocacy efforts across the country have drawn ample attention to the disease in general and treatment options at highly specialized cancer centers, in particular.

In this regard, the concept of repurposing drugs is incorporating many different aspects of PDAC with as many angles to tackle the disease and potentially epitomizes very fast transfer times from bench to bedside due to formerly acquired FDA approval status. Apart from trying out approved drugs based on basic, translational, and clinical observations as outlined above in Section General strategies toward drug repurposing as well as 5.1 Case studies of the past, we believe that in the present and future the field will experience much more sophisticated approaches. Among those, we only briefly describe ML and AI (Section Current development and case studies) that could lead to new (rapid) advances taking into account gigantic datasets of PDAC in general, of the patient and his disease in particular as well as the drug alone or in combination. Moreover, in the future, we believe that drug repurposing not only means to take into account previously mentioned factors but will incorporate preoperative testing and consideration of molecular tumor boards (Section Future directions). Last, peri-/postoperative priming of the tumor and its microenvironment, respectively, e.g., by respecting microbiotic colonization of tumor-bearing and tumor-neighboring organs as well as preventive antitumorigenic lifestyle changes could lead to new heights in PDAC treatment (Section Future directions of drug repurposing in PDAC).

Overall, drugs with FDA-approval harbor many advantages. In theory as shown in Fig. 9.1, they can be quickly translated into clinical use with comparatively little additional costs. Furthermore, repurposed drugs typically have very well-described side effects that—in comparison to classic chemotherapeutic regimens—appear negligible. Last, PDAC patients might benefit enormously from directed compounds, e.g., beta-blockers for high expressers of catecholamine receptors, autophagy modulators for highly autophagy-dependent cancers or CSC-specific therapies for tumors enriched with this powerful subpopulation.

Acknowledgments

We want to apologize to all authors whose work could not be cited due to space restrictions/limitations in this chapter.

References

[1] Stewart B, Wild CP. World cancer report 2014. 2019.

[2] Quante AS, Ming C, Rottmann M, Engel J, Boeck S, Heinemann V, et al. Projections of cancer incidence and cancer-related deaths in Germany by 2020 and 2030. Cancer Med 2016;5:2649–56. https://doi.org/10.1002/cam4.767.

[3] Rahib L, Smith BD, Aizenberg R, Rosenzweig AB, Fleshman JM, Matrisian LM. Projecting cancer incidence and deaths to 2030: the unexpected burden of thyroid, liver, and pancreas cancers in the United States. Cancer Res 2014;74:2913–21. https://doi.org/10.1158/0008-5472.CAN-14-0155.

[4] Malvezzi M, Carioli G, Bertuccio P, Rosso T, Boffetta P, Levi F, et al. European cancer mortality predictions for the year 2016 with focus on leukaemias. Ann Oncol 2016;27:725–31. https://doi.org/10.1093/annonc/mdw022.

[5] Siegel RL, Miller KD, Jemal A. Cancer statistics. CA Cancer J Clin 2018;68:7–30. https://doi.org/10.3322/caac.21442.

[6] Chan KKW, Guo H, Cheng S, Beca JM, Misner RR, Isaranuwatchai W, et al. Real-world outcomes of FOLFIRINOX vs gemcitabine and nab-paclitaxel in advanced pancreatic cancer: a population-based propensity score-weighted analysis. Cancer Med 2019; 68:7. https://doi.org/10.1002/cam4.2705.

[7] Nevala-Plagemann C, Hidalgo M, Garrido-Laguna I. From state-of-the-art treatments to novel therapies for advanced-stage pancreatic cancer. Nat Rev Clin Oncol 2019:1–16. https://doi.org/10.1038/s41571-019-0281-6.

[8] Vasen H, Ibrahim I, Ponce CG, Slater EP, Matthäi E, Carrato A, et al. Benefit of surveillance for pancreatic cancer in high-risk individuals: outcome of long-term prospective follow-up studies from three European expert centers. J Clin Oncol 2016;34. https://doi.org/10.1200/JCO.2015.64.0730. 2010–2019.

[9] Nuzhat Z, Kinhal V, Sharma S, Rice GE, Joshi V, Salomon C. Tumour-derived exosomes as a signature of pancreatic cancer – liquid biopsies as indicators of tumour progression. Oncotarget 2017;8:17279–91. https://doi.org/10.18632/oncotarget.13973.

[10] DiMasi JA, Grabowski HG, Hansen RW. Innovation in the pharmaceutical industry: new estimates of R&D costs. J Health Econ 2016;47:20–33. https://doi.org/10.1016/j.jhealeco.2016.01.012.

[11] Gupta SC, Sung B, Prasad S, Webb LJ, Aggarwal BB. Cancer drug discovery by repurposing: teaching new tricks to old dogs. Trends Pharmacol Sci 2013; 34:508–17. https://doi.org/10.1016/j.tips.2013.06.005.

[12] Gelijns AC, Rosenberg N, Moskowitz AJ. Capturing the unexpected benefits of medical research. N Engl J Med 1998;339:693–8. https://doi.org/10.1056/NEJM199809033391010.

[13] Boguski MS, Mandl KD, Sukhatme VP. Drug discovery. Repurposing with a difference. Science 2009;324: 1394–5. https://doi.org/10.1126/science.1169920.

[14] Fishman MC, Porter JA. Pharmaceuticals: a new grammar for drug discovery. Nature 2005;437: 491–3. https://doi.org/10.1038/437491a.

[15] Powell CE, Slater IH. Blocking of inhibitory adrenergic receptors by a dichloro analog of isoproterenol. J Pharmacol Exp Ther 1958;122:480–8.

[16] Black JW, Duncan WA, Shanks RG. Comparison of some properties of pronethalol and propranolol. Br J Pharmacol Chemother 1965;25:577–91. https://doi.org/10.1111/j.1476-5381.1965.tb01782.x.

[17] P.T.P.I.N.-P.. Anxiolytics not acting at the benzodiazepine receptor: beta blockers. Elsevier; 1992. https://doi.org/10.1016/0278-5846(92)90004-X [n.d.].

[18] Clevers H, Tuveson DA. Organoid models for cancer research. Annu Rev Cancer Biol 2019;3:223–34. https://doi.org/10.1146/annurev-cancerbio-030518-055702.

[19] Ideno N, Yamaguchi H, Okumura T, Huang J, Brun MJ, Ho ML, et al. A pipeline for rapidly generating genetically engineered mouse models of pancreatic cancer using in vivo CRISPR-Cas9-mediated somatic recombination. Lab Invest 2019. https://doi.org/10.1038/s41374-018-0171-z.

[20] Perales-Patón J, Piñeiro-Yañez E, Tejero H, López-Casas PP, Hidalgo M, Gómez-López G, et al. Pancreas cancer precision treatment using avatar mice from a bioinformatics perspective. Public Health Genomics 2017;20:81–91. https://doi.org/10.1159/000479812.

[21] Renz BW, Takahashi R, Tanaka T, Macchini M, Hayakawa Y, Dantes Z, et al. β2 adrenergic-neurotrophin feedforward loop promotes pancreatic cancer. Cancer Cell 2018;33:75–90. https://doi.org/10.1016/j.ccell.2017.11.007. e7.

[22] Doan TL, Pollastri M, Walters MA, I GGAR. The future of drug repositioning: old drugs, new opportunities. Annual Reports in Medicinal Chemistry 2011;Volume 46. Elsevier Inc. http://doi.org/10.1016/B978-0-12-386009-5.00004-7. ISSN: 0065-7743.

[23] Lancet WM. Thalidomide and congenital abnormalities, Softenon.Nl. (n.d.). 1961. https://doi.org/10.1016/S0140-6736(61)90927-8.

[24] Daniel Vasella MD, Slater R. Magic cancer bullet. Harper Collins; 2003.

[25] Goldstein I, Lue TF, Padma-Nathan H, Rosen RC, Steers WD, Wicker PA. Oral sildenafil in the treatment of erectile dysfunction. Sildenafil Study Group. N Engl J Med 1998;338:1397–404. https://doi.org/10.1056/NEJM199805143382001.

[26] Renz B, D'Haese J, Werner J, Westphalen C, Ilmer M. Repurposing established compounds to target pancreatic cancer stem cells (CSCs). Med Sci 2017;5:14. https://doi.org/10.3390/medsci5020014.

[27] Ilmer M, Westphalen CB, Nieß H, D'Haese JG, Angele MK, Werner J, et al. Repurposed drugs in pancreatic ductal adenocarcinoma: an update. Cancer J 2019;25:134–8. https://doi.org/10.1097/PPO.0000000000000372.

[28] Seufferlein T, Bachet JB, Van Cutsem E, Rougier P, Group OBOTEGW. Pancreatic adenocarcinoma: ESMO–ESDO Clinical Practice Guidelines for diagnosis, treatment and follow-up. Ann Oncol 2012;23:vii33–40. https://doi.org/10.1093/annonc/mds224.

[29] Conroy T, Desseigne F, Ychou M, Bouché O, Guimbaud R, Bécouarn Y, et al. FOLFIRINOX versus gemcitabine for metastatic pancreatic cancer. N Engl J Med 2011;364:1817–25. https://doi.org/10.1056/NEJMoa1011923.

[30] Hsueh C-T. Pancreatic cancer: current standards, research updates and future directions. J Gastrointest Oncol 2011;2:123–5. https://doi.org/10.3978/j.issn.2078-6891.2011.037.

[31] Conroy T, Hammel P, Hebbar M, Ben Abdelghani M, Wei AC, Raoul J-L, et al. FOLFIRINOX or gemcitabine as adjuvant therapy for pancreatic cancer. N Engl J Med 2018;379:2395–406. https://doi.org/10.1056/NEJMoa1809775.

[32] Valent P, Bonnet D, de Maria R, Lapidot T, Copland M, Melo JV, et al. Cancer stem cell definitions and terminology: the devil is in the details. Nat Rev Cancer 2012. https://doi.org/10.1038/nrc3368.

[33] Ilmer M, Horst D. Pancreatic CSCs and microenvironment. Genes Cancer 2015:365. https://doi.org/10.18632/genesandcancer.80.

[34] Ilmer M, Boiles AR, Regel I, Yokoi K, Michalski CW, Wistuba II, et al. RSPO2 enhances canonical Wnt signaling to confer stemness-associated traits to susceptible pancreatic cancer cells. Cancer Res 2015. https://doi.org/10.1158/0008-5472.CAN-14-1327.

[35] Lonardo E, Hermann PC, Mueller M-T, Huber S, Balic A, Miranda-Lorenzo I, et al. Nodal/activin signaling drives self-renewal and tumorigenicity of pancreatic cancer stem cells and provides a target for combined drug therapy. Cell Stem Cell 2011;9:433–46. https://doi.org/10.1016/j.stem.2011.10.001.

[36] Cioffi M, Trabulo SM, Sanchez-Ripoll Y, Miranda-Lorenzo I, Lonardo E, Dorado J, et al. The miR-17-92 cluster counteracts quiescence and chemoresistance in a distinct subpopulation of pancreatic cancer stem cells. Gut 2015;64:1936–48. https://doi.org/10.1136/gutjnl-2014-308470.

[37] Wang Z, Li Y, Kong D, Banerjee S, Ahmad A, Azmi AS, et al. Acquisition of epithelial-mesenchymal transition phenotype of gemcitabine-resistant pancreatic cancer cells is linked with activation of the notch signaling pathway. Cancer Res 2009;69:2400–7. https://doi.org/10.1158/0008-5472.CAN-08-4312.

[38] Guinot A, Oeztuerk-Winder F, Ventura J-J. miR-17-92/p38α dysregulation enhances Wnt signal and selects Lgr6+ cancer stem cell like cells during human lung adenocarcinoma progression. Cancer Res 2016; 76. https://doi.org/10.1158/0008-5472.CAN-15-3302. canres.3302.2015–4022.

[39] Guardavaccaro D, Clevers H. Wnt/{beta}-Catenin and MAPK signaling: allies and enemies in different battlefields. Sci Signal 2012;5:pe15. https://doi.org/10.1126/scisignal.2002921.

[40] Neth P, Ries C, Karow M, Egea V, Ilmer M. The Wnt signal transduction pathway in stem cells and cancer cells: influence on cellular invasion. Stem Cell Rev 2007;3:18–29. https://doi.org/10.1007/s12015-007-0001-y.

[41] Sato T, Vries RG, Snippert HJ, van de Wetering M, Barker N, Stange DE, et al. Single Lgr5 stem cells build crypt-villus structures in vitro without a mesenchymal niche. Nature 2009;459:262–5. https://doi.org/10.1038/nature07935.

[42] Kobayashi S, Yamada-Okabe H, Suzuki M, Natori O, Kato A, Matsubara K, et al. LGR5-Positive colon cancer stem cells interconvert with drug-resistant LGR5-negative cells and are capable of tumor reconstitution. Stem Cell 2012;30:2631–44. https://doi.org/10.1002/stem.1257.

[43] Horst D, Chen J, Morikawa T, Ogino S, Kirchner T, Shivdasani RA. Differential WNT activity in colorectal cancer confers limited tumorigenic potential and is regulated by MAPK signaling. Cancer Res 2012;72:1547–56. https://doi.org/10.1158/0008-5472.CAN-11-3222.

[44] Vermeulen L, De Sousa E Melo F, van der Heijden M, Cameron K, de Jong JH, Borovski T, et al. Wnt activity defines colon cancer stem cells and is regulated by the microenvironment. Nat Cell Biol 2010;12:468–76. https://doi.org/10.1038/ncb2048.

[45] Cairo S, Armengol C, De Reyničs A, Wei Y, Thomas E, Renard CA, et al. Hepatic stem-like phenotype and interplay of Wnt/[beta]-catenin and Myc signaling in aggressive childhood liver cancer. Cancer Cell 2008;14:471–84.

[46] Lu D, Choi MY, Yu J, Castro JE, Kipps TJ, Carson DA. Salinomycin inhibits Wnt signaling and selectively induces apoptosis in chronic lymphocytic leukemia cells. Proc Natl Acad Sci USA 2011;108:13253–7. https://doi.org/10.1073/pnas.1110431108.

[47] Berger M, Neth O, Ilmer M, Garnier A, Salinas-Martín MV, de Agustín Asencio JC, et al. Hepatoblastoma cells express truncated neurokinin-1 receptor and can be growth inhibited by aprepitant in vitro and in vivo. J Hepatol 2014;60:985–94. https://doi.org/10.1016/j.jhep.2013.12.024.

[48] Garnier AS, Vykoukal J, Hubertus J, Alt E, von Schweinitz D, Kappler R, et al. Targeting the neurokinin-1 receptor inhibits growth of human colon cancer cells. Int J Oncol 2015. https://doi.org/10.3892/ijo.2015.3016.

[49] Ilmer M, Garnier A, Vykoukal J, Alt E, von Schweinitz D, Kappler R, et al. Targeting the neurokinin-1 receptor compromises canonical Wnt signaling in hepatoblastoma. Mol Cancer Ther 2015; 14:2712–21. https://doi.org/10.1158/1535-7163.MCT-15-0206.

[50] Zhang N, Wei P, Gong A, Chiu W-T, Lee H-T, Colman H, et al. FoxM1 promotes β-catenin nuclear localization and controls Wnt target-gene expression and glioma tumorigenesis. Cancer Cell 2011;20: 427–42.

[51] Cui J, Shi M, Xie D, Wei D, Jia Z, Zheng S, et al. FOXM1 promotes the warburg effect and pancreatic cancer progression via transactivation of LDHA expression. Clin Cancer Res 2014;20:2595–606. https://doi.org/10.1158/1078-0432.CCR-13-2407.

[52] Blaj C, Bringmann A, Schmidt EM, Urbischek M, Lamprecht S, Fröhlich T, et al. ADNP is a therapeutically inducible repressor of WNT signaling in colorectal cancer. Clin Cancer Res 2016. https://doi.org/10.1158/1078-0432.CCR-16-1604.

[53] Kim SK, Kim H, Lee D-H, Kim T-S, Kim T, Chung C, et al. Reversing the intractable nature of pancreatic cancer by selectively targeting ALDH-high, therapy-resistant cancer cells. PloS One 2013;8:e78130. https://doi.org/10.1371/journal.pone.0078130.

[54] Han D, Wu G, Chang C, Zhu F, Xiao Y, Li Q, et al. Disulfiram inhibits TGF-β-induced epithelial-mesenchymal transition and stem-like features in breast cancer via ERK/NF-κB/Snail pathway. Oncotarget 2015;6:40907–19. https://doi.org/10.18632/oncotarget.5723.

[55] Cong J, Wang Y, Zhang X, Zhang N, Liu L, Soukup K, et al. A novel chemoradiation targeting stem and non-stem pancreatic cancer cells by repurposing disulfiram. Cancer Lett 2017;409:9–19. https://doi.org/10.1016/j.canlet.2017.08.028.

[56] Porporato PE. Understanding cachexia as a cancer metabolism syndrome. Oncogenesis 2016;5. https://doi.org/10.1038/oncsis.2016.3. e200.

[57] Ponnurangam S, Dandawate PR, Dhar A, Tawfik OW, Parab RR, Mishra PD, et al. Quinomycin A targets Notch signaling pathway in pancreatic cancer stem cells. Oncotarget 2016;7:3217–32. https://doi.org/10.18632/oncotarget.6560.

[58] Mani SA, Guo W, Liao M-J, Eaton EN, Ayyanan A, Zhou AY, et al. The epithelial-mesenchymal transition generates cells with properties of stem cells. Cell 2008; 133:704–15. https://doi.org/10.1016/j.cell.2008.03.027.

[59] Wellner U, Schubert J, Burk UC, Schmalhofer O, Zhu F, Sonntag A, et al. The EMT-activator ZEB1 promotes tumorigenicity by repressing stemness-inhibiting microRNAs. Nat Cell Biol 2009;11: 1487–95. https://doi.org/10.1038/ncb1998.

[60] Meidhof S, Brabletz S, Lehmann W, Preca BT, Mock K, Ruh M, et al. ZEB1-associated drug resistance in cancer cells is reversed by the class I HDAC inhibitor mocetinostat. EMBO Mol Med 2015;7:831–47. https://doi.org/10.15252/emmm.201404396.

[61] Bhutia YD, Hung SW, Patel B, Lovin D, Govindarajan R. CNT1 expression influences proliferation and chemosensitivity in drug-resistant pancreatic cancer cells. Cancer Res 2011;71:1825–35. https://doi.org/10.1158/0008-5472.CAN-10-2736.

[62] Zheng X, Carstens JL, Kim J, Scheible M, Kaye J, Sugimoto H, et al. Epithelial-to-mesenchymal transition is dispensable for metastasis but induces chemoresistance in pancreatic cancer. Nature 2015;527: 525–30. https://doi.org/10.1038/nature16064.

[63] Hermann PC, Huber SL, Herrler T, Aicher A, Ellwart JW, Guba M, et al. Distinct populations of cancer stem cells determine tumor growth and metastatic activity in human pancreatic cancer. Cell Stem Cell 2007;1:313–23. https://doi.org/10.1016/j.stem.2007.06.002.

[64] Lee Y, Kim SJ, Park HD, Park EH, Huang SM, Jeon SB, et al. PAUF functions in the metastasis of human pancreatic cancer cells and upregulates CXCR4 expression. PubMed – NCBI Oncogene 2009;29: 56–67. https://doi.org/10.1038/onc.2009.298.

[65] Singh S, Srivastava SK, Bhardwaj A, Owen LB, Singh AP. CXCL12-CXCR4 signalling axis confers gemcitabine resistance to pancreatic cancer cells: a novel target for therapy. Br J Cancer 2010;103: 1671–9. https://doi.org/10.1038/sj.bjc.6605968.

[66] Feig C, Jones JO, Kraman M, Wells RJB, Deonarine A, Chan DS, et al. Targeting CXCL12 from FAP-expressing carcinoma-associated fibroblasts synergizes with anti-PD-L1 immunotherapy in pancreatic cancer. Proc Natl Acad Sci USA 2013;110:20212–7. https://doi.org/10.1073/pnas.1320318110.

[67] Whatcott CJ, Han H, Posner RG, Hostetter G, Von Hoff DD. Targeting the tumor microenvironment in cancer: why hyaluronidase deserves a second look. Cancer Discov 2011;1:291–6. https://doi.org/10.1158/2159-8290.CD-11-0136.

[68] Provenzano PP, Cuevas C, Chang AE, Goel VK, Von Hoff DD, Hingorani SR. Enzymatic targeting of the stroma ablates physical barriers to treatment of pancreatic ductal adenocarcinoma. Cancer Cell 2012; 21:418–29. https://doi.org/10.1016/j.ccr.2012.01.007.

[69] Ramanathan RK, Abbruzzese J, chemotherapy TDC. A randomized phase II study of PX-12, an inhibitor of thioredoxin in patients with advanced cancer of the pancreas following progression after a gemcitabine. Springer; 2011 [n.d.].

[70] Hingorani SR. Intercepting cancer communiques: exosomes as heralds of malignancy. Cancer Cell 2015;28: 151–3. https://doi.org/10.1016/j.ccell.2015.07.015.

[71] Von Hoff DD, Ervin T, Arena FP, Chiorean EG, Infante J, Moore M, et al. Increased survival in pancreatic cancer with nab-paclitaxel plus gemcitabine. N Engl J Med 2013;369:1691–703. https://doi.org/10.1056/NEJMoa1304369.

[72] Olson P, Hanahan D. Breaching the cancer fortress. Science 2009;324:1400–1. https://doi.org/10.1126/science.1175940.

[73] Rhim AD, Oberstein PE, Thomas DH, Mirek ET, Palermo CF, Sastra SA, et al. Stromal elements act to restrain, rather than support, pancreatic ductal adenocarcinoma. Cancer Cell 2014;25:735–47. https://doi.org/10.1016/j.ccr.2014.04.021.

[74] Kruger S, Ilmer M, Kobold S, Cadilha BL, Endres S, Ormanns S, et al. Advances in cancer immunotherapy 2019 – latest trends. J Exp Clin Cancer Res 2019;38: 268. https://doi.org/10.1186/s13046-019-1266-0.

[75] Chauhan VP, Martin JD, Liu H, Lacorre DA, Jain SR, Kozin SV, et al. Angiotensin inhibition enhances drug delivery and potentiates chemotherapy by decompressing tumour blood vessels. Nat Comms 2013;4:2516. https://doi.org/10.1038/ncomms3516.

[76] Olive KP, Jacobetz MA, Davidson CJ, Gopinathan A, McIntyre D, Honess D, et al. Inhibition of hedgehog signaling enhances delivery of chemotherapy in a mouse model of pancreatic cancer. Science 2009;324:1457–61. https://doi.org/10.1126/science.1171362.

[77] Diop-Frimpong B, Chauhan VP, Krane S, Boucher Y, Jain RK. Losartan inhibits collagen I synthesis and improves the distribution and efficacy of nanotherapeutics in tumors. Proc Natl Acad Sci USA 2011;108: 2909–14. https://doi.org/10.1073/pnas.1018892108.

[78] Liu H, Naxerova K, Pinter M, Incio J, Lee H, Shigeta K, et al. Use of angiotensin system inhibitors is associated with immune activation and longer survival in nonmetastatic pancreatic ductal adenocarcinoma. Clin Cancer Res 2017;23:5959–69. https://doi.org/10.1158/1078-0432.CCR-17-0256.

[79] Murphy JE, Wo JY, Ryan DP, Jiang W, Yeap BY, Drapek LC, et al. Total neoadjuvant therapy with FOLFIRINOX followed by individualized chemoradiotherapy for borderline resectable pancreatic adenocarcinoma: a phase 2 clinical trial. JAMA Oncol 2018;4:963–9. https://doi.org/10.1001/jamaoncol.2018.0329.

[80] di Magliano MP, Logsdon CD. Roles for KRAS in pancreatic tumor development and progression. Gastroenterology 2013;144:1220–9. https://doi.org/10.1053/j.gastro.2013.01.071.

[81] Bailey P, Chang DK, Nones K, Johns AL, Patch A-M, Gingras M-C, et al. Genomic analyses identify molecular subtypes of pancreatic cancer. Nature 2016:1–19. https://doi.org/10.1038/nature16965.

[82] Hardie R-A, van Dam E, Cowley M, Han T-L, Balaban S, Pajic M, et al. Mitochondrial mutations and metabolic adaptation in pancreatic cancer. Cancer Metab 2017;5:2. https://doi.org/10.1186/s40170-017-0164-1.

[83] Daemen A, Peterson D, Sahu N, McCord R, Du X, Liu B, et al. Metabolite profiling stratifies pancreatic ductal adenocarcinomas into subtypes with distinct sensitivities to metabolic inhibitors. Proc Natl Acad Sci USA 2015;112. https://doi.org/10.1073/pnas.1501605112. E4410–7.

[84] King JK, Hahn BH. Systemic lupus erythematosus: modern strategies for management: a moving target. Best Pract Res Clin Rheumatol 2007;21:971–87. https://doi.org/10.1016/j.berh.2007.09.002.

[85] Takeda K, Kaisho T, Akira S. Toll-like receptors. Annu Rev Immunol 2003;21:335–76. https://doi.org/10.1146/annurev.immunol.21.120601.141126.

[86] Fujii S, Mitsunaga S, Yamazaki M, Hasebe T, Ishii G, Kojima M, et al. Autophagy is activated in pancreatic cancer cells and correlates with poor patient outcome. Cancer Sci 2008;99:1813–9. https://doi.org/10.1111/j.1349-7006.2008.00893.x.

[87] Yang S, Wang X, Contino G, Liesa M, Sahin E, Ying H, et al. Pancreatic cancers require autophagy for tumor growth. Genes Dev 2011;25:717–29. https://doi.org/10.1101/gad.2016111.

[88] Sousa CM, Biancur DE, Wang X, Halbrook CJ, Sherman MH, Zhang L, et al. Pancreatic stellate cells support tumour metabolism through autophagic alanine secretion. Nature 2016;536:479–83. https://doi.org/10.1038/nature19084.

[89] Yang A, Herter-Sprie G, Zhang H, Lin EY, Biancur D, Wang X, et al. Autophagy sustains pancreatic cancer growth through both cell-autonomous and nonautonomous mechanisms. Cancer Discov 2018;8:276–87. https://doi.org/10.1158/2159-8290.CD-17-0952.

[90] Karasic TB, O'Hara MH, Loaiza-Bonilla A, Reiss KA, Teitelbaum UR, Borazanci E, et al. Effect of gemcitabine and nab-paclitaxel with or without hydroxychloroquine on patients with advanced pancreatic cancer: a phase 2 randomized clinical trial. JAMA Oncol 2019;5:993–8. https://doi.org/10.1001/jamaoncol.2019.0684.

[91] Blum R, Kloog Y. Metabolism addiction in pancreatic cancer. Cell Death Dis 2014;5. https://doi.org/10.1038/cddis.2014.38. e1065.

[92] Balic A, Sørensen MD, Trabulo SM, Sainz B. Chloroquine targets pancreatic cancer stem cells via inhibition of CXCR4 and hedgehog signaling. Mole Cancer 2014.

[93] Kirpichnikov D, McFarlane SI, Sowers JR. Metformin: an update. Ann Intern Med 2002;137:25–33. https://doi.org/10.7326/0003-4819-137-1-200207020-00009.

[94] Wang Y, An H, Liu T, Qin C, Sesaki H, Guo S, et al. Metformin improves mitochondrial respiratory activity through activation of AMPK. Cell Rep 2019;29:1511–23. https://doi.org/10.1016/j.celrep.2019.09.070. e5.

[95] Hanahan D, Weinberg RA. Hallmarks of cancer: the next generation. Cell 2011;144:646–74. https://doi.org/10.1016/j.cell.2011.02.013.

[96] Feig C, Gopinathan A, Neesse A, Chan DS, Cook N, Tuveson DA. The pancreas cancer microenvironment. Clin Cancer Res 2012;18(16). https://doi.org/10.1158/1078-0432.CCR-11-3114. Clincancerres.Aacrjournals.org.

[97] Gukovsky I, Li N, Todoric J, Gukovskaya A, Karin M. Inflammation, autophagy, and obesity: common features in the pathogenesis of pancreatitis and pancreatic cancer. Gastroenterology 2013;144:1199–209. https://doi.org/10.1053/j.gastro.2013.02.007. e4.

[98] Alberti KGMM, Zimmet P, Shaw J. Metabolic syndrome—a new world-wide definition. A consensus statement from the International Diabetes Federation. Diabet Med 2006;23:469–80. https://doi.org/10.1111/j.1464-5491.2006.01858.x.

[99] Levi Z, Rottenberg Y, Twig G, Katz L, Leiba A, Derazne E, et al. Adolescent overweight and obesity and the risk for pancreatic cancer among men and women: a nationwide study of 1.79 million Israeli adolescents. Cancer 2018;125:118–26. https://doi.org/10.1002/cncr.31764.

[100] Pothuraju R, Rachagani S, Junker WM, Chaudhary S, Saraswathi V, Kaur S, et al. Pancreatic cancer associated with obesity and diabetes: an alternative approach for its targeting. J Exp Clin Cancer Res 2018;37:319. https://doi.org/10.1186/s13046-018-0963-4.

[101] Matsubara S, Ding Q, Miyazaki Y, Kuwahata T, Tsukasa K, Takao S. mTOR plays critical roles in pancreatic cancer stem cells through specific and stemness-related functions. Sci Rep 2013;3:3230. https://doi.org/10.1038/srep03230.

[102] Lonardo E, Cioffi M, Sancho P, Sanchez-Ripoll Y, Trabulo SM, Dorado J, et al. Metformin targets the metabolic achilles heel of human pancreatic cancer stem cells. PloS One 2013;8. https://doi.org/10.1371/journal.pone.0076518. e76518.

[103] Sancho P, Burgos-Ramos E, Tavera A, Bou Kheir T, Jagust P, Schoenhals M, et al. MYC/PGC-1α balance determines the metabolic phenotype and plasticity of pancreatic cancer stem cells. Cell Metabol 2015;22:590–605. https://doi.org/10.1016/j.cmet.2015.08.015.

[104] Jagust P, de Luxán-Delgado B, Parejo-Alonso B, Sancho P. Metabolism-based therapeutic strategies targeting cancer stem cells. Front Pharmacol 2019;10:203. https://doi.org/10.3389/fphar.2019.00203.

[105] Yue W, Yang CS, DiPaola RS, Tan X-L. Repurposing of metformin and aspirin by targeting AMPK-mTOR and inflammation for pancreatic cancer prevention and treatment. Cancer Prev Res 2014;7:388–97. https://doi.org/10.1158/1940-6207.CAPR-13-0337.

[106] Norman AW. Sunlight, season, skin pigmentation, vitamin D, and 25-hydroxyvitamin D: integral components of the vitamin D endocrine system. Am J Clin Nutr 1998;67:1108–10. https://doi.org/10.1093/ajcn/67.6.1108.

[107] Vuolo L, di somma C, Faggiano A, Colao A. Vitamin D and cancer. Front Endocrinol 2012;3. https://doi.org/10.3389/fendo.2012.00058.

[108] Ma J, Stampfer MJ, Gann PH, Hough HL, Giovannucci E, Kelsey KT, et al. Vitamin D receptor polymorphisms, circulating vitamin D metabolites, and risk of prostate cancer in United States physicians. Cancer Epidemiol Biomarkers Prev 1998;7:385–90.

[109] Lambrey G, N'Guyen TM, Garabedian M, Sebert JL, de Fremont JF, Marie P, et al. Possible link between changes in plasma 24,25-dihydroxyvitamin D and healing of bone resorption in dialysis osteodystrophy. Metab Bone Dis Relat Res 1982;4:25–30. https://doi.org/10.1016/0221-8747(82)90005-4.

[110] Feskanich D, Ma J, Fuchs CS, Kirkner GJ, Hankinson SE, Hollis BW, et al. Plasma vitamin D metabolites and risk of colorectal cancer in women. Cancer Epidemiol Biomarkers Prev 2004;13:1502–8.

[111] Ben Fradj MK, Gargouri MM, Hammami MB, Ben Rhouma S, Kallel A, Jemaa R, et al. Bladder cancer is associated with low plasma 25-hydroxyvitamin D concentrations in Tunisian population. Nutr Cancer 2016;68:208–13. https://doi.org/10.1080/01635581.2016.1134598.

[112] Ong JS, Cuellar-Partida G, Lu Y, Australian Ovarian Cancer S, Fasching PA, Hein A, et al. Association of vitamin D levels and risk of ovarian cancer: a Mendelian randomization study. Int J Epidemiol 2016;45:1619–30. https://doi.org/10.1093/ije/dyw207.

[113] Okazaki R, Ozono K, Fukumoto S, Inoue D, Yamauchi M, Minagawa M, et al. Assessment criteria for vitamin D deficiency/insufficiency in Japan – proposal by an expert panel supported by Research Program of Intractable Diseases, Ministry of Health, Labour and Welfare, Japan, the Japanese Society for Bone and Mineral Research and the Japan Endocrine Society [Opinion]. Endocr J 2017;64:1–6. https://doi.org/10.1507/endocrj.EJ16-0548.

[114] Holick MF, Binkley NC, Bischoff-Ferrari HA, Gordon CM, Hanley DA, Heaney RP, et al. Evaluation, treatment, and prevention of vitamin D deficiency: an Endocrine Society clinical practice guideline. J Clin Endocrinol Metab 2011;96:1911–30. https://doi.org/10.1210/jc.2011-0385.

[115] Grant WB. An estimate of premature cancer mortality in the US due to inadequate doses of solar ultraviolet-B radiation. Cancer 2002;94:1867–75. https://doi.org/10.1002/cncr.10427.

[116] Neale RE, Youlden DR, Krnjacki L, Kimlin MG, van der Pols JC. Latitude variation in pancreatic cancer mortality in Australia. Pancreas 2009;38:387–90. https://doi.org/10.1097/MPA.0b013e31819975f4.

[117] Mohr SB, Garland CF, Gorham ED, Grant WB, Garland FC. Ultraviolet B irradiance and vitamin D status are inversely associated with incidence rates of pancreatic cancer worldwide. Pancreas 2010;39:669–74. https://doi.org/10.1097/MPA.0b013e3181ce654d.

[118] Tran B, Whiteman DC, Webb PM, Fritschi L, Fawcett J, Risch HA, et al. Association between ultraviolet radiation, skin sun sensitivity and risk of pancreatic cancer. Cancer Epidemiol 2013;37:886–92. https://doi.org/10.1016/j.canep.2013.08.013.

[119] Garland CF, Cuomo RE, Gorham ED, Zeng K, Mohr SB. Cloud cover-adjusted ultraviolet B irradiance and pancreatic cancer incidence in 172 countries. J Steroid Biochem Mol Biol 2016;155:257–63. https://doi.org/10.1016/j.jsbmb.2015.04.004.

[120] Genkinger JM, Wang M, Li R, Albanes D, Anderson KE, Bernstein L, et al. Dairy products and pancreatic cancer risk: a pooled analysis of 14 cohort studies. Ann Oncol 2014;25:1106–15. https://doi.org/10.1093/annonc/mdu019.

[121] Waterhouse M, Risch HA, Bosetti C, Anderson KE, Petersen GM, Bamlet WR, et al. Vitamin D and pancreatic cancer: a pooled analysis from the pancreatic cancer case-control consortium. Ann Oncol 2015;26:1776–83. https://doi.org/10.1093/annonc/mdv236.

[122] Liu Y, Wang XJ, Sun XJ, Lu SN, Liu S. Vitamin intake and pancreatic cancer risk reduction: a meta-analysis of observational studies. Medicine 2018;97. https://doi.org/10.1097/md.0000000000010114.

[123] Stolzenberg-Solomon RZ, Jacobs EJ, Arslan AA, Qi D, Patel AV, Helzlsouer KJ, et al. Circulating 25-hydroxyvitamin D and risk of pancreatic cancer: cohort consortium vitamin D pooling project of rarer cancers. Am J Epidemiol 2010;172:81–93. https://doi.org/10.1093/aje/kwq120.

[124] Wolpin BM, Ng K, Bao Y, Kraft P, Stampfer MJ, Michaud DS, et al. Plasma 25-hydroxyvitamin D and risk of pancreatic cancer. Cancer Epidemiol Biomarkers Prev 2012;21:82–91. https://doi.org/10.1158/1055-9965.EPI-11-0836.

[125] Mohamed AA, Aref AM, Talima SM, Elshimy RAA, Gerges SS, Meghed M, et al. Association of serum level of vitamin D and VDR polymorphism Fok1 with the risk or survival of pancreatic cancer in Egyptian population. Indian J Cancer 2019;56:130–4. https://doi.org/10.4103/ijc.IJC_299_18.

[126] van Duijnhoven FJB, Jenab M, Hveem K, Siersema PD, Fedirko V, Duell EJ, et al. Circulating concentrations of vitamin D in relation to pancreatic cancer risk in European populations. Int J Cancer 2018;142:1189–201. https://doi.org/10.1002/ijc.31146.

[127] Wang KW, Dong M, Sheng WW, Liu QF, Yu DY, Dong Q, et al. Expression of vitamin D receptor as a potential prognostic factor and therapeutic target in pancreatic cancer. Histopathology 2015;67:386–97. https://doi.org/10.1111/his.12663.

[128] Cho M, Peddi PF, Ding K, Chen L, Thomas D, Wang J, et al. Vitamin D deficiency and prognostics among patients with pancreatic adenocarcinoma. J Transl Med 2013;11:206. https://doi.org/10.1186/1479-5876-11-206.

[129] Mukai Y, Yamada D, Eguchi H, Iwagami Y, Asaoka T, Noda T, et al. Vitamin D supplementation is a promising therapy for pancreatic ductal adenocarcinoma in conjunction with current chemoradiation therapy. Ann Surg Oncol 2018;25:1868–79. https://doi.org/10.1245/s10434-018-6431-8.

[130] Van Loon K, Owzar K, Jiang C, Kindler HL, Mulcahy MF, Niedzwiecki D, et al. 25-Hydroxyvitamin D levels and survival in advanced pancreatic cancer: findings from CALGB 80303 (Alliance). J Natl Cancer Inst (Bethesda) 2014;106. https://doi.org/10.1093/jnci/dju185.

[131] Yuan C, Wolpin BM. Pancreatic cancer survival: plasma levels of 25-hydroxyvitamin D and smoking reply. J Clin Oncol 2017;35. https://doi.org/10.1200/jco.2016.70.9261.

[132] Zhang X, Huang XZ, Chen WJ, Wu J, Chen Y, Wu CC, et al. Plasma 25-hydroxyvitamin D levels, vitamin D intake, and pancreatic cancer risk or mortality: a meta-analysis. Oncotarget 2017;8:64395–406. https://doi.org/10.18632/oncotarget.18888.

[133] Pettersson F, Colston KW, Dalgleish AG. Differential and antagonistic effects of 9-cis-retinoic acid and vitamin D analogues on pancreatic cancer cells in vitro. Br J Cancer 2000;83:239–45. https://doi.org/10.1054/bjoc.2000.1281.

[134] Schwartz GG, Eads D, Rao A, Cramer SD, Willingham MC, Chen TC, et al. Pancreatic cancer cells express 25-hydroxyvitamin D-1 alpha-hydroxylase and their proliferation is inhibited by the prohormone 25-hydroxyvitamin D3. Carcinogenesis 2004;25:1015–26. https://doi.org/10.1093/carcin/bgh086.

[135] Zugmaier G, Jager R, Grage B, Gottardis MM, Havemann K, Knabbe C. Growth-inhibitory effects of vitamin D analogues and retinoids on human pancreatic cancer cells. Br J Cancer 1996;73:1341–6. https://doi.org/10.1038/bjc.1996.256.

[136] Kawa S, Yoshizawa K, Tokoo M, Imai H, Oguchi H, Kiyosawa K, et al. Inhibitory effect of 220-oxa-1,25-dihydroxyvitamin D3 on the proliferation of pancreatic cancer cell lines. Gastroenterology 1996;110: 1605–13. https://doi.org/10.1053/gast.1996.v110.pm8613068.

[137] Naturecom [n.d.]. http://www.nature.com/ncb/journal/v13/n7/pdf/ncb2257.pdf [Accessed 6 August 2011].

[138] Ohlsson B, Albrechtsson E, Axelson J. Vitamins A and D but not E and K decreased the cell number in human pancreatic cancer cell lines. Scand J Gastroenterol 2004;39:882–5. https://doi.org/10.1080/00365520410006701.

[139] Persons KS, Eddy VJ, Chadid S, Deoliveira R, Saha AK, Ray R. Anti-growth effect of 1,25-dihydroxyvitamin D3-3-bromoacetate alone or in combination with 5-amino-imidazole-4-carboxamide-1-beta-4-ribofuranoside in pancreatic cancer cells. Anticancer Res 2010;30:1875–80.

[140] Li ZW, Guo JL, Xie KP, Zheng SJ. Vitamin D receptor signaling and pancreatic cancer cell EMT. Curr Pharmaceut Des 2015;21:1262–7. https://doi.org/10.2174/1381612821666141211151138.

[141] Kawa S, Nikaido T, Aoki Y, Zhai Y, Kumagai T, Furihata K, et al. Vitamin D analogues up-regulate p21 and p27 during growth inhibition of pancreatic cancer cell lines. Br J Cancer 1997;76:884–9. https://doi.org/10.1038/bjc.1997.479.

[142] Kawa S, Yoshizawa K, Nikaido T, Kiyosawa K. Inhibitory effect of 22-oxa-1,25-dihydroxyvitamin D3, maxacalcitol, on the proliferation of pancreatic cancer cell lines. J Steroid Biochem Mol Biol 2005; 97:173–7. https://doi.org/10.1016/j.jsbmb.2005.06.021.

[143] Mouratidis PX, Dalgleish AG, Colston KW. Investigation of the mechanisms by which EB1089 abrogates apoptosis induced by 9-cis retinoic acid in pancreatic cancer cells. Pancreas 2006;32:93–100. https://doi.org/10.1097/01.mpa.0000191648.47667.4f.

[144] Schwartz GG, Eads D, Naczki C, Northrup S, Chen T, Koumenis C. 19-nor-1 alpha,25-dihydroxyvitamin D2 (paricalcitol) inhibits the proliferation of human pancreatic cancer cells in vitro and in vivo. Cancer Biol Ther 2008;7:430–6. https://doi.org/10.4161/cbt.7.3.5418.

[145] Kanemaru M, Maehara N, Chijiiwa K. Antiproliferative effect of 1alpha,25-dihydroxyvitamin D3 involves upregulation of cyclin-dependent kinase inhibitor p21 in human pancreatic cancer cells. Hepato-Gastroenterology 2013;60:1199–205. https://doi.org/10.5754/hge11073.

[146] Chiang KC, Yeh CN, Hsu JT, Yeh TS, Jan YY, Wu CT, et al. Evaluation of the potential therapeutic role of a new generation of vitamin D analog, MART-10, in human pancreatic cancer cells in vitro and in vivo. Cell Cycle 2013;12:1316–25. https://doi.org/10.4161/cc.24445.

[147] Bruggemann LW, Queiroz KC, Zamani K, van Straaten A, Spek CA, Bijlsma MF. Assessing the

efficacy of the hedgehog pathway inhibitor vitamin D3 in a murine xenograft model for pancreatic cancer. Cancer Biol Ther 2010;10:79–88. https://doi.org/10.4161/cbt.10.1.12165.

[148] Albrechtsson E, Jonsson T, Moller S, Hoglund M, Ohlsson B, Axelson J. Vitamin D receptor is expressed in pancreatic cancer cells and a vitamin D3 analogue decreases cell number. Pancreatology 2003;3:41–6. https://doi.org/10.1159/000069149.

[149] Yu WD, Ma Y, Flynn G, Muindi JR, Kong RX, Trump DL, et al. Calcitriol enhances gemcitabine anti-tumor activity in vitro and in vivo by promoting apoptosis in a human pancreatic carcinoma model system. Cell Cycle 2010;9:3022–9. https://doi.org/10.4161/cc.9.15.12381.

[150] Sherman MH, Yu RT, Engle DD, Ding N, Atkins AR, Tiriac H, et al. Vitamin D receptor-mediated stromal reprogramming suppresses pancreatitis and enhances pancreatic cancer therapy. Cell 2014;159:80–93. https://doi.org/10.1016/j.cell.2014.08.007.

[151] Evans TR, Colston KW, Lofts FJ, Cunningham D, Anthoney DA, Gogas H, et al. A phase II trial of the vitamin D analogue Seocalcitol (EB1089) in patients with inoperable pancreatic cancer. Br J Cancer 2002; 86:680–5. https://doi.org/10.1038/sj.bjc.6600162.

[152] Barreto SG, Ramadwar MR, Shukla PJ, Shrikhande SV. Vitamin D3 in operable periampullary and pancreatic cancer: perioperative outcomes in a pilot study assessing safety. Pancreas 2008;36: 315–7. https://doi.org/10.1097/MPA.0b013e31815ac573.

[153] Blanke CD, Beer TM, Todd K, Mori M, Stone M, Lopez C. Phase II study of calcitriol-enhanced docetaxel in patients with previously untreated metastatic or locally advanced pancreatic cancer. Invest New Drugs 2009;27:374–8. https://doi.org/10.1007/s10637-008-9184-6.

[154] Borazanci EH, Jameson G, Korn RL, Caldwell L, Ansaldo K, Hendrickson K, et al. A Phase II pilot trial of nivolumab (N) plus albumin bound paclitaxel (AP) plus paricalcitol (P) plus cisplatin (C) plus gemcitabine (G) (NAPPCG) in patients with previously untreated metastatic pancreatic ductal adenocarcinoma (PDAC). Cancer Res 2019;79. https://doi.org/10.1158/1538-7445.Am2019-ct152.

[155] Colston KW, James SY, Ofori-Kuragu EA, Binderup L, Grant AG. Vitamin D receptors and anti-proliferative effects of vitamin D derivatives in human pancreatic carcinoma cells in vivo and in vitro. Br J Cancer 1997; 76:1017–20. https://doi.org/10.1038/bjc.1997.501.

[156] Jandaghi P, Najafabadi HS, Bauer AS, Papadakis AI, Fassan M, Hall A, et al. Expression of DRD2 is increased in human pancreatic ductal adenocarcinoma and inhibitors slow tumor growth in mice. Gastroenterology 2016;151:1218–31. https://doi.org/10.1053/j.gastro.2016.08.040.

[157] Fiorillo M, Lamb R, Tanowitz HB, Mutti L, Krstic-Demonacos M, Cappello AR, et al. Repurposing atovaquone: targeting mitochondrial complex III and OXPHOS to eradicate cancer stem cells. Oncotarget 2016;5.

[158] Ashton TM, Fokas E, Kunz-Schughart LA, Folkes LK, Anbalagan S, Huether M, et al. The anti-malarial atovaquone increases radiosensitivity by alleviating tumour hypoxia. Nat Comms 2016;7:12308. https://doi.org/10.1038/ncomms12308.

[159] Yang S, Kimmelman AC. A critical role for autophagy in pancreatic cancer. Autophagy 2011;7:912–3.

[160] Tremblay MR, Lescarbeau A, Grogan MJ, Tan E, Lin G, Austad BC, et al. Discovery of a potent and orally active hedgehog pathway antagonist (IPI-926). J Med Chem 2009;52:4400–18. https://doi.org/10.1021/jm900305z.

[161] Robarge KD, Brunton SA, Castanedo GM, Cui Y, Dina MS, Goldsmith R, et al. GDC-0449-a potent inhibitor of the hedgehog pathway. Bioorg Med Chem Lett 2009;19:5576–81. https://doi.org/10.1016/j.bmcl.2009.08.049.

[162] Sekulic A, Migden MR, Oro AE, Dirix L, Lewis KD, Hainsworth JD, et al. Efficacy and safety of vismodegib in advanced basal-cell carcinoma. N Engl J Med 2012;366:2171–9. https://doi.org/10.1056/NEJMoa1113713.

[163] Fu J, Rodova M, Nanta R, Meeker D, Van Veldhuizen PJ, Srivastava RK, et al. NPV-LDE-225 (Erismodegib) inhibits epithelial mesenchymal transition and self-renewal of glioblastoma initiating cells by regulating miR-21, miR-128, and miR-200. Neuro Oncol 2013;15:691–706. https://doi.org/10.1093/neuonc/not011.

[164] Nanta R, Kumar D, Meeker D, Rodova M, Van Veldhuizen PJ, Shankar S, et al. NVP-LDE-225 (Erismodegib) inhibits epithelial-mesenchymal transition and human prostate cancer stem cell growth in NOD/SCID IL2Rγ null mice by regulating Bmi-1 and microRNA-128. Oncogenesis 2013;2. https://doi.org/10.1038/oncsis.2013.5. e42–e42.

[165] D'Amato C, Rosa R, Marciano R, D'Amato V, Formisano L, Nappi L, et al. Inhibition of Hedgehog signalling by NVP-LDE225 (Erismodegib) interferes with growth and invasion of human renal cell carcinoma cells. Br J Cancer 2014;111:1168–79. https://doi.org/10.1038/bjc.2014.421.

[166] Garrido-Laguna I, Hidalgo M. Pancreatic cancer: from state-of-the-art treatments to promising novel therapies. Nat Rev Clin Oncol 2015;12:319–34. https://doi.org/10.1038/nrclinonc.2015.53.
[167] Arnold SA, Rivera LB, Carbon JG, Toombs JE, Chang C-L, Bradshaw AD, et al. Losartan slows pancreatic tumor progression and extends survival of SPARC-null mice by abrogating aberrant TGFβ activation. PloS One 2012;7. https://doi.org/10.1371/journal.pone.0031384. e31384.
[168] Škrtić M, Sriskanthadevan S, Jhas B, Gebbia M, Wang X, Wang Z, et al. Inhibition of mitochondrial translation as a therapeutic strategy for human acute myeloid leukemia. Cancer Cell 2011;20:674–88. https://doi.org/10.1016/j.ccr.2011.10.015.
[169] Lamb R, Ozsvari B, Lisanti CL, Tanowitz HB, Howell A, Martinez-Outschoorn UE, et al. Antibiotics that target mitochondria effectively eradicate cancer stem cells, across multiple tumor types: treating cancer like an infectious disease. Oncotarget 2015;6: 4569–84. https://doi.org/10.18632/oncotarget.3174.
[170] Piñeiro-Yañez E, Reboiro-Jato M, Gómez-López G, Perales-Patón J, Troulé K, Rodríguez JM, et al. PanDrugs: a novel method to prioritize anticancer drug treatments according to individual genomic data. Genome Med 2018;10:41. https://doi.org/10.1186/s13073-018-0546-1.
[171] Datta A, Kim H, McGee L, Johnson AE, Talwar S, Marugan J, et al. High-throughput screening identified selective inhibitors of exosome biogenesis and secretion: a drug repurposing strategy for advanced cancer. Sci Rep 2018;8:8161. https://doi.org/10.1038/s41598-018-26411-7.
[172] Risch HA, Lu L, Streicher SA, Wang J, Zhang W, Ni Q, et al. Aspirin use and reduced risk of pancreatic cancer. Cancer Epidemiol Biomarkers Prev 2016;26: 68–74. https://doi.org/10.1158/1055-9965.EPI-16-0508.
[173] Patrono C, Garcia Rodriguez LA, Landolfi R, Baigent C. Low-dose aspirin for the prevention of atherothrombosis. N Engl J Med 2005;353:2373–83. https://doi.org/10.1056/NEJMra052717.
[174] Langley RE, Burdett S, Tierney JF, Cafferty F, Parmar MK, Venning G. Aspirin and cancer: has aspirin been overlooked as an adjuvant therapy? Br J Cancer 2011;105:1107–13. https://doi.org/10.1038/bjc.2011.289.
[175] Easty GC, Easty DM. Prostaglandins and cancer. Cancer Treat Rev. 1976;3:217–25. https://doi.org/10.1016/s0305-7372(76)80011-4.
[176] Levine L. Arachidonic acid transformation and tumor production. Adv Cancer Res 1981;35:49–79. https://doi.org/10.1016/s0065-230x(08)60908-2.
[177] Molina MA, Sitja-Arnau M, Lemoine MG, Frazier ML, Sinicrope FA. Increased cyclooxygenase-2 expression in human pancreatic carcinomas and cell lines: growth inhibition by nonsteroidal anti-inflammatory drugs. Cancer Res 1999;59:4356–62.
[178] Chulada PC, Thompson MB, Mahler JF, Doyle CM, Gaul BW, Lee C, et al. Genetic disruption of Ptgs-1, as well as Ptgs-2, reduces intestinal tumorigenesis in Min mice. Cancer Res 2000;60:4705–8.
[179] Headley MB, Bins A, Nip A, Roberts EW, Looney MR, Gerard A, et al. Visualization of immediate immune responses to pioneer metastatic cells in the lung. Nature 2016;531:513–7. https://doi.org/10.1038/nature16985.
[180] Gay LJ, Felding-Habermann B. Contribution of platelets to tumour metastasis. Nat Rev Cancer 2011;11: 123–34. https://doi.org/10.1038/nrc3004.
[181] Gasic GJ, Gasic TB, Stewart CC. Antimetastatic effects associated with platelet reduction. Proc Natl Acad Sci USA 1968;61:46–52. https://doi.org/10.1073/pnas.61.1.46.
[182] Stark LA, Reid K, Sansom OJ, Din FV, Guichard S, Mayer I, et al. Aspirin activates the NF-kappaB signalling pathway and induces apoptosis in intestinal neoplasia in two in vivo models of human colorectal cancer. Carcinogenesis 2007;28:968–76. https://doi.org/10.1093/carcin/bgl220.
[183] Borthwick GM, Johnson AS, Partington M, Burn J, Wilson R, Arthur HM. Therapeutic levels of aspirin and salicylate directly inhibit a model of angiogenesis through a Cox-independent mechanism. Faseb J 2006; 20. https://doi.org/10.1096/fj.06-5987com. 2009–2016.
[184] Baron JA, Cole BF, Sandler RS, Haile RW, Ahnen D, Bresalier R, et al. A randomized trial of aspirin to prevent colorectal adenomas. N Engl J Med 2003;348: 891–9. https://doi.org/10.1056/NEJMoa021735.
[185] Rothwell PM, Fowkes FG, Belch JF, Ogawa H, Warlow CP, Meade TW. Effect of daily aspirin on long-term risk of death due to cancer: analysis of individual patient data from randomised trials. Lancet 2011;377:31–41. https://doi.org/10.1016/S0140-6736(10)62110-1.
[186] Burn J, Gerdes AM, Macrae F, Mecklin JP, Moeslein G, Olschwang S, et al. Long-term effect of aspirin on cancer risk in carriers of hereditary colorectal cancer: an analysis from the CAPP2 randomised controlled trial. Lancet 2011;378:2081–7. https://doi.org/10.1016/S0140-6736(11)61049-0.

[187] Chan AT, Ogino S, Fuchs CS. Aspirin use and survival after diagnosis of colorectal cancer. Jama 2009;302: 649–58. https://doi.org/10.1001/jama.2009.1112.

[188] Bosetti C, Gallus S, La Vecchia C. Aspirin and cancer risk: an updated quantitative review to 2005. Cancer Causes Control 2006;17:871–88. https://doi.org/10.1007/s10552-006-0033-7.

[189] Rothwell PM, Wilson M, Price JF, Belch JF, Meade TW, Mehta Z. Effect of daily aspirin on risk of cancer metastasis: a study of incident cancers during randomised controlled trials. Lancet 2012;379: 1591–601. https://doi.org/10.1016/S0140-6736(12)60209-8.

[190] Lebeau B, Chastang C, Muir JF, Vincent J, Massin F, Fabre C. No effect of an antiaggregant treatment with aspirin in small cell lung cancer treated with CCAVP16 chemotherapy. Results from a randomized clinical trial of 303 patients. The "Petites Cellules" Group Cancer 1993;71:1741–5. https://doi.org/10.1002/1097-0142(19930301)71:5<1741::aid-cncr2820710507>3.0.co;2-q.

[191] Creagan ET, Twito DI, Johansson SL, Schaid DJ, Johnson PS, Flaum MA, et al. A randomized prospective assessment of recombinant leukocyte A human interferon with or without aspirin in advanced renal adenocarcinoma. J Clin Oncol 1991;9:2104–9. https://doi.org/10.1200/JCO.1991.9.12.2104.

[192] Khalaf N, Yuan C, Hamada T, Cao Y, Babic A, Morales-Oyarvide V, et al. Regular use of aspirin or non-aspirin nonsteroidal anti-inflammatory drugs is not associated with risk of incident pancreatic cancer in two large cohort studies. Gastroenterology 2018; 154:1380–90. https://doi.org/10.1053/j.gastro.2017.12.001. e5.

[193] Zhang Y, Liu L, Fan P, Bauer N, Gladkich J, Ryschich E, et al. Aspirin counteracts cancer stem cell features, desmoplasia and gemcitabine resistance in pancreatic cancer. Oncotarget 2015;6:9999–10015. https://doi.org/10.18632/oncotarget.3171.

[194] Plassmeier L, Knoop R, Waldmann J, Kesselring R, Buchholz M, Fichtner-Feigl S, et al. Aspirin prolongs survival and reduces the number of Foxp3+ regulatory T cells in a genetically engineered mouse model of pancreatic cancer. Langenbeck's Arch Surg 2013;398: 989–96. https://doi.org/10.1007/s00423-013-1105-2.

[195] Weddle DL, Tithoff P, Williams M, Schuller HM. Beta-adrenergic growth regulation of human cancer cell lines derived from pancreatic ductal carcinomas. Carcinogenesis 2001;22:473–9. https://doi.org/10.1093/carcin/22.3.473.

[196] Guo K, Ma Q, Li J, Wang Z, Shan T, Li W, et al. Interaction of the sympathetic nerve with pancreatic cancer cells promotes perineural invasion through the activation of STAT3 signaling. Mol Cancer Ther 2013;12:264–73. https://doi.org/10.1158/1535-7163.MCT-12-0809.

[197] Zhang D, Ma Q-Y, Hu H-T, Zhang M. β2-adrenergic antagonists suppress pancreatic cancer cell invasion by inhibiting CREB, NFκB and AP-1. Cancer Biol Ther 2010;10:19–29. https://doi.org/10.4161/cbt.10.1.11944.

[198] Kim-Fuchs C, Le CP, Pimentel MA, Shackleford D, Ferrari D, Angst E, et al. Chronic stress accelerates pancreatic cancer growth and invasion: a critical role for beta-adrenergic signaling in the pancreatic microenvironment, Brain Behav. Immun 2014;40:40–7. https://doi.org/10.1016/j.bbi.2014.02.019.

[199] Jansen L, Hoffmeister M, Arndt V, Chang-Claude J, Brenner H. Stage-specific associations between beta blocker use and prognosis after colorectal cancer. Cancer 2014;120:1178–86. https://doi.org/10.1002/cncr.28546.

[200] Clark KL, Loscalzo M, Trask PC, Zabora J, Philip EJ. Psychological distress in patients with pancreatic cancer—an understudied group. Psycho Oncol 2010; 19:1313–20. https://doi.org/10.1002/pon.1697.

[201] Batty GD, Russ TC, Stamatakis E, Kivimäki M. Psychological distress in relation to site specific cancer mortality: pooling of unpublished data from 16 prospective cohort studies. BMJ 2017;356:j108. https://doi.org/10.1136/bmj.j108.

[202] Barron TI, Connolly RM, Sharp L, Bennett K, Visvanathan K. Beta blockers and breast cancer mortality: a population- based study. J Clin Oncol 2011;29: 2635–44. https://doi.org/10.1200/JCO.2010.33.5422.

[203] Watkins JL, Thaker PH, Nick AM, Ramondetta LM, Kumar S, Urbauer DL, et al. Clinical impact of selective and nonselective beta-blockers on survival in patients with ovarian cancer. Cancer 2015;121:3444–51. https://doi.org/10.1002/cncr.29392.

[204] Lemeshow S, Sørensen HT, Phillips G, Yang EV, Antonsen S, Riis AH, et al. β-Blockers and survival among Danish patients with malignant melanoma: a population-based cohort study. Cancer Epidemiol Biomarkers Prev 2011;20:2273–9. https://doi.org/10.1158/1055-9965.EPI-11-0249.

[205] Beg MS, Gupta A, Sher D, Ali S, Khan S, Gao A, et al. Impact of concurrent medication use on pancreatic cancer survival-SEER-medicare analysis. Am J Clin Oncol 2018;41:766–71. https://doi.org/10.1097/COC.0000000000000359.

[206] Udumyan R, Montgomery S, Fang F, Almroth H, Valdimarsdottir U, Ekbom A, et al. Beta-blocker drug use and survival among patients with pancreatic adenocarcinoma. Cancer Res 2017;77:3700–7. https://doi.org/10.1158/0008-5472.CAN-17-0108.

[207] Thaker PH, Han LY, Kamat AA, Arevalo JM, Takahashi R, Lu C, et al. Chronic stress promotes tumor growth and angiogenesis in a mouse model of ovarian carcinoma. Nat Med 2006;12:939–44. https://doi.org/10.1038/nm1447.

[208] Magnon C, Hall SJ, Lin J, Xue X, Gerber L, Freedland SJ, et al. Autonomic nerve development contributes to prostate cancer progression. Science 2013;341. https://doi.org/10.1126/science.1236361. 1236361–1236361.

[209] Partecke LI, Speerforck S, Käding A, Seubert F, Kühn S, Lorenz E, et al. Chronic stress increases experimental pancreatic cancer growth, reduces survival and can be antagonised by beta-adrenergic receptor blockade. Pancreatology 2016;16:423–33. https://doi.org/10.1016/j.pan.2016.03.005.

[210] Madden KS, Szpunar MJ, Brown EB. β-Adrenergic receptors (β-AR) regulate VEGF and IL-6 production by divergent pathways in high β-AR-expressing breast cancer cell lines. Breast Cancer Res Treat 2011;130: 747–58. https://doi.org/10.1007/s10549-011-1348-y.

[211] Pérez Piñero C, Bruzzone A, Sarappa MG, Castillo LF, Lüthy IA. Involvement of α2- and β2-adrenoceptors on breast cancer cell proliferation and tumour growth regulation. Br J Pharmacol 2012;166:721–36. https://doi.org/10.1111/j.1476-5381.2011.01791.x.

[212] Renz BW, Takahashi R, Tanaka T, Macchini M, Hayakawa Y, Dantes Z, et al. β2 Adrenergic-neurotrophin feedforward loop promotes pancreatic cancer. Cancer Cell 2017. https://doi.org/10.1016/j.ccell.2017.11.007.

[213] Sloan EK, Priceman SJ, Cox BF, Yu S, Pimentel MA, Tangkanangnukul V, et al. The sympathetic nervous system induces a metastatic switch in primary breast cancer. Cancer Res 2010;70:7042–52. https://doi.org/10.1158/0008-5472.CAN-10-0522.

[214] Krall JA, Reinhardt F, Mercury OA, Pattabiraman DR, Brooks MW, Dougan M, et al. The systemic response to surgery triggers the outgrowth of distant immune-controlled tumors in mouse models of dormancy. Sci Transl Med 2018;10:eaan3464. https://doi.org/10.1126/scitranslmed.aan3464.

[215] Monje M. Settling a nervous stomach: the neural regulation of enteric cancer. Cancer Cell 2017;31:1–2. https://doi.org/10.1016/j.ccell.2016.12.008.

[216] Kiberstis PA. Cancer and nerves: a tuf(t) partnership. Science 2017;355:144–5. https://doi.org/10.1126/science.355.6321.144-d.

[217] Lindsay TH, Jonas BM, Sevcik MA, Kubota K, Halvorson KG, Ghilardi JR, et al. Pancreatic cancer pain and its correlation with changes in tumor vasculature, macrophage infiltration, neuronal innervation, body weight and disease progression. Pain 2005;119: 233–46. https://doi.org/10.1016/j.pain.2005.10.019.

[218] Ayala GE, Dai H, Powell M, Li R, Ding Y, Wheeler TM, et al. Cancer-related axonogenesis and neurogenesis in prostate cancer. Clin Cancer Res 2008;14:7593–603. https://doi.org/10.1158/1078-0432.CCR-08-1164.

[219] Albo D, Akay CL, Marshall CL, Wilks JA, Verstovsek G, Liu H, et al. Neurogenesis in colorectal cancer is a marker of aggressive tumor behavior and poor outcomes. Cancer 2011;117:4834–45. https://doi.org/10.1002/cncr.26117.

[220] Zhao C-M, Hayakawa Y, Kodama Y, Muthupalani S, Westphalen CB, Andersen GT, et al. Denervation suppresses gastric tumorigenesis. Sci Transl Med 2014;6. https://doi.org/10.1126/scitranslmed.3009569. 250ra115.

[221] Mattingly RR, Sorisky A, Brann MR, Macara IG. Muscarinic receptors transform NIH 3T3 cells through a Ras-dependent signalling pathway inhibited by the Ras-GTPase-activating protein SH3 domain. Mol Cell Biol 1994;14:7943–52. https://doi.org/10.1128/mcb.14.12.7943.

[222] Partecke LI, Käding A, Trung DN, Diedrich S, Sendler M, Weiss F, et al. Subdiaphragmatic vagotomy promotes tumor growth and reduces survival via TNFα in a murine pancreatic cancer model. Oncotarget 2017;8:22501–12. https://doi.org/10.18632/oncotarget.15019.

[223] Renz BW, Tanaka T, Sunagawa M, Takahashi R, Jiang Z, Macchini M, et al. Cholinergic signaling via muscarinic receptors directly and indirectly suppresses pancreatic tumorigenesis and cancer stemness. Cancer Discov 2018;8:1458–73. https://doi.org/10.1158/2159-8290.CD-18-0046.

[224] Chu DJ, Yao DE, Zhuang YF, Hong Y, Zhu XC, Fang ZR, et al. Azithromycin enhances the favorable results of paclitaxel and cisplatin in patients with advanced non-small cell lung cancer. Genet Mol Res 2014;13:2796–805. https://doi.org/10.4238/2014.April.14.8.

[225] Rangarajan P, Subramaniam D, Paul S, Kwatra D, Palaniyandi K, Islam S, et al. Crocetinic acid inhibits hedgehog signaling to inhibit pancreatic cancer stem cells. Oncotarget 2015;6:27661–73. https://doi.org/10.18632/oncotarget.4871.

[226] Lockhart NR, Waddell JA, Schrock NE. Itraconazole therapy in a pancreatic adenocarcinoma patient: a case report. - PubMed - NCBI. J Oncol Pharm Pract 2015;22:528–32. https://doi.org/10.1177/1078155215572931.

[227] Zhou H-M, Dong T-T, Wang L-L, Feng B, Zhao H-C, Fan X-K, et al. Suppression of colorectal cancer metastasis by nigericin through inhibition of epithelial-mesenchymal transition. World J Gastroenterol 2012; 18:2640–8. https://doi.org/10.3748/wjg.v18.i21.2640.

[228] Kozono S, Ohuchida K, Eguchi D, Ikenaga N, Fujiwara K, Cui L, et al. Pirfenidone inhibits pancreatic cancer desmoplasia by regulating stellate cells. Cancer Res 2013;73:2345–56. https://doi.org/10.1158/0008-5472.CAN-12-3180.

[229] Huang S, Ren X, Wang L, Zhang L, Wu X. Lung-cancer chemoprevention by induction of synthetic lethality in mutant KRAS premalignant cells in vitro and in vivo. Cancer Prev Res (Phila) 2011;4: 666–73. https://doi.org/10.1158/1940-6207.CAPR-10-0235.

[230] Najumudeen AK, Jaiswal A, Lectez B, Oetken-Lindholm C, Guzmán C, Siljamäki E, et al. Cancer stem cell drugs target K-ras signaling in a stemness context. Oncogene 2016. https://doi.org/10.1038/onc.2016.59.

[231] Turnbaugh PJ, Ley RE, Mahowald MA, Magrini V, Mardis ER, Gordon JI. An obesity-associated gut microbiome with increased capacity for energy harvest. Nature 2006;444:1027–31. https://doi.org/10.1038/nature05414.

[232] Bercik P, Denou E, Collins J, Jackson W, Lu J, Jury J, et al. The intestinal microbiota affect central levels of brain-derived neurotropic factor and behavior in mice. Gastroenterology 2011;141:599–609. https://doi.org/10.1053/j.gastro.2011.04.052. 609 e1–3.

[233] Yano JM, Yu K, Donaldson GP, Shastri GG, Ann P, Ma L, et al. Indigenous bacteria from the gut microbiota regulate host serotonin biosynthesis. Cell 2015; 161:264–76. https://doi.org/10.1016/j.cell.2015.02.047.

[234] Czesnikiewicz-Guzik M, Muller DN. Scientists on the Spot: salt, the microbiome, and cardiovascular diseases. Cardiovasc Res 2018;114:e72–3. https://doi.org/10.1093/cvr/cvy171.

[235] Costello SP, Hughes PA, Waters O, Bryant RV, Vincent AD, Blatchford P, et al. Effect of fecal microbiota transplantation on 8-week remission in patients with ulcerative colitis: a randomized clinical trial. Jama 2019;321:156–64. https://doi.org/10.1001/jama.2018.20046.

[236] Pevsner-Fischer M, Tuganbaev T, Meijer M, Zhang SH, Zeng ZR, Chen MH, et al. Role of the microbiome in non-gastrointestinal cancers. World J Clin Oncol 2016;7:200–13. https://doi.org/10.5306/wjco.v7.i2.200.

[237] Choi IJ, Kook MC, Kim YI, Cho SJ, Lee JY, Kim CG, et al. *Helicobacter pylori* therapy for the prevention of metachronous gastric cancer. N Engl J Med 2018; 378:1085–95. https://doi.org/10.1056/NEJMoa1708423.

[238] Rubinstein MR, Wang X, Liu W, Hao Y, Cai G, Han YW. Fusobacterium nucleatum promotes colorectal carcinogenesis by modulating E-cadherin/beta-catenin signaling via its FadA adhesin. Cell Host Microbe 2013;14:195–206. https://doi.org/10.1016/j.chom.2013.07.012.

[239] Liebowitz D. Epstein-Barr virus and a cellular signaling pathway in lymphomas from immunosuppressed patients. N Engl J Med 1998;338:1413–21. https://doi.org/10.1056/NEJM199805143382003.

[240] Schiller JT, Castellsague X, Garland SM. A review of clinical trials of human papillomavirus prophylactic vaccines. Vaccine 2012;30(Suppl. 5):F123–38. https://doi.org/10.1016/j.vaccine.2012.04.108.

[241] Pushalkar S, Hundeyin M, Daley D, Zambirinis CP, Kurz E, Mishra A, et al. The pancreatic cancer microbiome promotes oncogenesis by induction of innate and adaptive immune suppression. Cancer Discov 2018;8:403–16. https://doi.org/10.1158/2159-8290.CD-17-1134.

[242] Aykut B, Pushalkar S, Chen R, Li Q, Abengozar R, Kim JI, et al. The fungal mycobiome promotes pancreatic oncogenesis via activation of MBL. Nature 2019; 574:264–7. https://doi.org/10.1038/s41586-019-1608-2.

[243] Sethi V, Kurtom S, Tarique M, Lavania S, Malchiodi Z, Hellmund L, et al. Gut microbiota promotes tumor growth in mice by modulating immune response. Gastroenterology 2018;155:33–7. https://doi.org/10.1053/j.gastro.2018.04.001. e6.

[244] De Monte L, Reni M, Tassi E, Clavenna D, Papa I, Recalde H, et al. Intratumor T helper type 2 cell infiltrate correlates with cancer-associated fibroblast thymic stromal lymphopoietin production and reduced survival in pancreatic cancer. J Exp Med 2011;208:469–78. https://doi.org/10.1084/jem.20101876.

[245] McAllister F, Bailey JM, Alsina J, Nirschl CJ, Sharma R, Fan H, et al. Oncogenic Kras activates a hematopoietic-to-epithelial IL-17 signaling Axis in preinvasive pancreatic neoplasia. Cancer Cell 2014; 25:621–37. https://doi.org/10.1016/j.ccr.2014.03.014.

[246] Routy B, Le Chatelier E, Derosa L, Duong CPM, Alou MT, Daillere R, et al. Gut microbiome influences efficacy of PD-1-based immunotherapy against epithelial tumors. Science 2018;359:91–7. https://doi.org/10.1126/science.aan3706.

[247] Zhang J, Haines C, Watson AJM, Hart AR, Platt MJ, Pardoll DM, et al. Oral antibiotic use and risk of colorectal cancer in the United Kingdom, 1989–2012: a matched case-control study. Gut 2019;68:1971–8. https://doi.org/10.1136/gutjnl-2019-318593.

[248] Geller LT, Barzily-Rokni M, Danino T, Jonas OH, Shental N, Nejman D, et al. Potential role of intratumor bacteria in mediating tumor resistance to the chemotherapeutic drug gemcitabine. Science 2017; 357:1156–60. https://doi.org/10.1126/science.aah5043.

[249] Vande Voorde J, Sabuncuoglu S, Noppen S, Hofer A, Ranjbarian F, Fieuws S, et al. Nucleoside-catabolizing enzymes in mycoplasma-infected tumor cell cultures compromise the cytostatic activity of the anticancer drug gemcitabine. J Biol Chem 2014;289:13054–65. https://doi.org/10.1074/jbc.M114.558924.

[250] Wilcox MH, Gerding DN, Poxton IR, Kelly C, Nathan R, Birch T, et al. Bezlotoxumab for prevention of recurrent *Clostridium difficile* infection. N Engl J Med 2017;376:305–17. https://doi.org/10.1056/NEJMoa1602615.

[251] Cammarota G, Masucci L, Ianiro G, Bibbo S, Dinoi G, Costamagna G, et al. Randomised clinical trial: faecal microbiota transplantation by colonoscopy vs. vancomycin for the treatment of recurrent *Clostridium difficile* infection. Aliment Pharmacol Ther 2015;41: 835–43. https://doi.org/10.1111/apt.13144.

[252] Riquelme E, Zhang Y, Zhang L, Montiel M, Zoltan M, Dong W, et al. Tumor microbiome diversity and composition influence pancreatic cancer outcomes. Cell 2019;178:795–806. https://doi.org/10.1016/j.cell.2019.07.008. e12.

[253] O'Keefe SJ, Li JV, Lahti L, Ou J, Carbonero F, Mohammed K, et al. Fat, fibre and cancer risk in African Americans and rural Africans. Nat Comms 2015; 6:6342. https://doi.org/10.1038/ncomms7342.

[254] Scharlau D, Borowicki A, Habermann N, Hofmann T, Klenow S, Miene C, et al. Mechanisms of primary cancer prevention by butyrate and other products formed during gut flora-mediated fermentation of dietary fibre. Mutat Res 2009;682:39–53. https://doi.org/10.1016/j.mrrev.2009.04.001.

[255] Hopkins BD, Pauli C, Du X, Wang DG, Li X, Wu D, et al. Suppression of insulin feedback enhances the efficacy of PI3K inhibitors. Nature 2018;560:499–503. https://doi.org/10.1038/s41586-018-0343-4.

[256] Zahra A, Fath MA, Opat E, Mapuskar KA, Bhatia SK, Ma DC, et al. Consuming a ketogenic diet while receiving radiation and chemotherapy for locally advanced lung cancer and pancreatic cancer: the University of Iowa experience of two phase 1 clinical trials. Radiat Res 2017;187:743–54. https://doi.org/10.1667/RR14668.1.

[257] Koulouris AI, Luben R, Banim P, Hart AR. Dietary fiber and the risk of pancreatic cancer. Pancreas 2019; 48:121–5. https://doi.org/10.1097/MPA.0000000000001191.

[258] Molina-Montes E, Sánchez M-J, Buckland G, Bueno-de-Mesquita HBA, Weiderpass E, Amiano P, et al. Mediterranean diet and risk of pancreatic cancer in the European Prospective Investigation into Cancer and Nutrition cohort. Br J Cancer 2017;116:811–20. https://doi.org/10.1038/bjc.2017.14.

[259] Pahwa R, Singh A, Jialal I. Chronic inflammation. 2019. https://doi.org/10.1016/j.jamcollsurg.2015.12.009.

[260] Network TCGAR, Raphael BJ, Hruban RH, Aguirre AJ, Moffitt RA, Yeh JJ, et al. Integrated genomic characterization of pancreatic ductal adenocarcinoma. Cancer Cell 2017;32:185–203. https://doi.org/10.1016/j.ccell.2017.07.007. e13.

[261] Tiriac H, Belleau P, Engle DD, Plenker D, Deschênes A, Somerville TDD, et al. Organoid profiling identifies common responders to chemotherapy in pancreatic cancer. Cancer Discov 2018;8:1112–29. https://doi.org/10.1158/2159-8290.CD-18-0349.

[262] Garnier A, Ilmer M, Becker K, Häberle B, Schweinitz DV, Kappler R, et al. Truncated neurokinin-1 receptor is an ubiquitous antitumor target in hepatoblastoma, and its expression is independent of tumor biology and stage. Oncol Lett 2016;11:870–8. https://doi.org/10.3892/ol.2015.3951.

[263] Garnier A, Ilmer M, Kappler R, Berger M. Therapeutic innovations for targeting hepatoblastoma. Anticancer Res 2016;36:5577–92.

CHAPTER

10

Animal models and in vivo investigations for drug repurposing in lung cancer

Hsuen-Wen Kate Chang[1], *Vincent H.S. Chang*[2,3]

[1]Laboratory Animal Center, Taipei Medical University, Taipei, Taiwan; [2]Department of Physiology, School of Medicine, College of Medicine, Taipei Medical University, Taipei, Taiwan; [3]The Ph.D. Program for Translational Medicine, College of Medical Science and Technology, Taipei Medical University, Taipei, Taiwan

OUTLINE

Introduction

Lung cancer is one of the most common and serious types of cancer, which causes the highest cancer mortality among men and women worldwide. The two major types of lung cancers are small cell lung cancer (SCLC) and non-small cell lung cancer (NSCLC), based on histological differentiation. NSCLCs are further divided into squamous cell carcinomas (SCCs), pulmonary adenocarcinomas (ADC), and large cell carcinomas. Lung ADC is the most prevalent form of NSCLC [1]. Lung cancer has a dismal prognosis of 15%, mainly attributed to ineffective early detection strategies and lack of therapeutic options for metastatic diseases [2]. This has spurred efforts for the development of molecularly targeted therapies.

Drug Repurposing in Cancer Therapy
https://doi.org/10.1016/B978-0-12-819668-7.00010-5

Drug repurposing relates to determining new targets for existing drugs and identifying new indications for known diseases. In addition to being time- and cost-efficient, drug repurposing offers a more favorable risk−return tradeoff than other available drug development strategies. Because the existing drugs have already been tested in terms of safety, dosage, and toxicity, they can often enter clinical trials much more rapidly than newly developed drugs [3]. Computational drug repurposing is deemed an effective alternative method for identifying novel connections between diseases and existing drugs [4]. The increase in drug−target information and the advances in systems pharmacology approaches have led to an increase in the success of in silico drug repurposing.

As summarized by Jegga et al. approaches for computational drug repurposing can be broadly categorized into knowledge-based and signature-based approaches [5]. Knowledge-based drug repurposing usually involves utilizing available information on genome-wide association studies and omics data such as genetic markers, structures, targets, and pathways to predict prospective disease mechanisms [6,7].

Signature-based drug repurposing involves comparing the drug-treated gene expression profile with its disease counterpart to construct a detailed map of connections between diseases and drug actions. In particular, large-scale chemical genomics databases, such as the Connectivity Map or the National Institutes of Health Library of Integrated Network-Based Cellular Signatures program's highly expanded chemical genomics data set, provide abundant information on the modes of action of drugs, which are reflected in the transcriptomic responses to chemical perturbation. Additionally, transcriptome-level expression profiles of approximately 20,000 genes have been computationally inferred using 1000 landmark genes (L1000) [8].

A literature-based approach was also used to identify licensed noncancer drugs with published evidence of anticancer activity. The Repurposing Drugs in Oncology project is an ongoing collaborative project that has focused exclusively on the potential use of licensed noncancer medications as sources of new cancer therapeutics [9].

By utilizing these computer-aided drug repurposing strategies, abundant amounts of data can be obtained. However, successful translation of repurposed drugs to high efficacy for its new indications (i.e., lung cancer) requires a detailed investigation of the underlying biology. To achieve the goal, target validation should be performed through intervention studies in human lung cancer cell lines. Because in vitro cell culture studies cannot fully mimic the complexity of lung carcinogenesis in vivo, developing mouse lung cancer models will provide insight into lung tumorigenesis. In the next section, we summarize related mouse models and their applications in repurposed lung cancer treatments.

Animal models and in vivo applications in drug repurposing

The importance of drug repurposing lies in the high costs and the prolonged time from target selection to regulatory approval of traditional drug development. Cell-based studies have limitations in terms of fully mimicking the complexity of the onset and development of tumorigenesis in living animals. In vivo evaluation is a key component in the drug discovery process, suggesting that evaluations based on small animal model offer a convenient platform to perform drug repurposing. Testing in an in vivo system allows the evaluation of multiple signaling mechanisms and tissue cross talk that cannot be studied in a cell-based system. This is still the only available platform for safety and efficacy studies with a high biomedical applicability to human use. A small molecule identified in an in vivo system has a higher likelihood of having bioactivity and clinical relevance.

Mouse models for human lung cancer

On the basis of histological differentiation, there are two major types of lung cancer: SCLC and NSCLC. NSCLCs are further divided into SCCs, ADC, and large cell carcinomas. Among them, lung ADC is the most prevalent form of NSCLC [1,10]. Mouse models for lung cancer are a valuable tool not only for understanding basic lung tumor biology but also for the development and validation of new tumor intervention strategies [11]. To this end, various lung tumor mouse models have been established, which resemble the different human lung cancer types in terms of genetic alterations and tumor cell characteristics.

Genetically engineered mouse models

The progress in whole genome analysis provides extensive genetic information on human lung cancer. The mouse is an ideal lung cancer model because of its anatomic and physiologic similarities to humans, and the fact that its genome can be manipulated allows for a genome-scale metabolic reconstruction of lung cancer. The evolution of human lung cancer mouse models was summarized by Dutt and Wong. Examples of such models are ectopic expression models, knock-out/knock-in models, inducible bitransgenic models, conditional bitransgenic mouse models using Cre/LoxP systems, and compound conditional models [12]. The activation of oncogenes and inactivation of tumor-suppressor genes are the events underlying tumorigenesis. Meuwissen and Berns summarized the genetic aberrations observed most frequently in NSCLC and SCLC [11].

Over the past decade, a number of genetic alterations have been described in NSCLC. The most commonly mutated genes in human NSCLC are TP53, KRAS, STK11, epidermal growth factor receptor (EGFR), NF1, EPHA3, CDNK2A, ERBB4, FGFR4, and INHBA [13]. Genetic mutations identified in humans have been introduced in mouse models, including KRAS, BRAF, EGFR, LKB1, RAC1, NF-kB, and TP53. KRAS mutations are identified in approximately 20%–25% of lung ADC in the West and in approximately 10%–15% in Asia. Globally, KRAS mutant tumors constitute the most frequent potentially targetable molecular subtype of NSCLC [14]. The codons 12, 13, and 61 of KRAS are mutated to an oncogenic form thought to limit its GTPase activity [15]. Several genetic mouse models of lung cancer have relied on initiation by oncogenic KRAS. Although some repurposed drugs have demonstrated considerable specificity toward the inhibition of mutant KRAS tumors in vitro and provide proof of concept of direct KRAS inhibition, long-term efficacy and toxicity limit their viability [16]. Because approximately 50% of humans with NSCLC possess alterations in the P53 tumor suppressor gene, models with both KRAS and P53 mutations closely resemble invasive and metastatic human ADC [17,18]. The types of mouse models engineered for human lung cancer are summarized by De Seranno et al. [19].

Kwon and Berns summarized the use of these genetic abnormalities for the development of mouse models for NSCLC [10] (Table 10.1).

Xenograft models

Xenograft models are primarily used to examine tumor response to therapy in vivo prior to translation into clinical trials. Lung cancer cell lines have contributed substantially to lung cancer translational research and biomedical discovery. Hundreds of cell lines have been established. These cell lines have been utilized worldwide in the scientific community and for drug screening. Lung cancer cell lines have contributed substantially to understanding the molecular biology of lung cancer and translating these findings to clinical applications. With the homogeneity of cell types and pathological differentiation, lung cancer cell lines have been widely used for cancer research, drug repurposing, and screening for lung cancer [20,21].

TABLE 10.1 Mouse models for NSCLC [10].

Mouse mutant	Tumor induction	Phenotype	References
LSL-KrasG12D endogenous control	Sporadic infection of lung cells with adeno-cre virus	Adenomas and adenocarcinomas long latency	[89]
KrasG12D LA1 and LA2 in wt or p53 deficiency	Spontaneous lung tumor development due to sporadic switching of LA allele	Adenomas and adenocarcinomas; a variety of tumor types	[106]
Tet-op-KrasG12D in wt or p53 and P19Arf deficiency	CCSP-rtTA transgene, treatment with doxycycline. Transgene directs rtTA expression in alveolar type II cells	Fast tumor growth, accelerated in p53 and Ink4A/Arf def. background	[78]
LSL-KrasG12D	Tamoxifen Cre-ERT2 knockins in SPC and CC10 (AT2 and Clara cells)	Adenomas and adenocarcinomas with SPC-Cre	[105]
$Trp53^{F2-10/F2-10}$; $Rb1^{F19/F19}$	Ad5-CMV-cre	Adenomas and adenocarcinomas short latency	[94]
LSL-KrasG12D; p531ax/lox	Sporadic switching of lung cells with lenti-cre virus	Accelerated tumor development; metastasis; role forNkx2-l and Hmga2	[104]
PTENlox/lox	Clara cell–specific CCSP-Cre.	No tumors	[89]
LSL-KrasG12D;	Clara cell–specific CCSP-Cre.	Accelerated tumor development	[107]
PTENlox/lox		Metastasis	[88]
LSL-KrasG12D; Lkb1lox/lox	Ad5-CMV-cre	Strongly augmented tumor growth and metastasis, both adenocarcinomas and squamous cell carcinomas	[91]
LSL-KrasG12Vgeo	Cre-ERT2 (RERT-ert) + tamoxifen	Adenomas and adenocarcinomas; not all KrasV12-expressing cells proliferate	[86]
LL-BrafV600E	Ad5-CMV-cre	Adenomas and rarely progress to adenocarcinoma	[75]
Inducible cRaf mutant	Clara and aalveolar type II cell–specific expression	Only expression in alveolar type II cells gives rise to macroscopic tumors; deinduction causes reversion	[71]
TRE-Egfr L858R/T790M/Del exon 19 mutants	CCSP-rtTA doxycycline inducible	Adenocarcinomas; T790M and L858R/T790M double mutant show less aggressive growth	[97,99]
Tet-op-PIK3CA H1047R; CCSP-rtTA	CCSP-rtTA doxycycline inducible	Adenocarcinoma with bronchioalveolar features	[76]

All the aforementioned murine lung tumor models developed pulmonary adenocarcinomas, which were used for the investigation of intervention strategies.

The xenograft transplantation of tumor cell lines into immunocompromized mice provides an environment for the involvement of stromal, vascular, and inflammatory components in tumorigenesis. However, an important limitation of this approach is that xenograft lung tumor models do not behave in the same way as actual lung tumors because they are not located in the environment of origin.

Establishing patient-derived xenograft (PDX) mouse models refers to the direct transfer of fresh tumor samples into immunodeficient mice. Human tumor cells may be transplanted into a mouse model subcutaneously, intraperitoneally, or orthotopically. Primary cells are more useful for mechanism study than stable cell lines. Multiple studies have reported that PDXs retain gene profiles and tumor heterogeneity after several transfers of tumors from mouse to mouse [22]. Orthotopic murine models involve implantation of tumor cell lines or patient-derived cells into animal organs or tissues that match the tumor histotype. The orthotopic models provide a tumor microenvironment in which tumor cells develop with similar biological and metastatic properties to clinical cases, thereby leading to more reliable translation. Richmond and Su summarized the advantages and disadvantages of xenografted and genetically engineered mice models in human cancer research [23] (Fig. 10.1).

Other transgenic models

Growing evidence indicates that viruses play a critical role in cancer development. Viruses can cause cellular transformation through the expression of viral oncogenes, through genomic integration to alter the activity of cellular protooncogenes or tumor suppressor genes, or by inducing inflammation that promotes oncogene activity [24]. Over the last few decades, human papillomavirus (HPV) has been reported in lung cancer tissues. However, the effect of viruses on host NSCLC tumor cells has not been investigated [24–26].

Various transgenic models have been developed through manipulation of genetic lesions found in human lung cancer into the mouse germline or pulmonary tissues and by introducing viral oncogenes to a specific subset of lung epithelial cells. Although continuous attempts to manipulate the mouse genome have enabled the reproduction of human lung cancers in mice with more accuracy, targeted gene-modulated models cannot fully represent the disease-related systemic scenarios related to lung cancer.

An estimated 15%–25% of human cancers may have a viral etiology, and two viruses in particular, HPV and Jaagsiekte sheep retrovirus (JSRV), were thought to play roles in the pathogenesis of human lung cancer [27]. Carraresi et al. generated transgenic mice expression of HPV-16 E6/E7 genes under the control of the murine keratin 5 gene promoter, resulting in lung ADC approximately 6 months postbirth [28]. Ovine pulmonary ADC is a naturally occurring lung cancer in sheep caused by retrovirus infection and has several features in common with human ADC, including its histopathology and cell signaling pathways [29,30]. The authors characterized a JSRV envelope protein transgenic mouse model. This spontaneous lung ADC model exhibits a metastatic phenotype and is used for drug repurposing [31,32].

In vivo investigations on drug repurposing in lung cancer

In 2018, we published a paper on the in vivo investigation of drug repurposing in lung ADC by comparing the transcriptome profiles obtained from a well-documented mouse lung cancer model [32]. We used the LINCS L1000 cellular signature bioinformatics approach to identify clinically approved candidate drugs to treat ADC. By using this approach, we identified the mTOR inhibitor temsirolimus, which has been approved by the FDA as a potential therapeutic agent for renal cell carcinoma.

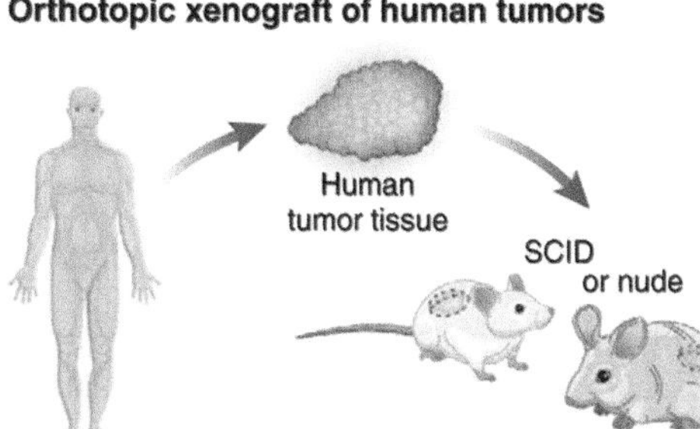

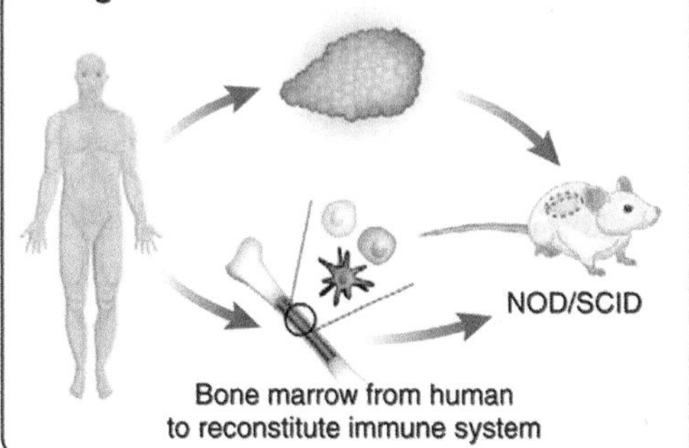

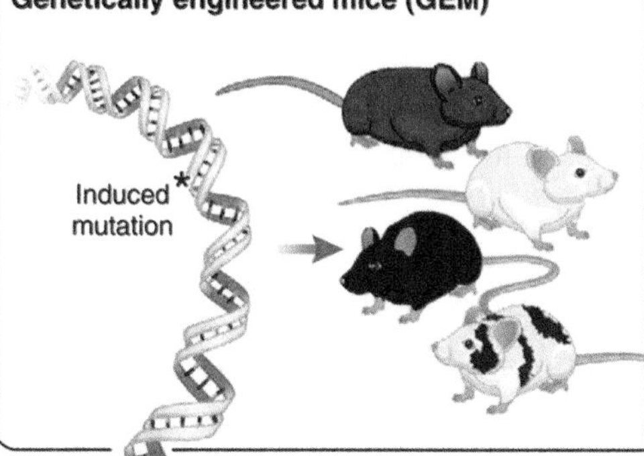

FIGURE 10.1 Types of murine model for studying human cancers [23].

In our mouse model studies, we designed concurrent and sequential administration of temsirolimus with either low or high doses of chemotherapy. In the concurrent schedule, administration of low-dose chemotherapy (C: cisplatin and G: gemcitabine) and temsirolimus (T + C + G) demonstrated greater inhibition of tumor growth compared with low-dose chemotherapy alone (C + G). In the sequential schedule, temsirolimus alone was administrated weekly for 3 weeks prior to the administration of high-dose chemotherapy (3 mg/kg cisplatin and 30 mg/kg gemcitabine) in the following weeks. The inhibition of tumor growth was less pronounced in this schedule. These findings indicate that concurrent administration of low-dose chemotherapy and temsirolimus is more effective in suppressing lung tumor growth, which may be advantageous in reducing the cytotoxicity caused by standard chemotherapy.

However, mTOR inhibitor administration was associated with pulmonary toxicity, as are numerous other anticancer agents [33]. Appropriate chemotherapeutic strategy management and clinical pulmonary symptom diagnosis should be considered when administering mTOR inhibitors. Our study results suggest that the combination of low-dose chemotherapy and temsirolimus treatment may be beneficial for the treatment of lung ADC; this warrants further investigation.

Repurposed therapies in lung cancer

The repurposing of existing therapies provides an opportunity to capitalize on well-characterized drugs that could synergize with existing therapies for lung cancer. The goal of this strategy is to deliver an effective drug with a favorable toxicity profile at a reduced cost. Various preclinical murine models have been adapted to demonstrate the efficacy of these agents and are being processed in prospective trials. Murray et al. and Saxena et al. have outlined various repurposed drugs, comprising metformin, statins, itraconazole, β-blockers, and nonsteroidal antiinflammatory drugs, which have been clinically assessed for treatment of NSCL [34]. Murray and Levy [35] have reviewed several repurposed drugs that have been clinically assessed in NSCLC in the ways of mechanisms of action, preclinical evidence, and clinical studies. Most in vivo investigations on drug repurposing in lung cancer have used xenografted or genetically engineered mouse models. Since the early mouse-in-mouse isograft models used for drug screening during the 1960 and 1970s, the conceptual targets of cancer treatment have progressed from actively dividing cells to oncogenic signaling and immune checkpoints. The preclinical models and cancer therapies have coevolved accordingly [36].

Previous literature has repurposed metformin hydrochloride, an oral biguanide used for treating type 2 diabetes, as an antineoplastic agent [37,38]. The anticancer effects of metformin were firstly reported by Evans et al. showing reduced risk of cancer in type 2 diabetes patients taking metformin [39]. Metformin has then been tested as a single agent or combined with chemotherapeutic agents in many different lung cancer cells. These in vitro data indicated numerous anticancer properties including increased apoptosis, reduced cell proliferation, and increased radiosensitization and are synergistic with chemotherapy, as summarized in Tables 10.2 and 10.3.

The in vivo efficacy of metformin has also been tested on various lung cancer murine models, which are either xenografted with lung cancer cells or induced tumorigenesis with carcinogens. These results showed promising preclinical benefit as listed in Table 10.4.

A single-center Phase II trial reported by Arrieta et al. [40] has demonstrated that the addition of metformin to a standard EGFR-tyrosine kinase inhibitor (TKI) treatment in patients with *EGFR*-mutated NSCLC improved progression-free and overall survival, without

TABLE 10.2 In vitro studies of metformin in lung cancer: as a single agent [37].

Cell type	Dose and duration	Findings	Mechanism	References
RERF-LC-AI (SCC), I A-5 (LCC), WA-hT (SCLC), A 549 (ADC)	0.3–20 mM met for 1–72 h RERF-LC-AI, $IC_{50} = 6$ mM A549. $IC_{50} = 1$ mM IA-5, $IC_{50} = 5$ mM WA-hT, $IC_{50} = 2$ mM	↓ cell proliferation ↑ apoptosis ↓ colony formation	↑ G0/G1 cell cycle arrest	[34]
Calu-1 (NSCLC), Calu-6 (ADC)	0.3–5 mM met for 0–72 h Calu-1 $IC_{50} = 16$ mM Calu-6 $IC_{50} = 18$ mM	↓ cell proliferation	↓ phosphorylation of IGF-IR substrates Akt and FOXO3a ↑ AMPK phosphorylation in calu-1 cells	[35]
A549 (ADC)	1–10 mM met for 24 h	↓ HO–1 mRNA and protein expression ↓ Nrf2 expression	↓ phosphorylation of Raf and ERK1/2	[36]
PC9 (ADC)	1–32 mM met for 0–72 h $-IC_{50} = 3.5$ mM	↓ proliferation		[37]
A549 (ADC), H1975 (ADC)	5–50 μM met for 24–72 h	↑ cytotoxicity ↓cellular TP and ERCC1 expression	↓ MEK1/2-ERK1/2 protein levels	[38]
A549 (ADC)	1–10 mM met for 24–72 h $IC_{50} = 5$ mM	↓ proliferation ↑apoptosis	↓ Akt levels reducing mTOR activation	[39]
H1299 (ADC). GLC82 (ADC), H1975 (ADC), CALU-3 (ADC), CALU-3 GEF-R (ADC), H460 (LCC), A549 (ADC)	0.1–20 mM met for 72 h $IC_{50} = 2$–2.25 mM for all cells except A549 and H460 $Ic_{50} > 20$ mM	↓ proliferation ↑ apoptosis	↑ MAPK activation	[40]
Calu-1 (NSCLC)	0.0375–10 mM met for, 6, 24,0.48 h	↑ apoptosis ↓ glucose uptake	↓ Hexokinase-II activity	[44]
A549 (ADC)	0.5–8 mM met for 24 h $-IC_{50} = 4$ Mm	↓ proliferation ↑ apoptosis	↑ G1 cell cycle arrest ↓ p38 MAPK phosphorylation	[47]
A549 (ADC)	10 mM met for 0–24 h	↓ Bmi–1 ↑ miR–15a, miR–128, miR–192 and miR–194	↑ phosphorylation of AMPK and expression of LKB1	[50]
TKI-sensitive PC-9 (ADC), TKI-resistant PC-9GR (ADC)	5 mM met for 48 h	↓ expression of markers of pulmonary fibrosis	↓ expression of α-actin and COL1A1 ↓ expression of pSMAD2, pSMAD3, pSTAT3, pAKT and dpERK1/2 ↓ TGF-β levels and activation.	[51]

TABLE 10.2 In vitro studies of metformin in lung cancer: as a single agent [37].—cont'd

Cell type	Dose and duration	Findings	Mechanism	References
H522 (ADC), H2342 (ADC), H2405 (ADC), A549 (ADC), SPC-A-1 (ADC), SW900(SCC), H1869(SCC), SK-MES-1(SCC), H661(LCC), H1299(ADC)	1.25–5 mM met for 7 days A549 IC50 = 7.97 mM SK-MES–1 IC_{50} = 13.36 mM	↓ proliferative Activities ↓ cancer stemness of A549 cells ↓ levels of stem cell markers	↓ levels of NLK, nanog, c-myc, and KLF4	[52]
A549 (ADC)	5–50 mM met for 24,48 or 72 h 24 h, IC_{50} = 3.5 mM 48 h, IC_{50} = 8 mM 72 h = 20 mM	↑ apoptosis ↑ cell cycle arrest at G_o-G_1 phase	↓ Bcl–2 protein levels ↑ Bax protein expression	[53]
H460 (LCC), H1299 (ADC)	5,10 or 20 mM met for 24, 48 or 72 h H460 24 h, IC_{50} > 20 mM H460 48 h, IC_{50} = 20 mM H460 72 h, IC_{50} = 10 mM H1299 24 h, IC_{50} > 20 mM H1299 48 h, IC_{50} = 20 mM H1299 72 h, IC_{50} = 20 mM	↓ proliferation ↑ apoptosis ↑ cell cycle arrest at G_o-G_1 phase	↑ AMPK phosphorylation ↓ mTOR and p70S6k phosphorylation	[38]

AC, Adenocarcinoma; *Akt*, Protein kinase B; *AMPK*, AMP-activated protein kinase; *Bax*, Bcl-2-like- protein 4; *Bcl-2*, B-cell lymphoma 2; *COL1A1*, Collagen type 1 alpha 1; *ERCC1*, excision repair cross-complementation 1; *ERK*, Extracellular regulated kinase; *FOXO3a*, Forkhead box O3; *IC_{50}*, half maximal inhibitory concentration; *IGF-1R*, Insulin-Like growth factor 1 receptor; *KLF4*, Kruppel-like factor 4; *LCC*, Large cell carcinoma; *LKB1*, Liver kinase B1; *MEK*, Mitogen-activated protein kinase kinase; *mTOR*, Mammalian target of rapamycin; *Nanog*, Homeobox protein Nanog; *NLK*, Nemo like kinase; *NSCLC*, Non-small cell lung cancer; *p70S6K*, Ribosomal protein S6 kinase beta-1; *Raf*, Rapidly accelerated fibrosarcoma; *SCC*, Squamous cell carcinoma; *SCLC*, Small cell lung cancer; *SMAD2*, SMAD family member 2; *SMAD3*, SMAD family member 3; *TGF-β*, Transforming growth factor β.

TABLE 10.3 In vitro studies of metformin in lung cancer: combined with chemotherapeutic agent [37].

Cell type	Dose and duration	Findings	Mechanism	Reference
H520 (SCC), H1703 (SCC)	0.1–1 mM met alone or in combination with 1–10 μM gefitinib for 24 h	↑ cytotoxicity and growth inhibition by gefitinib	↓ gefitinib-induced expression of MSH2 expression	[41]
PC-9 (ADC), PC-9GR (ADC), H1650-M3 (ADC)	5 mM met alone or in combination with either 1–16 μM gefitinib or 2.5–20 μM erlotinib for 48 h	Resensitized EGFR-TK1 resistant human lung cancer cells ↓ EMT in TK1 resistant human lung cancer cells	↑ E-cadherin expression ↓ Vimentin and SNAIL expression	[54]
H2228 (ADC), H3122 (ADC)	5 mM met alone or in combination with 400 nM crizotinib for 48 h	↓ cell proliferation ↓ apoptosis ↓ crizotinib sensitivity ↓ crizotinib resistance ↓ tumor invasion	↓ IGF1–R signaling. ↓ phosphorylation of mTOR, p70S6K, and S6	[55]

(*Continued*)

TABLE 10.3 In vitro studies of metformin in lung cancer: combined with chemotherapeutic agent [37].—cont'd

Cell type	Dose and duration	Findings	Mechanism	Reference
SW1271 (SCLC), H2347 (ADC)	10 mM met alone or in combination with 30 n M trametinib for 72h SW1271 IC_{50} = 29.9 mM H2347 IC_{50} = 6.79 mM	Combination is effective for treatment of NRAS mutant lung carcinomas	↓ cell viability ↓ activity of MAPK and PI3K/Akt/mTOR signaling pathways.	[56]
A549 (ADC), H460 (LCC)	0.5–2 mM met alone or in combination with 0.25–6 μM sorafenib, for up to 10 days	↓ proliferation	↓ AMPK activation	[57]
H358, Calu-3 (ADC), HI299 (ADC), H1975 (ADC)	2 mM met alone or in combination with 0.01 μM or 1 nM selumetinib for 72 h H358 IC_{50} = 1.5 mM Calu–3 IC_{50} = 1 mM H1299 IC_{50} = 1.5 mM H1975 IC_{50} = 2.5 mM	↓ proliferation ↓ apoptosis ↓ GLI1 transcriptional activity	↓ production of MMP–2 and MMP–9 by reducing NF-κB	[58]
AS_2 (ADC)	2.5–2.5 mM met alone or in combination with 2.5 μM cisplatin for 24–72 K	↓ secretion of VEGF ↓ cisplatin cytotoxicity ↓ cisplatin induced ROS production	↓ STAT3 pathway	[59]
H460 (LCC)	15.145–60.58 mM met alone or in combination with 0.0995–0.199 mM cisplatin or 0.0926–0.1852 mM etoposide IC_{50} = 60.58 mM	↓ proliferation	↓ metabolic viability	[60]
H1650 (ADC), H1703 (SCC)	0.1–1 mM met alone or in combination with 0.1–1 μM paclitaxel for 24 h	↓ cytotoxic effect of paclitaxel ↓ ERCC1 expression	↓ p38 MAPK phosphorylation	[61]
H1299 (ADC), H1650 (ADC)	5 mM met alone or in combination with 20 nM ciglitazone for 24 h	↓ growth ↑ apoptosis	↓ PDKl expression and promoter activity	[62]
HCC4006 (ADC), NC1-H1975 (ADC), HCC95 (SCC), NCI-H2122 (ADC), NCF-H3122(ADC)	0.01–10 mM met alone or in combination with 50–5 μM salinomycin for 48 or 72 h HCC4006 ~IC_{50} = 2 mM H1975~IC_{50} = 5 mM HCC95 ~IC_{50} = 5 mM H2122~IC_{50} > 1 mM H3122 ~IC_{50} = 5 mM	↓ cell death	↓ EGFR signaling ↓ Akt, EGFR 1/2, mTOR, and p70S6K activity	[63]
A549 (ADC), HCC4Q06 (ADC)	1–10 mM met alone or in combination with 0.1–1 μM salinomycin +10 ng/mL TGF-β for 48 h A549 ~IC_{50} = 20 mM HCC4006, IC_{50} = 20 mM	↑ TGF-β induced EMT ↓ cell migration	↑ E-cadherin expression	[64]

TABLE 10.3 In vitro studies of metformin in lung cancer: combined with chemotherapeutic agent [37].—cont'd

Cell type	Dose and duration	Findings	Mechanism	Reference
A549 (ADC), SPC-A-1 (ADC), H1975 (ADC), SK-MES-1 (SCC), H520 (SCC), PC-9 (ADC)	3 mM met alone or in combination with 100 ng/mL figitumumab for 6–48 h	↓ proliferation	↓ PI3K/Akt signaling pathways ↓ MEK/ERK signaling pathways ↓ IGF-1 receptor	[65]
A549 (ADC)	1–4 mM met alone for 12 h or in combination with 200 ng/mL TRAIL protein for 2 h	↓ apoptosis	↓ c-FI IP ↓ p62protein levels	[66]
PC9 (ADC), A549 (ADC),	20 μM met alone or in combination with 20 μg/mL β-element for 24 h	↓ cell growth	↓ Akt phosphorylation ↓ DNMT1	[67]
A549 (ADC), H1299 (ADC), SK-MES1 (SCC)	2.5–5 mM met alone or in combination with ionizing radiation for 72 h A549 ~ IC_{50} = 75 μM H1299 ~ IC_{50} = 25μM SK-MES1 ~ IC_{50} = 25 μM	↓ proliferation ↑ radiosensitization	↑ ATM-A M PK-P53 pathway activity ↓ Akt, mTOR, 4EBP1 pathways ↑ G_1 cell cycle arrest ↑ apoptosis	[68]

4EBP1, Eukaryotic translation initial factor 4E-binding protein l_r; *AC*, Adenocarcinoma; *Akt*, Protein kinase B; *AM PR*, AMP-activated protein kinase; *c-FLIP*, CASP-S and FADD-like apoptosis regulator; *DNMT1*, DNA methytransferase 1; EGFR, Epidermal growth factor receptor; *EMT*, Epithelial to mesenchymal transition; *ERCC1*, excision repair cross-complementation 1; *ERK*, Extracellular regulated kinase; *IC5U*, half maximal inhibitory concentration; *IGF-1R*, Insulin-like growth factor 1 receptor; *LCC*, Large cell carcinoma; *MAPK*, Mitogen activated protein kinase; *MEK*, Mitogen-activated protein kinase kinase; *MMP-2*, Matrix metalloproteinase-2; *MMP-9*, Matrix metallopeptidase-9; *MSH2*, MutS 2; *mTOR*, Mammalian target of rapamycin; *NF-κB*, Nuclear factor kappa-light chain-enhancer of activated B cells; *NRAS*, Neuroblastoma RAS viral oncogene homolog; *NSCLC*, Non-small cell lung cancer; *p53*, Tumor protein p53; *p62*, Nucleosome p62; *p7US6K*, Ribosomal protein S6K kinase beta-1; *PDK1*, Pyruvate dehydrogenase lipoamide kinase isozyme 1; *PI3K*, Phosphoinositide-3-kinase; *ROS*, Reactive oxygen species; *SCC*, Squamous cell carcinoma; *SCLC*, Small cell lung cancer; *SNAIL*, Zinc finger protein SN All; *STAT3*, Signal transducer and activator of transcription 3; *TGF-p*, Transforming growth factor fl; *TK1*, Tyrosine kinase inhibitor; *VFGF*, Vascular endothelial growth factor.

TABLE 10.4 In vivo studies of metformin in lung cancer [37].

Animal model	Dose and duration	Findings	Mechanism	References
7-week-old female Balb/c mice inoculated subcutaneously with A549 or A431 cells	Drinking water +250 mg/kg/day metformin for 21 days	↑ apoptosis ↓ growth of K-ras mutant tumors	↓ Akt levels ↓ mTOR activation	[39]
8-week-old LID mice were given 3 weekly injections of NNK (tobacco carcinogen)	Drinking water + 5 mg/mL metformin	↓ lung tumorigenesis	IGF-1–independent mechanism ↓ phosphorylation of multiple RTKs	[70]
6-week-old nude mice inoculated with HCC827–pSB388 cells	Drinking water +250 mg/kg body weight metformin, 2 days before tumor inoculation—till the mice were sacrificed	↓ growth and distant metastases ↓ IL-6 induced EMT	↓ STAT3 phosphorylation	[71]

(*Continued*)

TABLE 10.4 In vivo studies of metformin in lung cancer [37].—cont'd

Animal model	Dose and duration	Findings	Mechanism	References
4- to 6-week-old female Balb/c mice were injected subcutaneously with H1299 or CALU-3 GEF-R cells	Drinking water +200 mg/mL metformin +150 mg/kg/day gefitinib for 35 days	↓ proliferation ↑ apoptosis	↓ EGFR phosphorylation ↑ MAPK activation	[40]
6-week-old female Balb/cA-nu mice were inoculated with PC-9GR or PC-9 cells subcutaneously into the back next to the left forelimb	Drinking water +1 mg/mL metformin alone or in combination with 250 mg/L gefitinib for 30 days	↓ tumor growth in xenografts with TKI-resistant cancer cells	↓ IL-6 secretion and expression ↓ IL-6 signaling action ↓ EMT	[55]
Male SD rats with average weight of 200 g. Maintained under chloral hydrate anesthesia (500 mg/kg)	Gefitinib 200 mg/kg administered orally once/day for 3 days before animals received a single intratracheal administration of bleomycin (5 mg/kg). Gefitinib and metformin (300 mg/kg) was then continued once every 2 days for the following 21 days	↓ exacerbation of bleomycin-induced pulmonary fibrosis by gefitinib	↓ α-actin and COL1A1 ↓ expression of pSMAD2, pSMAD3, pSTAT3, pAKT, and ERK1/2 ↓ TGF-β levels and activation	[51]
6- to 8-week-old balb/c nude mice were subcutaneously injected with AS2 cells into the flanks	Drinking water with or without 500 mg/kg metformin with or without 4 mg/kg cisplatin	↓ xenograft growth ↓ cisplatin-induced ROS production and autocrine IL-6 secretion	↓ STAT3 pathway	[60]
Balb/c nude mice were subcutaneously injected with A549 cells into their right flanks	40 mg/kg/day metformin alone or in combination with 5 mg/kg/day cisplatin	↓ tumor size	↓ Bcl-2 protein levels ↑ box protein expression	[53]
7-week-old immunodeficient Balb/c female nude mice injected subcutaneously	Drinking water + 400 mg/kg/day body weight metformin + 30 mg/kg/day Sorafenib for 40 days	↓ proliferation	↑ apoptosis phosphorylation of AMPK ↑ inhibition of downstream mTOR signaling	[58]

TABLE 10.4 In vivo studies of metformin in lung cancer [37].—cont'd

Animal model	Dose and duration	Findings	Mechanism	References
with A549 cells into the right posterior flanks	Given either agent alone or in combination			
4- to 6-week-old nude mice bearing H1299 or H1975 cells that were grown subcutaneously	Drinking water +200 mg/mL metformin + 25 mg/kg selumetinib for 35 days	Expression changes of mesenchymal proteins, SNAIL, and vimentin ↓ tumor metastatic behavior	↓ production of MMP−2 and MMP-9 by reducing NF-kB	[59]
5-week-old balb/c-nude mice were subcutaneously injected with A549 or H1299 cells into the right flank	Drinking water+300 mg/kg body weight per day metformin till euthanasia. Xenografts were subjected to 0 Gy or 10 Gy IR while under gaseous anesthesia	↓ xenograft growth	↑ ATM-AMPK-P53 pathway ↓ Akt-mtOR-4EBPl pathways in tumors ↓ angiogenesis ↑ expression of apoptosis markers	[69]

1L-6, Interleukin-6; *4EBP1*, Eukaryotic translation initial factor 4E-binding protein 1; *Akt*, Protein kinase B; *AMPK*, AMP-activated protein kinase; *Bax*, Be I-2-1ike protein 4; *Bcl-2*, B-cell lymphoma 2; *COL1A1*, Collagen type 1 alpha 1; *EGFR*, Epidermal growth factor receptor; *EMT*, Epithelial to mesenchymal transition; *ERK*, Extracellular regulated kinase; *IGF-1*, Insulin-like growth factor 1; *K-Ras*, V-Ki-ras2 Kirsten rat sarcoma viral oncogene homolog; *MAPK*, Mitogen activated protein kinase; *MMP-2*, Matrix metalloproteinase-2; *MMP-9*, Matrix metallopeptidase-9; *mTOR*, Mammalian target of rapamycin; *NF-κB*, Nuclear factor kappa-light chain-enhancer of activated B cells; *p53*, Tumor protein p53; *ROS*, Reactive oxygen species; *RTK*, Receptor tyrosine kinase; *SMAD2*, SMAD family member 2; *SMAD3*, SMAD family member 3; *SNAIL*, Zinc finger protein SNAI1; *STAT3*, Signal transduced and activator of transcription 3; *TGF-β*, Transforming growth factor β; *TKI*, Tyrosine kinase inhibitor.

significantly increasing adverse events. In this Phase II trial of patients with EGFR-mutated lung ADC, the addition of metformin increased survival, warranting the design of a larger phase III study (Table 10.5).

Approximately 10%−15% of NSCLC harbor-activating mutations in EGFR have been recognized as a key predictor of therapeutic sensitivity to EGFR TKIs. Among them, approximately 10%−12% of EGFR-mutant NSCLC

TABLE 10.5 Representative clinically relevant mouse trials in cancer therapies [36].

Trial design	Cancer type	Model type	Engineered drivers	Drugs/ Treatment	Significance	Relevant publications
Preclinical	Hematopoietic (APL)	GEM	PML-RARα fusion PLZF-RARα fusion	Retinoic acid	Demonstrated the efficacy of retinoic acid plus As_2O_3 in specific APL subtypes, validated in clinic	[46,95]
Preclinical	Pancreas (neuroendocrine)	GEM	RIP1-Tag2	Sunitinib	Demonstrated the efficacy of sunitinib plus imatinib, validated in clinic. FDA approved for pancreatic cancer treatment in 2011	[96,98]

(*Continued*)

TABLE 10.5 Representative clinically relevant mouse trials in cancer therapies [36].—cont'd

Trial design	Cancer type	Model type	Engineered drivers	Drugs/ Treatment	Significance	Relevant publications
Preclinical	Medulloblastoma	GEM	Ptc1^{+}/− P53−/−	GDC-0449 (SMO inhibitor)	Demonstrated the efficacy of an Shh pathway small molecule inhibitor, validated in clinic	[100,101]
Preclinical	Pancreas (neuroendocrine)	GEM	RIP1-Tag2	Erlotinib rapamycin	Demonstrated efficacy of combining drugs targeting EGFR and mTOR	[74]
Coclinical	Pancreas (PDA)	GEM	LSL-Kras^{G12D} LSL-Trp53^{R172H} Pdx-1-cre	Gemcitabine Nab-paclitaxel	Provided mechanistic insight into clinical cooperation between gemcitabine and nab-paclitaxel	[82,85]
Coclinical	Pancreas (PDA)	GEM	LSL-Kras^{G12D} LSL-Trp53^{R172H} Pdx-1-cre	CD40 monoclonal antibody gemcitabine	Demonstrated that targeting stroma was effective in treatment of metastatic PDA	[48]
Coclinical	Lung (NSCLC)	GEM	KRAS^{G12D} P53$^{fl/fl}$ Lkb1$^{fl/fl}$	Selumetinib Docetaxel	Validation of improved response of adding selumetinib to docetaxel treatment	[72,90]
Coclinical	Lung (NSCLC)	GEM	EML4-ALK fusion	Crizotinib Docetaxel Pemetrexed	GEM model predicted clinical outcome of drug combinations	[73,93]
Coclinical	Various sarcomas	PDX	N/A	Various chemotherapies	PDX testing predicted clinical outcome of drug combinations	[102]
Postclinical	Ovarian (SEOC)	GDA; PDX	RB/p53-deficient BRCA1/2-deficient	Olaparib Cisplatin	Validation of treatment efficacy in BRCA mutant tumors in both GDA and PDX models	[92,103]
Postclinical	Pancreas (neuroendocrine)	GDA	RIP1 -Tag2	Anti-VEGFR1 and anti-VEGFR2 antibodies	Identification of mechanisms of resistance to antiangiogenic therapies	[49]

TABLE 10.5 Representative clinically relevant mouse trials in cancer therapies [36].—cont'd

Trial design	Cancer type	Model type	Engineered drivers	Drugs/ Treatment	Significance	Relevant publications
Biomarker	Lung (NSCLC)	GEM; Carcinogen-induced	Various models	N/A	Used in-depth quantitative MS-based proteomics to profile plasma proteins	[87]
Biomarker	Pancreas (PDA)	GEM	$Kras^{G12D}$ $Ink4a/Arf^{fl/fl}$ Pdx-1-cne	N/A	Used in-depth proteomic analyses to identify candidate markers applicable to human cancer	[77]

tumors have an in-frame insertion within exon 20 of EGFR and are general resistant to EGFR TKIs. It is now clear that not all activating EGFR mutations are inherently sensitive to TKIs. Inframe EGFR exon 20 insertions are associated with de novo resistance to current clinically available TKIs. In addition, 90% of HER2 mutations in NSCLC are exon 20 mutations, and approximately 3% of patients with NSCLC harbor HER2 mutations. Together, EGFR and HER2 exon 20 mutations are found in approximately 4% of all patients with NSCLC [41–43]. In a series of preclinical studies, engineered lung tumor cell lines and PDX models of NSCLC patients with EGFR and/or HER2 exon 20 insertions were tested for the therapeutic efficacy of different generations of TKIs, followed by the usage of GEM mouse models of NSCLC driven by an EGFR or HER2 exon 20 insertion [42,43]. The clinical trials of TKLs tested in EGFR exon 20 insertion NSCLC patients are listed in Table 10.6.

TABLE 10.6 Key clinical trials in EGFR exon 20 insertion positive NSCLC [41].

Inhibitor(s)	Target(s)	Clinical trial ID(s)	Key results	References
Gefitinib/Erlotinib	EGFR	Retrospective analysis of clinical studies	<3 months PFS 8%–27% RR	[52], [53]
Dacomitinib	EGFR/HER2/ HER4	NCT00225121	PR for 1 patient with D770delinsGY	[60]
Afatinib	EGFR/HER2/ HER4	NCT00525148 NCT00949650 NCT01121393	8.7% RR, 27 months PFS	[58]
Neratinib	EGFR/HER2/ HER4	NCT00266877	0% RR	[54]
Osimertinib	EGFR T790M	NCT03414814	Ongoing	[66–68]
Poziotinib	EGFR/HER2	NCT03066206	Ongoing, 64% RR	[31]
Cetuximab + erlotinib	EGFR	NCT00895362	D770>GY patient with 3.5 years PFS	[81]
Cetuximab + afatinib	EGFR	NCT03727724	Preliminary report, 3 out of 4 ex20ins patients with PR, 5.4 months PFS	[82]

(Continued)

TABLE 10.6 Key clinical trials in EGFR exon 20 insertion positive NSCLC [41].—cont'd

Inhibitor(s)	Target(s)	Clinical trial ID(s)	Key results	References
Luminespib	Hsp90	NCT01854034	17% RR, 2.9 months PFS	[87]
Tarloxotinib	EGFR	–	Preclinical inhibition of ex20ins EGFR	[89]
TAK-788	EGFR/HER2 ex 20 ins	NCT02716116	Ongoing, preliminary antitumor activity reported	[83]
TAS6417	EGFR ex 20 ins	–	Preclinical inhibition of ex20ins EGFR	[84]
Compound 1A	EGFR/HER2 ex 20 ins	–	Preclinical inhibition of ex20ins EGFR	[85]

Details for trials with NCT numbers can be accessed on https://clinicaltrials.gov/.
PFS progression-free survival, *PR* partial response, *RR* response rate, *ex20ins* exon 20 insertion.

Conclusion

Drug repurposing provides various advantages over new drug development. First and foremost, the risk of failure in Phase I and II clinical trials for safety and efficacy is greatly reduced. Second, it can reduce the time span for drug development. Third, in general, less investment is required, although this varies considerably depending on the stage and process of development of the repurposing candidate. Finally, repurposed drugs may reveal new targets and pathways that can be further exploited [44]. Although repurposed therapies have demonstrated promise in NSCLC, their efficacy requires further validation [35]. Although large amounts of preclinical data are available, which emphasize the antineoplastic properties of these drugs, this availability has not consistently translated to clinical utility. The most successful examples of drug repurposing thus far have not involved a systemic approach but relied on retrospective clinical experience (e.g., sildenafil citrate for erectile dysfunction and thalidomide for erythema nodosum leprosum) [44].

In this chapter, we review the animal models that have been used for drug repurposing for lung cancer. The most commonly used murine lung cancer models in preclinical studies and trials are genetically engineered mouse models followed by xenograft models. The technology required to produce these models is sophisticated, and the underlying mechanisms are well-documented. However, despite their promise, none of the repurposed drugs are approved for clinical use in lung cancer, and most fail due to a lack of efficacy. The indication is that current preclinical methods are limited in their predictive power, given the systemic complexity of human lung tumors.

It is noted that the divergence between clinical trial results and real-world outcomes is largely unknown for many cancer types. A study reviewed by Cramer-van der Welle et al., which aims to assess the efficacy–effectiveness gap in systemic treatment for metastatic NSCLC, showed that patients treated in real-world practice have a nearly one-quarter shorter survival than those in clinical trials. The constant pattern of reduced effectiveness shows that the existence of this gap is a general phenomenon irrespective of the type of systemic treatment regimen, which may partly be explained by patients' performance status, earlier discontinuation, and fewer subsequent lines of treatment [45].

In our previous studies, we provided a mouse lung cancer model with spontaneous lung tumors that was well-characterized in terms of the histopathology, transcriptome profile, and molecular signatures in human lung ADC. We then used this mouse model and the LINCS L1000 database to repurpose existing drugs for lung ADC. By using this approach, we identified the mTOR inhibitor temsirolimus, which had been approved by the FDA for treating renal cell carcinoma, as a potential therapeutic agent. Our study demonstrated that a combination of low-dose chemotherapy and temsirolimus treatment was more effective in inhibiting tumor growth than a doublet chemotherapy regimen in the lung tumor mouse model. Our study results also suggest that the combination of low-dose chemotherapy and temsirolimus treatment may be beneficial for the treatment of lung ADC, which warrants further investigation [31,32].

In summary, the repurposing of approved noncancer medications to treat lung cancer presents a unique opportunity to improve outcomes by delivering an effective drug at lower costs with manageable toxicity. Several such agents have demonstrated antineoplastic activity and are being studied in clinical trials. We now have a wealth of model systems; each system has specific and unique benefits as predictive models for future preclinical studies. Systematic comparisons of several mouse models improve the predictive power for outcomes in humans and thus improve outcomes for repurposed therapies for lung cancer.

References

[1] Teng X-D. World Health Organization classification of tumours, pathology and genetics of tumours of the lung. Zhonghua bing li xue za zhi = Chinese J pathol 2005;34(8):544–6.

[2] Julian R, Molina PY, Cassivi SD, Schild SE, Adjei AA. Non–small cell lung cancer: epidemiology, risk factors, treatment, and survivorship. Mayo Clin Proc 2008;83(5):584–94.

[3] Ashburn TT, Thor KB. Drug repositioning: identifying and developing new uses for existing drugs. Nat Rev Drug Discov 2004;3(8):673–83.

[4] Hurle MR, et al. Computational drug repositioning: from data to therapeutics. Clin Pharmacol Ther 2013;93(4):335–41.

[5] Yella JK, et al. Changing trends in computational drug repositioning, vol. 11. Basel, Switzerland): Pharmaceuticals; 2018. 2.

[6] Xue H, et al. Review of drug repositioning approaches and resources. Int J Biol Sci 2018;14(10):1232–44.

[7] Pan Pantziarka LM. Omics-driven drug repurposing as a source of innovative therapies in rare cancers. Expert Opinion on Orphan Drugs 2018;6(9):513–7.

[8] Vempati UD, et al. Metadata standard and data exchange specifications to describe, model, and integrate complex and diverse high-throughput screening data from the library of integrated network-based cellular signatures (LINCS). J Biomol Screen 2014;19(5):803–16.

[9] Pan Pantziarka CV, Vidula Sukhatme RC, Crispino S, Gyawali B, Ilse Rooman AMVN, Meheus L, Sukhatme VP, Gauthier B. ReDO_DB: the repurposing drugs in oncology database. Ecancermedicalscience 2018;12:886.

[10] Kwon MC, Berns A. Mouse models for lung cancer. Molecular Oncology 2013;7(2):165–77.

[11] Meuwissen R, Berns A. Mouse models for human lung cancer. Genes Dev 2005;19(6):643–64.

[12] Dutt A, Wong KK. Mouse models of lung cancer. Clin Canc Res 2006;12(14 Pt 2):4396s–402s.

[13] Ding L, et al. Somatic mutations affect key pathways in lung adenocarcinoma. Nature 2008;455(7216):1069–75.

[14] Ferrer I, et al. KRAS-Mutant non-small cell lung cancer: from biology to therapy. Lung Canc 2018;124:53–64.

[15] Campbell SL, et al. Increasing complexity of Ras signaling. Oncogene 1998;17(11):1395–413. Reviews.

[16] Roman M, et al. KRAS oncogene in non-small cell lung cancer: clinical perspectives on the treatment of an old target. Mol Canc 2018;17(1):33.

[17] Kim CF, et al. Mouse models of human non-small-cell lung cancer: raising the bar. Cold Spring Harbor Symp Quant Biol 2005;70:241–50.

[18] DuPage M, Dooley AL, Jacks T. Conditional mouse lung cancer models using adenoviral or lentiviral delivery of Cre recombinase. Nat Protoc 2009;4(7):1064–72.

[19] de Seranno S, Meuwissen R. Progress and applications of mouse models for human lung cancer. Eur Respir J 2010;35(2):426–43.

[20] Gazdar AF, et al. Lung cancer cell lines as tools for biomedical discovery and research. J Natl Cancer Inst 2010;102(17):1310–21.

[21] Gazdar AF, Gao B, Minna JD. Lung cancer cell lines: Useless artifacts or invaluable tools for medical science? Lung Canc 2010;68(3):309–18.

[22] CONG XU XL, PIXU LIU, MAN LI, FUWEN LUO. Patient-derived xenograft mouse models: a high fidelity tool for individualized medicine (Review). Oncology Lett. 2019;17:3–10.

[23] Richmond A, Su Y. Mouse xenograft models vs GEM models for human cancer therapeutics. Dis Models Mech 2008;1(2–3):78–82.

[24] Kim Y, Pierce CM, Robinson LA. Impact of viral presence in tumor on gene expression in non-small cell lung cancer. BMC Canc 2018;18(1):843.

[25] Kim J-MBaEH. Human papillomavirus infection and risk of lung cancer in never-smokers and women: an 'adaptive' meta-analysis. Epidemiol Health 2015;37: e2015052.

[26] Giuliani L, et al. Human papillomavirus infections in lung cancer. Detection of E6 and E7 transcripts and review of the literature. Anticancer Res 2007;27(4C): 2697–704.

[27] Sun S, Schiller JH, Gazdar AF. Lung cancer in never smokers–a different disease. Nat Rev Canc 2007; 7(10):778–90.

[28] Carraresi L, et al. Thymic hyperplasia and lung carcinomas in a line of mice transgenic for keratin 5-driven HPV16 E6/E7 oncogenes. Oncogene 2001;20(56): 8148–53.

[29] Youssef G, et al. Ovine pulmonary adenocarcinoma: a large animal model for human lung cancer. ILAR J 2015;56(1):99–115.

[30] Gray ME, et al. Ovine pulmonary adenocarcinoma: a unique model to improve lung cancer research. Frontiers Oncol 2019;9:335.

[31] Chang H-W, et al. Therapeutic effect of repurposed temsirolimus in lung adenocarcinoma model. Front Pharmacol 2018;9:778.

[32] Chang HW, et al. Characterization of a transgenic mouse model exhibiting spontaneous lung adenocarcinomas with a metastatic phenotype. PLoS ONE 2017;12(4):e0175586.

[33] Duran I, et al. Characterisation of the lung toxicity of the cell cycle inhibitor temsirolimus. Eur J Canc 2006; 42(12):1875–80.

[34] Saxena A, et al. Therapeutic effects of repurposed therapies in non-small cell lung cancer: what is old is new again. Oncol 2015;20(8):934–45.

[35] Murray JC, Levy B. Repurposed drugs trials by cancer type: lung cancer. Canc J 2019;25(2):127–33.

[36] Day C-P, Merlino G, Van Dyke T. Preclinical mouse cancer models: a maze of opportunities and challenges. Cell 2015;163(1):39–53.

[37] Yousef M, Tsiani E. Metformin in lung cancer: review of in vitro and in vivo animal studies. Cancers 2017; 9(5).

[38] Guo Q, Liu Z, Jiang L, Liu M, Ma J, Yang C, Han L, Nan K, Liang X. Metformin inhibits growth of human non–small cell lung cancer cells via liver kinase B-1–independent activation of adenosine monophosphate–activated protein kinase. Mol Med Rep 2016;13(3): 2590–6.

[39] Evans JMM, et al. Metformin and reduced risk of cancer in diabetic patients. Br Med J 2005;330(7503): 1304–5.

[40] Arrieta O, Barrón F. Effect of metformin plus tyrosine kinase inhibitors compared with tyrosine kinase inhibitors Alone in patients with epidermal growth factor receptor–mutated lung adenocarcinoma A phase 2 randomized clinical trial. JAMA Oncol 2019;5(11): 1–11.

[41] Vyse S, Huang PH. Targeting EGFR exon 20 insertion mutations in non-small cell lung cancer. Signal Trans Target Therapy 2019;4:5.

[42] Robichaux JP, et al. Mechanisms and clinical activity of an EGFR and HER2 exon 20-selective kinase inhibitor in non-small cell lung cancer. Nat Med 2018;24(5): 638–46.

[43] Arcila ME, et al. Prevalence, clinicopathologic associations, and molecular spectrum of ERBB2 (HER2) tyrosine kinase mutations in lung adenocarcinomas. Clin Canc Res 2012;18(18):4910–8.

[44] Pushpakom S, et al. Drug repurposing: progress, challenges and recommendations. Nat Rev Drug Discov 2019;18(1):41–58.

[45] Cramer-van der Welle CM, et al. Systematic evaluation of the efficacy-effectiveness gap of systemic treatments in metastatic nonsmall cell lung cancer. Eur Respir J 2018;52(6).

[46] Ablain J, de The′ H. Retinoic acid signaling in cancer: The parable of acute promyelocytic leukemia. Int. J. Cancer 2014;135:2262–72.

[47] Wang Y, Lin B, Wu J, Zhang H, Wu B. Metformin inhibits the proliferation of A549/CDDP cells by activating p38 mitogen-activated protein kinase. Oncol. Lett. 2014;8:1269–74.

[48] Beatty GL, Torigian DA, Chiorean EG, Saboury B, Brothers A, Alavi A, Troxel AB, Sun W, Teitelbaum UR, Vonderheide RH, O'Dwyer PJ. A phase I study of an agonist CD40 monoclonal antibody (CP-870,893) in combination with gemcitabine in patients with advanced pancreatic ductal adenocarcinoma. Clin. Cancer Res. 2013;19:6286–95.

[49] Casanovas O, Hicklin DJ, Bergers G, Hanahan D. Drug resistance by evasion of antiangiogenic targeting of VEGF signaling in late-stage pancreatic islet tumors. Cancer Cell 2005;8:299–309.

[50] Huang D, He X, Zou J, Guo P, Jiang S, Lv N, Alekseyev Y, Luo L, Luo Z. Negative regulation of Bmi-1 by AMPK and implication in cancer progression. Oncotarget 2015;7:6188–200.

[51] Li L, Wang H, Kunlin L, Kejun Z, Caiyu L, Rui H, Conghua L, Yubo W, Hengyi C, Fenfen S, et al. Metformin attenuates gefitinib-induced exacerbation of pulmonary fibrosis by inhibition of TGF-_ signaling pathway. Oncotarget 2015;6:43605–19.

[52] Dong S, Zeng L, Liu Z, Li R, Zou Y, Li Z, Ge C, Lai Z, Xue Y, Yang J, et al. NLK functions to maintain proliferation and stemness of NSCLC and is a target of metformin. J. Hematol. Oncol. J. Hematol. Oncol. 2015;8:120.

[53] Wang J, Gao Q, Wang D, Wang Z, Hu C. Metformin inhibits growth of lung adenocarcinoma cells by inducing apoptosis via the mitochondria-mediated pathway. Oncol. Lett. 2015;10:1343–9.

[54] Li L, Han R, Xiao H, Lin C, Wang Y, Liu H, Li K, Chen H, Sun F, Yang Z, et al. Metformin sensitizes EGFR-TKI–resistant human lung cancer cells in vitro and in vivo through inhibition of IL-6 signaling and EMT reversal. Clin. Cancer Res. 2014;20:2714–26.

[55] Li L, Wang Y, Peng T, Zhang K, Lin C, Han R, Lu C, He Y. Metformin restores crizotinib sensitivityin crizotinib-resistant human lung cancer cells through inhibition of IGF1-R signaling pathway. Oncotarget 2016;7:34442–52.

[56] Vujic I, Sanlorenzo M, Posch C, Esteve-Puig R, Yen AJ, Kwong A, Tsumura A, Murphy R, Rappersberger K, Ortiz-Urda S. Metformin and trametinib have synergistic effects on cell viability and tumor growth in NRAS mutant cancer. Oncotarget 2015;6:969–78.

[57] Groenendijk FH, Mellema WW, van der Burg E, Schut E, Hauptmann M, Horlings HM, Willems SM, van den Heuvel MM, Jonkers J, Smit EF, et al. Sorafenib synergizes with metformin in NSCLC through AMPK pathway activation. Int. J. Cancer 2015;136:1434–44.

[58] Della Corte CM, Ciaramella V, Mauro CD, Castellone MD, Papaccio F, Fasano M, Sasso FC, Martinelli E, Troiani T, De Vita F, et al. Metformin increases antitumor activity of MEK inhibitors through GLI1 downregulation in LKB1 positive human NSCLC cancer cells. Oncotarget 2015;7:4265–78.

[59] Lin C-C, Yeh H-H, Huang W-L, Yan J-J, Lai W-W, Su W-P, Chen HHW, Su W-C. Metformin enhances cisplatin cytotoxicity by suppressing signal transducer and activator of transcription–3 activity independently of the liver kinase B1–AMP-activated protein kinase pathway. Am. J. Respir. Cell Mol. Biol. 2013;49:241–50.

[60] Teixeira SF, dos Santos Guimarães I, Madeira KP, Daltoé RD, Silva IV, Rangel LBA. Metformin synergistically enhances antiproliferative effects of cisplatin and etoposide in NCI-H460 human lung cancer cells. J. Bras. Pneumol. 2013;39:644–9.

[61] Tseng S-C, Huang Y-C, Chen H-J, Chiu H-C, Huang Y-J, Wo T-Y, Weng S-H, Lin Y-W. Metformin-mediated downregulation of p38 mitogen-activated protein kinase-dependent excision repair cross-complementing 1 decreases DNA repair capacity and sensitizes human lung cancer cells to paclitaxel. Biochem. Pharmacol. 2013;85:583–94.

[62] Hann SS, Tang Q, Zheng F, Zhao S, Chen J, Wang Z. Repression of phosphoinositide-dependent protein kinase 1 expression by ciglitazone via Egr-1 represents a new approach for inhibition of lung cancer cell growth. Mol. Cancer 2014;13:149.

[63] Xiao Z, Sperl B, Ullrich A, Knyazev P. Metformin and salinomycin as the best combination for the eradication of NSCLC monolayer cells and their alveospheres (cancer stem cells) irrespective of EGFR, KRAS, EML4/ALK and LKB1 status. Oncotarget 2014;5: 12877–90.

[64] Koeck S, Amann A, Huber JM, Gamerith G, Hilbe W, Zwierzina H. The impact of metformin and alinomycin on transforming growth factor _-induced epithelial-to-mesenchymal transition in non-small cell lung cancer cell lines. Oncol. Lett. 2016;11:2946–52.

[65] Cao H, Dong W, Qu X, Shen H, Xu J, Zhu L, Liu Q, Du J. Metformin enhances the therapy effects of anti-IGF-1R mAb figitumumab to NSCLC. Sci. Rep. 2016;6:31072.

[66] Nazim UM, Moon J-H, Lee J-H, Lee Y-J, Seol J-W, Eo S-K, Lee J-H, Park S-Y. Activation of autophagy flux by metformin downregulates cellular FLICE-like inhibitory protein and enhances TRAILinduced apoptosis. Oncotarget 2016;7:23468–81.

[67] Zhao S, Wu J, Zheng F, Tang Q, Yang L, Li L, Wu W, Hann SS. _-elemene inhibited expression of DNA methyltransferase 1 through activation of ERK1/2 and AMPK_ signalling pathways in human lung cancer cells: The role of Sp1. J. Cell. Mol. Med. 2015;19: 630–41.

[68] Storozhuk Y, Hopmans SN, Sanli T, Barron C, Tsiani E, Cutz J-C, Pond G, Wright J, Singh G, Tsakiridis T. Metformin inhibits growth and enhances radiation response of non-small cell lung cancer (NSCLC) through ATM and AMPK. Br. J. Cancer 2013;108:2021–32.

[69] Quinn BJ, Dallos M, Kitagawa H, Kunnumakkara AB, Memmott RM, Hollander MC, Gills JJ, Dennis PA. Inhibition of lung tumorigenesis by metformin is associated with decreased plasma IGF-I and diminished receptor tyrosine kinase signaling. Cancer Prev. Res. 2013;6:801–10.

[70] Zhao Z, Cheng X, Wang Y, Han R, Li L, Xiang T, He L, Long H, Zhu B, He Y. Metformin Inhibits the IL-6-Induced Epithelial-Mesenchymal Transition and Lung Adenocarcinoma Growth and Metastasis. PLoS ONE 2014;9. e95884.

[71] Ceteci F, Xu J, Ceteci S, Zanucco E, Thakur C, Rapp UR. Conditional expression of oncogenic C-RAF in mouse pulmonary epithelial cells reveals

differential tumorigenesis and induction of autophagy leading to tumor regression. Neoplasia 2011; 13:1005–18.

[72] Chen Z, Cheng K, Walton Z, Wang Y, Ebi H, Shimamura T, Liu Y, Tupper T, Ouyang J, Li J, et al. A murine lung cancer co-clinical trial identifies genetic modifiers of therapeutic response. Nature 2012;483: 613–7.

[73] Chen H, YaoW Chu Q, et al. Synergistic effects of metformin in combination with EGFR-TKI in the treatment of patients with advanced non–small cell lung cancer and type 2 diabetes. Cancer Lett 2015;369: 97–102.

[74] Chiu CW, Nozawa H, Hanahan D. Survival benefit with proapoptotic molecular and pathologic responses from dual targeting of mammalian target of rapamycin and epidermal growth factor receptor in a preclinical model of pancreatic neuroendocrine carcinogenesis. J. Clin. Oncol. 2010;28:4425–33.

[75] Dankort D, Filenova E, Collado M, Serrano M, Jones K, McMahon M. A new mouse model to explore the initiation, progression, and therapy of BRAFV600E-induced lung tumors. Genes Dev 2007; 21:379–84.

[76] Engelman JA, Chen L, Tan X, Crosby K, Guimaraes AR, Upadhyay R, Maira M, et al. Effective use of PI3K and MEK inhibitors to treat mutant Kras G12D and PIK3CA H1047R murine lung cancers. Nat. Med. 2008;14:1351–6. https://doi.org/10.1038/nm.1890.

[77] Faca VM, Song KS, Wang H, Zhang Q, Krasnoselsky AL, Newcomb LF, Plentz RR, Gurumurthy S, Redston MS, Pitteri SJ, et al. A mouse to human search for plasma proteome changes associated with pancreatic tumor development. PLoS Med 2008;5:e123.

[78] Fisher GH, Wellen SL, Klimstra D, Lenczowski JM, Tichelaar JW, Lizak MJ, Whitsett JA, Koretsky A, Varmus HE. Induction and apoptotic regression of lung adenocarcinomas by regulation of a K-Ras transgene in the presence and absence of tumor suppressor genes. Genes Dev 2001;15:3249–62.

[79] Wheler JJ, et al. Combining erlotinib and cetuximab is associated with activity in patients with non-small cell lung cancer (including squamous cell carcinomas) and wild-type EGFR or resistant mutations. Mol. Cancer Ther. 2013. https://doi.org/10.1158/1535-7163.MCT-12-1208.

[80] van Veggel B, et al. Afatinib and cetuximab in four patients with EGFR Exon 20 insertion–positive advanced NSCLC. J. Thorac. Oncol. 2018;13:1222–6.

[81] Estrada-Bernal A, et al. Abstract A157: Antitumor activity of tarloxotinib, a hypoxia-activated EGFR TKI, in patient-derived lung cancer cell lines harboring EGFR exon 20 insertions. EGFR/Her2 2018;17. A157–A157.

[82] Frese KK, Neesse A, Cook N, Bapiro TE, Lolkema MP, Jodrell DI, Tuveson DA. nab-Paclitaxel potentiates gemcitabine activity by reducing cytidine deaminase levels in a mouse model of pancreatic cancer. Cancer Discov 2012;2:260–9.

[83] Doebele RC, et al. First report of safety, PK, and preliminary antitumor activity of the oral EGFR/HER2 exon 20 inhibitor TAK-788 (AP32788) in non–small cell lung cancer (NSCLC). J. Clin. Oncol. 2018;36(15 (suppl)). 9015–9015.

[84] Hasako S, et al. TAS6417, a novel EGFR inhibitor targeting exon 20 insertion mutations. Mol. Cancer Ther. 2018;17:1648–58.

[85] Goldstein D, El-Maraghi RH, Hammel P, Heinemann V, Kunzmann V, Sastre J, Scheithauer W, Siena S, Tabernero J, Teixeira L, et al. nab-Paclitaxel plus gemcitabine for metastatic pancreatic cancer: long-term survival from a phase III trial. J. Natl. Cancer Inst. 2015;107.

[86] Guerra C, Mijimolle N, Dhawahir A, Dubus P, Barradas M, Serrano M, Campuzano V, Barbacid M. Tumor induction by an endogenous K-ras oncogene is highly dependent on cellular context. Cancer Cell 2003;4:111–20.

[87] Hanash S, Taguchi A. Application of proteomics to cancer early detection. Cancer J 2011;17:423–8.

[88] Iwanaga K, Yang Y, Raso MG, Ma L, Hanna AE, Thilaganathan N, Moghaddam S, Evans CM, Li H, Cai W-W, et al. Pten inactivation accelerates oncogenic K-rasinitiated tumorigenesis in a mouse model of lung cancer. Cancer Res. 2008;68:1119–27.

[89] Jackson EL, Willis N, Mercer K, Bronson RT, Crowley D, Montoya R, Jacks T, Tuveson DA. Analysis of lung tumor initiation and progression using conditional expression of oncogenic K-ras. Genes Dev. 2001;15:3243–8.

[90] Jänne PA, et al. Dacomitinib as first-line treatment in patients with clinically or molecularly selected advanced non-small-cell lung cancer: a multicentre, openlabel, phase 2 trial. Lancet Oncol 2014;15:1433–41.

[91] Ji H, Ramsey MR, Hayes DN, Fan C, McNamara K, Kozlowski P, Torrice C, Wu MC, Shimamura T, Perera SA, et al. LKB1 modulates lung cancer differentiation and metastasis. Nature 2007;448:807–10.

[92] Kortmann U, McAlpine JN, Xue H, Guan J, Ha G, Tully S, Shafait S, Lau A, Cranston AN, O'Connor MJ, et al. Tumor growth inhibition by olaparib in BRCA2 germline-mutated patient-derived ovarian cancer tissue xenografts. Clin. Cancer Res. 2011;17:783–91.

[93] Lunardi A, Pandolfi PP. A co-clinical platform to accelerate cancer treatment optimization. Trends Mol. Med. 2015;21:1–5.

[94] Meuwissen R, Linn SC, van der Valk M, Mooi WJ, Berns A. Mouse model for lung tumorigenesis through Cre/lox controlled sporadic activation of the K-Ras oncogene. Oncogene 2001;20:6551–8.

[95] Pandolfi PP. Oncogenes and tumor suppressors in the molecular pathogenesis of acute promyelocytic leukemia. Hum. Mol. Genet. 2001;10:769–75.

[96] Pietras K, Hanahan D. A multitargeted, metronomic, and maximum-tolerated dose "chemo-switch" regimen is antiangiogenic, producing objective responses and survival benefit in a mouse model of cancer. J. Clin. Oncol. 2005;23:939–52.

[97] Politi K, Zakowski MF, Fan P-D, Schonfeld EA, Pao W, Varmus HE. Lung adenocarcinomas induced in mice by mutant EGF receptors found in human lung cancers respond to a tyrosine kinase inhibitor or to down-regulation of the receptors. Genes Dev. 2006;20:1496–510.

[98] Raymond E, Dahan L, Raoul J-L, Bang Y-J, Borbath I, Lombard-Bohas C, Valle J, Metrakos P, Smith D, Vinik A, et al. Sunitinib malate for the treatment of pancreatic neuroendocrine tumors. N. Engl. J. Med. 2011;364:501–13.

[99] Regales L, Balak MN, Gong Y, Politi K, Sawai A, Le C, Koutcher JA, Solit DB, Rosen N, Zakowski MF, Pao W. Development of new mouse lung tumor models expressing EGFR T790M mutants associated with clinical resistance to kinase inhibitors. PLoS ONE 2007;2(8):e810.

[100] Romer JT, Kimura H, Magdaleno S, Sasai K, Fuller C, Baines H, Connelly M, Stewart CF, Gould S, Rubin LL, Curran T. Suppression of the Shh pathway using a small molecule inhibitor eliminates medulloblastoma in Ptc1(+/-)p53(-/-) mice. Cancer Cell 2004;6:229–40.

[101] Rudin CM, Hann CL, Laterra J, Yauch RL, Callahan CA, Fu L, Holcomb T, Stinson J, Gould SE, Coleman B, et al. Treatment of medulloblastoma with hedgehog pathway inhibitor GDC-0449. N. Engl. J. Med. 2009;361:1173–8.

[102] Stebbing J, Paz K, Schwartz GK, Wexler LH, Maki R, Pollock RE, Morris R, Cohen R, Shankar A, Blackman G, et al. Patient-derived xenografts for individualized care in advanced sarcoma. Cancer 2014; 120:2006–15.

[103] Szabova L, Bupp S, Kamal M, Householder D.B, Hernandez L, Schlomer J.J, Baran M.L, Yi M, Stephens R.M, Annunziata C.M, Martin P.L, Van Dyke T.A, Weaver Ohler Z, Difilippantonio S. Pathway-specific engineered mouse allograft models functionally recapitulate human serous epithelial ovarian cancer. PLoS ONE. 2014;9(4):e95649. https://doi.org/10.1371/journal.pone.0095649. eCollection 2014.

[104] Winslow MM, Dayton TL, Verhaak RGW, Kim-Kiselak C, Snyder EL, Feldser DM, Hubbard DD, DuPage MJ, Whittaker CA, Hoersch S, et al. Suppression of lung adenocarcinoma progression by Nkx2-1. Nature 2011;473:101–4.

[105] Xu X, Rock JR, Lu Y, Futtner C, Schwab B, Guinney J, Hogan BLM, Onaitis MW. Evidence for type II cells as cells of origin of K-Ras-induced distal lung adenocarcinoma. Proc. Natl. Acad. Sci. USA 2012; 109:4910–5.

[106] Johnson L, Mercer K, Greenbaum D, Bronson RT, Crowley D, Tuveson DA, Jacks T. Somatic activation of the K-ras oncogene causes early onset lung cancer in mice. Nature 2001;410(6832):1111–6. https://doi.org/10.1038/35074129.

[107] Li H, Cho SN, Evans CM, Dickey BF, Jeong J-W, Demayo FJ. Cre mediated recombination in mouse clara cells. Genesis 2008;46:300–7.

CHAPTER

11

Identification of chemosensitizers by drug repurposing to enhance the efficacy of cancer therapy

Ge Yan, Thomas Efferth

Department of Pharmaceutical Biology, Institute of Pharmacy and Biochemistry, Johannes Gutenberg University, Mainz, Germany

OUTLINE

Introduction

The development of drug resistance, which eventually leads to therapy failure and tumor relapse, emerges as a major challenge in cancer treatment. Clinic drug resistance is featured by the resistance toward a broad spectrum of drugs, which are structurally and functionally unrelated, i.e., multidrug resistance (MDR). Both intrinsic and acquired drug resistance are attributed to at least one of the following mechanisms: (a) altered expression of drug efflux transporters; (b) sequestration of drugs mediated by acidic cellular vesicles [1]; (c) enhanced drug biotransformation and metabolism [2,3]; (d) impairment of apoptosis through DNA

https://doi.org/10.1016/B978-0-12-819668-7.00011-7

repair augment [4], checkpoint activation [5] or cellular stress harness [6,7]; (e) stemness induction and epithelial–mesenchymal transition (EMT) [8]; (f) altered epigenomics by methylation, acetylation, and microRNA (miRNA) regulation [9]; and (g) mutations of drug targets [10] (Fig. 11.1).

In order to combat MDR, chemosensitizers aiming to improve the efficacy of standard cytotoxic regimens have been proposed and developed accordingly. In principle, drugs reversing the resistance mechanisms, except for targets mutations, are able to restore cellular sensitivity toward chemotherapies (Fig. 11.1): (1) drugs restraining efflux transporters can enhance intracellular dose of chemotherapies; (2) drugs disrupting acidic cellular vesicles can prevent vesicle sequestration of cytotoxic agents and thereby enhance the accessibility of chemotherapies to their targets; (3) inhibitors of metabolic enzymes responsible for biotransformation can elevate the concentration of therapeutically effective drugs; (4) drugs exhibiting concomitant effects on cellular process including oncogenic signaling, cell cycle, DNA repair, stress handling, or energy supply can augment the apoptosis induced by anticancer agents; (5) drugs reversing EMT process can sensitize tumors toward treatment; and (6) miRNAs negatively regulating prosurvival gene expression can promote apoptosis induced by anticancer agents. The first three types of chemosensitizers aiming at enhancing the therapeutic concentration of anticancer agents are referred to as quantitative chemosensitizers in this chapter, while the latter three types aiming at improving the killing effect of anticancer agents are referred to as qualitative chemosensitizers.

Noticeably, numerous drugs already existing on the market as clinically approved medications for a broad range of indications outside oncology exhibit potent potential to restore cellular response through abovementioned mechanisms. Considering the high failure rates and costs during conventional drug development, repurposing of these drugs to enhance tumor response toward chemotherapies is considered to be a more efficient and economical strategy. In this chapter, clinically approved drugs with chemosensitizing properties and their functional mechanisms will be discussed in depth.

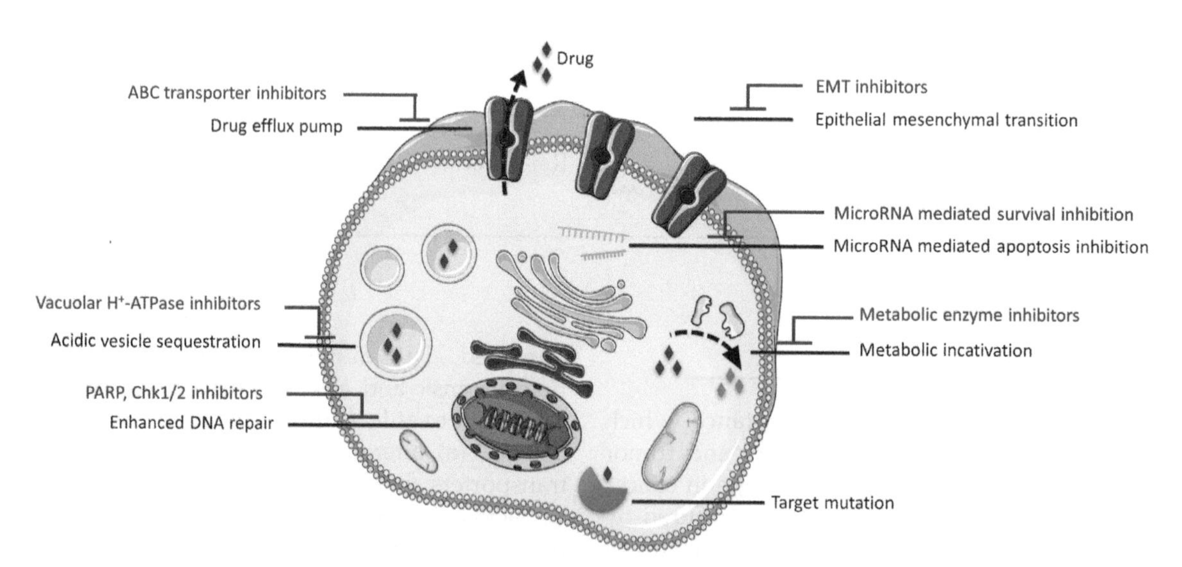

FIGURE 11.1 Resistance mechanisms (black) and reversing strategies (red).

Quantitative chemosensitizers

ATP-binding cassette transporters inhibitors

Insufficient drug accumulation in tumor cells due to overexpression of energy-dependent efflux transporters accounts for a large proportion of MDR cases. These drug transporters belong to the ATP-binding cassette (ABC) transporter family and are characterized by the capacity of binding and extruding drugs against the concentration gradient at the expense of ATP hydrolysis. A total of 49 members of the family were found in the human genome, which are divided into seven subfamilies designated from A to G [11–13]. In this chapter, we concentered on the most prominent ABC transporters, i.e., P-glycoprotein (P-gp, also known as MDR1 or ABCB1), MDR-associated protein family members (MRP, also known as ABCC), and breast cancer–resistance protein (BCRP, also known as ABCG2, MXR, or ABC-P) (Fig. 11.2). They confer resistance toward a broad range of anticancer agents (Table 11.1) through conformational change. Taking P-gp as an example, P-gp adopts an inward-face conformation showing a turned V-like structure, which forms the main substrate binding cavity at the inner leaflet of the membrane [18]. Upon binding to its substrates at the top of the turned "V" structure, each of the two NBDs recruit one ATP molecule. Accompanied by ATP hydrolysis, P-gp undergoes conformational change, which eventually opens the inner membrane and allows access of the bound substrates to the periplasm. Similar to P-gp, the substrate transportation of MRP depends on conformation alteration upon hydrolysis of ATP molecules [19]. However, the concrete process in BCRP has not been demonstrated. Due to the structural specificity of BCRP, many studies have indicated that BCRP requires homodimerization [20] or higher-order oligomerization [21].

P-glycoprotein

P-gp encoded by the *MDR1* gene was among the first discovered ABC transporters. P-gp contains approximately 1280 amino acids with a molecular mass of ~170 kDa, which presents a tandemly duplicated structure with two intracellular nucleotide-binding domains (NBDs) and two hydrophobic six-helices transmembrane domains (TMDs) [13] (Fig. 11.2A). P-gp transports a broad range of substrates from small molecules such as organic cations to macromolecules such as proteins, most of which are electrically neutral or slightly positive lipophilic molecules. These substrates varied in structure and functionality. Regarding anticancer agents, it excretes doxorubicin, daunorubicin, vincristine, vinblastine, paclitaxel, and docetaxel, etc. (Table 11.1) [17].

Considering the significant clinic relevance of P-gp as a key mediator of MDR, extensive studies have been devoted to identify potent and specific inhibitors by drug repurposing (Table 11.2). These investigated inhibitors were initially designed for antihypertension, mental disorder control, antiinflammation, and antiinfection. Mechanically speaking, these drugs are normally antagonists against ion channels, enzymes, adrenergic receptors, histamine receptors, hormone receptors, etc. However, not all inhibitors have equal potency due to different capability to form complex with P-gp drug-binding regions [22].

Multidrug resistance–associated protein

The human MRP family composes another important branch of ABC transporters. Nine members of the family have been identified. MRP1, also known as ABCC1 or GS-X pump, was the first discovered member of MRPs in the early 1990s [23,24]. MRP1 with a molecular mass of 190 kDa is ubiquitously expressed in the body, specifically in polarized epithelial cells. Compared to the structure of P-gp, MRPs have an additional membrane-spanning

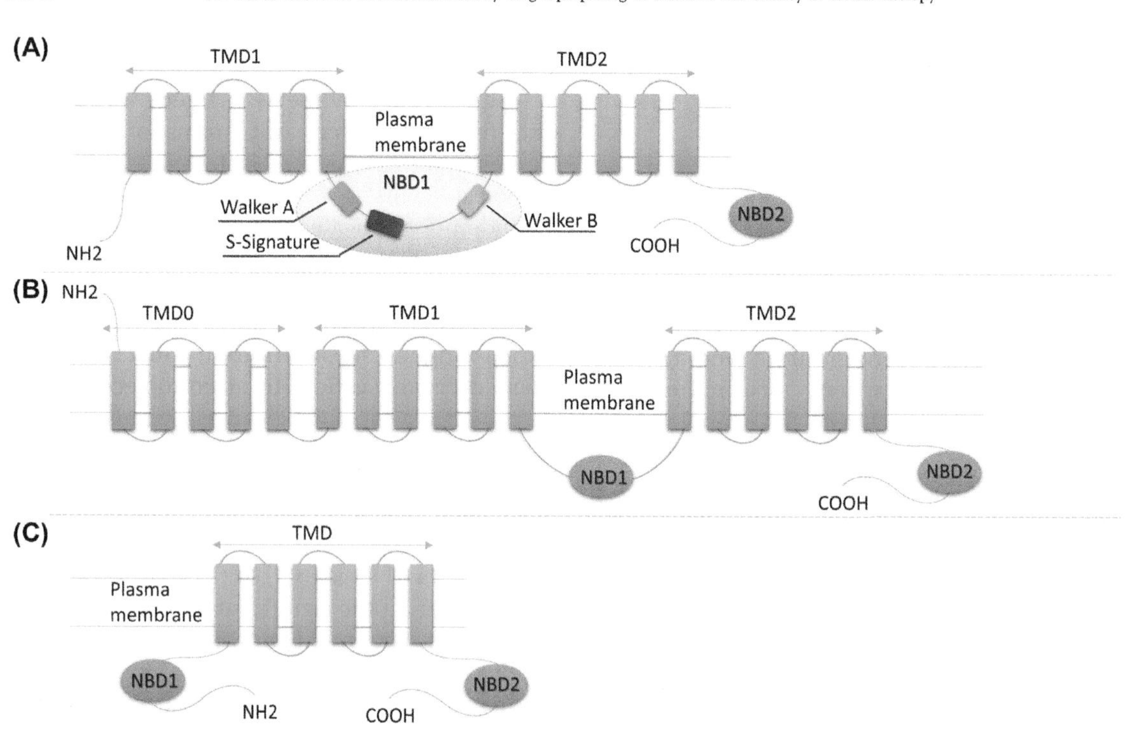

FIGURE 11.2 Topological structure of typical ABC transporters. *NBD*, nucleotide-binding domain; *TMD*, transmembrane domain. Each NBD domain comprises the Walker A/B motifs and S-signature motif. Only the first NBD is shown in detail, while the rest NBDs are shown in simplified pattern. (A) Model structure of P-gp. (B) Model structure of MRP1. (C) Model structure of BCRP.

N-terminal region (TMD0) or an intracellular loop (Fig. 11.2B). MRP4, MRP5, MRP8, and MRP9 lack the third membrane-spanning domain but possess the additional loop. By comparison, MRP1, MRP2, MRP3, MRP6, and MRP7 contain both additional segments. The most recent discovered ABCC13 gene encodes a pseudoprotein belonging to ABCC transporter family [25]. Unlike P-gp, all members can excrete lipophilic anions including anticancer agents. The resistance profiles among MRPs are partially overlapping but not identical, e.g., anthracyclines, *vinca* alkaloids, epipodophyllotoxins, methotrexate, and camptothecins are substrates of MRP1. Cisplatin is a substrate for MRP2, but not for MRP1 [26]. There is no cross-resistance to taxanes, which are an important component in P-gp resistance profile (Table 11.1). Although there is considerate overlap between the resistance profiles of P-gp and MRP1, the substrates of MRP1 are mostly lipophilic anions and modified with glutathione, glucosylation, sulfation, or glucuronylation. Among these modifications, glutathione conjugation and glucosylation represent typical products of phase II drug metabolism. In addition to the conjugation, MRP1 can cotransport free glutathione and xenobiotics (e.g., etoposide) or only transport anionic substrate (e.g., estrone-3-sulfate), in the presence of free glutathione but without actually transporting glutathione.

Some specific compounds including MK571 and reversan [27] have been developed to target MRP1. Under the scope of drug repurposing, many drugs inhibiting P-gp also have an impact

TABLE 11.1 Resistance to anticancer agents mediated by three principal ABC transporters involved in MDR.

Drug classification		Resistance-mediating proteins		
		P-gp	MRP1	BCRP
Topoisomerase I inhibitors	Irinotecan/SN-38	R	R	R
	Topotecan	R	R	R
	Methotrexate/MTX		R	R
Topoisomerase II inhibitors/anthracyclines	Doxorubicin	R	R	R
	Daunorubicin	R		R
	Epirubicin	R	R	R
	Mitoxantrone	R	R	R
	Bisantrene	R		R
Topoisomerase II inhibitors/epipodophyllotoxins	Etoposide/VP-16	R	R	
	Teniposide/VM-26	R	R	
Vinca alkaloids	Vinblastine	R	R	
	Vincristine	R	R	
	Vinorelbine	R		
	Catharanthine	R		
Taxanes	Docetaxel	R		
	Paclitaxel	R		
Antibiotics	Actinomycin D	R		
	Mitomycin	R		
Tyrosine kinase inhibitors	Imatinib	R	R	R
	Gefitinib		R	R
CDK inhibitor	Flavopiridol/alvocidib			R
Others	Rhodamine 123	R		R

R indicates resistance conferred by the protein toward a certain clinic drug. The drugs are compiled from elegant studies [14–17].

on MRP1. These drugs are broadly used as antiarrhythmic agent (verapamil), calcium channel blocker (nicardipine), calmodulin antagonist (cyclosporine A), nonsteroidal antiinflammatory drug (indomethacin), tyrosine kinase inhibitor (imatinib), HIV therapy (abacavir), or antibiotic (penicillin G) [28]. In addition, the uricosuric agents probenecid [29] and sulfinpyrazone, the antiedema agent furosemide, the anticoagulant agent dipyridamole [30] and a veterinarian antibiotic difloxacin [31] also inhibit MRP1 [32].

Breast cancer–resistance protein

BCRP was initially identified as an approximately 75 kDa plasma membrane protein in a

TABLE 11.2 Approved drugs as P-gp inhibitors.

Target	Drug	Mechanism	Purpose
Receptor	Carvedilol	Adrenergic receptor inhibitor	Congestive heart failure (CHF)
	Doxazosin	Adrenergic receptor inhibitor	Antihypertension
	Prazosin	Adrenergic receptor inhibitor	Antihypertension
	Yohimbine	Adrenergic receptor inhibitor	Antiimpotence
	Amoxapine	Dopamine receptor inhibitor	Mental disorder
	Bromocriptine	Dopamine receptor inhibitor	Mental disorder
	Chlorpromazine	Dopamine receptor inhibitor	Mental disorder
	Fluoxetine	Dopamine receptor inhibitor	Mental disorder
	Haloperidol	Dopamine receptor inhibitor	Mental disorder
	Loxapine	Dopamine receptor inhibitor	Mental disorder
	Paroxetine	Dopamine receptor inhibitor	Mental disorder
	Prochlorperazine	Dopamine receptor inhibitor	Mental disorder
	Sertraline	Dopamine receptor inhibitor	Mental disorder
	Bicalutamide	Nonsteroidal agents	Hormone regulation
	Medroxyprogesterone acetate	Steroidal agents	Hormone regulation
	Megestrol acetate	Steroidal agents	Hormone regulation
	Mifepristone	Steroidal agents	Hormone regulation
	Progesterone	Steroidal agents	Hormone regulation
	Tamoxifen	Steroidal agents	Hormone regulation
	Toremifene	Steroidal agents	Hormone regulation
	Erlotinib	Tyrosine kinase receptor inhibitor	Anticancer
	Gefitinib	Tyrosine kinase receptor inhibitor	Anticancer
	Lapatinib	Tyrosine kinase receptor inhibitor	Anticancer
	Astemizole[a]	Histamine receptor antagonist	Antihistamine
	Azelastine	Histamine receptor antagonist	Antihistamine
	Benzquinamide	Histamine receptor antagonist	Antihistamine
	Terfenadine[a]	Histamine receptor antagonist	Antihistamine
	Methylprednisolone	Glucocorticoid receptor agonist	Antiinflammation
	Nicotine	Nicotinic cholinergic receptor agonist	Nicotine withdrawal
	Sirolimus	mTOR inhibitor	Immunosuppressants
	Midazolam	GABA receptor potentiator	Anesthetic
Enzyme	Bafilomycin	H+-ATPase (vacuolar) inhibitor	Antibiotic
	Concanamycin A	H+-ATPase (vacuolar) inhibitor	Antibiotic

TABLE 11.2 Approved drugs as P-gp inhibitors.—cont'd

Target	Drug	Mechanism	Purpose
	Ibuprofen	Inhibitor of cyclooxygenase	Antiinflammation
	Indomethacin	Inhibitor of cyclooxygenase	Antiinflammation
	Econazole	Inhibits cytochrome P450–dependent enzymes	Antifungal
	Itraconazole	Inhibits cytochrome P450–dependent enzymes	Antifungal
	Ketoconazole	Inhibits cytochrome P450–dependent enzymes	Antifungal
	Caffeine	Phosphodiesterase inhibitor	CNS stimulant
	Dipyridamole	Phosphodiesterase inhibitor	Antiplatelet aggregation
	Pentoxifylline	Phosphodiesterase inhibitor	Chronic occlusive arterial
	Vardenafil	Phosphodiesterase inhibitor	Erectile dysfunction
	Ritonavir	HIV protease inhibitor	HIV/AIDS
	Atorvastatin	HMG-CoA reductase	Antidyslipidemias
	Disulfiram	Aldehyde dehydrogenase inhibitor	Alcohol deterrent
Ion channel	Amiodarone	Calcium channel blocker	Antianginal and antiarrhythmia
	Bepridil[a]	Calcium channel blocker	Antianginal
	Diltiazem	Calcium channel blocker	Antihypertension
	Emopamil	Calcium channel blocker	Renal injury
	Felodipine	Calcium channel blocker	Antihypertension
	Isradipine	Calcium channel blocker	Antihypertension
	Mibefradil[a]	Calcium channel blocker	Antihypertension
	Nicardipine	Calcium channel blocker	Antianginal
	Nifedipine	Calcium channel blocker	Antihypertension
	Nimodipine	Calcium channel blocker	Antihypertension
	Nitrendipine	Calcium channel blocker	Antihypertension
	Verapamil	Calcium channel blocker	Antiarrhythmia
	Reserpine	Calcium channel blocker	Antihypertension
	Cinchocaine	Sodium channel blocker	Anesthetic
	Lidocaine	Sodium channel blocker	Anesthetic
	Propafenone	Sodium channel blocker	Antiarrhythmia
	Propofol	Sodium channel blocker	Anesthetic
	Quinidine	Sodium channel blocker	Antiarrhythmia

(Continued)

TABLE 11.2 Approved drugs as P-gp inhibitors.—cont'd

Target	Drug	Mechanism	Purpose
P-gp–specific inhibitors	Biricodar	P-gp/MRP1 inhibitor	MDR
	Dofequidar	P-gp inhibitor	MDR
	Elacridar	P-gp inhibitor	MDR
	Laniquidar	P-gp inhibitor	MDR
	Tariquidar	P-gp inhibitor	MDR
	Tesmilifene	P-gp inhibitor	MDR
	Valspodar	P-gp inhibitor	MDR
	Zosuquidar	P-gp inhibitor	MDR
Other	Cyclosporin A	Calmodulin antagonists	Immunosuppressants
	trans-Flupenthixol[a]	Calmodulin antagonists	Mental disorder
	Fluphenazine	Calmodulin antagonists	Mental disorder
	Perphenazine	Calmodulin antagonists	Mental disorder
	Prochlorperazine	Calmodulin antagonists	Mental disorder
	Tacrolimus	Calmodulin antagonists	Immunosuppressants
	Trifluoperazine	Calmodulin antagonists	Mental disorder
	Cefoperazone	Penicillin-binding protein inhibitor	Antibiotic
	Ceftriaxone	Penicillin-binding protein inhibitor	Antibiotic
	Azithromycin	Interfering with 50S ribosomal subunit	Antibiotic
	Clarithromycin	Interfering with 50S ribosomal subunit	Antibiotic
	Erythromycin	Interfering with 50S ribosomal subunit	Antibiotic
	Tetracycline	Interfering with 50S ribosomal subunit	Antibiotic
	Curcumin	Scavenger of oxygen species	Antiinflammation
	Beta carotene	Scavenger of oxygen species	Antioxidation
	Tetrabenazine	Vesicular monoamine transporter inhibitor	Hyperkinetic disorders
	Brefeldin A	Not completely established	Antibiotic
	Colchicine	Not completely established	Antiinflammation
	Metronidazole	Not completely established	Antiinfection
	Mitotane	Not completely established	Anticancer
	Quinine	Not completely established	Antimalarial
	Valinomycin	Not completely established	Antibiotic
	Epigallocatechin gallate	Not completely established	Antihypertension

The drugs are compiled from elegant studies [85–91].

[a] *indicates the drug has been withdrawn from clinic.*

human multidrug-resistant cell line MCF-7/AdrVp in 1998 [33]. Unlike P-gp and the MRP family members, BCRP is termed as half-transporter due to the fact that it only has one region with one six-helices TMD and a single NBD composing 655 amino acids and it requires homodimerization or multimerization for transport activity [34] (Fig. 11.2C).

The BCRP is capable of expelling a large range of xenobiotics, including anthracyclines (e.g., doxorubicin), topoisomerase inhibitors (e.g., mitoxantrone), and tyrosine kinase inhibitors (e.g., imatinib) (Table 11.1). It can export both hydrophobic substrates such as mitoxantrone and hydrophilic conjugated organic anions. Moreover, mutated BCRP potentially increases drug resistance [35,36]. Multiple binding sites of BCRP are suggested according to the type of substrates. The drug-binding sites in BCRP with doxorubicin, prazosin, and daunorubicin cluster in one region. Rhodamine 123 and methotrexate bind in possibly overlapping but not identical regions [37].

Fumitremorgin C, which is initially designed to treat fungi infection, exhibits highly selective activity against BCRP, but no or negligible inhibition for P-gp and MRP1. Chemotherapies (ortataxel, tamoxifen, and daunomycin), anti-HIV therapies (amprenavir, atazanavir, lopinavir, nelfinavir, and saquinavir), HMG-CoA reductase inhibitors (atorvastatin, cerivastatin, fluvastatin, pitavastatin, and rosuvastatin), and a couple of tyrosine kinase inhibitors including gefitinib and imatinib, as well as reserpine, also inhibit BCRP [38].

Acidic vesicle inhibitors

Increasing evidence indicates that acidic vesicles sequestration impedes the effective accessibility of basic drugs, e.g., doxorubicin, vincristine [39] and sunitinib [40], to their intracellular targets, and high pH gradient between cytoplasm (alkaline) and the lumen of intracellular vesicles (acidic) contributes to MDR [1,41]. These vesicles involves lysosomes, endocytic compartments, secretory vesicles, and multivesicular bodies [42]. A turnover of acidic vesicles and normalization of pH gradient through inhibition of vacuolar H^+-ATPase is a smart strategy to reverse MDR [43–45]. Proton pump inhibitors originally indicated for reduction of gastric acid represent a class of drugs suitable for the purpose [46,47] (Table 11.3).

Metabolic enzyme inhibitors

Metabolic inactivation of drugs *in vivo* attenuates therapeutic effect. For instance, bleomycin could be inactivated by a cysteine protease [48]. Carbonyl reductase leads to resistance toward daunorubicin [49]. Glutathione detoxification is highly associated with chemoresistance [50]. Autoinduction of cytochrome P450 isoenzymes (CYPs) speeds up the clearance of active drugs [51]. These studies raise the rationale to inhibit

TABLE 11.3 Acidic vesicle inhibitors with chemosensitizer property.

Drug	Original indication	Restore sensitivity to
Concanamycin A	V-ATPase inhibitor	Daunomycin, doxorubicin, and epirubicin
Bafilomycin A	V-ATPase inhibitor	Cisplatin
Pantoprazole	Proton pump inhibitor	Cisplatin, 5-fluorouracil, vinblastine
Omeprazole	Proton pump inhibitor	Cisplatin, 5-fluorouracil, vinblastine, methotrexate, doxorubicin
Esomeprazole	Proton pump inhibitor	Cisplatin, 5-fluorouracil, vinblastine
Lansoprazole	Proton pump inhibitor	Paclitaxel

The drugs are compiled from elegant studies [44–46,92,93].

biotransformation of chemotherapies. Actually, some clinic drugs targeting these enzymes like CYP3A4 inhibitors ketoconazole and CsA have improved standard chemotherapies activity [52,53]. However, metabolic enzymes play a pivotal role in detoxification, and long-term inhibition of their activity probably gives rise to severe side effects. Temporary inhibition may allow the balance between safety and efficacy.

Qualitative chemosensitizers

Cellular process regulators

In addition to broad-spectrum chemosensitizers as ABC transporter inhibitors, many clinical drugs exhibiting concomitant effects on specific cellular regulation including oncogenic signaling, cell cycle, DNA repair, stress handling, or energy supply are also reported to sensitize resistant cell lines toward a certain type of therapy. Many of these drugs are also applied as monotherapy in cancer treatment. However, their novel role as chemosensitizers potentially expends their original application scope.

Inhibition of oncogenic regulatory molecules such as protein kinases including epidermal growth factor receptor 1/2, breakpoint cluster region–Abelson murine leukemia viral oncogene homolog, vascular endothelial growth factor receptor, anaplastic lymphoma kinase, Janus kinase 2, fms-like tyrosine kinase 3, serine/threonine-protein kinase B-Raf, Bruton's tyrosine kinase [54], and protein kinase C [55] sensitizes tumors to toxic regimens. The mechanisms regarding their effect on reversing MDR have not been fully addressed but are commonly believed to function through attenuation of prosurvival signaling, promotion of apoptosis, and interactions with drug transporters.

Besides, poly(ADP-ribose) polymerase (PARP) sensation as well as checkpoint kinase 1/2 (Chk1/Chk2) activation in response to DNA-damaging agents or radiation therapy allows the cancerous cells to evade expected apoptosis [56,57]. Thereby, small molecules such as PARP inhibitors (AG014699, AG14361, CEP-6800, and some alkaloids) or Chk1/Chk2 inhibitors (UCN-01 and CEP-3891) potentiate systemic anticancer drugs including mitomycin C, temozolomide, doxorubicin, cisplatin, 5-fluorouracil, doxorubicin, and gemcitabine [56,58–61]. Repression of prosurvival modules in response to stress such as autophagy by thioridazine, apigenin, chloroquine, and hydroxychloroquine [62–66] and heat shock proteins induction by ganetespib, luminespib, and tanespimycin [67–70] has proven to be another promising way to improve drug response. Gypenoside promotes apoptosis induced by 5-fluorouracil through activation of p53 [71]. Lastly, total ATP depletion by glycolysis inhibitors such as 2-deoxy-D-glucose [72] and 3-bromopyruvate [73] fundamentally cut off the power required by prosurvival signaling and thereby promote cell death.

EMT inhibitors

EMT initiated by cytokines and growth factors allows cells to acquire motility as well as plasticity. In carcinoma, the transition imparts aggressiveness and resistance through altered cell behavior, gene expression, and chromosome instability [74]. It has been reported that tumors harboring high proportion of mesenchymal cells exhibited less sensitivity toward treatment [75]. During EMT process, the cell proliferative phenotype is switched to invasive phenotype, which gives rise to resistance toward treatment targeting at rapid division [76]. Besides, evidence showed that E-cadherin downregulation in EMT was associated poor response to cetuximab [77]. EMT-mediated resistance is also attributed to elevated ABC transporters expression, cancer stem cell transformation, and long noncoding RNAs [78].

Drugs repressing EMT such as salinomycin, curcumin, metformin, and histone deacetylase

TABLE 11.4 EMT inhibitors with chemosensitizer property.

Drug	Original indication	Restore sensitivity to
Curcumin	Antiinflammation	5-Fluorouracil
Salinomycin	Antibiotic	Doxorubicin
Metformin	Antidiabetes	Gefitinib
Mocetinostat	HDAC inhibitor	Docetaxel, gemcitabine
Entinostat	HDAC inhibitor	Erlotinib
Vorinostat	HDAC inhibitor	Exmestane
Daurinoline	EMT inhibition	Paclitaxel

The drugs were compiled from elegant studies[8,94-99].

inhibitors [8,79] (Table 11.4) can be repurposed to sensitize tumors toward common chemotherapies.

Conclusions and perspectives

Different biological mechanisms conferring drug resistance have been discussed in this chapter. Instead of being independent, these mechanisms are often found to interlink with each other. For instance, the autoinduction of P450 enzymes activity is often accompanied by the upregulation of P-gp [80]. During the process of EMT, specific transcription factors such as Twist, Snail, and Slug also upregulate ABC transporters [81]. Extracellular vesicles incorporating drug efflux pumps, miRNAs, and long noncoding RNAs can transfer resistance among cells [82].

Multiple strategies to overcome drug resistance and enhance drug efficacy have been proposed accordingly. In order to efficiently repurpose existing drugs, these strategies should provide an optimized and synergistical solution in combating MDR. However, some strategies should be cautiously applied. For instance, metabolic enzymes repression needs a second thought since the purpose of biotransformation is detoxification. The possible risks should be strictly evaluated. As to a single chemosensitizer, it can function through more than one way. For instance, vacuolar H^+-ATPase inhibitors can play a dual role in inhibiting P-gp and vesicle sequestration. On the other hand, different chemosensitizers with their own targets can work together as cocktail therapy. However, the drug interactions and complex molecular networks may lead to unpredictable outcomes. In this case, the chemosensitizers can only serve the patients better if applied in a personalized manner considering the specific symptoms copresented by one patient. Generally speaking, the development of already approved drugs as chemosensitizers for MDR presenting tumors may shorten the time and the costs, since preclinical data on safety and toxicity are already available. However, there is a big caveat. These drugs have been approved for diseases other than cancer. If they are repurposed for cancer treatment, the initial indications may appear as side effects. This might be one reason why none P-gp inhibitor passed clinical phase III as of yet [83].

Nevertheless, the drug repurposing concept has great potentials. The task for the future is to identify drugs for repurposing with sufficient specificity and tolerable side effects. Rather than identifying chemosensitizers among existing drugs by chance, unintended side effects, or off-label use experience, a comprehensive understanding of genomics and molecular networks involved in cancer and other diseases under the help of high-throughput technologies allows the scientists to have a better chance in successful drug repurposing. High-quality big data originated from chemical structure database and biological molecular networks represent valuable resource for chemosensitizer identification with computational approach [84]. By exploiting the information, scientists can efficiently compare the structural similarity and functional mechanisms with molecular profiles of existing chemosensitizers with other potential drugs to discover other possibilities.

Acknowledgments

A Ph.D. stipend of the Chinese Scholarship Council to G.Y. is gratefully acknowledged.

References

[1] Millot C, Millot JM, Morjani H, Desplaces A, Manfait M. Characterization of acidic vesicles in multidrug-resistant and sensitive cancer cells by acridine orange staining and confocal microspectrofluorometry. J Histochem Cytochem 1997;45(9):1255–64. https://doi.org/10.1177/002215549704500909.

[2] Rochat B, Morsman JM, Murray GI, Figg WD, McLeod HL. Human CYP1B1 and anticancer agent metabolism: mechanism for tumor-specific drug inactivation? J Pharmacol Exp Therapeut 2001;296(2):537–41.

[3] Sebti SM, Jani JP, Mistry JS, Gorelik E, Lazo JS. Metabolic inactivation: a mechanism of human tumor resistance to bleomycin. Cancer Res 1991;51(1):227–32.

[4] Sakthivel KM, Hariharan S. Regulatory players of DNA damage repair mechanisms. Role in cancer chemoresistance. Biomed pharmacother 2017;93:1238–45. https://doi.org/10.1016/j.biopha.2017.07.035.

[5] Bao S, Wu Q, McLendon RE, Hao Y, Shi Q, Hjelmeland AB, et al. Glioma stem cells promote radioresistance by preferential activation of the DNA damage response. Nature 2006;444(7120):756–60. https://doi.org/10.1038/nature05236.

[6] Dharmaraja AT. Role of reactive oxygen species (ROS) in therapeutics and drug resistance in cancer and bacteria. J Med Chem 2017;60(8):3221–40. https://doi.org/10.1021/acs.jmedchem.6b01243.

[7] Okon IS, Zou M-H. Mitochondrial ROS and cancer drug resistance. Implications for therapy. Pharmacol Res 2015;100:170–4. https://doi.org/10.1016/j.phrs.2015.06.013.

[8] Du B, Shim JS. Targeting epithelial-mesenchymal transition (EMT) to overcome drug resistance in cancer. Molecules 2016;21(7). https://doi.org/10.3390/molecules21070965.

[9] Hu Q, Baeg GH. Role of epigenome in tumorigenesis and drug resistance. Food Chem Toxicol 2017;109(Pt 1):663–8. https://doi.org/10.1016/j.fct.2017.07.022.

[10] Siegfried Z, Karni R. The role of alternative splicing in cancer drug resistance. Curr Opin Genet Dev 2018;48:16–21. https://doi.org/10.1016/j.gde.2017.10.001.

[11] Efferth T. The human ATP-binding cassette transporter genes: from the bench to the bedside. Curr Mol Med 2001;1(1):45–65.

[12] Efferth T, Volm M. Multiple resistance to carcinogens and xenobiotics. P-glycoproteins as universal detoxifiers. Arch Toxicol 2017;91(7):2515–38. https://doi.org/10.1007/s00204-017-1938-5.

[13] Gillet J-P, Efferth T, Remacle J. Chemotherapy-induced resistance by ATP-binding cassette transporter genes. Biochim Biophys Acta 2007;1775(2):237–62. https://doi.org/10.1016/j.bbcan.2007.05.002.

[14] Doyle LA, Ross DD. Multidrug resistance mediated by the breast cancer resistance protein BCRP (ABCG2). Oncogene 2003;22(47):7340–58. https://doi.org/10.1038/sj.onc.1206938.

[15] Fojo T, Menefee M. Mechanisms of multidrug resistance. The potential role of microtubule-stabilizing agents. Ann Oncol 2007;18(Suppl. 5):v3–8. https://doi.org/10.1093/annonc/mdm172.

[16] Hipfner DR, Deeley RG, Cole SPC. Structural, mechanistic and clinical aspects of MRP1. Biochim Biophys Acta Biomembr 1999;1461(2):359–76. https://doi.org/10.1016/S0005-2736(99)00168-6.

[17] Zhou S-F. Structure, function and regulation of P-glycoprotein and its clinical relevance in drug disposition. Xenobiotica 2008;38(7–8):802–32. https://doi.org/10.1080/00498250701867889.

[18] Aller SG, Yu J, Ward A, Weng Y, Chittaboina S, Zhuo R, et al. Structure of P-glycoprotein reveals a molecular basis for poly-specific drug binding. Science 2009;323(5922):1718–22. https://doi.org/10.1126/science.1168750.

[19] Cole SPC. Multidrug resistance protein 1 (MRP1, ABCC1), a "multitasking" ATP-binding cassette (ABC) transporter. J Biol Chem 2014;289(45):30880–8. https://doi.org/10.1074/jbc.R114.609248.

[20] Kage K, Tsukahara S, Sugiyama T, Asada S, Ishikawa E, Tsuruo T, Sugimoto Y. Dominant-negative inhibition of breast cancer resistance protein as drug efflux pump through the inhibition of S-S dependent homodimerization. Int J Cancer 2002;97(5):626–30.

[21] Bhatia A, Schäfer H-J, Hrycyna CA. Oligomerization of the human ABC transporter ABCG2. Evaluation of the native protein and chimeric dimers. Biochemistry 2005;44(32):10893–904. https://doi.org/10.1021/bi0503807.

[22] Bentz J, O'Connor MP, Bednarczyk D, Coleman J, Lee C, Palm J, et al. Variability in P-glycoprotein inhibitory potency (IC_{50}) using various in vitro experimental systems. Implications for universal digoxin drug-drug interaction risk assessment decision criteria. Drug Metabol Dispos 2013;41(7):1347–66. https://doi.org/10.1124/dmd.112.050500.

[23] Cole SP, Bhardwaj G, Gerlach JH, Mackie JE, Grant CE, Almquist KC, et al. Overexpression of a transporter gene in a multidrug-resistant human lung cancer cell line. Science 1992;258(5088):1650–4. New York, N.Y.

[24] Krishnamachary N, Center MS. The MRP gene associated with a non-P-glycoprotein multidrug resistance encodes a 190-kDa membrane bound glycoprotein. Cancer Res 1993;53(16):3658–61.

[25] Yabuuchi H, Takayanagi S-I, Yoshinaga K, Taniguchi N, Aburatani H, Ishikawa T. ABCC13, an unusual truncated ABC transporter, is highly expressed in fetal human liver. Biochem Biophys Res Commun 2002;299(3):410–7. https://doi.org/10.1016/S0006-291X(02)02658-X.

[26] Kruh GD, Belinsky MG. The MRP family of drug efflux pumps. Oncogene 2003;22(47):7537–52. https://doi.org/10.1038/sj.onc.1206953.

[27] Burkhart CA, Watt F, Murray J, Pajic M, Prokvolit A, Xue C, et al. Small-molecule multidrug resistance-associated protein 1 inhibitor reversan increases the therapeutic index of chemotherapy in mouse models of neuroblastoma. Cancer Res 2009;69(16):6573–80. https://doi.org/10.1158/0008-5472.can-09-1075.

[28] Stefan SM, Wiese M. Small-molecule inhibitors of multidrug resistance-associated protein 1 and related processes. A historic approach and recent advances. Med Res Rev 2019;39(1):176–264. https://doi.org/10.1002/med.21510.

[29] Sirotnak FM, Wendel HG, Bornmann WG, Tong WP, Miller VA, Scher HI, Kris MG. Co-administration of probenecid, an inhibitor of a cMOAT/MRP-like plasma membrane ATPase, greatly enhanced the efficacy of a new 10-deazaaminopterin against human solid tumors in vivo. Clin Cancer Res 2000;6(9):3705–12.

[30] Janneh O, Jones E, Chandler B, Owen A, Khoo SH. Inhibition of P-glycoprotein and multidrug resistance-associated proteins modulates the intracellular concentration of lopinavir in cultured CD4 T cells and primary human lymphocytes. J Antimicrob Chemother 2007;60(5):987–93. https://doi.org/10.1093/jac/dkm353.

[31] Norris MD, Madafiglio J, Gilbert J, Marshall GM, Haber M. Reversal of multidrug resistance-associated protein-mediated drug resistance in cultured human neuroblastoma cells by the quinolone antibiotic difloxacin. Med Pediatr Oncol 2001;36(1):177–80. https://doi.org/10.1002/1096-911X(20010101)36:1<177::AID-MPO1042>3.0. CO;2-Q.

[32] Bakos É, Evers R, Sinkó E, Váradi A, Borst P, Sarkadi B. Interactions of the human multidrug resistance proteins MRP1 and MRP2 with organic anions. Mol Pharmacol 2000;57(4):760–8. https://doi.org/10.1124/mol.57.4.760.

[33] Doyle LA, Yang W, Abruzzo LV, Krogmann T, Gao Y, Rishi AK, Ross DD. A multidrug resistance transporter from human MCF-7 breast cancer cells. Proc Natl Acad Sci USA 1998;95(26):15665–70.

[34] Nakanishi T, Doyle LA, Hassel B, Wei Y, Bauer KS, Wu S, et al. Functional characterization of human breast cancer resistance protein (BCRP, ABCG2) expressed in the oocytes of *Xenopus laevis*. Mol Pharmacol 2003;64(6):1452–62. https://doi.org/10.1124/mol.64.6.1452.

[35] Eddabra L, Wenner T, El Btaouri, Hassan, Baranek T, Madoulet C, Cornillet-Lefebvre P, Morjani H. Arginine 482 to glycine mutation in ABCG2/BCRP increases etoposide transport and resistance to the drug in HEK-293 cells. Oncol Rep 2012;27(1):232–7. https://doi.org/10.3892/or.2011.1468.

[36] Honjo Y, Hrycyna CA, Yan QW, Medina-Pérez WY, Robey RW, van de Laar A, et al. Acquired mutations in the MXR/BCRP/ABCP gene alter substrate specificity in MXR/BCRP/ABCP-overexpressing cells. Cancer Res 2001;61(18):6635–9.

[37] Clark R, Kerr ID, Callaghan R. Multiple drugbinding sites on the R482G isoform of the ABCG2 transporter. Br J Pharmacol 2006;149(5):506–15. https://doi.org/10.1038/sj.bjp.0706904.

[38] Gameiro M, Silva R, Rocha-Pereira C, Carmo H, Carvalho F, Bastos Mde L, Remião F. Cellular models and in vitro assays for the screening of modulators of P-gp, MRP1 and BCRP. Molecules 2017;22(4). https://doi.org/10.3390/molecules22040600.

[39] Zhitomirsky B, Assaraf YG. Lysosomal sequestration of hydrophobic weak base chemotherapeutics triggers lysosomal biogenesis and lysosome-dependent cancer multidrug resistance. Oncotarget 2015;6(2):1143–56. https://doi.org/10.18632/oncotarget.2732.

[40] Gotink KJ, Broxterman HJ, Labots M, Haas RR de, Dekker H, Honeywell RJ, et al. Lysosomal sequestration of sunitinib: a novel mechanism of drug resistance. Clin Cancer Res 2011;17(23):7337–46. https://doi.org/10.1158/1078-0432.ccr-11-1667.

[41] Zhitomirsky B, Assaraf YG. The role of cytoplasmic-to-lysosomal pH gradient in hydrophobic weak base drug sequestration in lysosomes. Can Cell Microenviron 2015. https://doi.org/10.14800/ccm.807.

[42] Kaufmann AM, Krise JP. Lysosomal sequestration of amine-containing drugs: analysis and therapeutic implications. J Pharmaceut Sci 2007;96(4):729–46. https://doi.org/10.1002/jps.20792.

[43] Fais S, Milito Ade, You H, Qin W. Targeting vacuolar H+-ATPases as a new strategy against cancer. Cancer Res 2007;67(22):10627–30. https://doi.org/10.1158/0008-5472.can-07-1805.

[44] Kulshrestha A, Katara GK, Ibrahim SA, Riehl V, Sahoo M, Dolan J, et al. Targeting V-ATPase isoform restores cisplatin activity in resistant ovarian cancer. Inhibition of autophagy, endosome function, and ERK/MEK pathway. J Oncol 2019;2019(7):1–15. https://doi.org/10.1155/2019/2343876.

[45] Ouar Z, Bens M, Vignes C, Paulais M, Pringel C, Fleury J, et al. Inhibitors of vacuolar H+-ATPase impair the preferential accumulation of daunomycin in lysosomes and reverse the resistance to anthracyclines in drug-resistant renal epithelial cells. Biochem J 2003;370(Pt 1):185–93. https://doi.org/10.1042/bj20021411.

[46] Chen M, Zou X, Luo H, Cao J, Zhang X, Zhang B, Liu W. Effects and mechanisms of proton pump inhibitors as a novel chemosensitizer on human gastric adenocarcinoma (SGC7901) cells. Cell Biol Int 2009; 33(9):1008–19. https://doi.org/10.1016/j.cellbi.2009.05.004.

[47] Spugnini EP, Citro G, Fais S. Proton pump inhibitors as anti vacuolar-ATPases drugs: a novel anticancer strategy. J Exp Clin Cancer Res 2010;29:44. https://doi.org/10.1186/1756-9966-29-44.

[48] Schwartz DR, Homanics GE, Hoyt DG, Klein E, Abernethy J, Lazo JS. The neutral cysteine protease bleomycin hydrolase is essential for epidermal integrity and bleomycin resistance. Proc Natl Acad Sci U S A 1999;96(8):4680–5. https://doi.org/10.1073/pnas.96.8.4680.

[49] Ax W, Soldan M, Koch L, Maser E. Development of daunorubicin resistance in tumour cells by induction of carbonyl reduction. Biochem Pharmacol 2000;59(3): 293–300. https://doi.org/10.1016/s0006-2952(99)00322-6.

[50] Townsend DM, Tew KD. The role of glutathione-S-transferase in anti-cancer drug resistance. Oncogene 2003;22(47):7369–75. https://doi.org/10.1038/sj.onc.1206940.

[51] Rochat B. Role of cytochrome P450 activity in the fate of anticancer agents and in drug resistance: focus on tamoxifen, paclitaxel and imatinib metabolism. Clin Pharmacokinet 2005;44(4):349–66. https://doi.org/10.2165/00003088-200544040-00002.

[52] Kruijtzer CMF, Beijnen JH, Schellens JHM. Improvement of oral drug treatment by temporary inhibition of drug transporters and/or cytochrome P450 in the gastrointestinal tract and liver: an overview. Oncologist 2002;7(6):516–30. https://doi.org/10.1634/theoncologist.7-6-516.

[53] Piska K, Koczurkiewicz P, Wnuk D, Karnas E, Bucki A, Wójcik-Pszczoła, Katarzyna, et al. Synergistic anticancer activity of doxorubicin and piperlongumine on DU-145 prostate cancer cells - the involvement of carbonyl reductase 1 inhibition. Chem Biol Interact 2019;300:40–8. https://doi.org/10.1016/j.cbi.2019.01.003.

[54] Chen Z, Yang D, editors. Protein kinase inhibitors as sensitizing agents for chemotherapy. Elsevier; 2019.

[55] Fine RL, Chambers TC, Sachs CW. P-glycoprotein, multidrug resistance and protein kinase C. Stem Cell 1996;14(1):47–55. https://doi.org/10.1002/stem.140047.

[56] Luo Y, Leverson JD. New opportunities in chemosensitization and radiosensitization: modulating the DNA-damage response. Expet Rev Anticancer Ther 2005; 5(2):333–42. https://doi.org/10.1586/14737140.5.2.333.

[57] Zhou B-BS, Bartek J. Targeting the checkpoint kinases: chemosensitization versus chemoprotection. Nat Rev Cancer 2004;4(3):216–25. https://doi.org/10.1038/nrc1296.

[58] Ghelli Luserna Di Rorà A, Iacobucci I, Imbrogno E, Papayannidis C, Derenzini E, Ferrari A, et al. Prexasertib, a Chk1/Chk2 inhibitor, increases the effectiveness of conventional therapy in B-/T- cell progenitor acute lymphoblastic leukemia. Oncotarget 2016;7(33): 53377–91. https://doi.org/10.18632/oncotarget.10535.

[59] Maugeri-Saccà M, Bartucci M, de Maria R. Checkpoint kinase 1 inhibitors for potentiating systemic anticancer therapy. Cancer Treat Rev 2013;39(5):525–33. https://doi.org/10.1016/j.ctrv.2012.10.007.

[60] Mon MT, Yodkeeree S, Punfa W, Pompimon W, Limtrakul P. Alkaloids from stephania venosa as chemo-sensitizers in SKOV3 ovarian cancer cells via akt/NF-κB signaling. Chem Pharmaceut Bull 2018; 66(2):162–9. https://doi.org/10.1248/cpb.c17-00687.

[61] Parsels LA, Morgan MA, Tanska DM, Parsels JD, Palmer BD, Booth RJ, et al. Gemcitabine sensitization by checkpoint kinase 1 inhibition correlates with inhibition of a Rad51 DNA damage response in pancreatic cancer cells. Mol Cancer Therapeut 2009;8(1):45–54. https://doi.org/10.1158/1535-7163.mct-08-0662.

[62] Cudjoe EK, Lauren Kyte S, Saleh T, Landry JW, Gewirtz DA. Autophagy inhibition and chemosensitization in cancer therapy. In: Daniel Johnson E, editor. Targeting cell survival pathways to enhance response to chemotherapy. Elsevier; 2019. p. 259–73.

[63] Gao A-M, Zhang X-Y, Hu J-N, Ke Z-P. Apigenin sensitizes hepatocellular carcinoma cells to doxorubic through regulating miR-520b/ATG7 axis. Chem Biol Interact 2018;280:45–50. https://doi.org/10.1016/j.cbi.2017.11.020.

[64] Johannessen T-C, Hasan-Olive MM, Zhu H, Denisova O, Grudic A, Latif MA, et al. Thioridazine inhibits autophagy and sensitizes glioblastoma cells to temozolomide. Int J Cancer 2019;144(7):1735–45. https://doi.org/10.1002/ijc.31912.

[65] Johnson ED, editor. Targeting cell survival pathways to enhance response to chemotherapy. Elsevier; 2019.

[66] Li Y, Cao F, Li M, Li P, Yu Y, Xiang L, et al. Hydroxychloroquine induced lung cancer suppression by enhancing chemo-sensitization and promoting the transition of M2-TAMs to M1-like macrophages. J Exp Clin Cancer Res 2018;37(1):259. https://doi.org/10.1186/s13046-018-0938-5.

[67] Guttmann DM, Koumenis C. The heat shock proteins as targets for radiosensitization and chemosensitization in cancer. Cancer Biol Ther 2011;12(12):1023–31. https://doi.org/10.4161/cbt.12.12.18374.

[68] He S, Smith DL, Sequeira M, Sang J, Bates RC, Proia DA. The HSP90 inhibitor ganetespib has chemosensitizer and radiosensitizer activity in colorectal cancer. Invest New Drugs 2014;32(4):577–86. https://doi.org/10.1007/s10637-014-0095-4.

[69] Kryeziu K, Bruun J, Guren TK, Sveen A, Lothe RA. Combination therapies with HSP90 inhibitors against colorectal cancer. Biochim Biophys Acta Rev Cancer 2019;1871(2):240–7. https://doi.org/10.1016/j.bbcan.2019.01.002.

[70] Lee SH, Lee EJ, Min KH, Hur GY, Lee SH, Lee SY, et al. Quercetin enhances chemosensitivity to gemcitabine in lung cancer cells by inhibiting heat shock protein 70 expression. Clin Lung Cancer 2015;16(6):e235–43. https://doi.org/10.1016/j.cllc.2015.05.006.

[71] Kong L, Wang X, Zhang K, Yuan W, Yang Q, Fan J, et al. Gypenosides synergistically enhances the antitumor effect of 5-fluorouracil on colorectal cancer in vitro and in vivo: a role for oxidative stress-mediated DNA damage and p53 activation. PLoS One 2015;10(9):e0137888. https://doi.org/10.1371/journal.pone.0137888.

[72] Gupta S, Mathur R, Dwarakanath BS. The glycolytic inhibitor 2-deoxy-D-glucose enhances the efficacy of etoposide in ehrlich ascites tumor-bearing mice. Cancer Biol Ther 2005;4(1):87–94. https://doi.org/10.4161/cbt.4.1.1381.

[73] Yadav S, Pandey SK, Kumar A, Kujur PK, Singh RP, Singh SM. Antitumor and chemosensitizing action of 3-bromopyruvate: implication of deregulated metabolism. Chem Biol Interact 2017;270:73–89. https://doi.org/10.1016/j.cbi.2017.04.015.

[74] Derynck R, Weinberg RA. EMT and cancer: more than meets the eye. Dev Cell 2019;49(3):313–6. https://doi.org/10.1016/j.devcel.2019.04.026.

[75] Katsuno Y, Meyer DS, Zhang Z, Shokat KM, Akhurst RJ, Miyazono K, Derynck R. Chronic TGF-β exposure drives stabilized EMT, tumor stemness, and cancer drug resistance with vulnerability to bitopic mTOR inhibition. Sci Signal 2019;12(570). https://doi.org/10.1126/scisignal.aau8544.

[76] Voulgari A, Pintzas A. Epithelial-mesenchymal transition in cancer metastasis: mechanisms, markers and strategies to overcome drug resistance in the clinic. Biochim Biophys Acta 2009;1796(2):75–90. https://doi.org/10.1016/j.bbcan.2009.03.002.

[77] Black PC, Brown GA, Inamoto T, Shrader M, Arora A, Siefker-Radtke AO, et al. Sensitivity to epidermal growth factor receptor inhibitor requires E-cadherin expression in urothelial carcinoma cells. Clin Cancer Res 2008;14(5):1478–86. https://doi.org/10.1158/1078-0432.ccr-07-1593.

[78] Heery R, Finn SP, Cuffe S, Gray SG. Long non-coding RNAs: key regulators of epithelial-mesenchymal transition, tumour drug resistance and cancer stem cells. Cancers 2017;9(4). https://doi.org/10.3390/cancers9040038.

[79] Singh A, Settleman J. EMT, cancer stem cells and drug resistance: an emerging axis of evil in the war on cancer. Oncogene 2010;29:4741 [EP -.].

[80] Burt RK, Thorgeirsson SS. Coinduction of MDR-1 multidrug-resistance and cytochrome P-450 genes in rat liver by xenobiotics. J Natl Cancer Inst 1988; 80(17):1383–6. https://doi.org/10.1093/jnci/80.17.1383.

[81] Saxena M, Stephens MA, Pathak H, Rangarajan A. Transcription factors that mediate epithelial-mesenchymal transition lead to multidrug resistance by upregulating ABC transporters. Cell Death Dis 2011;2:e179. https://doi.org/10.1038/cddis.2011.61.

[82] Sousa D, Lima RT, Vasconcelos MH. Intercellular transfer of cancer drug resistance traits by extracellular vesicles. Trends Mol Med 2015;21(10):595–608. https://doi.org/10.1016/j.molmed.2015.08.002.

[83] Amiri-Kordestani L, Fojo T. Why do phase III clinical trials in oncology fail so often? J Natl Cancer Inst 2012;104(8):568–9. https://doi.org/10.1093/jnci/djs180.

[84] Dudley JT, Deshpande T, Butte AJ. Exploiting drug-disease relationships for computational drug repositioning. Brief Bioinf 2011;12(4):303–11. https://doi.org/10.1093/bib/bbr013.

[85] Ford JM, Hait WN. Pharmacologic circumvention of multidrug resistance. Cytotechnology 1993;12(1–3):171–212.

[86] Hu Y, Qin X, Cao H, Yu S, Feng J. Reversal effects of local anesthetics on P-glycoprotein-mediated cancer multidrug resistance. Anti Cancer Drugs 2017;28(3):243–9. https://doi.org/10.1097/CAD.0000000000000455.

[87] La X, Zhang L, Li Z, Li H, Yang Y. (-)-Epigallocatechin gallate (EGCG) enhances the sensitivity of colorectal cancer cells to 5-FU by inhibiting GRP78/NF-κB/miR-155-5p/MDR1 pathway. J Agric Food Chem 2019;67(9):2510–8. https://doi.org/10.1021/acs.jafc.8b06665.

[88] Palmeira A, Sousa E, Vasconcelos MH, Pinto MM. Three decades of P-gp inhibitors. Skimming through several generations and scaffolds. Comput Mater Continua (CMC) 2012;19(13):1946–2025. https://doi.org/10.2174/092986712800167392.

[89] Saeed M, Zeino M, Kadioglu O, Volm M, Efferth T. Overcoming of P-glycoprotein-mediated multidrug resistance of tumors in vivo by drug combinations. Synergy 2014;1(1):44–58. https://doi.org/10.1016/j.synres.2014.07.002.

[90] Seo E-J, Sugimoto Y, Greten HJ, Efferth T. Repurposing of bromocriptine for cancer therapy. Front Pharmacol 2018;9:1030. https://doi.org/10.3389/fphar.2018.01030.

[91] Zeino M, Paulsen MS, Zehl M, Urban E, Kopp B, Efferth T. Identification of new P-glycoprotein inhibitors derived from cardiotonic steroids. Biochem Pharmacol 2015;93(1):11–24. https://doi.org/10.1016/j.bcp.2014.10.009.

[92] Lindner K, Borchardt C, Schöpp M, Bürgers A, Stock C, Hussey DJ, et al. Proton pump inhibitors (PPIs) impact on tumour cell survival, metastatic potential and chemotherapy resistance, and affect expression of resistance-relevant miRNAs in esophageal cancer. J Exp Clin Cancer Res 2014;33:73. https://doi.org/10.1186/s13046-014-0073-x.

[93] Luciani F, Spada M, de Milito A, Molinari A, Rivoltini L, Montinaro A, et al. Effect of proton pump inhibitor pretreatment on resistance of solid tumors to cytotoxic drugs. J Natl Cancer Inst 2004;96(22):1702–13. https://doi.org/10.1093/jnci/djh305.

[94] Li D-D, Qin X-C, Yang Y, Chu H-X, Li R-L, Ma L-X, et al. Daurinoline suppressed the migration and invasion of chemo-resistant human non-small cell lung cancer cells by reversing EMT and Notch-1 and sensitized the cells to Taxol. Environ Toxicol Pharmacol 2019;66:109–15. https://doi.org/10.1016/j.etap.2018.12.005.

[95] Li L, Han R, Xiao H, Lin C, Wang Y, Liu H, et al. Metformin sensitizes EGFR-TKI-resistant human lung cancer cells in vitro and in vivo through inhibition of IL-6 signaling and EMT reversal. Clin Cancer Res 2014;20(10):2714–26. https://doi.org/10.1158/1078-0432.CCR-13-2613.

[96] Meidhof S, Brabletz S, Lehmann W, Preca B-T, Mock K, Ruh M, et al. ZEB1-associated drug resistance in cancer cells is reversed by the class I HDAC inhibitor mocetinostat. EMBO Mol Med 2015;7(6):831–47. https://doi.org/10.15252/emmm.201404396.

[97] Ronnekleiv-Kelly SM, Sharma A, Ahuja N. Epigenetic therapy and chemosensitization in solid malignancy. Cancer Treat Rev 2017;55:200–8. https://doi.org/10.1016/j.ctrv.2017.03.008.

[98] Toden S, Okugawa Y, Jascur T, Wodarz D, Komarova NL, Buhrmann C, et al. Curcumin mediates chemosensitization to 5-fluorouracil through miRNA-induced suppression of epithelial-to-mesenchymal transition in chemoresistant colorectal cancer. Carcinogenesis 2015;36(3):355–67. https://doi.org/10.1093/carcin/bgv006.

[99] Zhou Y, Liang C, Xue F, Chen W, Zhi X, Feng X, et al. Salinomycin decreases doxorubicin resistance in hepatocellular carcinoma cells by inhibiting the β-catenin/TCF complex association via FOXO3a activation. Oncotarget 2015;6(12):10350–65. https://doi.org/10.18632/oncotarget.3585.

CHAPTER

12

Drugs repurposed to potentiate immunotherapy for cancer treatment

Kenneth K.W. To[1], *William C.S. Cho*[2]

[1]School of Pharmacy, Faculty of Medicine, The Chinese University of Hong Kong, Hong Kong SAR, China; [2]Department of Clinical Oncology, Queen Elizabeth Hospital, Hong Kong SAR, China

OUTLINE

Drug Repurposing in Cancer Therapy
https://doi.org/10.1016/B978-0-12-819668-7.00012-9

Introduction

Cancer is a leading cause of death worldwide. Given the unmet need for more effective anticancer therapy, intense efforts have been invested in searching for better anticancer drugs in a more efficient manner. Pharmaceutical companies and academic investigators alike have become increasingly interested in finding new uses of the existing drugs, a process referred to as drug repurposing or repositioning, to treat cancer. The existing drugs might have been used in the clinic for different indications. In other cases, the drug candidates might have been halted from further development due to insufficient efficacy in their original intended indications or marketing considerations. Since the safety profile and pharmacokinetic properties of approved drugs have already been well studied in clinical trials, repurposing a drug promises faster access of drugs to patients and it can save time and money. It has been estimated that the classical de novo drug discovery process generally takes about 14 years and US$2.5 billion to approve and launch a new drug starting from hit selection in vitro [1].

Drug repurposing has been largely a serendipitous process when an off-target effect of a drug was identified for a new medical indication. In recent years, in silico predictive tools and high-throughput screening methods have been developed to facilitate drug repurposing process. Various molecular docking- [2] and omics-based [3] approaches have been applied for drug repurposing. Drugs having similar chemical structures and biological activities have been evaluated to find out whether they may have similar clinical indications. Similarity of protein structures, at local ligand binding sites, has been exploited to identify drugs to target diseases beyond their original indications. A few electronic resources are invaluable for this approach, which include Protein Data Bank [4], Protein-binding Sites (ProBis) [5], and Protein–Ligand Interaction Profiler [6], and DrugPredict [3]. Moreover, databases including the Connectivity Map [7] and the Library of Integrated Network-based Cellular Signatures [8,9] have been used to identify drugs with similar transcriptional signature for drug repurposing. More recently, a few databases, including Drug Repurposing Hub [10], Drug Target Commons [11], and Open Targets [12], have been established to integrate various computational approaches to enable the search for repurposed drugs using more comprehensive information. Two most recently developed databases, repoDB [13] and repurposeDB [14], also incorporate information about clinical results of drug repurposing. Schneider et al. recently reported a comprehensive visual analytics tool, called ClinOmicsTrail^{bc}, that analyzes and visualizes clinical biomarkers, genomics/epigenomics, and transcriptomics datasets to facilitate a holistic assessment of the use of targeted drugs, drug candidates for repurposing, and immunotherapeutic agents in the treatment of breast cancer [15].

Promise of cancer immunotherapy

Immune checkpoint blockade with anticytotoxic T lymphocyte-associated antigen 4 (CTLA-4) (ipilimumab and tremelimumab), anti-programmed cell death receptor (PD-1) (nivolumab and pembrolizumab), or anti–PD-ligand (PD-L1) (duralumab, atezolizumab, and avelumab) monoclonal antibodies are shifting cancer therapy paradigm, which induce durable tumor responses and overall survival benefit in a wide variety of cancer types. While the CTLA-4 blockade may be associated with immune-related adverse events (irAEs) (such as colitis, hepatitis, dermatitis, endocrinopathies, and neuropathies), the PD-1 blockade is more tolerated by the patients with minimal irAE.

PD-1 is an inhibitory receptor expressed on activated T cells, B cells, and natural killer cells, which normally function to blunt the immune response. PD-1 is engaged by its major ligand PD-L1, which are expressed in tumor cells and infiltrating immune cells, to suppress the T cell–mediated cancer killing effect. Anti–PD-1/PD-L1 antibodies work by binding to inhibitory PD-1 receptor on tumor-reactive T cells and PD-L1 on tumor cells, respectively, thereby disrupting the PD-1:PD-L1 interaction and reactivating the antitumor T cell–mediated cell cytotoxicity. Clinical benefit from anti-PD-1/PD-L1 therapy is associated with high tumor mutational load, high levels of pretreatment tumor-infiltrating T cells, and high expression of pretreatment PD-L1 on tumor cells and tumor-infiltrating immune cells.

Despite enormous clinical success achieved by immune checkpoint inhibitors, both primary and acquired resistance are preventing cancer patients from responding to these immunotherapeutic agents or having a durable disease control. Cancer resistance to immunotherapy can be mediated by both tumor-intrinsic and tumor-extrinsic mechanisms. A few comprehensive reviews on the resistance mechanisms to cancer immunotherapy can be found in recent publications [16–18]. Tumor-intrinsic factors include T-cell exhaustion, loss of antigen protein expression, and absence of antigen presentation. On the other hand, absence of T cells with tumor antigen–specific T-cell receptors and presence of immunosuppressive cells (i.e., regulatory T cells, myeloid-derived suppressor cells (MDSCs), and tumor-associated macrophages [TAMs]) in the tumor microenvironment (TME) represent the major tumor-extrinsic factors leading to resistance [17].

New opportunities for drug repurposing in cancer immunotherapy

Fueled by the clinical success of immune checkpoint inhibitors (anti-CTLA-4, anti-PD-1, and anti-PD-L1 antibodies) in ever-growing number of tumor types, there has been a resurgence of research interest in immunological approaches to treat cancer. Nevertheless, not all tumors respond to immune checkpoint therapy. In order to maximize the clinical efficacy of cancer immunotherapy, additional approaches for enhancing tumor immunity are urgently needed. Tumors may not respond well to immunotherapy due to either tumor-intrinsic or tumor-extrinsic factors as described above.

Drugs originally approved for noncancer indications have been combined with checkpoint inhibitors to boost antitumor immunity [19–21]. It is noteworthy that these repurposed drugs do not need to be directly cytotoxic themselves, thus imposing less toxicity issues to normal tissues. On the other hand, anticancer drugs not initially intended for cancer immunotherapy have been used to modulate the immune system so that these drugs form part of the antitumor immune cocktail (sometimes referred to as "soft repurposing") [22,23]. The repurposed drugs act by either exerting immunostimulatory activities or abolishing immunosuppressive TME (Fig. 12.1). For example, the angiotensin II receptor blocker (ARB) (valsartan) has been shown to increase the population of the tumor antigen gp70-specific T cells [24]. Metformin was reported to exert multiple effect to remodel the TME, including (i) protection of $CD8^+$ T cells from apoptosis [25], (ii) degradation of PD-L1 [26], and (iii) reduction of intratumoral hypoxia [20], which collectively preserve functionality of antitumor immune cells and reverse the immunosuppressive TME. Cyclophosphamide [22] and sunitinib [23], which are clinically

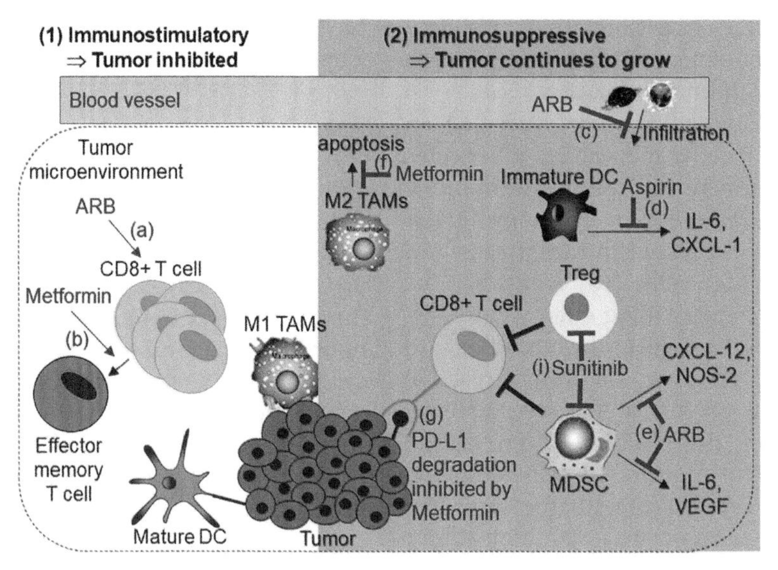

FIGURE 12.1 Mechanisms of repurposed drugs to combat resistance to immunotherapy. Repurposed drugs act by either (1) exerting immunostimulatory activities or (2) abolishing immunosuppressive tumor microenvironment. (1) (a) Valsartan (an angiotensin II receptor blocker; ARB) increases the population of gp70-specific T cells but does not increase the total number of T cells. (b) Metformin induces the effective memory T cell (TEM) phenotype, which stimulates immune attack against tumor cells. (2) (c) ARB inhibits infiltration of immunosuppressive cancer-associated fibroblast (CAF) and M2 TAMs. (d) Aspirin inhibits production of immunosuppressive factors such as IL-6 and CXCL-1 from dendritic cells and macrophages. (e) ARB inhibits production of immunosuppressive factors from myeloid-derived suppressor cells (MDSC) (e.g., IL-6 and VEGF) and CAF (e.g., CXCL-12 and NOS-2). Metformin exerts multiple actions to remodel tumor microenvironment, including (f) protection of $CD8^+$ T cells from apoptosis, (g) degradation of PD-L1, and (h) reduction of intratumoral hypoxia, which collectively preserve functionality of antitumor immune cells and reverse the immunosuppressive tumor microenvironment. (i) Sunitinib suppresses regulatory T cells (Treg) and depletes MDSC to enhance antigen-specific immune responses and tumor eradication.

approved anticancer drugs, have been shown to reduce the abundance of immunosuppressive Treg and/or MDSCs. These drugs represent untapped opportunities for new and affordable treatment options for enhancing the efficacy of immunotherapy that can be readily evaluated. More emerging and promising drug candidates for repurposing are described according to their mechanisms of action in the next section.

Promising examples of repurposed drug candidates for cancer immunotherapy

Small molecules for reversal of T-cell exhaustion

Cytotoxic T lymphocytes (CTLs) play a central role in mediating immune surveillance to recognize and remove unwanted virus-infected cells and malignant tumor cells in our body. As a self-protective mechanism, viruses and tumors are able to upregulate several inhibitory checkpoint receptors on the surfaces of CTLs to counteract the host immune surveillance activity (which is also commonly described as T-cell exhaustion). The checkpoint inhibitory therapies (anti-PD-1 or anti-CTLA-4 monoclonal antibodies) neutralize the inhibitory receptors such as PD-1 or CTLA-4 on exhausted T cells, thereby restoring their effector immune responses [27,28]. However, responses in many cancer patients are limited due to insufficient restoration of T-cell function [17]. Active research is underway to discover additional targets and pharmacologic agents to overcome the limitations of the current immune checkpoint blockade [29].

To this end, it has been recently shown that low molecular weight therapeutics can complement or replace existing immune checkpoint blockade biologics (comprehensively reviewed in Ref. [30]). Conventional chemotherapy, targeted therapies, and radiation therapy have been shown to induce antitumor immunity. Well-known examples include anthracycline-class cytotoxic chemotherapeutic drugs (e.g., doxorubicin, which induces immunogenic cell death [ICD] and block immunosuppressive pathways) and vascular endothelial growth factor (VEGF) inhibitors (e.g., bevacizumab, which increases the numbers of intratumoral cytotoxic T cells and reduce accumulation of immunosuppressive Treg cells) [30]. Therefore, rational combination of these traditional treatment modalities with immunotherapy has been shown to increase the rate of complete and durable clinical response in cancer patients.

A few recent studies have identified new T cell–modifying drugs by phenotypic screening of chemical libraries [31–34]. These investigations have a drawback as they relied on artificial activation of T cells from naïve mice via antibody stimulation with CD3/CD28 molecules rather than antigen-experienced T cells exhibiting dysfunctional effector responses. To this end, functional exhaustion of virus-specific T cells was first described in mice infected with the clone 13 (CL13) variant of lymphocytic choriomeningitis virus (LCMV) [27]. CL13 leads to sustained expression of inhibitory receptors (including PD-1) and immunosuppressive cytokine interleukin-10 (IL-10), suboptimal CD4, and CD8 T-cell activity and persistent viral infection (reviewed in Ref. [35]). Most recently, a high-throughput screening platform has been developed, which utilized an in vivo LCMV-CL13 model to identify small molecules that reverse T-cell exhaustion [36]. In this study, C57BL/6 mice were infected with LCMV-CL13. Virus-specific CD8+ T cells lost their capacity to express IFN-γ gradually, which closely mimic the immunosuppressive environment occurring in vivo during T-cell exhaustion. As a result, IFN-γ was used as a disease-linked biomarker to indicate T-cell dysfunction. A total of 19 positive hits was identified from the ReFRAME drug repurposing compound library (~12,000 repurposed molecules), including known and novel immunomodulatory compounds, that restores cytokine production and enhances the proliferation of exhausted T cells [36]. Since the ReFRAME compound library comprises mostly clinically evaluated drugs, translation of lead compounds to patient use is expected to be more streamlined.

Drugs navigating metabolic pathways to enhance antitumor immunity

Another major mechanism contributing to unresponsiveness to cancer immunotherapy is the predominance of immunosuppressive TME that limits reinvigoration of antitumor immunity. In the TME, cancer cells are generally depriving nutrients from T cells and redirect glucose and amino acids for their own advantage. To this end, various metabolic machineries are known to regulate the behavior of immune cells in response to the nutrient deprivation in the TME. In particular, tumor-infiltrating immune cells often experience metabolic stress due to the dysregulated metabolic activity of tumors, thus impairing antitumor immune responses. Therefore, the repurposing of drugs capable of targeting cancer metabolism may potentiate cancer immunotherapy by metabolic reprogramming the TME. A few excellent reviews on this topic have been published recently [37–39]. Fig. 12.2 illustrates a few promising approaches to restore the metabolic fitness of T cells in TME and to enhance the efficacy of cancer immunotherapy. A summary of recent clinical trials investigating the combination of immune checkpoint inhibitors and drugs modulating metabolic pathways are listed in Table 12.1.

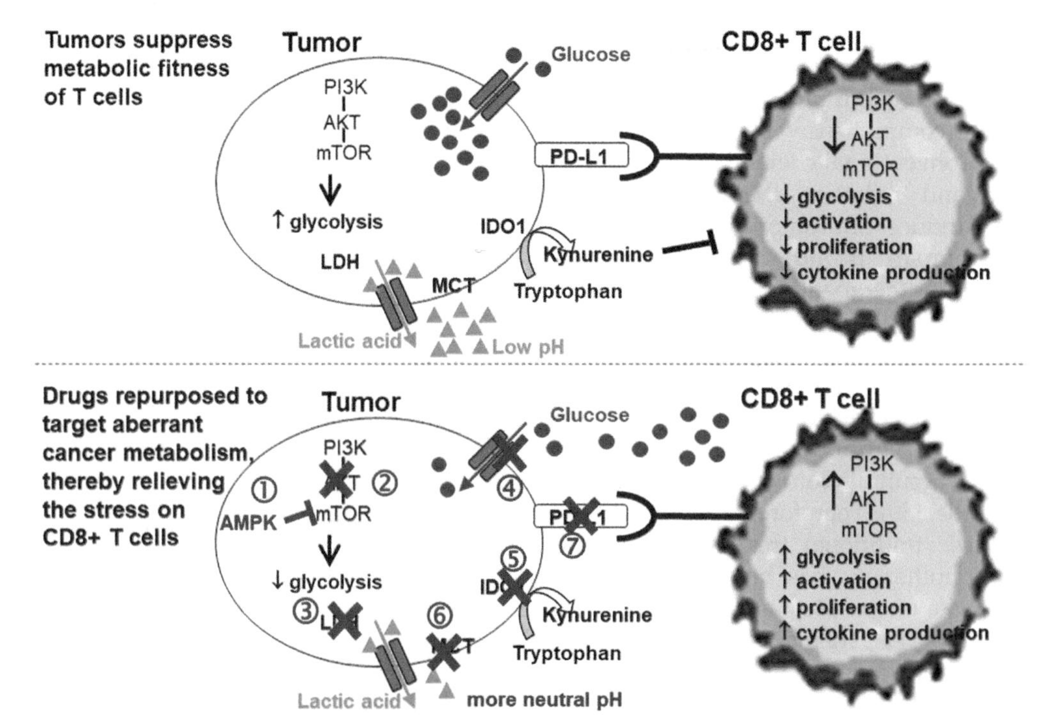

FIGURE 12.2 Therapeutic interventions that may be used to target aberrant cancer metabolism and restore the metabolic fitness of CD8+ T cells to boost antitumor immunity. Upper panel: Tumors suppress the metabolic fitness of T cells by (i) depleting glucose and amino acids in TME; (ii) increasing lactate concentration and acidity in TME; and (iii) suppressing glycolysis of T cells, subsequently inhibiting their activation and proliferation. Lower panel: Various repurposed drugs have been investigated to relieve the metabolic stresses imposed by tumors on T cells through one of the following mechanisms: ① activating AMPK (metformin); ② inhibiting mTOR (rapamycin analogues); ③ inhibition of LDH (galloflavin); ④ inhibiting glycolysis (dichloroacetate); ⑤ inhibiting amino acid metabolism (acivicin, azaserine); ⑥ inhibiting MCT (lenalidomide, pomalidomide); and ⑦ degrading PD-L1 (metformin).

Targeting glucose metabolism to enhance cancer immunotherapy

Cancer is notorious for its metabolic abnormalities. The well-known Warburg effect suggests that cancer cells preferentially adopt aerobic glycolysis, instead of the more efficient oxidative phosphorylation, as the major mode of glucose metabolism to provide energy. As a consequence, high level of lactic acid is generated and it acidifies the TME. It has been suggested that lactic acid production in the extracellular milieu is responsible for the immunosuppressive properties of TME [40], partially by inhibiting proliferation of CTLs [41–43]. Novel therapeutic approaches have been investigated to target lactate production by inhibiting lactate dehydrogenase (LDH) and monocarboxylate transporter (MCT) to tamper the acidic TME [44]. LDH is the enzyme catalyzing the conversion of pyruvate to lactate, thereby impairing glycolysis and proliferation of cancer cells [45]. However, while LDH inhibition by galloflavin was found to reduce lactate levels [46], it was also shown to reduce IFN-γ production by T cells [47]. Therefore, differential effects of LDH inhibitors on cancer and immune cells should be considered when they combined with immunotherapy. Apart from the inhibition of LDH, the lactate transporters MCT1-4 could also be inhibited to prevent the formation of

TABLE 12.1 Clinical trials investigating combination of immune checkpoint inhibitors with drugs navigating metabolic pathways.

Drug affecting metabolic pathway	Immune checkpoint inhibitor	Cancer type	Phase	ClinicalTrials.gov registration #
Aspirin (COX-1/COX-2 inhibitor)	Pembrolizumab (anti−PD-1 antibody)	Recurrent or metastatic HNSCC	I	NCT03245489
Aspirin	BAT1306 (anti−PD-1 antibody)	Advanced-stage MSI-H/dMMR cancers	II	NCT03638297
Celecoxib (COX-2 inhibitor)	BAT1306 (anti−PD-1 antibody)	Advanced-stage MSI-H/dMMR cancers	II	NCT03638297
Grapiprant (EP4 prostaglandin receptor antagonist)	Pembrolizumab	Advanced-stage CRC	I	NCT03658772
Grapiprant	Pembrolizumab	NSCLC	I/II	NCT03696212
Indoximod (IDO1/IDO2 inhibitor)	Pembrolizumab or nivolumab	Advanced-stage melanoma	II/III	NCT03301636
Indoximod	Ipilimumab (anti−CTLA-4 antibody), nivolumab, or pembrolizumab	Metastatic melanoma	I/II	NCT02073123
Metformin (antihyperglycemic drug)	Nivolumab (anti−PD-1 antibody)	Metastatic NSCLC	II	NCT03048500
Metformin	Pembrolizumab	Advanced stage melanoma	I	NCT03311308

COX, cyclooxygenase; *CRC*, colorectal cancer; *EP4*, prostaglandin E2 receptor; *HNSCC*, head and neck squamous cell carcinoma; *IDO*, indoleamine 2,3-dioxygenase; *MSI-H/dMMR*, microsatellite instability high and/or deficient mismatch repair; *NSCLC*, non−small-cell lung cancer.

acidic TME [48]. Recently, a few clinically approved drugs, including thalidomide, lenalidomide, and pomalidomide, have been identified as new MCT inhibitors [49]. Interestingly, lenalidomide was also found to enhance IL-2 and IFN-γ secretion from T cells [50], thus suggesting the dual role of lenalidomide in suppressing cancer proliferation but activating T-cell function. A nonsteroidal antiinflammatory drug (NSAID) (diclofenac) has also been investigated for its effects on lactate secretion and transport [51,52]. In a glioma model, diclofenac was found to reduce the abundance of infiltrating Tregs and lactic acid concentration in the TME and inhibit tumor growth [51,52]. It was therefore proposed to combine diclofenac and cancer immunotherapeutic agents for better efficacy.

Interestingly, the neutralization of TME by buffering the lactic acid with bicarbonate has also been proposed to enhance the therapeutic outcome of cancer immunotherapy [53]. Orally administered bicarbonate was reported to inhibit tumor growth when combined with anti−PD-1 immunotherapy in melanoma, and it also improved survival when combined with adoptive T-cell transfer [54]. The proton pump inhibitor (esomeprazole) has also been shown to neutralize the pH of TME and subsequently potentiate the anticancer effects by CTLs and NK cells [55,56].

Antihyperglycemic drugs have been repurposed to target the metabolic pathways of cancer cells and/or to reverse metabolic defects within the antitumor immune populations to potentiate cancer immunotherapy. Metformin is a widely prescribed antihyperglycemic drug with multiple mechanisms of action [57]. Preclinical investigations have demonstrated beneficial combinations of metformin with cancer immunotherapy [20,26]. The drug combination was shown to reduce accumulation of MDSCs in tumors but increase proliferation and cytokine secretion from CD8+ T cells infiltrated in tumors. In a mouse melanoma model, metformin was found to inhibit tumor cell metabolism (via oxidative phosphorylation and glycolysis) but enhance CD8+ T-cell metabolism (via oxidative phosphorylation) and cytokine production [20]. On the other hand, metformin was also reported to directly interact with the PD-L1/PD-1 axis to interfere with CTL immunity. It has been shown that metformin could phosphorylate PD-L1 (at Ser195) by activating AMPK, thereby leading to aberrant glycosylation and subsequent degradation in the endoplasmic reticulum [26]. The finding suggests that combination of metformin and anti-CTLA-4 monoclonal antibody could provide a dual immune checkpoint blockade to prevent immune escape [26]. This may represent a more economical, tolerable, and effective treatment strategy other than the current regimen of anti-PD1/anti-CTLA-4 combination. A few ongoing clinical trials are investigating the combination of metformin and immune checkpoint inhibitors (NCT03048500: nivolumab–metformin combination in NSCLC; NCT03311308: pembrolizumab–metformin combination in advanced melanoma; UMIN registration number 000028405: nivolumab–metformin combination in NSCLC).

Emerging evidence suggests that inhibition of the PI3K–AKT–mTOR pathway could suppress the glycolytic metabolism and sensitize tumor cells to chemotherapy [58,59]. The analogs of rapamycin, which inhibits the mTOR signaling, have been approved for treating breast, pancreatic, and renal cancers [60–62]. Interestingly, rapamycin was found to mediate the opposite effects on T cells where it broadens Tregs and cytotoxic memory T cells but decreases Teff proliferation [63]. It has been recently reported in glioblastoma that the combination of rapamycin and immunotherapy augmented cytotoxic and memory T-cell functions [64], thereby enhancing the therapeutic outcome.

Targeting amino acid catabolism to potentiate cancer immunotherapy

A few amino acids, including L-arginine, tryptophan, and glutamine, are crucial for supporting tumor progression and immunity [65]. In particular, tryptophan was known to support an oncogenic signature and maintain the immunosuppressive phenotype in various cancer types [66]. To this end, indoleamine 2,3-dioxygenase (IDO) is an intracellular monomeric enzyme that controls tryptophan breakdown into kynurenine to mediate the immunosuppressive effect. Consistently, it has been reported that the genetic silencing of IDO enhanced antitumor immunity in metastatic liver cancer [67], potentiated cytotoxic T-cell function, and decreased abundance of Treg cells [68]. Therefore, IDO is considered a useful target for drug discovery in cancer immunotherapy [69,70]. To this end, imatinib has been shown to improve antitumor immunity by activating T-effector cells and suppressing Tregs in an IDO-dependent manner [71]. The combination of imatinib and anti–CTLA-4 monoclonal antibody is currently under clinical investigation in gastrointestinal stromal tumors [72]. Glutamine is another critical amino acid important for supporting cancer cell metabolism and the rapidly dividing T cells. A few glutamine analogs, including acivicin, azaserine, and 6-diazo-5-oxo-L-norleucine, have been shown to impair the activity of enzymes utilizing glutamine in various tumor models [73].

Nonsteroidal antiinflammatory drugs

NSAIDs are clinically approved for use as antipyretic, analgesic, and antiinflammatory agents. Cyclooxygenase (COX) is an enzyme catalyzing the conversion of arachidonic acid to prostaglandins, thromboxanes, and prostacyclins, which play a key role in inflammation, temperature set-point regulation, and platelet adhesion. NSAIDs primarily work by inhibiting COX enzymes. There are two COX enzymes, namely COX-1 and COX-2. COX-1 is constitutively expressed in the body, whereas COX-2 is inducible upon inflammation. Most NSAIDs are nonselective and they inhibit both COX-1 and COX-2.

Besides the classical physiological roles of COX enzymes, recent studies revealed that they are also important in modulating antitumor immunity. PGE2 is a prostaglandin that has been reported to suppress tumor antigen-presenting dendritic cells and inhibit T-cell activation [74]. It also promotes propagation of immunosuppressive Tregs, MDSC, and M2 macrophages [75]. PGE2 is also known to prevent the T helper 1 (Th1) to T helper 2 (Th2) transition, thus inhibiting the formation of cytotoxic T cell and promoting cancer proliferation [74]. In addition, the downregulation of COX-2 in tumors could upregulate CXCL9 and CXCL10 to promote CD8+ T-cell infiltration and enhance interferon (INF) signaling [76]. NSAIDs is therefore hypothesized to sensitize tumor cells to immunotherapy (Fig. 12.3).

In fact, aspirin has been combined with anti–PD-1 monoclonal antibody to produce a synergistic antitumor immune response in a mouse CT26 colorectal cancer model [76]. It is noteworthy that the combination of NSAIDs and cancer immunotherapy may only benefit a subset of patient population (Fig. 12.4). In a recent large prospective cohort study, the adjuvant use of aspirin was found to be more strongly associated with improved survival in patients bearing tumor with low PD-L1 expression [19].

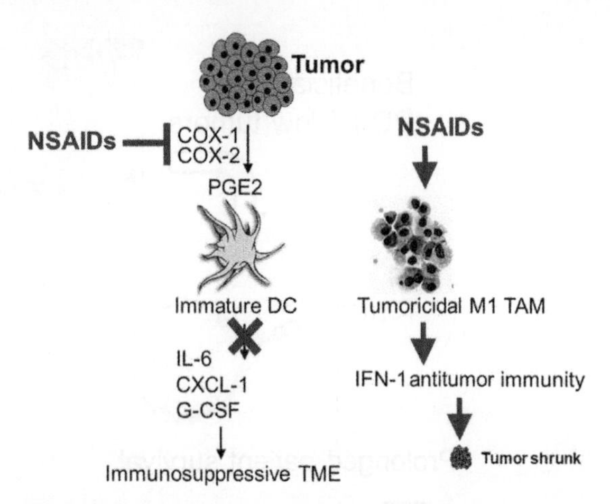

FIGURE 12.3 Mechanisms of NSAIDs to potentiate anticancer immunotherapy. NSAIDs inhibit cyclooxygenase (COX) enzymes and synthesis of prostaglandin E2 (PGE2), thus suppressing the release of several immunosuppressive factors including IL-6 and CXCL-1 from immature dendritic cells (DC) in the immune infiltrates. NSAIDs also activate tumoricidal M1 tumor–associated macrophage (TAM) to provide type-I interferon (IFN-1)–based antitumor immunity.

Therefore, the use of a reliable predictive biomarker for patient selection would be critical to ensure beneficial patient outcome. The COX-2 inhibitor (celecoxib) was also found to promote the conversion of the immunosuppressive M2 to the tumoricidal M1 TAM phenotype, potentially reversing immunotherapy resistance [77]. A few clinical trials are currently ongoing to investigate the combination of aspirin and anti–PD-1/anti–CTLA-4 immunotherapy in several cancer types (NCT02659384, NCT03245489, and NCT03396952).

Drugs modulating RANK–RANKL to enhance cancer immunotherapy

Receptor activator of nuclear factor kappa B (RANK) and its ligand (RANKL) belong to the tumor necrosis factor superfamily [78]. While the RANK–RANKL system was first discovered to regulate osteoclast function and bone

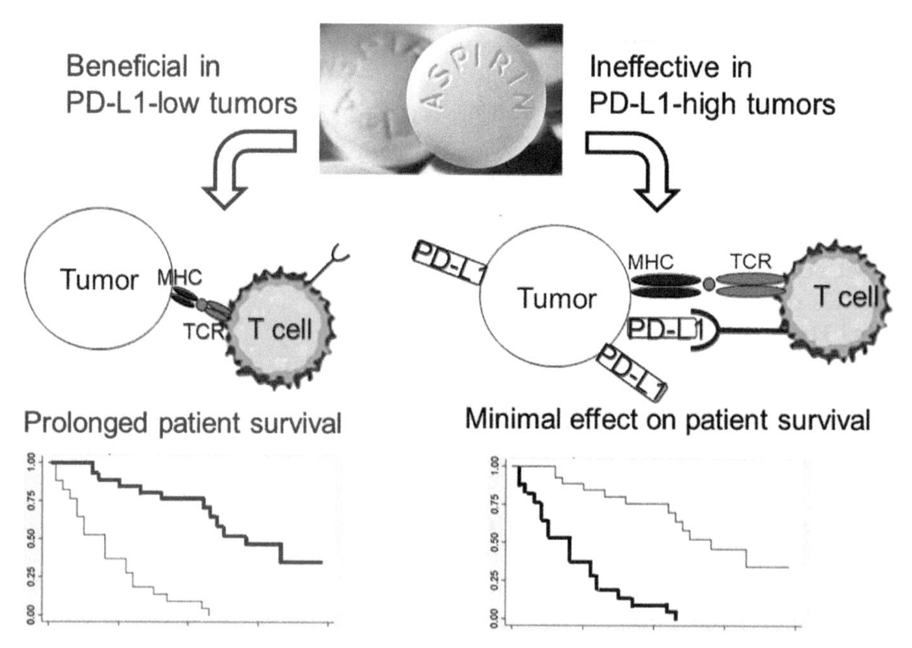

FIGURE 12.4 Aspirin only produces patient survival benefit in colorectal cancer patients bearing tumor with low PD-L1 level.

remodeling in bone homeostasis [78,79], it is also recognized to play important roles in dendritic cell survival and function, M1 macrophage activation, and T-cell activation [80,81]. Binding of RANKL to RANK activates numerous signal transduction pathways, including the production of nuclear factor kappa B, to regulate the cross talk between inflammation and cancer development [82]. RANKL has been proposed to promote the negative selection of tumor-specific T cells by inducing the expression of self-antigens, which are shared with tumors in the thymus, thereby promoting the tolerance to tumor antigens and tumor immune evasion [83]. RANK is highly expressed in TAMs, whereas RANKL is highly expressed in tumor-infiltrating T cells [78,84,85]. RANK–RANKL signaling in M2 TAMs has been shown to produce chemokines that recruit immunosuppressive Tregs to the TME [78,84,85].

With accumulating knowledge about the modulatory role of RANK–RANKL signaling in the immune system, there has been research interest in investigating the repurposing of RANKL inhibitors to improve response to immune checkpoint inhibitors in the treatment of cancer [86]. Denosumab is a RANKL inhibitor first approved in 2010 for the prevention of skeletal related events in patients with advanced malignancies involving the bone. Recent clinical data suggest that combination of denosumab and immune checkpoint inhibitors could act synergistically to improve response rate (melanoma and NSCLC) and prolong survival (NSCLC) [85,87,88]. A few more clinical trials are underway to investigate the combination of denosumab and PD-1 inhibitors (Table 12.2), including (i) KEYPAD trial (phase II, NCT03280667) evaluating the combination of denosumab and pembrolizumab in patients with VEGF receptor tyrosine kinase inhibitor–refractory clear renal cell carcinoma; and (ii) CHARLI trial (phase Ib/II, NCT03161756) investigating ipilimumab-nivolumab-denosumab and nivolumab-denosumab in patients with unresectable stage III/IV melanoma.

TABLE 12.2 Clinical trials investigating combination of immune checkpoint inhibitors with drugs inhibiting the RANK–RANKL system.

Cancer type	Drug inhibiting RANK–RANKL	Immune checkpoint inhibitor	Primary endpoint	ClinicalTrials.gov registration #
Melanoma (unresectable, stage III/IV)	Denosumab	Nivolumab or ipilimumab-nivolumab	PFS; grade 3–4 immune-related adverse reactions	NCT03161756 (CHARLI)—Phase Ib/II
Melanoma (stage III/IV; cutaneous melanoma)	Denosumab	Nivolumab or pembrolizumab	Antitumor effect as represented by change in density of TILs in tumor tissue	NCT03620019—Phase II
NSCLC (stage IV with bone metastases)	Denosumab	Nivolumab	ORR (according to PD-L1 expression rate)	NCT03669523 (DENIVOS)—Phase II
Renal cell carcinoma (unresectable, metastatic clear cell)	Denosumab	Pembrolizumab	ORR	NCT03280667 (KEYPAD)—Phase II

ORR, objective response rate; *PFS*, progression free survival; *TILs*, tumor-infiltrating lymphocytes.

Repurposing antiestrogens for cancer immunotherapy

Subpopulations of immature myeloid lineage cells frequently arise during tumor progression. These cells are known as the MDSCs, which promote tumor progression and metastasis [89]. MDSCs also inhibit the adaptive immune response and therefore interfere with therapy with immune checkpoint blockade. It has been recently demonstrated that MDSCs express ERα and that estradiol signaling through ERα regulates MDSC expansion in various cancer types [90]. Importantly, depletion of estrogen was found to retard tumor progression by reducing MDSCs regardless of the actual ER status of the tumors. Therefore, antiestrogen therapy may be combined with immune checkpoint inhibitors to provide better cancer immunotherapy by eliminating MDSCs [91]. Since the beneficial effect is independent of the ERα status of the tumor, this novel approach may be used in cancer type other than breast cancer (e.g., lung cancer and melanoma) [92]. While the composition of the TME in different tumor type can be considerably heterogeneous, MDSC expansion and recruitment is generally observed in various tumor types. Therefore, the novel approach of targeting MDSCs will be highly desirable because it is expected to modulate the immunosuppressive TME and enhance cancer immunotherapy [91].

Antiestrogens can be classified into one of the following categories: (i) selective estrogen receptor modulators (SERMs), e.g., tamoxifen, clomifene, and raloxifene; (ii) ER silent antagonist and selective estrogen receptor degrader, e.g., fulvestrant; and (iii) aromatase inhibitors that interfere with ER ligand synthesis, e.g., anastrozole. Since these classes of antiestrogens have different mechanisms, their effectiveness in MDSC inhibition in the context of different tumor type still warrants further investigation.

Ospemifene is a recently approved SERM. Compared with other SERMs such as tamoxifen, it has relatively weaker estrogenic/antiestrogenic effects [93]. Indeed, it is devoid of activity in multidrug resistant breast cancer [94]. Currently, ospemifene is indicated for the treatment of postmenopausal women with dyspareunia and vulvovaginal atrophy [95]. Because of the weak hormonal effect of ospemifene, its effect on anticancer immunity is not mediated by the aforementioned ERα-dependent inhibition of MDSCs. It has been recently reported that ospemifene stimulated T cells through PI3K and calmodulin signaling pathways [96]. When used in combination with an antigen-specific peptide cancer vaccine, ospemifene was shown to enhance antigen-specific immune response and increase CTL activity in tumor-bearing mice [96].

Drugs inhibiting the renin–angiotensin system to reprogram TME and overcome resistance to immunotherapy

The renin–angiotensin system (RAS) is a critical regulator of blood pressure and fluid balance. When a reduced renal blood flow is detected by the kidney, renin is released into the circulation to promote the production of angiotensin II to induce vasoconstriction and fluid retention. Numerous drugs, including the angiotensin-converting enzyme inhibitors and ARB, have been approved to manipulate the RAS in the management of hypertension, heart failure, and diabetic nephropathy.

It has been recently reported that angiotensin II was produced by hypoxic cancer cells to create an immunosuppressive TME by facilitating infiltration of fibroblasts and other inflammatory cells [21]. Therefore, inhibitors targeting the RAS are hypothesized to reprogram the TME for the circumvention of resistance to cancer immunotherapy [21]. In a murine colorectal cancer model, candesartan (a clinically approved ARB) was found to enhance the antitumor immune response to both anti–PD-1 and anti–CTLA-4 checkpoint immunotherapy by blocking angiotensin II signaling and promoting infiltration of CD8+ T cells [97]. Another widely used ARB, valsartan, was also shown to abrogate the immunosuppressive TME and potentiate the CD8+ T cell–mediated antitumor response to anti–PD-1 monoclonal antibody in a murine colon cancer model [24]. Valsartan was found to suppress the production of a few immunosuppressive factors including IL-6 and VEGF in MDSCs and macrophage, thereby promoting antitumor immune response [24].

Repurposing rotavirus vaccines to overcome resistance to immune checkpoint blockade

Intratumoral injections of oncolytic viruses, toll-like receptor (TLR) agonists, or stimulator of INF gene agonists are currently in active clinical investigation with an aim to stimulate "pattern recognition receptors (PRRs)" for the priming of antitumor immunity by converting noninfiltrated "cold" tumors into immune cell–infiltrated "hot" tumors [98,99]. Pediatric cancers are generally not responsive to immune checkpoint blockade because of low T-cell infiltrates, high myeloid cell infiltrates, and usually low mutational load [100,101]. For pediatric cancer patients, the commercially available antiinfectious disease vaccines represent an alternative source of PRR agonists that may be used to potentiate the effect of immune checkpoint inhibitors. In a recent study by Shekarian et al., the TLR stimulatory activity of various viral- and bacterial-based vaccine was evaluated [102]. Interestingly, rotavirus vaccines were found to possess both immunostimulatory and oncolytic properties. They can also directly kill cancer cells in vitro, showing features consistent with ICD. In several immunocompetent murine tumor models, intratumoral injection of both live and heat/UV-inactivated rotavirus was found to generate synergistic anticancer effects with anti-CTLA4/anti–PD-L1 antibodies. The enhanced anticancer effect was likely due to the upregulation of the double-stranded RNA receptor retinoic acid-induced gene 1. Since rotavirus vaccines are routinely used in pediatric and adult populations, the intratumoral injection of rotavirus vaccines is expected to be translated to oncology clinic quickly to enhance cancer immunotherapy.

Recent advances of repurposing nanoparticle-based drug delivery systems to potentiate cancer immunotherapy and to alleviate side effects

Multiple immune checkpoint inhibitors have been combined in clinical settings to improve the efficacy of cancer immunotherapy. However, the combination therapies also lead to more irAEs. Moreover, many immune checkpoint inhibitors administered systemically have limited tumor specificity, which could lead to off-target adverse effects due to immune reactions against normal tissues [103]. The adverse reactions can range from relatively minor conditions, including skin redness, to more severe aliments such as pneumonitis, colitis, and endocrinopathies [104]. Effective strategies are needed to increase the efficacy of cancer immunotherapy by minimizing off-target adverse effects.

Nanoparticle (NP)-based drug delivery systems are under intensive investigation to achieve targeted delivery of various anticancer agents, including small molecule drugs, siRNAs, DNAs, and monoclonal antibodies, to tumor sites [105–108]. Optimization of NPs in terms of size, shape, surface charges, and hydrophobicity has been shown to successfully improve the delivery of immunotherapeutic agents to tumor and/or lymph nodes [109,110]. Moreover, targeted delivery of NPs to immune cells could be achieved by chemical or physical modification of immune system–targeting ligands that specifically interact with overexpressed receptors on the surface of the target tumor or immune cells [30,111,112]. Therefore, the repurposing of existing NP-based formulations has been used to deliver ICD-inducing cytotoxic drugs, cytokine-related immune modulators, or adjuvants to target tumor sites and/or TME. In fact, a few NP formulations related to cancer immunotherapy is currently in clinical investigation. A few representative examples are CYT004-MelQbG10 (A-type CpG with virus-like NP) for treating malignant melanoma (phase IIa, NCT00306566), Lipovaxin-MM (liposomal formulation of IFN-γ) for treating malignant melanoma (phase I, NCT01052142), and Oncoquest-L (liposomal formulation of IL-2) for treating lymphoma (phase II, NCT02194751).

NP-based formulations for delivering immunogenic cell death–inducing cytotoxic drugs

To enhance the limited clinical efficacy of immune checkpoint inhibitors, it has been proposed that additional synergistic treatment modalities have to be coadministered to induce ICD of tumors [113]. Interestingly, a number of classical cytotoxic drugs, including doxorubicin, 5-fluorouracil, gemcitabine, mitoxantrone, oxaliplatin, and paclitaxel, are known to elicit an immune response to activate apoptosis [114–117]. However, classical cytotoxic anticancer drugs are lethal to both cancer and normal cells, thus leading to high off-target toxicity in normal tissues and immune cells. Therefore, the simple combination of cancer immunotherapy and the conventional cytotoxic drugs is expected to cause severe systemic toxicity and immune suppression. To this end, the tumor-specific delivery of cytotoxic drugs facilities tumor immunogenicity of antigens in various tumor sites and enhances ICD to potentiate cancer immunotherapy [118–120]. Table 12.3 summarizes recent examples of such NP-based delivery systems of cytotoxic drugs to enhance ICD in the TME.

Interestingly, the encapsulation of cytotoxic anticancer drugs in NPs has been shown to eliminate the undesirable immune side effects of peptides and liposomes making up the NPs [118,121]. Liposomes, modified with the cyclic RGD peptide, are known to induce acute systemic anaphylaxis, IgG immune complex–triggered complement activation, and cytokine release. Reduction of the cyclic RGD peptide

TABLE 12.3 NP-based delivery systems for tumor targeting delivery of ICD-inducing cytotoxic drugs and/or their combination with immune modulators.

ICD-inducing cytotoxic drugs	NP-based formulation	Mechanism to enhance cancer immunotherapy	References
Docetaxel	NPs made up with ApoAI mimetic peptide and phospholipids	- Induced ICD and recruited cytotoxic T lymphocytes in TME - Evaluated in glioblastoma	[137]
Doxorubicin	Polymer–lipid manganese dioxide NP	- Attenuated hypoxia and acidosis - Remodeled TME and boost antitumor T cell activity - Promoted macrophage phenotype polarization from M1 to M2 type - Evaluated in breast cancer	[138]
Doxorubicin	Nano-sized prodrug consisting of doxorubicin, MMP cleavable peptide (CPLGLAGG), and HA	- Induced the release of ICD-associated molecules (HMGB1, IFN, and TNF-α) in TME to activate T cells - Upregulated PD-L1; when combined with anti–PD-1 antibody, they elicited strong antitumor immune response mediated by robust TILs - Evaluated in melanoma	[139]
Oxaliplatin (in combination with IDO pathway inhibitor—indoximod)	Mesoporous silica NP	- Induced ICD and recruited cytotoxic T lymphocytes in TME - Promoted the release of HMGB1 and ATP from tumor cells to generate immunogenic stimuli to the antigen presenting DC - Increased cell surface–exposed calreticulin on the dying cell surfaces in TME, subsequently leading to enhancement of ICD effect by indoximod - Evaluated in pancreatic cancer	[140]
Paclitaxel	Drugs loaded in nanoformulation exosomes isolated from M1 macrophages	- Exosomes are representative endogenous NPs that contain various immune-modulating cytokines derived from cells - Paclitaxel induced ICD - Proinflammatory cytokines enhanced anticancer immunity	[141]
Paclitaxel	TME-activated binary cooperative NP	- Acidity-induced cleavage of PEG NP shell and glutathione-mediated linker allowed tumor/TME-specific targeting - The enhanced accumulation of self-assembled cytotoxic prodrugs in TME triggered ICD of tumor cells - The delivery of TME-activated prodrug to tumor cells increased immunogenicity of tumor cells without affecting normal immune cells	[142]

ApoAI, apolipoprotein-I; *EPR*, enhanced permeability and retention; *HA*, hyaluronic acid; *HMGB1*, high mobility group box 1; *ICD*, immunogenic cell death; *IDO*, indoleamine 2,3-dioxygenase; *IFN*, interferon; *MMP*, matrix metalloproteinase; *NP*, nanoparticle; *PEG*, polyethylene glycol; *TNF-α*, tumor necrosis factor-α; *TILs*, tumor-infiltrating lymphocytes; *TME*, tumor microenvironment.

content or injection doses could not resolve the acute systemic toxicity. Lu et al. reported that the encapsulation of doxorubicin in cyclic RGD peptide–modified liposomes eliminated the systemic immune response by inhibiting immunotoxicity and antibody overproduction [121].

NP-based formulations for delivering cytokine-related immune modulators

Cytokines, INFs, ILs, and chemokines are cytokines that have been widely used as immunomodulatory agents for cancer therapy. IFN-α is a clinically approved cytokine used for the treatment of leukemia. Recombinant IL-2 has also been developed for cancer immunotherapy [122]. However, the short half-lives and limited stability of these cytokines have posed great hurdle for their clinical use. To this end, NP-based delivery systems have been used to overcome limitations of cytokine-based therapy including short half-life, autoimmune attack, and inflammatory immune reaction [123–125]. Moreover, NPs have also been developed to facilitate the use of cytokines as immune-modulating agents in anticancer treatment. Cytokines can be readily encapsulated or chemically conjugated to NPs to prevent their degradation by enzymes in vivo. Conjugation of cytokines to the surface of NPs also allows their specific delivery to target receptors on cancer cell surface in a controlled release manner [126,127]. Moreover, NP-based delivery systems also improve the targeting efficiency and pharmacokinetics/pharmacodynamics of the encapsulated or conjugated cytokines in vivo [125]. In recent years, multifaceted NPs have been developed, which contain two or more immune modulating agents to improve the selective delivery and targeting ability to target cancer and immune cells with efficacy [125,128,129]. Table 12.4 summarizes NP-based formulations developed recently for delivering cytokines to enhance anticancer response.

NP-based formulations for delivering adjuvants to enhance anticancer immunity

Adjuvants are used in cancer immunotherapy to activate the APCs and enhance the immune response [130]. However, only limited adjuvants have been approved for human use due to unwanted side effects. In recent years, NP-based formulations have been repurposed to deliver adjuvants to enhance their immunogenicity and reduce toxicity. Moreover, NP-based delivery systems have also been used to codeliver adjuvant and antigen to channel the immune response to Th1 or Th2 [131]. Aluminum salts (alum), lipopolysaccharide, CpG oligodeoxynucleotides, layered double hydroxide, and polyinosinic:polycytidylic acid (poly I:C) are commonly used adjuvants in NP formulations for cancer immunotherapy [132,133]. Table 12.5 summarizes representative NP formulations for the codelivery of adjuvant and antigen to enhance antitumor immunity.

Challenges and perspective

There is a remarkable growing interest in adopting the drug repurposing approach for cancer therapy in recent years. It is encouraging to note that the major pharmaceutical regulatory authorities (including Food and Drug Administration in the United States and European Medicines Agency in Europe) have already launched drug repurposing programs to identify new therapeutic uses for existing pipeline medications developed by the pharmaceutical industry [134]. A streamlined strategy to identify potential repurposed drugs and an appropriate platform for preclinical and clinical investigation are highly needed to realize the full potential of drug repurposing for cancer therapy. Some repurposed drug candidates may show good efficacy, but only at a dose remarkably higher

TABLE 12.4 NP-based delivery systems for tumor targeting delivery of cytokines to enhance cancer immunotherapy.

Cytokine-related immune modulatory agent	NP-based formulation	Mechanism to enhance cancer immunotherapy	References
IL-2	Hydroxyethyl starch nanocapsules	- Compared with free IL-2, the NP formulation increased the binding affinity to IL-2 receptor - The nanocapsules are significantly absorbed by activated $CD4^{+}CD25^{+}$ T cells compared with naïve $CD4^{+}CD25^{-}$ T cells, thus enhancing activated T-cell proliferation	[126]
TRAIL	Lipid nanocarriers	- The expression of TRAIL on the surface of the lipid nanocarriers effectively enhanced the proapoptotic activity of extrinsic TRAIL pathway and increased the activity of apoptosis-inducing caspases	[127]
Peptide loaded with MHC and anti-CD19	Multifaceted NPs to facilitate simultaneous targeting of lymphocytes and cancer cells	- Stimulated T cells to kill tumor cells by coating a single NP with tumor cell–binding antibody (α-human CD19) and loaded antigen-specific T cell–binding peptide (MHC-binding peptide)	[128]
Calreticulin	Multifaceted NPs to simultaneously target cancer cell–specific receptors and signaling phagocytosis of macrophages	- Specifically targeted cancer cells overexpressing human EGFR-2 and strongly stimulated professional APCs by calreticulin, which is a protein-inducing phagocytosis - Phagocytosis by APCs was triggered by multifaceted NPs, which are selectively bound to cancer cells to cause innate and adaptive immunity	[143]
M1 macrophage markers and mRNAs of proinflammatory factors	M1 macrophage–derived nanovesicles	- M1 macrophage–derived nanovesicles effectively encapsulate M1 macrophage markers and mRNAs of proinflammatory factors and replicated the function of M1 macrophages during immunotherapy - Efficiently polarized M2 tumor–associated macrophages to antitumor M1 type macrophage - Enhanced the secretion of antitumor cytokines and suppressed tumor growth	[144]
Cancer cell–specific biomarker	Tumor cell–derived nanovesicles	- The tumor cell–derived nanovesicle was composed of various proteins in the tumor cell membrane, which acts as tumor antigens and a pathogen ligand to facilitate antigen recognition by APCs	[145]
Anti–4-1BB antibody	Dual-targeting NP to simultaneously stimulate immune cells and block immune checkpoints on cancer cells	- The multifaceted NPs were coated with two types of antibodies (anti-4-1BB and anti-PD-L1) - Anti-4-1BB induced tumor-targeted $CD8^{+}$ T cells to increase secretion of cytokines including IFN-γ	[129]

TABLE 12.4 NP-based delivery systems for tumor targeting delivery of cytokines to enhance cancer immunotherapy.—cont'd

Cytokine-related immune modulatory agent	NP-based formulation	Mechanism to enhance cancer immunotherapy	References
		- Anti–PD-L1 blocked the immunosuppressive pathway of cancer cells to prevent immune evasion from cytotoxic T cells - Evaluated in melanoma	
TRAIL	Multifaceted NPs coated with TRAIL and E-selectin adhesion molecules	- E-selectin on NP surface specifically binds to leukocyte membrane; therefore, TRAIL is circulated in the blood by attaching to white blood cells and can evade renal clearance - The approach mimics the cytotoxic activity of tumor-targeted T cells to eradicate metastatic cancer cells from the blood circulation - Evaluated in colon carcinoma	[146]

APC, antigen-presenting cell; *IL-2*, interleukin-2; *MHC*, major histocompatibility complex; *TRAIL*, tumor necrosis factor related apoptosis-inducing ligand.

TABLE 12.5 NP-based delivery systems for codelivery of adjuvant and antigen to APCs for cancer immunotherapy.

Adjuvant	NP-based formulation	Mechanism to enhance cancer immunotherapy	References
TLR3 agonist	Synthetic vaccine NPs—composed of two NPs (one carrying the adjuvant TLR3 agonist [poly I:C]) and the other carrying the model tumor antigen (ovalbumin)	- Poly I:C is a double-stranded RNA, and it interacts with TLR3 expressing on leukocyte membranes - NPs are preferentially taken up by APCs to enhance the secretion of type I IFN-α and IFN-β and proinflammatory cytokines - Evaluated in lymphoma cells	[147]
Albumin-binding adjuvant (AlbiCpG)	Albumin-binding vaccine (AlbiVax)—consisting of AlbiCpG and AlbiAg (antigen self-assembled with endogenous albumin in vivo)	- AlbiVax-based nanocomplexes are efficiently delivered to lymph nodes and induced antigen-specific T-cell response in mice - Evaluated in lymphoma cells	[130]
CpG	Synthetic high-density lipoprotein (sHDL) as the nanocarrier	- sHDL could deliver multiepitope antigens that induced broad T-cell responses to potentiate tumor immunotherapy - Evaluated in melanoma cells	[148,149]

APC, antigen-presenting cell; *NP*, nanoparticle; *TLR3*, toll-like receptor 3.

than that needed for their original indications. Although the adverse effect profile of repurposed drugs had generally been established, there are still risks of developing toxicity especially at high doses.

It is challenging to identify the "right" repurposed drug to target a specific oncogenic pathway or to treat a specific population of cancer patients. In the era of personalized medicine, the use of an appropriate biomarker for patient selection in clinical trials becomes critical. It is noteworthy that efficacy of numerous drug repurposing approaches in cancer therapy was demonstrated only in a subset of patients [17]. A representative example is the differential enhancement of patient survival in cancer patients bearing tumor with low PD-L1 expression upon the adjuvant use of aspirin [19]. On the other hand, patients with certain characteristics may also receive additional benefits. In prostate cancer, genome-scale metabolic modeling has been applied to offer insights into cancer metabolism and help identify potential biomarkers and drug targets [135]. The induction of severe irAEs by immune checkpoint inhibitors in some patients has raised safety concerns. It will be important to develop tools for reliable prediction of normal tissue toxicity in patients receiving the combined treatment of immunotherapy and repurposed drugs [136]. To this end, in-depth preclinical mechanistic studies, biomarker selection, and perhaps also prospective clinical studies will be needed for further exploration.

References

[1] Nosengo N. Can you teach old drugs new tricks? Nature 2016;534:314–6.

[2] Sohraby F, Bagheri M, Aryapour H. Performing an in silico repurposing of existing drugs by combining virtual screening and molecular dynamics simulation. Methods Mol Biol 2019;1903:23–43.

[3] Nagaraj AB, Wang QQ, Joseph P, Zheng C, Chen Y, Kovalenko O, et al. Using a novel computational drug-repositioning approach (DrugPredict) to rapidly identify potent drug candidates for cancer treatment. Oncogene 2018;37:403–14.

[4] Rose PW, Prlic A, Altunkaya A, Bi C, Bradley AR, Christie CH, et al. The RCSB protein data bank: integrative view of protein, gene and 3D structural information. Nucleic Acids Res 2017;45: D271–81.

[5] Konc J, Janezic D. ProBiS-ligands: a web server for prediction of ligands by examination of protein binding sites. Nucleic Acids Res 2014;42(Web Server issue):W215–20.

[6] Salentin S, Schreiber S, Haupt VJ, Adasme MF, Schroeder M. PLIP: fully automated protein-ligand interaction profiler. Nucleic Acids Res 2015;43(W1): W443–7.

[7] Lamb J, Crawford ED, Peck D, Modell JW, Blat IC, Wrobel MJ, et al. The Connectivity Map: using gene-expression signatures to connect small molecules, genes, and disease. Science 2006;313: 1929–35.

[8] Keenan AB, Jenkins SL, Jagodnik KM, Koplev S, He E, Torre D, et al. The library of integrated network-based cellular signatures NIH program: system-level cataloging of human cells response to pertubations. Cell Syst 2018;6:13–24.

[9] Koleti A, Terryn R, Stathias V, Chung C, Cooper DJ, Turner JP, et al. Data portal for the Library of Integrated Network-based Cellular Signatures (LINCS) program: integrated access to diverse large-scale cellular perturbation response data. Nucleic Acids Res 2018;46(D1):D558–66.

[10] Corsello SM, Bittker JA, Liu Z, Gould J, McCarren P, Hirschman JE, et al. The Drug Repurposing Hub: a next-generation drug library and information resource. Nat Med 2017;23:405–8.

[11] Tang J, Tanoli ZU, Ravikumar B, Alam Z, Rebane A, Vaha-Koskela M, et al. Drug target commons: a community effort to build a consensus knowledge base for drug-target interactions. Cell Chem Biol 2018;25: 224–9.

[12] Khaladkar M, Koscielny G, Hasan S, Agarwal P, Dunham I, Rajpal D, et al. Uncovering novel repositioning opportunities using the Open Targets platform. Drug Discov Today 2017;22:1800–7.

[13] Brown AS, Patel CJ. A standard database for drug repositioning. Sci Data 2017;4:170029.

[14] Shameer K, Glicksberg BS, Hodos R, Johnson KW, Badgeley MA, Readhead B, et al. Systematic analyses of drugs and disease indications in RepurposeDB reveal pharmacological, biological and epidemiological factors influencing drug repositioning. Brief Bioinform 2018;19:656–78.

[15] Schneider L, Kehl T, Thedinga K, Grammes NL, Backes C, Mohr C, et al. ClinOmicsTrailbc: a visual analytics tool for breast cancer treatment stratification. Bioinformatics April 30, 2019. https://doi.org/10.1093/bioinformatics/btz302. pii: btz302.

[16] Pitt JM, Vetizou M, Daillere R, Roberti MP, Yamazaki T, Routy B, et al. Resistance mechanisms to immune-checkpoint blockade in cancer: tumor-intrinsic and −extrinsic factors. Immunity 2016;44:1255−69.

[17] Sharma P, Hu-Lieskovan S, Wargo JA, Ribas A. Primary, adaptive, and acquired resistance to cancer immunotherapy. Cell 2017;168:707−23.

[18] Jenkins RW, Barbie DA, Flaherty KT. Mechanisms of resistance to immune checkpoint inhibitors. Br J Cancer 2018;118:9−16.

[19] Hamada T, Cao Y, Qian ZR, Masugi Y, Nowak JA, Yang J, et al. Aspirin use and colorectal cancer survival according to tumor CD274 (Programmed cell death 1 ligand 1) expression status. J Clin Oncol 2017;35:1836−44.

[20] Scharping NE, Menk AV, Whetstone RD, Zeng X, Delgoffe GM. Efficacy of PD-1 blockade is potentiated by metformin-induced reduction of tumor hypoxia. Cancer Immunol Res 2017;5:9−16.

[21] Pinter M, Jain RK. Targeting the renin-angiotensin system to improve cancer treatment: implications for immunotherapy. Sci Transl Med 2017;9. eaan5616.

[22] Scurr M, Pembroke T, Bloom A, Roberts D, Thomson A, Smart K, et al. Low-dose cyclophosphamide induces antitumor T-cell responses, which associate with survival in metastatic colorectal cancer. Clin Cancer Res 2017;23:6771−80.

[23] Chen HM, Ma G, Gildener-Leapman N, Eisenstein S, Coakley BA, Ozao J, et al. Myeloid-derived suppressor cells as an immune parameter in patients with concurrent sunitinib and stereotactic body radiotherapy. Clin Cancer Res 2015;21:4073−85.

[24] Nakamura K, Yaguchi T, Ohmura G, Kobayashi A, Kawamura N, Iwata T, et al. Involvement of local renin-angiotensin system in immunosuppression of tumor microenvironment. Cancer Sci 2018;109:54−64.

[25] Eikawa S, Nishida M, Mizukami S, Yamazaki C, Nakayama E, Udono H. Immune-mediated antitumor effect by type 2 diabetes drug, metformin. Proc Natl Acad Sci USA 2015;112:1809−14.

[26] Cha JH, Yang WH, Xia W, Wei Y, Chan LC, Lim SO, et al. Metformin promotes antitumor immunity via endoplasmic-reticulum-associated degradation of PD-L1. Mol Cell 2018;71:606−620.e7.

[27] Barber DL, Wherry EJ, Masopust D, Zhu B, Allison JP, Sharpe AH, et al. Restoring function in exhausted CD8 T cells during chronic viral infection. Nature 2006;439:682−7.

[28] Brooks DG, McGavern DB, Oldstone MB. Reprogramming of antiviral T cells prevents inactivation and restores T cell activity during persistent viral infection. J Clin Invest 2006;116:1675−85.

[29] Baumeister SH, Freeman GJ, Dranoff G, Sharpe AH. Coinhibitory pathways in immunotherapy for cancer. Annu Rev Immunol 2016;34:539−73.

[30] Gotwals P, Camercon S, Cipolletta D, Cremasco V, Crystal A, Hewes B, Mueller B, Quaratino S, Sabatos-Peyton C, Petruzzelli L, et al. Prospects for combining targeted and conventional cancer therapy with immunotherapy. Nat Rev Cancer 2017;17:286−301.

[31] Chen EW, Brzostek J, Gascoigne NRJ, Rybakin V. Development of a screening strategy for new modulators of T cell receptor signaling and T cell activation. Sci Rep 2018;8:10046.

[32] Chen EW, Ke CY, Brzostek J, Gascoigne NRJ, Rybakin V. Identification of mediators of T-cell receptor signaling via the screening of chemical inhibitor libraries. J Vis Exp 2019;143:e58946.

[33] Deng J, Wang ES, Jenkins RW, Li S, Dries R, Yates K, et al. CDK4/6 inhibition augments antitumor immunity by enhancing T-cell activation. Cancer Discov 2018;8:216−33.

[34] Fouda A, Tahsini M, Khodayarian F, Al-Nafisah F, Rafei M. A fluorescence-based lymphocyte assay suitable for high-throughput screening of small molecules. J Vis Exp 2017;121:55199.

[35] Hashimoto M, Kamphorst AO, Im SJ, Kissick HT, Pillai RN, Ramalingam SS, et al. CD8 T cell exhaustion in chronic infection and cancer: opportunities for interventions. Annu Rev Med 2018;69:301−18.

[36] Marro BS, Zak J, Zavareh RB, Teijaro JR, Lairson LL, Oldstone MBA. Discovery of small molecules for the reversal of T cell exhaustion. Cell Rep 2019;29:3293−302.

[37] Li X, Wenes M, Romero P, Huang SC, Fendt SM, Ho PC. Navigating metabolic pathways to enhance antitumor immunity and immunotherapy. Nat Rev Clin Oncol 2019;16:425−41.

[38] Shevhenko I, Bazhin AV. Metabolic checkpoints: novel avenues for immunotherapy of cancer. Front Immunol 2018;9:1816.

[39] Kouidhi S, Ben Ayed F, Benammar Elgaaied A. Targeting tumor metabolism: a new challenge to improve immunotherapy. Front Immunol 2018;9:353.

[40] McCarty MF, Whitaker J. Manipulating tumor acidification as a cancer treatment strategy. Altern Med Rev 2010;15:264−72.

[41] Nakagawa Y, Negishi Y, Shimizu M, Takahashi M, Ichikawa M, Takahashi H. Effects of extracellular pH and hypoxia on the function and development

of antigen-specific cytotoxic T lymphocytes. Immunol Lett 2015;167:72–86.

[42] Bosticardo M, Ariotti S, Losanna G, Bernabei P, Forni G, Novelli F. Biased activation of human T lymphocytes due to low extracellular pH is antagonized by B7/CD28 costimulation. Eur J Immunol 2001;31: 2829–38.

[43] Romero-Garcia S, Moreno-Altamirano MMB, Prado-Garcia H, Sanchez-Garcia FJ. Lactate contribution to the tumor microenvironment: mechanisms, effects on immune cells and therapeutic relevance. Front Immunol 2016;7:52.

[44] Peppicelli S, Bianchini F, Calorini L. Extracellular acidity, a <reappreciated> trait of tumor environment driving malignancy: perspectives in diagnosis and therapy. Cancer Metastasis Rev 2014;33:823–32.

[45] Le A, Cooper CR, Gouw AM, Dinavahi R, Maitra A, Deck KM, et al. Inhibition of lactate dehydrogenase A induces oxidative stress and inhibits tumor progression. Proc Natl Acad Sci USA 2010;107:2037–42.

[46] Doherty JR, Cleveland JL. Targeting lactate metabolism for cancer therapeutics. J Clin Invest 2013; 123:3685–92.

[47] Huber V, Camisaschi C, Berzi A, Ferro S, Lugini L, Triulzi T, et al. Cancer acidity: an ultimate frontier of tumor immune escape and a novel target of immunomodulation. Semin Cancer Biol 2017; 43(Suppl. C):74–89.

[48] Hong CS, Graham NA, Gu W, Espindola Camacho C, Mah V, Maresh EL, et al. MCT1 modulates cancer cell pyruvate export and growth of tumors that co-express MCT1 and MCT4. Cell Rep 2016;14:1590–601.

[49] Eichner R, Heider M, Fernandez-Saiz V, van Bebber F, Garz A-K, Lemeer S, et al. Immunomodulatory drugs disrupt the cereblon-CD147-MCT1 axis to exert antitumor activity and teratogenicity. Nat Med 2016;22: 735–43.

[50] Gorgun G, Calabrese E, Soydan E, Hideshima T, Perrone G, Bandi M, et al. Immunomodulatory effects of lenalidomide and pomalidomide on interaction of tumor and bone marrow accessory cells in multiple myeloma. Blood 2010;116:3227–37.

[51] Chirasani SR, Leukel Pm Gottfried E, Hochrein J, Stadler K, Neumann B, et al. Diclofenac inhibits lactate formation and efficiently counteracts local immune suppression in a murine glioma model. Int J Cancer 2013;132:843–53.

[52] Pantziarka P, Sukhatme V, Bouche G, Meheus L, Sukhatme VP. Repurposing drugs in oncology (ReDO) – diclofenac as an anti-cancer agent. Ecancermedicalscience 2016:10.

[53] Choi SYC, Collins CC, Gout PW, Wang Y. Cancer-generated lactic acid: a regulatory, immunosuppressive metabolite? J Pathol 2013;230:350–5.

[54] Pilon-Thomas S, Kodumudi KN, El-Kenawi AE, Russell S, Weber AM, Luddy K, et al. Neutralization of tumor acidity improves antitumor responses to immunotherapy. Cancer Res 2016;76:1381–90.

[55] Fais S. Proton pump inhibitor-induced tumor cell death by inhibition of a detoxification mechanism. J Intern Med 2010;267:515–25.

[56] Koltai T. Cancer: fundamentals behind pH targeting and the double-edged approach. OncoTargets Ther 2016;76:1381–90.

[57] Rena G, Hardie DG, Pearson ER. The mechanisms of action of metformin. Diabetologia 2017;60:1577–85.

[58] Chi KH, Wang YS, Huang YC, Chiang HC, Chi MS, Chi CH, et al. Simultaneous activation and inhibition of autophagy sensitizes cancer cells to chemotherapy. Oncotarget 2016;7:58075–88.

[59] Huang S, Yang ZJ, Yu C, Sinicrope FA. Inhibition of mTOR kinase by AZD8055 can antagonize chemotherapy-induced cell death through autophagy induction and down-regulation of p62/sequestosome 1. J Biol Chem 2011;286:40002–12.

[60] Woo S-U, Sangai T, Akcakanat A, Chen H, Wei C, Meric-Bernstam F. Vertical inhibition of the PI3K/Akt/mTOR pathway is synergistic in breast cancer. Oncogenesis 2017;6:e385.

[61] Deng F, Ma YX, Liang L, Zhang P, Feng J. The pro-apoptosis effect of sinomenine in renal carcinoma via inducing autophagy through inactivating PI3K/AKT/mTOR pathway. Biomed Pharmacother 2017; 13:1269–74.

[62] Schmidt KM, Hellerbrand C, Ruemmele P, Michalski CW, Kong B, Kroemer A, et al. Inhibition of mTORC2 component RICTOR impairs tumor growth in pancreatic cancer models. Oncotarget 2017;8:24491–505.

[63] Chi H. Regulation and function of mTOR signaling in T cell fate decision. Nat Rev Immunol 2012;12:325–38.

[64] Mineharu Y, Kamran N, Lowenstein PR, Castro MG. Blockade of mTOR signaling via rapamycin combined with immunotherapy augments anti-glioma cytotoxic and memory T-cell functions. Mol Cancer Ther 2014; 13:3024–36.

[65] Lukey MJ, Katt WP, Cerione RA. Targeting amino acid metabolism for cancer therapy. Drug Discov Today 2017;22:796–804.

[66] Van Baren N, Van den Eynde BJ. Tryptophan-degrading enzymes in tumoral immune resistance. Front Immunol 2015;6:34.

[67] Huang TT, Yen MC, Lin CC, Weng TY, Chen YL, Lin CM, et al. Skin delivery of short hairpin RNA of indoleamine 2,3-dioxygenase induces antitumor immunity against orthotropic and metastatic liver cancer. Cancer Sci 2011;102:2214–20.

[68] Yen MC, Lin CC, Chen YL, Huang SS, Yang HJ, Chang CP, et al. A novel cancer therapy by skin delivery of indoleamine 2,3-dioxygenase siRNA. Clin Cancer Res 2009;15:641–9.

[69] Platten M, von Knebel Doeberitz N, Oezen I, Wick W, Ochs K. Cancer immunotherapy by targeting IDO1/TDO and their downstream effectors. Front Immunol 2015;5:673.

[70] Zhai L, Spranger S, Binder DC, Gritsina G, Lauing KL, Giles FJ, et al. Molecular pathways: targeting Ido1 and other tryptophan dioxygenases for cancer immunotherapy. Clin Cancer Res 2015;21:5427–33.

[71] Balachandran VP, Cavnar MJ, Zeng S, Bamboat ZM, Ocuin LM, Obaid H, et al. Imatinib potentiates antitumor T cell responses in gastrointestinal stromal tumor through the inhibition of Ido. Nat Med 2011;17: 1094–100.

[72] Reilley MJ, Bailey A, Subbiah V, Janku F, Naing A, Falchook G, eta l. Phase I clinical trial of combination imatinib and ipilimumab in patients with advanced malignancies. J Immunother Cancer 2017;5:35.

[73] Ahluwalia GS, Grem JL, Hao Z, Cooney DA. Metabolism and action of amino acid analog anti-cancer agents. Pharmacol Ther 1990;46:243–71.

[74] Sharma S, Stolina M, Yang SC, Baratelli F, Lin JF, Atianzar K, et al. Tumor cyclooxygenase 2-dependent suppression of dendritic cell function. Clin Cancer Res 2003;9:961–8.

[75] Marzbani E, Inatsuka C, Lu H, Disis ML. The invisible arm of immunity in common cancer chemoprevention agents. Cancer Prev Res (Phila) 2013;6:764–73.

[76] Zelenay S, van der Veen AG, Bottcher JP, Snelgrove KJ, Rogers N, Acton SE, et al. Cyclooxygenase-dependent tumor growth through evasion of immunity. Cell 2015;162:1257–70.

[77] Nakanishi Y, Nakatsuji M, Seno H, Ishizu S, Akitake-Kawano R, Kanda K, et al. COX-2 inhibition alters the phenotype of tumor-associated macrophages from M2 to M1 in Apc Min/+ mouse polyps. Carcinogenesis 2011;32:1333–9.

[78] Gonzalez-Suarez E, Sanz-Moreno A. RANK as a therapeutic target in cancer. FEBS J 2016;283:2018–33.

[79] Hanada R, Hanada T, Sigl V, Schramek D, Penninger JM. RANKL/RNK-beyond bones. J Mol Med (Berl) 2011;89:647–56.

[80] Cheng ML, Fong L. Effects of RANKL-targeted therapy in immunity and cancer. Front Oncol 2014;3:329.

[81] Huang R, Wang X, Zhou Y, Xiao Y. RANKL-induced M1 macrophages are involved in bone formation. Bone Res 2017;5:17019.

[82] Xia Y, Shen S, Verma IM. NF-kappaB, an active player in human cancers. Cancer Immunol Res 2014;2: 823–30.

[83] Khan IS, Mouchess ML, Zhu ML, Conley B, Fasano KJ, Hou Y, et al. Enhancement of an antitumor immune response by transient blockade of central T cell tolerance. J Exp Med 2014;211:761–8.

[84] Renema N, Navet B, Heymann MF, Lezot F, Heymann D. RANK-RANKL signaling in cancer. Biosci Rep 2016;310:C663–72.

[85] Ahern E, Harjunpaa H, Barkauskas D, Allen S, Takeda K, Yagita H, et al. Co-administration of RANKL and CTLA4 antibodies enhances lymphocyte-mediated antitumor immunity in mice. Clin Cancer Res 2017;23:5789–801.

[86] de Groot AF, Appelman-Dijkstra NM, van der Burg SH, Kroep JR. The anti-tumor effect of RANKL inhibition in malignant solid tumors – a systematic review. Cancer Treat Rev 2018;62:18–28.

[87] Liede A, Hernandez RK, Wade SW, Bo R, Nussbaum NC, Ahern E, et al. An observational study of concomitant immunotherapies and denosumab in patients with advanced melanoma or lung cancer. Oncoimmunology 2018;7:e1480301.

[88] Smyth MJ, Yagita H, McArthur GA. Combination anti-CTLA-4 and anti-RANKL in metastatic melanoma. J Clin Oncol 2016;34:e104–6.

[89] Gabrilovich DI, Nagaraj S. Myeloid-derived suppressor cells as regulators of the immune system. Nat Rev Immunol 2009;9:162–74.

[90] Svoronos N, Perales-Puchalt A, Allegrezza MJ, Rutkowski MR, Payne KK, Tesone AJ, et al. Tumor cell-independent estrogen signaling drives disease progression through mobilization of myeloid-derived suppressor cells. Cancer Discov 2017;7:72–85.

[91] Welte T, Zhang XH, Rosen JM. Repurposing antiestrogens for tumor immunotherapy. Cancer Discov 2017; 7:17–9.

[92] Sharma P, Allison JP. Immune checkpoint targeting in cancer therapy: toward combination strategies with curative potential. Cell 2015;161:205–14.

[93] Kangas L. Biochemical and pharmacological effects of toremifene metabolites. Cancer Chemother Pharmacol 1990;27:8–12.

[94] Wiebe V, Koester S, Lindberg M, Emshoff V, Baker J, Wurz G, et al. Toremifene and its metabolites enhance doxorubicin accumulation in estrogen receptor negative multidrug resistant human breast cancer cells. Invest New Drugs 1992;10:63–71.

[95] DeGregorio MW, Zerbe RL, Wurz GT. Ospemifene: a first-in-class, non-hormonal selective estrogen receptor modulator approved for the treatment of dyspareunia associated with vulvar and vaginal atrophy. Steroids 2014;90:82–93.

[96] Kao CJ, Wurz GT, Lin YC, Vang DP, Phong B, DeGregorio MW. Repurposing ospemifene for potentiating an antigen-specific immune response. Menopause 2017;24:437–51.

[97] Xie G, Cheng T, Lin J, Zhang L, Zheng J, Liu Y, et al. Local angiotensin II contributes to tumor resistance to

checkpoint immunotherapy. J Immunother Cancer 2018;6:88.

[98] Marabelle A, Tselikas L, de Baere T, Houot R. Intratumoral immunotherapy: using the tumor as the remedy. Ann Oncol 2017;28:xii33–43.

[99] Aznar MA, Tinari N, Rullan AJ, Sanchez-Paulete AR, Rodriguez-Ruiz ME, Melero I. Intratumoral delivery of immunotherapy-act locally, think globally. J Immunol 2017;198:31–9.

[100] Merchant MS, Wright M, Baird K, Wexler LH, Rodriguez-Galindo C, Bernstein D, et al. Phase I clinical trial of ipilimumab in pediatric patients with advanced solid tumors. Clin Cancer Res 2015;22: 1364–70.

[101] Geoerger B, Bergeron C, Gore L, Sender L, Dunkel IJ, Herzog C, et al. Phase II study of ipilimumab in adolescents with unresectable stage III or IV malignant melanoma. Eur J Cancer 2017;86:358–63.

[102] Shekarian T, Sivado E, Jallas AC, Depil S, Kielbassa J, Janoueix-Lerosey I, et al. Repurposing rotavirus vaccines for intratumoral immunotherapy can overcome resistance to immune checkpoint blockade. Sci Transl Med 2019;11. eaat5025.

[103] Zimmer L, Goldinger SM, Hofmann L, Loquai C, Ugurel S, Thomas I, et al. Neurological, respiratory, musculoskeletal, cardiac and ocular side-effects of anti-PD-1 therapy. Eur J Cancer 2016;60:210–25.

[104] Winer A, Bodor JN, Borghaei H. Identifying and managing the adverse effects of immune checkpoint blockade. J Thorac Dis 2018;10:S480–9.

[105] Liu L, Chen Q, Ruan C, Chen X, Zhang Y, He X, et al. Platinum-based nanovectors engineered with immune-modulating adjuvant for inhibiting tumor growth and promoting immunity. Theranostics 2018;8:2974–87.

[106] Kong M, Tang J, Qiao Q, Wu T, Qi Y, Tan S, et al. Biodegradable hollow mesoporous silica nanoparticles for regulating tumor microenvironment and enhancing antitumor efficiency. Theranostics 2017;7: 3276–92.

[107] He J, Duan S, Yu X, Qian Z, Zhou S, Zhang Z, et al. Folate-modified chitosan nanoparticles containing the IP-10 gene enhance melanoma-specific cytotoxic CD8(+)CD28(+) T lymphocyte responses. Theranostics 2016;6:752–61.

[108] Liu X, Li Y, Sun X, Muftuoglu Y, Wang B, Yu T, et al. Powerful anti-colon cancer effect of modified nanoparticle-mediated IL-15 immunogene therapy through activation of the host immune system. Theranostics 2018;8:3490–503.

[109] Toy R, Roy K. Engineering nanoparticles to overcome barriers to immunotherapy. Bioeng Transl Med 2016; 1:47–62.

[110] Kumar S, Anselmom AC, Banerjee A, Zakrewsky M, Mitragotri S. Shape and size-dependent immune response to antigen-carrying nanoparticles. J Control Release 2015;220:141–8.

[111] Zhu YQ, Feijen J, Zhong ZY. Dual-targeted nanomedicines for enhanced tumor treatment. Nano Today 2018;18:65–85.

[112] Phung CD, Nguyen HT, Tran TH, Choi HG, Yong CS, Kim JO. Rational combination immunotherapeutic approaches for effective cancer treatment. J Control Release 2019;294:114–30.

[113] Galon J, Bruni D. Approaches to treat immune hot, altered and cold tumours with combination immunotherapies. Nat Rev Drug Discov 2019;18: 197–218.

[114] Pfirschke C, Engblom C, Rickelt S, Cortez-Retamozo V, Garris C, Pucci F, et al. Immunogenic chemotherapy sensitizes tumors to checkpoint blockade therapy. Immunity 2016;44:343–54.

[115] Casares N, Pequignot MO, Tesniere A, Ghiringhelli F, Roux S, Chaput N, et al. Caspase-dependent immunogenicity of doxorubicin-induced tumor cell death. J Exp Med 2005;202:1691–701.

[116] Bracci L, Schiavoni G, Sistigu A, Belardelli F. Immune-based mechanisms of cytotoxic chemotherapy: implications for the design of novel and rationale-based combined treatments against cancer. Cell Death Differ 2014;21:15–25.

[117] Brown JS, Sundar R, Lopez J. Combining DNA damaging therapeutics with immunotherapy: more haste, less speed. Br J Cancer 2018;118: 312–24.

[118] Gu ZL, Wang QJ, Shi YB, Huang Y, Zhang J, Zhang XK, et al. Nanotechnology-mediated immunochemotherapy combined with docetaxel and PD-L1 antibody increase therapeutic effects and decrease systemic toxicity. J Control Release 2018;286:369–80.

[119] Pusuluri A, Wu D, Mitragotri S. Immunological consequences of chemotherapy: single drugs, combination therapies and nanoparticle-based treatments. J Control Release 2019;305:130–54.

[120] Zhang B, Hu Y, Pang Z. Modulating the tumor microenvironment to enhance tumor nanomedicine delivery. Front Pharmacol 2017;8:952.

[121] Wang XY, Wang H, Jiang K, Zhang YY, Zhan CY, Ying M, et al. Liposomes with cyclic RGD peptide motif triggers acute immune response in mice. J Control Release 2019;293:201–14.

[122] Riley RS, June CH, Langer R, Mitchell MJ. Delivery technologies for cancer immunotherapy. Nat Rev Drug Discov 2019;18:175–96.

[123] Kedar E, Gur H, Babai I, Samira S, Even-Chen S, Barenholz Y. Delivery of cytokines by liposomes: hematopoietic and immunomodulatory activity of interleukin-2 encapsulated in conventional liposomes and in long-circulating liposomes. J Immunother 2000;23:131–45.

[124] Christian DA, Hunter CA. Particle-mediated delivery of cytokines for immunotherapy. Immunotherapy 2012;4:425–41.

[125] Guimaraes PPG, Gaglione S, Sewastianik T, Carrasco RD, Langer R, Mitchell MJ. Nanoparticles for immune cytokine TRAIL-based cancer therapy. ACS Nano 2018;12:912–31.

[126] Frick SU, Domogalla MP, Baier G, Wurm FR, Mailander V, Landfester K, et al. Interleukin-2 functionalized nanocapsules for T cell-based immunotherapy. ACS Nano 2016;10:9216–26.

[127] Nair PM, Flores H, Gogineni A, Marsters S, Lawrence DA, Kelley RF, et al. Enhancing the antitumor efficacy of a cell-surface death ligand by covalent membrane display. Proc Natl Acad Sci USA 2015; 112:5679–84.

[128] Schutz C, Varela JC, Perica K, Haupt C, Oelke M, Schneck JP. Antigen-specific T cell Redirectors: a nanoparticle based approach for redirecting T cells. Oncotarget 2016;7:68503–12.

[129] Kosmides AK, Sidhom JW, Fraser A, Bessell CA, Schneck JP. Dual targeting nanoparticle stimulates the immune system to inhibit tumor growth. ACS Nano 2017;11:5417–29.

[130] Zhu GZ, Lynn GM, Jacobson O, Chen K, Liu Y, Zhang HM, et al. Albumin/vaccine nanocomplexes that assemble in vivo for combination cancer immunotherapy. Nat Commun 2017;8.

[131] Wilson JT, Keller S, Manganiello MJ, Cheng C, Lee CC, Opara C, Convertine A, Stayton PS. pH-responsive nanoparticle vaccines for dual-delivery of antigens and immunostimulatory oligonucleotides. ACS Nano 2013;7:3912–25.

[132] Zhang F, Stephan SB, Ene CI, Smith TT, Holland EC, Stephan MT. Nanoparticles that reshape the tumor milieu create a therapeutic window for effective T-cell therapy in solid malignancies. Cancer Res 2018;78:3718–30.

[133] Sun B, Ji Z, Liao YP, Wang M, Wang X, Dong J, Chang CH, Li R, Zhang H, Nel AE, Xia T. Engineering an effective immune adjuvant by designed control of shape and crystallinity of aluminum oxyhydroxide nanoparticles. ACS Nano 2013;7:10834–49.

[134] Nowak-Sliwinska P, Scapozza L, Altaba ARI. Drug repurposing in oncology: compounds, pathways, phenotypes and computational approaches for colorectal cancer. Biochim Biophys Acta Rev Cancer 2019;1871:434–54.

[135] Turanli B, Zhang C, Kim W, Benfeitas R, Uhlen M, Arga KY, Mardinoglu A. Discovery of therapeutic agents for prostate cancer using genome-scale metabolic modeling and drug repositioning. EBioMedicine 2019;42:386–96.

[136] Wirsdorfer F, de Leve S, Jendrossek V. Combining radiotherapy and immunotherapy in lung cancer: can we expect limitations due to altered normal tissue toxicity? Int J Mol Sci 2018;20(1):E24.

[137] Kadiyala P, Li D, Nunez FM, Altshuler D, Doherty R, Kuai R, et al. High-density lipoprotein-mimicking nanodiscs for chemo-immunotherapy against glioblastoma multiforme. ACS Nano 2019;13:1365–84.

[138] Amini MA, Abbasi AZ, Cai P, Lip H, Gordijo CR, Li J, et al. Combining tumor microenvironment modulating nanoparticles with doxorubicin to enhance chemotherapeutic efficacy and boost antitumor immunity. J Natl Cancer Inst 2019;111:399–408.

[139] Gao F, Zhang C, Qiu WX, Dong X, Zheng DW, Wu W, et al. PD-1 blockade for improving the antitumor efficacy of polymer-doxorubicin nanoprodrug. Small 2018;14.

[140] Lu J, Liu X, Liao YP, Salazar F, Sun B, Jiang W, et al. Nano-enabled pancreas cancer immunotherapy using immunogenic cell death and reversing immunosuppression. Nat Commun 2017;8:1811.

[141] Wang PP, Wang HH, Huang QQ, Peng C, Yao L, Chen H, et al. Exosomes from M1-polarized macrophages enhance paclitaxel antitumor activity by activating macrophages-mediated inflammation. Theranostics 2019;9:1714–27.

[142] Feng B, Zhou FY, Hou B, Wang DG, Wang TT, Fu YL, et al. Binary cooperative prodrug nanoparticles improve immunotherapy by synergistically modulating immune tumor microenvironment. Adv Mater 2018;30.

[143] Yuan H, Jiang W, von Roemeling CA, Qie Y, Liu X, Chen Y, et al. Multivalent bi-specific nanobioconjugate engager for the targeted cancer immunotherapy. Nat Nanotechnol 2017;12:763–9.

[144] Choo YW, Kang M, Kim HY, Han J, Kang S, Lee JR, et al. Ma macrophage-derived nanovesicles potentiate the anticancer efficacy of immune checkpoint inhibitors. ACS Nano 2018;12:8977–93.

[145] Noh YW, Kim SY, Kim JE, Kim S, Ryu J, Kim I, et al. Multifaceted immunomodulatory nanoliposomes: reshaping tumors into vaccines for enhanced cancer immunotherapy. Adv Funct Mater 2017;27.

[146] Mitchell MJ, Wayne E, Rana K, Schaffer CB, King MR. TRAIL-coated leukocytes that kill cancer cells in the circulation. Proc Natl Acad Sci USA 2014;111:930–5.

[147] Kim SY, Noh YW, Kang TH, Kim JE, Kim S, Um SH, et al. Synthetic vaccine nanoparticles target to lymph node triggering enhanced innate and adaptive antitumor immunity. Biomaterials 2017;130:56–66.

[148] Kuai R, Ochyl LJ, Bahjat KS, Schwendeman A, Moon JJ. Designer vaccine nanodiscs for personalized cancer immunotherapy. Nat Mater 2017;16:489–96.

[149] Lu F, Mosley YC, Carmichael B, Brown DD, HogenEsch H. Formulation of aluminum hydroxide adjuvant with TLR agonists poly(I:C) and CpG enhances the magnitude and avidity of the humoral immune response. Vaccine 2019;37:1945–53.

CHAPTER

13

Nanoparticle-based formulation for drug repurposing in cancer treatment

Bei Cheng, Peisheng Xu

Department of Drug Discovery and Biomedical Sciences, College of Pharmacy, University of South Carolina, Columbia, SC, United States

OUTLINE

Introduction of nanomedicine

Nanomedicine was first entered the preclinical in the mid-1980s, and the first product is PEGylated adenosine deaminase enzyme that was approved by FDA in early 1990 [1]. Ever since then, scientists hold great passion in this area and many different types of nanoparticles (NPs) have been produced. Perhaps one of the most famous products is the Doxil (DOX liposomal), which appeared in the United States in 1995 for the treatment of patients with ovarian cancer and AIDS-related Kaposi's sarcoma after the failure of prior other therapy. It can prolong drug circulation time and avoid the reticuloendothelial system due to the PEGylation of the liposomes [2,3].

NPs can have a broad size distribution, ranging from larger than atoms to several hundreds of nanometers. The size of a NP has a

Drug Repurposing in Cancer Therapy
https://doi.org/10.1016/B978-0-12-819668-7.00013-0

significant impact on cellular interaction and in vivo pharmacokinetics, including cellular uptake, biodistribution, and circulation half-life [4]. NP can be generated from a variety of materials, such as lipid [5], polymer [6], protein constructs, carbon dots, and inorganic nanomaterials such as mesoporous silica NPs [7] and gold nanoclusters [8]. Nanomedicine has broad applications such as anatomical and functional imaging (CT, MRI, PET), diagnosis (detection of molecules cells, tissues), and therapy (mostly focus on cancer therapy) [9]. Superparamagnetic iron oxide NPs that are composed of nano-sized iron oxide cores coated with dextran shell have a strong T2 effect and can significantly improve the MRI sensitivity [10]. Quantum dots were used to develop a fluorescent polarization assay to identify the synthetic peptides' antigenicity. Quantum dots were first conjugated with different peptides and the corresponding antigenicity was measured by using the hepatitis B virus surface antigen as the target. This assay can be completed within a few minutes with high sensitivity and specificity to be 85.4% and 98.6%, respectively [11].

There are several advantages of NPs over conventional formulations. One of the major benefits of NPs is their high tumor accumulation, which can be achieved through either passive targeting or active targeting. The passive targeting of NPs is exploiting the unique tumor microenvironment. In tumor tissue, the absence of a supporting matrix for the vascular tissue intimates the formation of very leaky vessels and pores (100 nm–2 μm) formed by adjacent endothelial cells [12]. NPs are accumulated in tumor tissue by taking advantage of this leaky vasculature and poor drainage system, also called EPR (enhanced permeability and retention) effect. While the free drug enters tumor tissue by free diffusion, NPs can extravasate into the leaky tumor blood vessel and remain inside the tumor due to the insufficient lymphatic drainage system [13]. Vlerken et al. performed the in vivo biodistribution of free drug and the PLGA-based NPs in mice xenografts cancer model and observed a 6.5-fold increase of the NP accumulation in the tumor site compared with its free drug form [14]. Active targeting is employed to further improve tumor-selective localization of the NP. Conjugating targeting ligands is one of the most popular ways to achieve active targeting. The ligands on the surface of NPs can be recognized and internalized by its corresponding receptors expressed on the surface of cancer cells (e.g., different antibodies, antibody fragments) [15]. Besides higher tumor targeting, the encapsulation of therapeutic molecules in a NP can improve their solubility, bioavailability, and altered biodistribution and facilitate the cellular uptake.

Despite the advantages, there are several biological and technological challenges yet to be solved for the clinical translation of nanomedicine. One of the main biological challenges is the toxicity of nanomaterials. Other than a few polyethylene glycol (PEG) and PLGA-based polymer, sugar, or protein-based system, there are quite a few synthetic polymers that are eventually proved clinically. Fundamental research is needed to study how the different types of NPs are eliminated from the body. It was proposed that small NPs (smaller than 6 nm) undergo efficient urinary excretion [16,17]. Most of the NPs that cannot be cleared by the renal system will eventually be accumulated in the liver via the mononuclear phagocyte system (MPS). This hepatic processing and biliary excretion are usually slow from several hours to months, which raises the concern of chronic toxicity to the liver [18]. To overcome the in vivo degradation and toxicity of NP, ideally designed NP should be smaller enough in the circulation system to undergo efficient renal clearance and retain the features that are not favored by the MPS. The technological challenges associated with nanomedicine mainly focus on the scale-up synthesis of polymer and NP. To develop clinically translatable polymers, the synthesis steps should be reproducible and reliable. Nanomanufacturing also requires high-quality control over the size, size distribution, shape, morphology, surface charge, and drug loading efficiency of NP. The overall outlook of NPs is promising, as it is a revolutionary approach to addresses the many disadvantages associated with the free drugs and it is now being expanding to many diseases.

Combinational nanomedicine: two or more drugs in one particle

Drug combination includes two or more drugs in a single dosage form at a fixed dose, which needs to take into account the diverse pharmacokinetics, physiological varieties, and different drug dose regime. Cancer is intrinsic heterogenic, even to the same tumor of different patients or the different cells within the same tumors. Frequently, the intrinsic or acquired cancer drug resistance to a single chemotherapeutics due to multiple-drug resistance, apoptosis suppression, or enhancing DNA repairing is the main reason for a quick cancer relapse or incurability [19]. Therefore, the addition of combinational therapy to chemotherapy regimen is particularly beneficial as different drugs can target different pathways or genes that can greatly reduce the number of cancer cells that survived the treatment and significantly delayed the cancer recurrence or even eradicate it. Combinational nanomedicine is providing even greater advantages for cancer treatment and offering superior therapeutic outcomes to the current drug cocktail therapy. Especially in the field of immunotherapy, compared with conventual therapies or cancer nanomedicine, the combinational nanoimmunotherapy substantially improves the patient's overall survival and long-term memory responses (Fig. 13.1) [20,21].

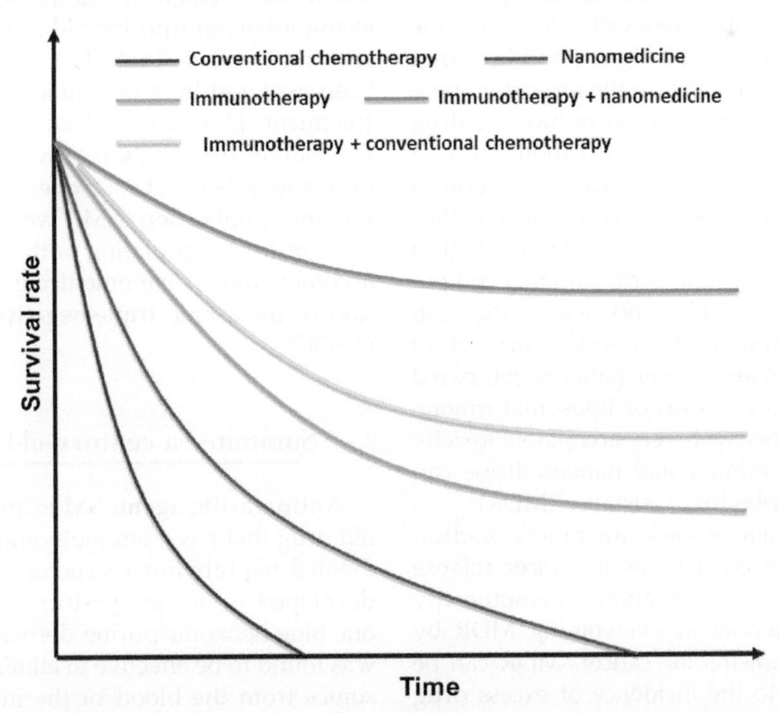

FIGURE 13.1 Potential clinical benefits of combination nanoimmunotherapy.

By taking advantage of NPs, drugs can be delivered simultaneously to the target of interests and be maintained at the optimum drug ratio. Different drugs can be loaded at desired molar ratios and released in a controlled manner based on nature of the nanomaterials' properties despite the physicochemical property difference of the individual drugs. CPX-351 is a liposome formulation encapsulating cytarabine and daunorubicin at the 5:1 molar ratio. CPX-351 exhibited a prolonged $t_{1/2}$ of 40.4 h for cytarabine and 31.5 h for daunorubicin, markable different than their nonliposomal formulation (1–3 h for cytarabine and 18.5 h for daunorubicin). It has been used in several Phase I and II clinical trials for treating acute myeloid leukemia and myelodysplastic syndrome and was exhibited an increased overall patient survival rate. After the IV infusion, the optimized dosage was well maintained for 24 h [22]. Secondly, combinational nanomedicine can enhance tumor therapeutic efficacy. One advantage of nanomedicine versus conventional medication is EPR effect. Therefore, the bioavailability of the payloads in tumor will be higher, and less drug dose was required to achieve the same therapeutic effect, which in turn would reduce the drug dose and inhibit side effects. Irinotecan is a potent anticancer drug, but its application is hindered by its severe side effects of severe diarrhea, and even life-threatening hematological and GI toxicity [23]. The combinational of liposomal irinotecan, 5-FU and leucovorin can significantly improve the overall survival of metastatic pancreatic cancer patients compared with 5-FU and leucovorin or liposomal irinotecan alone, together with very acceptable toxicity [24]. Thirdly, combinational nanomedicine can overcome multiple-drug resistance (MDR).

MDR is a major obstacle for cancer eradication and plays a crucial role in cancer relapse and metastasis. Combination chemotherapy holds great potential in overcoming MDR by targeting mechanisms for cancer. MDR can be developed due to the incidence of excess drug efflux, increase DNA repair, malfunctional checkpoints, and activation of prosurvival pathways [25]. Some nanomaterials themselves can function as P-gp inhibitors in nature, including the polysaccharides, PEG, and various Pluronic-based polymers [26]. Pluronic P85 has been widely used in inhibiting P-gp in brain endothelial cells by increasing ATP depletion, inducing membrane fluidization, and inhibiting P-gp ATPase activity [27]. NPs can also reverse the MDR by increasing cellular uptake, thus bypassing the efflux pumps [28]. By codelivering doxorubicin (DOX) and paclitaxel in a cross-linking NP, the cellular uptake of coloading NPs in 4T1 cells was significantly increased compared with free drug combination or single drug–loaded NPs.

Repurposing drugs with nanomedicine in cancer therapy: Repurposing old medicines in cancer therapy is a cost-efficient way and is now gathering its momentum. The discovery of new pharmacological targets has extended the life of some centenary-old medicines [29]. Some examples of repurposing old drugs in nanomedicine are listed in Table 13.1. Suramin (SM) has been explored in many clinical trials for cancer treatment. However, SM was withdrawn from the market due to its narrow therapeutic window and side effects. To overcome the obstacle for the application SM, we introduced the concept of encapsulating both SM and DOX as a combinational nanomedicine for the prevention of metastatic triple-negative breast cancer (TNBC).

Suramin—a century-old medicine

Antiparasitic agent: SM is an over 100-year-old drug that has a unique symmetrical polysulfonated naphthylurea structure. It is a product developed from the dyestuff industry. In 1906, one blue benzopurpurine derivate, trypan blue, was found to be effective in eliminating trypanosomes from the blood of the infected animals.

TABLE 13.1 List of NPs of repurposed drugs in cancer therapy.

Drug category	Drug	New mechanism of actions/indication	Types of NP	Disease model	References
CNS	Perphenazine	Disrupting cancer cell cholesterol homeostasis	Liposome	Xenografted melanoma tumor	[30]
Iron deficiency anemia	Ferumoxytol	MRI of brain tumors, accumulation in tumor lesions	Inorganic	Patients with advanced solid tumor	[31]
NSAID	Celecoxib	Decrease levels of key cyclin	Liposome	Xenografted melanoma tumor	[32]
Antidiabetic	Metformin	Decrease glucose metabolites	Polymer	Patient-derived orthotopic pancreatic tumor	[33]
Alcohol aversion	Disulfiram	Degrades to diethyldithiocarbamate and forms a copper complex	Polymer	Xenografted ovary tumor	[34]
Metabolic bone diseases	Bisphosphonate	Cytotoxic effects	Metal-organic	Xenografted lung and prostate tumor	[35]

However, due to its easiness of staining, Bayer research groups sought to investigate the colorless version of this compound. After extensive optimizing over 1000 naphthalene urea, Bayer finally discovered the Germanin in 1917 and later changed its name to SM [36]. The presence of six negatively charged sulfonate groups allowed SM to bind with many proteins with basic amino acids, and as a result, it serves a multitarget physiological pharmaceutical agent. Thus, SM interacts with all kinds of proteins, including albumin globulins, fibrinogen, and histones, and inhibits a wide variety of enzymes, e.g., hyaluronidase, urease, hexokinase, fumarase, and succinic dehydrogenase [37].

Primarily as an **antiparasitic agent**, it is used to treat African human trypanosomiasis caused by two subspecies of *Trypanosoma brucei, T.b. rhodesiense* and *T.b. gambiense*, and now being listed as a WHO essential medicine.

Antiretroviral agent: In 1980s, SM received much attention in the field of antiretroviral as it inhibited reverse transcriptase. It is a potent inhibitor of the reverse transcriptase of retroviruses, including the HTLV-III/LAV reverse transcriptase. It exhibits potent anti-HIV activities alone or together with other antiretroviral drugs [37,38]. Never entering the clinical practice in treating HIV, there are many reports following the possible functions of SM in the field of treating HIV.

Treatment in autism spectrum disorder: Autism spectrum disorder shows the symptoms of communication and language difficulties, repetitive behaviors, and the inability to socialize. A group of researchers recently reported that a low dose of SM could effectively treat children with autism spectrum disorder [39].

Antineoplastic agent: Until 1900, it has been gradually used for cancer treatment. Researchers showed significant interest in applying it for various types of cancers, including prostate cancer, breast cancer, lung cancer, and bladder cancer alone. It is a highly unspecific inhibitor that was reported to inhibit various growth factors. Most likely owing to its structural similarity to glycosaminoglycans, it unspecifically binds to the a/b FGF (acidic/basic fibroblast growth factors) at the site where the polysulfated drug binding is located [40]. Besides this charge–charge interaction, it has also been reported that SM binding to hFGF1 is partially through hydrophobic interactions at the site of L14, C16, L133, and L135 in hFGF1. It blocked the interaction between hFGF1 and FGFR2 D2 [41]. FGFs and their receptors play crucial roles in many fundamental processes from embryogenesis to adult life, such as proliferation, differentiation, migration, angiogenesis, and wound healing [42–44]. Like many other mitogens, FGFs act like a double-edged sword and deregulate the signaling, thereby causing many types of human cancers, including lung, breast, ovarian, and prostate cancers. SM can inhibit the angiogenesis process by binding with FGFs and deactivating its downstreaming signals.

The pharmacokinetics and pharmacology of SM

It has been reported that SM entered the cells in different pathways depending on the cell type. It can enter cells by the ionic interactions with the membrane component or through the pinocytosis process with the endpoint at the endosomal/lysosomal. In DU145 and LNCap prostate cancer cells, it showed a pattern of trypsin insensitive, low-affinity cell surface binding pattern. Analysis of the exocytosis of SM in DU145 showed that 64% drug was effluxed from a shallow endosome compartment with a $t_{1/2}$ of 3.15 min and 31% of the drug from a deep endosome compartment with a $t_{1/2}$ of 433 min [45].

SM has a very high level of albumin binding ratio over 99%, leading to the $t_{1/2}$ distribution half-life of 2 h and $t_{1/2}$ elimination over 20 days with high deviation among different patients. What's more, the SM/albumin complex is very poorly internalized into cells in contrast to the albumin itself [45]. The undesirable pharmacokinetic characteristics as well as its poor cell permeability led to a narrow therapeutic window and dose-related toxicity. Due to its poor oral bioavailability, SM is primarily administrated intravenously. SM is excreted in the urine in its unchanged form.

The toxicity of the SM is dose-dependent. The toxicity is strongly associated with serum SM level or the cumulative dose [46]. At a high dose, it can act as a direct cytotoxic medicine. The early clinical trials tend to use very high dose of SM, typically exceeding 200 μM. At this concentration, it is cytotoxic or cytostatic to a variety of tumor cells as well as normal cells. It induced severe side effects such as proteinuria, renal dysfunction, and the most troublesome sensory motor neuropathy. Even at a concentration higher than 50 μM, it caused cell arresting at G1 phase [47].

Formulating SM in NP: Inspired by the benefits of combination therapy of SM with chemotherapeutic agents, we thought that formulating SM with chemotherapeutic agents into NPs would be a good idea. It offers the following benefits: (1) Increasing the cell permeability of SM: The SM–albumin complex was existed as over 99% of its parent form, showing very low cell permeability. As a result, a further increase of SM concentration was required in all the clinical trials. By formulating SM into NP, it can enter cells more efficiently through endocytosis. (2) Control the pharmacokinetics of SM and other chemotherapeutics: Different pharmacokinetic profiles of different drugs often undermine the therapeutic effect, especially when highly different physicochemical properties drugs were combined [48]. SM has a unique pharmacokinetic if administrated alone. It has a very long $t_{1/2}$. The typical chemotherapeutics usually have entirely different pharmacokinetics. For example, the DOX used in our design has a $t_{1/2}$ of 10 min in the initial phase and 30 h in the terminal phase, which is significantly different from that of the SM [49]. If merely by mixing these two drugs, it would be hard to achieve simultaneously effect at the targeted concentration for DOX on DNA repair and SM on FGF inhibition. NPs have the advantages of controlled release of drugs, efficient entering cells, and also inhibiting the SM interfering with albumin.

There are a few other reports about SM-based NPs formulation for treating various diseases. SM was loaded into hyperbranched poly(amino-ester) (HBPAE) through electrostatic adsorption and conjugated with alendronate as the bone-targeting moiety. The resultant SM-HBPAE-ALE has a spherical size of 65 nm and exhibited high potency in inhibiting the proliferation of diseased fibrous dysplasia cells [50]. A serial of different molecule weight of heparin–SM conjugates formed nanocomplex with polycationic protamine–PEG system via ionic interactions. It was found that heparin (molecular weight of 10 kDa)–SM conjugate had the size around 100 nm, and it allowed for the accumulation at the tumor site for up to 48 h. The NP formulation was found to be more efficient in inhibiting tumor progression compared with the heparin–SM conjugate itself [51]. Besides the therapeutic purposes, SM was also used as an interlayer linker to facilitate multiple layers loading of albumin on the surface of silica NPs [52]. However, none of these reports have extensively investigated the possible role of SM in the NP formulation to treat cancer.

Improve therapy efficacy: SM has been frequently combined with chemotherapeutic agents in treating various cancers, including breast cancer, prostate cancer, lung cancer, and brain cancer. It has been combined with gemcitabine, DOX, docetaxel, paclitaxel, and cisplatin in different clinical trials as summarized in Table 13.2. Recent clinical trials prefer to use SM as a chemo/radiation sensitizer at a nontoxic dose that can effectively inhibit FGFs instead of as a cytotoxic agent. It is usually used at the targeted concentration of 10–50 μM, which is well-tolerated [53,54]. Interestingly, it has been reported that SM can produce chemosensitization at nontoxic doses. SM at the concentration of 10–50 μM significantly enhanced the therapeutic efficacy of chemotherapy (DOX, paclitaxel, docetaxel, mitomycin). Some of the combinational therapy even showed improved clinical efficiency. Hormone-refractory prostate cancer patients treated with hydrocortisone and SM showed moderate palliative benefits and a delay in disease progression [55]. Besides the clinical trials, several preclinical studies observed the synergistic effect in a mice model with naïve A549 tumor that SM can enhance the antitumor effect when combined with docetaxel [56]. It could also improve the mitomycin antitumor activity in a bladder cancer xenograft mouse model and histocultures of human bladder tumors from bladder cancer patients at the concentration of 20 μM [57]. Other than the antitumor effect, it was also reported that SM overcomes the FGF-induced drug resistance at the concentration of 1–17 μM in human prostate PC3 cancer cells in the presence of DOX [58].

Reduce renal toxicity: SM can induce renal proximal tubule cell outgrowth, scattering, and proliferation and thus reverse the chemotherapeutical agent–induced renal injury, including reduced tubular necrosis [59,60]. Approximately 30% of patients administrated cisplatin suffered from renal injury. Treatment of SM followed by cisplatin after 24 h can partially recover kidney function in several rodent models without sacrificing the chemotherapeutical effect of cisplatin [59]. It is also a potent antifibrotic agent that may have potential for patients with fibrotic kidney disease. Administration of SM reduces serum creatinine level, ammoniates proteinuria, inhibits renal fibroblasts activation, and prevents the progression of renal fibrosis and glomerulosclerosis in chronic kidney disease [61].

Reduce cardiotoxicity: SM can inhibit enterovirus 71–caused cardiopulmonary complication by binding of the naphthalenedisulfonic acid to the viral capsid [62].

TABLE 13.2 Clinical trials using suramin as an antineoplastic agent in cancer treatment.

Condition or disease	NCT number	Drug	Stages	Status
Prostate cancer	NCT00002723	Suramin	Phase III	Completed, no result
Recurrent bladder cancer	NCT00006476	Suramin	Phase I	Completed, no result
Non-small cell lung cancer	NCT01038752	Suramin/docetaxel/carboplatin	Phase II	Terminated, insufficient data
Recurrent breast cancer stage IIIB breast cancer stage IV breast cancer	NCT00054028	Suramin/paclitaxel	Phase I&II	Complete response and partial response
Prostate cancer	NCT00002881	Suramin/flutamide/goserelin acetate/leuprolide acetate	Phase III	Completed, no result
Recurrent non-small cell lung cancer stage IIIB non-small Ce	NCT00006929	Suramin/carboplatin/paclitaxel	Phase II	Completed, no result
Recurrent non-small cell lung cancer stage IIIB non-small Ce	NCT00066768	Suramin/docetaxel/ gemcitabine hydrochloride	Phase I	Completed, no result
Recurrent primary brain tumor	NCT00002639	Suramin	Phase II	Completed, no result
Advanced solid tumors	NCT00003038	Suramin/doxorubicin hydrochloride	Phase I	Completed, no result
Recurrent renal cell carcinoma stage IV renal cell cancer	NCT00083109	Suramin/fluorouracil	Phase I and II	Completed, no result
Prostatic neoplasm	NCT00001266	Suramin/leuprolide/flutamide	Phase II	Completed, no result
Carcinoma, non-small cell lung	NCT01671332	Suramin/docetaxel	Phase II	Completed, result submitted
Brain and central nervous system tumors	NCT00004073	Suramin/radiation therapy	Phase II	Overall survival not significantly improved; toxicity with high dose
Multiple myeloma and plasma cell neoplasm	NCT00002652	Suramin	Phase II	Completed, no result
Bladder carcinoma	NCT00001381	Suramin	Phase I	Completed, no result
Stage III or stage IV adrenocortical carcinoma	NCT00002921	Suramin/hydrocortisone	Phase II	Terminated; well tolerated but no response
Chemotherapy-induced polyneuropathy	NCT02871284	Suramin/chemotherapy	Not applicable	Active, not recruiting

SM **act as a gelator in the dosage form:** Inspired by the classical hydrogel formation of chitosan/tripolyphosphate (TPP), it is a small molecule with five negative charges. At an acidic condition, TPP polyions can align along the c-axil to increase the orientation of chitosan sheet and interact with the positively charged amine groups of chitosan to form NPs. Since SM is small molecular with six negative charges, we postulated it could form NP with glycol chitosan by acting as an anionic compartment through gelating with cationic counterpart as shown in Fig. 13.2 [63]. We thoroughly investigated all the possible factors that would affect the NPs formation in terms of SM concentration, pH of the solution, SM loading, and DOX loading. Surprisingly, the concentration of SM used during NP fabrication was found to have the greatest impact on the size and the PDI of NPs. When the concentration of SM increased to 1 mg/mL, the final NPs have a size of 600 nm with a PDI of 0.25, significantly larger than the ideal size of 180 nm (Fig. 13.3). The pH of the buffer is expected to impact NP size significantly.

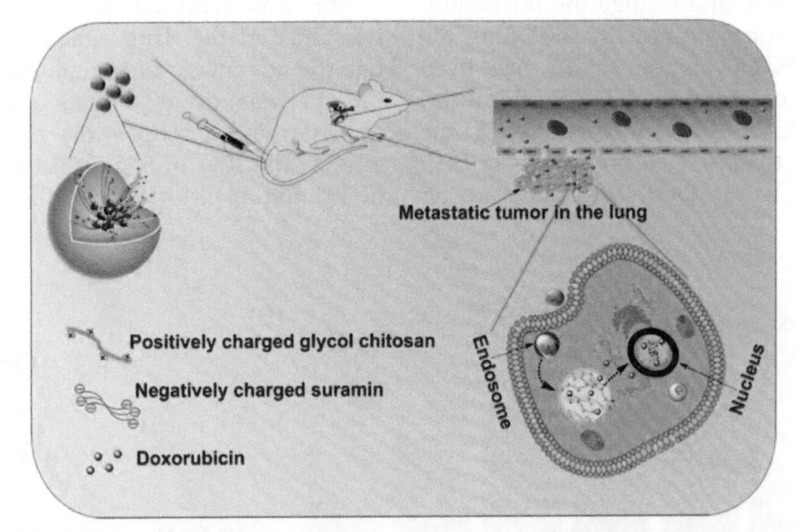

FIGURE 13.2 Scheme of the assembly of GCS-SM/DOX NP and its action in treating breast cancer lung metastasis [63]. *Copyright © 2019 Elsevier B.V.*

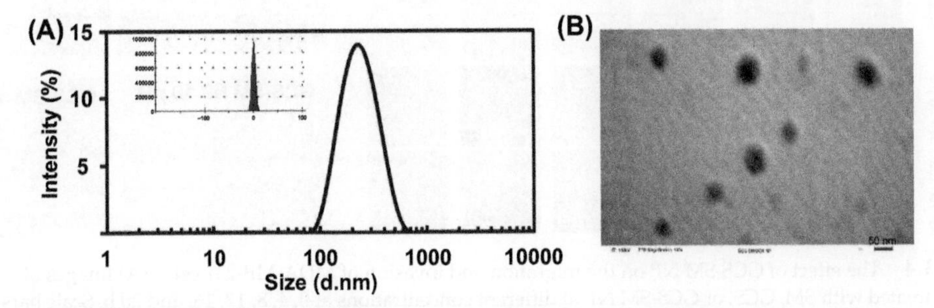

FIGURE 13.3 Characterization of GCS-SM/DOX NP. (A) Hydrodynamic size distribution and surface charge (inserted). (B) Transmission electron microscopy images of GCS-SM/DOX NPs with 100 K magnification. The scale bar is 50 nm [63]. *Copyright © 2019 Elsevier B.V.*

The NPs formed by electrostatic interaction generally were affected by the pH of the buffer. The primary amine in glycol chitosan has a pKa of 6.5, which means at pH 6.5, 50% of amine was positively protonated. We varied the pH from 6 to 7.4, and we found the optimized pH of the buffer when preparing the NPs is at 7.0 with ionic strength at 20 mM.

Synergy effect of SM **and DOX** in vitro: Our combination NP system provides a solid foundation for these two drugs to maximize their synergy effect in treating triple-negative metastatic breast cancer. On the one hand, SM was used to bind to a/b FGF and change the microenvironment of the tumor, especially inhibiting the angiogenesis. FGFR1 amplification has been found in 15% of patients with breast cancer, and targeting FGFRs serves a crucial role in the anti-angiogenic treatment of breast cancer [64]. On the other hand, DOX was focused on interfering with the cancer cell cycle and killing cancer cells. In the MDA-MB-231 human breast cancer cell line, we found out that SM itself at the dose of 40 and 100 μM was not as effective as its NP form in inhibiting cell migration in the wound healing assay and cell invasion assay (Fig. 13.4). It may be related that SM has much less cell internalization compared with its NP counterpart. In the tube formation assay, however, the free SM is more effective in inhibiting tube formation compared with its NP form in terms of tube length as well as the number of the junctions (Fig. 13.5).

SM was reported to suppress the cross-resistance of the drug against DOX. Since NP has the advantage of maintaining an optimum ratio between drugs, it offers us the opportunity to determine a ratio while not inducing side effects. The IC_{50} of SM or SM NPs has an IC_{50} of 300 and 63 μM in MDA-MB-231 after treated

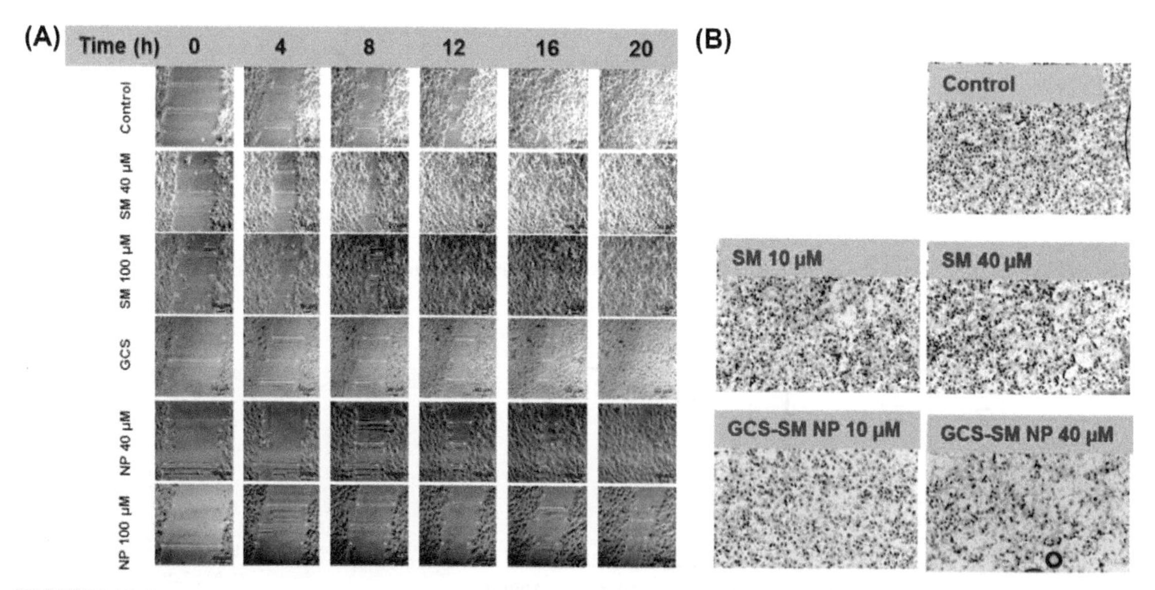

FIGURE 13.4 The effect of GCS-SM NP on the migration and invasion of MDA-MB-231 cells. (A) Images of cell migration of cells after treated with SM, GCS, or GCS-SM NP at different concentrations at 0, 4, 8, 12, 16, and 20 h. Scale bars are 150 μm. (B) Representative images of MDA-MB-231 cells in the invasion assay [63]. *Copyright © 2019 Elsevier B.V.*

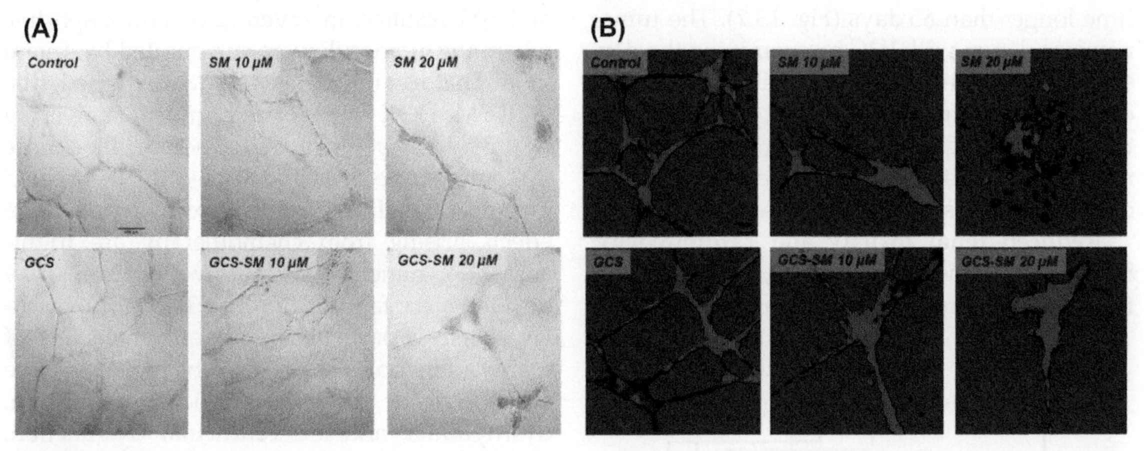

FIGURE 13.5 The in vitro HUVEC tube formation assay. (A) Photographs of representative tubes formed in different treatments. Magnification: 10X. (B) Confocal microscopy images of representative tubes formed in different treatments [63]. *Copyright © 2019 Elsevier B.V.*

with a/b FGF. To reduce the cytotoxicity of SM, we limited the concentration of SM at 10, 20, and 100 μM. Cytotoxicity of DOX was greatly enhanced by the addition of SM at all concentrations, but a further increase of SM from 20 to 100 μM did not additionally boost its effect (Fig. 13.6). Moreover, it was revealed that the GCS-SM/DOX NPs exhibited stronger cell-killing effects than free DOX when SM/DOX is at the ratio from 3.5 to 20.

Synergy effect of suramin and DOX in vivo

A TNBC metastatic mouse model was adopted to evaluate the effectiveness of the GCS-SM/DOX NP on inhibiting the metastasis of TNBC and affecting the survival time. It was found that DOX and SM combination significantly improved the median survival time from 64.5 to 81.5 days. Furthermore, GCS-SM/DOX NP showed a median survival

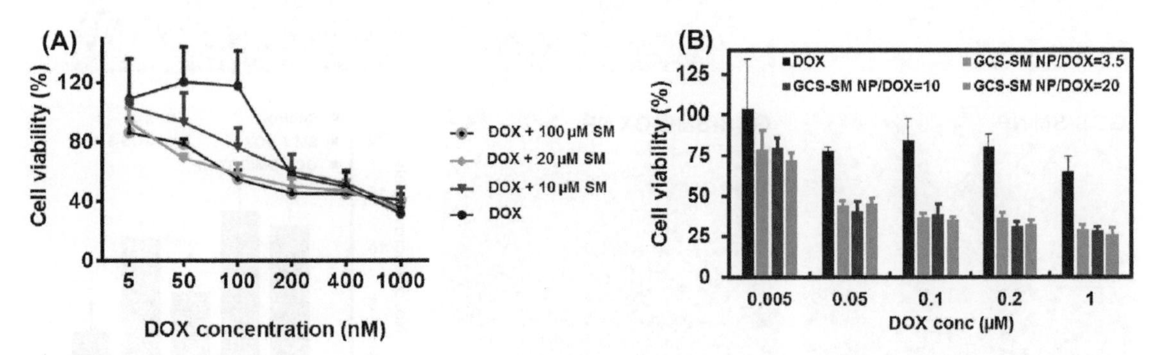

FIGURE 13.6 Cytotoxicity of DOX and suramin/DOX combination. (A) Cell viability of DOX in combination with different concentrations of suramin in cells pretreated with a/b FGF. (B) Cell viability of combinational NPs at different ratios [63]. *Copyright © 2019 Elsevier B.V.*

time longer than 85 days (Fig. 13.7). The tumor neovasculature with IHC by examining the density of CD31-stained microvessels in the tumor of the xenograft mouse model exhibited that the GCS-SM/DOX NP and the free drug combination can significantly decrease the formation of new blood vessels in tumor tissues (Fig. 13.8).

Reduced renal toxicity and cardiotoxicity: Surprisingly, histological analysis of kidney revealed that free drug combination of DOX and SM resulted in severe glomerulonephritis, where the glomeruli were surrounded by severe inflammatory infiltrates. As we expected, the GCS-SM/DOX NP–related group showed a healthy kidney histologic structure, like those in control, free SM, and GCS-SM NP–treated groups (Fig. 13.9A). The cardiovascular side effects arising from chemotherapy are frightening. A patient that survived cancer may develop heart failure after that. The cardiovascular system is particularly prone to the action of many antineoplastic drugs, which may induce vasospastic or ischemia, arterial hypertension, dysrhythmia, and left ventricular dysfunction, leading to heart failure, myocarditis, and pericarditis [65]. Some chemotherapy agents cause heart muscle to weaken soon after chemotherapy begins, and some can happen years after the chemotherapy. Chemotherapy agents are known to have cardiotoxicity, including cisplatin, DOX and other anthracyclines, fluorouracil 5-FU, IL-2, mitomycin, mitoxantrone, and sunitinib.

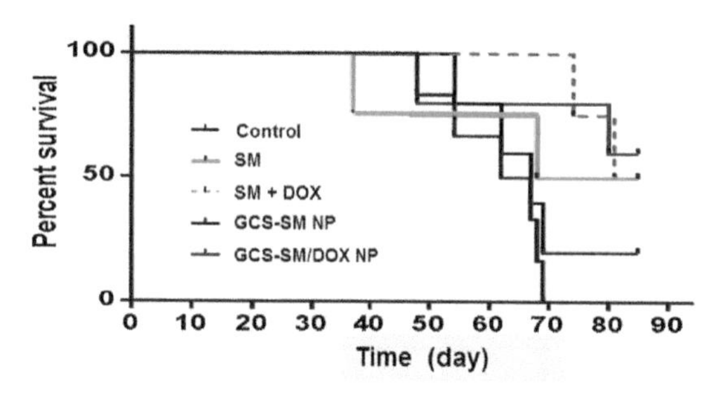

FIGURE 13.7 Survival curves of mice receiving different treatments [63]. *Copyright © 2019 Elsevier B.V.*

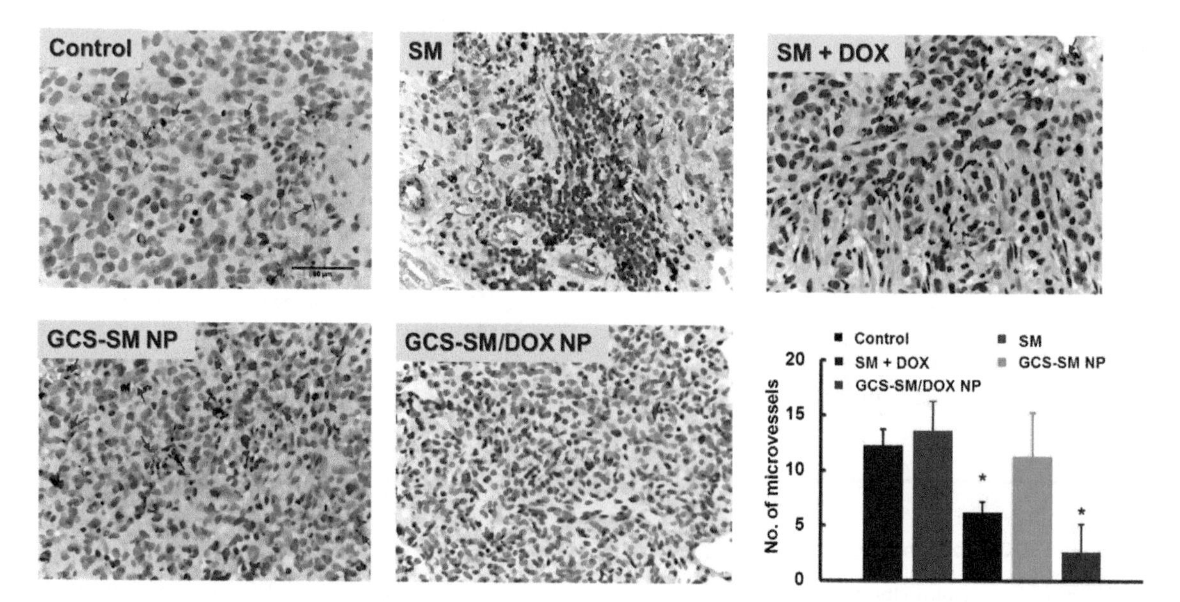

FIGURE 13.8 IHC detection of CD31 in tumor tissues of different treatment groups. IHC staining (CD31) of tumor blood vessels (brown), microvessels (*red arrow*). Scale bars are 50 μm. The tumors were cut into 5 μm thick sections, stained with CD31, and imaged (10 fields per section) (*$P < .05$ compared with control) [63]. *Copyright © 2019 Elsevier B.V.*

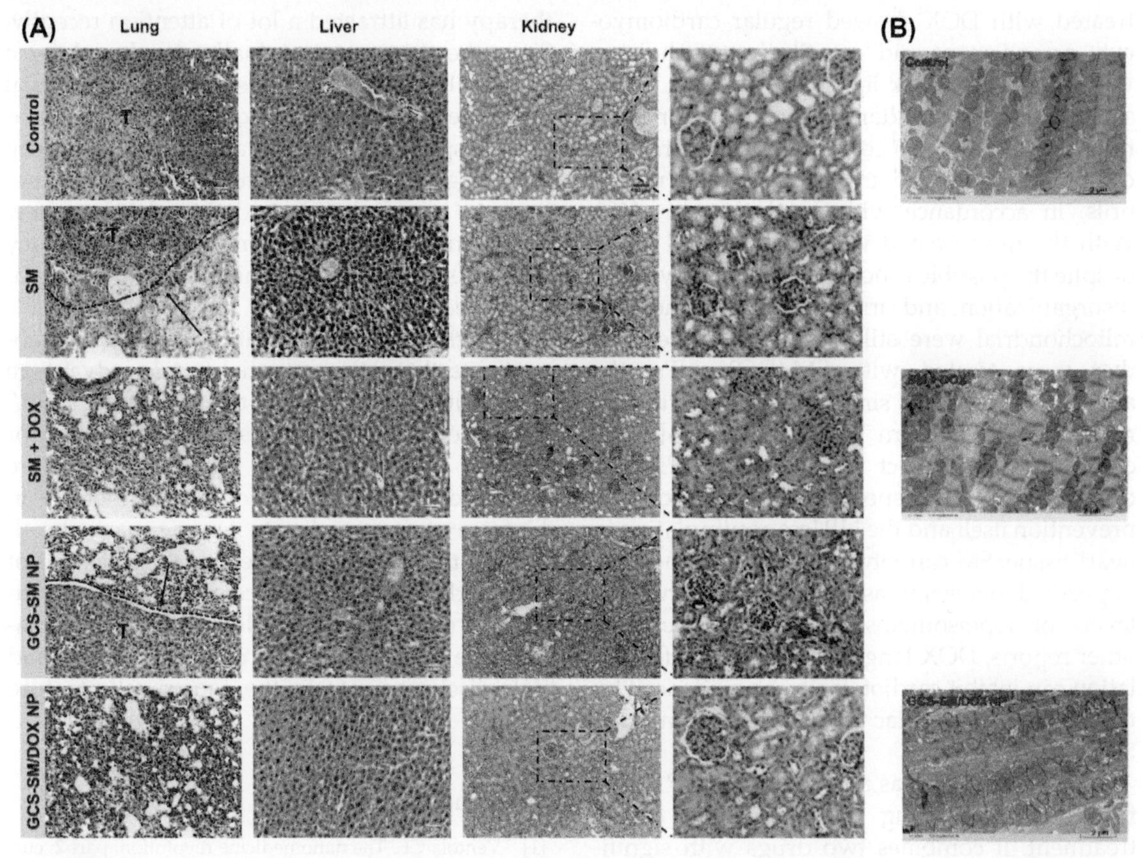

FIGURE 13.9 Histology analysis and TEM analysis of organs from different treatment groups. (A) H&E stained images of lungs, livers, and kidneys. *Black arrows* indicate the tumor (T) area in the lung. Green (dark gray in print version) arrows indicate the glomerulonephritis areas in the kidney. (B) Transmittance electron microscopy images of heart tissues from the control, free drug combination, and GCS-SM/DOX NP treatment groups [63]. *Copyright © 2019 Elsevier B.V.*

For TNBC, anthracyclines and taxanes are the two major classes of drugs among the first-line drugs for breast cancer. Anthracyclines were reported to lead to cardiovascular problems in up to 40% patients [65]. DOX is commonly used in the clinical practice among anthracyclines, and it is usually administrated through intravenous infusion. The action of DOX is mainly through interference with DNA replication at several levels, showing damaging effects in every phase of the cell cycle. However, without exception, it has very severe dose-depending cardiotoxic effects, primarily resulting from iron accumulation in the mitochondrial and production of ROS [66]. The DOX-dependent cardiotoxic often results in irregular aligned mitochondrial and reduced perivascular fibrosis, which can be observed under transmittance electron microscopy (TEM). The ultrastructural of normal myocardium could be evaluated by TEM. Pronounced ultrastructural changes were reported in the ultrathin myocardial sections obtained from the DOX group, with the signature of myofibrillogenesis, z-disc degradation, and chromatin disintegration in the nuclei [67,68]. The control mice not treated with DOX showed regular cardiomyocyte arrangement and myofibrils morphology, with the typical dark-light band pattern of the sarcomeres and abundant spherical or elongated mitochondrial, which contained packed, regular cristae well-arranged in rows between myofibrils, in accordance with another report [69]. With the mice treated with free SM and DOX, despite the possible function of SM, the myofibril disorganization and irregular arrangement of mitochondrial were still present. For the mice that were treated with SM/DOX NPs, the arrangement was similar to the typical mitochondrial pattern (Fig. 13.9B) [63]. The cardioprotective effect of SM/DOX NPs may originate from two aspects, from the SM cardioprevention itself and the NP less accumulation in heart tissue. SM can inhibit the catalytic activity of purified topoisomerase by the reduction of the levels of topoisomerase. In accordance with other reports, DOX long-circulating nanoformulation can inhibit cardiotoxicity, as evidenced by the unchanged cardiac myocyte injury marker troponin-T level [70].

This study serves as an example of NP-based formulation for drug repurposing in cancer treatment. It combines two drugs with significantly different pharmacokinetics into a single NP and delivers them with high efficiency. GCS-SM/DOX NP exhibited a better overall survival rate for mice with metastatic breast cancer than NP alone and reduced the side effect of DOX and SM. To make the study clinically translatable, a detailed pharmacokinetic study for both drugs is necessary to verify the assumption that both drugs are reaching the target of interests at the optimum concentration.

Outlook of repurposing with nanomedicine in cancer treatment

Benefited from the advantages of validated clinical safety, lower failure risk, less investment, and shortened drug development time frame repurposing, a old drug for a new disease therapy has attracted a lot of attention recently. Since the drug was originally developed for a specifical disease, the chemical and physical properties of the compound has been optimized for that specific clinical situation during the drug screening and/or preformulation step. Therefore, to repurpose an existing drug, a new formulation is needed to maximize the efficacy for cancer therapy while minimizing it potential side effects, especially for the application of a drug with narrow therapeutic window. As nanomedicines have the potential to take advantage of the tumor blood capillary–associated "EPR" effect to enrich the repurposed drug in the tumor site, we are expecting to see more and more drugs being repurposed for cancer therapy in the form of nanomedicine. Among many developed nanomedicines, there is a special type of nanomedicine that can tracelessly release conjugated drug in the tumor deserving more attention due to its free of premature release and high discharge rate in the cancer cells feature [71].

References

[1] Ventola CL. The nanomedicine revolution: part 2: current and future clinical applications. P T 2012;37(10):582–91.

[2] Bulbake U, Doppalapudi S, Kommineni N, Khan W. Liposomal formulations in clinical use: an updated review. Pharmaceutics 2017;9(2).

[3] Barenholz Y. Doxil(R)–the first FDA-approved nano-drug: lessons learned. J Contr Release 2012;160(2):117–34.

[4] Hoshyar N, Gray S, Han H, Bao G. The effect of nanoparticle size on in vivo pharmacokinetics and cellular interaction. Nanomedicine 2016;11(6):673–92.

[5] Liu Y, Cao ZT, Xu CF, Lu ZD, Luo YL, Wang J. Optimization of lipid-assisted nanoparticle for disturbing neutrophils-related inflammation. Biomaterials 2018;172:92–104.

[6] Bei C, Bindu T, Remant KC, Peisheng X. Dual secured nano-melittin for the safe and effective eradication of cancer cells. J Mater Chem B 2015;3(1):25–9.

[7] Cheng B, He H, Huang T, Berr SS, He J, Fan D, Zhang J, Xu P. Gold nanosphere gated mesoporous silica nanoparticle responsive to near-infrared light and redox potential as a theranostic platform for cancer therapy. J Biomed Nanotechnol 2016;12(3):435–49.

[8] Mao W, Kim HS, Son YJ, Kim SR, Yoo HS. Doxorubicin encapsulated clicked gold nanoparticle clusters exhibiting tumor-specific disassembly for enhanced tumor localization and computerized tomographic imaging. J Contr Release 2018;269:52–62.

[9] Pelaz B, Alexiou C, Alvarez-Puebla RA, Alves F, Andrews AM, Ashraf S, Balogh LP, Ballerini L, Bestetti A, Brendel C, Bosi S, Carril M, Chan WC, Chen C, Chen X, Chen X, Cheng Z, Cui D, Du J, Dullin C, Escudero A, Feliu N, Gao M, George M, Gogotsi Y, Grunweller A, Gu Z, Halas NJ, Hampp N, Hartmann RK, Hersam MC, Hunziker P, Jian J, Jiang X, Jungebluth P, Kadhiresan P, Kataoka K, Khademhosseini A, Kopecek J, Kotov NA, Krug HF, Lee DS, Lehr CM, Leong KW, Liang XJ, Ling Lim M, Liz-Marzan LM, Ma X, Macchiarini P, Meng H, Mohwald H, Mulvaney P, Nel AE, Nie S, Nordlander P, Okano T, Oliveira J, Park TH, Penner RM, Prato M, Puntes V, Rotello VM, Samarakoon A, Schaak RE, Shen Y, Sjoqvist S, Skirtach AG, Soliman MG, Stevens MM, Sung HW, Tang BZ, Tietze R, Udugama BN, VanEpps JS, Weil T, Weiss PS, Willner I, Wu Y, Yang L, Yue Z, Zhang Q, Zhang Q, Zhang XE, Zhao Y, Zhou X, Parak WJ. Diverse applications of nanomedicine. ACS Nano 2017;11(3):2313–81.

[10] Wei H, Bruns OT, Kaul MG, Hansen EC, Barch M, Wisniowska A, Chen O, Chen Y, Li N, Okada S, Cordero JM, Heine M, Farrar CT, Montana DM, Adam G, Ittrich H, Jasanoff A, Nielsen P, Bawendi MG. Exceedingly small iron oxide nanoparticles as positive MRI contrast agents. Proc Natl Acad Sci USA 2017;114(9):2325–30.

[11] Meng Z, Song R, Chen Y, Zhu Y, Tian Y, Li D, Cui D. Rapid screening and identification of dominant B cell epitopes of HBV surface antigen by quantum dot-based fluorescence polarization assay. Nanoscale Res Lett 2013;8(1):118.

[12] Kalyane D, Raval N, Maheshwari R, Tambe V, Kalia K, Tekade RK. Employment of enhanced permeability and retention effect (EPR): nanoparticle-based precision tools for targeting of therapeutic and diagnostic agent in cancer. Mater Sci Eng C Mater Biol Appl 2019;98:1252–76.

[13] Peer D, Karp JM, Hong S, Farokhzad OC, Margalit R, Langer R. Nanocarriers as an emerging platform for cancer therapy. Nat Nanotechnol 2007;2(12):751–60.

[14] van Vlerken LE, Duan Z, Little SR, Seiden MV, Amiji MM. Biodistribution and pharmacokinetic analysis of Paclitaxel and ceramide administered in multifunctional polymer-blend nanoparticles in drug resistant breast cancer model. Mol Pharm 2008;5(4):516–26.

[15] Rosenblum D, Joshi N, Tao W, Karp JM, Peer D. Progress and challenges towards targeted delivery of cancer therapeutics. Nat Commun 2018;9(1):1410.

[16] Choi HS, Liu W, Misra P, Tanaka E, Zimmer JP, Itty Ipe B, Bawendi MG, Frangioni JV. Renal clearance of quantum dots. Nat Biotechnol 2007;25(10):1165–70.

[17] Du B, Yu M, Zheng J. Transport and interactions of nanoparticles in the kidneys. Nature Reviews Materials 2018;3(10):358–74.

[18] Zhang YN, Poon W, Tavares AJ, McGilvray ID, Chan WCW. Nanoparticle-liver interactions: cellular uptake and hepatobiliary elimination. J Contr Release 2016;240:332–48.

[19] Mansoori B, Mohammadi A, Davudian S, Shirjang S, Baradaran B. The different mechanisms of cancer drug resistance: a brief review. Adv Pharmaceut Bull 2017;7(3):339–48.

[20] Nam J, Son S, Park KS, Zou W, Shea LD, Moon JJ. Cancer nanomedicine for combination cancer immunotherapy. Nature Rev Mater 2019;4(6):398–414.

[21] Dai W, Wang X, Song G, Liu T, He B, Zhang H, Wang X, Zhang Q. Combination antitumor therapy with targeted dual-nanomedicines. Adv Drug Deliv Rev 2017;115:23–45.

[22] Feldman EJ, Lancet JE, Kolitz JE, Ritchie EK, Roboz GJ, List AF, Allen SL, Asatiani E, Mayer LD, Swenson C, Louie AC. First-in-man study of CPX-351: a liposomal carrier containing cytarabine and daunorubicin in a fixed 5:1 molar ratio for the treatment of relapsed and refractory acute myeloid leukemia. J Clin Oncol 2011;29(8):979–85.

[23] Fujita K, Kubota Y, Ishida H, Sasaki Y. Irinotecan, a key chemotherapeutic drug for metastatic colorectal cancer. World J Gastroenterol 2015;21(43):12234–48.

[24] Wang-Gillam A, Hubner RA, Siveke JT, Von Hoff DD, Belanger B, de Jong FA, Mirakhur B, Chen LT. NAPOLI-1 phase 3 study of liposomal irinotecan in metastatic pancreatic cancer: final overall survival analysis and characteristics of long-term survivors. Eur J Cancer 2019;108:78–87.

[25] Holohan C, Van Schaeybroeck S, Longley DB, Johnston PG. Cancer drug resistance: an evolving paradigm. Nat Rev Cancer 2013;13(10):714–26.

[26] Werle M. Natural and synthetic polymers as inhibitors of drug efflux pumps. Pharm Res 2008;25(3):500–11.

[27] Kabanov AV, Batrakova EV, Miller DW. Pluronic block copolymers as modulators of drug efflux transporter activity in the blood-brain barrier. Adv Drug Deliv Rev 2003;55(1):151–64.

[28] Liu Y, Fang J, Joo KI, Wong MK, Wang P. Codelivery of chemotherapeutics via crosslinked multilamellar liposomal vesicles to overcome multidrug resistance in tumor. PLoS One 2014;9(10):e110611.

[29] Shah RR, Stonier PD. Repurposing old drugs in oncology: opportunities with clinical and regulatory challenges ahead. J Clin Pharm Therapeut 2019;44(1): 6–22.

[30] Kuzu OF, Gowda R, Noory MA, Robertson GP. Modulating cancer cell survival by targeting intracellular cholesterol transport. Br J Cancer 2017;117(4): 513–24.

[31] Ramanathan RK, Korn RL, Raghunand N, Sachdev JC, Newbold RG, Jameson G, Fetterly GJ, Prey J, Klinz SG, Kim J, Cain J, Hendriks BS, Drummond DC, Bayever E, Fitzgerald JB. Correlation between ferumoxytol uptake in tumor lesions by MRI and response to nanoliposomal irinotecan in patients with advanced solid tumors: a pilot study. Clin Cancer Res 2017;23(14):3638–48.

[32] Gowda R, Kardos G, Sharma A, Singh S, Robertson GP. Nanoparticle-based celecoxib and plumbagin for the synergistic treatment of melanoma. Mol Cancer Therapeut 2017;16(3):440–52.

[33] Elgogary A, Xu Q, Poore B, Alt J, Zimmermann SC, Zhao L, Fu J, Chen B, Xia S, Liu Y, Neisser M, Nguyen C, Lee R, Park JK, Reyes J, Hartung T, Rojas C, Rais R, Tsukamoto T, Semenza GL, Hanes J, Slusher BS, Le A. Combination therapy with BPTES nanoparticles and metformin targets the metabolic heterogeneity of pancreatic cancer. Proc Natl Acad Sci USA 2016;113(36):E5328–36.

[34] He H, Markoutsa E, Li J, Xu P. Repurposing disulfiram for cancer therapy via targeted nanotechnology through enhanced tumor mass penetration and disassembly. Acta Biomaterialia 2018;68:113–24.

[35] Au KM, Satterlee A, Min Y, Tian X, Kim YS, Caster JM, Zhang L, Zhang T, Huang L, Wang AZ. Folate-targeted pH-responsive calcium zoledronate nanoscale metal-organic frameworks: turning a bone antiresorptive agent into an anticancer therapeutic. Biomaterials 2016;82:178–93.

[36] Steverding D. The development of drugs for treatment of sleeping sickness: a historical review. Parasites Vectors 2010;3(1):15.

[37] De Clercq E. Suramin in the treatment of AIDS: mechanism of action. Antivir Res 1987;7(1):1–10.

[38] Tan S, Li JQ, Cheng H, Li Z, Lan Y, Zhang TT, Yang ZC, Li W, Qi T, Qiu YR, Chen Z, Li L, Liu SW. The antiparasitic drug suramin potently inhibits formation of seminal amyloid fibrils and their interaction with HIV-1. J Biol Chem 2019;294:13740–54.

[39] Naviaux RK, Curtis B, Li K, Naviaux JC, Bright AT, Reiner GE, Westerfield M, Goh S, Alaynick WA, Wang L, Capparelli EV, Adams C, Sun J, Jain S, He F, Arellano DA, Mash LE, Chukoskie L, Lincoln A, Townsend J. Low-dose suramin in autism spectrum disorder: a small, phase I/II, randomized clinical trial. Ann Clin Transl Neurol 2017;4(7):491–505.

[40] Middaugh CR, Mach H, Burke CJ, Volkin DB, Dabora JM, Tsai PK, Bruner MW, Ryan JA, Marfia KE. Nature of the interaction of growth factors with suramin. Biochemistry 1992;31(37):9016–24.

[41] Wu ZS, Liu CF, Fu B, Chou RH, Yu C. Suramin blocks interaction between human FGF1 and FGFR2 D2 domain and reduces downstream signaling activity. Biochem Biophys Res Commun 2016;477(4):861–7.

[42] Turner N, Grose R. Fibroblast growth factor signalling: from development to cancer. Nat Rev Cancer 2010; 10(2):116–29.

[43] Holland EC, Varmus HE. Basic fibroblast growth factor induces cell migration and proliferation after glia-specific gene transfer in mice. Proc Natl Acad Sci USA 1998;95(3):1218–23.

[44] Rogelj S, Klagsbrun M, Atzmon R, Kurokawa M, Haimovitz A, Fuks Z, Vlodavsky I. Basic fibroblast growth factor is an extracellular matrix component required for supporting the proliferation of vascular endothelial cells and the differentiation of PC12 cells. J Cell Biol 1989;109(2):823–31.

[45] Stein CA, Khan TM, Khaled Z, Tonkinson JL. Cell surface binding and cellular internalization properties of suramin, a novel antineoplastic agent. Clin Cancer Res 1995;1(5):509–17.

[46] Arlt W, Reincke M, Siekmann L, Winkelmann W, Allolio B. Suramin in adrenocortical cancer: limited efficacy and serious toxicity. Clin Endocrinol 1994;41(3): 299–307.

[47] Palayoor ST, Bump EA, Teicher BA, Coleman CN. Apoptosis and clonogenic cell death in PC3 human prostate cancer cells after treatment with gamma radiation and suramin. Radiat Res 1997;148(2):105–14.

[48] Zhang M, Liu E, Cui Y, Huang Y. Nanotechnology-based combination therapy for overcoming multidrug-resistant cancer. Cancer Biol Med 2017; 14(3):212–27.

[49] Greene RF, Collins JM, Jenkins JF, Speyer JL, Myers CE. Plasma pharmacokinetics of adriamycin and adriamycinol: implications for the design of in vitro experiments and treatment protocols. Cancer Res 1983; 43(7):3417–21.

[50] Lv M, Li X, Huang Y, Wang N, Zhu X, Sun J. Inhibition of fibrous dysplasia via blocking Gsalpha with suramin sodium loaded with an alendronate-conjugated polymeric drug delivery system. Biomater Sci 2016;4(7): 1113–22.

[51] Park J, Hwang SR, Choi JU, Alam F, Byun Y. Self-assembled nanocomplex of PEGylated protamine and heparin-suramin conjugate for accumulation at the tumor site. Int J Pharm 2018;535(1–2):38–46.

[52] Chou HC, Chiu SJ, Hu TM. LbL assembly of albumin on nitric oxide-releasing silica nanoparticles using suramin, a polyanion drug, as an interlayer linker. Biomacromolecules 2015;16(8):2288–95.

[53] Lustberg MB, Pant S, Ruppert AS, Shen T, Wei Y, Chen L, Brenner L, Shiels D, Jensen RR, Berger M, Mrozek E, Ramaswamy B, Grever M, Au JL, Wientjes MG, Shapiro CL. Phase I/II trial of non-cytotoxic suramin in combination with weekly paclitaxel in metastatic breast cancer treated with prior taxanes. Cancer Chemother Pharmacol 2012;70(1): 49–56.

[54] Lam ET, Au JL, Otterson GA, Guillaume Wientjes M, Chen L, Shen T, Wei Y, Li X, Bekaii-Saab T, Murgo AJ, Jensen RR, Grever M, Villalona-Calero MA. Phase I trial of non-cytotoxic suramin as a modulator of docetaxel and gemcitabine therapy in previously treated patients with non-small cell lung cancer. Cancer Chemother Pharmacol 2010;66(6): 1019–29.

[55] Small EJ, Meyer M, Marshall ME, Reyno LM, Meyers FJ, Natale RB, Lenehan PF, Chen L, Slichenmyer WJ, Eisenberger M. Suramin therapy for patients with symptomatic hormone-refractory prostate cancer: results of a randomized phase III trial comparing suramin plus hydrocortisone to placebo plus hydrocortisone. J Clin Oncol 2000;18(7):1440–50.

[56] Lu Z, Wientjes TS, Au JL. Nontoxic suramin treatments enhance docetaxel activity in chemotherapy-pretreated non-small cell lung xenograft tumors. Pharm Res 2005; 22(7):1069–78.

[57] Xin Y, Lyness G, Chen D, Song S, Wientjes MG, Au JL. Low dose suramin as a chemosensitizer of bladder cancer to mitomycin C. J Urol 2005;174(1):322–7.

[58] Zhang Y, Song S, Yang F, Au JL, Wientjes MG. Nontoxic doses of suramin enhance activity of doxorubicin in prostate tumors. J Pharmacol Exp Therapeut 2001;299(2):426–33.

[59] Dupre TV, Doll MA, Shah PP, Sharp CN, Kiefer A, Scherzer MT, Saurabh K, Saforo D, Siow D, Casson L, Arteel GE, Jenson AB, Megyesi J, Schnellmann RG, Beverly LJ, Siskind LJ. Suramin protects from cisplatin-induced acute kidney injury. Am J Physiol Ren Physiol 2016;310(3):F248–58.

[60] Zhuang S, Schnellmann RG. Suramin promotes proliferation and scattering of renal epithelial cells. J Pharmacol Exp Therapeut 2005;314(1):383–90.

[61] Liu N, Tolbert E, Pang M, Ponnusamy M, Yan H, Zhuang S. Suramin inhibits renal fibrosis in chronic kidney disease. J Am Soc Nephrol 2011;22(6):1064–75.

[62] Ren P, Zou G, Bailly B, Xu S, Zeng M, Chen X, Shen L, Zhang Y, Guillon P, Arenzana-Seisdedos F, Buchy P, Li J, von Itzstein M, Li Q, Altmeyer R. The approved pediatric drug suramin identified as a clinical candidate for the treatment of EV71 infection-suramin inhibits EV71 infection in vitro and in vivo. Emerg Microb Infect 2014;3(9):e62.

[63] Cheng B, Gao F, Maissy E, Xu PS. Repurposing suramin for the treatment of breast cancer lung metastasis with glycol chitosan-based nanoparticles. Acta Biomaterialia 2019;84:378–90.

[64] Golfmann K, Meder L, Koker M, Volz C, Borchmann S, Tharun L, Dietlein F, Malchers F, Florin A, Buttner R, Rosen N, Rodrik-Outmezguine V, Hallek M, Ullrich RT. Synergistic anti-angiogenic treatment effects by dual FGFR1 and VEGFR1 inhibition in FGFR1-amplified breast cancer. Oncogene 2018; 37(42):5682–93.

[65] Cuomo A, Rodolico A, Galdieri A, Russo M, Campi G, Franco R, Bruno D, Aran L, Carannante A, Attanasio U, Tocchetti CG, Varricchi G, Mercurio V. Heart failure and cancer: mechanisms of old and new cardiotoxic drugs in cancer patients. Card Fail Rev 2019;5(2): 112–8.

[66] Salvatorelli E, Menna P, Chello M, Covino E, Minotti G. Modeling human myocardium exposure to doxorubicin defines the risk of heart failure from low-dose doxorubicin. J Pharmacol Exp Therapeut 2017;362(2): 263–70.

[67] Ichikawa Y, Ghanefar M, Bayeva M, Wu R, Khechaduri A, Naga Prasad SV, Mutharasan RK, Naik TJ, Ardehali H. Cardiotoxicity of doxorubicin is mediated through mitochondrial iron accumulation. J Clin Invest 2014;124(2):617–30.

[68] Lodi M, Priksz D, Fulop GA, Bodi B, Gyongyosi A, Nagy L, Kovacs A, Kertesz AB, Kocsis J, Edes I, Csanadi Z, Czuriga I, Kisvarday Z, Juhasz B, Lekli I, Bai P, Toth A, Papp Z, Czuriga D. Advantages of prophylactic versus conventionally scheduled heart failure therapy in an experimental model of doxorubicin-induced cardiomyopathy. J Transl Med 2019;17(1):229.

[69] Ferreiro SF, Vilarino N, Carrera C, Louzao MC, Cantalapiedra AG, Santamarina G, Cifuentes JM, Vieira AC, Botana LM. Subacute cardiovascular toxicity of the marine phycotoxin azaspiracid-1 in rats. Toxicol Sci 2016;151(1):104–14.

[70] Maksimenko A, Dosio F, Mougin J, Ferrero A, Wack S, Reddy LH, Weyn AA, Lepeltier E, Bourgaux C, Stella B, Cattel L, Couvreur P. A unique squalenoylated and nonpegylated doxorubicin nanomedicine with systemic long-circulating properties and anticancer activity. Proc Natl Acad Sci USA 2014;111(2):E217–26.

[71] Sui BL, Cheng C, Wang MM, Hopkins E, Xu PS. Heterotargeted nanococktail with traceless linkers for eradicating cancer. Adv Funct Mater 2019;29:1906433.

CHAPTER

14

Nanotechnological approaches in cancer: the role of celecoxib and disulfiram

João Basso[1,2,*], *Maria Mendes*[1,2,3,*], *Ana Fortuna*[1,4], *Rui Vitorino*[5], *João Sousa*[1,2], *Alberto Pais*[2], *Carla Vitorino*[1,2,3]

[1]Faculty of Pharmacy, University of Coimbra, Azinhaga de Santa Comba, Coimbra, Portugal; [2]Coimbra Chemistry Center, Department of Chemistry, University of Coimbra, Coimbra, Portugal; [3]Centre for Neurosciences and Cell Biology (CNC), University of Coimbra, Rua Larga, Faculty of Medicine, Coimbra, Portugal; [4]CIBIT/ICNAS–Coimbra Institute for Biomedical Imaging and Translational Research, University of Coimbra, Coimbra, Portugal; [5]Department of Medical Sciences and Institute of Biomedicine–iBiMED, University of Aveiro, Aveiro, Portugal

OUTLINE

* These authors contributed equally to this work and should be regarded as cofirst authors.

Drug Repurposing in Cancer Therapy
https://doi.org/10.1016/B978-0-12-819668-7.00014-2

Introduction

The limited success of new drug approvals and the long development process of novel medicines are some of the problems found in the drug discovery field. Thus, the use of already approved drugs, originally developed for a specific therapeutic indication, in the treatment of other diseases is being increasingly studied [1]. Herein, the pharmaceutical industry has found a niche and a possible solution for some issues, including the reduction of costs and duration of the stages in drug development associated to product approval (Fig. 14.1). The redirection of "old drugs" has prompted advantages regarding the mechanism of action, identification of molecular targets, and pharmacological properties, including pharmacokinetics, pharmacodynamics, posology, toxicological profile, and drug–drug interactions, since most of the aforementioned parameters are already well established. Indeed, the main regulatory agencies have developed rescue programs of drugs that failed for a determined disease but may have a promising potential on other conditions [2]. Drug repurposing can also be applied to combination therapy aiming at targeting different signaling pathways or receptors. In this case, drug dosage regimens may be tailored in order to minimize side effects while maintaining or potentiating drug efficacy. Preclinical studies regarding drug synergism or antagonism should be conducted, in order to prove biological efficacy at an early stage. In addition, the pharmacokinetic profile of the coadministration should also be assessed at least in vivo, as there may be potential clinical implications between drugs.

Despite having a higher probability of success, drug repurposing does not come without failure risk. Lack of clinical efficacy, organization within the industry, intellectual and legal barriers, and regulatory hurdles may impair the development of novel medicines based on old drugs (for additional details, please see Ref. [3]). Focusing on clinical efficacy, the use of animal models does not always correlate well with clinical practice. As an example, topiramate has been suggested to have a positive impact on inflammatory bowel disease due to gene expression signatures and a rodent model but failed to prove its benefit on a cohort study [4]. In fact, animal models can lack quality and/or validation. Furthermore, basic research is not always reproducible between research groups, thus creating an accuracy gap in knowledge transfer. A rudimental understanding of the target behavior in humans is also pointed as a common cause for translational failure [5]. In order to reduce the probability of failure in clinical trials, the use of patient-derived xenograft (PDX) and genetically engineered mouse models is being encouraged, as they are more truthful representation of the pathologies. Overall, PDX models retain the main characteristics of the donor and are a stronger predictor of clinical outcomes than the conventional cell lines [6].

Cases of a successful reposition include thalidomide (teratogenic for the fetus, currently used for refractory multiple myeloma), sildenafil (antianginal drug now used for the treatment of erectile dysfunction), exenatide (used in type II diabetes, now repurposed in the control of obesity issues), methotrexate (primarily developed for cancer, is also used for rheumatoid arthritis and psoriasis), and everolimus (an immunosuppressor drug used to avoid organ transplant rejection, is also used against neuroendocrine tumors of gastrointestinal (GI) or lung origin, or HER2-negative breast cancer) [2,7–11].

Oncology is one of the areas that has mostly benefited from drug repurposing, exhibiting favorable clinical trial outcomes [12,13]. The poor specificity and therapeutic response, which

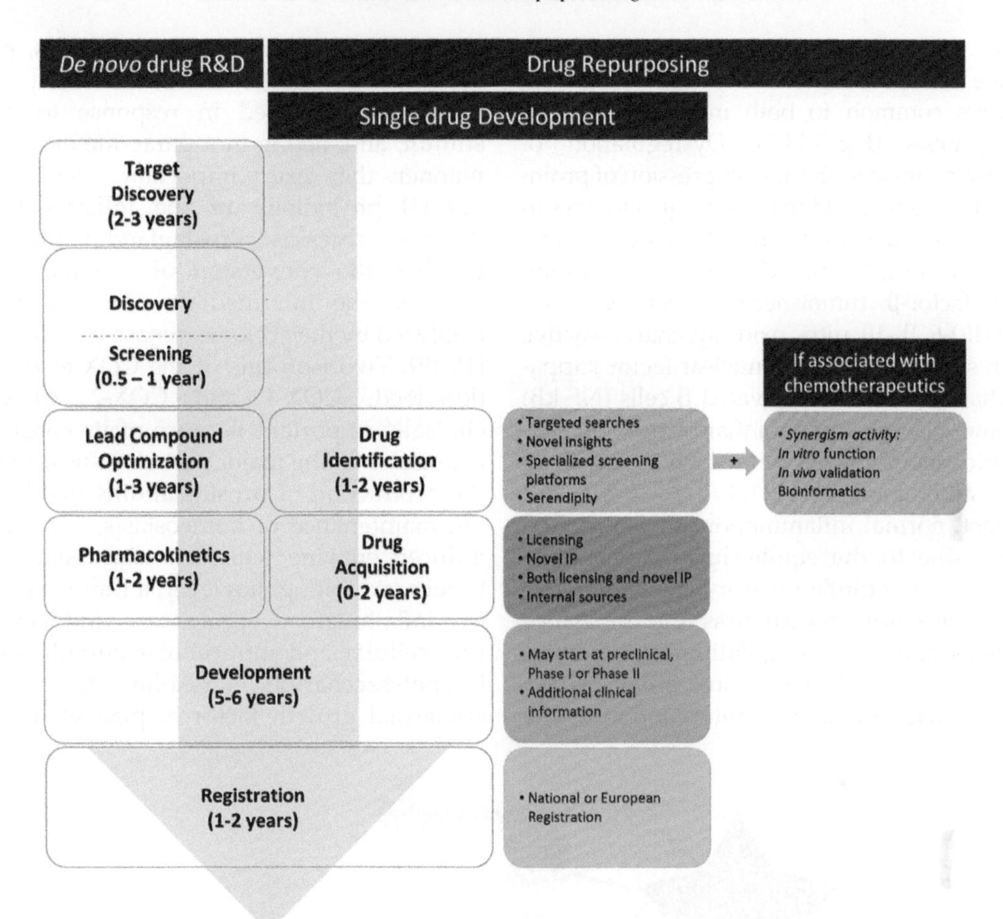

FIGURE 14.1 Infographic comparison of drug discovery and development between novel and repurposed drugs.

lead to dose-limiting side effects, impose the need to develop alternative therapies. Thus, the main groups of noncancer drugs that might offer effective treatment for cancer therapy are antidepressants, antipsychotic drugs, cardiovascular drugs, antimicrobiological agents, a drug for treating alcoholism, and nonsteroidal antiinflammatory drugs (NSAIDs) [8].

This review provides a comprehensive overview of current repurposing nanotechnological approaches specifically concerning celecoxib (CXB) and disulfiram (DSF) for the treatment of neoplastic diseases.

NSAIDs as a class of repurposed drugs

NSAIDs are one of the pharmacotherapeutic groups most commonly prescribed for the symptomatic treatment of pain and fever. NSAIDs have sparked interest in the treatment of cancer, as a class of drug repurposing, due to their antiinflammatory properties [14–16]. Indeed, cancer is linked to chronic inflammation. Therefore, the use of antiinflammatory drugs seems to play an important role in the treatment and prevention of cancer [14]. NSAIDs may exert anticancer effects due to their ability to induce apoptosis, inhibit angiogenesis, and enhance cellular immune responses, which are signaling pathways common to both inflammation and carcinogenesis (Fig. 14.2). Dysregulation of signaling pathways, aberrant expression of proinflammatory genes, and the release of cytokines in the tumor microenvironment (TME) are potential therapeutic targets for NSAIDs. Transforming growth factor-β, tumor-necrosis factor-α, interleukin (IL)-6, IL-10, nitric oxide synthase, reactive oxygen species (ROS), and nuclear factor kappa-light-chain-enhancer of activated B cells (NF-kB) are some examples of proinflammatory mediators endogenously produced, which are involved in TME and inflammation [17].

In fact, normal inflammation is a controlled process, due to the equilibrium between the production of antiinflammatory and proinflammatory cytokines. Nevertheless, the imbalance of factors and/or dysregulation of signaling pathways may lead to a chronic inflammation process that demands intervention. The prostanoids, including prostaglandins, prostacyclins, and thromboxanes, are a family of lipid mediators produced in response to diverse stimuli, and, acting in a paracrine or autocrine manner, they exert important roles either in normal physiology of the inflammation or disease processes. Prostaglandin production requires the conversion of arachidonic acid (AA) to the intermediate prostaglandin H2 catalyzed by the cyclooxygenase (COX) enzyme [18,19]. Two isoforms of the COX enzyme are described, COX-1 and COX-2. COX-1 is clinically important because of its constitutive expression in the major part of tissues, allowing the production of prostaglandins that helps in the maintenance of homeostasis; COX-2 is an induced enzyme with low expression in few tissues, exhibiting, however, a rapid expression in inflammatory processes, inducible by extracellular and intracellular stimuli, such as lipopolysaccharide, forskolin, IL-1, TNF-α, epidermal growth factor α, platelet-activating

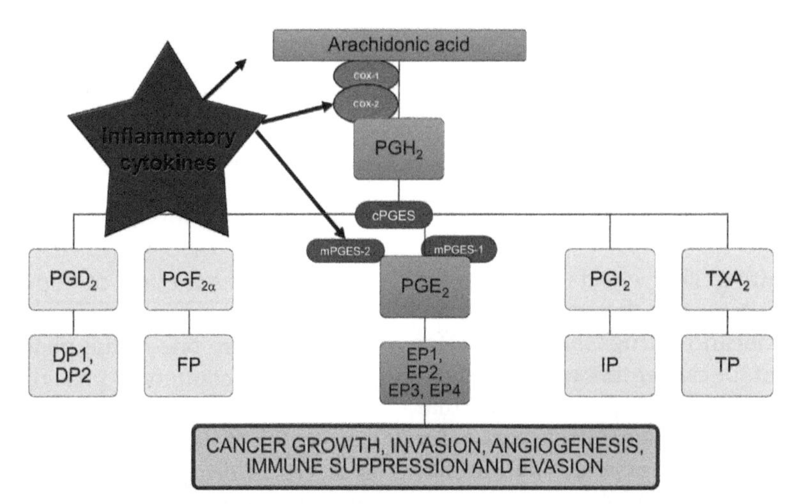

FIGURE 14.2 Cyclooxygenase-2 (COX-2) signaling pathway. Inflammatory cytokines are the stimuli for the induction of COX-2, which converts the arachidonic acid to the intermediate molecule prostaglandin H_2 (PGH_2). PGH_2 is subsequently converted to five different prostanoids: prostaglandin D_2 (PGD_2), prostaglandin E_2 (PGE_2, which have four distinct receptors: epoxygenases [EP]1, EP2, EP3, and EP4), PGF2α (identification of a single F prostanoid receptor [FP]), prostacyclin (PGI_2, the I prostanoid [IP] receptor), and thromboxane A_2 (TXA_2, T prostanoid [TP] receptor). The conversion of PGH_2 into PGE_2 is induced by three isozymes (membrane-associated PGE synthase—mPGES-1, mPGES-2, and cytosolic—cPGES). There are two types of G protein–coupled receptors (GPCRs), DP1 and DP2, that are activated by PGD_2 [22].

factor, interferon-γ, and endothelin [20,21]. Thus, the understanding of these mechanisms of action, regulation, and function has enabled the design and synthesis of COX inhibitors.

Depending on the generation of NSAIDs, these can be selective or not in the blocking of COX or prostaglandin endoperoxide H synthase. The first generation of NSAIDs, which includes aspirin, indomethacin, diclofenac, and sulindac, inhibits both COX-1 and COX-2 activity, while the second generation of NSAIDs includes, for example, CXB and rofecoxib, and it selectively inhibits COX-2 activity. Preclinical studies reflect the importance of targeting COX-2 in cancer, as it is highly expressed in several signaling pathways involved in invasion, proliferation, and angiogenesis [23]. Among NSAIDs, the second generation stands out due to its selective COX-2 inhibition. Therefore, in repurposing strategies, the use of CXB, rofecoxib, valdecoxib, etoricoxib, or lumiracoxib is the natural and first choice. However, note that only CXB is approved by the Food and Drug Administration (FDA), as the remaining drugs were either withdrawn or not approved in the United States due to safety concerns. The potential approval of CXB for cancer treatment may be easily accomplished, if supported by solid clinical data, rather than the approval of other coxibs.

Celecoxib: physicochemical, pharmacokinetic, and pharmacodynamic parameters

CXB was the first approved drug belonging to the class of selective COX-2 inhibitors, approved by the FDA) in 1998. CXB inhibits the synthesis of prostaglandins by the inhibition of COX-2 in humans, having demonstrated efficiency to treat inflammatory diseases, such as rheumatoid arthritis and osteoarthritis, ankylosing spondylitis, acute pain, and primary dysmenorrhea. CXB is available in four dosage forms, respectively, 50, 100, 200, and 400 mg capsules. The dosage is different depending on the type of disease [24,25]. The dependence is related to the amount of AA, i.e., a higher amount of AA leads to increased prostaglandin production and, subsequently, a higher dose of NSAIDs is required for their effectiveness.

CXB physicochemical properties are a relevant information for the understanding of its pharmacokinetics and pharmacodynamics. Structurally, CXB is a diarylsubstituted pyrazole compound, also known as benzene sulphonamide. Chemically, it is termed as 4-[5-(4-methylphenyl)-3-trifluoromethyl-1H-pyrazoyl-1-yl] (Fig. 14.3) having an empirical formula of $H_{14}F_3N_3O_2S$. It exhibits their action by specifically inhibiting COX-2 isoenzyme with a 5- to 50-fold selectivity. The sulphonamide moiety is the main group that defines its COX-2 selectivity and antiinflammatory activity. CXB is described in both crystalline and amorphous forms, wherein the crystalline form evidences less bioavailability than the amorphous one. CXB belongs to class II of Biopharmaceutical Classification System (BCS), characterized by low solubility and high permeability. Its poor water solubility is confirmed in several studies (c. 4.3 μg/mL) (Table 14.1) [26]. CXB is a drug that respects

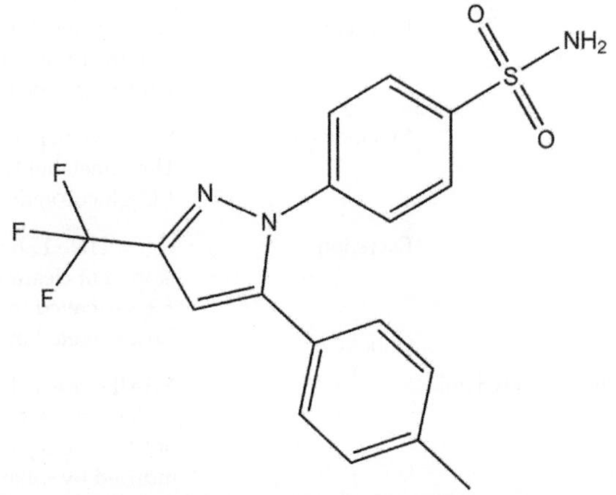

FIGURE 14.3 Chemical structure of celecoxib.

TABLE 14.1 Physicochemical, toxicological, pharmacokinetic, and pharmacodynamic properties of celecoxib (Human clinical data, unless specified as otherwise; retrieved from Refs. [28–31] or generated in Chemicalize, developed by Chemaxon).

Physicochemical	IUPAC name	4-[5-(4-methylphenyl)-3-(trifluoromethyl)-1H-pyrazol-1-yl]benzene-1-sulfonamide
	Traditional name	Celecoxib
	SMILES	CC1=CC=C(C=C1)C1=CC(=NN1C1=CC=C(C=C1)S(N) (=O)=O)C(F)(F)F
	InChI	InChI = 1 S/C17H14F3N3O2S/c1-11-2-4-12(5-3-11)15-10-16(17(18,19)20)22-23(15)13-6-8-14(9-7-13)26(21,24)25/h2-10H,1H3,(H2,21,24,25)
	Molecular formula	$C_{17}H_{14}F_3N_3O_2S$
	Molar mass	381.37 g/mol
	Melting point	158.0°C
	Water solubility	4.3 mg/mL
	Octanol/water partition coefficient	3.53
	Dissociation constant	11.1
	Density	1.43 g/cm^3
	Lipinski's rule of five	Yes
	Topological polar surface area	77.98 Å^2
	Molar refractivity	92.23 cm^3/mol
Toxicological	Oral LD_{50} (both in rat and dog)	>2000 mg/kg
	Oral TD_{LO}	5.71 mg/kg
	Side effects	Risk of cardiovascular events
Pharmacokinetic	Absorption	T_{max} = 2–4 h
	Distribution	High protein bound (>97%) to albumin Volume of distribution = 5.7–7.1 L/kg, higher when compared to other NSAIDs, which may relate to CXB lipophilic nature
	Metabolism	Extensive hepatic metabolism—cytochrome P450 (CYP) 2C9 Three metabolites identified in plasma: hydroxycelecoxib, carboxycelecoxib, and 1-O-glucuronide
	Excretion	$t_{1/2}$ = 11.2–15.6 h Apparent clearance (CL/F) ~ 30 L/h <2% excreted in urine 2.6% excreted in feces
Pharmacodynamic		NSAIDs are a class of antiinflammatory drugs that inhibit both types of cyclooxygenases (COX-1 and COX-2). CXB is a selective noncompetitive inhibitor of COX-2 enzyme. COX-2 is expressed densely in inflamed tissues where it is induced by inflammatory mediators. Thus, CXB inhibits expression of COX-2 and reduces the synthesis of metabolites, such as PGE_2, PGI_2, TXA_2, PGD_2, and PGF_2, leading to an improvement of pain and inflammation.

LD50, lethal dose of 50% (one half) of a group; $t_{1/2}$, elimination half-life; TD_{LO}, toxic dose low; T_{max}, time to achieve the maximum concentration.

the Lipinski's "rule of five," with good absorption and permeation in biological systems. It is a hydrophobic compound, with a log P of 3.5 [26], and weakly acidic with a pKa of 11.1, ascribed to the ionization of the primary amine functional group.

The pharmacokinetic and pharmacodynamic properties of the CXB are summarized in Table 14.1. This information is crucial for a rational newer repurposing. CXB is administered orally (once or twice a day, depending on the therapeutic indication) and quickly absorbed, exhibiting a maximum peak serum concentration between 2 and 4 h. CXB metabolism occurs essentially by the cytochrome P450 2C9 in the liver (97%), being described three metabolites, namely hydroxycelecoxib, carboxycelecoxib, and 1-O-glucuronide, with renal and fecal excretion. The unchanged form of CXB is also excreted in urine and feces in a low extent (<3%). CXB is used for antiinflammatory and analgesic effects by blocking the synthesis of different inflammatory prostanoids, including PGs and thromboxanes. These products are the end of fatty acid metabolism produced by COX enzymatic activity [27]. These mediators are crucial for pathological and physiological processes, like pain, inflammation, glaucoma, osteoporosis, cardiovascular diseases, and cancer. Thus, the production of PGs is dependent on the accessibility of AA from the cellular phospholipids—secretory or cytoplasmic phospholipases. PG synthesis is stimulated by inflammatory cytokines and the consequent release of AA. Thus, the activation of COX-1 (encoded by $PTGS_1$) and COX-2 (encoded by $PTGS_2$) lead to the synthesis of prostanoids. COX converts AA in PGG_2, following reduction of PGG_2 to PGH_2. The latter is converted into the active metabolites PGH_2, PGD_2, $PGF_2\alpha$, prostacyclin (PGI_2), and thromboxane (TXA_2). These molecules interact with specific prostanoid G protein–coupled receptors, which mediate the physiological responses including blood pressure regulation, fever, inflammation, and GI protection.

Role of CXB in cancer

CXB has been studied in the oncological field, due to its potential anticancer properties. The overexpression of COX-2 is chronically found in several steps of carcinogenesis, which leads to higher levels of prostaglandin in neoplastic tissues. This is evident in different cancers like brain, colon, lung, and breast. The level of COX-2 is directly proportional to the degree of cancer invasiveness. Considering the TEM, the presence of inflammatory cells is a stimulus for the tumorigenesis, which contributes to the augment of COX-2 enzymes, following the activation of different mechanisms involved in cancer progression (Table 14.2). The COX-2 release is stimulated by inflammatory cytokines (e.g., oncogenes or tumor promoters) to synthesize PGE_2, which promotes tumor cell proliferation, angiogenesis, and migration by activation/production of vascular endothelial growth factor. PGE2 inhibits apoptosis, inducing the protooncogene Bcl-2, epidermal growth factor receptor, or activation of PI3kinase/Akt and mitogen-activated protein kinase (MAPK) pathways. In Table 14.2, the different antitumor mechanisms where CXB may effectively act are highlighted.

Glioblastoma (GB) is one of the most aggressive central nervous system tumors, due to its invasive nature and genetic and epigenetic variability which, consequently, reflects in resistance to currently used forms of therapy [17]. Temozolomide, a drug used as the first line of GB treatment, has shown limited efficacy. New drugs are urgently needed, but the process of developing novel compounds able to reach clinical application is highly time-consuming and expensive. So, focus on drug repositioning has growth, and the results seem to be promising [8]. Human malignant glioma cells show constitutively prominent levels of COX-2, with several unregulated transcription factors [32]. The recurrent tumors, including GB, have high expression of COX-2, mPGES-1, and cPGES. All these three therapeutic targets are relevant

TABLE 14.2 The role of COX-2 in cancer progression and the proposed celecoxib molecular mechanism against cancer progression [38–44].

Possible role of COX-2 in cancer progression and respective mediators		Proposed molecular mechanism of celecoxib against cancer progression	
Cell proliferation	EGFR VEGF	Inhibition of cell proliferation	Increase p21 Increase p27 Decrease expression of cyclins Decrease β-catenin Activation of Akt/survivin and Akt/ID3 pathway
Migration		Induction of apoptosis	Decrease PDK1/Akt Activation of caspases and CHOP Decrease sarcoplasmic/ER calcium ATPase
Angiogenesis		Inhibition of angiogenesis	Decrease EGFR Decrease VEGF Decrease MMP-2/9
Metastasis		Inhibition of metastasis	Decrease Sp1
Apoptosis inhibition	Bcl-2 NF-kB PUMA BH3	Inhibition tumor growth	Decrease carbonic anhydrases
Metastasis	MMP-2/9		
Invasion			
Immunosuppression	IL-12 IL-10		

Akt/PKB, protein kinase B; *Bcl-2*, B-cell lymphoma 2; *EGFR*, epidermal growth factor receptor; *ER*, endoplasmic reticulum; *IL*, interleukin; *MMP*, matrix metalloproteinase; *NF-κB*, factor nuclear kappa B; *PDK1*, phosphoinositide-dependent kinase-1; *VEGF*, vascular endothelial growth factor.

for CXB action. Indeed, the mechanism of action previously detailed of CXB may be repurposed to block cell proliferation and invasion, angiogenesis, and inflammation pathways [33–36]. Additionally, COX-2 has been described as playing a major role in glioma resistance and progression [37].

The ideal of chemotherapy is to inhibit the progression of abnormal cells with no impact on the functionality of normal cells. Given the selectivity of CXB to inhibit COX-2, its use as a potential chemotherapeutic drug is promising, due to the less toxicity to other tissues, with safety and efficacy parameters for long-term treatment.

Clinical development of CXB in cancer

CXB has demonstrated antitumor clinical efficacy in many types of cancer, as displayed in Table 14.3. A search from the "ClinicalTrials.gov" database with "CXB" and "cancer" keywords showed around 236 studies [45]. Several types of cancer are addressed, including GB, lung, breast, bladder, colorectal, advanced head and neck, metastases, or recurrent tumors. In most clinical studies, CXB is coadministrated with other chemotherapeutics (NCT02574728, NCT02885974, NCT02054104, and NCT00504660) or with biological therapies

TABLE 14.3 Clinical trials in cancer considering celecoxib as anticancer drug.

Clinical trial number	Treatment (route of administration)	Cancer	Status
NCT03710876	✓ Adenovirus-delivered interferon alpha-2b ✓ Celecoxib ✓ Gemcitabine (intrapleural administration)	Malignant pleural mesothelioma	Active (phase III)
NCT02574728	✓ Sirolimus (per os) ✓ Etoposide (per os) ✓ Celecoxib (per os) ✓ Cyclophosphamide (per os)	Solid or central nervous system tumors, recurrent or refractory	Active (phase II)
NCT02885974	✓ Celecoxib (per os) ✓ Gemcitabine hydrochloride (IV) ✓ Cisplatin (IV)	Bladder cancer before surgery	Active (phase I)
NCT01356290	✓ Bevacizumab (IV) ✓ Combination (oral) with o thalidomide o celecoxib o fenofibrate o etoposide o cyclophosphamide	Medulloblastoma	Active (phase II)
NCT04162873	✓ Celecoxib (per os)	Advanced Head and Neck Cancer	Active (phase II)
NCT04081389	✓ Celecoxib (per os) ✓ Recombinant interferon alfa-2b (IV) ✓ Rintatolimod (IV) + Standard chemotherapy (paclitaxel, doxorubicin, and cyclophosphamide) (IV/per os)	Early Stage Triple-Negative Breast Cancer	Active (phase I)
NCT03599453	✓ Celecoxib (per os) ✓ Recombinant interferon alfa-2b (IV) ✓ Rintatolimod (IV) ✓ Pembrolizumab (IV) + Standard chemotherapy (paclitaxel, doxorubicin, and cyclophosphamide) (IV/per os)	Metastatic triple-negative breast Cancer	Active (phase II)
NCT03403634	✓ Celecoxib (per os) ✓ Recombinant interferon alfa-2b (IV) ✓ Rintatolimod (IV)	Colorectal cancer	Active (phase IIA)
NCT02432378	✓ Celecoxib (per os) ✓ Recombinant interferon alfa-2b (IV) ✓ Cisplatin (IV) ✓ Rintatolimod (IV)	Recurrent ovarian cancer	Active (phase I/II)

(Continued)

TABLE 14.3 Clinical trials in cancer considering celecoxib as anticancer drug.—cont'd

Clinical trial number	Treatment (route of administration)	Cancer	Status
NCT04093323	✔ Celecoxib (per os) ✔ Polarized dendritic cell vaccine (intradermally) ✔ Interferon alpha-2 (IV) ✔ rintalolimid (IV)	HLA-A2+ refractory melanoma	Not yet recruiting (phase II)
NCT02615574	✔ Celecoxib (per os) ✔ Polarized dendritic cell vaccine (intradermally) ✔ Interferon Alpha-2 (intranasally) ✔ Rintalolimid (IV)	Refractory metastatic colorectal cancer	Approved (phase II)
NCT02054104	✔ Celecoxib (per os) ✔ Metronomic cyclophosphamide (per os) ✔ Cell lysate vaccine (subcutaneous)	Lungs, esophagus, pleura, or mediastinum	Temporarily closed to accrual
NCT00045591	✔ Celecoxib (per os)	Metastatic or recurrent breast cancer	Terminated (phase II)
NCT00504660	✔ Celecoxib (per os) ✔ 6-Thioguanine (per os) ✔ Capecitabine (per os) ✔ Lomustine (per os) ✔ Temozolomide (per os)	Anaplastic glioma	Completed (phase II)
NCT00047281	✔ Celecoxib (per os) ✔ Thalidomide (per os) + Combination chemotherapy	Relapsed or refractory malignant glioma	Completed (phase II)
NCT00056082	✔ Celecoxib (per os)	Breast cancer in at-risk women	Completed (phase II)

IV, intravenous.

(NCT03710876, NCT01356290, NCT04081389, NCT03599453, NCT03403634, or NCT02432378). The latter therapeutic regimen allows to stimulate the immune system through different pathways and arrest the progression of the tumor cells. Only a small fraction of clinical trials make use of solely CXB, since the better outcomes are obtained when the drug is employed as adjuvant therapy.

Drug delivery nanosystems containing CXB

CXB, as COX-2 specific drug, holds a potential therapeutic benefit in cancer treatment. Despite the few expression of COX-2 in the body, the long-term use of CXB could be toxic for GI and cardiovascular systems. Its interference in the synthesis of prostaglandins may result in side effects for both systems. Thus, bearing in mind the CXB physicochemical properties and envisioning an improvement in targeting effect to the tumor tissues, different drug delivery nanosystems (DDNSs) have been developed. Moreover, the improvement in dissolution rate (note that CXB is a BCS class II drug), and consequently in the absorption profile, so as to enhance the bioavailability and the therapeutic effect is pointed out as main reasons for using nanotechnological approaches, Several DDNS have been investigated for CXB encapsulation, including micelles, liposomes, cyclodextrins, mesoporous silica NPs, silk fibroin NPs, polymeric NPs, and quantum dots [46–48]. Apart from cancer, CXB nanoparticles have also been developed for other diseases, encompassing ischemia, rheumatoid arthritis, edema, gastric irritancy, impaired gastric motility, and cancer.

In general, favorable outcomes have been retrieved from CXB nanoformulations. Details are provided in Table 14.4, with particular emphasis on the different mechanisms of action that support cancer CXB repurposing.

Disulfiram: physicochemical, pharmacokinetic, and pharmacodynamics parameter

The discovery of the therapeutic properties of DSF (Fig. 14.4) goes back to the beginning of the 19th century, when two Danish physicians attempted to evaluate its potential anthelminthic properties and became ill, while concomitantly consuming alcohol beverages at a cocktail party. Further studies on DSF laid the foundation for its use in alcohol dependences [73]. Despite being a nontoxic drug, DSF has been abandoning the clinical practice over the last decades due to a low therapy adhesion by the patients. In fact, DSF is responsible for the 5- to 10-fold increase in blood acetaldehyde concentration in the presence of alcohol, thus leading to a severe symptomatic reaction characterized by intense throbbing in the head and neck, respiratory difficulties, nausea, vomiting, sweating, hypotension, and blurred vision, among others. The appearance of symptoms occurs rapidly, only 5–10 min after the consumption of alcohol. Moreover, the food or medicinal products containing alcohol, after-shave lotions, or perfumes can also trigger DSF-alcohol reaction symptoms [73]. The use of DSF as a therapeutic approach against alcohol addiction is then being replaced by naltrexone and acamprosate.

DSF, due to the formation of various active metabolites, is a classical inhibitor of the aldehyde dehydrogenase 1 and 2, the enzymes that catalyze the oxidation of acetaldehyde to acetic acid. However, not only DSF is an unstable drug, but it also suffers a rapid biotransformation once it reaches the bloodstream and the liver (Fig. 14.5).

A simplified overview of physicochemical, toxicological, pharmacokinetic, and pharmacodynamic properties is presented in Table 14.5.

Role of DSF in cancer

Following an oral administration, DSF is rapidly converted to diethyldithiocarbamic acid (DDC) in the stomach. Without the presence of copper, DDC decomposes in diethylamine (DEA) and CS_2. Copper ions existing in the stomach are responsible for the formation of the complex copper (II) diethyldithiocarbamate ($Cu(DDC)_2$), the principal compound that forms the basis of DSF anticancer therapy. Interestingly, not only higher levels of copper are found in tumors than in healthy tissues but also higher copper levels may be directly correlated with cancer progression [77]. On the contrary, zinc, selenium, and iron are diminished in pancreatic cancer tumors [78]. Note that the formation of DDC complexes can also occur with other metallic ions, such as zinc and nickel, although such compounds may present different anticancer activities over tumor cells [79].

TABLE 14.4 List of celecoxib delivery nanosystems for several types of cancer.

Nanostructure	Surface modification strategy	Drug(s)	Therapeutic indication	Mechanism of action	Studies in vitro/in vivo	References
Liposomes	Protein transduction domain (PTD) peptide: Gly-Arg-Lys-Lys-Arg-Arg-Gln-Arg-Arg-Arg-Cys-Gly-NH_2	Celecoxib Epirubicin	• Invasive breast cancer	• Induction of apoptosis by activation of caspase-8— and caspase-3—signaling pathways. • Activation of the proapoptotic protein Bax. • Suppression of the antiapoptotic protein Mcl-1.	✓ Cytotoxicity studies: The combinatory effect between CXB plus epirubicin had a significantly stronger cytotoxic effect than both drugs alone. ✓ Transport across BBB: The results showed a higher transport ability for the targeted drug-loaded liposomes across the BBB. ✓ In vivo: After intravenous administration, the results showed that the mice treated with targeted epirubicin plus celecoxib liposomes had the longest survival time, 30 days.	[49,50]
	Methoxy polyethylene glycol distearoyl ethanolamine (MPEG-DSPE2000)	Celecoxib	• Colon cancer	–	✓ In vitro: Release studies showed that PEGylated liposomes promoted a slow CXB release, when compared to other formulations due to delayed drug diffusion from the formulation. ✓ Stability studies: The storage at 4°C improved formulation stability.	[51]
	–	Celecoxib Genistein	• Prostate cancer	• Induction of apoptosis by COX pathway. • Suppression of GSH synthesis, which could lead to the high ROS production in cells. The cleaved caspase-3 expression was significantly increased in PC-3 cells exposed.	✓ Prostate cancer cell viability: NP formulation is more effective in PC-3 cells than LNCaP, due to the higher express of Glut-1 receptors in PC-3 cells than LNCaP cells. ✓ Cell migration assay: The combination of CXB and genistein (CGL) may lead to higher long-term suppression of PC-3 cells in vitro than the CXB-liposome and genestein-liposome formulations. ✓ ROS generation assay: The combination of both drugs, CGL formulation, induces a threefold increase in ROS production. These results are in agreement with the lower % of cell viability associated to CGL formulation.	[52]

–	Celecoxib Plumbagin	• Melanoma	• CelePlum-777 leads to a small effect on the signaling of other pathways, such as AKT and cPLA2 signaling pathways; induced caspase-3/7 activity. • Plum inhibits COX-2 signaling, which in turn increases COX-2 levels. CXB + Plum decreased the protein levels of pSTAT3 (Y705) and influences the levels of certain cyclins (cyclin A2, E1, and E2; cyclins B1, D1, and H); increase in cleave caspase-3 and PARP protein levels in cultured cells.	✓ In vitro cytotoxicity study: ○ Normal cells were less sensitive to the free drugs than melanoma cells. ○ CXB + Plum had an additive killing effect. ○ IC_{50} CelePlum-777 > IC_{50} CXB-liposomes/pulmonary liposomes ~ IC_{50} empty liposomes. ✓ Melanoma tumor growth: ○ CelePlum-777 in /saline did not change the tumor inhibitory efficacy. ○ CelePlum-777 or liposomes containing CXB and plum led to similar levels of tumor inhibition. ○ CelePlum-777 led to 50% decrease in proliferating cells and vascular development. ✓ CelePlum-777 displayed a decrease in the proliferative potential of melanoma cells, when compared to the individual agents, suggesting a synergistic effect on tumor development.	[53]
–	Letrozole Celecoxib	• Breast cancer	• Tumor growth biomarkers: ○ Inhibition of aromatase level in mammary tumors: PC-NCs < free CXB/LTZ. ○ Level of VEGF: PC-NCs < free CXB/LTZ. ○ Level of caspase-3 in mammary tumors: PC-NCs > free CXB/LTZ ○ NF-κB level: PC-NCs > free CXB/LTZ.	✓ Serum stability and hemocompatibility: Particle size of MRP-NCs was increased once mixed with serum after 2 h, followed by stabilization after 6 h. ✓ In vitro cytotoxicity study: IC_{50}PC-NCs > IC_{50} free drugs. ✓ In vivo pharmacokinetics: PC-NCs could extend the half-life of LTZ, due to their reduced uptake by RES. ✓ In vivo antitumor efficacy: ○ Tumor volume: PC-NCs < combined free LTZ/CXB. ✓ Histopathological study: PC-NCs displayed higher % necrosis than free drug combination.	[54]
NBD peptide Hyaluronic acid	Curcumin Celecoxib		• TN-CCLP and HA/TN-CCLP were most efficient in	✓ In vitro cytotoxicity: CXB + CUR had synergistic effect (CI = 0.33); Tan + CXB + CUR prompted additive effects. IC_{50} TN-	[55]

Continued

TABLE 14.4 List of celecoxib delivery nanosystems for several types of cancer.—cont'd

Nanostructure	Surface modification strategy	Drug(s)	Therapeutic indication	Mechanism of action	Studies in vitro/in vivo	References
			• Metastatic inflammatory breast cancer	downregulating the expressions of NF-κB and IL-6.	CCLP < IC_{50} HA/TN-CCLP < IC_{50} TN-LP < IC_{50} CCLP, which suggest that the combination of CXB + CUR and TN could exert the addictive tumor cytotoxicity. ✓ In vitro cellular uptake and uptake inhibition: TN-LP > HA/TN-LP, thus TN had a higher impact on cellular uptake, however, and was nonselective for both abnormal cells and normal ones. ✓ In vitro antimigration and anti-inflammation studies: TNs-CCLP and HA/TN-CCLP showed the more significant inhibitory effects. ✓ In vivo and ex vivo biodistribution studies: HA/TN-LP was detected in the tumor after 2 h postadministration and reached a maximum at 12 h. ✓ In vivo antitumor efficacy: CCLP < TN-LP < TN-CCLP < HA/TN-CCLP. The presence of HA and TN led to a longer circulation time and, consequently, a higher tumor accumulation. ○ The histopathology of tumor tissues when administrated HA/TN-CCLP formulations showed severe necrosis. ○ HA/TN-CCLP showed the maximum in vivo inhibition for inflammatory factors (NF-κB, IL-6 and TNF-α). ✓ In vivo antimetastasis efficacy: TN-CCLP < HA/TN-CCLP showed a decrease in the area and number of tumors; the absence of macrophages and macroscopic metastatic nodules was observed in the HA/TN-CCLP treatment. This behavior could be justified by the effective improvement in the tumor inflammatory microenvironment and additional blockade of tumor cell metastasis.	

Nanostructured lipid carriers	PEGylated	Celecoxib	• Breast cancer • Acute promyelocytic leukemia	CXB-SLN in comparison to all experimental groups: • Induction of apoptosis through the activation of caspase-3. • Inhibition of Bcl-2, P-AKT, COX-2, MK, and drug efflux protein levels. • Inhibition of drug efflux proteins, indirectly proceeded through p-AKT/AKT pathway.	Cytotoxicity studies: IC_{50}CXB-solid lipid nanoparticles (SLNs) > IC_{50}CXB-nanoemulsions (NE) > IC_{50}CXB-nanostructured lipid carriers (NLCs) > IC_{50}CXB ✔ Apoptotic and antiapoptotic protein levels: CXB induced the decrease in COX-2, MK, Bcl-2, ABCG2 and P-gp levels, and p-AKT/AKT ratio in comparison to the formulations.	[56]
	–	Celecoxib Docetaxel (DOC)	• Non-small cells lung cancer	• CXB-NLC treatments decreased Bcl-2 and increased Bax, cleaved caspase-9, and cleaved caspase-3 expression. • CXB-NLC + DOC treatments increased the expression of Bax, cleaved caspase-9, cleaved caspase-3 proteins, and decreased the expression of Bcl-2, when compared to Cxb-NLC and doc alone. The combined treatments reduced VEGF expression and reduced survivin protein expression.	✔ *In vitro cytotoxicity study:* ○ IC_{50} CXB-NLC and DOC were time dependent. DOC and CXB-NLC combination showed CI values < 1, which suggest a moderate synergistic activity. ✔ Anticancer activity: Animals treated with CXB-NLC + DOC and DOC displayed a lower lung weight than untreated animals. ✔ TUNEL assay and immunohistochemistry (IHC) for cleaved caspase-3: CXB-NLC + DOC treatments displayed a significantly higher number of apoptotic cells than DOC alone. ✔ Proteomic analysis: Proteins S100A6 and S100P were downregulated in the CXB-NLC + DOC−treated lung tumors, showing the enhanced anticancer activity of the combination. Vimentin, associated to cancer invasion and metastasis, was drastically downregulated in CXB-NLC + DOC treated samples. ✔ NLC lung toxicity: Blank-/Cxb-NLC did not cause any pulmonary edema.	[57]

Continued

TABLE 14.4 List of celecoxib delivery nanosystems for several types of cancer.—cont'd

Nanostructure	Surface modification strategy	Drug(s)	Therapeutic indication	Mechanism of action	Studies in vitro/in vivo	References
	–	Celecoxib	• Lung cancer	–	✓ In vitro cytotoxicity of aerosolized: In vitro cytotoxicity results showed a direct relation between drug concentration versus exposure time. IC_{50} CXB-Sol > IC_{50} CXB-NLC. ✓ In vivo lung deposition and systemic pharmacokinetic studies: At 20 min of exposure, CXB lung concentrations were found to be constant. ○ Deposition of CXB-NLC < CXB-Sol. ○ Clearance of CXB-NLC < CXB-Sol. ✓ CXB-NLC modulated the pharmacokinetic behavior, showing significantly higher CXB plasma levels, being detected in plasma for up to 24 h. The peak plasma concentrations were reached after 4 h after administration.	[58]
Micelles	–	Paclitaxel Celecoxib	• Lung cancer	• Inhibition of CXCL12/CXCR4 axis, which has an important role in tumor stromal formation and angiogenesis. • Suppression of fibroblast proliferation. • Activation stimulated by FGF-2. • Induction of G1-S cell cycle arrest and apoptosis of TAF.	✓ In vivo studies: CXB treatment significantly decreased the microvessel density.	[59]
	Hyaluronic acid		• Breast cancer			[60]

Polymeric nanoparticles		Celecoxib Doxorubicin		• Free CXB ~ DOX/CXB mixture < loaded-HPPDC markedly downregulated the mRNA expression of MDR1 and reduced the protein expressions of both P-gp and COX-2. • CXB could reverse drug resistance in breast cancer by suppressing the P-gp and COX-2 expressions.	✔ Cellular uptake and intracellular location: In MCF-7 cells, the free DOX > HPPD > loaded-HPPDC, 8 and 24 h after incubation. In MCF-7/ADR cells, HPPD ~ loaded-HPPDC > free DOX, 8 and 24 h after incubation due to the effective circumvention of P-gp efflux mechanism. ✔ In vitro drug resistance: ○ MCF-7/ADR cells: IC_{50} DOX > IC_{50} CXB. IC_{50} DOX > IC_{50} HPPD ~ loaded-HPPDC. ○ HPPD exerted potent induction effect on the late apoptosis. ✔ In vivo biodistribution and tumor accumulation: HPPDC/Cy5.5 displayed a significant tumor targeting ability at 6 h. ✔ In vivo drug resistance: All the treatment groups, except free CXB, remarkedly inhibited the tumor growth. However, loaded-HPPDC had significantly higher inhibitory effect. Loaded-HPPDC showed high significant effect on the induction of cell apoptosis.	
	–	Celecoxib Naringin	• Lung cancer	• CXB induction apoptosis depends on the expression of Bax or Bak. • Naringin decreases the proapoptotic protein Bax.	✔ In vitro cytotoxicity study: IC_{50} value CXB $<$ IC_{50} value loaded-nanoparticle $<$ IC_{50} value narginin + CXB $<$ IC_{50} value narginin. ✔ In vivo studies: 30 min after administration, naringin and CXB displayed accumulation in lung tissues with systemic distribution to the other body organs.	[61]

Continued

TABLE 14.4 List of celecoxib delivery nanosystems for several types of cancer.—cont'd

Nanostructure	Surface modification strategy	Drug(s)	Therapeutic indication	Mechanism of action	Studies in vitro/in vivo	References
	–	Paclitaxel Celecoxib	• Breast cancer	• Decreases P-gp expression by PTX/CXB@LPNP. These results showed that the dual-drug delivery may effectively downregulate the P-gp expression. • Decreases IL-10 expression: PTX/CXB@LPNP > PTX@LPNP > free PTX.	✓ Cytotoxicity studies: Blank nanoparticles did not show cytotoxicity. IC_{50} value PTX/CXB@LPNP > IC_{50} value PTX@LPNP > IC_{50} value free PTX. ✓ Cellular uptake: PTX/CXB@LPNP had the highest intracellular drug concentration, due to the effective downregulation of P-gp by CXB. ✓ Apoptosis studies: PTX/CXB@LPNP > PTX@LPNP > PTX.	[62]
	RGD R8	Celecoxib Doxorubicin	• Breast cancer	• CXB enhanced the sensibility of cancer cells to chemotherapeutics because it can reduce the P-gp expression. • DOX/CXB@PNP showed a significant downregulation of P-gp expression.	✓ Cytotoxicity studies: Blank NPs did not exhibit inhibitory effect. IC_{50} value DOX/CXB@PNP > IC_{50} value DOX@NP ~ IC_{50} value DOX@PNP > IC_{50} value free DOX. ✓ Cellular uptake: The NPs displayed a higher accumulation of drug inside the cells. However, no significant differences were noted between NPs with or without targeting.	[63]
	Alendronate (ALE)	Celecoxib Doxorubicin (DOX)	• Bone metastasis	–	✓ Cytotoxicity studies: Blank polymeric NPs and ALE-blank-polymeric NPs were not toxic. ○ IC_{50} value free CXB > IC_{50} value free DOX > IC_{50} value DOX + CXB > IC_{50} value PLCA-DOX/CXB ~ IC_{50} value ALE-polymeric NPs-DOX/CXB ✓ Targeting of NPs: Ale-polymeric NPs bound a higher amount of CA^{2+} than polymeric NPs. The introduction of drugs did not change the affinity of NPs.	[64]

	Peptide linker (MMP-2-sensitive)	Paclitaxel Celecoxib	• Fibrosarcoma	• SN-PTX/CXB, which induces a higher apoptosis, showed a decrease in cleaved caspase-3, expression of Bcl-2 protein, and the activation of COX-2/PGE_2 pathway.	✓ Inhibition of PGE_2 secretion: Polymeric NPs were tested in different cells (HT-1080, RAW 264.7, and L-929 cells). In all the cells, the higher amount of CXB leads to a decreased of PGE_2 level. ✓ Cytotoxicity studies: IC_{50} value MMP-2-sensitive nanospheres (SNs) CXB/PTX < IC_{50} value SN-PTX < IC_{50} value SN-PTX + PGE_2. CXB + PTX had a synergistic effect (CI < 1). ✓ In vivo synergistic anticancer activity: SN-CXB/PTX had a higher accumulation in tumor. The liver was the other organ with a high amount of NPs. ○ Tumor volume: SN-CXB/PTX < SN-PTX < InN-CXB/PTX < SN-CXB. ✓ SN-CXB, SN-PTX, and InN-CXB/PTX displayed a medium level of cancer cell density and apoptosis.	[65]
	PEG	Celecoxib Brefeldin a (BFA)	• Metastatic breast cancer	• Inhibition of COX-2, which decreases the expression level of metastasis-associated proteins (MMP-9 and VEGF)	✓ Cytotoxicity studies: Blank NPs did not show cytotoxicity. ○ IC_{50} value free CXB > IC_{50} value free BFA > IC_{50} value free CXB/BFA > IC_{50} value CBNPs. ✓ Cell migration and invasion: Inhibition of cell migration CXB/BFA NPs > CXB/BFA.	[66]
Protein nanoparticles (K237-HSA-DC)	Peptide (K237(NH_2-LHHQYHHYYMTH-COOH)	Celecoxib Doxorubicin (DOX)	• Lung cancer	• CXB + DOX inhibited the metabolism. It is associated to a decrease of expression levels of both GLUT-1 and hexokinase. CXB can effectively enhance the inhibitory impact of DOX on energy metabolism and GSH biosynthesis. • K237-HSA-DC downregulated the expression of hexokinase and GLUT-1 and GSH biosynthesis–related proteinases.	✓ In vitro cytotoxicity study: CXB could not inhibit cell proliferation. DOX + CXB showed a synergistic inhibitory effects. ✓ Cellular uptake: K237-HSA-DC > HSA-DC. The incubation with sorafenib (inhibitor of VEGFR-2) showed a decrease in the K237-HAS-DC uptake. ✓ In vivo pharmacokinetic and biodistribution: 2 h after injection, K237-HAS-DC had higher accumulation in the tumor. The free drugs were completely eliminated in 2 h. ✓ Therapeutic efficacy: K237-HSA-DC > DOX + CXB > DOX. K237-HSA-DC were more effective in inducing cell apoptosis.	[67]

Continued

TABLE 14.4 List of celecoxib delivery nanosystems for several types of cancer.—cont'd

Nanostructure	Surface modification strategy	Drug(s)	Therapeutic indication	Mechanism of action	Studies in vitro/in vivo	References
Dendrimer	Biotin	Celecoxib Fmoc-L-Leucine	• Glioblastoma (U-118 MG) • Squamous cell carcinoma (SCC-15)	–	✓ COX-2 expression: The level was about 2 and 2.5 times higher in glioblastoma and squamous carcinoma. ✓ Biotin uptake: The results did not show differences between cancer cells and normal cells. ✓ Cytotoxicity studies: Dendrimers $G3^{1B16C15L}$ conjugated with both drugs showed the highest cytotoxicity.	[33]
Mesoporous silica nanoparticles + β-cyclodextrin (MSCPs)	Buthionine sulfoximine (BSO)	Celecoxib Doxorubicin (DOX)	• Liver cancer	• Free DOX and DOX@MSBPs: Increased mRNA and protein levels of Oct-3/4, Nanog, and Notch-3 in HepG2. • DOX@MSCPs suppressed the PGE_2 production by CXB. • CXB suppressed P-gp upregulation.	✓ Cytotoxicity studies: IC_{50} value DOX@MSCPs < IC_{50} value DOX@MSBPs < IC_{50} value DOX. The higher cytotoxicity from DOX@MSCPs is attributed to the presence of CXB. ✓ In vivo biocompatible: The free DOX displayed a severe damage on the major organs. DOX@MSCPs showed a good biocompatibility. ✓ Antitumor activity: A high concentration of CXB was found in tumor mass in mice treated with MSCPs or DOX@MSCPs. The NPs had a higher accumulation on lungs. ○ Tumor reduction: DOX@MSCPs > DOX@ MSBPs > free DOX.	[68]
Polymer/inorganic hybrid nanoparticles (BSO/CXB@BNP)	Buthionine sulfoximine (BSO)	Celecoxib Doxorubicin (DOX)	• Multiple drug resistance (MDR) • Breast cancer	• Production of IL-10 in drug-resistant cells: BSO/CXB@BNP treatment resulted in smaller IL-10 concentration.	✓ Cytotoxicity studies: IC_{50} value DOX@BNP < IC_{50} value DOX@NP ~ IC_{50} value free DOX; DOX@BNP showed the highest cell inhibition effects due to the presence of biotin moieties for tumor cell targeting. ○ CXB@BNP treatment led to a decrease in P-gp level. ○ BSO/CXB@BNP showed the highest efficiency in downregulation of P-gp. ○ Reducing the intracellular GSH: BSO/CXB@BNP > BSO@BNP > free BSO > free CXB. ✓ Cellular uptake: Free DOX < DOX@NP < DOX@BNP < CXB@BNP, DOX@BNP < BSO@BNP, DOX@BNP < BSO/CXB@BNP, free DOX < BSO/CXB@BNP, DOX@BNP	[69]

Halloysite nanotubes (HNTs)	pH-responsive hydroxypropyl methylcellulose acetate succinate polymer	Celecoxib Atorvastatin calcium (ATV)	• Colon cancer	–	✔ Cytotoxicity studies: IC_{50} value HNT@HF ~ HNTs. ○ The cytotoxicity effect of NPs was pH-dependent. At pH 6.5, the HNT-ATV@HF-CEL displayed a weaker inhibition effect. ✔ Drug permeation: HNT-ATV@HF-CEL > free drugs.	[70]
Quantum dots	Cationic gelatin, specific to matrix metalloproteinase (MMP-2)	Celecoxib Rapamycin (RAP)	• Breast cancer	• Generation of reactive oxygen species causing mitochondrial dysfunction by changing MMP. • Induction of cell apoptosis by release of apoptotic factors and cytochrome c. • Reduction of p-AKT protein level, which is responsible for regulation of tumor genesis (CS-NCS < G-CS-NCs < G-QDs-CS-NCs).	✔ In vitro serum stability: Increased NP size when previously mixed with 10% FBS. After 6 h of incubation, the PS drastically decreased. ✔ In vitro cytotoxicity study: ○ Blank NCs did not show any significant cytotoxicity. ○ IC_{50} CXB + RAP > IC_{50} free drugs alone. ○ IC_{50} QDs-G-CS-NCs < IC_{50} CXB/RAP solution. ✔ Cellular uptake study: electrostatic interaction and interaction with CD44 increased the penetration of particles into the cells. ✔ In vivo studies: ○ Antitumor activity: Blank NC < CS-NCs < G-CS-NC < G-QDs-CS-NCs. ○ G-QDs—-CS-NCs showed a significant higher necrosis level than positive control group. ✔ G-QDs-CS-NCs were effectively localized in the tumor tissue and with a lower fluorescence intensity in liver and kidney tissues.	[71]

Continued

TABLE 14.4 List of celecoxib delivery nanosystems for several types of cancer.—cont'd

Nanostructure	Surface modification strategy	Drug(s)	Therapeutic indication	Mechanism of action	Studies in vitro/in vivo	References
	Lactoferrin (LF)	Celecoxib Honokiol (HNK)	• Breast cancer	• NCs showed a strong antiangiogenic and apoptotic effect by interfering with reduction of VEGF-1, increase of caspase-3 levels, and reduction of p-AKT protein level. • CXB decreased the intratumoral production of IL-1, IL-6, and COX-2 and led to a reduction of protumor angiogenic and inflammatory microenvironment.	✓ In vitro serum stability: The particle size increased with 10% FBS, which can be explained by the shaping of protein corona. ✓ In vitro cytotoxicity study: ○ Blank nanocapsules (NCs) did not show any cytotoxicity ○ Synergistic cytotoxic between both drugs (CXB/HNK) ○ IC_{50} value Quantum dots (QDs)-LF-NCs > IC_{50} value LF-CS–NCs > IC_{50} value CS–NCs > IC_{50} value CXB/HNK solution. ✓ Cellular uptake study: The presence of LF lead to a higher uptake NCs. ✓ In vivo antitumor efficacy: Blank NCs < CXB/HNK solution < CS–NCs < LF-CS-NCs < QDs-LF-CS-NCs. ✓ Immunogenicity: LF stimulus anti-LF antibodies.	[72]

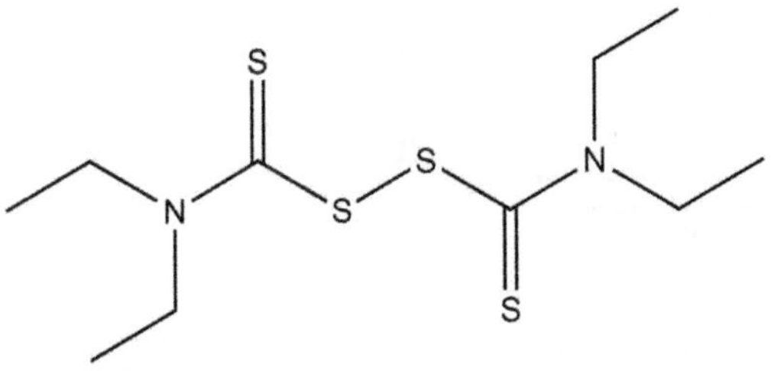

FIGURE 14.4 Chemical structure of disulfiram.

The several mechanisms with which DSF induces cancer cell death suggest an incredible versatile behavior of this drug. It is believed DSF acts on at least 19 different targets/pathways that reduce (or contribute to the reduction of) the viability of a tumor cell. A highly reported cell death mechanism concerns the depletion of intracellular glutathione (GSH) with consequent production of ROS and the activation of the MAPK pathway. Induction of apoptosis (modification in cellular levels of Bax [proapoptotic] and Bcl-2 [antiapoptotic]) is also verified [80,81].

Cells exposed to $Cu(DDC)_2$ present a similar phenotype as those treated with inhibitors of the proteasome, which includes the cytosolic accumulation of polyubiquitinated proteins, most likely by ubiquitin E3 ligase inhibition. This led to a categorization of DSF as a proteasome inhibitor [82,83]. $Cu(DDC)_2$ also targets the zinc (II)-binding thiolate of NPL4 proteins, thus leading to their aggregation. Consequently, P97 segregase enzymes are inactivated, causing an accumulation of misfolded proteins and cell death via the p97-NPL4-UFDI pathway [84,85]. The upregulation of RECK by DSF was also observed in vitro. The downregulation of this gene results in tumor proliferation, invasion, angiogenesis, and metastasis [86–88]. $Cu(DDC)_2$ also induced cell death by paraptosis, a nonapoptotic event that is independent of caspases [89]. Therefore, DSF may be useful against drug-resistant cancer cells due to apoptotic defects. Other pathways with relevance in drug-resistant cancers include a cytotoxic effect over cancer stem cells, which are often resistant to conventional chemotherapy, play a major role in cancer relapse, and may have their overexpressed ALDH levels targeted by DSF [90,91]. In parallel, DSF has showed an inhibitory effect over the activation of NF-kB, a drug resistance marker and transcription factor that promotes, under a hypoxic environment, cell proliferation, invasion, and migration [92].

Drug resistance (for example, of doxorubicin or paclitaxel) to chemotherapy is also associated to the overexpression of several membrane transporters, including P-glycoprotein (P-gp) and multidrug resistance protein 1, that promote the cellular efflux of chemotherapeutics, that is, the drug transport from the cytosol to the extracellular environment. DSF is able to covalently modify cysteine motifs of these transporters in two different locations, the drug-binding and the ATP site [93,94].

DSF also inhibits the antichemotherapeutic activity of DNA methyltransferases and glutathione S-transferase P1 [95,96]). Regarding GB, DSF has showed a strong inhibition of O6-methylguanine DNA methyltransferase (MGMT), a critical enzyme in temozolomide resistance [97]. Copper-DSF complex adopts a planar conformation with ability to inhibit DNA replication enzymes (e.g., DNA topoisomerases) and impairing cellular division (via reduction of cyclin-dependent kinase 1 and polo-like kinase 1 expression levels with G2 arrest) [98,99]. Matrix metalloproteinases and copper/zinc superoxide dismutase may also be inhibited due to the chelation of these two ions [100–102].

Clinical development of DSF in cancer

The promising data gathered from fundamental research have prompted the use of DSF against cancer, usually in combination with copper, in several clinical trials (Table 14.6). Associations with chemotherapy, namely, cisplatin,

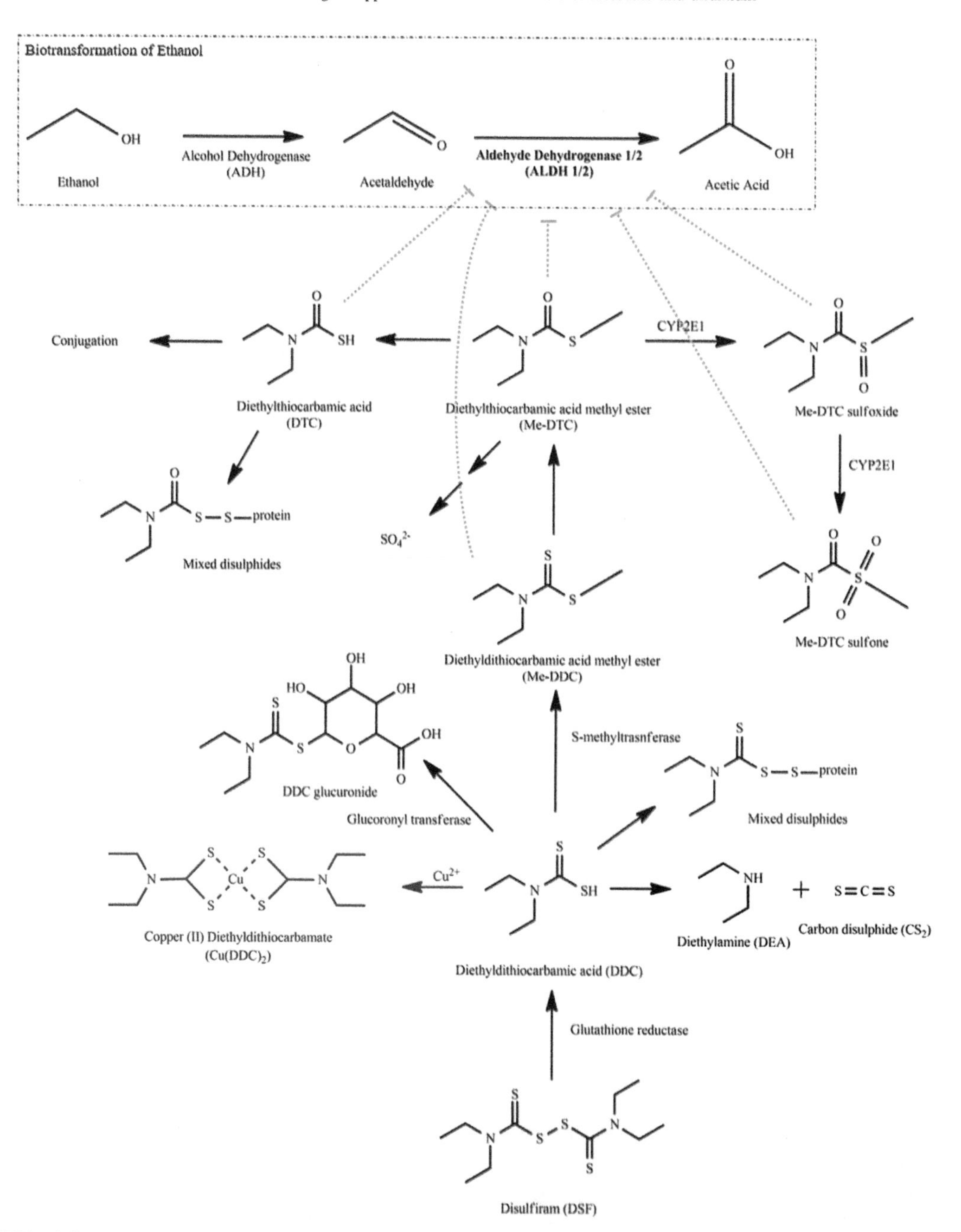

FIGURE 14.5 Metabolic pathway of disulfiram (DSF), an inhibitor of both aldehyde dehydrogenase 1 and 2 (ALDH 1/2). *Green dotted arrows* indicate the inhibition of ALDH by the active metabolites of DSF. Blue structures represent the decomposition of diethyldithiocarbamic acid (DDC) in the absence of copper or enzymes. The red structure characterizes the complex formed between DDC and Cu^{2+}, copper (II) diethyldithiocarbamate ($Cu(DDC)_2$), recognized by its anticancer properties [74,75].

TABLE 14.5 Physicochemical, toxicological, pharmacokinetic, and pharmacodynamic properties of disulfiram (Human clinical data, unless specified as otherwise; retrieved from Refs. [73,75,76] or generated in Chemicalize, developed by Chemaxon).

Physicochemical	Name	Disulfiram, tetraethylthiuram disulfide, Antabuse
	IUPAC name	N,N-diethyl[(diethylcarbamothioyl)disulfanyl]carbothioamide
	SMILES	CCN(CC)C(=S)SSC(=S)N(CC)CC
	InChI	InChI = 1 S/C10H20N2S4/c1-5-11(6-2)9(13)15-16-10(14)12(7-3)8-4/h5-8H2,1-4H3
	Molecular formula	$C_{10}H_{20}N_2S_4$
	Elemental composition	C (40.51%), H (6.8%), N (9.45%), S (43.25%)
	Molar mass	296.52 g/mol
	Melting point	69–71°C
	Lipinski's rule of five	Yes
	pKa	– (nonionizable)
	Isoelectric point	– (nonionizable)
	logP	4.16 (pH independent)
	Water solubility	0.2 mg/mL
	BCS class	II (low solubility, high permeability)
	Topological polar surface area	6.48 Å^2
Toxicological	Oral DL_{50} (mouse)	1980 mg/kg
	Intraperitoneal LD_{50} (mouse)	75 mg/kg
	Intraperitoneal LD_{50} (rat)	248 mg/kg
	Subcutaneous LD_{50} (mouse)	2600 mg/kg
	Side effects	DSF is generally nontoxic. However, it has been reported with the appearance of urticaria and acneform eruptions, headache, tremor, dizziness, restlessness, metallic/garlic-like taste, gastrointestinal disorders, psychosis, peripheral neuropathies, and ketosis.
Pharmacokinetic	Absorption	Generally above 80%
	Distribution	Uniformly distributed throughout the body. Approximately 96% binds to plasma albumin.
	Metabolism	DSF is rapidly metabolized to diethyldithiocarbamate and the latter conjugated with glucuronic acid in the liver. Oxidation to sulfate and decomposition to carbon disulphide (CS_2) and diethylamine (DEA) also occurs.
	Excretion	Up to 20% intact in feces. 65% eliminated by the kidneys (mainly as diethyldithiocarbamic acid glucuronide or inorganic sulfate). 15% eliminated by the lungs (CS_2).

Continued

TABLE 14.5 Physicochemical, toxicological, pharmacokinetic, and pharmacodynamic properties of disulfiram (Human clinical data, unless specified as otherwise; retrieved from Refs. [73,75,76] or generated in Chemicalize, developed by Chemaxon).—cont'd

Pharmacodynamic	DSF is a prodrug of at least five active metabolites that act as a suicide substrates of the hepatic aldehyde dehydrogenase (ALDH). Consequently, the irreversible inactivation of this enzyme, both in cytosol and mitochondria, blocks the metabolism of consumed alcohol and increases acetaldehyde concentration. DSF-alcohol reaction symptoms include intense throbbing in the head and neck, respiratory difficulties, nausea, vomiting, sweating, hypotension, and blurred vision among others. Alcohol sensitization may occur up to 14 days after the injection of DSF tablets, due to the slow renovation of ALDH.

TABLE 14.6 Clinical trials in cancer considering disulfiram as anticancer drug.

Clinical trial number	Treatment (route of administration)	Cancer	Status
NCT03323346	DSF (per os)	Female breast neoplasm	Recruiting (phase II)
	Copper (per os)	Metastatic breast cancer	
NCT03950830	DSF (per os)	Germ cell tumor	Recruiting (phase II)
	Cisplatin (IV)		
NCT01907165	Radiation	Glioblastoma	Completed (phase I)
	Temozolomide (per os)		
	DSF (per os)		
	Copper (per os)		
NCT03363659	Radiation	Glioblastoma	Recruiting (phase I)
	Temozolomide (per os)		
	DSF (per os)		
	Copper (per os)		
NCT01777919	Radiation	Glioblastoma	Unknown (phase II)
	Temozolomide (per os)		
	DSF (per os)		
	Copper (per os)		
NCT03151772	DSF (per os)	Glioblastoma	Recruiting (phase I)
	Copper (per os)		
	Metformin (per os)		

TABLE 14.6 Clinical trials in cancer considering disulfiram as anticancer drug.—cont'd

Clinical trial number	Treatment (route of administration)	Cancer	Status
NCT02715609	Radiation Temozolomide (per os) DSF (per os) Copper (per os)	Glioblastoma multiforme	Recruiting (phase I, II)
NCT02678975	DSF (per os) Copper (per os) Lomustine (per os) Procarbazine (per os) Vincristine (IV) Temozolomide (per os)	Glioma Glioblastoma	Recruiting (phase II, III)
NCT03034135	Temozolomide (per os) DSF (per os) Copper (per os)	Recurrent glioblastoma	Completed (phase II)
NCT02770378	Temozolomide (per os) Aprepitant (per os) Minocycline (per os) Disulfiram (per os) Celecoxib (per os) Sertraline (per os) Captopril (per os) Itraconazole (per os) Ritonavir (per os) Auranofin (per os)	Recurrent glioblastoma	Active, not recruiting (phase I/II)
NCT00742911	DSF (per os) Copper (per os)	Liver metastasis	Completed (phase I)
NCT02101008	DSF (per os) Chelated zinc (per os)	Melanoma	Completed (phase II)
NCT00571116	DSF (per os) Arsenic trioxide (IV)	Metastatic melanoma	Terminated (phase I)
NCT00256230	DSF (per os)	Stage IV melanoma	Completed (phase I, II)

(Continued)

TABLE 14.6 Clinical trials in cancer considering disulfiram as anticancer drug.—cont'd

Clinical trial number	Treatment (route of administration)	Cancer	Status
NCT02671890	DSF (per os) Gemcitabine (IV)	Metastatic pancreatic adenocarcinoma	Recruiting (phase I)
		Recurrent pancreatic carcinoma	
		Solid neoplasm	
		Stage IV pancreatic cancer	
NCT03714555	Nab-paclitaxel (IV)	Metastatic pancreatic cancer	Recruiting (Phase II)
	Gemcitabine (IV)		
	FOLFIRINOX (folinic acid, 5-fluorouracil, irinotecan, oxaliplatin) (IV)		
	Disulfiram (per os)		
	Copper (ns)		
NCT00312819	DSF (ns) Cisplatin (ns)	Non-small cell lung cancer	Completed (phase II, III)
NCT01118741	DSF (per os)	Prostate cancer	Completed (phase II)
NCT02963051	DSF (per os) Copper (per os/IV)	Prostate cancer	Active, not recruiting (phase Ib)

IV, intravenous; *ns*, not specified.

temozolomide, lomustine, procarbazine, vincristine, arsenic trioxide, paclitaxel, gemcitabine, 5-fluorouracil, irinotecan, and oxaliplatin, have also been explored, aiming at simultaneously boosting the therapy efficacy while reducing toxicity and circumventing drug resistance.

Although the use of DSF appears to extend to different types of cancer, including breast, pancreatic, lung, skin, hepatic, and prostate cancer, the use of this drug in GB clearly stands out, as it counts for more approximately 40% of the number of the trials.

Nonetheless, available results of clinical trials proved to be unsuccessful and disappointing, with no clinical evidence of a strong contribution of DSF to the anticancer activity of chemotherapeutic regimens. Potential causes underlying the lack of efficacy in clinic are the weak stability of DSF in the stomach (rapid decomposition into CS_2 and DEA), and in the bloodstream (extremely low half-life time), and the inability to form $Cu(DDC)_2$ at significant concentrations at the tumor. In addition, as DSF possesses a low aqueous solubility that limits its intravenous administration, the administration of DSF is currently limited to the enteral route. Therefore, up to date, all clinical trials regarding the use of DSF are based on the administration of DSF tablets (generally, Antabuse), the only DSF dosage form approved by FDA and other Medicine Agencies. In addition, the separate administration of DSF and copper may desynchronize the availability of both entities to tumor tissues, thus reducing the targeting of DSF and the efficacy of the clinical trial in increasing the median overall survival of cancer patients.

Drug delivery nanosystems containing DSF or DSF derivatives

The use of nanotechnological approaches for cancer therapy keeps thriving in fundamental research, paving a path for a successful translation to the clinic. Several nanosystems were already developed and explored in vitro and in in vivo models of different types of neoplastic diseases, which are presented in Table 14.7.

Up to date, there are more than 30 different nanosystems developed using DSF or its copper complex with application in cancer. Taking into consideration the type of nanocarrier, they may be categorized in liposomes, nanostructured lipid carriers and nanocapsules, lipid emulsions, polymeric micelles and nanoparticles, albumin, or silica nanoparticles and nanococrystals. Furthermore, within each category, they may vary in size, composition, production method, and targeting strategy to the tumor tissue.

However, all nanosystems possess some common characteristics: their size is usually not superior to 200 nm (for cellular internalization, administration, and clearance purposes), they promote a controlled release of DFS (thus increasing its stability and pharmacokinetic performance), and they all showed promising anticancer results in vitro and/or in vivo. Considering the importance of copper for a proper anticancer activity, oral supplementation, coadministration, or codelivery of this ion is not uncommon, thus counting for approximately 22% of all cases.

Furthermore, almost 20% of the reported nanosystems include the administration of the active metabolite of DSF, $Cu(DDC)_2$, or sodium diethyldithiocarbamate coadministered with copper.

In spite of the significant and selective cytotoxicity over tumor cells, the use of DSF-associated nanotechnology also extends to overcoming single- or multidrug resistance. In fact, c. 22% of the designed platforms aim at reversing chemotherapy resistance in breast, colorectal, prostate, or lung cancer, which is justified by the aforementioned wide versatility of anticancer mechanisms of DFS.

It should be noted that the use of nanosystems containing DSF is not restricted to cancer. In fact, DSF or its derivatives have also been explored in ocular pathologies, namely, in cataracts, glaucoma, and uveitis [136–146]. Among these, hydroxypropyl-substituted β-cyclodextrins have been intensively explored. Their lipophilic core aim at complexing with DSF or derivatives, whereas the hydrophilic surface increasing drug solubility [147,148]. Moreover, common polymers used in ophthalmic preparations, such as hydroxypropyl methylcellulose, help increasing drug solubility by increasing the stability constants of the inclusions [141].

Conclusions

Drug repurposing has received increasing attention both from the pharmaceutical industry and the public sector/academia as a faster and cheaper strategy for expanding the arsenal of anticancer drugs. A wide variety of drugs clinically used to treat noncancerous diseases also interfere with malignancy-associated pathways. Repurposed drugs have a key advantage: they already have an approved status for clinical use by regulatory authorities. Moreover, they are, for the most part, inexpensive and their side effect and safety profiles are well characterized. Nonetheless, the efficiency in drug repositioning still needs to be addressed carefully, as clinical results do not often corroborate in vitro or in vivo study results.

CXB and DSF have been intensively explored within repurposing, in multiple anticancer approaches. However, their promising activity has yet to be successfully translated to clinical practice. Additional technological strategies may be required, with drug encapsulation being an encouraging approach to improve their pharmacokinetic performance and consequent increase in efficacy and safety.

TABLE 14.7 Drug delivery nanosystems with disulfiram or its copper complex with application in cancer.

Nanostructure	Surface modification strategy	Drug(s)	Therapeutic indication	Mechanism of action	In vitro/in vivo studies	References
Liposomes	Hyaluronic acid	$Cu(DDC)_2$	Pancreatic cancer stem cells (CSCs)	ROS-mediated anticancer activity	Improved ability to decrease proliferation of patient derived CSCs (vs. free DSF and copper complex). Sphere formation capability was decreased only by liposomes.	[103]
	–	DSF Copper	Breast cancer stem cells	Blockage of nuclear factor-κB (NFkB) activation	In vitro inhibition of the hypoxia induced NFkB pathway promoted a lack of stemness and of chemotherapy resistance. Liposomes strongly inhibit xenograft development in vivo.	[104]
	–	$Cu(DDC)_2$	Glioblastoma	–	Improved blood circulation of the complex. 25% increase in overall survival in an F98 glioma rat model. 45% of tumor burden reduction in a murine model.	[105]
	PEGylation	DSF	Colorectal cancer	–	Reversion of chemoresistance (5-fluorouracil and paclitaxel) at nanomolar concentration.	[106]
Nanostructured lipid carriers	PEGylation with vitamin E-TPGS	DSF	Breast cancer	–	Reduction of cell viability with nanomolar concentrations. Decrease in tumor volume and approximately 48% of growth inhibition in 4T1 murine models, compared to free DSF (9%) and unPEGylated particles (29%).	[107]
	PEGylation with biotin-PEG_{2000}-DSPE	DSF Copper supplementation	Breast cancer	–	Strong cytotoxicity in the presence of copper over 4T1 cells. Tumor accumulation with a significant impairment in tumor growth.	[108]

Lipid emulsion	–	DSF	–	–	Type of lecithin impacts DSF stability.	[109]
	–	DSF	Breast cancer	–	DSF formulations proved to be more effective at lower concentrations than an intravenous paclitaxel solution, while oral administration of DSF did not show an antitumor effect.	[110]
Lipid nanocapsules	PEGylation with pH-triggered TAT peptide	DSF Copper supplementation	Liver cancer	–	DSF induced cytotoxicity over HepG2 cells in a copper-dependent way. Improved pharmacokinetics and tumor accumulation (vs. free DSF/unmodified nanocapsules).	[111]
Polymeric micelles	pH-sensitive polymer derivative	DSF Doxorubicin	Multidrug-resistant breast cancer	Inhibition of P-glycoprotein (DSF) Induction of apoptosis in targeted tumor cells (doxorubicin)	Temporal drug release (fast DSF release inhibited P-glycoprotein while a triggered controlled release of doxorubicin impaired tumor growth). Significant tumor impairment with no severe systemic toxicity.	[112]
	pH-sensitive polymer derivative	DSF Paclitaxel	Multidrug-resistant breast cancer	Inhibition of P-glycoprotein (DSF)	Triggerable micelles showed an enhanced uptake by cancer cells. Increased cytotoxicity over tumor cells.	[113]
	Redox-sensitive behavior	DSF	Lung cancer and metastasis	Apoptosis induction (early and late stage)	Marked inhibition of cell proliferation, colony formation, cell invasion, and tube formation of HMEC-1 cells. Improved pharmacokinetic profile, with pronounced tumor accumulation. Significant inhibition of tumor growth and prevention of metastasis.	[114]
	–	DSF Copper supplementation	Liver cancer	–	Improved biodistribution and marked tumor growth inhibition (vs. 5-fluorouracil).	[115]
	PEGylation	$Cu(DDC)_2$	Drug-resistant prostate cancer	Induction of parapoptosis (caspase-independent cell death)	Remarkable serum stability for 72 h. Reversion of drug resistance.	[89]

Continued

TABLE 14.7 Drug delivery nanosystems with disulfiram or its copper complex with application in cancer.—cont'd

Nanostructure	Surface modification strategy	Drug(s)	Therapeutic indication	Mechanism of action	In vitro/in vivo studies	References
Polymeric nanoparticles	PEGylation	$Cu(DDC)_2$	Non-small cell lung cancer	ROS production and apoptosis induction	Superior anticancer activity in mice when compared to the administration of the free copper complex.	[116]
	Redox sensitive polymer Lactobionic acid	Sodium diethyldithiocarbamate Copper supplementation	Metastatic ovarian cancer	Proteasome inhibition	Increased uptake due to the targeting of b-D-galactose receptors by lactobionic acid. Strong anticancer efficacy on a tumor-bearing mice model of metastatic cancer.	[117]
	PEGylation	DSF	Breast cancer	Apoptosis	Increase in plasma concentration of DSF up to 24 h (vs. 1 h for free DSF). Reduction of approximately 43% in tumor growth in 24 days. Enhanced synergistic activity in delaying tumor growth with PEGylated copper oleate liposomes (combinatory treatment, approximately 50% at 14 days).	[118,119]
	PEGylation Folate	DSF	Breast cancer	ROS production and apoptosis induction	Intravenous administration decreases tumor growth rate while prolonging overall survival in a murine model.	[120]
	PEGylation Folate	DSF	Breast cancer	Inhibition of caspase-3 activity, cell cycle arrest in G0/G1 and S phases, induction of apoptosis	The impairment in tumor growth proved to be dependent on the route of administration (intravenous > intratumoral > intraperitoneal = no treatment).	[121]
	–	DSF	Hepatocellular carcinoma	Morphological changes in cell nuclei with DNA fragmentation, G0/G1 cell cycle arrest, apoptosis activation	Reduced cell proliferation with superior stability.	[122]
	PEGylation	DSF Doxorubicin	Breast cancer	–	Synergistic cytotoxicity effect between both drugs. Improved intracellular accumulation in tumors, with consequent reduction in tumor growth rate.	[123]

–	DSF	Hepatocellular carcinoma	–	Polysorbate 80 showed a better ability to stabilize, surface modify, and control DSF release than Pluronic 188. Less pronounced toxicity over HEP3bB cells, which was correlated with the controlled release of DSF.	[124]
–	DSF	–	–	DSF and polymer (PLGA) interactions were found to be strongly influenced by the surfactant. DSF increased rigidity and crystallinity of the nanoparticles.	[125]
–	DSF Copper supplementation	Liver cancer cell–like stem cells	Induction of apoptosis	Marked synergistic cytotoxicity between DSF and 5-fluorouracil or sorafenib. Eradication of migration, invasion, and sphere-forming and clonogenic abilities. Impairment in tumor growth and metastasis development in animal model.	[126]
–	DSF	Non-small cell lung cancer	–	Strong cytotoxic effect (nanomolar range). Type of PLGA does not impact the characteristics of the nanoparticles (except for drug release). Production method (sonication time) strongly impacts the cytotoxicity of the nanosystem.	[127]
–	DSF	Hepatocellular carcinoma	Inhibition of the proteasome, metalloproteinase 9, and COX-2 with consequent apoptosis triggering	Significant antimetastatic and cytotoxic ability over cancer cells was due to DSF, while blank nanoparticles proved to have a neutral activity.	[128]
PEGylation	DSF	Glioblastoma	Production of ROS Reduction of mitochondrial membrane potential Activation of MAPK pathway Induction of apoptosis	Selective tumor accumulation for more than 24 h. Strong impairment in tumor growth due to the encapsulation and passive delivery of DSF.	[96]

Continued

TABLE 14.7 Drug delivery nanosystems with disulfiram or its copper complex with application in cancer.—cont'd

Nanostructure	Surface modification strategy	Drug(s)	Therapeutic indication	Mechanism of action	In vitro/in vivo studies	References
	PEGylation	DSF	Melanoma	–	Selective toxicity over melanoma cells (while inducing minimal death over normal fibroblasts and adipose derived stem cells).	[129]
	PEGylation	DSF	Hepatocellular carcinoma	–	Enhanced pharmacokinetic profile. Impairment in tumor growth (similar to the administration of 5-fluorouracil) while DSF solution showed negligible performance.	[130]
Nanococrystals	–	DSF Paclitaxel	Drug-resistant lung carcinoma	DSF modulates multidrug resistance gene-1 and inactivates P-glycoprotein G2/M cell cycle phase arrest Activation of apoptosis	Increase in cell uptake by cancer cells (14-fold) and tumor (7-fold). Reversion of paclitaxel resistance (due to DSF activity). Strong cytotoxic activity (when compared to paclitaxel nanocrystals). Effective reduction in tumor volume (12-fold) when compared to saline.	[131,132]
Albumin nanoparticles	Mannose	$Cu(DDC)_2$ Regorafenib	Drug-resistant colon cancer	Caspase-3 activation (apoptosis) Production of ROS Autophagy induction (increase in LC3 levels)	Reversion of paclitaxel and regorafenib resistance. High tumor accumulation and penetration. Remarkable impairment in tumor growth.	[133]
Silica nanoparticles	PEGylation pH-triggered release of copper	DSF Copper	Breast cancer	Production of ROS Induction of apoptosis	Intracellular release of copper and DSF in tumor (due to lower pH). Marked reduction in tumor growth rate (while saline and DSF alone proved to be ineffective).	[134]
Soy nanoparticles	–	DSF	Breast cancer	–	Improved cytotoxicity of the loaded nanosystem, as well as efficient time-dependent cell uptake by cancer cells.	[135]

Acknowledgments

The authors acknowledge Fundação para a Ciência e a Tecnologia (FCT), the Portuguese Agency for Scientific Research, for financial support through the Research Project no. 016648 (Ref. POCI-01-0145-FEDER-016648), the project PEst-UID/NEU/04539/2013, and COMPETE (Ref. POCI-01-0145-FEDER-007440). The Coimbra Chemistry Centre is supported by FCT, through the Project PEst-OE/QUI/UI0313/2014 and POCI-01-0145-FEDER-007630. João Basso and Maria Mendes acknowledge the Ph.D. research grants SFRH/BD/149138/2019 and SFRH/BD/133996/2017 assigned by FCT. RV also acknowledges FCT for the financial support of the projects IF/00286/2015, iBiMED (UID/BIM/04501/2019 and POCI-01-0145-FEDER-007628), and UnIC (UID/IC/00051/2019).

References

[1] Nowak-Sliwinska P, Scapozza L, I Altaba AR. Drug repurposing in oncology: compounds, pathways, phenotypes and computational approaches for colorectal cancer. Biochim Biophys Acta Rev Cancer 2019;1871(2):434–54.

[2] Polamreddy P, Gattu N. The drug repurposing landscape from 2012 to 2017: evolution, challenges, and possible solutions. Drug Discov Today 2019;24:789–95.

[3] Pushpakom S, Iorio F, Eyers PA, Escott KJ, Hopper S, Wells A, et al. Drug repurposing: progress, challenges and recommendations. Nat Rev Drug Discov 2019;18:41–58.

[4] Crockett SD, Schectman R, Stürmer T, Kappelman MD. Topiramate use does not reduce flares of inflammatory bowel disease. Dig Dis Sci 2014;59:1535–43.

[5] Perry CJ, Lawrence AJ. Hurdles in basic science translation. Front Pharmacol 2017;8:478.

[6] Hidalgo M, Amant F, Biankin AV, Budinská E, Byrne AT, Caldas C, et al. Patient-derived xenograft models: an emerging platform for translational cancer research. Cancer Discov 2014;4:998–1013.

[7] Zheng W, Sun W, Simeonov A. Drug repurposing screens and synergistic drug-combinations for infectious diseases. Br J Pharmacol 2018;175:181–91.

[8] Basso J, Miranda A, Sousa J, Pais A, Vitorino C. Repurposing drugs for glioblastoma: from bench to bedside. Cancer Lett 2018;428:173–83. https://doi.org/10.1016/j.canlet.2018.04.039.

[9] Langedijk J, Mantel-teeuwisse AK, Slijkerman DS. Drug repositioning and repurposing: terminology and definitions in literature. Drug Discov Today 2015;20. https://doi.org/10.1016/j.drudis.2015.05.001.

[10] Parsons CG. CNS repurposing-Potential new uses for old drugs: examples of screens for Alzheimer's disease, Parkinson's disease and spasticity. Neuropharmacology 2019;147:4–10.

[11] Shoaib M, Amjad Kamal M, Mohd Danish Rizvi S. Repurposed drugs as potential therapeutic candidates for the management of Alzheimer's disease. Curr Drug Metabol 2017;18:842–52.

[12] ClinicalTrials.gov. Clinical Trials – Celecoxib; 2019.

[13] Antoszczak M, Markowska A, Markowska J, Huczyński A. Old wine in new bottles: drug repurposing in oncology. Eur J Pharmacol 2020;866:172784.

[14] Shebl FM, Hsing AW, Park Y, Hollenbeck AR, Chu LW, Meyer TE, et al. Non-steroidal anti-inflammatory drugs use is associated with reduced risk of inflammation-associated cancers: NIH-AARP study. PLoS One 2014;9:e114633.

[15] Todoric J, Antonucci L, Karin M. Targeting inflammation in cancer prevention and therapy. Cancer Prev Res 2016;9:895–905.

[16] Wong RSY. Role of nonsteroidal anti-inflammatory drugs (NSAIDs) in cancer prevention and cancer promotion. Adv Pharmacol Sci 2019;2019.

[17] Mendes M, Sousa J, Pais A, Vitorino C. Targeted theranostic nanoparticles for brain tumor treatment. Pharmaceutics 2018;10:181.

[18] Fitzpatrick FA. Cyclooxygenase enzymes: regulation and function. Curr Pharmaceut Des 2004;10:577–88.

[19] Hla T, Bishop-Bailey D, Liu CH, Schaefers HJ, Trifan OC. Cyclooxygenase-1 and-2 isoenzymes. Int J Biochem Cell Biol 1999;31:551–7.

[20] Garavito RM, DeWitt DL. The cyclooxygenase isoforms: structural insights into the conversion of arachidonic acid to prostaglandins. Biochim Biophys Acta Mol Cell Biol Lipids 1999;1441:278–87.

[21] Kam PCA, See AU. Cyclo-oxygenase isoenzymes: physiological and pharmacological role. Anaesthesia 2000;55:442–9.

[22] Smyth EM, Grosser T, Wang M, Yu Y, Fitzgerald GA. Prostanoids in health and disease. 2009. p. 423–8. https://doi.org/10.1194/jlr.R800094-JLR200.

[23] Xu X-C. COX-2 inhibitors in cancer treatment and prevention, a recent development. Anticancer Drugs 2002;13:127–37. https://doi.org/10.1097/00001813-200202000-00003.

[24] Shin S. Safety of celecoxib versus traditional nonsteroidal anti-inflammatory drugs in older patients with arthritis. J Pain Res 2018;11:3211.

[25] Nissen SE, Yeomans ND, Solomon DH, Lüscher TF, Libby P, Husni ME, et al. Cardiovascular safety of celecoxib, naproxen, or ibuprofen for arthritis. N Engl J Med 2016;375:2519–29.

[26] PubChem. Characteristics of clecoxib. n.d.

[27] Domiati S, Ghoneim A. Celecoxib for the right person at the right dose and right time: an updated overview. Springer Sci Rev 2015;3:137–40.
[28] Celecoxib pathway, pharmacodynamics overview | PharmGKB. n.d.
[29] DrugBank. Celecoxib 2020. https://www.drugbank.ca/drugs/DB00482.
[30] Davies NM, McLachlan AJ, Day RO, Williams KM. Clinical pharmacokinetics and pharmacodynamics of celecoxib. Clin Pharmacokinet 2000;38:225–42.
[31] PubChem Celecoxib 2018. https://pubchem.ncbi.nlm.nih.gov/compound/2662.
[32] Shono T, Tofilon PJ, Bruner JM, Owolabi O, Lang FF. Cyclooxygenase-2 expression in human gliomas: prognostic significance and molecular correlations. Cancer Res 2001;61:4375–81.
[33] Uram Ł, Filipowicz A, Misiorek M, Pieńkowska N, Markowicz J, Wałajtys-Rode E, et al. Biotinylated PAMAM G3 dendrimer conjugated with celecoxib and/or Fmoc-l-Leucine and its cytotoxicity for normal and cancer human cell lines. Eur J Pharmaceut Sci 2018;124:1–9.
[34] Vera M, Barcia E, Negro S, Marcianes P, Garcia-Garcia L, Slowing K, et al. New celecoxib multiparticulate systems to improve glioblastoma treatment. Int J Pharm 2014;473:518–27.
[35] Sato A, Mizobuchi Y, Nakajima K, Shono K, Fujihara T, Kageji T, et al. Blocking COX-2 induces apoptosis and inhibits cell proliferation via the Akt/survivin-and Akt/ID3 pathway in low-grade-glioma. J Neuro Oncol 2017;132:231–8.
[36] Li Z, Chang C, Wang L, Zhang P, Shu HG. Cyclooxygenase-2 induction by amino acid deprivation requires p38 mitogen-activated protein kinase in human glioma cells. Cancer Invest 2017;0:1–11. https://doi.org/10.1080/07357907.2017.1292517.
[37] Oliver L, Olivier C, Vallette FM. Prostaglandin E 2 plays a major role in glioma resistance and progression. Transl Cancer Res 2016;5:1073–7. https://doi.org/10.21037/tcr.2016.11.20.
[38] Gong L, Thorn CF, Bertagnolli MM, Grosser T, Altman RB, Klein TE. Celecoxib pathways: pharmacokinetics and pharmacodynamics. Pharmacogenet Genom 2012;22:310.
[39] Vosooghi M, Amini M. The discovery and development of cyclooxygenase-2 inhibitors as potential anticancer therapies. Expet Opin Drug Discov 2014;9:255–67.
[40] Liu R, Xu K-P, Tan G-S. Cyclooxygenase-2 inhibitors in lung cancer treatment: bench to bed. Eur J Pharmacol 2015;769:127–33.
[41] Zhang Z, Chen F, Shang L. Advances in antitumor effects of NSAIDs. Cancer Manag Res 2018;10:4631.
[42] Zhang P, He D, Song E, Jiang M, Song Y. Celecoxib enhances the sensitivity of non-small-cell lung cancer cells to radiation-induced apoptosis through downregulation of the Akt/mTOR signaling pathway and COX-2 expression. PLoS One 2019;14.
[43] Chiang S-L, Velmurugan BK, Chung C-M, Lin S-H, Wang Z-H, Hua C-H, et al. Preventive effect of celecoxib use against cancer progression and occurrence of oral squamous cell carcinoma. Sci Rep 2017;7:6235.
[44] Sato A, Mizobuchi Y, Nakajima K, Shono K, Fujihara T, Kageji T. Blocking COX-2 induces apoptosis and inhibits cell proliferation via the Akt/survivin- and Akt/ID3 pathway in low-grade-glioma. J Neuro Oncol 2017;0. https://doi.org/10.1007/s11060-017-2380-5.
[45] ClinicalTrials.gov. Celecoxib – Clinical Trials; 2020.
[46] Bao Z, Zhou Y, Lei L, Zhang R, Song Q, Li X, et al. A facile strategy to generate high drug payload celecoxib micelles for enhanced corneal permeability. J Biomed Nanotechnol 2019;15:822–9.
[47] Crivelli B, Bari E, Perteghella S, Catenacci L, Sorrenti M, Mocchi M, et al. Silk fibroin nanoparticles for celecoxib and curcumin delivery: ROS-scavenging and anti-inflammatory activities in an in vitro model of osteoarthritis. Eur J Pharm Biopharm 2019;137:37–45.
[48] Jansook P, Kulsirachote P, Asasutjarit R, Loftsson T. Development of celecoxib eye drop solution and microsuspension: a comparative investigation of binary and ternary cyclodextrin complexes. Carbohydr Polym 2019;225:115209.
[49] Ju R-J, Zeng F, Liu L, Mu L-M, Xie H-J, Zhao Y, et al. Destruction of vasculogenic mimicry channels by targeting epirubicin plus celecoxib liposomes in treatment of brain glioma. Int J Nanomed 2016;11:1131.
[50] Ju R-J, Li X-T, Shi J-F, Li X-Y, Sun M-G, Zeng F, et al. Liposomes, modified with PTDHIV-1 peptide, containing epirubicin and celecoxib, to target vasculogenic mimicry channels in invasive breast cancer. Biomaterials 2014;35:7610–21.
[51] Dave V, Gupta A, Singh P, Gupta C, Sadhu V, Reddy KR. Synthesis and characterization of celecoxib loaded PEGylated liposome nanoparticles for biomedical applications. Nano-Struct Nano-Objects 2019;18:100288.
[52] Tian J, Guo F, Chen Y, Li Y, Yu B, Li Y. Nanoliposomal formulation encapsulating celecoxib and genistein inhibiting COX-2 pathway and Glut-1 receptors to prevent prostate cancer cell proliferation. Cancer Lett 2019;448:1–10.
[53] Gowda R, Kardos G, Sharma A, Singh S, Robertson GP. Nanoparticle-based celecoxib and plumbagin for the synergistic treatment of melanoma. Mol Cancer Ther 2017;16:440–52.

[54] Elzoghby AO, Mostafa SK, Helmy MW, Eldemellawy MA. Multi-reservoir phospholipid shell encapsulating protamine nanocapsules for Co-delivery of Letrozole and Celecoxib in breast cancer therapy. Pharm Res 2017. https://doi.org/10.1007/s11095-017-2207-2.

[55] Sun Y, Li X, Zhang L, Liu X, Jiang B, Long Z, et al. Cell permeable NBD peptide-modified liposomes by hyaluronic acid coating for the synergistic targeted therapy of metastatic inflammatory breast cancer. Mol Pharm 2019;16:1140–55.

[56] Üner M, Yener G, Ergüven M. Design of colloidal drug carriers of celecoxib for use in treatment of breast cancer and leukemia. Mater Sci Eng C 2019:109874.

[57] Patel AR, Chougule MB, Townley I, Patlolla R, Wang G, Singh M. Efficacy of aerosolized celecoxib encapsulated nanostructured lipid carrier in non-small cell lung cancer in combination with docetaxel. Pharm Res (N Y) 2013;30:1435–46.

[58] Patlolla RR, Chougule M, Patel AR, Jackson T, Tata PNV, Singh M. Formulation, characterization and pulmonary deposition of nebulized celecoxib encapsulated nanostructured lipid carriers. J Contr Release 2010;144:233–41.

[59] Zhang B, Jin K, Jiang T, Wang L, Shen S, Luo Z, et al. Celecoxib normalizes the tumor microenvironment and enhances small nanotherapeutics delivery to A549 tumors in nude mice. Sci Rep 2017;7:10071.

[60] Zhang S, Guo N, Wan G, Zhang T, Li C, Wang Y, et al. pH and redox dual-responsive nanoparticles based on disulfide-containing poly (β-amino ester) for combining chemotherapy and COX-2 inhibitor to overcome drug resistance in breast cancer. J Nanobiotechnol 2019;17:1–17.

[61] Said-Elbahr R, Nasr M, Alhnan MA, Taha I, Sammour O. Nebulizable colloidal nanoparticles co-encapsulating a COX-2 inhibitor and a herbal compound for treatment of lung cancer. Eur J Pharm Biopharm 2016;103:1–12.

[62] Zeng S-Q, Chen Y-Z, Chen Y, Liu H. Lipid–polymer hybrid nanoparticles for synergistic drug delivery to overcome cancer drug resistance. New J Chem 2017; 41:1518–25.

[63] Wu J-L, He X-Y, Liu B-Y, Gong M-Q, Zhuo R-X, Cheng S-X. Fusion peptide functionalized hybrid nanoparticles for synergistic drug delivery to reverse cancer drug resistance. J Mater Chem B 2017;5: 4697–704.

[64] Kozlu S, Sahin A, Ultav G, Yerlikaya F, Calis S, Capan Y. Development and in vitro evaluation of doxorubicin and celecoxib co-loaded bone targeted nanoparticles. J Drug Deliv Sci Technol 2018;45: 213–9.

[65] Huang J, Xu Y, Xiao H, Xiao Z, Guo Y, Cheng D, et al. Core-Shell-distinct nanodrug showing on-demand sequential drug release to act on multiple cell types for synergistic anticancer therapy. ACS Nano 2019; 13(6):7036–49.

[66] Yu R-Y, Xing L, Cui P-F, Qiao J-B, He Y-J, Chang X, et al. Regulating the Golgi apparatus by co-delivery of a COX-2 inhibitor and Brefeldin A for suppression of tumor metastasis. Biomater Sci 2018;6:2144–55.

[67] Shi L, Xu L, Wu C, Xue B, Jin X, Yang J, et al. Celecoxib-induced self-assembly of smart albumin-doxorubicin conjugate for enhanced cancer therapy. ACS Appl Mater Interfaces 2018;10:8555–65.

[68] Liu J, Chang B, Li Q, Xu L, Liu X, Wang G, et al. Redox-Responsive dual drug delivery nanosystem suppresses cancer repopulation by abrogating doxorubicin-promoted cancer stemness, metastasis, and drug resistance. Adv Sci 2019;6:1801987.

[69] Wu C, Gong M-Q, Liu B-Y, Zhuo R-X, Cheng S-X. Co-delivery of multiple drug resistance inhibitors by polymer/inorganic hybrid nanoparticles to effectively reverse cancer drug resistance. Colloids Surf B Biointerfaces 2017;149:250–9.

[70] Li W, Liu D, Zhang H, Correia A, Mäkilä E, Salonen J, et al. Microfluidic assembly of a nano-in-micro dual drug delivery platform composed of halloysite nanotubes and a pH-responsive polymer for colon cancer therapy. Acta Biomater 2017;48:238–46.

[71] AbdElhamid AS, Helmy MW, Ebrahim SM, Bahey-El-Din M, Zayed DG, Zein El Dein EA, et al. Layer-by-layer gelatin/chondroitin quantum dots-based nanotheranostics: combined rapamycin/celecoxib delivery and cancer imaging. Nanomedicine 2018;13.

[72] AbdElhamid AS, Zayed DG, Helmy MW, Ebrahim SM, Bahey-El-Din M, Zein-El-Dein EA, et al. Lactoferrin-tagged quantum dots-based theranostic nanocapsules for combined COX-2 inhibitor/herbal therapy of breast cancer. Nanomedicine 2018; 13:2637–56.

[73] Brunton LL, Hilal-Dandan R, Knollmann BC. Chapter 23: ethanol. Goodman Gilman's pharmacol. Basis ther.. McGraw-Hill Education; 2018.

[74] Petersen EN. The pharmacology and toxicology of disulfiram and its metabolites. Acta Psychiatr Scand 1992;86:7–13.

[75] Johansson B. A review of the pharmacokinetics and pharmacodynamics of disulfiram and its metabolites. Acta Psychiatr Scand 1992;86:15–26.

[76] National Center for Biotechnology Information. PubChem Database. Disulfiram, CID=3117. n.d. https://pubchem.ncbi.nlm.nih.gov/compound/Disulfiram [Accessed 3 January 2020].

[77] Gupte A, Mumper RJ. Elevated copper and oxidative stress in cancer cells as a target for cancer treatment. Cancer Treat Rev 2009;35:32–46.

[78] Fabris C, Farini R, Del Favero G, Gurrieri G, Piccoli A, Sturniolo GC, et al. Copper, zinc and copper/zinc ratio in chronic pancreatitis and pancreatic cancer. Clin Biochem 1985;18:373–5.

[79] Cvek B, Milacic V, Taraba J, Dou QP. Ni(II), Cu(II), and Zn(II) diethyldithiocarbamate complexes show various activities against the proteasome in breast cancer cells. J Med Chem 2008;51:6256–8. https://doi.org/10.1021/jm8007807.

[80] Viola-Rhenals M, Patel KR, Jaimes-Santamaria L, Wu G, Liu J, Dou QP. Recent advances in Antabuse (disulfiram): the importance of its metal-binding ability to its anticancer activity. Curr Med Chem 2018;25:506–24.

[81] Yip NC, Fombon IS, Liu P, Brown S, Kannappan V, Armesilla AL, et al. Disulfiram modulated ROS–MAPK and NFκB pathways and targeted breast cancer cells with cancer stem cell-like properties. Br J Cancer 2011;104:1564.

[82] Lövborg H, Öberg F, Rickardson L, Gullbo J, Nygren P, Larsson R. Inhibition of proteasome activity, nuclear factor-KB translocation and cell survival by the antialcoholism drug disulfiram. Int J Cancer 2006;118:1577–80.

[83] Chen D, Cui QC, Yang H, Dou QP. Disulfiram, a clinically used anti-alcoholism drug and copper-binding agent, induces apoptotic cell death in breast cancer cultures and xenografts via inhibition of the proteasome activity. Cancer Res 2006;66:10425–33.

[84] Skrott Z, Mistrik M, Andersen KK, Friis S, Majera D, Gursky J, et al. Alcohol-abuse drug disulfiram targets cancer via p97 segregase adaptor NPL4. Nature 2017;552:194.

[85] Xu L, Xu J, Zhu J, Yao Z, Yu N, Deng W, et al. Universal anticancer Cu (DTC) 2 discriminates between thiols and zinc (II) thiolates oxidatively. Angew Chem 2019;131:6131–4.

[86] Takahashi C, Sheng Z, Horan TP, Kitayama H, Maki M, Hitomi K, et al. Regulation of matrix metalloproteinase-9 and inhibition of tumor invasion by the membrane-anchored glycoprotein RECK. Proc Natl Acad Sci U S A 1998;95:13221–6.

[87] Noda M, Oh J, Takahashi R, Kondo S, Kitayama H, Takahashi C. RECK: a novel suppressor of malignancy linking oncogenic signaling to extracellular matrix remodeling. Cancer Metastasis Rev 2003;22:167–75.

[88] Murai R, Yoshida Y, Muraguchi T, Nishimoto E, Morioka Y, Kitayama H, et al. A novel screen using the Reck tumor suppressor gene promoter detects both conventional and metastasis-suppressing anticancer drugs. Oncotarget 2010;1:252.

[89] Chen W, Yang W, Chen P, Huang Y, Li F. Disulfiram copper nanoparticles prepared with a stabilized metal ion ligand complex method for treating drug-resistant prostate cancers. ACS Appl Mater Interfaces 2018;10:41118–28.

[90] Liu P, Kumar IS, Brown S, Kannappan V, Tawari PE, Tang JZ, et al. Disulfiram targets cancer stem-like cells and reverses resistance and cross-resistance in acquired paclitaxel-resistant triple-negative breast cancer cells. Br J Cancer 2013;109:1876.

[91] Liu P, Brown S, Goktug T, Channathodiyil P, Kannappan V, Hugnot JP, et al. Cytotoxic effect of disulfiram/copper on human glioblastoma cell lines and ALDH-positive cancer-stem-like cells. Br J Cancer 2012;107:1488–97. https://doi.org/10.1038/bjc.2012.442.

[92] Cvek B. Targeting malignancies with disulfiram (Antabuse): multidrug resistance, angiogenesis, and proteasome. Curr Cancer Drug Targets 2011;11:332–7.

[93] Loo TW, Bartlett MC, Clarke DM. Disulfiram metabolites permanently inactivate the human multidrug resistance P-glycoprotein. Mol Pharm 2004;1:426–33.

[94] Sauna ZE, Peng X-H, Nandigama K, Tekle S, Ambudkar SV. The molecular basis of the action of disulfiram as a modulator of the multidrug resistance-linked ATP binding cassette transporters MDR1 (ABCB1) and MRP1 (ABCC1). Mol Pharmacol 2004;65:675–84.

[95] Lin J, Haffner MC, Zhang Y, Lee BH, Brennen WN, Britton J, et al. Disulfiram is a DNA demethylating agent and inhibits prostate cancer cell growth. Prostate 2011;71:333–43.

[96] Rao Madala SRP H, Ali-Osman F, Zhang R, Srivenugopal KS. Brain-and brain tumor-penetrating disulfiram nanoparticles: sequence of cytotoxic events and efficacy in human glioma cell lines and intracranial xenografts. Oncotarget 2018;9:3459.

[97] Paranjpe A, Zhang R, Ali-Osman F, Bobustuc GC, Srivenugopal KS. Disulfiram is a direct and potent inhibitor of human O6-methylguanine-DNA methyltransferase (MGMT) in brain tumor cells and mouse brain and markedly increases the alkylating DNA damage. Carcinogenesis 2014;35:692–702. https://doi.org/10.1093/carcin/bgt366.

[98] Yakisich JS, Å S, Eneroth P, Cruz M. Disulfiram is a potent in vitro inhibitor of DNA topoisomerases. Biochem Biophys Res Commun 2001;289:586–90.

[99] Tesson M, Anselmi G, Bell C, Mairs R. Cell cycle specific radiosensitisation by the disulfiram and copper complex. Oncotarget 2017;8:65900.

[100] Marikovsky M, Nevo N, Vadai E, Harris-Cerruti C. Cu/Zn superoxide dismutase plays a role in angiogenesis. Int J Cancer 2002;97:34–41.

[101] Li Y, Fu S-Y, Wang L-H, Wang F-Y, Wang N-N, Cao Q, et al. Copper improves the anti-angiogenic activity of disulfiram through the EGFR/Src/VEGF pathway in gliomas. Cancer Lett 2015;369:86–96.

[102] Goto K, Arai J, Stephanou A, Kato N. Novel therapeutic features of disulfiram against hepatocellular carcinoma cells with inhibitory effects on a disintegrin and metalloproteinase 10. Oncotarget 2018;9:18821.

[103] Marengo A, Forciniti S, Dando I, Dalla Pozza E, Stella B, Tsapis N, et al. Pancreatic cancer stem cell proliferation is strongly inhibited by diethyldithiocarbamate-copper complex loaded into hyaluronic acid decorated liposomes. Biochim Biophys Acta Gen Subj 2019;1863:61–72.

[104] Liu P, Wang Z, Brown S, Kannappan V, Tawari PE, Jiang W, et al. Liposome encapsulated Disulfiram inhibits NFκB pathway and targets breast cancer stem cells in vitro and in vivo. Oncotarget 2014;5:7471.

[105] Wehbe M, Anantha M, Shi M, Leung AW, Dragowska WH, Sanche L, et al. Development and optimization of an injectable formulation of copper diethyldithiocarbamate, an active anticancer agent. Int J Nanomed 2017;12:4129.

[106] Najlah M, Said Suliman A, Tolaymat I, Kurusamy S, Kannappan V, Elhissi A, et al. Development of injectable PEGylated liposome encapsulating disulfiram for colorectal cancer treatment. Pharmaceutics 2019;11:610.

[107] Banerjee P, Geng T, Mahanty A, Li T, Zong L, Wang B. Integrating the drug, disulfiram into the vitamin E-TPGS-modified PEGylated nanostructured lipid carriers to synergize its repurposing for anti-cancer therapy of solid tumors. Int J Pharm 2019;557:374–89.

[108] Liu X, Chu H, Cui N, Wang T, Dong S, Cui S, et al. In vitro and in vivo evaluation of biotin-mediated PEGylated nanostructured lipid as carrier of disulfiram coupled with copper ion. J Drug Deliv Sci Technol 2019;51:651–61.

[109] Chen X, Zhang L, Hu X, Lin X, Zhang Y, Tang X. Formulation and preparation of a stable intravenous disulfiram-loaded lipid emulsion. Eur J Lipid Sci Technol 2015;117:869–78.

[110] Li H, Liu B, Ao H, Fu J, Wang Y, Feng Y, et al. Soybean lecithin stabilizes disulfiram nanosuspensions with a high drug-loading content: remarkably improved antitumor efficacy. J Nanobiotechnol 2020;18:1–11.

[111] Zhang L, Tian B, Li Y, Lei T, Meng J, Yang L, et al. A copper-mediated disulfiram-loaded pH-triggered PEG-shedding TAT peptide-modified lipid nanocapsules for use in tumor therapy. ACS Appl Mater Interfaces 2015;7:25147–61.

[112] Duan X, Xiao J, Yin Q, Zhang Z, Yu H, Mao S, et al. Smart pH-sensitive and temporal-controlled polymeric micelles for effective combination therapy of doxorubicin and disulfiram. ACS Nano 2013;7:5858–69.

[113] Huo Q, Zhu J, Niu Y, Shi H, Gong Y, Li Y, et al. pH-triggered surface charge-switchable polymer micelles for the co-delivery of paclitaxel/disulfiram and overcoming multidrug resistance in cancer. Int J Nanomed 2017;12:8631.

[114] Duan X, Xiao J, Yin Q, Zhang Z, Yu H, Mao S, et al. Multi-targeted inhibition of tumor growth and lung metastasis by redox-sensitive shell crosslinked micelles loading disulfiram. Nanotechnology 2014;25:125102.

[115] Miao L, Su J, Zhuo X, Luo L, Kong Y, Gou J, et al. mPEG5k-b-PLGA2k/PCL3. 4k/MCT mixed micelles as carriers of disulfiram for improving plasma stability and antitumor effect in vivo. Mol Pharm 2018;15:1556–64.

[116] Peng X, Pan Q, Zhang B, Wan S, Li S, Luo K, et al. Highly stable, coordinated polymeric nanoparticles loading copper (II) diethyldithiocarbamate for combinational chemo/chemodynamic therapy of cancer. Biomacromolecules 2019;20(6):2372–83.

[117] He H, Markoutsa E, Li J, Xu P. Repurposing disulfiram for cancer therapy via targeted nanotechnology through enhanced tumor mass penetration and disassembly. Acta Biomater 2018;68:113–24.

[118] Song W, Tang Z, Lei T, Wen X, Wang G, Zhang D, et al. Stable loading and delivery of disulfiram with mPEG-PLGA/PCL mixed nanoparticles for tumor therapy. Nanomed Nanotechnol Biol Med 2016;12:377–86.

[119] Zhou L, Yang L, Yang C, Liu Y, Chen Q, Pan W, et al. Membrane loaded copper oleate pegylated liposome combined with disulfiram for improving synergistic antitumor effect in vivo. Pharm Res (N Y) 2018;35:147.

[120] Fasehee H, Dinarvand R, Ghavamzadeh A, Esfandyari-Manesh M, Moradian H, Faghihi S, et al. Delivery of disulfiram into breast cancer cells using folate-receptor-targeted PLGA-PEG nanoparticles: in vitro and in vivo investigations. J Nanobiotechnol 2016;14:32.

[121] Fasehee H, Zarrinrad G, Tavangar SM, Ghaffari SH, Faghihi S. The inhibitory effect of disulfiram encapsulated PLGA NPs on tumor growth: different administration routes. Mater Sci Eng C 2016;63:587–95.

[122] Hoda M, Pajaniradje S, Shakya G, Mohankumar K, Rajagopalan R. Anti-proliferative and apoptosis-triggering potential of disulfiram and disulfiram-loaded polysorbate 80-stabilized PLGA nanoparticles on hepatocellular carcinoma Hep3B cell line. Nanomed Nanotechnol Biol Med 2016;12:1641–50.

[123] Tao X, Gou J, Zhang Q, Tan X, Ren T, Yao Q, et al. Synergistic breast tumor cell killing achieved by intracellular co-delivery of doxorubicin and disulfiram via core–shell–corona nanoparticles. Biomater Sci 2018; 6:1869–81.

[124] Hoda M, Sufi SA, Shakya G, Kumar KM, Rajagopalan R. Influence of stabilizers on the production of disulfiram-loaded poly (lactic-co-glycolic acid) nanoparticles and their anticancer potential. Ther Deliv 2015;6:17–25.

[125] Hoda M, Sufi SA, Cavuturu B, Rajagopalan R. Stabilizers influence drug–polymer interactions and physicochemical properties of disulfiram-loaded poly-lactide-co-glycolide nanoparticles. Futur Sci OA 2017;4:FSO263.

[126] Wang Z, Tan J, McConville C, Kannappan V, Tawari PE, Brown J, et al. Poly lactic-co-glycolic acid controlled delivery of disulfiram to target liver cancer stem-like cells. Nanomed Nanotechnol Biol Med 2017; 13:641–57.

[127] Najlah M, Ahmed Z, Iqbal M, Wang Z, Tawari P, Wang W, et al. Development and characterisation of disulfiram-loaded PLGA nanoparticles for the treatment of non-small cell lung cancer. Eur J Pharm Biopharm 2017;112:224–33.

[128] Hoda M, Cavuturu BM, Iqbal S, Shakya G, Rajagopalan R. Disulfiram and disulfiram-loaded poly-[lactide-co-glycolic acid] nanoparticles modulate metastatic markers and proteasomal activity in hepatocarcinoma Hep3b cell line. Eur J Nanomed 2017;9: 127–38.

[129] Zhou D, Gao Y, Xu Q, Meng Z, Greiser U, Wang W. Anticancer drug disulfiram for in situ RAFT polymerization: controlled polymerization, multifacet self-assembly, and efficient drug delivery. ACS Macro Lett 2016;5:1266–72.

[130] Zhuo X, Lei T, Miao L, Chu W, Li X, Luo L, et al. Disulfiram-loaded mixed nanoparticles with high drug-loading and plasma stability by reducing the core crystallinity for intravenous delivery. J Colloid Interface Sci 2018;529:34–43.

[131] Mohammad IS, He W, Yin L. A smart paclitaxel-disulfiram nanococrystals for efficient MDR reversal and enhanced apoptosis. Pharm Res (N Y) 2018;35:77.

[132] Mohammad IS, Teng C, Chaurasiya B, Yin L, Wu C, He W. Drug-delivering-drug approach-based codelivery of paclitaxel and disulfiram for treating multidrug-resistant cancer. Int J Pharm 2019;557: 304–13.

[133] Zhao P, Yin W, Wu A, Tang Y, Wang J, Pan Z, et al. Dual-targeting to cancer cells and M2 macrophages via biomimetic delivery of mannosylated albumin nanoparticles for drug-resistant cancer therapy. Adv Funct Mater 2017;27:1700403.

[134] Wu W, Yu L, Jiang Q, Huo M, Lin H, Wang L, et al. Enhanced tumor-specific disulfiram chemotherapy by in situ Cu^{2+} chelation-initiated nontoxicity-to-toxicity transition. J Am Chem Soc 2019;141: 11531–9.

[135] Farooq MA, Li L, Parveen A, Wang B. Globular protein stabilized nanoparticles for delivery of disulfiram: fabrication, characterization, in vitro toxicity, and cellular uptake. RSC Adv 2020;10: 133–44.

[136] Ito Y, Cai H, Koizumi Y, HORI R, TERAO M, KIMURA T, et al. Effects of lipid composition on the transcorneal penetration of liposomes containing disulfiram, a potential anti-cataract agent, in the rabbit. Biol Pharm Bull 2000;23:327–33.

[137] Liu C, Lan Q, He W, Nie C, Zhang C, Xu T, et al. Octa-arginine modified lipid emulsions as a potential ocular delivery system for disulfiram: a study of the corneal permeation, transcorneal mechanism and anti-cataract effect. Colloids Surf B Biointerfaces 2017;160:305–14.

[138] Buckiová D, Ranjan S, Newman TA, Johnston AH, Sood R, Kinnunen PKJ, et al. Minimally invasive drug delivery to the cochlea through application of nanoparticles to the round window membrane. Nanomedicine 2012;7:1339–54.

[139] Nagai N, Yoshioka C, Mano Y, Tnabe W, Ito Y, Okamoto N, et al. A nanoparticle formulation of disulfiram prolongs corneal residence time of the drug and reduces intraocular pressure. Exp Eye Res 2015;132:115–23.

[140] Wang S, Li D, Ito Y, Nabekura T, Wang S, Zhang J, et al. Bioavailability and anticataract effects of a topical ocular drug delivery system containing disulfiram and hydroxypropyl-beta-cyclodextrin on selenite-treated rats. Curr Eye Res 2004;29:51–8.

[141] Wang S, Li D, Ito Y, Liu X, Zhang J, Wu C. An ocular drug delivery system containing zinc diethyldithiocarbamate and HPβCD inclusion complex-corneal permeability, anti-cataract effects and mechanism studies. J Pharm Pharmacol 2004;56:1251–7.

[142] Nagai N, Takeda M, Ito Y, Takeuchi N, Kamei A. Delay in ICR/f rat lens opacification by the instillation of eye drops containing disulfiram and hydroxypropyl-β-cyclodextrin inclusion complex. Biol Pharm Bull 2007;30:1529–34.

[143] Wang S, Jang T, Wang Z. Permeability and anticataract effects of a topical ocular drug delivery system of disulfiram. Int Conf Biomed Eng Informatics 2008;1: 596–600. IEEE.

[144] Ito Y, Nagai N, Shimomura Y. Reduction in intraocular pressure by the instillation of eye drops containing disulfiram included with 2-hydroxypropyl-β-cyclodextrin in rabbit. Biol Pharm Bull 2010;33:1574–8.

[145] Kanai K, Ito Y, Nagai N, Itoh N, Hori Y, Chikazawa S, et al. Effects of instillation of eyedrops containing disulfiram and hydroxypropyl-β-cyclodextrin inclusion complex on endotoxin-induced uveitis in rats. Curr Eye Res 2012;37:124–31.

[146] Nagai N, Mano Y, Ito Y. An ophthalmic formulation of disulfiram nanoparticles prolongs drug residence time in lens. Biol Pharm Bull 2016;39:1881–7.

[147] Müller BW, Brauns U. Solubilization of drugs by modified β-cyclodextrins. Int J Pharm 1985;26:77–88.

[148] Kristinsson JK, Fridriksdóttir H, Thorisdottir S, Sigurdardottir AM, Stefansson E, Loftsson T. Dexamethasone-cyclodextrin-polymer co-complexes in aqueous eye drops. Aqueous humor pharmacokinetics in humans. Invest Ophthalmol Vis Sci 1996;37: 1199–203.

CHAPTER

15

Clinical trials on combination of repurposed drugs and anticancer therapies

Süreyya Ölgen

Faculty of Pharmacy, Biruni University, Istanbul, Zeytinburnu, Turkey

OUTLINE

This book section provides information about the use of existing repurposed drugs in cancer treatment and their clinical applications.

Developing the same drug which has previously been designed for a certain disorder, to serve a new therapeutic application for the treatment of a different disorder with a different pharmacological activity is called drug repositioning or drug repurposing [1]. It is the act of taking a drug intended to treat one patient population and demonstrating its efficacy in the treatment of completely different group of

Drug Repurposing in Cancer Therapy
https://doi.org/10.1016/B978-0-12-819668-7.00015-4

patient. Drug repurposing can be considered as recycling in its most basic level.

Drug repurposing studies are relatively new and can be promising for rapid clinical impact at a lower cost than de novo drug development. Drug repurposing studies started in the early 1990s, but the most successes have been established in last decades. According to the search of SciFinder database (was seen in November 26, 2019), 2141 references are found related with drug repurposing, and among them, 244 are directly related with drug repurposed for anticancer activity. Serendipity was the starting point of drug repurposing that plays important role in identification of new uses of old drugs in clinic. The fact that many drugs have been prescribed without approval in clinical use has played an important role in these inventions.

The way of drug repurposing studies

Drug repositioning opportunities evolve from observations, discussions, and other collaborations, including the purposeful development of platforms for drug identification, which identify the potential targets and allow the accessing of compounds. Drug repurposing candidates can be obtained from drugs in clinical development that have failed to demonstrate efficacy for a particular indication during Phase II or III trials but have no major safety concerns and drugs that have been discontinued for commercial reasons and drugs patents are close to expiry in market, and from nonreleased drug candidates. Three main approaches based on computational, biological experimental, and mixed are generally used for drug repurposing studies. Several strategies involved the drug repurposing studies that include various in silico methods, genomic, high-throughput screening technologies, and literature mining. These screening methods also provide opportunities for the exploitation of the open-source model (Table 15.1). DrugBank, the Potential Drug Target Database, Therapeutic Target Database, and Super Target represents the open sources that provide targets and drugs, including protein and active-site structures, association with related diseases, biological functions, and signaling pathways. Compound-specific databases are provided by PubChem, ChEMBL, and ChemSpider, the US FDA's Electronic Orange Book's Discontinued Drug Products List, and ID (Investigational Drug) Map [2].

Several open-source models permit sharing the data, resources, compounds, clinical molecules, and small libraries, as well as screening platforms in the search for new indications of old drugs or failed candidates. As a result of successful findings of drug repurposing, new data banks have been obtained to the creation of open sources in this field. An online Drug Repurposing Hub (www.broadinstitute.org/repurposing) is created, which contains the

TABLE 15.1 Open-source models for drug repurposing studies.

Compound-specific databases	Data obtained from
• Pubchem • ChEMBL • ChemSpider • ID (Investigational Drug) Map • US FDA's Electronic Orange Book's Discontinued Drug Products List • On-line Drug Repurposing Hub (www.broadinstitute.org/repurposing)	• Resources • Compounds • Molecules in clinical studies • Small libraries • Screening platforms in the search for new indications of old drugs or failed candidates

detailed annotation for each compounds and is designed to rapidly identify drugs for evaluation in disease models [3].

Conventional drug discovery and development process generally takes 10–15 years that includes five stages: discovery and preclinical, safety review, clinical research, FDA review, and FDA postmarket safety monitoring [4]. Drug repurposing process takes 9–10 years, and it has four stage as compound identification, compound acquisition, development, and FDA postmarket safety monitoring (Fig. 15.1).

Approaches of drug repurposing

Several rational approaches to identification of drug repurposing candidates are used. Drug repurposing approaches include various relationships such as drug–disease, drug–drug, or drug–target relationships (Fig. 15.2). Experimental data related to disease (e.g., omics data collected from patients) or knowledge on how drugs modulate phenotypes related to disease (e.g., known from their side effects) are utilized in disease focused approaches. Drug focus is based on the single drug that interacts with multiple targets and related with approved molecules for particular indication that can help to identify active compounds that were originally developed for different indications. When primary and/or secondary targets of compounds are known and often involved in several biological processes that targets relevant to one disease or biological process, it is used to find new indications [5].

Identification of interaction with multiple targets related with several diseases can arise serendipitously during screening studies and observation of some side effects [6]. Deep understanding of genomics and molecular pathways involved in human diseases and drug activity and using this information are important to finding new mechanisms, dosing levels, routes of administration, or innovative targets [7].

Source for drug repurposing

Mechanism action of disease, clinical development for new indication, and the original indication can be carried out simultaneously to identify drug for repurposing aim [8].

At the beginning, generic drugs were preferred as sources of repositioning target. These drugs are safer, easy, and cheap to obtain for clinical trials because their original patents have been expired. Another source of drug repurposing studies are off-label uses of drugs by physicians for the treatment of diseases. FDA-approved drugs can be repurposed for their potential new therapeutic applications. Failed drugs in clinical studies is the another source for drug repurposing studies that are

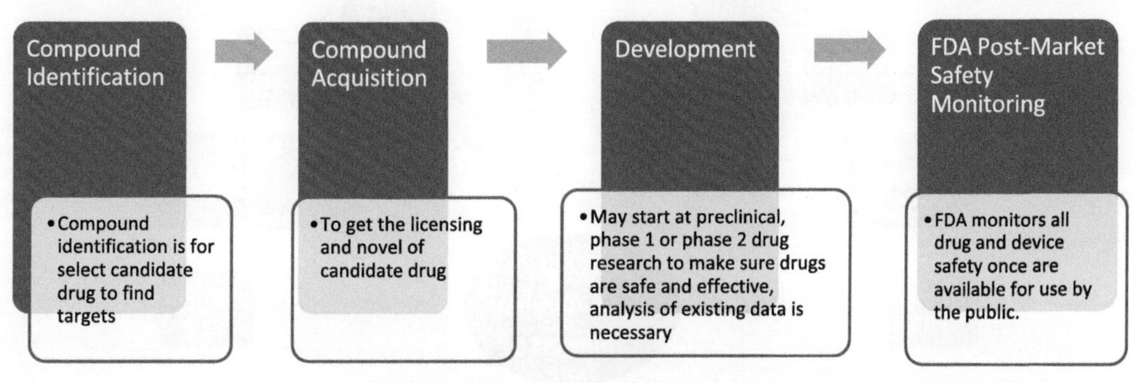

FIGURE 15.1 Drug repurposing process.

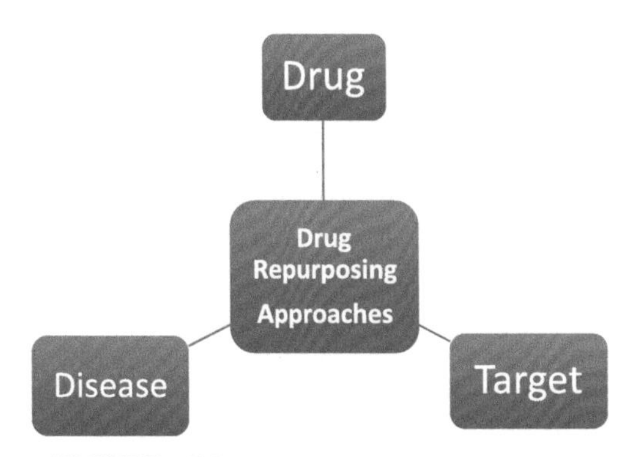

FIGURE 15.2 Drug repurposing approaches.

safe in humans and can be used for some other activities. All publicly available data have been constituted from the information of a social network that allows to obtain all relationships between drugs, molecular pathways, genes, and other biological suitable entities [9]. The other sources of drug repurposing are shown in Fig. 15.3.

Advantages and disadvantages of drug repurposing

Drug repurposing discovery has some advantages that lowers time-consuming efforts and paper works for licensing and eliminates scientific barriers for proving its safety and efficacy (Table 15.2). Furthermore, when it is approved, their integration into healthcare will be very fast. Effective and ideal candidate for drug repurposing should not need further chemical optimization, and some part of clinical studies should have been done earlier. Their ADMET properties, safety, and clinical efficacy should also be appropriate, and they should be effective for the new indication in the same concentration act for the original disease. If reformulation is needed for new clinical use, safety and ADME studies should not be necessary [10]. Since the safety, efficacy, and toxicity of an existing drug have been extensively studied and passed all clinical tests in Phase I, Phase II, and Phase III, safety of repurposed drugs has already been confirmed. All data for repurposed drugs are already exist, this saves time and money that provides hope to patients whose treatment conditions are costly for conventional drug development. The most favorite side of drug repurposing is reducing the time for research and development. Moreover, the approval rate of these drugs on market takes approximately 30%, while this rate is ~10% for new drug applications. As a result, repurposed drugs can enter the pipeline at the efficacy stage, thus significantly decreasing the failure rate probability

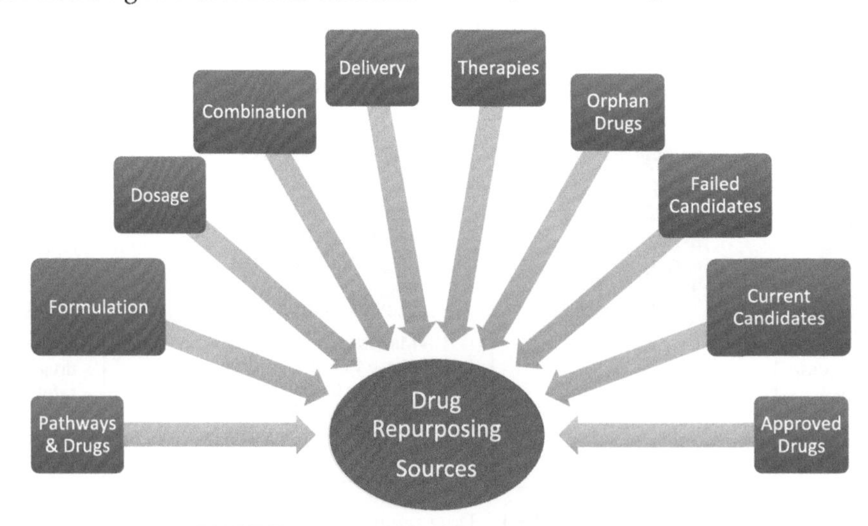

FIGURE 15.3 Potential drug repurposing sources.

TABLE 15.2 Advantages and disadvantages of drug repurposing studies.

Advantages	Disadvantages
• Lowers time-consuming efforts and paper works for licensing	• It still needs a drug development phase to fully analyze data from preclinical and clinical studies
• Eliminates scientific barriers for proving its safety and efficacy	• For the generic compounds, there is less option to define intellectual properties (IP)
• Integration into healthcare will be very fast	• Much higher level of evidence is required to repurpose of on-label compounds
• No need further chemical optimization	• On-label drug repurposing is needed longer and more costly clinical trials
• For new clinical use, safety and ADME studies should not be necessary	• On-label drug repurposing requires advance level of industry experience
• All data for repurposed drugs are already existing, this saves time and Money	• On-patent drug, current regulations for drug repurposing are very strict.
• Advantage to find new treatment option for orphan, rare, and neglected diseases	• On-patent drugs, the product is used in agreement with the approved product label is necessary
	• Off-patent or generic drugs, the new indication should be found, and it should have clinical benefits

and increasing the chances for a successful launch [4,11]. On the other hand, the chemistry, manufacturing process, quality profile, and pharmacokinetic and pharmacodynamic profile of repurposed drugs are already proved. Repurposing drugs have already a wealth of accessible data that provide a high level of safety for the pharmacokinetics, toxicities, bioavailability, dosing, and protocols of the agents. Repurposed drugs bring an advantage of a reduced failure rate as they have already been tested for safety [12]. As well as finding new treatment options for orphan, rare and neglected diseases, providing therapeutic efficacy which did not exist previously is also an advantage for drug repurposing studies.

Drug repurposing has some risks because it still needs a drug development phase. The biological activity, pharmacological parameters, and clinical observations need to be fully analyzed to limit failure in late-stage clinical trials or after marketing. For the generic compounds, there is less option to define IP, current formulations or delivery method may not be appropriate, and drug can be taken by patients outside of clinical studies. If on-label compounds are tried to be repurposed, it may require a much higher level of evidence, which could be even higher to secure reimbursement. It may also need longer and more costly clinical trials. Moreover, advanced level of industry experience is required. When off-label compounds are tried to be repurposed, efficacy may be lower, and monitoring the safety is difficult [13,14]. In various cases, a drug may show potent activity for a new indication, but if higher dosage is applied than first approved one, this results in possible adverse effects relating to toxicity. Patent exclusivity is another problem in limiting the potential of repurposing a drug. In the United States and the European Patent Convention, a patent protection time is determined as 20years. It is normal that a pharmaceutical company will apply for a patent prior to commencing clinical trials. When repurposing on-patent drug, current regulations for drug repurposing are very strict. This can be a deterrent for repurposing drugs for some research groups. On-patent drugs can be screened for new therapeutic use without the need of chemistry and manufacturing applications. However, approval of the original owners providing that the product is used in agreement with the approved product label is necessary. In the case of off-patent or generic drugs, the new indication should be found and it should have clinical benefits. These indications and new claim must not have been previously published. Another difficulty for research is the requirement of a new formulation to have market priority with the proposed new indication [6].

Advantages and disadvantages of drug repurposing studies are summarized in Table 15.2.

Anticancer drug repurposing

Repositioned anticancer drugs have a potential to be an excellent strategy for future anticancer drug development [15]. Recent modern anticancer drugs show better response in clinical studies, but the side effects and poor quality of life still obstacle and cause discontinuation of drugs, dose reduction, and emergence of drug resistance [16]. For instance; kinase inhibitors target oncogenes or oncoproteins that show high selectivity for tumor cells and prevent tumor progression and reduce side effects. However, most kinase inhibitors often have some pitfalls and lead to drug resistance [17].

The starting point for development of cancer therapeutics from noncancer drugs is that different diseases share with common molecular pathways and targets in the cell. This fact has been discovered through progress in genomics, proteomics, and informatics technologies. Advances of analytical tools allow researchers to screen large numbers of existing drugs against a particular disease target at the same time [18].

Drug repositioning has been applied in anticancer drug discovery due to the combination of great need for new anticancer drugs and the availability of a wide variety of cell- and target-based screening assays. With recent successful clinical introduction of noncancer drugs for cancer treatment, drug repositioning became a powerful choice to discover and develop novel anticancer drug candidates from the existing drug data. With the increasing knowledge about the pharmacological properties and identification of new targets, the chemical structure of the old drugs emerges as a great treasure for new cancer drug discovery. Although the several repurposed drugs are being on the market, the therapeutic potential of a large number of noncancer drugs is limited during their repositioning due to various problems including the lack of efficacy and intellectual property. Increasing knowledge about the pharmacological properties of old drugs may have further applications to utilizing them as a drug or using their scaffold for repurposing as cancer therapeutics. Since repurposing drugs may have potential applications with other promising biological targets, it is believed that the scaffold of these drugs will be beneficial for new drug design. The scaffold repurposing may be better tools to develop novel logical properties through the generation of novel chemotypes. The previous SARs and the known pharmacokinetics parameters of old drugs may provide to obtain more efficient drugs instead of a random chemical classification. The scaffold can be optimized by rational design to eliminate the known unwanted properties for increasing the success rate of further development [19].

By using high-throughput technologies, a large number of genetic and genomic data were obtained by using next-generation sequencing (NGS) and were generated a large volume of genetic and genomic data from several national and international cancer genome projects. These international or national cancer genome projects have identified many cancer-driven genes. Moreover, the cancer genomics data pioneer for the revolution of a novel oncology drug discovery from candidate target or gene studies toward targeting clinically relevant driver alterations or molecular features for the development of precision cancer therapy. For example, among the several selective kinases, axitinib inhibits BRC-ABL1 and it is more effective for treating tumors that have the clinically relevant driver mutations. However, millions of cancer patients are administrating various anticancer medications without their benefits due to both somatically acquired and inherited mutations. Although recent advances in NGS technologies enable the identification of various genetic or epigenetic alterations (e.g., clinically relevant driver mutations) with clinical practice, it is still necessary to develop more efficient and selective target for particularly genetic profiles or molecular features. Ideal candidates for repositioning can contribute to the therapeutically unmet need for more efficient anticancer agents, including drugs that selectively target cancer stem cells [20].

In this review, the recent advances in the development of novel small molecules for cancer therapy by their structure to repurposing with highlighted examples are summarized. The relevant strategies, advantages, challenges, and future research directions associated with this approach are also discussed. Clinical trials on combination of repurposed drugs and anticancer therapies are explained in pharmacological classification.

Central nervous system drugs

Studies on antipsychotic drugs such as valproic acid (VPA), phenothiazines (PTZs), selective serotonin reuptake inhibitors, tricyclic antidepressants, and monoamine oxidase inhibitors demonstrated several anticancer activities. The details of these studies were reported in some recent literatures [21].

The first example of these studies is the reporting of sedative thalidomide as an oncology drug (Fig. 15.4). Thalidomide was first used to treat for morning sickness in pregnant women. It was quickly withdrawn from the market due its teratogenic properties. Later studies showed that thalidomide, especially its *R*-isomer, exhibited potent antiangiogenic properties, and it would considered to use it as an anticancer agent [22]. Thalidomide shows its mechanism of action in six different way; it affects DNA replication, transcription, synthesis, and function of growth factors and integrins, angiogenesis, chondrogenesis, and cell death [23]. Understanding the adverse effects of thalidomide prompted scientist to investigate the therapeutic efficacy in multiple myeloma (MM). The response rate of treatment with thalidomide alone in MM and combination with dexamethasone resulted a 32% and >60%, respectively, which was quite effective for the treatment of MM. Structurally similar compound lenalidomide combination with dexamethasone was another promising treatment option in MM. Lenalidomide launches growth arrest and apoptosis of drug-resistant MM cells. It also inhibits to binding of MM cells to bone marrow and stromal cells and angiogenesis and modulates cytokine secretion. Combination of lenalidomide with dexamethasone enhances the response of the survival and prolongs the time. It was approved by FDA for therapy of MM patients in 2006 [24]. Another study showed that triple combination of lenalidomide–dexamethasone and the proteasome inhibitor bortezomib (Fig. 15.4) extended the life of MM patients by few years [25]. This combination was also found highly tolerable and is highly effective therapeutic against newly diagnosed MM [26].

Investigation of mechanisms of action of thalidomide and other structurally similar drugs such as lenalidomide, pomalidomide, and temozolomide (Fig. 15.4) demonstrated that this mechanism was associated with the upregulation of p21 and induction of apoptosis and direct affection in proliferation of MM cell lines and patient's specimens [27]. Pomalidomide alone was not very effective against MM. However, it showed synergistic effects when combined with dexamethasone. This combination was found safe and effective in relapsing and

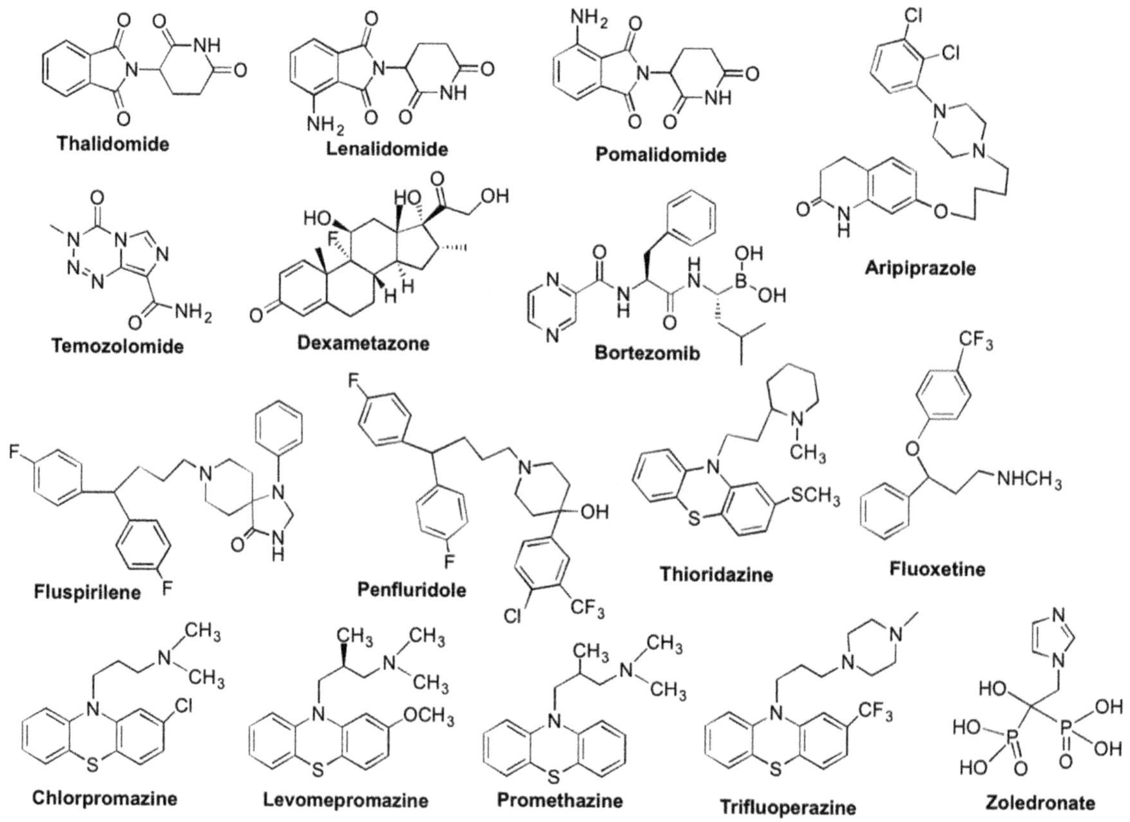

FIGURE 15.4 Chemical structures of thalidomide analogs and their combination drugs effective in anticancer therapies.

refractory of patients [28]. It was shown that temozolomide was highly effective in combination regimen for the treatment of metastatic melanoma [29]. A combination of thalidomide and temozolomide is currently under clinical Phase I/II studies for brain metastatic melanoma [30]. It was also found that thalidomide exhibited antiangiogenic effects and inhibits the processing of mRNA encoding peptide molecules including tumor necrosis factor-alpha (TNF-α) and the angiogenic factor, vascular endothelial growth factor (VEGF). As a result of these findings, thalidomide was investigated against several cancer types such as renal carcinoma and prostate cancer (PCa) [31,32]. Nevertheless, additional clinical studies are necessary to clarify the role of thalidomide in the therapy of PCa [33]. It was also shown that thalidomide inhibits nuclear factor kappa light chain of activated B cells (NF-kB), which is the reason of inflammation, survival, proliferation, invasion, and metastasis of tumors [34].

Some clinical studies of thalidomide combination therapy with anticancer agents have been done for the treatment of PCa patients. The aim of one of the studies was to evaluate clinical efficacy of thalidomide combination with capecitabine and cyclophosphamide against advanced castrate-resistant prostate cancer (CRPC). The results were found promising with excellent safety profile for patients with advances CRPC [35]. In other Phase I study, the potential efficacy and safety profile of thalidomide and cyclophosphamide in patients with hormone refractory prostate cancer (HRPC), previously treated with docetaxel-based regimens, was shown. It was suggested that Phase II clinical trials were needed to evaluate clinical efficacy of this regimen in HRPC [36]. In addition, a Phase I study was conducted to determine the feasibility and safety of combination regimen of gemcitabine with interferon, thalidomide, and capecitabine in patient with metastatic renal cell carcinoma (MRCC) in order to establish a Phase II trial to evaluate antitumor activity. The main aim of this study was to determine the effectiveness of adding gemcitabine to this combination. Weak responses obtained from this regimen demonstrated that adding gemcitabine to the three-drug regimen was not successful. Further investigation on this combination without the gemcitabine was recommended as a reasonable treatment option [37].

The clinical potency of the combination of weekly intravenous (IV) gemcitabine with continuous infusion fluorouracil (5-FU) and daily oral thalidomide in patients with MRCC was studied. During the application of therapy, several vascular toxicities such as deep venous thromboembolism (VTE), vein thrombosis (DTV), and pulmonary embolization were observed. In this study, the addition of thalidomide to gemcitabine and 5-FU was found ineffective, and it was recommended that further efforts need to be directed toward improving our knowledge of the incidence and pathophysiology of this potentially lethal adverse event [38].

The efficacy and safety of zoledronate (Fig. 15.4), thalidomide, and interferon treatment for renal carcinoma and bone metastases was studied in 15 patients. The results show that the therapeutic benefit of this combination was minimally enhanced. Nevertheless, the clinical efficacy of bone-targeted therapy might be improved by selecting appropriate patients with bone disease [39].

Fluspirilene (Fig. 15.4) is currently used as antipsychotic drug and it also demonstrated antiproliferative effect in human hepatocellular. The studies showed that oral fluspirilene was used to treat human hepatocellular carcinoma (HCC) compared to 5-fluorouracil. It was shown that it exhibited anticancer effect in human hepatoma HepG2 and Huh7 cells by inhibition of cyclin-dependent kinase 2 (CDK2). Combination therapy with fluspirilene and 5-fluorouracil produced the highest therapeutic effect. Since fluspirilene has been used safely for a long time as therapeutic agents and valuable results received proving it a potential anti-HCC agent, it may immediately enter in clinical therapy. Having being used with HCC could also be combined with other chemotherapeutic drugs for treatment [40].

It was determined that penfluridol (Fig. 15.4), which is clinically used for schizophrenia, had antitumor activities in non-small cell lung cancer (NSCLC) and acute myeloid leukemia (AML) cell lines. It was shown that penfluridol inhibited the viability and motility of NSCLC cells in vitro and in vivo. It induced nonapoptotic cell death by blocking autophagic flux and accumulation of autophagosome-related protein, light chain 3 (LC3) B-II, in HCC827 and A549 NSCLC cells. It was also observed that penfluridol used in patients with lung tumors expressing high LC3B had longer overall and disease-free survival times. Mechanistically, it was determined that penfluridol acted as inducer of endoplasmic reticulum (ER) stress and mitogen-activated protein kinase (MAPK) p38 activation [41]. In another study, it was shown that penfluridol suppressed the cell viability of AML cells with FLT3-WT and FLT3-ITD depending on its concentration. It was found that penfluridol induced apoptosis by activation of protein phosphatase 2A (PP2A) that suppresses AKt and MAPK activities and also augmented the reactive oxygen species (ROS) levels. In clinical studies, it was observed that patients with AML expressing high PP2A had essential prognoses and combination with penfluridol with an autophagy inhibitor that may become a favorable treatment for AML patient [42].

PTZs are old drugs used for the treatment of psychotic diseases as well as chemotherapy-induced emesis. Recently, they have been intensively studied as anticancer agents to targeting various signaling pathways. It has been shown that PTZs possess cytotoxic activities in tumor cell lines. However, there are some confusion about the cell death mechanism of PTZs [43]. Other PTZ derivatives with different chemical structures such as chlorpromazine (CPZ), levomepromazine, promethazine, trifluoperazine, and thioridazine (TRDZ) were studied against several leukemic cell lines. These drugs showed strong cytotoxic effect and antiproliferative activity against leukemic cells in clinically relevant doses (up to 20 μM). Among them, TRDZ was found the strongest apoptotic agent. It is considered that the activity mechanism of PTZs is associated with inhibition of mitochondrial DNA polymerase in treated leukemic cells and decreasing of ATP production, which are crucial events for the viability of cancer cells [44].

PTZs were studied for potential therapy for lung cancer (LC), which is the most prevalent and deadly malignancies worldwide. It was found that PTZs decreased cell viability and induced cell death preferentially in small cell lung carcinoma over NSCLC cell lines [45].

Antipsychotic drug CPZ (Fig. 15.4) was identified as a potential treatment agent for colorectal cancer (CRC) that effectively inhibited tumor growth and induced apoptosis in CRC cell in a p53-dependent manner. In addition, it was reported that CPZ induced the degradation of sirtuin 1 (SIRT1), which is associated with downstream of JNK and JNK which is suppressed by SIRT1 downregulation. In clinical studies, it was found that high SIRT1 expression and poor outcome in CRC patient was significant. Therefore, it was concluded that SIRT1 was an attractive therapeutic target and treatment agent for CRC [46]. Moreover, CPZ was reported as anticancer agent for the treatment of breast cancer by suppression of stemness properties including mammosphere formation and stemness-related gene expressions in breast cancer cells. Since CPZ targets the breast cancer stem cells (SSCs), it can be used for breast cancer associated with therapeutic resistance and metastasis [47].

Antipsychotic aripiprazole (Fig. 15.4) was reported to have a partial dopamine agonist effect alone or in combination with other chemotherapeutic agents to inhibit the growth in serum-cultured cancer cells and cancer stem cells. Furthermore, it was found that it inhibited sphere formation, as well as stem cell marker expression of cancer stem cells. The results were considered that aripiprazole might be used as an anticancer stem cell drug [48].

Antidepressant drug fluoxetine (Fig. 15.4) was studied for the treatment of colon cancer. It was shown that fluoxetine induced p53-independent apoptosis by altering mitochondrial membrane potential that leaded to DNA fragmentation, depending on concentration used. Furthermore, it induced the cell death with high selectivity for colon, breast, and ovarian carcinoma cells compared with normal cells. These results provided new findings about the cytotoxic effects of fluoxetine and demonstrated a new potential use for cancer treatment [49].

NSAID drugs (coxibs)

Acetylsalicylic acid (aspirin, Fig. 15.5) is the most commonly used pain killer, fever reducer, and preventer for heart attack and stroke. Acetylsalicylic acid was investigated for cancer therapy 46 years ago by Gasic et al. [50]. In this study, it was presented that platelet reduction by neuraminidase in tumor-bearing mice was reduced 50% in lung metastases, significantly reduced the metastases in tumor-bearing mice,

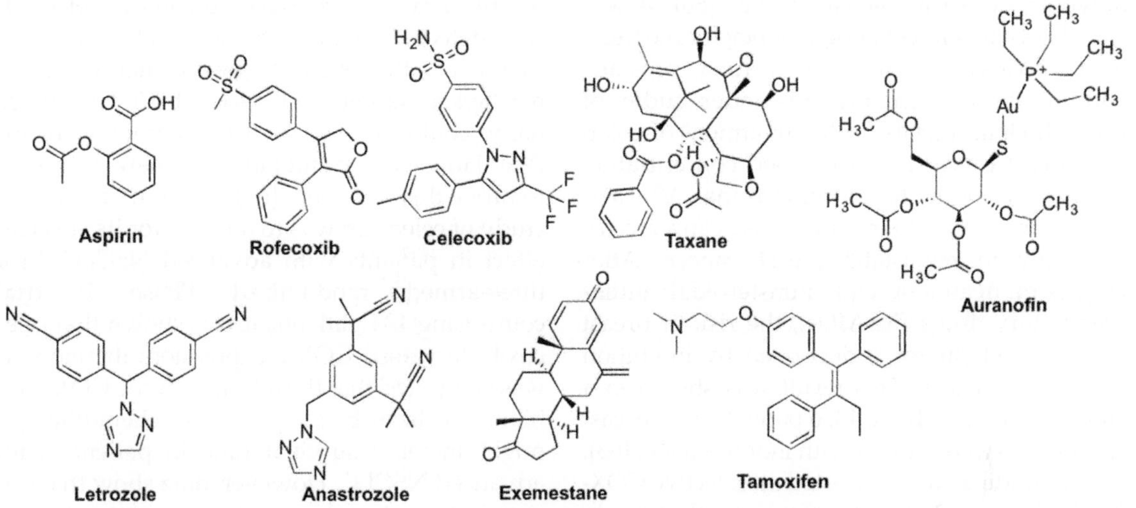

FIGURE 15.5 Chemical structures of coxibs and their combination drugs effective in anticancer therapies.

and inhibited the platelet formation [51]. It was also shown that lower dose of acetylsalicylic acid (75 mg) reduced the deaths of several cancers such as gastrointestinal, esophageal, pancreatic, brain, lung, and colorectal [51,52]. Aspirin and its metabolites were studied on several targets, such as adenosine monophosphate (AMP)−activated protein kinase (AMPK), cyclin-dependent kinase, heparanase, and histone to reveal the therapeutic effect on cancer. Combination of aspirin with other cancer therapies, such as sorafenib, radiotherapy, chemotherapy, and photothermal therapy, may be considered as promising therapy. However, more clinical studies are required to prove this statement [53].

Cyclooxygenase-2 (COX-2) is an enzyme that regulates the responses in various proinflammatory agents including cytokines, tumor promoters, and mitogens, which causes several physiological responses such as inflammation, pain, glaucoma, asthma, tumor, and osteoarthritis [54]. Overexpression of COX-2 has been determined in several cancers such as colorectal, breast, pancreatic, lung, and hematological (Burkitt's lymphoma and leukemia). COX-2 regulates cell proliferation and apoptosis mainly in solid tumors and hematological malignancies. COX-2 also causes the programmed cell death that enhances the efficacy of anticancer therapies [55]. Many studies demonstrated the evidences for overexpression of COX-2 in the role of cell proliferation, apoptosis, angiogenesis, and multidrug resistance (MDR), and therefore, it has been considered that they might be beneficial in several cancer types [56].

Celecoxib (Fig. 15.5) was developed against treatment of osteoarthritis and rheumatoid arthritis [57]. It selectively inhibits COX-2 enzyme and blocks the synthesis of several inflammatory prostanoids, therefore it exhibits its antiinflammatory and analgesic activities [58]. Celecoxib exhibits chemopreventive activities via the inhibition of COX-2 and COX-2−independent targets such as nuclear NF-kB, AKT8 virus oncogene cellular homolog (AKT is a serine/threonine kinase), glycogen synthase kinase (GSK) 3β, and β-catenin. It is also an important inhibitor of apoptosis protein and the B-cell lymphoma (BCL)-2 families, which are essential for several types of cancers [59]. Celecoxib was first used for the treatment of colorectal polyps and familial adenomatous polyposis (FAP) in clinical studies [60]. It was also shown that it had good potency for the treatment of cancers such as colon, lung, prostate, and breast [61]. Currently, it is under studies of Phase II clinical trials for the treatment of colon and breast cancer [62]. Increased concentration of prostaglandin E2, thromboxane A2, and COX-2 mRNA is found in breast cancer cells, especially in metastatic breast cancer. After long-term treatment with nonsteroidal antiinflammatory drugs (NSAIDs), the risk of breast cancer development is decreased by inhibition of such mediators. This result was shown in a clinical study involving 4876 patients with breast cancer who were treated with indomethacin [63]. Similar findings were reported on selective COX-2 inhibitors in 2006. Therefore, another study was conducted to show the risk of breast cancer in woman, and it was reported to be reduced by as much as 71% when the patients use celecoxib and rofecoxib (Fig. 15.5) [64]. Similarly, it was confirmed that aspirin had protection against cancer and decreased the development of breast cancer [65]. An antiangiogenic mechanism of celecoxib was reported, which includes apoptosis, cell cycle arrest, regulation of angiogenesis, and invasion and induces stress of ER [66]. In another study, it was found that it decreased the expression of vascular endothelial cell growth factor and matrix metalloproteinase-9 (MMP-9) in cancer tissues and cell lines [18]. Known anticancer agents, for instance, taxanes and aromatase inhibitors such as letrozole, anastrazole, and exemestane (Fig. 15.5), have been used in combination with COX-2 inhibitors that increased the efficacy of monotherapies [67]. A study of randomized Phase II study of celecoxib in combination with exemestane showed that no significant benefits were obtained from patients. However, further evaluation in large patient population was found important to warrant the results [68]. A double-blind clinical trial was conducted on 73 patients, and 64 were randomly assigned and included in the intention-to-treat analysis. No treatment differences were obtained between the primary or secondary outcomes. Celecoxib was detected in prostate tissue of patients in treatment; this result showed that celecoxib reached its target, and celecoxib, 400 mg twice daily, could be used as a preventive agent for PCa. However, additional studies are necessary to prove this statement [69]. The aim of a clinical study of celecoxib was to analyze for its survival effect in patients with advanced NSCLC. In a three-armed randomized Phase II trial comprising 134 patients, it was shown that celecoxib decreased COX-2 expression; therefore, it was suggested that the addition of the COX-2 inhibitor celecoxib to palliative chemotherapy might increase survival time in patients with advanced NSCLC. However, data show that celecoxib did not prolong survival in patients with advanced NSCLC [70]. Celecoxib was added to combination of cisplatin, gemcitabine, navelbine, and docetaxel in order to evaluate the efficacy and safety in treatment of 44 naive NSCLC patients to determine the benefiting effect of celecoxib by molecular analysis. After treatment of this combination, the 1-year survival rate was found as 68%. It was also reported that this combination was well tolerated and slightly active against COX-2 IHC–positive patients. However, it was suggested that further trials needed to confirm these results [71]. Combination of celecoxib with gemcitabine was studied in 25 patients with advanced or metastatic pancreatic cancer to determine the effects on survival, disease progression, and tolerability. It was shown that this therapy had no significant improvement in patients with advanced pancreatic cancer. The suggestive correlation between VGEF and patient survival was explained. It was also presented that higher dose of celecoxib needed to observe significant antitumor activity [72]. The clinical study of low-dose (100 mg twice daily) celecoxib combination with weekly cisplatin was applied to determine the safety in locally advanced undifferentiated nasopharyngeal carcinoma. Although this combination was found highly effective against locally advanced undifferentiated nasopharyngeal carcinoma, the response rates, toxicity parameters, and overall survival were not found statistically significant [73]. Phase II clinical trial was conducted to evaluate the efficacy of celecoxib in treating 32 patients with progressive metastatic differentiated thyroid cancer (DTC). The results showed that using 400 mg of celecoxib orally twice per day failed to halt DTC in most patients [74].

A gold compound auranofin (Fig. 15.5) had a long history for the treatment of rheumatoid arthritis. But nowadays auranofin is rarely used in clinical therapy for rheumatoid arthritis. Recently, it was reported that auranofin inhibited the activity of thioredoxin redceructase (TrxR), an enzyme of thioredoxin system that is important for maintaining the intracellular redox state. Inhibition of TrxR causes the increase activity of oxidative stress and induces apoptosis, which is associated with tumor progression in breast cancer, ovarian cancer, and LCs. In addition, auranofin exhibited preclinical efficacy in chronic lymphocytic leukemia (CLL) cells. Auranofin also displayed synergistic lethality with heme oxygenase-1 and glutamate-cysteine ligase inhibitory activity against CLL cells by eliminating apoptosis resistance. Recent studies showed that auranofin inhibited proteasome activity at the level of the proteasome-associated deubiquitinases and proteasome processing in cells. It was found that it caused mitochondrial dysfunction and induced the oxidative stress that is explained as the dominant mechanisms underlying its cytotoxicity. Auranofin also has good pharmacokinetic properties and is orally bioavailable. It is sufficient to elicit toxicity to cancer cells in low doses and has neoplastic effects in vivo. All these findings showed that there is a possibility to use auranofin as anticancer agent [75–77].

Anticonvulsant (antiepileptic) drugs

1-methyl-1-cyclohexanecarboxylic acid (Fig. 15.6) was studied for its anticancer activities. Obtained activity of this drug inspired scientist to evaluate another anticonvulsant/antimigraine compound, VPA (Fig. 15.6), for its anticancer efficacy. In clinical trials, VPA was originally studied for neurodisorders or other diseases, such as drug addiction and amyotrophic lateral sclerosis [78]. VPA exhibits its pharmacological activity upon several mechanisms that involve histone deacetylases (HDACs), GSK3α and GSKβ, protein kinase B (is also known as Akt), and the extracellular signal–regulated kinase (ERK) phosphoinositol pathways. It plays a role on the activities of the

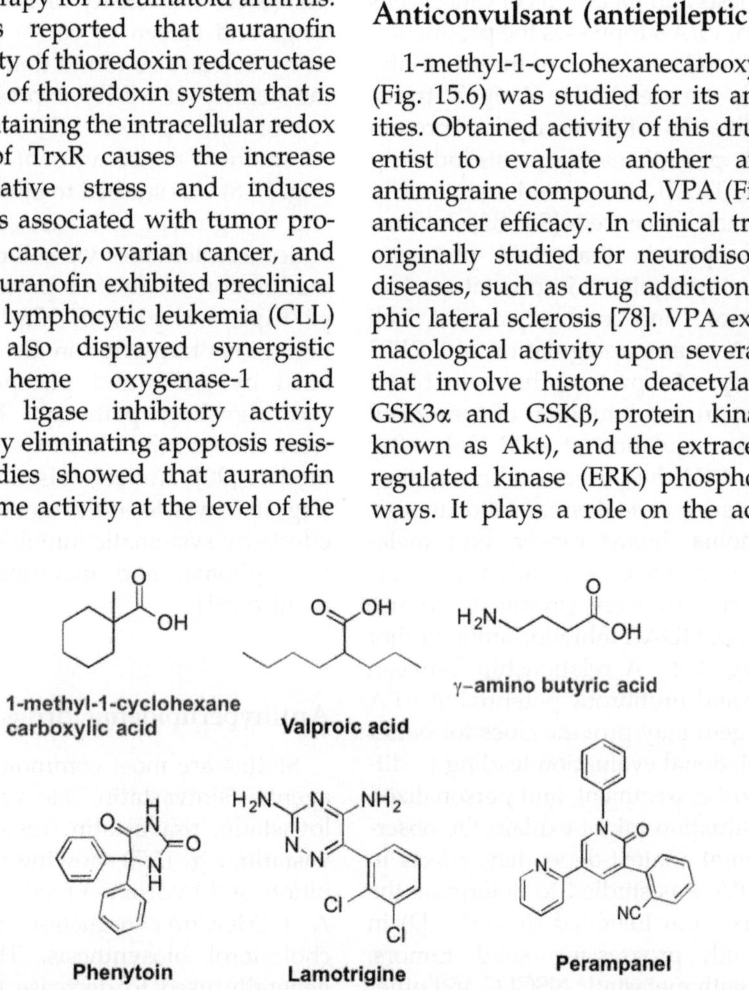

FIGURE 15.6 Chemical structures of anticonvulsant drugs effective in anticancer therapies.

tricarboxylic acid cycle, oxidative phosphorylation system, and γ-amino butyric acid (Fig. 15.6) [79]. The first anticancer activity of VPA was identified in human leukemia cells, and later, it was tested for the treatment of solid and hematopoietic malignancies in clinical trials alone or in combination with other anticancer compounds [80,81]. Inducing the aberrant transcription of genes by HDACs regulates several cellular functions such as cell proliferation, cell cycle regulation, and apoptosis. Studies showed that VPA inhibited HDAC prevents survival, invasion, agiogenesis and metastatis of cancel cells [82]. In addition, VPA suppresses the production of interleukin-6, and as a result of these activity, TNF-α increases the acetylation of signal transducer and activators of transcription protein (STAT-1). VPA potentiates apoptosis-inducing ligand (Apo2L/TRAIL)–mediated cytotoxicity in cultured thoracic cancer cells [83–85]. A transmembrane glycoprotein β-amyloid precursor protein (APP) is upregulated in prostate, colon, pancreatic tumor, and oral squamous cell carcinoma. VPA causes upregulation of GRP78 (a major ER chaperone protein) that is involved in APP maturation and inhibition of tumor cell growth by downregulation of APP [86]. After these findings, VPA has been studied against several cancers such as leukemia, solid tumors, SCLCs, lymphoma, breast cancer, and melanoma, prostate, thoracic, and advanced sarcomas. These studies were promising to use VPA as first place HDAC inhibitor among other anticancer drugs [78]. A relationship between antiangiogenic and profibrotic potential of VPA as anticancer agent may provide clues for better potential translational evaluation leading to differential diagnostic, treatment, and personalized therapy. This situation might explain the observation of different context-dependent effects in patient [87]. VPA was studied to determine the safety and maximum tolerated dose (MTD) in 26 patients with progressing solid tumors. In one patient with metastatic NSCLC and other with metastatic CRC having 3 and 5 months disease, respectively, no progression was observed. After these findings, it was concluded that further studies in larger group of patients were needed to investigate the effectivity of VPA alone or in combination with other anticancer drug [88].

A voltage-gated Na^+ channel–blocker phenytoin (Fig. 15.6) is an antiepileptic drug that was found to inhibit the migration and invasion of metastatic MDA-MB-231 (an epithelial, human breast cancer cell line) cells in in vitro studies. Animal studies showed that phenytoin decreased the density of metastatic cells in the lungs and spleen of 66.3% and 92.4%, respectively. It was also found that phenytoin reduced the density of MMP9-expressing cells and the rate of tumor growth [89].

Another anticonvulsant drug lamotrigine (Fig. 15.6) was studied to inhibit the proliferation and cell growth of breast cells. It exhibited its anticancer activities via cell cycle arrest and the modulation of related proteins (cyclin D1, E, p27Kip1, and p21waf1/Cip1), genes of FoxO3, ubiquitous transcription factor negatively regulated by AKT, and phosphoinosite-3-kinase/Akt signaling pathway. These mechanisms were shown in in vitro experiments and animal models [90]. Anticonvulsant agent perampanel (Fig. 15.6) was found to mediate antitumorigenic effects by systematic inhibition of cell proliferation, glioma, and metastasis cell growth in in vitro [91].

Antihyperlipidemic drugs (statins)

Statins are most common antihiperlipidemic agents (simvastatin, atorvastatin, fluvastatin, lovastatin, pravastatin, rosuvastatin, and pitavastatin, Fig. 15.7) showing their effects by inhibition of 3-hydroxy-3-methylglutaryl-coenzyme A (HMGCoA) reductase that catalyzes the cholesterol biosynthesis. Therefore, they are generally used to decrease the level of cholesterol for the treatment of hypercholesteremia.

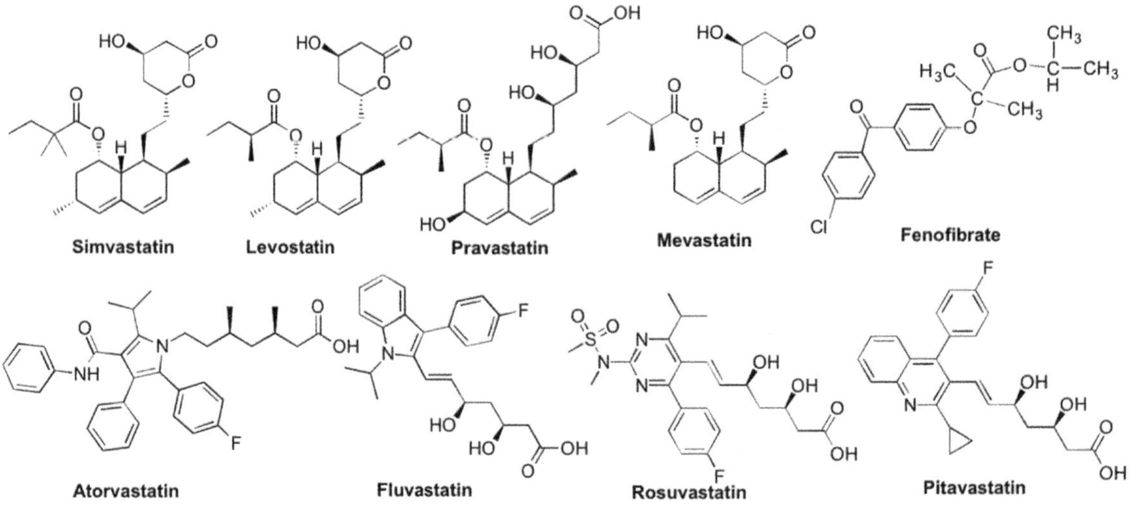

FIGURE 15.7 Chemical structures of antihyperlipidemic drugs effective in anticancer therapies.

In recent clinical studies, it was demonstrated that statins exhibited promising anticancer activities [92]. Statins induce cell death in cancer cells at higher concentrations than their effective concentrations for hypercholesteremia treatment, but the optimal drug concentration for cancer treatment is not known [93]. Besides inhibition of HMG-CoA mechanism, it was found that statins inhibited cell proliferation, promote cell apoptosis and tumor cell differentiation, and modulate tumor microenvironment. Although several studies showed the chemotherapeutic effects of statins in many types of cancers, the potency of statins should be confirmed and validated in clinical studies [94]. It was also shown that statins lowered the risk of many types of cancers such as breast, stomach, liver, prostate, pancreatic, lung, melanoma, hematological malignancies, and esophageal adenocarcinoma [95]. Simvastatin, mevastatin, lovastatin, and pravastatin (Fig. 15.7) are reported as anticancer agent against chronic myeloid leukemia (CML) cells via TNF-induced apoptosis [96]. The effects of statins on cell cycle of YT-INDY natural killer cell leukemia cell line were studied, and the results demonstrated that simvastatin, fluvastatin, and lovastatin decreased the cellular proliferation [97]. It was also found that statins have therapeutic benefits for leukemia therapy [98]. The use of statins in patients with glioblastoma is considered as a new challenge for the possible anticancer roles of these drugs. But current analysis and recent evidence from literature showed that the results were doubtful, and the need to continue further investigation was not necessary [99].

It was reported that simvastatin particularly reduced the action of CRC [100,101]. Another study demonstrated that simvastatin used as a preventive agent for protecting against death from PCa. However, longer clinical analysis and experimental studies on the biological mechanisms are required to prove the therapy with statins against PCa [102]. A Phase II study was applied to determine the clinical benefit of adding simvastatin to afatinib treatment comparison with afatinib monotherapy in nonadenocarcinomatous non-small cell lung cancer (NA-NSCLC). The main idea of this study comes from that the combination of statins with EGFR-blocking agents may be effective against specific tumor types. It was found that the addition of

simvastatin to afatinib did not enhance efficacy of afatinib alone used for treatment of NA-NSCLC patients [103]. HMGCoA inhibitor simvastatin was added to XELIRI (capecitabine–irinotecan) and FOLFIRI (leucovorin–fluorouracil–irinotecan) combination to evaluate clinical benefits to patients against pretreated metastatic CRC in a randomized Phase III clinical trial. It was found that adding low-dose simvastatin (40 mg) to XELIRI/FOLFIRI combination did not improve the response of the therapy [104]. Pravastatin was combined with other anticancer drugs idarubicin and cytarabine. The results showed that this combination was active against AML in Phase I clinical studies [105].

A lipid-lowering agent fenofibrate (Fig. 15.7) and a natural antioxidant agent salicin were investigated as possible anticancer agents against breast and pancreatic cancer models. Both drugs exhibited in vivo anticancer activities by decreasing tumor weight and volume through activation of the caspase 3/7 apoptotic pathway [106]. Fenofibrate also shows its anticancer effects by apoptosis, cell cycle arrest, inhibition of invasion, and migration. In vivo experimental results confirmed that fenofibrate exerted positive effects against tumor growth in high doses (200 mg/kg). In addition to breast and pancreatic cancer, it was found effective against LC, liver, glioma, and prostate cancer cell lines. Fenofibrate is considered as an adjuvant drug in cancer treatment, which will be used in combination therapy in future clinical studies [107].

Antidiabetic drugs (biguanides)

It was considered a drug that is used for prevention of a metabolic syndrome and might also be used in preventive and therapeutic oncology. The current aim of this approach is to improve the survival of patients and limit cancer progression. Since an epidemic of diabetes and obesity is a characteristic factor to increase the incidence of certain cancers, it was considered that antidiabetic drugs in combination with other anticancer agents might have success in clinical therapy [108]. Therefore, biguanide was studied as effective drugs for glioblastoma (GBM). Chloride intracellular channel 1 (CLIC1) is the relevant molecular factor controlling the aggressive behavior of GMB. CLIC1 controls several physiological cell functions, and its abnormal expression induces tumor development via cell proliferation, invasion, and metastasis. Recently, it was reported that biguanides selectively inhibit CLIC1 activity that could be used to develop novel drugs with a strong efficacy against glioblastoma stem cells [109].

Metformin (Fig. 15.8) is a therapeutic agent for diabetes treatment that has a biguanide structure. It was shown that it decreased cancer cell growth and proliferation via the inhibitory mechanism of AMPK on the mammalian target of rapamycin (mTOR) pathway [110]. Insulin-like growth factor 1 (IGF-1) provides the occurrence of several cancer types. It was shown that metformin reduced the IGF-1 and circulation level of insulin. It was reported that metformin inhibited proliferation and growth of the certain types of cancers over than 100 clinical studies [111]. Studies showed that metformin exhibited its anticancer effect by lowering circulating glucose and insulin level that slows tumor proliferation and reduces the energy consumption into the neoplastic cells. In addition, it was demonstrated that metformin blocked invasion of tumor cells by inhibiting MMP-9 and gelatinase-2 activation [112].

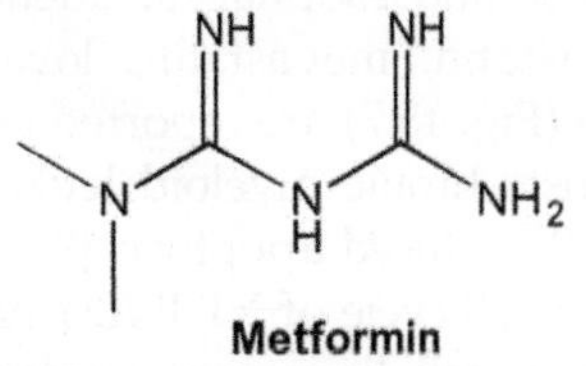

NH NH
N N NH₂
H
Metformin

FIGURE 15.8 Chemical structures of antidiabetic drug, metformin.

It was also shown that metformin reduced cancer incidence and mortality depending on the dose in many preclinical and clinical studies of it. A study showed that activation of AMPK played a role in the antitumor activity of metformin in breast cancer cells [113–115]. During the clinical studies, it was shown that metformin therapy of nondiabetic women with breast cancer diminished the number of Ki67-positive cancer cells (a proliferative marker) and also gene expression of molecules associated with mTOR and AMPK pathways [116]. Another study demonstrated that small dose metformin treatment decreased the marker of CRC by 40% in nondiabetic patient [117]. The incidence of gastric cancer in diabetic patients was reduced by daily treatment of metformin [118,119]. The results obtained from five clinical studies with considerable numbers of patients and evaluation of data from 422 articles showed that long-term use of metformin decreased the risk of gastric cancer. However, additional well-established trials are necessary to confirm these findings [120]. In another study, obese mice with thyroid cancer were studied using preclinical model of ThrbPV/PVPten ± mice (a metastatic thyroid cancer model) that fed with high-fed diet. This study showed that metformin could be used for the treatment of patients with thyroid cancer [121]. Metformin is also studied for its therapeutic use for prostate, endometrial, and pancreatic cancers in clinical trials [122].

In preclinical and clinical studies, it was shown that metformin had antitumor activity in nondiabetic postmenopausal women with ER-positive breast cancer. Metformin show a synergy with neoadjuvant letrozole in ER-positive breast cancer patients similar to that of the mTOR inhibitor, everolimus [123]. A study demonstrated that metformin effected the hormone receptor–positive (HR+) luminal cells in the normal murine mammary gland at a clinically relevant dose. In this study, it was shown that it decreased oxygen consumption rate, increased the length of cell cycle of luminal cells, and slightly lowered the circulating of estrogen in clinical doses. It did not make any change in circulating sex hormone–binding globulin (SHBG) levels in postmenopausal women, but it showed a modest increase in circulating SHBG and estrogen levels in premenopausal polycystic ovarian syndrome patients. Therefore, it was considered that metformin had a potential to prevent and treat breast cancer [124]. A Phase II trial was conducted to claim the hypothesize that could had better outcomes to adding metformin in combination of everolimus and exemestane for the treatment of obese patients with metastatic, HR+, HER2-negative breast cancer. It has been reported that this combination was safe and had moderate clinical benefit [125].

Metformin is weakly cationic compound and targets the mitochondria to induce cytotoxic effects in tumor cells. It was hypothesized that attaching a positively charged lipophilic substituent might increase its mitochondria-targeting potential and enhance the antitumor activity, including aggressive cancers like pancreatic ductal adenocarcinoma [126]. To prove this aim, more than 100 clinical trials of metformin and combination with anticancer drugs in oncology studies were conducted to confirm possible clinic utility of metformin. A study showed that metformin did not improve survival of patients with advanced pancreatic cancer. In addition, it was presented that there was no benefit of the addition of metformin to the combination of gemcitabine and erlotinib in the treatment of advanced pancreatic cancer [127–129].

Phase I clinical study was conducted to evaluate clinical benefits of immune checkpoint inhibitors (ICIs) and metformin combination for treatment of NSCLC patients. A significant benefit was observed in terms of response rate from this combination although small sample size had been used. It is suggested that larger retrospective and prospective studies should be run to understand the potential benefit of

metformin in combination with ICIs. Dose escalation Phase I studies is also recommended for determination of optimal metformin dose with antitumor properties [130].

Antimalarial drugs

A relationship between anticancer and antimalarial activities has been discovered in the frame of diagnostics, drug research, treatment, prevention, and epidemiology [131–134]. Different chemical structure–bearing antimalarial drugs such as artemisinins (ARSs), 4-aminoquinolines, 8-aminoquinolines, and synthetic peroxides have exhibited anticancer activities in various cancer cell lines [135–140]. They act as sensitizers against drug-resistant tumor cell lines or act as synergistic agents with other anticancer drugs. Several mechanisms have been proposed to explain the exact mode of action of antimalarial drugs in cancer. However, the mechanism is still needed to be explored in many ways, including induction of autophagy, interference with angiogenesis, and induction of apoptosis. It was reported that antimalarial agent ARS and its derivatives such as dihydroartemisinin (DHA), artesunate, artemether, and arteether (Fig. 15.9) exhibited anticancer activity [141]. They demonstrated their mechanisms of action via several mechanisms, such as oxidative stress response by reactive oxygen species and nitric oxide, DNA damage and repair, various cell death modes (apoptosis, autophagy, ferroptosis, necrosis, necroptosis, oncosis), inhibition of angiogenesis and tumor-related signal transduction pathways (e.g., Wnt/catenin pathway, AMPK pathway, metastatic pathways, and others), and inhibition of signal transducers (NF-B, MYC/MAX, AP-1, CREBP, mTOR etc.) [142]. The endoperoxide

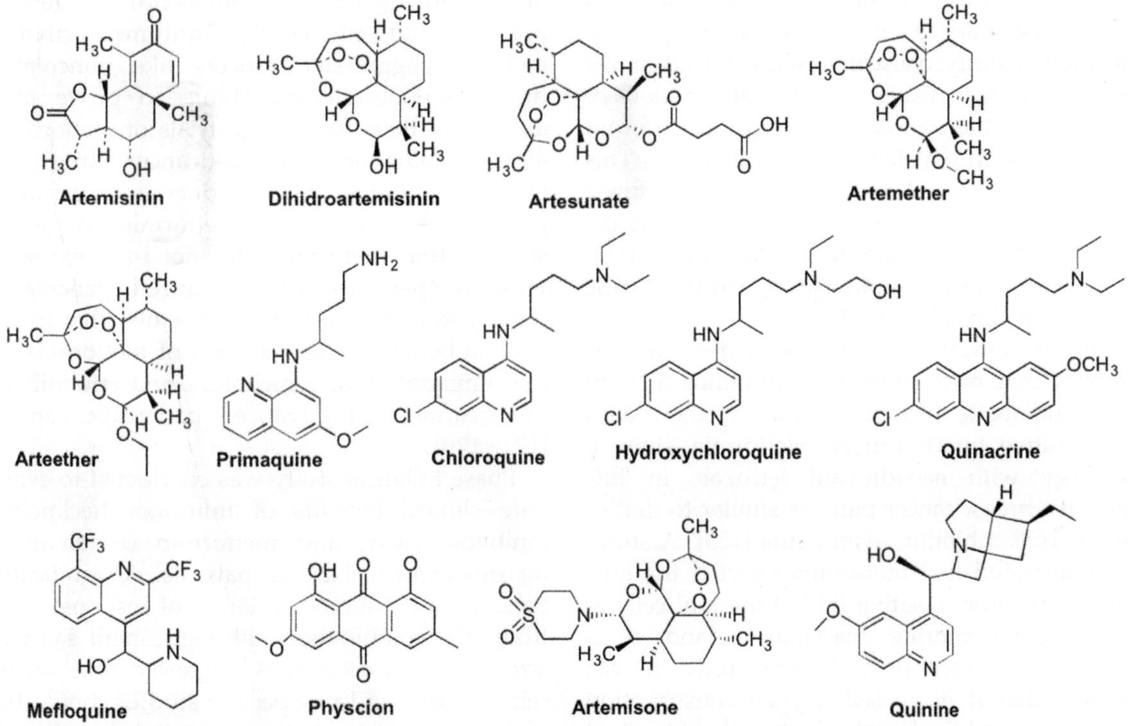

FIGURE 15.9 Chemical structures of antimalarial drugs effective as potential anticancer therapies.

structure of ARSs is necessary to obtain both antimalarial and anticancer activities. The endoperoxide bond is activated by the reduced heme or ferrous ion, generates carbon-centered radicals, and acts as alkylating agents. Although endoperoxide moiety is not completely responsible for anticancer activity, the absence of the endoperoxide moiety significantly reduces the cytotoxicity [143,144]. This finding prompt to consider some other mechanisms such as apoptosis that might be involved the anticancer activity of ARS derivatives [145,146]. ARS shows its cytotoxic activity by blocking cytokines, as well as inhibiting tumor invasion, migration, and metastasis [147,148]. The antineoplastic activity of ARS is controlled by other factors such as calcium metabolism, ER stress, and the expression of translationally controlled tumor protein [149]. ARS also exhibited its antitumor effects via inhibition of tumor angiogenesis, which depends on the mRNA expression of angiogenesis-related genes [150]. It was found that the nature of the linker in dimers of ARS had an important role in anticancer activity [151]. Activated ARS exposes highly alkylating carbon-centered radicals and reactive oxygen species (ROS) [145]. ROS generation may contribute to the selective action of ARS on cancer cells by DNA damage, promotion of apoptosis, growth arrest, and reduction of angiogenesis [146]. A wealth of publications with several case report and pilot Phase I and Phase II clinical trials of ARS type of drugs with their anticancer activities were appeared during the past two decades. However, more extensive clinical Phase II and III clinical trials should be done to provide more compelling evidence to prove suitability of ARS and its derivatives for oncology use [142].

Among the dimeric and trimeric ARS derivatives, a series of C-10 nonacetal dimers has been reported as antitumor compounds and found more effective than doxorubicin (DOX). It was found that the semisynthetic derivative DHA demonstrated antineoplastic activity against pancreatic, leukemic, osteosarcoma, and lung cancer cells. Combination of small-molecule 6-phosphogluconate dehydrogenase inhibitor physcion (Fig. 15.9) and DHA presented a novel leukemia treatment without inducing hemolysis. It was found that combined therapy synergistically reduced tumor growth in xenograft nude mice injected with human K562 leukemia cells and cell viability of primary leukemia cells from human patients [152]. Artemisone (Fig. 15.9) also demonstrated better anticancer properties than ARS [153]. Arsenuate (Fig. 15.9) showed its cytotoxic effects on artesunate-treated cells that made oncosis-like microscopic changes at subcellular structures and artesunate-treated HeLa cells by activation of ROS generation [154]. It was shown that artesunate inhibited leukemia, colon, melanoma, breast, ovarian, prostate, neuronal, and renal cancer cells. Although lower dose of artesunate kills the oncosis-like cells, higher concentration inducing apoptosis was reported to be needed [155]. It was presented that artesunate had a broad range of anticancer activities including antiangiogenic and immunomodulatory effects as well as reduction of postsurgical distant metastases [156]. Phase I study of artesunate was conducted to determine the MTD and dose-limiting toxicities (DLTs) in patient with solid tumor malignancies. Treatment of the MTD of IV artesuanate at 18 mg/kg dose was found as tolerable, and modest clinical activity was observed in pretreated population [157].

Potent antimalarial agents' aminoquinoline derivatives were investigated for their anticancer effects as well. It was reported that an 8-aminoquinoline derivative primaquine (Fig. 15.9) exhibited significant cytostatic activity against different cancer cell lines and showed high selectivity toward breast adenocarcinoma MCF-7 cell line [158–164].

It was shown that other 8-aminoquinoline derivative, mefloquine (Fig. 15.9), stimulated growth arrest and apoptosis of CRC cells in mice and exhibited antitumor action through inhibition of tumor NF-κB signaling pathway. It was found that mefloquine was a NF-JB inhibitor and induced growth arrest and apoptosis of CRC cells in cell culture assay and in mice. Since mefloquine exerts an antitumor activity and it might be used as an anti-CRC agent in clinic as a single drug or in combination with another anticancer drug such as DOX [165].

Synthetic 4-aminoquinoline derivatives of quinine, chloroquine (CQ), and hydroxychloroquine (HCQ) (Fig. 15.9) are used as antimalarial agents. Anticancer properties affecting the toll-like receptor 9, p53, and angiogenesis in cancer cells have been reported. They are especially found to be useful in combination with conventional anticancer drugs for the treatment of different cancers [166]. In addition, it was reported that CQ showed anticancer activity via inhibition of BCL-2 proteins family, and CQ could be potentiated the antitumor activity of chemotherapeutic drugs such as DOX. CQ was reported as a promising antitumor compound for HCC treatment via inducing extrinsic and intrinsic apoptotic pathway [167].

Some valuable clinical findings were obtained by using CQ or HCQ as autophagy inhibitors in the treatment of cancers. Autography inhibition is an orchestrated hemostatic process to eliminate unwanted proteins and damaged organelles. Autography inhibition enhances the responsiveness to MAPK, JNK1, ERK, and p38kinase pathway targeted therapies in different cancer types. A metaanalysis was run to explore autophagy inhibitor properties of CQ and HCQ. It was found that both drugs can significantly enhance the relative risk of overall response rate (ORR) and 6-month progression-free survival (PFS) rate. Combination with gemcitabine yielded the best ORR, and applying a combination of autophagy inhibitor, temozolomide, and radiation resulted the best 6-month PFS rate. It was also reported that autophagy inhibitor could cause the best survival benefit in glioblastoma patients. Evaluating of these results showed that further studies should have been done to prove its efficacy and safety profile comparison with different autophagy inhibitor base on the combination therapy [168].

Quinacrine (QC, Fig. 15.9) has been used as an antimalarial drug and as an antibiotic. It was reported that it inhibited tumorigenesis in endometrial cancer (EC) in in vitro and in vivo EC mouse model. QC was combined with cisplatin, carboplatin, or paclitaxel studied against in vitro chemoresistant cell line and the highest level of synergism was observed. QC and carboplatin/paclitaxel combination significantly augmented the antiproliferative ability of these agents. However, QC and these combinations did not delay tumor growth. This preclinical study supports QC's efficacy as an adjunct to standard chemotherapy in the treatment of EC. It was proven that the maintenance therapy with QC is superior to combination treatment as it resulted in long-term stabilization of disease and further prolongation of overall survival compared to combination. The results showed that further investigation of the QC role in a Phase I/II clinical trial investigating its combination with platinum-based chemotherapy for patients with chemoresistant EC will be beneficial in clinical therapy [169].

Antiviral drugs

Anticancer activities of antiviral agents have been investigated. Acyclovir used against herpes simplex virus infections (Fig. 15.10) was reported to decrease the growth and proliferation rate of cells and regulate the apoptosis-associated cytokine, caspase-3. It was also found that acyclovir inhibited tumor invasion and ability for colony formation and suppressed the growth of breast cancer cells. These results showed that acyclovir had good potency to suppress secondary tumor formation. But more research is necessary to clarify the effects of biochemical mechanisms of acyclovir on potential anticancer activities [170].

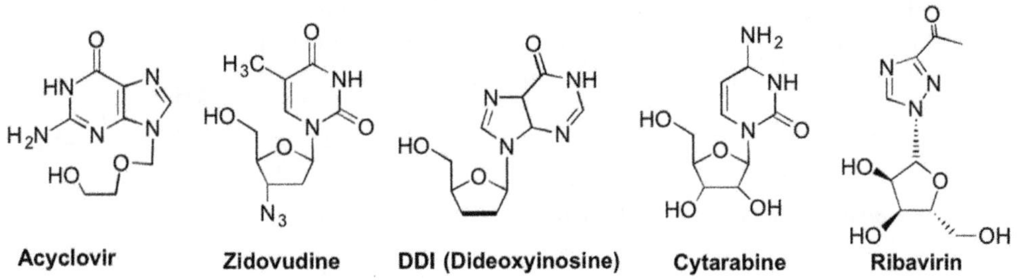

FIGURE 15.10 Chemical structures of the antiviral drugs effective in anticancer therapies.

Human immunodeficiency virus (HIV) aspartyl protease inhibitors are generally used in highly active antiretroviral therapy that shows antiangiogenic and antitumor effects [171]. Zidovudine (azidothymidine [AZT] in Fig. 15.10) is a thymidine analogue, which was first isolated from *Cryptotethia crypta* and approved for the treatment HIV infection by inhibiting HIV reverse transcriptase [172]. Zidovudine was reported to be effective against several types of cancers such as leukemia, lymphoma, Kaposi sarcoma, and pancreatic cancer in 1960s. Zidovudine was also found effective against gemcitabine-resistant pancreatic cancer. It was reported as inhibitor of epithelial–mesenchymal transition (EMT)–like phenotype cells and inhibitor of activation of Akt-GSK3β-Snail pathway. It exhibited that combination therapy of gemcitabine and zidovudine effectively restrains chemoresistance-specific signaling [173]. A study showed that low-dose AZT and dideoxyinosine (Fig. 15.10) decreased tumor cell growth in vivo. Therapeutic dose of AZT for the treatment of HIV was found effective for the treatment of solid cancers. It was shown that low-dose dual reverse transcriptase inhibitor treatment caused a significant telomeres shortening, DNA damage, and apoptosis of tumor cells [174].

Cytarabine (Fig. 15.10) is approved by FDA for treatment of several cancer types such as AML, ALL, and NHLs. Recently, it was reported for treatment of meningeal leukemia, lymphomas, and recurrent embryonal brain tumors. It is also a potent inhibitor of Kaposi sarcoma associated herpes virus (KSVH)–induced primary infused lymphoma (PEL) by controlling cell cycle arrest and apoptosis. Cytarabine induced cellular cytotoxicity via inhibition of host DNA and RNA syntheses and inhibited KSHV lytic replication program by preventing virion production. Cytarabine also induced the major latent protein LANA (latency-associated nuclear antigen) of KSVH. As a result of all those findings, cytarabine was identified as an ideal candidate for repurposing for PEL therapy and decided to evaluate it in advanced clinical trials [175].

It was reported that antiviral agent ribavirin decreased the elevated level of EIF4E protein in most cases of infant acute lymphoblastic leukemia (ALL). It was reported that antiviral agent ribavirin (Fig. 15.10) decreased the elevated level of EIF4E protein in most cases of infant ALL and divided infant ALL cells on bone marrow stromal cells at clinical dose to inhibit proliferation. This result suggested that ribavirin might be a promising agent for ALL treatment [176].

Antimicrobial drugs

Tropodithietic acid (TDA, Fig. 15.11) is a marine source of product, which has been shown to be a potent antibacterial activity. Based on the structural similarity to polyether anticancer agents, cytotoxicity of TDA's has been examined and found that it exhibited potent anticancer activities. It was also reported that TDA harbor lethal and growth-inhibitory activities against several cancer cell lines [177].

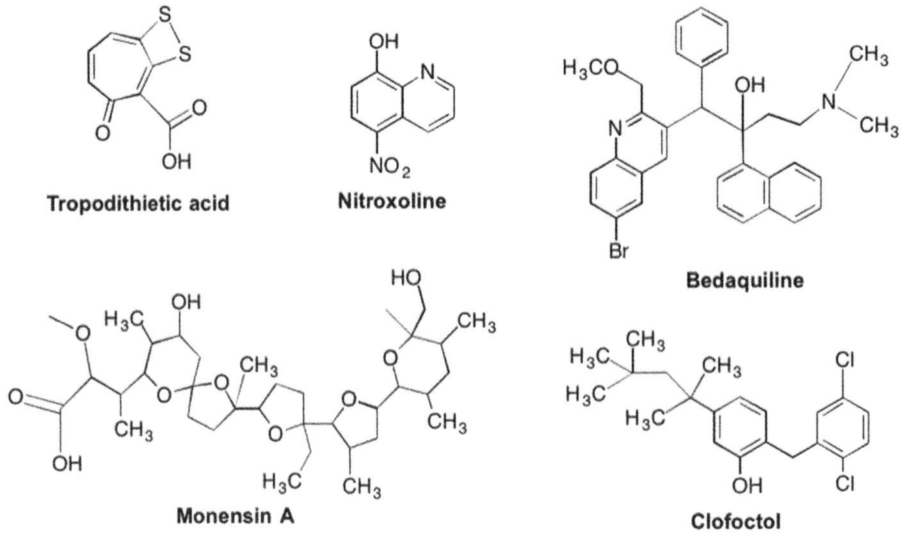

FIGURE 15.11 Chemical structures of antimicrobial drugs effective as potential anticancer therapies.

A wide-spectrum antibacterial agent monensin A (Fig. 15.11) is a polyether ionophore antibiotic and was found that it exhibited in vitro antiproliferative activity against cancer cells at micromolar concentrations [178].

Bedaquiline (Fig. 15.11) is a diarylquinoline compound which is approved for the treatment of multidrug-resistant pulmonary tuberculosis by inhibiting the bacterial ATP synthase. It was hypothesized that bedaquiline might target the mitochondrial ATP synthase, leading to mitochondrial dysfunction and ATP depletion, which prompted scientist to search anticancer activity of bedaquiline. It was found that its anticancer activity directed against cancer stem-like cells, and more specifically, it inhibited mitochondrial oxygen consumption, induced oxidative stress in MCF7 breast cells, and blocked the propagation and expansion of MCF7 at 1 μM of an IC50 value [179].

Urinary antibiotic nitroxoline (Fig. 15.11) was identified to disrupt the interaction between the first bromodomain-containing proteins and acetylated H4 peptide. Bromodomain and extraterminal domain (BET) family of proteins that can specifically bind acetylated lysine residues in histones serve as chromatin-targeting modules that decipher the histone acetylation code. BET proteins play a crucial role in regulating gene transcription through epigenetic interactions between bromodomains and acetylated histones during cellular proliferation and differentiation processes, and therefore, BET is believed to be a promising drug target for therapeutic intervention of cancer. Nitroxoline is defined as a selective BET inhibitor, and it was shown as an anticancer agents against MLL leukemia, one of the BET-related diseases, and effectively inhibited the proliferation of MLL leukemia cells by inducing cell cycle arrest and apoptosis [180]. Nitroxoline has recently been investigated for its inhibition of angiogenesis, inducing apoptosis, and blocking cancer cell invasion. Clinical studies demonstrated that the routine nitroxoline administration for therapy of urinary system infections also sufficiently treated urological cancer within two to four-fold higher doses. In addition, it was shown that nitroxoline sulfate metabolite in urine effectively inhibited proliferation of cancer cell. Due to adequate feasibility of

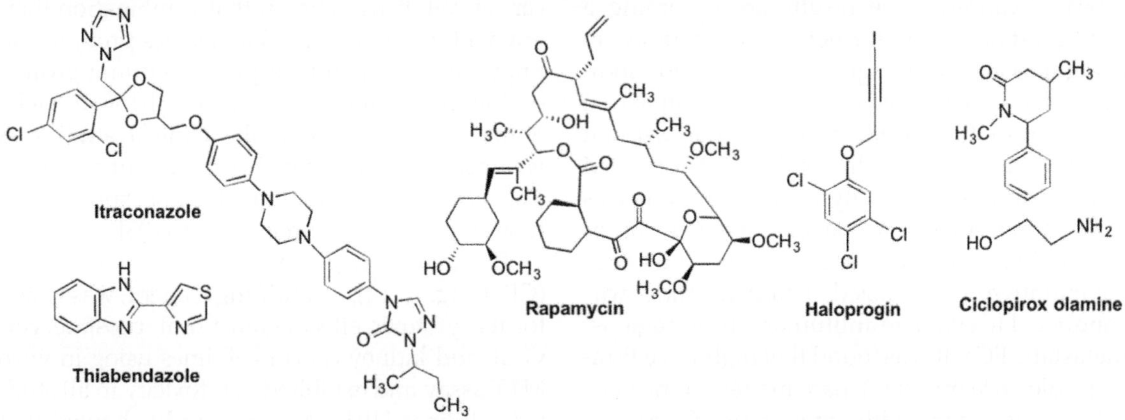

FIGURE 15.12 Chemical structures of antifungal drugs effective as potential anticancer therapies.

nitroxoline to use for urological cancer treatment, it has been approved to enter into a Phase II clinical trials in China for bladder cancer treatment [181].

Antibacterial clofoctol (Fig. 15.11) was reported effective for the treatment of PCa that inhibits cell proliferation. Clofoctol activates all three unfolded protein-response pathways including inositol requiring enzyme 1, double-stranded RNA-activated PK-like ER kinase, and activating transcription factor 6. Therefore, clofoctol was evaluated in human clinical studies for the treatment of PCa, and these findings suggested that clofoctol had a therapeutic value against prostate and other cancers [182].

Antifungal drugs

It is offered that scientist should consider cancer not only as a metabolic disease but also a metabolic parasite. A number of studies in the literature including in in vitro, in vivo, and clinical studies identified that antiparasitic and antifungal medications could be useful in cancer therapy [183].

Antifungal drugs are repurposed as a results of repurposing drugs in oncology project for cancer treatment. Among them, triazole ring–bearing antifungal agent itraconazole (Fig. 15.12) was reported as a potent antiangiogenic compound that inhibited cell signals, induced the autophagic growth arrest, reversed MDR in cancer and inhibited angiogenesis and Hedgehog (Hh)-signaling pathway [184,185]. In clinical trials, itraconazole was found to be beneficial for the treatment of leukemia, ovarian, breast, and pancreatic cancers. Combination chemotherapy with itraconazole was studied for non-small cell lung (combination with pemetrexed, bevacizumab, and ramucirumab) and skin (combination with vismodegib and sonidegib) cancers. Itraconazole enhanced the anticancer effect of other drugs in combinations [186,187]. Antiangiogenetic activity of itraconazole was explained through inhibition of the mTOR signaling that is due to direct inhibition of the lysosomal protein Niemann-Pick type C protein 1. This unique mechanism of itraconazole could potentially bring several clinical advantages over other mTOR inhibitors. It has nonimmunosuppressive properties, it has two distinct targets that decreases the likelihood of developing drug resistance, it specifically targets endothelial cells rather than cancer cells and decreases the occurrence of resistance-causing mutations, and it minimizes the risk of side effects. Since itraconazole is also well-tolerated in most patients and it is efficient in numerous types of cancer, these results are becoming a strong rationale for further clinical studies to using it as anticancer agents [188]. A study indicated that itraconazole had a modest impact on serum prostate–specific antigen levels without affecting circulating androgen levels that proved to have antitumor activity for PCa [189]. It was also shown that itraconazole delayed tumor growth in murine PCa xenograft models. A noncomparative, randomized, Phase II study was conducted to evaluate antitumor efficacy against metastatic PCa. It was found that high-dose Iitraconazole (600 mg/day) had modest antitumor activity in men with metastatic C-reactive protein prostate cancer (CRPC) [190]. The effectiveness of itraconazole in combination with other drugs to affecting cell survival, increasing the drug efficacy, and overcoming to drug resistance has been studied. However, it was found that itraconazole should be used longer time periods at various stages of cancers, in tumors associated with drug resistance and in other malignancies known to be affected by the Hh pathway and angiogenesis [185]. After findings potent antiangiogenic activity and enhanced cytotoxic efficacy of itraconazole in LC models, and exploratory clinical study was performed in order to determine the efficacy of itraconazole with cytotoxic chemotherapy in the treatment of advanced LC. Itraconazole was found well be tolerated in combination with pemetrexed, and the results have found consistency with preclinical data [191].

An antifungal agent rapamycin (Fig. 15.12) was first isolated from *Streptomyces hygroscopicus* and was identified as cell growth inhibitor in tumor cell lines. This inhibition was found to be depending on the inhibition of mTOR signaling pathway that would provide more clues to develop novel rapamycin analogues for treating breast and other cancers [192].

A combination of an experimental anticancer drug RAPTA-T 9, which contains ruthenium and an antifungal agent haloprogin (Fig. 15.12) was found synergistically induce the killing of cancer cell. It was shown that combination therapy with these two small molecules provided an improved therapeutic response in tumor growth of highly aggressive syngeneic B16F10 melanoma tumor model in the clinical studies. It was also reported that this treatment might be also efficient to inhibit metastatic spread and treat to metastasis of melanoma [193].

The anticancer potential of ciclopirox olamine (CPO, Fig. 15.12), an antifungal agent, was tested for its cytotoxic effect against oral, prostate, cervical, and kidney cancer cell lines using in vitro MTT assay and exhibited cytotoxicity in all studied cell lines [194]. Another study showed that CPO decreased cell growth and viability of malignant leukemia, myeloma, and solid tumor cell lines as well as primarily AML patients by inhibition of iron-dependent enzyme ribonucleotide reductase [195]. CPO alone and its combination with gemcitabine was evaluated in human pancreatic cell lines such as BxPC-3, Panc-1, and MIApaCa-2 and in humanized xenograft mouse models. Preclinical pharmacodynamic activity of CPO was examined for cell proliferation and clonogenic growth potential. It was found that these inhibitory effects were related with induction of reactive oxygen species (ROS), which were strongly associated with reduced BCL-xL and survivin level and activation of caspases. These results have shown that antitumor efficacy of CPO was superior and combination with gemcitabine was even more effective in pancreatic patient [196]. An oral formulation of CPO was studied in human Phase I study of relapsed or refractory hematological malignancies at 5–80 mg/m^2 daily dose for 5 days in 21-day treatment cycles. It was found that CPO was well tolerated in patients without DLT and displayed sustained pharmacodynamic activity [197].

Thiabendazole (TBZ, Fig. 15.12), an orally active antifungal drug, was studied for its vascular disrupt and antiangiogenic effects. Later studies showed that TBZ slowed tumor growth and decreased vascular density in animal tests; thus, it was considered that it might be chosen as a potential complementary therapeutic use in combination with current antiangiogenic therapies [198].

Antihelmintic drugs

Among the antihelminthic drugs niclosamide (Fig. 15.13) was reported as a potential antitumor agent. It was found that niclosamide and its ethanolamine salt (NEN) increased gene expression in HCC, disrupted interaction between CDC37 and HSP90 in HCC cells, and reduced the viability of HCC cells. It was also shown that NEN reduced the expression of proteins in STAT3, AKT−mTOR, and EGFR−Ras−Raf signaling pathways in mice. This result suggested that it might be considered for the treatment against HCC [199]. It was also demonstrated that niclosamide inhibited the expression of CIP2A and reactivated the tumor suppressor PP2A in NSCLC cells in addition to inhibition of cell proliferation, colony formation, and tumor sphere formation and inducing the mitochondrial dysfunction through increases mitochondrial production in NSCLC cells [200]. Furthermore, it was found that niclosamide induced mitochondrial dysfunctions via inhibiting mitochondrial respiration, complex I activity, and ATP generation, which led to oxidative stress. Niclosamide also increased the ROS; this mechanism of action suggested that it was through inhibition of proliferation and induction of apoptosis in cervical cancer cells. It was also shown that niclosamide played an essential role of mitochondrial metabolism and mTOR signaling in cervical cancer. These results indicated that niclosamide was a potential candidate for cervical cancer treatment as a single agent or might be used it as combination with standard chemotherapeutic agents [201].

Anticancer activity of a combination of niclosamide and sorafenib on renal cell carcinoma (RCC) was investigated. Niclosamide inhibited the expression of C-MYC and E2F1 while inducing the expression of PTEN gene in RCC cells. This effect was also being synergized with the suppression of RCC cell proliferation by sorafenib. These results suggested that niclosamide might be used as a potent agent against RCC [202]. Cytotoxic effect and migratory ability of niclosamide in urological cancer cell lines were measured by using MTT and scratch migration assay, respectively. Analysis of apoptosis and cell cycle changes showed that niclosamide has good ability in inhibition proliferation and migration. For this reason, it may have a potential to use it as a novel alternative treatment agent for bladder cancer [203].

Niclosamide was reported as antiadrenocortical carcinoma (ACC) agent by testing it on ACC cell lines, BD140A, SW-13, and NCI-H295R. It inhibited cellular proliferation in all tested cell lines by inducing caspase-dependent apoptosis and G1 cell cycle arrest. It also decreased cellular migration in addition to expression of b-catenin and reduced the level of mediators of EMT, such as N-cadherin and vimentin. These results suggested that niclosamide therapy was a good option in clinical trial for ACC who did not respond to standard therapy [204].

Mebendazole (MBZ, Fig. 15.13) was evaluated against several models of different types of cancer in preclinical studies. It was shown that

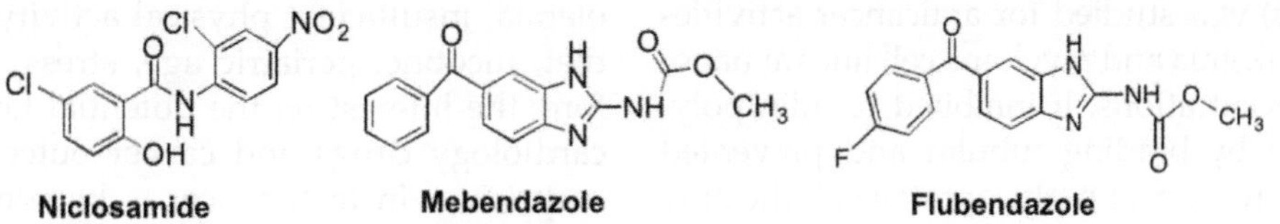

FIGURE 15.13 Chemical structures of antihelminthic drugs effective as potential anticancer therapies.

MBZ inhibited LC cells and chemoresistant melanoma cells and diminished growth of breast, ovary, colon carcinomas, and osteosarcoma with IC_{50} values in the range of 0.1–0.8 μM. In addition, MBZ presented inhibition in mice with inoculated H460 NSCLC cells. In preclinical studies, anticancer activities have been determined in adrenocortical cancer models [205]. It was shown that MBZ inhibited Hh signaling pathway in many types of cancer and therefore slowed the growth of Hh-driven human medulloblastoma cells at clinically applicable concentrations. MBZ was also combined with Hh inhibitor vismodegib, and this combination resulted in additive Hh signaling inhibition. These results suggested that MBZ could be safely administered and repurposed as a prospective therapeutic agent for many tumors [206]. It was found that MBZ decreased COX-2 expression, blood vessel formation, and VEGFR phosphorylation. It worked synergistically with antiinflammatory agent sulindac to reduce overexpression of MYC, BCL-2, and various proinflammatory cytokines. A combination of sulindac and MBZ was found effective against the reducing of FAP in patient with colon cancer [207]. Animal studies showed that MBZ was a promising drug for Glioblastoma Multiforme (GBM) therapy. MBZ disrupted microtubule formation in GBM cells, and in vitro activity studies presented a correlation with reduced tubulin polymerization. It was also shown that MBZ significantly extended survival time in mouse glioma models. These results have been found promising to clinical application of MBZ in therapy of GMB patient [208].

Another broad-spectrum benzimidazole-containing antihelmintic drug flubendazole (Fig. 15.13) was studied for anticancer activities against leukemia and myeloma cell lines at nanomolar concentrations. It inhibited tubulin polymerization by binding tubulin and prevented to overexpression of p-glycoproteins, indicating that flubendazole could overcome the resistance. A study was shown that combination of flubendazole and vinblastine reduced the viability of OCI-AML2 cells. All these findings identified flubendazole as a novel microtubule inhibitor for the treatment of leukemia and myeloma [209]. Signal transducer and activator of transcription 3 (STAT3) factor plays important role the occurrence of 70% of human cancers. When STAT3 signaling pathway is triggered by cancer cells, this has been implicated in the autophagic process. In vitro studies shown that flubendazole exerted potent antitumor activity by blocking the IL-6–induced nuclear translocation of STAT3 in human colorectal endometrium (CRC) cell lines and in the nude mouse model. In addition, flubendazole was found to reduce the expression of P-mTOR, P62, and BCL-2 and upregulate the beclin1 and LC3-I/II, which are major autophagy-related genes that induced potent cell apoptosis in CRC cells. It was also reported that flubendazole displayed a synergistic effect with the chemotherapeutic agent 5-fluorouracil in the treatment of CRC [210].

Cardiovascular drugs

Drug repurposing of approved cardiovascular drugs, such as aspirin, β-blockers, angiotensin-converting enzyme inhibitors, angiotensin II receptor blockers, cardiac glycosides, and statins, has been studied regarding of their antitumor activities. Although not all clinical trials have not present expected results, several experimental studies have shown acceptable properties [211].

Cardiovascular diseases and cancer share a common etiology, including obesity, metabolic abnormalities, hyperglycemia, hypercholesterolemia, insufficient physical activity, improper diet, nicotine, geriatric age, stress, etc. Therefore, the interest on the potential link between cardiology drugs and cancer outcomes is not surprising. In fact, cancer is known to induce the profound malfunction of typical cardiovascular-regulating systems, including the renin–angiotensin system, sympathetic nervous system, and coagulation cascade. For this reason, some cardiology drugs such as angiotensin-converting enzyme inhibitors, angiotensin receptor blockers, β-blockers, statins, and heparins have been studied for their repurposing potential in preclinical and clinical trials. All of them have been shown to attenuate cancer development through different mechanisms. The renin–angiotensin system inhibitors primarily reduce the inflammation, angiogenesis, and immunosuppression, β-blockers suppress migration and metastasis, heparins decreases metastasis, and statins effect cell growth, apoptosis, migration, and angiogenesis. Although there are strong indications for repurposing use of these drugs, clinical data are not currently sufficient. Notwithstanding the assumptions presented above, it is also important to consider that the inconsistency of clinical data exists for each discussed class of drugs. Therefore, further investigations in this context are necessary [212].

Carvedilol (Fig. 15.14) was investigated for its inhibitory effect on EGF-induced malignant transformation of JB6Pþ cells, which is a skin cell model used to study tumor detection. Carvedilol failed to promote anchorage-independent growth of JB6Pþ cells, and it dependently inhibited EGF-induced malignant transformation of JB6Pþ cells at nontoxic concentrations; these results suggested that carvedilol has chemopreventive activity against skin cancer but might not be an effective treatment of established tumors. On the basis of this finding, carvedilol is considered as a very promising candidate for future clinical trials of skin cancer prevention [213].

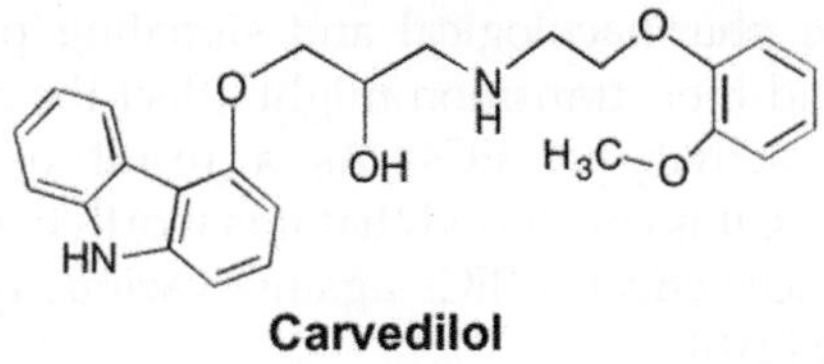

FIGURE 15.14 Chemical structures of antihypertensive drug carvedilol effective as potential anticancer therapies.

Miscellaneous drugs repurposed for anticancer activities

An antialcoholism drug disulfiram (Fig. 15.15) alone or in combination with anticancer drugs has been identified as a promising agent against various human cancers such as breast, cervical, colorectal, lung, melanoma, prostate, glioblastoma, liver, myeloma, and leukemia. These results have been confirmed in both preclinical and clinical studies. Disulfiram is converted to ethyldithiocarbamate (EtDTC) in the human body, and a bis (diethyldithiocarbamate)-copper (II) complex ($Cu(EtDTC)_2$) is formed. This complex inhibits the cellular proteasome by targeting the 19S rather than 20S proteasome. It also targets MDR, angiogenesis, invasion, and anticancer activity as a DNA demethylation agent or inhibitor of RING-finger E3 ubiquitin ligases. In preclinical studies, it was found that (Cu(EtDTC)2) complex of disulfiriam acts like a proteasome inhibitor which indudes oxidateive stress, reduces nuclear factor kappa binding (NFkB) activity and enhances the sensitivity of cancer cells to chemotherpeutic agents [214].

EtDTC has been confirmed for use in the treatment of high-risk breast cancer patients in a Phase II clinical trial. Another study showed that a combination of oral zinc gluconate and disulfiram induced the clinical remission of hepatic metastasis in patient with ocular melanoma. This complex was also found promising against liver cancer. With all these results, it was concluded that more clinical studies were required to prove the potential of disulfiram as an anticancer therapy in clinical studies [215].

Anticancer activity of IV anesthetic drug propofol (Fig. 15.15) alone and its combination with BCR–ABL tyrosine kinase inhibitors (TKIs) have been studied against in CML cell lines, patient progenitor cells, and mouse xenograft model. It was shown that propofol alone inhibited proliferation and induced the apoptosis in in vitro cell culture system and in in vivo xenograft model. Furthermore, it was found effective in

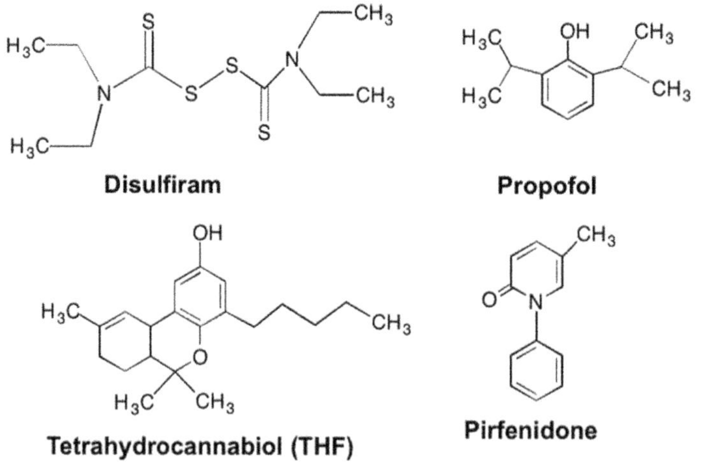

Disulfiram

Propofol

Tetrahydrocannabiol (THF)

Pirfenidone

FIGURE 15.15 Chemical structures of miscellaneous drugs effective as potential anticancer therapies.

inducing apoptosis and inhibited colony formation in CML CD34 progenitor cells than normal bone marrow counterparts. Since it was shown that propofol suppressed the phosphorylation of Akt, mTOR, S6, and 4EBP1 in K562 in several studies, it is suggested that propofol augmented BCR–ABL TKI's inhibitory effects [216].

Antifibrotic drug pirfenidone (Fig. 15.15) has been studied to be repurposed as anticancer drugs because of the activation of fibroblasts in inflammatory conditions has similar characteristics as cancer-associated fibroblasts (CAFs) that contribute actively to the malignant phenotype. To claim this hypothesis, the effect of pirfenidone alone and in combination with cisplatin on human patient-derived CAF and NSCLC cell lines was tested. The impact of cell death in tumor growth in a mouse model was determined. Although the mechanisms was uncertain, pirfenidone–cisplatin combination has been found effective against tumor-promoting CAFs and tumor cell death. The results demonstrated that the combination of cisplatin and pirfenidone was effective in preclinical studies for NSCLC, and it might be a new therapeutic option for patient [217].

The endocannabinoids (ECs) are the active component of marijuana (*Cannabis sativa*), which have significant psychoactive effects. Tetrahydrocannabinol (Fig. 15.15), is the chemical responsible for most of marijuana's psychological effects. It relieves chronic pain, induces an increase in appetite, alleviates nausea, and eliminates anxiety. ECs have been studied for anticancer activity in the last decade, and endocannabinoid system became an attractive new targets for the treatment of various cancer subtypes. It was reported that the ECs played dual role in both tumorigenesis and inhibition of tumor growth and metastatic spread that has transformed the cannabinoid receptors CB1R and CB2R, which are the members of a large family of membrane proteins called G protein–coupled receptors. These receptor-containing heteromers have been reported in cancer cells. These protein complexes provide unique pharmacological and signaling properties, and their transition might affect the antitumoral activity of ECs. As a result of these findings, it is concluded that it is worth to search anticancer effect of ECs against various type of cancer [218].

Drug repurposing studies for personalized medicine

Although, many biology-based precision drugs are available, which neutralize aberrant molecular pathways in cancer, molecular heterogeneity and the lack of reliable diagnostic biomarkers for many drugs make failure for chosen treatment of cancer in many individuals. Single-cell sequencing technologies are used for tumor characterization and heterogeneity, at the genetic, epigenetic, transcriptomic level, and provide increasing knowledge of cancer progression and treatment [219]. NGS technologies give genetic information about cancer cells to clarify the map of intratumor heterogeneity and enable to identify true genetic biomarkers that correlate with aggressive tumors. Furthermore, these studies help to understand the prognosis of disease and identify better drugs for specific patient. Many high-throughput sequencing projects are set up to identify genetic and epigenetic mutations, diagnosis, prognosis, and the potential efficacy of various drugs taken earlier by cancer patients. In this context, anticancer and nonanticancer drugs have been recently screened for genetic analyses to find the most effective drug that might be used for cancer treatment in clinic. As a result of these studies, a few commercial services of these strategies in a number of cancer clinics, such as Foundation Medicine, was applied [220]. In silico studies are used for identification of the effects of a single mutation on several pathways and design-specific inhibitors against tumor triggers that enable to reach specifically to targeted tumor cells without affecting the normal tissues. Recent innovative studies such as big data analysis for prediction to cancer cell mutations and use number of aberration data against various known drugs with molecular targets and their combinations were employed to regimen of cancer patients [221]. This approach was used to predict possible regimen for four MM patient's cells, and it was shown that the method was effective [222]. An integrative network analysis was used to detect mutated genes, and it was found that only 33.3% of the current cancer drugs would be potentially effective. In addition to this study, a big data analyses were applied to evaluate the chance of treatment of these cancer cells by potential drugs was found 66.7% [223]. To eliminate the difficulties for the integration of NGS with clinically validated biomarkers, the Database of Evidence for Precision Oncology (DEPO) was used for the druggability of genomic, transcriptomic, and proteomic biomarkers. A pan cancer cohort of 6570 tumors was used to identify tumors with potentially druggable biomarkers consisting of drug-associated mutations, mRNA expression outliers, and protein/phosphoprotein expression outliers identified by DEPO and found that 3% were druggable based on FDA-approved in specific cancer types. Overall, the results suggested that this analysis platform was efficient to detect multiomics alterations as biomarkers of druggability and helps current efforts to treat oncology patients. This study is important to describe the possibility of druggable alternatives and interpretation of clinical studies in cancer treatment [224].

A novel targeted RNA NGS (t/RNA-NGS) technique was used to detect the analysis of biological pathway activities involved in tumor behavior in many cancer types (e.g., tyrosine kinase signaling, angiogenesis signaling, immune response, metabolism). Quantitative measurement of transcript levels and splice variants of hundreds of genes identified known glioma-associated molecular aberrations and aberrant expression levels of actionable genes and mutations with nonassociated glioma to provide the possibility of drug repurposing for individual patients. Thus, t/RNA-NGS was reported as an efficient method for repurposing of drugs in an individualized manner [225].

The rate of nontreatable and metastatic thyroid cancers of follicular origins was reported as 5%. To understand the treatment of this type of thyroid cancer patients and further understanding of the tumor's genetic information, a NGS technology was applied to interrogate 740 mutational hotspots in 46 oncogenes. The studies provided to determine 21 mutations and 11 oncogenes in the 22 fatal thyroid cancer samples and brought new insights to future targeted therapy for these patients. According to obtained results, dual therapies targeting both the oncogenes and tumor suppressor genes, as those reported in HCC and melanoma, and 44 combination therapies inhibiting both the BRAF and MEK pathways could be potential promising personalized treatment choices for 45 patients who are fatal thyroid cancer [226].

Using NGS technologies and early biomarkers of therapeutic responses, a serial changes of plasma-circulating tumor DNA (ctDNA) in 41 metastatic colorectal cancer (mCRC) patients receiving first-line chemotherapies was examined and tested its association with treatment outcomes according to radiological assessments. In this study, somatic mutations in 50 cancer-related genes in ctDNA before each of the first four treatment cycles was detected and mutations in 95.7% of pretreatment ctDNA samples were observed. This data suggested that ctDNA mutations could be detected in a high proportion of treatment-naive mCRC patients via NGS. Early changes in the ctDNA levels showed the potential of later radiologic responses. In this study, it was concluded that ctDNA monitoring might be integrated with imaging to assess responses to anticancer treatment in mCRC [227].

Recent progress of NGS technology allows the identification of important variants and structural changes in DNA and RNA in cancer patients' samples and correlation of specific variants and/or structural changes with actionable therapeutics known to inhibit these variants. To efficiently use this technology, IMPACT web portal was created to connect molecular profiles of tumors to approved drugs, investigational therapeutics, and pharmacogenetics-associated drugs. This portal contains 776 drugs connected to 1326 target genes and 435 target variants, fusion, and copy number alterations. This portal can be used into 3 levels to search for various genetic alterations and connect them for actionable therapeutics. Scientist can use this portal to find approved drugs with variant-specific information and gene-level information, drugs currently in oncology clinical trials, and pharmacogenetics associations between approved drugs and genes. This portal is useful to link between the actionable? therapeutics for translational and drug repurposing research. It is also used to query genes and variants to approved and investigational drugs. Thus, it is considered as a valuable database for personalized medicine and drug repurposing [228].

Although NGS and big data analyses entered into clinical practice quite fast, and tumor-specific drug repurposing studies, further efforts are necessary to generalize or widely adopt these approaches [229]. These progresses will allow to using small molecule drugs, which includes drug repurposing inherently and associate big data analyses and continue to evolve in the forthcoming years' further fine-tuning application of these technologies for cancer treatment.

Conclusion

As explained at the beginning of this chapter, there are many advantages and disadvantages of drug reconstruction which are summarized in Table 15.2. Drug repurposing treatment has come a considerable progress in many type of cancers today. Different chemical entities have been used to identify alternative treatment strategies for several cancers. Table 15.3 summarized which type of repurposed drugs is used for their anticancer properties in different cancers. As seen, many different variety of drugs

TABLE 15.3 Repurposed drugs for their novel therapeutic indications in cancer.

Cancer Types	Drugs	Primary Indication of Drugs	References
Bladder	Nitroxolin	Antimicrobial	[182]
	Niclosamide	Antihelminthic	[204]
Breast	Clorpromazine	Antipsychotic	[47]
	Celecoxib	NSAID	[62–65]
	Valproic acid	Antiepileptic	[78]
	Phenytoin	Antiepileptic	[89]
	Metformin	Antidiabetic	[113–116: 123–125]
	Acyclovir	Antiviral	[171]
	Disulfiram	Antialcoholism	[216]
	Artesunate	Antimalarial	[156]
	Rapamycin	Antifungal	[193]
Cervical	Niclosamide	Antihelminthic	[202]
Colorectal	Chlorpromazine	Antipsychotic	[46]
	Valproic acid	Antiepileptic	[88]
	Simvastatin	Antihiperlipidemic	[100, 101, 104]
	Metformin	Antidiabetic	[117]
	Mefloquine	Antimalarial	[166]
	Flubendazole	Antihelminthic	[211]
	Artesunate	Antimalarial	[156]
	Mebendazole	Antihelminthic	[208]
Endometrial	Metformin	Antidiabetic	[122]
	Quinacrine	Antimalarial	[170]
Gastric	Metformin	Antidiabetic	[118–120]
Glioblastoma	Biguanides	Antidiabetic	[109]
	Temozolomide	Antimalarial	[169]
	Mebendazole	Antihelminthic	[209]
Hepatocellular carcinoma	Fluspirilen	Antipsychotic	[40]
	Chloroquine	Antimalarial	[168]
	Niclosamide	Antihelminthic	[200]
Leukemia	Penfluridol	Antipsychotic	[41]
	Auranofin	Antireumatoid	[75–77]
	Valproic acid	Antiepileptic	[78, 80, 81]
	Statins (simvastatin, mevastatin, lovastatin, and pravastatin)	Antihiperlipidemic	[96–98, 105]
	Dihydroartemisinin	Antimalarial	[153, 154]
	Artesunate	Antimalarial	[155, 156]
	Zidovudine	Antiviral	[174]
	Ribavirin	Antiviral	[177]
	Nitroxoline	Antimicrobial	[181]
	Itraconazole	Antifungal	[187, 188]
	Ciclopirox olamine	Antifungal	[196]

(*Continued*)

TABLE 15.3 Repurposed drugs for their novel therapeutic indications in cancer.—cont'd

Cancer Types	Drugs	Primary Indication of Drugs	References
	Flubendazole	Antihelminthic	[210]
	Disulfiram	Antialcoholism	[215]
	Propofol	Anesthetic	[217]
Lung	Penfluridol	Antipsychotic	[41]
	Phenpthiazines	Antipsychotic	[45]
	Aspirin	NSAI	[51, 52]
	Celecoxib	NSAI	[61]
	Auranofin	Antireumatoid	[75-77]
	Valproic acid	Antiepileptic	[78]
	Phenytoin	Antiepileptic	[89]
	Statins	Antihiperlipidemic	[95–97, 103]
	Fenofibrate	Antihiperlipidemic	[107]
	Dihydroartemisinin	Antimalarial	[153]
	Itraconazole	Antifungal	[187, 188, 192]
	Mebendazole	Antihelminthic	[206]
	Disulfiram	Antialcoholism	[215]
Lymphoma	Celecoxib	NSAI	[59]
Melanoma	Valproic acid	Antiepileptic	[78]
	Statins	Antihiperlipidemic	[95]
	Artemisone	Antimalarial	[156]
	Haloprogin	Antifungal	[194]
	Mebendazole	Antihelminthic	[206]
	Disulfiram	Antialcoholism	[215, 216]
	Artesunate	Antimalarial	[156]
Osteosarcoma	Dihydroartemisinin	Antimalarial	[153]
Overian	Artesunate	Antimalarial	[156]
Pancreatic	Metformin	Antidiabetic	[126–129]
	Zidovudine	Antiviral	[174]
	Itraconazole	Antifungal	[187, 188]
	Ciclopirox olamine	Antifungal	[197]
	Celecoxib	NSAI	[72]
	Fenofibrate	Antihiperlipidemic	[107]
Prostate	Ciclopirox olamine	Antifungal	[195]
	Celecoxib	NSAI	[96]
	Simvastatin	Antihiperlipidemic	[102]
	Fenofibrate	Antihiperlipidemic	[107]
	Metformin	Antidiabetic	[122]
	Clofoctol	Antimicrobial	[183]
	Itraconazole	Antifungal	[190, 191]
Renal	Artesunate	Antimalarial	[156]
	Ciclopirox olamine	Antifungal	[195]
Skin	Carvedilol	Cardivascular	[214]

TABLE 15.3 Repurposed drugs for their novel therapeutic indications in cancer.—cont'd

Cancer Types	Drugs	Primary Indication of Drugs	References
Solid tumors	Valproic acid	Antiepileptic	[88]
	Zidovudine	Antiviral	[175]
	Perampanel	Anticonvulsant	[91]
Thoracic	Valproic acid	Antiepileptic	[83–85]
Tyroid	Celecoxib	NSAI	[74]
	Metformin	Antidiabetic	[121]
Urological	Nitroxoline	Antimicrobial	[182]
	Niclosamide	Antihelminthic	[204]

were used for repurposing. In fact, further investigations are needed in order to provide more efficient and safe therapeutic options, and this is mentioned after studies in many literature.

Although the most effective results obtained in oncological drug repurposing with combination studies promise to create synergistically or additively therapeutic benefits to minimize drug resistance, it can equally produce unwanted side effects. If complications from combination therapies create further problems for patients in terms of more severe clinical symptoms and toxicities that badly impact and alter the patient's life expectancy, this also brings financial difficulties and lowers the patient's quality of life. The most of studied combination summarized in this chapter showed synergicity, success, and slight to moderate benefits without any toxic and adverse reactions, but some serious side effects and lack of effectiveness were reported. For instance; therapeutic benefit was minimal during the treatment of renal cancer with zoledronate, thalidomide, and interferon combination therapy [39]. A combination of gemcitabine, 5-FU, and thalidomide resulted lethal in vascular toxicities [38]. Another big problem is to conduct efficient clinical studies. Many of the prospective clinical studies have been poorly designed because of taking account into some or partial factors identified in preclinical studies, which considered that compound would be effective. The heterogeneity of cancers, studies on small number of patient in clinical trials, and unsuitable drug doses are the main obstacles. As a matter of fact, in some studies, it was reported that well-designed and detailed clinical studies were needed in order to proove the use of repurposed drugs and combinations that were provided certain statements for anticancer therapy.

In fact, lack of the drug–drug interaction studies is the weakest part of drug repurposing studies in current efforts, and well-designed prospective studies related with interactions between the chemical entities are needed for novel indications in oncology. The scaffold repurposing is an important tool to novel potential applications for cancer target and can be rationally designed to eliminate unwanted physicochemical properties to increase the success of therapy. Already studied SAR and pharmacokinetic parameters will be beneficial to aim of these studies. Obviously, the most important advantage is that pharmacokinetic and toxicity data are available to confirm efficacy and safety and this lowers to cost of studies, but it should remember that current drug repurposing strategies are not completely enough for novel therapeutic benefits in cancer treatment. Further investigation on drug repurposing studies undoubtedly will be beneficial to discover new therapeutic applications by using old drugs; however, medicinal chemist should find novel strategies to combine this method with other de novo drug development methods that allow to use multiple heterogeneous information sources such as machine learning and big data analysis.

References

[1] Mucke HAM. Therapeutic drug repurposing, repositioning and rescue-Part I: overview. Drug Discov World Winter 2014;15:49–62.

[2] Allarakhia M. Open-source approaches for the repurposing of existing or failed candidate drugs: learning from and applying the lessons across diseases. Drug Des Dev Ther 2013;7:753–66.

[3] Corsello SM, Bittker JA, Liu Z, Gould J, McCarren P, Hirschman JE, Johnston SE, Vrcic A, Wong B, Khan M, Asiedu J, Narayan R, Mader CC, Subramanian A, Golub TR. The drug repurposing hub: a next-generation drug library and information resource. Nat Med 2017;23(4):405–8.

[4] Xue H, Li J, Xie H, Wang Y. Review of drug repositioning approaches and resources. Int J Biol Sci 2018; 14(10):1232–44.

[5] Khanapure A, Chuki P, De Sousa A. Drug repositioning: old drugs for new indications. Indian J Appl Res 2014;4(8):462–6.

[6] McCabe B, Liberante F, Mills KI. Repurposing medicinal compounds for blood cancer treatment. Ann Hematol 2015;94:1267–76.

[7] Würth R, Thellung S, Bajetto A, Mazzanti M, Florio T, Barbieri F. Drug repositioning opprtunities for cancer therapy: novel molecular targets for known compounds. Drug Discov Today 2016;21(1):190–9.

[8] Ashburn TT, Thor KB. Drug repositioning: identifying and developing new uses for existing drugs. Nature 2004;3:673–83.

[9] Nosengo N. New tricks for old drugs. Nature 2016; 534:314–6.

[10] Wilkinson GF, Pritchard K. In vitro screening for drug repositioning. J Biomol Screen 2015;20(2):167–79.

[11] Hernandez JJ, Pryszlak M, Smith L, Yanchus C, Kurji N, Shahani VM, Molinski SV. Giving drugs a second chance: overcoming regulatory and financial hurdles in repurposing approved drugs as cancer therapeutics. Front Oncol 2017;7:273.

[12] Parmar M, Panchal S. Drug Reposition of non-cancer drugs for cancer treatments through pharmacovigilance approach-repurposing drugs in oncology. Asian J Pharmaceut Clin Res 2019;12(2):310–4.

[13] Naylor S, Schonfeld JM. Therapeutic drug repurposing, repositioning and rescue. Drug Discov World Winter 2014;15:49–62.

[14] Ding X. Drug repositioning needs a rethink. Nature 2016;535:355.

[15] Shim JS, Liu JO. Recent advances in drug repositioning for the discovery of new anticancer drugs. Int J Biol Sci 2014;10(7):654–63.

[16] Das AK. Anticancer effect of antimalarial artemisinin compounds. Ann Med Health Sci Res 2015;5(2):93–102.

[17] Cheng F, Zhao J, Fooksa M, Zhao Z. A network-based drug repositioning infrastructure for precision cancer medicine through targeting signi cantly mutated genes in the human cancer genomes. J Am Med Inf Assoc 2016;23(4):681–91.

[18] Gupta SC, Sung B, Prasad S, Webb LJ, Aggarwal BB. Cancer drug discovery by repurposing: teaching new tricks to old dogs. Trends Pharmacol Sci 2013; 34(9):508–17.

[19] Chen H, Wu J, Gao Y, Chen H, Zhou J. Scaffold repurposing of old drugs towards new cancer drug discovery. Curr Top Med Chem 2016;6(19). 2107-2014.

[20] Cheng F, Hong H, Yang S, Wei Y. Individualized network-based drug repositioning infrastructure for precision oncology in the panomics era. Briefings Bioinf 2017;18(4):682–97.

[21] Huang J, Zhao D, Liu Z, Liu F. Repurposing psychiatric drugs as anti-cancer agents. Cancer Lett 2018;419: 257e265.

[22] Zhou S, Wang F, Hsieh T-C, Wu JW, Wu E. Thalidomide-A notorious sedative to a wonder anticancer drug. Curr Med Chem 2013;20(33):4102–8.

[23] Stephens TD, Fillmore BJ. Hypothesis: thalidomide embryopathy-proposed mechanism of action. Teratology 2000;61(3):189–95.

[24] Hideshima T, Raje N, Richardson PG, Anderson KC. A review of lenalidomide in combination with dexamethasone for the treatment of multiple myeloma. Therapeut Clin Risk Manag 2008;4(1):129–36.

[25] Galustian C, Dalgleish A. Lenalidomide: a novel anticancer drug with multiple modalities. Expet Opin Pharmacother 2009;10(1):125–33.

[26] Licht JD, Shortt J, Johnstone R. From anecdote to targeted therapy: the curious case of thalidomide in multiple myeloma. Cancer Cell 2014;25:9–11.

[27] Richardson PG, Weller E, Lonial S, Jakubowiak AJ, Jagannath S, Raje NS, Avigan DE, Xie W, Ghobrial IM, Schlossman RL, Mazumder A, Munshi NC, Vesole DH, Joyce R, Kaufman JL, Doss D, Warren DL, Lunde LE, Kaster S, DeLaney C, Hideshima T, Mitsiades CS, Knight R, Esseltine D-L, Anderson KC. Lenalidomide, bortezomib, and dexamethasone combination therapy in patients with newly diagnosed multiple myeloma. Blood 2010;116(5):679–86.

[28] Ríos-Tamayo R, Martín-García A, Alarcón-Payer C, Sánchez-Rodríguez D, María del Valle Díaz de la Guardia A, Collado CG, Morales AJ, Chacón MJ, Barrera JC. Pomalidomide in the treatment of multiple myeloma: design, development and place in therapy. Drug Des Dev Ther 2017;11:2399–408.

[29] Agarwala SS, Kirkwood JM. Temozolomide, a novel alkylating agent with activity in the central nervous system, may improve the treatment of advanced metastatic melanoma. Oncologist 2000;5:144–51.

[30] Hwu WJ. New approaches in the treatment of metastatic melanoma: thalidomide and temozolomide. Oncology 2000;14(12):25–8.

[31] Soape MP, Verma R, Payne JD, Wachtel M, Hardwicke F, Cobos E. Treatment of hepatic epithelioid hemangioendothelioma: finding uses for thalidomide in a new era of medicine. Case Rep Gastrointest Med 2015:4. Article ID 326795.

[32] Eisen T, Boshoff C, Mak I, Sapunar F, Vaughan MM, Pyle L, Johnston SRD, Ahern R, Smith IE, Gore ME. Continuous low dose thalidomide: a phase II study in advanced melanoma, renal cell, ovarian and breast cancer. Br J Cancer 2000;82(4):812–7.

[33] Hwang C, Heath EI. Angiogenesis inhibitors in the treatment of prostate cancer. J Hematol Oncol 2010; 3(26):1–12.

[34] Keifer JA. Inhibition of NF-kappa B activity by thalidomide through suppression of IKappaB kinase activity. J Biol Chem 2001;276(25):22382–7.

[35] Meng L-J, Wang J, Fan W-F, Pu X-L, Liu F-Y, Yang M. Evaluation of oral chemotherapy with capecitabine and cyclophosphamide plus thalidomide and prednisone in prostate cancer patients. J Cancer Res Clin Oncol 2011;138(2):333–9.

[36] Di Lorenzo G, Autorino R, De Laurentiis M, Forestieri V, Romano C, Prudente A, De Placido S. Thalidomide in combination with oral daily cyclophosphamide in patients with pretreated hormone refractory prostate cancer: A phase I clinical trial. Cancer Biol Ther 2007;6(3):313–7.

[37] Amato RJ, Khan M. A phase I clinical trial of low-dose interferon-α-2A, thalidomide plus gemcitabine and capecitabine for patients with progressive metastatic renal cell carcinoma. Cancer Chemother Pharmacol 2007;61(6):1069–73.

[38] Desai AA, Vogelzang NJ, I Rini B, Ansari R, Krauss S, Stadler WM. A high rate of venous thromboembolism in a multi-institutional Phase II trial of weekly intravenous gemcitabine with continuous infusion fluorouracil and daily thalidomide in patients with metastatic renal cell carcinoma. Cancer 2002;95(8):1629–36.

[39] Tannir N, Jonasch E, Pagliaro LC, Mathew P, Siefker-Radtke A, Rhines L, Tu S-M. Pilot trial of bone-targeted therapy with zoledronate, thalidomide, and interferon-γ for metastatic renal cell carcinoma. Cancer 2006;107(3):497–505.

[40] Shi X-N, Li H, Yao H, Liu X, Li L, Leung K-S. In silico identification and in vitro and in vivo validation of anti-psychotic drug fluspirilene as a potential CDK2 inhibitor and a candidate anti-cancer drug. PLoS One 2015;10(7). e0132072.

[41] Hung W-Y, Chang J-H, Cheng Y, Cheng G-Z, Huang H-C, Hsiao M, Chung C-L, Lee W-J, Chien M-H. Autophagosome accumulation-mediated ATP energy deprivation induced by penfluridol triggers nonapoptotic cell death of lung cancer via activating unfolded protein response. Cell Death Dis 2019;10:538.

[42] Wu S-Y, Wu S-Y, Wen Y-C, Lee W-J, Ku C-C, Yang Y-C, Chien M-H, Chow J-M, Yang S-F. Penfluridol triggers cytoprotective autophagy and cellularapoptosis through ROS induction and activation of the PP2A-modulated MAPK pathway in acute myeloid leukemia with different FLT3 statuses. J Biomed Sci 2019; 26(1):63.

[43] Wu C-H, Bai L-Y, Tsai M-H, Chu P-C, Chiu C-F, Chen MY, Chiu S-J, Chiang J-H, Weng J-R. Pharmacological exploitation of the phenothiazine antipsychotics to develop novel antitumor agents-A drug repurposing strategy. Sci Rep 2016;6:27540.

[44] Zhelev Z, Ohba H, Bakalova R, Hadjimitova V, Ishikawa M, Shinohara Y, Baba Y. Phenothiazines suppress proliferation and induce apoptosis in cultured leukemic cells without any influence on the viability of normal lymphocytes. Cancer Chemother Pharmacol 2004;53(3):267–75.

[45] Zong D, Zielinska-Chomej K, Juntti T, Mork B, Lewensohn R, Haag P, Viktorsson K. Harnessing the lysosome-dependent antitumor activity of phenothiazines in human small cell lung cancer. Cell Death Dis 2014;5:e1111.

[46] Lee W-Y, Lee W-T, Cheng C-H, Chen K-C, Chou C-M, Cheng H-W, Ho M-N, Lin C-W, Cheng C-H, Chou C-M, Lin C-W, Chung C-H, Sun M-S. Re-positioning antipsychotic chlorpromazine for treating colorectal cancer by inhibiting sirtuin 1. Oncotarget 2015;6(29): 27580–95.

[47] Yang C-E, Lee W-Y, Cheng H-W, Chung C-H, Mi F-L, Lin C-W. Antipsychotic chlorpromazine suppresses YAP signaling, stemness properties, and drug resistance in breast cancer cells. Chem Biol Interact 2019; 312:30621–5.

[48] Suzuki S, Okada M, Kuramoto K, Takeda H, Sakaki H, Watarai H, Sanomachi T, Seino S, Yoshioka T, Kitanaka C. Aripiprazole, an antipsychotic and partial dopamine agonist, inhibits cancer stem cells and reverses chemoresistance. Anticancer Res 2016;36:5153–62.

[49] Marcinkute M, Afshinjavid S, Fatokun AA, Javid FA. Fluoxetine selectively induces p53-independent apoptosis in human T colorectal cancer cells. Eur J Pharmacol 2019;857:172441.
[50] Gasic GJ, Gasic TB, Murphy S. Anti-metastatic effect of aspirin. Lancet 1972;300:932–3.
[51] Rothwell PM, Fowkes FG, Belch JF, Ogawa H, Warlow CP, Meade TW. Effect of daily aspirin on long-term risk of death due to cancer: analysis of individual patient data from randomised trials. Lancet 2011;377:31–41.
[52] Din FV, Theodoratou E, Farrington SM, Tenesa A, Barnetson RA, Cetnarskyi R, Stark L, Porteous ME, Campbell H, Dunlop MG. Effect of aspirin and NSAIDs on risk and survival from colorectal cancer. Gut 2010;59:1670–9.
[53] Hua H, Zhang H, Kong Q, Wang J, Jiang Y. Complex roles of the old drug aspirin in cancer chemoprevention and therapy. Med Res Rev 2018;39(1): 114–45.
[54] Turini ME, DuBois RN. Cyclooxygenase-2: a therapeutic target. Annu Rev Med 2002;53:35–57.
[55] Sobolewski C, Cerella C, Dicato M, Ghibelli L, Diederich M. The role of cyclooxygenase-2 in cell proliferation and cell death in human malignancies. Int J Cell Biol 2010;1–21. ID:21558.
[56] Mohsen V, Mohsen A. The discovery and development of cyclooxygenase-2 inhibitors as potential anticancer therapies. Expert Opin Drug Discov 2014;9(3): 255–67.
[57] Lanas A. Clinical experience with cyclooxygenase-2 inhibitors. Rheumatology 2002;41(Suppl. 1):16–22.
[58] Fitz Gerald GA, Patrono C. The coxibs, selective inhibitors of cyclooxygenase-2. N Engl J Med 2001;345: 433–42.
[59] Jendrossek V. Targeting apoptosis pathways by celecoxib in cancer. Cancer Lett 2011;332:313–24.
[60] Steinbach G, Lynch PM, Phillips RK, Wallace MH, Hawk E, Gordon GB, Wakabayashi N, Saunsers B, Shen Y, Fujimura T, Su LK, Levin B, Godio L, Patterson S, Rodriguez-Bigas MA, Jester SL, King KL, Schumacher M, Abbruzzese J, DuBois RN, Hittelman WN, Zimmerman S, Sherman JW, Kelloff G. The effect of celecoxib, a cyclooxygenase-2 inhibitor in familial adenomatous polyposis. N Engl J Med 2000;342:1946–52.
[61] Schonthal AH. Direct non-cyclooxygenase-2 targets of celecoxib and their potential relevance for cancer therapy. Br J Cancer 2007;97(11):1465–8.
[62] Harris RE. Cyclooxygenase-2 (COX-2) blockade in the chemoprevention of cancers of the colon, breast, prostate, and lung. Inflammopharmacology 2009;17: 55–67.
[63] Friedman GD, Ury HK. Initial screening for carcinogenicity of commonly used drugs. J Natl Cancer Inst 1980;65:723–33.
[64] Harris RE, Beebe-Donk J, Alshafie GA. Reduction in the risk of human breast cancer by selective cyclooxygenase-2 (COX-2) inhibitors. BMC Cancer 2006;6:27.
[65] Algra AM, Rothwell PM. Effects of regular aspirin on long-term cancer incidence and metastasis: a systematic comparison of evidence from observational studies versus randomised trials. Lancet Oncol 2012; 13:518–27.
[66] Gonga L, Thorna CF, Bertagnollic MM, Grosserd T, Altmana RB, Kleina TE. Celecoxib pathways: pharmacokinetics and pharmacodynamics. Pharmacogenetics Genom 2012;22(4):310–8.
[67] Regulski M, Regulska K, Prukała W, Piotrowska H, Stanisz B, Murias M. COX-2 inhibitors: a novel strategy in the management of breast cancer. Drug Discov Today 2016;21(4):598–615.
[68] Falandry C, Debled M, Bachelot T, Delozier T, Cretin J, Romestaing P, Mille D, You B, Mauriac L, Pujade-Lauraine E, Freyer G. Celecoxib and exemestane versus placebo and exemestane in postmenopausal metastatic breast cancer patients: a double-blind phase III GINECO study. Breast Cancer Res Treat 2009;116:501–8.
[69] Antonarakis ES, Heath EI, Walczak JR, Nelson WG, Fedor H, De Marzo AM, Carducci MA. Phase II, randomized, placebo-controlled trial of neoadjuvant celecoxib in men with clinically localized prostate cancer: evaluation of drug-specific biomarkers. J Clin Oncol 2009;27(30):4986–93.
[70] Koch A, Bergman B, Holmberg E, Sederholm C, Ek L, Kosieradzki J, Sörenson S. Effect of celecoxib on survival in patients with advanced non-small cell lung cancer: a double blind randomised clinical phase III trial (CYCLUS study) by the Swedish Lung Cancer Study Group. Eur J Cancer 2011;47(10):1546–55.
[71] Zhao J, Wang Z, Duan J, Guo Q, Bai H, Yang L, Wang J. A phase II clinical trial of celecoxib combined with platinum-based regimen as first-line chemotherapy for advanced non-small cell lung cancer patients with cyclooxygenase-2 positive expression. Chin J Cancer Res 2009;21(1):1–12.
[72] Dragovich T, Burris H, Loehrer P, Von Hoff DD, Chow S, Stratton S, Gordon M. Gemcitabine plus celecoxib in patients with advanced or metastatic pancreatic adenocarcinoma. Am J Clin Oncol 2008; 31(2):157–62.
[73] Mohamma-Dianpanah M, Shafizad A, Khademi B, Ansari M, Mosalaei A, Ghalaei R-S, Mosleh-Shirazi M. Efficacy and safety of concurrent

chemoradiation with weekly cisplatin ± low-dose celecoxib in locally advanced undifferentiated nasopharyngeal carcinoma: a phase II-III clinical trial. J Cancer Res Therapeut 2011;7(4):442.

[74] Mrozek E, Kloos RT, Ringel MD, Kresty L, Snider P, Arbogast D, Shah MH. Phase II study of celecoxib in metastatic differentiated thyroid carcinoma. J Clin Endocrinol Metabol 2006;91(6):2201–4.

[75] Onodera T, Momose I, Kawada M. Potential anticancer actisvity of auranofin. Chem Pharm Bull 2019;67:186–91.

[76] Fiskus W, Saba N, Shen M, Ghias M, Liu J, Das Gupta S, Chauhan L, Rao R, Gunewardena S, Schorno K, Austin CP, Maddocks K, Byrd J, Melnick A, Huang P, Wiestner A, Bhalla KN. Auranofin induces lethal oxidative and endoplasmic reticulum stress and exerts potent preclinical activity against chronic lymphocytic leukemia. Cancer Res 2014;74(9):2520–32.

[77] Zhang X, Selvaraju K, Saei AA, D'Arcy P, A Zubarev R, Arner ES, Linder S. Repurposing of auranofin: thioredoxin reductase remains a primary target of the drug. Biochimie 2019. https://doi.org/10.1016/j.biochi.2019.03.015.

[78] Chateauvieux S, Morceau F, Dicato M, Diederich M. Molecular and therapeutic potential and toxicity of valproic acid. J Biomed Biotechnol 2010:18. ID 479364.

[79] Kostrouchová M, Kostrouch Z, Kostrouchová M. Valproic acid, a molecular lead to multiple regulatory pathways. Folia Biol 2007;53:37–49.

[80] Blaheta RA, Michaelis M, Driever PH, Cinatl J. Evolving anticancer drug valproic acid: insights into the mechanism and clinical studies. Med Res Rev 2005;25:383–97.

[81] Raffoux E, Chaibi P, Dombret H, Degos L. Valproic acid and all-trans retinoic acid for the treatment of elderly patients with acute myeloid leukemia. Haematologica 2005;90:986–8.

[82] Michaelis M, Doerr HW, Cinatl J. Valproic acid as anti-cancer drug. Curr Pharmaceut Des 2007;13:3378–93.

[83] Abdul M, Hoosein N. Inhibition by anticonvulsants of prostate-specific antigen and interleukin-6 secretion by human prostate cancer cells. Anticancer Res 2001;21:2045–8.

[84] Kramer OH, Baus D, Knauer SK, Stein S, Jager E, Stauber RH, Grez M, Pfitzner E, Heinzel T. Acetylation of stat1 modulates NF-kappaB activity. Genes Dev 2006;20:473–85.

[85] Ziauddin MF, Yeow W-S, Maxhimer JB, Baras A, Chua A, Reddy RM, Tsai W, Cole GW, Schrump DS, Nguyen DM. Valproic acid, an antiepileptic drug with histone deacetylase inhibitory activity, potentiates the cytotoxic effect of Apo2L/TRAIL on cultured thoracic cancer cells through mitochondria-dependent caspase activation. Neoplasia 2006;8(6):446–57.

[86] Venkataramani V, Rossner C, Iffland L, Schweyer S, Tamboli IY, Walter J, Wirths O, Bayer TA. Histone deacetylase inhibitor valproic acid inhibits cancer cell proliferation via down-regulation of the zlzheimer amyloid precursor protein. J Biol Chem 2010;285(14):10678–89.

[87] Murugavel S, Bugyei-Twum A, Matkar PN, Al-Mubarak H, Chen HH, Adam M, Jain S, Narang T, Abdin RM, Qadura M, Connelly KA, Leong-Poi H, Singh KK. Valproic acid induces endothelial-to-mesenchymal transition-like phenotypic switching. Front Pharmacol 2018;9:737.

[88] Atmaca A, Al-Batran S-E, Maurer A, Neumann A, Heinzel T, Hentsch B, Jäger E. Valproic acid (VPA) in patients with refractory advanced cancer: a dose escalating phase I clinical trial. Br J Cancer 2007;97(2):177–82.

[89] Nelson M, Yang M, Dowle AA, Thomas JR, Brackenbury WJ. The sodium channel-blocking antiepileptic drug phenytoin inhibits breast tumour growth and metastasis. Mol Cancer 2015;14(13):1–7.

[90] Pellegrino M, Rizza P, Nigro A, Ceraldi R, Ricci E, Perrotta I, Saveria Aquila S, Lanzino M, Andò S, Morelli C, Sisci D. FoxO3a mediates the inhibitory effects of the antiepileptic drug lamotrigine on breast cancer growth. Mol Cancer Res 2018;9. https://doi.org/10.1158/1541-7786.

[91] Hornschemeyer J, Bergner C, Krause BJ. AMPA receptor antagonist perampanel affects glioblastoma cell growth and glutamate release in vitro. PLoS One 2019;14(2). e0211644.

[92] Papanagnou P, Stivarou T, Papageorgiou I, Papadopoulos G, Pappas A. Marketed drugs used for the management of hypercholesterolemia as anticancer armament. Onco Targets Ther 2017;10:4393–411.

[93] Burke LP, Kukoly CA. Statins induce lethal effects in acute myeloblastic lymphoma cells within 72 hours. Leuk Lymphoma 2008;49(2):322–30.

[94] Altwa AK. Statins are potential anticancerous agents. Oncol Rep 2015;33:1019–39.

[95] Chae YK, Yousaf M, Malecek M-K, Carneiro B, Chandra S, Kaplan J, Kalyan J, Sassano A, Platanias LC, Giles F. Statins as anti-cancer therapy; can we translate preclinical and epidemiologic data into clinical benefit? Discov Med 2015;20. http://www.discoverymedicine.com.

[96] Ahn KS, Sethi G, Aggarwal BB. Simvastatin, potentiates TNF-alpha-induced apoptosis through the down-regulation of NF-kappaB-dependent antiapoptotic gene products: role of Ikappabalpha kinase and TGF-beta-activated kinase-1. J Immunol 2007;178:2507–16.

[97] Crosbie J, Magnussen M, Dornbier R, Iannone A, Steele TA. Statins inhibit proliferation and cytotoxicity of a human leukemic natural killer cell line. Biomark Res 2013;1:33.

[98] Pradelli D, Soranna D, Zambon A, Catapano A, Mancia G, La Vecchia C, Corrao G. Statins use and the risk of all and subtype hematological malignancies: a meta-analysis of observational studies. Cancer Med 2015;4(5):770–80.

[99] Coogan PF, Smith J, Rosenberg L. Statin use and risk of colorectal cancer. J Natl Cancer Inst 2007;99: 32–40.

[100] Broughton T, Sington J, Beales ILP. Statin use is associated with a reduced incidence of colorectal cancer: a colonoscopy-controlled case-control study. BMC Gastroenterol 2012;12:36.

[101] Happold C, Gorlia T, Nabors LB, Erridge SC, Reardon DA, Hicking C, Picard M, Stupp R, Weller M. For the EORTC brain tumor group and on behalf of the CENTRIC and CORE clinical trial groups. Do statins, ACE inhibitors or sartans improve outcome in primary glioblastoma? J Neuro Oncol 2018;138(1):163–71.

[102] Chen YA, Lin Y-J, Lin C-L, Lin H-J, Wu H-S, Hsu H-Y, Sun Y-C, Wu H-Y, Lai C-H, Kao C-H. Simvastatin therapy for drug repositioning to reduce the risk of prostate cancer mortality in patients with hyperlipidemia. Front Pharmacol 2018;9:225.

[103] Lee Y, Lee K-Y, Lee G-K, Lee S-H, Lim K-Y, Joo J-J, Go Y-J, Lee J-S, Han J-Y. Randomized phase II study of afatinib plus simvastatin versus afatinib alone in previously treated patients with advanced nonadenocarcinomatous non-small cell lung cancer. Cancer Res Treat 2017;49(4):1001–11.

[104] Lim SH, Kim TW, Hong YS, Han S-W, Lee K-H, Kang HJ, Kang WK. A randomised, double-blind, placebo-controlled multi-centre phase III trial of XELIRI/FOLFIRI plus simvastatin for patients with metastatic colorectal cancer. Br J Cancer 2015; 113(10):1421–6.

[105] Shadman M, Mawad R, Dean C, Chen TL, Shannon-Dorcy K, Sandhu V, Hendrie PC, Scott BL, Walter RB, Becker PS, Pagel JM, Estey EH. Idarubicin, cytarabine, and pravastatin as induction therapy for untreated acute myeloid leukemia and high-risk myelodysplastic syndrome. Am J Hematol 2015;90(6): 483–6.

[106] Sabaa M, EL Fayoumi HM, Elshazly S, Youns M, Barakat W. Anticancer activity of salicin and fenofibrate. Naunyn Schmiedebergs Arch Pharmacol 2017;390(10):1061–71.

[107] Lian X, Wang G, Zhou H, Zheng Z, Fu Y, Cai L. Anticancer properties of fenofibrate: a pepurposing use. J Cancer 2018;9(9):1527–37.

[108] Berstein LM. Clinical usage of hypolipidemic and antidiabetic drugs in the prevention and treatment of cancer. Cancer Lett 2005;224(2):203–12.

[109] Barbieri F, Verduci I, Carlini V, Zona G, Pagano A, Mazzanti M, Florio T. Repurposed biguanide drugs in glioblastoma exert antiproliferative effects via the inhibition of intracellular chloride channel 1 activity. Front Oncol 2019;9:135.

[110] Kalender A, Selvaraj A, Kim SY, Gulati P, Brule S, Viollet B, Kemp BE, Bardeesy N, Dennis P, Schlager JJ, Marette A, Kozma SC, Thomas G. Metformin independent of AMPK, inhibits mTORC1 in A rag GTPase-dependent manner. Cell Metabol 2010; 11:390–401.

[111] Kasznicki J, Sliwinska A, Drzewoski J. Metformin in cancer prevention and therapy. Ann Transl Med 2014;2(6):57.

[112] Heckman-Stoddard BM, De Censi A, Sahasrabuddhe VV, Ford LG. Repurposing metformin for the prevention of cancer and cancer recurrence. Diabetologia 2017;60:1639–47.

[113] Bo S, Benso A, Durazzo M, Ghigo E. Metformin blocks progression of obesity-activated thyroid cancer in a mouse model. J Endocrinol Invest 2012;35(2):231–5.

[114] Bodmer M, Meier C, Krahenbühl S, Jick SS, Meier CR. Long-term metformin use is associated with decreased risk of breast cancer. Diabetes Care 2010; 33:1304–8.

[115] Jiralerspong S, Palla SL, Giordano SH, Meric-Bernstam F, Liedtke C, Barnett CM, Hsu L, Hung MC, Hortobagyi GN, Gonzalez-Angulo AM. Metformin and pathologic complete responses to neoadjuvant chemotherapy in diabetic patients with breast cancer. J Clin Oncol 2009;27:3297–302.

[116] Campagnoli C, Pasanisi P, Abbà C, Ambroggio S, Biglia N, Brucato T, Colombero R, Danese S, Donadio M. Effect of different doses of metformin on serum testosterone and insulin in non-diabetic women with breast cancer: a randomized study. Clin Breast Cancer 2012;12:175–82.

[117] Hosono K, Endo H, Takahashi H, Sugiyama M, Sakai E, Uchiyama T, Suzuki K, Lida H, Sakamoto Y, Yoneda K, Koide T, Tokoro C, Abe Y, Inamori M, Nakagama H, Nakajima A. Metformin suppresses solorectal aberrant crypt foci in a short-term clinical trial. Cancer Prev Res 2010;3:1077–83.

[118] Zhou XL, Xue W-H, Ding X-F, Li L-F, Dou M-M, Zhang W-J, Lv Z, Fan Z-R, Zhao J, Wang L-X. Association between metformin and the risk of gastric cancer in patients with type 2 diabetes mellitus: a meta-analysis of cohort studies. Oncotarget 2017;8(33):55622–31.

[119] Tseng C-H. Metformin reduces gastric cancer risk in patients with type 2 diabetes mellitus. Aging 2016; 8(8):1636–49.

[120] Li P, Zhang C, Gao P, Chen X, Ma B, Yu D, Song Y, Wang Z. Metformin use and its effect on gastric cancer in patients with type 2 diabetes: a systematic review of observational Studies. Oncol Lett 2018;15:1191−9.

[121] Park J, Kim WG, Zhao L, Enomoto K, Willingham M, Chen SY. Metformin blocks progression of obesity-activated thyroid cancer in a mouse model. Oncotarget 2016;7(23):34832−44.

[122] Quinn BJ, Kitagawa H, Memmott RM, Gills JJ, Dennis PA. Repositioning metformin for cancer prevention and treatment. Trends Endocrinol Metabol 2013;24:469−80.

[123] Kim J, Lim W, Kim E-K, Kim M-K, Paik N-S, Jeong SS, Han W. Phase II randomized trial of neoadjuvant metformin plus letrozole versus placebo plus letrozole for estrogen receptor positive postmenopausal breast cancer. BMC Cancer 2014;14(1):170−4.

[124] Shehata M, Kim H, Vellanki R, Waterhouse PD, Mahendralingam M, Casey AE, Koritzinsky M, Khokha R. Identifying the murine mammary cell target of metformin exposure. Commun Biol 2019;2: 192.

[125] Yam C, Esteva FJ, Patel MM, Raghavendra AS, Ueno NT, Moulder SL, Valero V. Efficacy and safety of the combination of metformin, everolimus and exemestane in overweight and obese postmenopausal patients with metastatic, hormone receptor-positive, HER2-negative breast cancer: a phase II study. Invest N Drugs 2019. https://doi.org/10.1007/s10637-018-0700-z.

[126] Cheng G, Zielonka J, Ouari O, Lopez M, McAllister D, Boyle K, Barrios CS, Weber JJ, Johnson BD, Hardy M, Dwinell MB, Kalyanaraman B. Mitochondria-targeted analogues of metformin exhibit enhanced antiproliferative and radiosensitizing effects in pancreatic cancer cells. Cancer Res 2016;76(13): 3904−15.

[127] Pollak M. Repurposing biguanides to target energy metabolism for cancer treatment. Nat Med 2014; 20(6):591−3.

[128] Kordes S, Pollak MN, Zwinderman AH, Mathôt RA, Weterman MJ, Beeker A, Punt CJ, Richel DJ, Wilmink JW. Metformin in patients with advanced pancreatic cancer: a double-blind, randomised, placebo-controlled phase 2 trial. Lancet Oncol 2015; 16(7):839−47.

[129] Papanagnou P, Stivarou T, Tsironi M. Unexploited antineoplastic effects of commercially available anti-diabetic drugs. Pharmaceuticals 2016;9:24.

[130] Afzal MZ, Dragnev K, Sarwar T, Shirai K. Clinical outcomes in non small-cell lung cancer patients receiving concurrent metformin and immune checkpoint inhibitors. Lung Cancer Manag 2019. https://doi.org/10.2217/lmt-2018-0016.

[131] Higginbotham S, Wong WR, Linington RG, Spadafora C, Iturrado L, Arnold AE. Sloth sair as a novel source of fungi with potent anti-parasitic, anti-cancer and anti-bacterial bioactivity. PLoS One 2014; 9:e84549.

[132] Johnston WT, Mutalima N, Sun D, Emmanuel B, Bhatia K, Aka P, Wu X, Borgstein E, Liomba GN, Kamiza S, Mkandawire N, Batumba M, Carpenter LM, Jaffe H, Molyneux EM, Goedert JJ, Soppet D, Newton R, Bulaiteye SM. Relationship between plasmodium falciparum malaria prevalence, genetic diversity and endemic burkitt lymphoma in Malawi. Sci Rep 2014;4:3741.

[133] Khan KH. DNA vaccines: roles against diseases. Germs 2013;3:26−35.

[134] Fedosov DA, Dao M, Suresh KS. Computational biorhelogy of human blood flow in health and disease. Ann Biomed Eng 2014;42:368−87.

[135] Huijsduijnen HR, Guy RK, Chibale K, Haynes RK, Peitz I, Kelter G, Phillips MA, Vennerstrom JL, Yuthavong Y, Wells TN. Anticancer properties of distinct antimalarial drug classes. PLoS One 2013;8: e82962.

[136] Keum K-C, Yoo N-C, Yoo W-M, Chang KK, Choon YN, Min YW. Anti-cancer composition composed of anti-cancer and anti-malarial drug. 2002. WO 2002013826 A1.

[137] Liu F, Shang Y, Chen S-Z. Chloroquine potentiates the anti-cancer effect of lidamycin on non-small cell lung cancer cells in vitro. Acta Pharmacol Sin 2014;35: 645−52.

[138] Kamal A, Aziz A, Shouman S, El-Demerdash E, Elgendy M, Abdel-Naim AB. Chloroquine as a promising adjuvant chemothreaphy together with sunitinib. Sci Proc 2014;1:e384.

[139] Ganguli A, Choudhury D, Datta S, Bhattacharya S, Chakrabarti G. Inhibition of autophagy by chloroquine potentiates synergistically anti-cancer property of artemisinin by promoting ROS dependent apoptosis. Biochimie 2014;107:338−49.

[140] Soo GW, Law JH, Kan E, Tan SY, Lim WY, Chay G, Bukhari NI, Segarra I. Differential effects of ketoconazole and primaquine on the pharmacokinetics and tissue distribution of imatinib in mice. Anti Cancer Drugs 2010;21:695−703.

[141] Wong YK, Xu C, Kalesh KA, He Y, Lin Q, Wong WSF, Shen HM, Wang J. Artemisinin as an anticancer drug: recent advances in target profiling and mechanisms of action. Med Res Rev 2017;37:1492−517.

[142] Efferth T. From ancient herb to modern drug: artemisia annua and artemisinin for cancer therapy. Semin Cancer Biol 2017;46:65–83.

[143] Olliaro PL, Haynes RK, Meunier B, Yuthavong Y. Possible modes of action of the artemisinin-type compounds. Trends Parasitol 2001;17:122–6.

[144] Zhang S, Gerhard GS. Heme mediates cytotoxicity from artemisinin and serves as a general anti-proliferation target. PLoS One 2009;4:e7472.

[145] Hamacher-Brady A, Stein HA, Turschner S, Toegel I, Mora R, Jennewein N, Efferth T, Eils R, Brady NR. Artesunate activates mitochondrial apoptosis in breast cancer cells via iron-catalyzed lysosomal reactive oxygen species production. J Biol Chem 2011; 286:6587–601.

[146] Mercer AE, Maggs JL, Sun XM, Cohen GM, Chadwick J, O'Neill PM, Park BK. Evidence for the involvement of carbon-centered radicals in the induction of apoptotic cell death by artemisinin compounds. J Biol Chem 2007;282(9):372–9382.

[147] Mercer AE, Copple IM, Maggs JL, O'Neill PM, Park BK. The role of heme and the mitochondrion in the chemical and molecular mechanisms of mammalian cell death induced by the artemisinin antimalarials. J Biol Chem 2011;286:987–96.

[148] Bostwick DG, Alexander EE, Singh R, Shan A, Qian J, Santella RM, Oberley LW, Yan T, Zhong W, Jiang X, Oberley TD. Antioxidant enzyme expression and reactive oxygen species damage in prostatic intraepithelial neoplasia and cancer. Cancer 2000;89:123–34.

[149] Lu JJ, Chen SM, Zhang XW, Ding J, Meng LH. The anti-cancer activity of dihydroartemisinin is associated with induction of iron-dependent endoplasmic reticulum stress in colorectal carcinoma HCT116 cells. Invest N Drugs 2011;29:1276–83.

[150] Anfosso L, Efferth T, Albini A, Pfeffer U. Microarray expression profiles of angiogenesis-related genes predict tumor cell response to artemisinins. Pharmacogenomics J 2006;6:269–78.

[151] Xhamla N, Naki T, Blessing AA. Quinoline-based hybrid compounds with antimalarial activity. Molecules 2017;22:2268–90.

[152] Elf S, Lin R, Xia S, Pan Y, Shan C, Wu S, Chen J. Targeting 6-phosphogluconate dehydrogenase in the oxidative PPP sensitizes leukemia cells to antimalarial agent dihydroartemisinin. Oncogene 2016;36(2):254–62.

[153] Lu YY, Chen TS, Qu JL, Pan WL, Sun L, Wei XB. Dihydroartemisinin (DHA) induces caspase-3-dependent apoptosis in human lung adenocarcinoma ASTC-a-1 cells. J Biomed Sci 2009;16:16.

[154] Zhou C, Pan W, Wang XP, Chen TS. Artesunate induces apoptosis via a bak-mediated caspase-independent intrinsic pathway in human lung adenocarcinoma cells. J Cell Physiol 2012;227:3778–86.

[155] Efferth T, Sauerbrey A, Olbrich A, Gebhart E, Rauch P, Weber HO, Hengstler JG, Halatsch ME, Volm M, Tew KD, Ross DD, Funk JO. Molecular modes of action of artesunate in tumor cell lines. Mol Pharmacol 2003;64:382–94.

[156] Augustin Y, Krishna S, Kumar D, Pantziarka P. The wisdom of crowds and the repurposing of artesunate as an anticancer drug. Ecancer 2015;9:ed50.

[157] Deeken JF, Wang H, Hartley M, Cheema AK, Smaglo B, Hwang JJ, Pishvaian MJ. A phase I study of intravenous artesunate in patients with advanced solid tumor malignancies. Cancer Chemother Pharmacol 2018;81(3):587–96.

[158] Džimbeg G, Zorc B, Kralj M, Ester K, Pavelić K, Balzarini J, De Clercq E, Mintas M. The novel primaquine derivatives of N-alkyl, cycloalkyl or aryl urea: synthesis, cytostatic and antiviral activity evaluations. Eur J Med Chem 2008;43:1180–7.

[159] Šimunović M, Perković I, Zorc B, Ester K, Kralj M, Hadjipavlou-Litina D, Pontiki E. Urea and carbamate derivatives of primaquine: synthesis, cytostatic and antioxidant activities. Bioorg Med Chem 2009;17: 5605–13.

[160] Perković I, Tršinar S, Žanetić J, Kralj M, Martin-Kleiner I, D Balzarini J, Hadjipavlou-Litina AM, Katsori B, Zorc B. Novel 1-acyl-4-substituted semicarbazide derivatives of primaquine-synthesis, cytostatic, antiviral and antioxidative studies. J Enzym Inhib Med Chem 2013;28:601–10.

[161] Pavić K, Perković I, Cindrić M, Pranjić M, Martin-Kleiner I, Kralj M, Schols D, Hadjipavlou-Litina D, Katsori A-M, Zorc B. Synthesis, biological evaluation, and quantitative structure-activity relationship analysis of new schiff bases of hydroxysemicarbazide as potential antitumor agents. Eur J Med Chem 2014; 86:502–14.

[162] Perković I, Antunović M, Marijanović I, Pavić K, Ester K, Kralj M, Vlainić J, Kosalec I, Schols D, Hadjipavlou-Litina D, Pontiki E, Zorc B. Novel urea and bis-urea primaquine derivatives with hydrocyphenyl or halogenphenyl substituents: synthesis and biological evaluation. Eur J Med Chem 2016;124: 622–36.

[163] Pavić K, Perković I, Gilja P, Kozlina F, Ester K, Kralj M, Schols D, Hadjipavlou-Litina D, Pontiki E, Zorc B. Design, Synthesis and biological evaluation of novel primaquine-cinnamic acid conjugates of the amide and acylsemicarbazide type. Molecules 2016; 21(12):1629–53.

[164] Pavić K, Perković I, Pospíšilová S, Machado M, Fontinha D, Prudêncio M, Jampilek J, Coffey A, Rimac H, Zorc B. Primaquine hybrids as promising antimycobacterial and antimalarial agents. J Med Chem 2018;143:769–79.

[165] Xu X, Wang J, Han K, Li S, Xu F, Yang Y. Antimalarial drug mefloquine inhibits nuclear factor kappa B signaling and induces apoptosis in colorectal cancer cells. Cancer Sci 2018;109:1220–9.

[166] Verbaanderd C, Maes H, Schaaf MB, Sukhatme VP, Pantziarka P, Sukhatme V, Agostinis P, Bouche G. Repurposing drugs in oncology (ReDO) chloroquine and hydroxychloroquine as anti-cancer agents. Ecancer 2017;11:781.

[167] Helmy SA, El-Mesery M, El-Karef A, Eissa LA, El Gayar AM. Chloroquine upregulates TRAIL/TRAILR2 expression and potentiates doxorubicin anti-tumor activity in thioacetamide-induced hepatocellular carcinoma model. Chem Biol Interact 2018; 279:84–94.

[168] Xu R, Ji Z, Xu C, Zhu J. Clinical value of using chloroquine or hydroxychloroquine as autophagy inhibitors in the treatment of cancers. Medicine 2018;97(46):e12912.

[169] Kalogera E, Roy D, Khurana A, Mondal S, Weaver AL, He X, Dowdy SC, Shridhar V. Quinacrine in endometrial cancer: repurposing an old antimalarial drug. Gynecol Oncol 2017;146:187–95.

[170] Shaimerdenova M, Karapina O, Mektepbayeva D, Alibek K, Akilbekova D. The effects of antiviral treatment on breast cancer cell line. Infect Agents Cancer 2017;12:18.

[171] Monini P, Sgadari C, Toschi E, Barillari G, Ensoli B. Antitumor effects of antiretroviral therapy. Nat Rev Cancer 2004;4:861–75.

[172] Mitsuya H, Weinhold KJ, Furman PA, St Clair MH, Lehrman SN, Gallo RC, Bolognesi D, Barry DW, Broder S. 3′-Azido-3′-deoxythymidine (BW A509U): an antiviral agent that inhibits the infectivity and cytopathic effect of human T-lymphotropic virus type III/lymphadenopathy-associated virus in vitro. Proc Natl Acad Sci USA 1985;82:7096–100.

[173] Namba T, Kodama R, Moritomo S, Hoshino T, Mizushima T. Zidovudine, an anti-viral drug, resensitizes gemcitabine-resistant pancreatic cancer cells to gemcitabine by inhibition of the Akt-GSK3β-snail pathway. Cell Death Dis 2015;6:e1795.

[174] Aschacher T, Sampl S, Käser L, Bernhard D, Spittler A, Holzmann K, Bergmann M. The combined use of known antiviral reverse transcriptase inhibitors AZT and DDI induce anticancer effects at low concentrations. Neoplasia 2012;14:44–53.

[175] Gruffaz M, Zhou S, Vasan K, Rushing T, Michael QL, Lu C, Gao S-J. Repurposing cytarabine for treating primary effusion lymphoma by targeting kaposi's sarcoma-associated herpesvirus latent and lytic replications. mBio 2018;9(3). e00756-18.

[176] Urtishak KA, Wang L-S, Culjkovic-Kraljacic B, Davenport JW, Porazzi P, Vincent TL, Felix CA. Targeting EIF4E signaling with ribavirin in infant acute lymphoblastic leukemia. Oncogene 2018;38(13): 2241–62.

[177] Wilson MZ, Wang R, Gitai Z, Seyedsayamdost MR. Mode of action and resistance studies unveil new roles for tropodithietic acid as an anticancer agent and the γ-glutamyl cycle as a proton sink. Proc Natl Acad Sci USA 2016;113(6):1630–5.

[178] Klejborowska G, Jedrzejczyk M, Stepczynska N, Maj E, Wietrzyk J, Huczynski A. Antiproliferative activity of ester derivatives of monensin A at the C-1 and C-26 positions. Chem Biol Drug Des 2019;94(4): 1859–64.

[179] Fiorillo M, Lamb R, Tanowitz HB, Cappello AR, Martinez-Outschoorn UE, Sotgia F, Lisanti MP. Bedaquiline, an FDA-approved antibiotic, inhibits mitochondrial function and potently blocks the proliferative expansion of stem-like cancer cells (CSCs). Aging 2016;8(8):1593–606.

[180] Jiang H, Xing J, Wang C, Zhang H, Yue L, Wan X, Chen W, Ding H, Xie Y, Tao H, Chen Z, Jiang H, Chen K, Chen S, Zheng M, Zhang Y, Luo C. Discovery of novel BET inhibitors by drug repurposing of nitroxoline and its analogues. Org Biomol Chem 2017;15: 9352–61.

[181] Zhang Q, Wang S, Yang D, Pan K, Li L, Yuang S. Preclinical pharmacodynamic evaluation of antibiotic nitroxoline for anticancer drug repurposing. Oncol Lett 2016;11:3265–72.

[182] Wang M, Shim JS, Li R-J, Dang Y, He Q, Das M, Liu JO. Identification of an old antibiotic clofoctol as a novel activator of unfolded protein response pathways and an inhibitor of prostate cancer. Br J Pharmacol 2014;171:4478–89.

[183] Guilford FT, Yu S. Antiparasitic and antifungal medications for targeting cancer cells. Literature review and case studies. Alternative Ther Health Med 2019; 25(4):26–31.

[184] Chong CR, Xu J, Lu J, Bhat S, Sullivan DJ, Liu JO. Inhibition of angiogenesis by the antifungal drug itraconazole. ACS Chem Biol 2017;2(4):263–70.

[185] Pounds R, Leonard S, Dawson C, Kehoe S. Repurposing itraconazole for the treatment of cancer (review). Oncol Lett 2017;14:2587–97.

[186] Pantziarka P, Sukhatme V, Bouche G, Meheus L, Sukhatme VP. Repurposing drugs in oncology (ReDO)-Itraconazole as an anti-cancer agent. Ecancer 2015;9:521.

[187] Tsubamoto H, Ueda T, Inoue K, Sakata K, Shibahara H, Sonoda T. Repurposing itraconazole as an anticancer agent (review). Oncol Lett 2017;14: 1240–6.
[188] Head SA, Shi WQ, Yang EJ, Nacev BA, Hong SY, Pasunooti KK, Liu JO. Simultaneous targeting of NPC1 and VDAC1 by itraconazole leads to synergistic inhibition of mTOR signaling and angiogenesis. ACS Chem Biol 2016;12(1):174–82.
[189] Lee M, Hong H, Kim W, Zhang L, Friedlander TW, Fong L, Lin AM, Small EJ, Wei XX, Rodvelt TJ, Miralda B, Stocksdale B, Ryan CJ, Aggarwal R. Itraconazole as a noncastrating treatment for biochemically recurrent prostate cancer: a phase 2 study. Clin Genitourin Cancer 2019;17(1):e92–6.
[190] Antonarakis ES, Heath EI, Smith DC, Rathkopf D, Blackford AL, Danila DC, King S, Forst A, Ajiboye AS, Zhao M, Mendonca J, Kachhap SK, Rudek MA, Carducci MA. Repurposing itraconazole as a treatment for advanced prostate cancer: a noncomparative randomized phase II trial in men with metastatic castration-resistant prostate cancer. Oncologist 2013;18:163–73.
[191] Rudin CM, Brahmer JR, Juergens RA, Hann CL, Ettinger DS, Sebree R, Liu JO. Phase 2 study of pemetrexed and itraconazole as second-line therapy for metastatic nonsquamous non-small-cell lung cancer. J Thorac Oncol 2013;8(5):619–23.
[192] Seto B. Rapamycin and mTOR: a serendipitous discovery and implications for breast cancer. Clin Transl Med 2012;1:29.
[193] Riedel T, Demaria O, Zava O, Joncic A, Gilliet M, Dyson PJ. Drug repurposing approach identifies a synergistic drug combination of an antifungal agent and an experimental organometallic drug for melanoma treatment. Mol Pharm 2017;15(1):116–26.
[194] Shaikh K, Pawar A, Aphale S, Moghe A. Effect of vesicular encapsulation on in-vitro cytotoxicity of ciclopirox olamine. Int J Drug Deliv 2012;4(2):139–46.
[195] Eberhard Y, McDermott SP, Wang X, Gronda M, Venugopal A, Wood TE, Hurren R, Datti A, Batey RA, Wrana J, Antholine WE, Dick JE, Schimmer AD. Chelation of intracellular iron with the antifungal agent ciclopirox olamine induces cell death in leukemia and myeloma cells. Blood First 2009;114(14):3064–73.
[196] Mihailidou C, Papakotoulas P, Papavassiliou AG, Karamouzis MV. Superior efficacy of the antifungal agent ciclopiroxolamine over gemcitabine in pancreatic cancer models. Oncotarget 2018;9(12):10360–74.
[197] Minden MD, Hogge DE, Weir SJ, Kasper J, Webster DA, Patton L, Schimmer AD. Oral ciclopirox olamine displays biological activity in a phase I study in patients with advanced hematologic malignancies. Am J Hematol 2014;89(4):363–8.
[198] Cha HJ, Byrom M, Mead PE, Ellington AD, Wallingford JB, Marcotte EM. Evolutionarily repurposed networks reveal the well-known antifungal drug thiabendazole to be a novel vascular disrupting agent. PLoS Biol 2012;10(8). e1001380.
[199] Chen B, Wei W, Ma L, Yang B, Gill RM, Chua MS, Butte AJ, So S. Computational discovery of niclosamide ethanolamine, a repurposed drug candidate that reduces growth of hepatocellular carcinoma cells in vitro and in mice by inhibiting CDC37 signaling. Gastroenterology 2017;152(8):2022–36.
[200] Kim M, Choe MH, Yoon YN, Ahn J, Yoo M, Jung K-Y, Kim J-S. Antihelminthic drug niclosamide inhibits CIP2A and reactivates tumor suppressor protein phosphatase 2A in non-small cell lung cancer cells. Biochem Pharmacol 2017;144:78–89.
[201] Chen L, Wang L, Shen H, Lin H, Li D. Anthelminthic drug niclosamide sensitizes the responsiveness of cervical cancer cells to paclitaxel via oxidative stress-mediated mTOR inhibition. Biochem Biophys Res Commun 2017;484:416e421.
[202] Yu X, Liu F, Zeng L, He F, Zhang R, Yan S, Zeng Z, Zhang W, Huang S, Zhang L, Dai Z, Wang X, Liu B, Haydon RC, Luu HH, Gan H, He T-C, Chen L, Shu Y, Zhao C, Wu X, Lei J, Yang C, Ji X, Wu K, Wu Y, An L, Gong C, Yuan C, Feng Y, Huang B, Liu W, Zhang B. Niclosamide exhibits potent anticancer activity and synergizes with sorafenib in human renal cell cancer cells. Cell Physiol Biochem 2018;47:957–71.
[203] Wu C-L, Chen C-L, Huang H-S, Yu D-S. A new niclosamide derivatives-B17 can inhibit urological cancers growth through apoptosis-related pathway. Cancer Med 2018;7:3945–54.
[204] Satoh K, Zhang L, Zhang Y, Chelluri R, Boufraqech M, Nilubol N, Patel D, Shen M, Kebebew E. Identification of niclosamide as a novel anticancer agent for adrenocortical carcinoma. Clin Cancer Res 2016;22(14): 3458–66.
[205] Pantziarka P, Sukhatme V, Bouche G, Meheus L, Sukhatme VP. Repurposing drugs in oncology (ReDO)-mebendazole as an anti-cancer agent. Ecancer 2014;8:443.
[206] Larsen AR, Chung JH, Bunz F, Bai R-Y, Borodovsky A, Riggins GI, Rudin CM. Repurposingthe antihelmintic mebendazole as a hedgehog inhibitor. Mol Cancer Therapeut 2015; 14(1):3–13.
[207] Williamson T, Bai R-Y, Riggins GJ, Staedtke V, Huso D, Riggins G. Mebendazoleand a non-steroidal anti-inflammatory combine to reduce tumor initiation

in a colon cancer preclinical model. J. Oncotarget 2016; 7(42):68571−84.

[208] Bai R-Y, Staedtke V, Aprhys CM, Gallia GL, Riggins GJ. Antiparasitic mebendazole shows survival benefit in 2 preclinical models of glioblastoma multiforme. Neuro Oncol 2011;13(9):974−82.

[209] Spagnuolo PA, Hu J, Hurren R, Wang X, Gronda M, Sukhai MA, Di Meo Ashley A, Boss J, Ashali I, Beheshti ZR, Fine N, Simpson CD, Sharmeen S, Rottapel R, Schimmer AD. The antihelmintic flubendazole inhibits microtubule function through a mechanism distinct from vinca alkaloids and displays preclinical activity in leukemia and myeloma. Blood 2010;115(23):4824−33.

[210] Lin S, Yang L, Yao Y, Xu L, Zhao C, Lin S, Yang L, Yao Y, Xu L, Xiang Y, Wang L, Zuo Z, Huang X, Zhao H. Flubendazole demonstrates valid antitumor effects by inhibiting STAT3 and activating autophagy. J Exp Clin Cancer Res 2019;38(1):293.

[211] Junichi Ishida I, Konishi M, Ebner N, Springer J. Repurposing of approved cardiovascular drugs. J Transl Med 2016;14:269.

[212] Regulska K, Regulski M, Karolak B, Michalak M, Murias M, Stanisz B. Beyond the boundaries of cardiology: still untapped anticancer properties of the cardiovascular system-related drugs. Pharmacol Res 2019. https://doi.org/10.1016/j.phrs.2019.104326.

[213] Chang A, Yeung S, Thakkar A, Huang KM, Liu MM, Kanassatega R-S, Parsa C, Orlando R, Jackson EK, Andresen BT, Huang Y. Prevention of skin carcinogenesis by the ß-blocker carvedilol. Cancer Prev Res 2014;8(1):27−36.

[214] Owunari GU, Minakiri SI. Disulfiram and copper gluconate in cancer chemotherapy; a review of the literature. Cancer Res J 2014;2(5):88−92.

[215] Skrott Z, Cvek B. Diethyldithiocarbamate complex with copper: the mechanism of action in cancer cells. Mini Rev Med Chem 2012;12:1184−92.

[216] Tan Z, Peng A, Xu J, Ouyang M. Propofol enhances BCR-ABL TKIs' inhibitory effects in chronic myeloid leukemia through Akt/mTOR suppression. BMC Anesthesiol 2017;17:132.

[217] Mediavilla-Varela M, Boateng K, Noyes D, Antonia SJ. The anti-fibrotic agent pirfenidone synergizes with cisplatin in killing tumor cells and cancer-associated fibroblasts. BMC Cancer 2016;16:176.

[218] Moreno E, Cavic M, Krivokuca A, Casado V, Canela E. The endocannabinoid system as a target in cancer diseases: are we there yet? Front Pharmacol 2019;10:339.

[219] Gonza lez-Silva L, Quevedo L, Varela I. Tumor functional heterogeneity unraveled by scRNA-seq technologies. Trends Cancer 2020;6(1):13−9.

[220] Ding X. Drug repositioning needs a rethink. Nature 2016;535:355.

[221] Yadav SS, Li J, Lavery HJ, Yadav KK, Tewari AK. Next-generation sequencing technology in prostate cancer diagnosis, prognosis, and personalized treatment. Urol Oncol 2015;33. 267.e1-13.

[222] Doudican NA, Kumar A, Singh NK, Nair PR, Lala DA, Basu K, Talawdekar AA, Sultana Z, Tiwari KK, Tyagi A, Abbasi T, Vali S, Vij R, Fiala M, King J, Perle M, Mazumder A. Personalization of cancer treatment using predictive simulation. J Transl Med 2015;13:43.

[223] Cheng F, Hong H, Yang S, Wei Y. Individualized network-based drug repositioning infrastructure for precision oncology in the panomic era. Briefings Bioinf 2017;18:682−97.

[224] Sengupta S, Sun SQ, Huang K-L, Oh C, Bailey MH, Varghese R, Wyczalkowski MA, Ning J, Tripathi P, McMichael JF, Johnson KJ, Kandoth C, Welch J, Ma C, Wendl MC, Payne SH, Fenyö D, Townsend RR, Dipersio JF, Chen F, Ding L. Integrative omics analyses broaden treatment targets in human cancer. Genome Med 2018;10:60.

[225] Lenting K, van den Heuvel CNAM, van Ewijk A, ElMelik D, de Boer R, Tindall E, Wei G, Kusters B, te Dorsthorst M, ter Laan M, Huynen MA, Leenders WP. Mapping actionable pathways and mutations in brain tumours using targeted RNA next generation sequencing. Acta Neuropathol Commun 2019;7:185.

[226] Lu J-Y, Cheng W-C, Chen K-Y, Lin C-C, Chang C-C, Kuo K-T, Chen P-L. Using Ion Torrent sequencing to study genetic mutation profiles of fatal thyroid cancers. J Formos Med Assoc 2018;117:488−96.

[227] Jia N, Sun Z, Gao X, Cheng Y, Zhou Y, Shen C, Chen W, Wang X, Shi R, Li N, Jand JZ, Bai C. Serial monitoring of circulating tumor DNA in patients with metastatic colorectal cancer to predict the therapeutic response. Front Genet 2019;10:470.

[228] Hintzsche JD, Yoo M, Kim J, Amato CM, Robinson WA, Tan AC. IMPACT web portal: oncology database integrating molecular profiles with actionable therapeutics. BMC Med Genom 2018;11(Suppl. 2):72−6.

[229] Hyman DM, Taylor BS, Baselga J. Implementing genome-driven oncology. Cell 2017;168:584−99.

Index

Note: 'Page numbers followed by "f" indicate figures and "t" indicates tables.'

Printed and bound by CPI Group (UK) Ltd, Croydon, CR0 4YY

Printed and bound by CPI Group (UK) Ltd, Croydon, CR0 4YY
25/09/2024
01037685-0005